Negative Emissions Technologies for Climate Change Mitigation

Negative Emissions Technologies for Climate Change Mitigation

Steve Rackley
Graham Andrews
Diarmaid Clery
Renaud De Richter
George Dowson
Pol Knops
Wei Li
Stephen Mccord
Tingzhen Ming
Adrienne Sewel
Peter Styring
Michael Tyka

ELSEVIER

Elsevier
Radarweg 29, PO Box 211, 1000 AE Amsterdam, Netherlands
The Boulevard, Langford Lane, Kidlington, Oxford OX5 1GB, United Kingdom
50 Hampshire Street, 5th Floor, Cambridge, MA 02139, United States

ISBN: 978-0-12-819663-2

For information on all Elsevier publications visit our website at https://www.elsevier.com/books-and-journals

Publisher: Candice G Janco
Acquisitions Editor: Jessica Mack
Editorial Project Manager: Lira Faurillo
Production Project Manager: R.Vijay Bharath
Cover Designer: Mark Rogers

Typeset by TNQ Technologies

for
our children's
children's
children

Contents

11. Afforestation and other land- and soil-based methods

Steve Rackley

12. Bioenergy with carbon capture and storage

Diarmaid Clery and Steve Rackley

13. Direct air capture

Peter Styring, George Dowson and Steve Rackley

17. Carbon dioxide utilization

Peter Styring, Stephen McCord and Steve Rackley

Preface

The start of the Industrial Revolution in the mid-1700s was a turning point in human civilization comparable in importance to the adoption of agricultural practices over 10,000 years earlier. Both had a dramatic impact on productivity—of the land in the first instance, and of labor in the second—and both had, and continue to have, profound impacts on the planet, shaping its landscapes and changing its climate. The Industrial Revolution was powered by the energy stored in cheap and widely available fossil carbon—initially coal, later oil and gas—and the unintended consequences of this exploitation have been a 50% increase in the carbon dioxide (CO_2) concentration in the atmosphere, a 25% increase in the acidity of the ocean, and an increase in global mean surface temperature that is rapidly approaching 1.5°C.

A growing realization of the potentially catastrophic consequences of continued greenhouse gas emissions led to the establishment in 1988 of the Intergovernmental Panel on Climate Change (IPCC), the 1990 publication of the IPCC's First Assessment Report, and, in 1992, the adoption of the UN Framework Convention on Climate Change (UNFCCC); yet half of the historical increase in atmospheric CO_2 has occurred since 1990, with CO_2 emissions almost doubling in the same period.

Rapid decarbonization of industry, accompanied by major changes in individual and societal behaviors and expectations, is essential if we are to avoid the worst consequences of climate change. But 30 years of inaction mean that these changes are no longer enough, and the removal of massive quantities of CO_2 from the atmosphere is now essential if we are to succeed in limiting warming to at most 2°C.

Developing the capability to remove gigatonnes of CO_2 per year by mid-century is an immense undertaking—in effect a Second Industrial Revolution—and the aim of this book is to contribute in small part to this endeavor, by providing a comprehensive overview of the technologies, as background for decision-makers and others currently involved in the field, and hopefully to inspire the coming generation of scientists and engineers to tackle this most pressing challenge.

Recent progress has been encouraging and is exemplified by the explosive growth in research and development work published in the last few years. It is clearly impossible to fully cover this wealth of scholarship in a single volume; instead the aim here has been to demonstrate the breadth of ongoing work with sufficient detail to encourage and guide further study. Climate-related papers are increasingly being published with open access, but where this is not the case, a paper's corresponding author will invariably respond to a personal request for a preprint, and contact details can usually be found on institutional websites.

When this book was conceived in 2018 the task of reviewing and summarizing negative emissions technologies (NETs) in an introductory text seemed a manageable task for a sole author, but by 2021, when the initial target date for the manuscript was being rescheduled for the second time, the rapid progress being made across the whole field made it clear that this was no longer possible. This work would not have been completed except as a collaborative effort, and I am deeply grateful to my coauthors, Graham Andrews, Diarmaid Clery, Reinaud de Richter, George Dowson, Pol Knops, Wei Li, Stephen McCord, Tingzhen Ming, Adrienne Sewel, Peter Styring, and Michael Tyka, for rising to the challenge. All who read and learn from this book are in your debt.

Several chapters have also benefited greatly from review by professionals engaged in various aspects of NETs development, and the climate crisis more generally. In this respect special thanks are due to Rob Bellamy, Stefano Caserini, Andrei Girz, David Hawkins, Duncan McLaren, Chris Pearce, Greg Rau, Omar Sadoon, Marius Sandru, and Adele Walker. Their critical input, generously provided, is reflected in these pages; responsibility for the remaining shortcomings, errors, and omissions remains with the authors. Comments, suggestion, and other feedback from readers will be most welcome and can be sent to nets4climate@gmail.com. A number of companies and publishers have kindly given permission for photographs and figures to be reproduced here. Any requests for further reuse should be directed to the cited copyright owners.

Once again it has been a pleasure to work with the team at Elsevier, and my thanks are due to Vijay Bharath, Lira Faurillo, Ivy Dawn Torre, Candice Janco, Jessica Mack, Marisa LaFleur, and Joseph Hayton.

The techniques and technologies described here may help us to avoid breaching the Paris Agreement goal of 2°C and perhaps, even still, the aspiration of 1.5°C. But only if we accelerate the transformation of our energy system and industrial landscape to deliver the deep and rapid cuts in CO_2 emissions that are essential to put us on a path toward that goal.

Steve Rackley
June 2023

Chapter 1

The climate challenge: climate change, mitigation, and negative emissions

Steve Rackley

The large-scale removal from the atmosphere of greenhouse gases, predominantly carbon dioxide, is a relatively recent addition to the range of options being implemented or under consideration to address climate change, having gained increased attention following the publication of the IPCC's Fifth Assessment Report (AR5, IPCC, 2014a,b,c) and the 2015 Paris Agreement.

Carbon dioxide (CO_2) can be removed from the atmosphere through a range of methods, collectively known as negative emissions technologies or NETs, many of which mimic or enhance the natural processes—such as photosynthesis, air:sea gas exchange, and rock weathering—that are responsible for turning over around a quarter of the atmospheric CO_2 inventory each year.

This chapter provides a brief historical introduction to the climate challenge and the emergence of NETs on the mitigation scene, before reviewing the part they will be called upon to play in the second half of this century if the Paris goal of limiting global warming to at most 2°C is to be achieved.

1.1 First steps in climate science

1.1.1 Carbonic acid and the Earth's temperature

The absorption of infrared radiation by carbon dioxide was first demonstrated experimentally in 1856 by American physicist Eunice Foote, who also made the connection with global temperature by noting that a higher atmospheric CO_2 concentration [CO_2] in the past would have necessarily resulted in a higher surface temperature (Foote, 1856). Evidence in the geological record for significant variations in Earth's temperature has been amassing during the 19th century, and a raft of competing theories, summarized by Arrenhius (1896), had been proposed to account for the apparent succession of "ice ages." Motivated by the question of whether the temperature changes occurring between glacial and "genial" periods of geological history could be attributed to the influence of changing "carbonic acid in the air," Swedish physicist Svante Arrhenius investigated the influence of [CO_2] on the global mean surface temperature. Using infrared observations of the moon, performed by American astronomer Samuel P. Langley, Arrhenius calculated that a 50% increase in [CO_2] would result in an increase in surface temperature of 3−4°C (Arrhenius, 1896). Given the relatively small total mass of CO_2 in the atmosphere compared with that locked in carbonate sediments, he concluded it was entirely possible that the required glacial-to-genial temperature swing could be due solely to [CO_2] variations.

Although we now know that cyclical changes in the Earth's orbital geometry have been the primary drivers of recent glaciations, with [CO_2] variation and other feedbacks being secondary causes (or possibly effects), Arrhenius' calculation stands as the point of origin of the scientific endeavor that has culminated in our understanding of the profound and potentially catastrophic impact that the combustion of fossil fuels is having on the Earth's climate.

The first attempt to measure the warming effect predicted by Arrhenius was undertaken by the British engineer and amateur meteorologist Guy S. Callendar in the mid-1930s. Callendar laboriously analyzed historical temperature measurements for 200 weather stations and concluded that a 0.3°C rise in mean land temperatures had occurred since the 1880s. He estimated that fossil fuel combustion was releasing some 4.3 billion tonnes of CO_2 ($GtCO_2$) annually to the atmosphere, with about 30% being taken up by the oceans, and that this had resulted in an increase in [CO_2] of 5%−6%

Negative Emissions Technologies for Climate Change Mitigation. https://doi.org/10.1016/B978-0-12-819663-2.00011-3

between 1900 and 1936 (Callendar, 1938), a value that he found was confirmed by the then sparsely available [CO_2] measurements. Callendar also updated the warming calculation and found an effect about half as large as that predicted 40 years earlier by Arrhenius, a difference he attributed to the improved infrared absorption measurements that had become available.

Arrhenius' work was the first attempt to calculate what is now known as the equilibrium climate sensitivity (ECS), which is defined as the increase in global mean surface temperature (GMST) following a doubling of [CO_2]. Given the simplicity of his calculation approach and the limitations of the then available infrared absorption data, it is remarkable that his result—around 6°C—was within a factor 2 of the currently accepted value. Callendar revised this estimate to ~2°C, and later estimates bracket his, for example, 1.8°C (Kaplan, 1960), 3.8°C (Plass, 1956). Today's best estimate of 3°C comes from the IPCC's AR6 (IPCC, 2021), which concluded that the ECS is likely (>67% likelihood) to be between 2.5 and 4.0°C, very unlikely (<10% likelihood) to be less than 1.5°C or greater than 5°C (National Academy of Sciences, 1979). However, recent paleoclimate analysis by Hansen and colleagues suggests that ECS is at least 4°C, and could be as high as 10°C after the full impact of slow feedbacks related to ice sheets and vegetation cover is felt (Hansen et al., 2022). Charles Keeling noted in 1970 that ECS calculations are very sensitive to the assumptions made about the response of clouds to a warming climate (Keeling, 1970), and cloud plus cloud feedback effects remained major factors in ECS uncertainty four decades later (Webb et al., 2013).

ECS and the related transient climate response (TCR; see Glossary), and the uncertainties in these factors, play an important role in determining the "carbon budget"—the amount of CO_2 that can be emitted to the atmosphere while keeping GMST below a given limit—and are further discussed below.

1.1.2 The CO_2 balance

During the 1950s, one aspect of Callendar's work that received great attention was the question of CO_2 transfer between the atmosphere, the terrestrial biosphere, and the oceans—the "CO_2 balance" as it was then called. Callendar had estimated that some 25%–30% of historical fossil fuel CO_2 emissions had been taken up by the oceans, on the assumption of an "equilibrium time" (what we would now call the ventilation time) of 2000–5000 years. Recent estimates put the age of the oldest deep waters, which occur in the North Pacific, at 1300–1400 years (Khatiwala et al., 2012), with relatively younger deep waters in the Atlantic and Southern oceans, while complete ventilation of the world's oceans is estimated to take around 5000 years (England, 1995).

Callendar's assessment was widely dismissed, with later studies suggesting that virtually all of the CO_2 from fossil fuel combustion must have quickly found its way into the ocean. Some 20 years after Callendar's initial estimate, Scripps oceanographer Roger Revelle concluded that the average residence time of a CO_2 molecule in the atmosphere before being dissolved into the sea was ~10 years and that, on this basis, all but 2.5% of industrial emissions from fossil fuel combustion up to that time would have been taken up by the oceans (Revelle and Suess, 1957). It was almost another 2 decades before the improved understanding of the coupling of the mixed layer and the deep ocean allowed Oeschger et al. to conclude that, depending on the degree of CO_2 fertilization of the biosphere (see Fig. 1.1), around 55%–60% of anthropogenic CO_2 had remained in the atmosphere, with 35%–40% being taken up by the ocean (Oeschger et al., 1975).

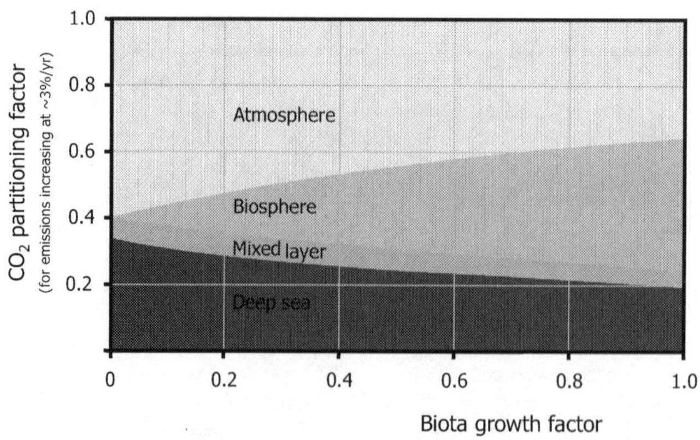

FIGURE 1.1 The CO_2 partitioning balance as a function of biota growth. *After Oeschger et al. (1975).*

Subsequent work, using more and more detailed ocean circulation models, has refined the estimate of the oceanic uptake to around 25% of cumulative fossil fuel plus land-use change CO_2 emissions over the industrial era (IPCC, 2021), and around 26% in the decade to 2020 (Friedlingstein et al., 2021). The IPCC's AR6 estimates that ~42% of total anthropogenic CO_2 emissions from 1750 to 2019 (1045 ± 18 $GtCO_2$ from a total of 2512 ± 275 $GtCO_2$) have remained in the atmosphere, with the balance (~34%) being taken up in the terrestrial biosphere.

Helped by the cancellation of errors arising from several simplifying assumptions and then unknown processes, Callendar's estimate had been right! Fig. 1.2 illustrates the carbon balance for the decade to 2020.

1.1.3 Signs of trouble in the greenhouse

Both Arrhenius and Callendar viewed the warming affect due to increasing [CO_2] from fossil fuel combustion in a positive light. Callendar calculated the polar shift of climatic zones out to the 22nd century and concluded that more favorable conditions for plant growth, both from increased temperature and CO_2 availability, would be beneficial to mankind. He predicted a mean [CO_2] and temperature increase for the 22nd century of 360 ppm and 0.6°C assuming a continuation of fossil fuel combustion at early 20th century levels, and could not have foreseen the explosive economic growth that started in mid-century.

In 1957, Revelle and Suess noted that "... human beings are now carrying out a large scale geophysical experiment of the kind that could not have happened in the past or be repeated in the future. Within a few centuries we are returning to the atmosphere and oceans the concentrated organic carbon stored in sedimentary rocks over hundreds of millions of years." This experiment was looked upon as an opportunity to gain insight into "the processes determining weather and climate," and they concluded that the anticipated 20%–40% increase in [CO_2] "in coming decades ... should certainly be adequate to allow a determination of the effects, if any, of changes in atmospheric carbon dioxide on weather and climate throughout the Earth."

The International Geophysical Year (1957–58) was an opportunity to begin establishing an accurate baseline against which future changes in atmospheric CO_2 could be measured, and, following his recruitment by Revelle to the Scripps Institute, Charles Keeling began the Mauna Loa CO_2 observations that bear his name. After a decade of observations, and with a strong trend established (Fig. 1.3), it was still not known what increase in CO_2 the Earth's atmosphere could accommodate without noticeable warming. In addressing the question whether emissions from fossil fuel use were changing man's environment, Keeling, in 1970, would say no more than caution that "[i]f the human race survives into the 21st century with the vast population increase that now seems inevitable, the people living then, along with their other troubles, may also face the threat of climatic change brought about by an uncontrolled increase in atmospheric CO_2 from fossil fuels" (Keeling, 1970).

Nevertheless, the Keeling curve's inexorable rise was ominous, and CO_2 emissions were recognized by some scientists (largely outside the peer-reviewed literature) as posing an environmental risk that could lead to "measurable and perhaps marked climate change" (Revelle et al., 1965), including the melting of glaciers and Arctic ice, sea level rise, and coastal flooding, as well as droughts in some regions.

Although there was a growing awareness that climate change could potentially have serious long-term consequences, it was impossible in the early 1970s for the scientific community to give any meaningful estimate of the magnitude or

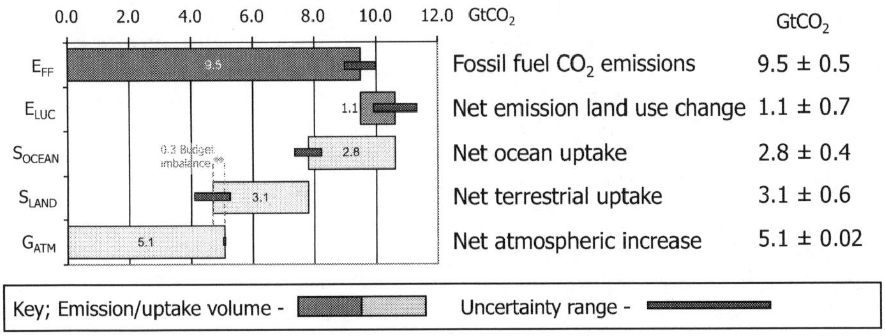

FIGURE 1.2 The CO_2 balance for the decade to 2020. *Data from Friedlingstein et al. (2021).*

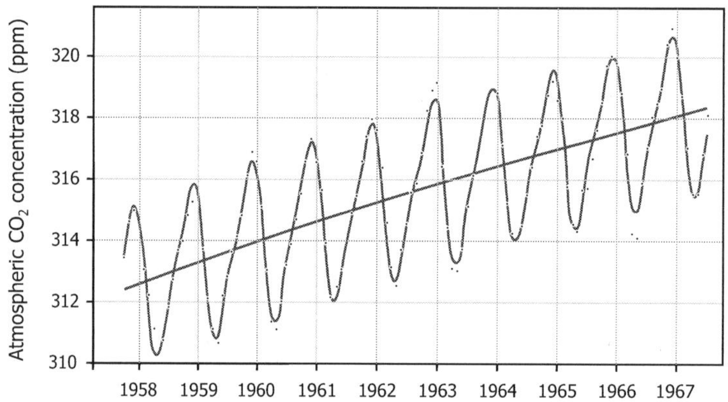

FIGURE 1.3 First decade of the Mauna Loa CO_2 record. *After Keeling (1970).*

timescale of the potential impacts. Getting that clarity required the investigation of numerous processes in the atmosphere, biosphere, and oceans, the acquisition of wide-ranging environmental data, and, perhaps most important, the integration of knowledge and data into computer models—ultimately encompassing the complete Earth climate system. By 1980, a mean surface temperature rise of 0.4°C had been observed over the previous century (Hansen et al., 1981), consistent with the increase in atmospheric CO_2 concentration, and indicating significant consequences for 21st-century climate, including shifting climatic zones leading to new drought-prone regions in North America and central Asia, and erosion of the west Antarctic ice sheet accelerating global sea level rise. These predictions added extra urgency to the need to improve Earth system models.

1.2 Earth system models—from pencils to petaFLOPs

Apart from the sheer willpower needed to carry through the "tedious calculations," Arrhenius' first estimate of ECS required no more than a slide rule and a year's supply of pencils and paper to complete. One can speculate that, working perhaps 15 h a week and performing around 10 arithmetic operations per minute, his work was equivalent to some 500,000 "floating point operations" at a computing speed of around 15 milliFLOPS (floating point operations per second). Almost 150 years later, the advent of digital computers, such as the University of Pennsylvania's ENIAC (Electronic Numerical Integrator And Computer), the University of Michigan's MIDAC (Michigan Digital Automatic Computer), and the Swedish Board for Computing Machinery's BESK (Binär Elektronisk SekvensKalkylator), allowed researchers from the late 1940s to perform these calculations thousands of times faster.

The pace of computing hardware developments far outstripped the very gradual accumulation of empirical data and understanding of the processes influencing climate, and early models of these processes often required extremely simplifying assumptions. One such example was the investigation of the atmosphere–biosphere–ocean carbon cycle developed by Erikson and Welander (1956) based on the simple analytical model shown in Fig. 1.4A.

The assumption that gross primary production (GPP—see Glossary) in the terrestrial biosphere was limited by CO_2 availability led Erikson and Welander to conclude, incorrectly, that there was a self-sustaining oscillation of CO_2 between the atmosphere and biosphere, with an amplitude of around 120 ppm and a period of several hundred years, the ocean playing a minor role in dampening these oscillations.

During the 1950s, the numerical simulation took over from analytical methods as the preferred tool for investigating atmospheric processes, primarily driven by the desire for improved weather forecasting. Numerical simulations use a two- or three-dimensional grid of points, with properties such as temperature, pressure, and humidity defined at each grid location at the start of the simulation. Physics (and, later, chemistry and biology) can then be used to predict how these properties will change at the end of a time step, with 3 hours being a typical time step for this generation of forecasting models. By the early 1960s, numerical simulation models focusing on atmospheric processes, with simplified treatment of the oceans and ice cover, could reproduce the main features of global atmospheric circulation.

As environmental data became increasingly available in the next decades, simulation models evolved to include additional physical, chemical, and biological processes in the atmosphere, on the land surface, and in the oceans, the

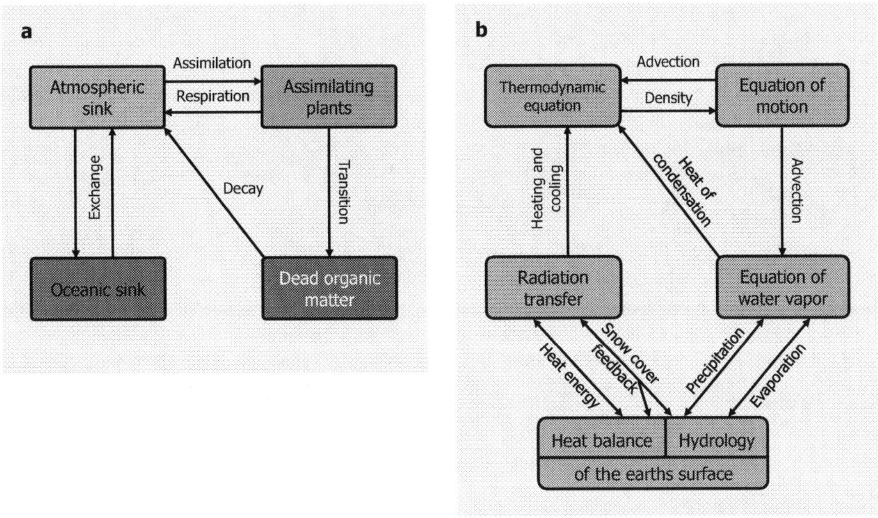

FIGURE 1.4 (A) analytical carbon cycle model of Erikssson and Welander (1956); (B) numerical hydrological cycle model of Manabe and Wetherald (1975).

incorporation of the hydrological cycle being an important example (Fig. 1.4B). Increasing process detail has been accompanied by a refinement of the simulation grid resolution from around 1000 km in 1980 to close to 10 km by 2020, enabled by increasing computing power (Fig. 1.5).

By the late 1980s, incorporating the impact of sulfate aerosol release from volcanic eruptions, modelers were able to achieve a reasonable match to the temperature history of a previous half-century and therefore had greater confidence in the predictions of the warming that would result if fossil fuel emissions were to continue unchecked.

The extent and distribution of clouds, and their response to warming, have been recognized since the very first steps as a critical factor in determining the warming effect of increasing [CO_2] (Plass, 1956), and even today, the greater part of the

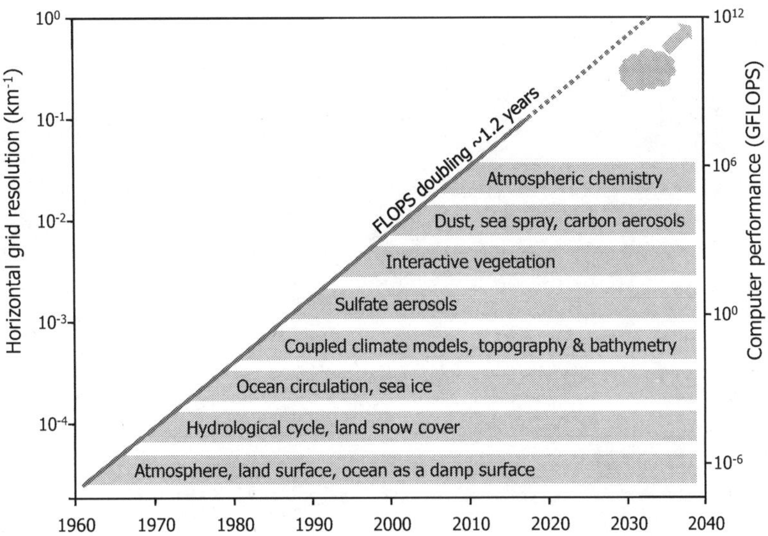

FIGURE 1.5 Climate model and computing power development.

uncertainty in the ECS range (3.0 ± 1.5°C) is due to differences in the modeling of low clouds (Schneider et al., 2017). Schneider et al. speculate that the problem of clouds may finally be solved by around 2060 if the trend of computer performance shown in Fig. 1.5 continues and climate model resolution can be reduced to ~10 m. As was the case for the weather forecasters of the 1950s who needed 24 h to compute and postprocess a 24-h forecast, "Computing advances may eventually deliver answers—but not before Earth's climate system itself has revealed its sensitivity in the experiment we are currently performing on it" (Schneider et al., 2017).

Fig. 1.6 shows schematically the components of a current generation Earth system model used to simulate the impact of CO_2 emissions on climate and ecosystems. Such models incorporate all the physical, chemical, and biological processes that are known to be relevant, and are constrained by a broad range of environmental data.

Integrated assessment models (IAMs) go one step further and use multicriteria optimization to investigate the economic and other impacts of responses such as emissions reduction, greenhouse gas (GHG) removal, and adaptation (Fig. 1.7), for example, to identify the lowest-cost path to avoid exceeding a certain warming limit. By modeling wider industrial and economic activity, IAMs are able to create scenario ranges for non-CO_2 GHG drivers of warming which, as discussed later, are an important input to the assessment of the remaining carbon budget to avoid exceeding any given warming limit.

An interesting recent development in the IAM field is the application of agent-based models, which provide an alternative and perhaps more realistic assessment of the economic impacts of climate change (see Lamperti et al., 2018 for a starting point in this field).

While IAMs are an essential tool, their outputs do not constitute an optimal goal toward which policymakers should aim. Model results based on simple optimization metrics will inevitably simplify or exclude many factors influencing the deployment of any given solution, such as evolving costs and achievable deployment rates, negative ecological and environmental impacts (many, as yet, poorly understood), as well as social and political challenges. IAMs must therefore be seen as one component in an iterative policymaking process, rather than as providing a settled objective basis for that process (Fuhrman et al., 2019; Workman et al., 2020). Of particular importance is the need to supplement IAMs with regional models of the implications of indicated pathways, as well as expert elicitations, scenario analyses, and other approaches, to ensure that a broad cross-disciplinary perspective is adopted when applying IAM outputs in decision-making (Gambhir et al., 2019).

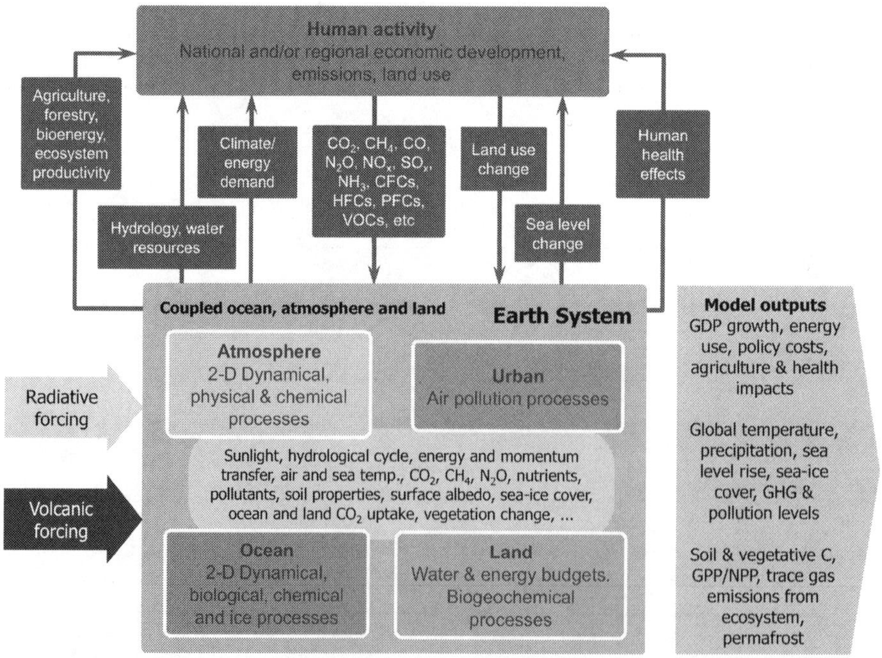

FIGURE 1.6 Integrated earth system model framework—MIT IGSM. *After Prin (2013).*

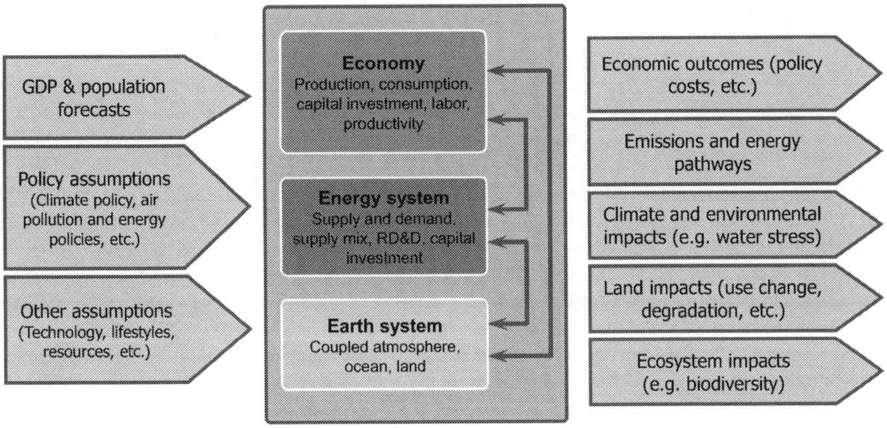

FIGURE 1.7 Schematic integrated assessment model (IAM) framework.

1.3 The long road to international consensus

As confidence in the predictive capabilities of Earth system models grew through the 1970 and 1980, the early warnings of Revelle, Keeling, and others of the potentially damaging impacts of global warming gained credibility and were repeated in increasingly strong terms, notably at the "Changing Atmosphere" conference held in Toronto in 1988, at which 300 scientists, policymakers, and corporate and environmental leaders from 48 countries called for CO_2 emissions to be reduced by 20% below 1988 levels by 2005 (in the event, emissions from fossil fuel combustion rose by 36% during that period).

In the same year, the Intergovernmental Panel on Climate Change (IPCC) was formed by the World Meteorological Organization and the United Nations Environment Program (UNEP), to assess and synthesize the "scientific, technical and socio-economic information relevant to understanding the scientific basis of risk of human-induced climate change, its potential impacts and options for adaptation and mitigation." Six years later, in 1994, the UN Framework Convention on Climate Change (UNFCCC) came into effect with the objective to "stabilize greenhouse gas concentrations in the atmosphere at a level that would prevent dangerous anthropogenic interference with the climate system," with the first Conference of Parties (COP1) being held in Berlin, in 1995.

Despite the unequivocal conclusion of the 1988 Toronto conference, the need to get governments as well as scientists to sign up to the text meant that the IPCC's Second Assessment Report (SAR or AR2), published in 1995, was only able to conclude that "[t]he body of statistical evidence … when examined in the context of our physical understanding of the climate system, now points to a discernible human influence on global climate." A further decade would pass before the first commitment to action came into force in 2005.

The Kyoto Protocol, agreed in 1997 at COP3, committed the "Annex 1" group of 37 industrialized countries and economies in transition to reduce their GHG emissions, with an aggregate reduction of around 5% from a base year (in most cases 1990) during the first commitment period from 2005 to 2012. In the event, a reduction of around 7% was achieved, although this was largely the result of economic decline following the breakup of the former Soviet block and the recession following the 2008 global financial crisis, rather than the overdelivery of concerted mitigation efforts (these target and actual figures include the US and Canada which, respectively, failed to ratify the Protocol and withdrew 2 weeks before the end of the first commitment period (Olivier et al., 2011)).

By the time the second commitment period of the Kyoto Protocol (2012–20) was agreed at COP18 (The "Doha Climate Gateway"), the scope of the agreement was limited to just 15% of global CO_2 emissions, since Canada, Japan, Russia, Belarus, Ukraine, and New Zealand had withdrawn, the United States had not ratified, and major emitting developing countries like China, India, and Brazil were still not subject to any reduction targets.

If nothing else, the growth in global CO_2 emissions from 22.7 $GtCO_2$ in 1990, the Kyoto Protocol base year, to 29.3 in 2005, 34.5 in 2012, and 40.8 $GtCO_2$ in 2021, reflects the diversity of interests and challenges that have had to be addressed and included in moving from the scientific consensus—already strong in 1988—to the global, intergovernmental alignment that was eventually achieved at the Paris COP21 in 2015, and that must initiate the emissions decline to net-zero in the next (very) few years.

1.4 The climate challenge—warming limits and the carbon budget

The Paris Agreement aims to hold the increase in the global average temperature to well below 2°C above preindustrial levels and to pursue efforts to limit the temperature increase to 1.5°C, recognizing that this would significantly reduce the risks and impacts of climate change. Parties to the agreement take "common but differentiated responsibility" for achieving this goal, by committing to increasingly ambitious "nationally determined contributions" (NDCs), to be progressively ramped up on a 5-year reporting cycle.

Taken together, the initial (2015) NDCs implied a GMST warming of 2.6−3.1°C by 2100 (Rogelj et al., 2016a), falling well short of achieving the 2°C limit, let alone 1.5°C, although updated NDCs submitted before COP26 in 2021 did increase the likelihood that these goals might still be achieved (Ou et al., 2021; Meinshausen et al., 2022). Framing the climate challenge in terms of a GMST limit is, nevertheless, an important development, as it provides the basis for establishing a carbon budget—the maximum quantity of CO_2 equivalent (CO_2e) emissions that, for a given likelihood, will avoid exceedance of the temperature limit.

The future trajectory of anthropogenic emissions of CO_2 and other GHGs will be determined by a wide range of social, economic, technological, and political factors, and many scenarios have been published by the IPCC and others as the basis for future climate change predictions. Fig. 1.8 shows a variety of such emissions scenarios, including the IPCC's SRES scenarios (IPCC, 2000) and the Representative Concentration Pathways (RCPs) published in 2011 and used in the IPCC's AR5 cycle (Van Vuuren et al., 2011).

The initial application of the SRES scenarios tended to focus on stabilizing [CO_2] at a certain level (450 ppm, 550 ppm, etc.) but since the recognition at COP15 (Copenhagen, 2009) that a 2°C GMST limit was needed to prevent dangerous climate system changes, emissions scenarios have increasingly been presented in terms of GMST rather than [CO_2] or [CO_2e].

In 2021, IPCC AR6 concluded that GMST rise by 2100, relative to the period 1850 to 1900, is very likely (>90% likelihood) to exceed 2°C for all shared socioeconomic pathways (SSPs; see O'Neill et al., 2017) that deliver less emissions mitigation than implied by RCP2.6 (IPCC, 2021). Warming is also expected to continue beyond 2100 for all RCP (and SSP equivalent) scenarios except RCP2.6.

In the analysis conducted for AR5, the "strong, consistent, almost linear relationship between cumulative CO_2 emissions and projected global temperature change to the year 2100," depicted in Fig. 1.9, was observed both for the RCPs and for the wider set of mitigation scenarios (IPCC, 2014a). This relationship, which is the net result of a number of carbon cycle and climate feedbacks (Rogelj et al., 2016), provides the basis for associating any given warming level to 2100 with a carbon budget—a cumulative CO_2 emissions limit that takes account of non-CO_2 GHGs. Stated in this way, a carbon budget concisely captures the scale of the challenge if we are to avoid dangerous climate change but, due to the complexity of the climate system and the uncertainties around non-CO_2 drivers, they cannot give a precise constraint on how much CO_2 can be emitted for a given likelihood of avoiding any specific warming limit.

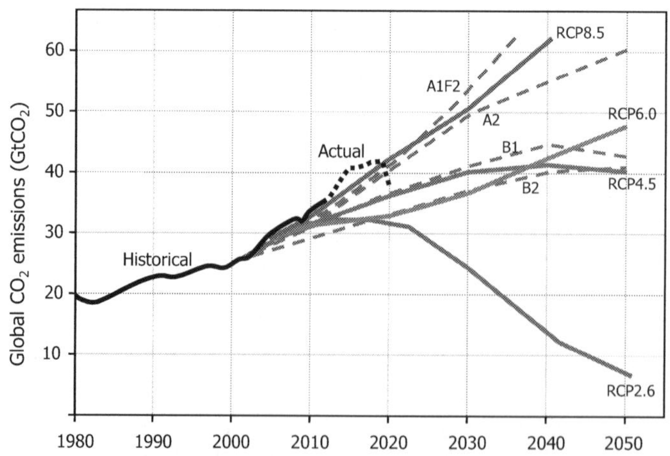

FIGURE 1.8 Future global CO_2 emissions scenarios to 2050. *After Peters et al. (2013).*

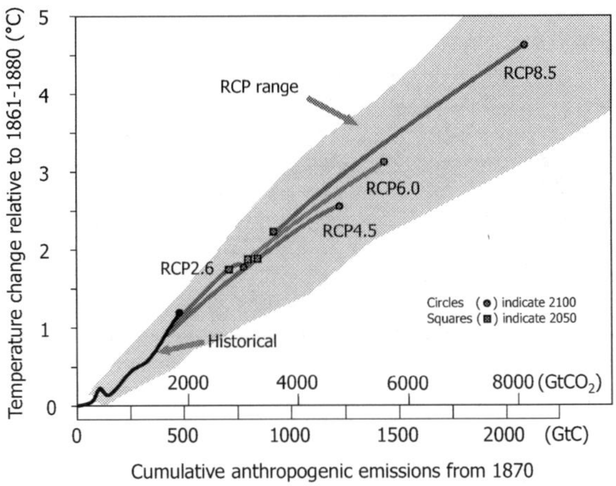

FIGURE 1.9 Global mean surface temperature rise versus cumulative CO_2 emissions. *After IPCC (2014b).*

Table 1.1 shows the cumulative emissions volumes consistent with the 1.5 and 2°C warming limits for a range of probabilities (IPCC, 2021).

The remaining budget (1150 $GtCO_2$) that gives >67% likelihood of staying below 2°C warming is around 28 years of annual emissions at recent levels (2015–20) and, assuming a linear decline, emissions would need to reach zero in the 2070s to stay within this budget. Although many different emissions pathways would result in the same cumulative volume, net emissions of all GHGs must reach zero at that point, and must remain at or below zero thereafter to avoid further warming. The following section briefly discusses the ways in which this dramatic transition might be achieved.

1.5 Responses to the climate challenge—getting to zero emissions

The fundamental drivers of rising GHG emissions since the start of the industrial revolution have been global population and GDP growth, coupled with a lack of strong economic or other incentives to drive industrial production toward higher energy and carbon efficiency. The various factors that relate these drivers to GHG emissions can be revealed using the Kaya identity (Peters et al., 2017);

$$C = G \times I_E \times I_C \tag{1.1}$$

where, G is the Gross Domestic Product (GDP) in constant prices, I_E is total primary energy use per unit GDP (i.e., the energy intensity of production), and I_C is the CO_2 emissions per unit energy use (i.e., the carbon intensity of primary energy). Here primary energy includes that derived from both fossil and nonfossil fuels. Population is not shown as a separate component; instead, the focus is on aggregated GDP. The carbon intensity of primary energy can be further decomposed as follows:

$$I_C = F_i \times F_s \tag{1.2}$$

TABLE 1.1 Cumulative CO_2 emission consistent with 1.5 and 2°C warming limits at different likelihood levels.

Net warming limit		<1.5°C			<2°C	
Likelihood of meeting the goal	>67%	>50%	>33%	>67%	>50%	>33%
Cumulative from 1850 ($GtCO_2$)	2790	2890	3040	3540	3740	4090
Remaining budget ($GtCO_2$)	400	500	650	1150	1350	1700

After IPCC (2021).

where F_i is the average carbon intensity of the fossil-fueled component of total primary energy and F_s is the share of fossil fuels in the primary energy supply (PES). These factors can be further broken down (Fig. 1.10) to provide a set of lower-level indicators that can be used to track the change needed to meet climate goals.

Approximate current values of the Level 2 and 3 factors are shown in Table 1.2.

Reducing emissions therefore requires a comprehensive transition in the global energy system, to reduce the carbon intensity of energy supply and the energy intensity of production, briefly reviewed in the following sections. This approach focuses on the energy system, which is currently responsible for over 70% of CO_2 emissions (Table 1.3).

Aside from energy production and end-use, similar transitions must substantially reduce GHG emissions from other major sources such as agriculture. By 2050, the world must be fed by zero emissions, sustainable agricultural system (Rockstrom et al., 2017), driven by financial mechanisms to incentivize carbon management in this sector, and by product labeling to encourage the necessary shift in consumer behavior. Energy labeling has been effective in nudging consumers toward more eco-friendly purchases, particularly for appliances and lighting, and carbon labeling could also have an important impact, for example, to encourage low-carbon food choices (Liu et al., 2016; Camilleri et al., 2019).

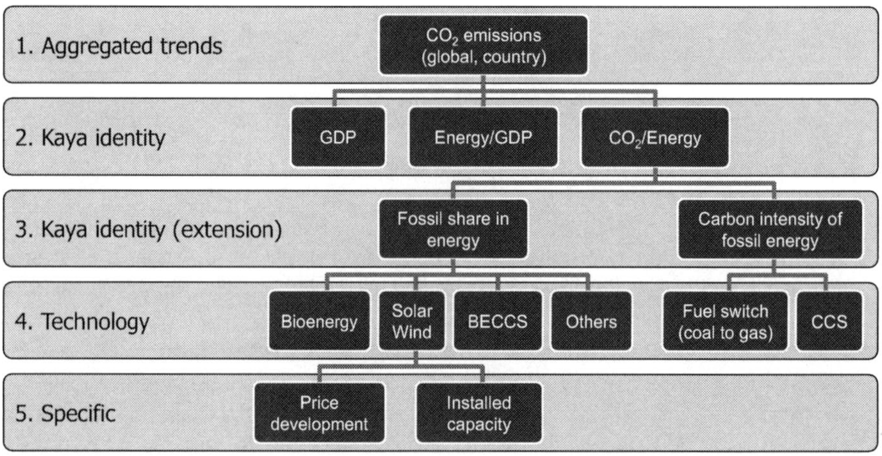

FIGURE 1.10 Hierarchy of indicators for tracking progress toward climate goals. *After Peters et al. (2017).*

TABLE 1.2 Approximate values of some Kaya identity factors.

Factor		Approximate current value	
Global GDP	G	$78\ 10^{12}$	US$ (nominal)
Total primary energy supply		$570\ 10^{9}$	GJ (570 EJ)
Energy intensity of production	I_E	7.3	MJ/US$
Global CO_2 emissions	C	38	$GtCO_2$
Carbon intensity of primary energy	I_C	67	$kgCO_2/GJ$
Carbon intensity of production	$I_E \times I_C$	0.49	$kgCO_2/\$$
Fossil share of primary energy	F_s	85	%
Carbon intensity of fossil energy	F_i	0.78	$kgCO_2/GJ$

Note; GDP expressed in terms of purchasing power parity is approximately $108 trillion, which, if used instead of nominal $, will reduce all "per dollar" factors by 28%. Final energy consumption is roughly 68% of primary supply, so all "per GJ" factors would be increased by 47% if expressed in terms of final energy consumption.

TABLE 1.3 Global CO_2 emissions; distribution by economic sector.

Sector	Emissions %	Subsector	Emissions %
Energy	73.2%	Energy use in industry	24.2%
		Energy use in buildings	17.5%
		Transport	16.2%
		Unallocated fuel combustion	7.8%
		Fugitive emissions	5.8%
		Agriculture and fisheries	1.7%
Agriculture, forestry and land use	18.4%	Livestock and manure	5.8%
		Agricultural soils'	4.1%
		Crop burning	3.5%
		Deforestation	2.2%
		Others	2.7%
Industry	5.2%	Cement	3.0%
		Chemicals	2.2%
Waste	3.2%	Landfills	1.9%
		Wastewater	1.3%

Data from ourworldindata.org/emissions-by-sector.

1.5.1 Carbon intensity of energy

Meeting the Paris GMST goals requires a rapid decrease in the carbon intensity of energy supply, as illustrated in Fig. 1.11 for a range of scenarios. The 1.5°C goal requires I_C to be halved before 2040, and no later than 2045 for the 2°C goal, and by around mid-century the global economy must be powered by carbon-free energy (Rockstrom et al., 2017).

The transformation of the global energy system that must deliver this reduction is under way, exemplified by the impressive growth in wind and solar renewable energy from 0.8% of PES in 2005 to 2.8% in 2015 and 5.7% in 2020 (Dale,

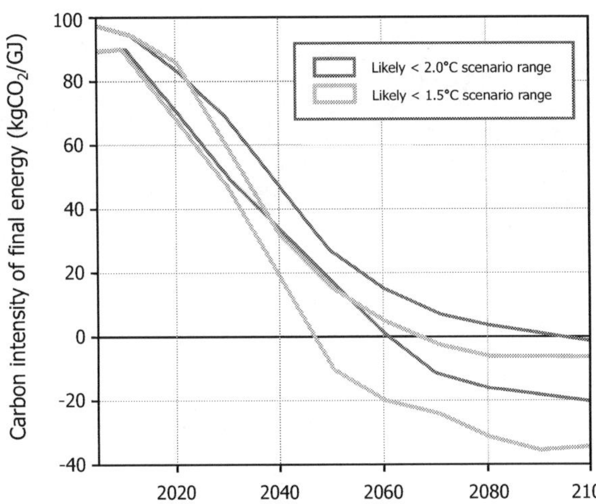

FIGURE 1.11 Decline in carbon intensity of final energy required to meet Paris GMST goals. *After Rogelj et al. (2015).*

2021), although further progress in this sector alone is unlikely to be sufficient to achieve the GSMT goals (Peters, 2016). Progress is required on many fronts, as shown in Table 1.4; rapid acceleration and, in many cases, a dramatic change in political will is needed during the 2020s if the window of opportunity to achieve the 1.5°C goal is to remain open.

Consider carbon pricing; the various forms of carbon pricing (e.g., tax, cap-and-trade) have been applied to some degree in many states, countries, and regions, and its importance as a driver of the energy transition, both to enable appropriate investment and to address issues of international equity and free-riding has been widely recognized (Nordhaus, 2015, 2020). However, this recognition has not prevented carbon pricing being used too often as a political football, to be picked up or kicked into touch at the whim of the party seeking or holding power. Of the 116 scenarios in the IPCC's AR5 assessment that achieve the 2°C goal, 76 assume a global carbon price was in place from 2010, and a further 24 scenarios assume this to have been the case from 2020 (Peters, 2016). Carbon prices will also have to increase significantly to achieve tighter warming targets. For example, IAM studies indicate that a 2–3 times higher carbon price is required in scenarios with a >67% chance of achieving the 1.5°C goal compared to those likely to achieve the 2°C goal (Rogelj et al., 2015).

Failure to agree on a global carbon pricing system, in whatever form, and to robustly implement it by the early 2020s significantly reduces the likelihood of delivering the Paris goals.

1.5.2 Energy intensity of production and other demand-side responses

Baseline scenarios (previously known as "business as usual" scenarios) in the IPCC's AR5 showed substantial growth in energy demand by 2050 in the key end-use sectors of industry, buildings, and transportation, which account for roughly 90% of final energy use, and for baseline CO_2 emissions from these sectors to grow by 50%–150% without substantial energy efficiency improvements (IPCC, 2014c).

As shown in Fig. 1.12, a reduction of over 50% has been seen in the US energy intensity of production over the last 4 decades, with a further threefold reduction forecast by 2050. Most scenarios that meet the 1.5°C goal assume both substantial demand reductions and also a decrease in the energy intensity of production at a faster pace than historically observed, as a result of stronger energy efficiency and other climate-driven policies (Rogelj et al., 2015).

Energy efficiency improvements across all sectors of the global economy are thus both necessary and likely to play a major part in achieving rapid emissions reduction, and many technical and structural mitigation measures are available, as summarized in Table 1.5.

TABLE 1.4 Actions to rapidly reduce the carbon intensity of energy supply.

I_C driver	Actions required
Reducing fossil share of energy supply (F_s)	Accelerate the elimination of fossil fuel subsidies ahead of the G7's 2025 target
	Establish global carbon pricing rising to \$400–\$500/tCO_2 by 2050.
	Double the renewable share of energy supply every 5–7 years (solar, wind, bioenergy, etc.)
	Develop storage and long-distance transmission systems to enable high renewables penetration
	Increase the use of low-, zero-carbon, and carbon-negative transportation fuels (electricity, natural gas, hydrogen, biofuels)
	Develop nonfossil-based fuels for shipping and aviation
Reducing carbon intensity of fossil-fueled energy supply (F_i)	Continue fuel switching from coal to gas, with rapid retirement of non-CCS capable coal capacity
	Increase efficiency of gas-powered generation plant (e.g., combined cycle, combined heat, and power)
	Accelerate deployment of CCS for remaining large fossil-fueled power generation plant

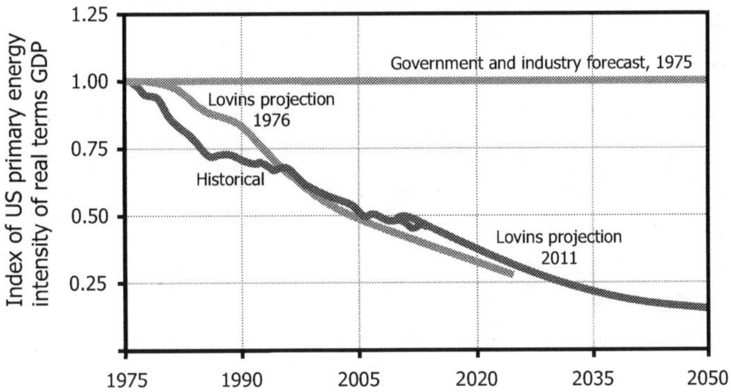

FIGURE 1.12 US primary energy intensity, 1975–2017, and projections. *After Lovins (2018).*

These changes often have significant cobenefits, such as in improved public health or productivity, and can in many cases be delivered at a lower cost than traditional systems. One such example, with very high energy saving and emissions reduction potential, is the application of integrated design processes in buildings—both new build and retrofit. Integrated design—bringing the skills and experience of mechanical and electrical engineers and other consultants to bear from the very beginning of the design process, and taking a full life-cycle view of the brief (i.e., building design, construction, and operation)—can result in energy savings in excess of 80% compared to a traditional baseline (Lovins, 2018). Integrative planning can also lead to large efficiency gains in other sectors, such as transport and industry.

Behavioral change is also needed to maximize the impact of technical and structural improvements. As Jamieson pointed out in the first paper to discuss the ethics of intentional climate modification (see Chapter 3), changing human behavior is often a more desirable response to a problem than intervening in physical systems. It is flexible, can adapt and if necessary reverse, through learning in the face of changing conditions (Jamieson, 1996).

Examples of behavioral change leading to reduced energy use include the following:

- Transportation: Journey avoidance, higher vehicle occupancy, use of public transportation
- Buildings: Thermostat setting, reduced appliance use, flexible use of shared space (hotdesking)
- Industry: Reduced demand for consumer products, including cars

Wide-ranging policies and incentives are needed to drive these technical, structural, and behavioral changes, including the following:

- Sectoral decarbonization targets, particularly at the national level
- Development and enforcement of stringent building codes defining minimum energy performance, at pace to avoid further lock-in
- Progressively strengthening minimum energy performance standards for domestic appliances and energy consuming devices
- Incentives for retrofit of high energy efficiency equipment, in both industrial and residential sectors
- Increasingly stringent fuel efficiency standards for the transport sector
- Grants and other economic instruments, information policies, and campaigns, including energy and carbon labeling, to inform and drive consumer choices
- Incentives to encourage journey avoidance and use of public transport.

Most integrated assessment studies show that the uptake of efficiency improvements, leading to reduced overall mitigation costs, will be most effectively driven by a combination of carbon pricing and other policies, including minimum performance standards/codes (Mundaca et al., 2018).

1.5.3 Closing the emissions gap

The foregoing has been, of necessity, no more than a cursory look at the key areas for mitigating the baseline growth in emissions that would otherwise occur over the coming decades, and extensive discussion of these and other opportunities can be found in the IPCCs assessment reports (IPCC, 2014a,c, 2021).

TABLE 1.5 Major technical efficiency and structural mitigation measures by sector.

	Measures to reduce energy intensity of production	
Sector	**Technical**	**Structural**
Transport	Fuel-efficient engine and vehicle design Advanced propulsion systems Use of lighter vehicle construction materials	Shift from light-duty vehicles to public transit, cycling/walking Shift from aviation and heavy-duty vehicles to rail Improved freight logistics Improved infrastructure planning
Buildings	Improved device efficiency, including advanced biomass stoves and lighting High-performance boilers and appliances Interconnectivity for demand leveling (smart appliances) Use of heat pumps, including ground or air loop systems	Integrated design approach Low- or zero-energy buildings Building automation and controls (e.g., smart heating controls) Urban planning District heating/cooling and CHP Smart meters/grids
Industry	Use of most energy-efficient systems (furnaces, boilers, electric motors) Improved electronic control systems Waste heat recovery Material efficiency; reduce losses and recycle/reuse waste material	Improve product-service efficiency More intensive use of products Greater durability for longer product life

After IPCC (2014c), Table TS.3.

Nevertheless, it is clear that comprehensive decarbonization of the energy system and improvements in end-use energy efficiency are essential if the Paris goals are to be achieved, and must proceed at an accelerating pace in the coming years. The IPCC's AR5 concluded already in 2014 that carbon dioxide removal (CDR, or negative emissions) would also be required to meet the 2°C limit if (as has been the case) stringent emissions reduction was further delayed (IPCC, 2014c). Since the NDCs pledged under the Paris Agreement do not foresee a peak in emissions before 2030 (Fig. 1.13) that "further delay" seems likely to be at least a decade.

Of the 400 scenarios reviewed for AR5 that were consistent with the 2°C limit, 86% relied on negative emissions, while the remainder assumed emissions had already peaked around 2010 (IPCC, 2014c). Going further, to achieve the 1.5°C warming limit, global CO_2 emissions must reach net-zero by around mid-century, with residual CO_2 emissions from "hard to decarbonize" sectors being offset by active measures to remove CO_2 from the atmosphere (Fig. 1.14).

The scale of negative emissions required to meet the 2°C climate goal has been estimated by the US National Academy of Sciences, Engineering and Medicine to be ~10 $GtCO_2$/yr globally by 2050 and ~20 $GtCO_2$/yr globally by 2100 (NASEM, 2019)—a scale equivalent to between 25% and 50% of current emissions. The potential for the various negative

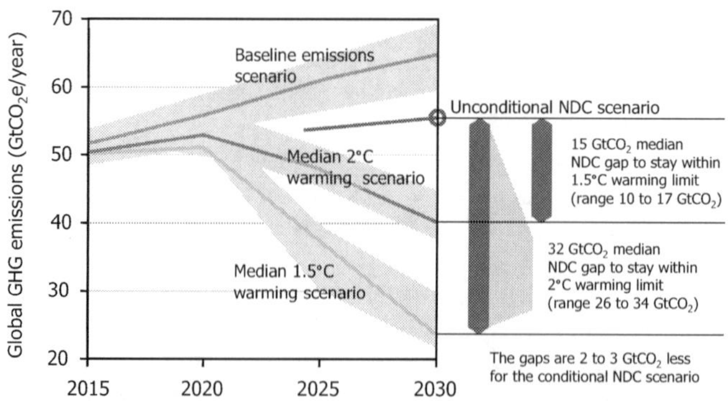

FIGURE 1.13 Nationally determined contributions gap ("The Emissions Gap") to meet Paris goals. *After UNEP (2018).*

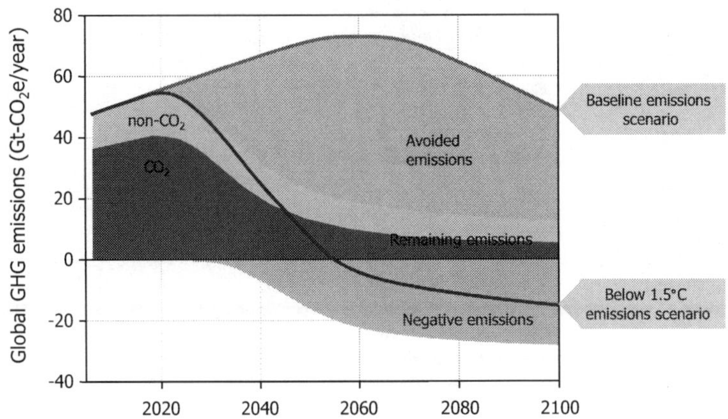

FIGURE 1.14 Schematic emissions scenario consistent with the 1.5°C warming limit.

emissions techniques and technologies to contribute to this need, and the shared and unique technical, social, and other challenges that each will face are discussed in the following chapters.

Although so-called deep mitigation pathways have been proposed that reduce the need for negative emissions, for example, through accelerated renewable energy growth and lifestyle changes (van Vuuren et al., 2018), recent history leaves little room for doubt that negative emissions, using a variety of techniques and technologies, deployed at a global scale, will be an essential part of the solution. Alongside these methods, other techniques to counter warming, such as stratospheric aerosol injection and marine cloud brightening, remain the subject of ongoing discussion (Lawrence et al., 2018), and in some cases are seeing renewed R&D activity (Irvine et al., 2019; Shyur et al., 2019). The optimal response to the climate challenge may ultimately include these techniques, alongside negative emissions and the stringent emissions reductions that are essential in all scenarios (Belaia, 2019).

1.6 Conclusion

The foregoing discussion has highlighted the delays, indecision, and vacillation that have occurred since the 1970s, in order to provide a realistic, perhaps somewhat skeptical, vantage point from which to consider the likely course and progress of international and national level efforts toward the Paris goals in the coming decades.

It is clear that emissions pathways consistent with the 1.5°C goal require net-zero CO_2 emissions globally by mid-century, and for global emissions to remain net negative through the second half of this century. This requires two transformations. Firstly, the carbon intensity of the energy and industrial landscape must be reduced to a level that is as low as reasonably practicable, consistent with achieving a low or lowest-cost overall outcome. Secondly, NETs must be developed and deployed at a global scale, creating a new industrial sector of a similar scale to the current fossil fuel sector.

While there is some degree of flexibility in choosing between mitigation options in pursuit of the 2°C goal, this flexibility is virtually eliminated in the more stringent case (Rogelj et al., 2015); these transformations are therefore both essential. Only if the first is successful will the second enable us to achieve the Paris goals, and avoid increasingly dangerous climate change. Against any emissions pathway resembling a "Business as usual" scenario, the impact of NETs would pale into insignificance. In the short term, ambitious emissions mitigation commitments must be made and implemented, alongside R&D and early deployment efforts to ensure that NETs are matured and available for rapid upscaling well before 2050.

References and resources

References

Arrhenius, S., 1896. On the influence of carbonic acid in the air upon the temperature of the ground. Philosoph. Magaz. J. Sci. 41 (251), 237–276.

Belaia, M., 2019. Optimal Climate Strategy with Mitigation, Carbon Removal, and Solar Geoengineering. arXiv preprint arXiv:1903.02043.

Callendar, G.S., 1938. The artificial production of carbon dioxide and its influence on temperature. Q. J. R. Meteorol. Soc. 64 (275), 223–240.
Camilleri, A.R., Larrick, R.P., Hossain, S., Patino-Echeverri, D., 2019. Consumers underestimate the emissions associated with food but are aided by labels. Nat. Clim. Change 9 (1), 53.
Dale, S., 2021. BP Statistical Review of World Energy. British Petroleum Company.
England, M.H., 1995. The age of water and ventilation timescales in a global ocean model. J. Phys. Oceanogr. 25 (11), 2756–2777.
Eriksson, E., Welander, P., 1956. On a mathematical model of the carbon cycle in nature. Tellus 8 (2), 154–175.
Foote, E., 1856. Circumstances affecting the heat of the sun's rays. Am. J. Sci. Arts 22 (66), 382–383.
Friedlingstein, P., Jones, M.W., O'Sullivan, M., et al., 2021. Global carbon budget 2021. Earth Syst. Sci. Data Discuss. 1–191.
Fuhrman, J., McJeon, H., Doney, S., Shobe, W., Clarens, A.F., 2019. From zero to hero? Why integrated assessment modeling of negative emissions technologies is hard and how we can do better. Front. Clim. 1, 11.
Gambhir, A., Butnar, I., Li, P.H., Smith, P., Strachan, N., 2019. A review of criticisms of integrated assessment models and proposed approaches to address these, through the lens of BECCS. Energies 12 (9), 1747.
Hansen, J., Johnson, D., Lacis, A., et al., 1981. Climate impact of increasing atmospheric carbon dioxide. Science 213 (4511), 957–966.
Hansen, J.E., Sato, M., Simons, L., et al., 2022. Global warming in the pipeline. arXiv preprint arXiv:2212.04474.
IPCC, 2000. In: Nakicenovic, N., et al. (Eds.), Special Report on Emissions Scenarios (SRES): A Special Report of Working Group III of the Intergovernmental Panel on Climate Change. IPCC, Geneva, Switzerland.
IPCC, 2014a. In: Core Writing Team, Pachauri, R.K., Meyer, L.A. (Eds.), Climate Change 2014: Synthesis Report. Contribution of Working Groups I, II and III to the Fifth Assessment Report of the Intergovernmental Panel on Climate Change. IPCC, Geneva, Switzerland.
IPCC, 2014b. In: Stocker, T., et al. (Eds.), Climate Change 2014: The Physical Science Basis. Working Group I Contribution to the Fifth Assessment Report of the Intergovernmental Panel on Climate Change. IPCC, Geneva, Switzerland.
IPCC, 2014c. In: Edenhofer, O.,R., et al. (Eds.), Climate Change 2014: Mitigation of Climate Change. Working Group III Contribution to the Fifth Assessment Report of the Intergovernmental Panel on Climate Change. IPCC, Geneva, Switzerland.
IPCC, 2021. In: Masson-Delmotte, V., Zhai, P., Pirani, A., et al. (Eds.), Climate Change 2021: The Physical Science Basis. Contribution of Working Group I to the Sixth Assessment Report of the Intergovernmental Panel on Climate Change. IPCC, Geneva, Switzerland.
Irvine, P., Emanuel, K., He, J., Horowitz, L.W., Vecchi, G., Keith, D., 2019. Halving warming with idealized solar geoengineering moderates key climate hazards. Nat. Clim. Change 1.
Jamieson, D., 1996. Ethics and intentional climate change. Clim. Change 33 (3), 323–336.
Kaplan, L.D., 1960. The influence of carbon dioxide variations on the atmospheric heat balance. Tellus 12 (2), 204–208.
Keeling, C.D., 1970. Is carbon dioxide from fossil fuel changing man's environment? Proc. Am. Phil. Soc. 114 (1), 10–17.
Khatiwala, S., Primeau, F., Holzer, M., 2012. Ventilation of the deep ocean constrained with tracer observations and implications for radiocarbon estimates of ideal mean age. Earth Planet Sci. Lett. 325, 116–125.
Lamperti, F., Dosi, G., Napoletano, M., Roventini, A., Sapio, A., 2018. Faraway, so close: coupled climate and economic dynamics in an agent-based integrated assessment model. Ecol. Econ. 150, 315–339.
Lawrence, M.G., Schäfer, S., Muri, H.J., et al., 2018. Evaluating climate geoengineering proposals in the context of the Paris Agreement temperature goals. Nat. Commun. 9 (1), 1–19.
Liu, T., Wang, Q., Su, B., 2016. A review of carbon labeling: standards, implementation, and impact. Renew. Sustain. Energy Rev. 53, 68–79.
Lovins, A.B., 2018. How big is the energy efficiency resource? Environ. Res. Lett. 13 (9), 090401.
Manabe, S., Wetherald, R.T., 1975. The effects of doubling the CO_2 concentration on the climate of a general circulation model. J. Atmos. Sci. 32 (1), 3–15.
Meinshausen, M., Lewis, J., McGlade, C., et al., 2022. Realization of Paris Agreement pledges may limit warming just below 2° C. Nature 604 (7905), 304–309.
Mundaca, L., Ürge-Vorsatz, D., Wilson, C., 2018. Demand-side approaches for limiting global warming to 1.5°C. Energy Effici. 2018, 1–20. https://doi.org/10.1007/s12053-018-9722-9.
NAS (National Academy of Sciences), 1979. Carbon Dioxide and Climate: A Scientific Assessment: Report of an Ad Hoc Study Group on Carbon Dioxide and Climate. National Academy of Sciences, Climate Research Board, Washington, DC.
NASEM (National Academies of Sciences, Engineering, and Medicine), 2019. Negative Emissions Technologies and Reliable Sequestration: A Research Agenda. The National Academies Press, Washington, DC. https://doi.org/10.17226/25259.
Nordhaus, W., 2015. Climate clubs: overcoming free-riding in international climate policy. Am. Econ. Rev. 105 (4), 1339–1370.
Nordhaus, W., 2020. The climate club: how to fix a failing global effort. Foreign Aff. 99, 10.
Oeschger, H., Siegenthaler, U., Schotterer, U., Gugelmann, A., 1975. A box diffusion model to study the carbon dioxide exchange in nature. Tellus 27 (2), 168–192.
Olivier, J.G., Janssens-Maenhout, G., Peters, J.A.H.W., Wilson, J., 2011. Long-term Trend in Global CO_2 Emissions: 2011 Report. PBL Netherlands Environmental Assessment Agency, The Hague, Netherlands.
Ou, Y., Iyer, G., Clarke, L., et al., 2021. Can updated climate pledges limit warming well below 2°C? Science 374 (6568), 693–695.
O'Neill, B.C., Kriegler, E., Ebi, K.L., et al., 2017. The roads ahead: narratives for shared socioeconomic pathways describing world futures in the 21st century. Global Environ. Change 42, 169–180.
Peters, G.P., Andrew, R.M., Boden, T., et al., 2013. The challenge to keep global warming below 2°C. Nat. Clim. Change 3 (1), 4–6.
Peters, G.P., 2016. The 'best available science' to inform 1.5°C policy choices. Nat. Clim. Change 6 (7), 646.

Peters, G.P., Andrew, R.M., Canadell, J.G., et al., 2017. Key indicators to track current progress and future ambition of the Paris Agreement. Nat. Clim. Change 7 (2), 118–122.

Plass, G.N., 1956. The carbon dioxide theory of climatic change. Tellus 8 (2), 140–154.

Prinn, R.G., 2013. Development and application of earth system models. Proc. Natl. Acad. Sci. U. S. A. 110 (Suppl. 1), 3673–3680.

Revelle, R., Suess, H.E., 1957. Carbon dioxide exchange between atmosphere and ocean and the question of an increase of atmospheric CO_2 during the past decades. Tellus 9 (1), 18–27.

Revelle, R., Broecker, W., Craig, H., Kneeling, C.D., Smagorinsky, J., 1965. Restoring the Quality of Our Environment: Report of the Environmental Pollution Panel. President's Science Advisory Committee. The White House, p. 317.

Rockström, J., Gaffney, O., Rogelj, J., Meinshausen, M., Nakicenovic, N., Schellnhuber, H.J., 2017. A roadmap for rapid decarbonization. Science 355 (6331), 1269–1271.

Rogelj, J., Luderer, G., Pietzcker, R.C., et al., 2015. Energy system transformations for limiting end-of-century warming to below 1.5°C. Nat. Clim. Change 5 (6), 519–527.

Rogelj, J., Schaeffer, M., Friedlingstein, P., et al., 2016. Differences between carbon budget estimates unravelled. Nat. Clim. Change 6 (3), 245–252.

Schneider, T., Teixeira, J., Bretherton, C.S., et al., 2017. Climate goals and computing the future of clouds. Nat. Clim. Change 7 (1), 3–5.

Shyur, Y., Dykema, J.A., Keith, D., Keutsch, F.N., 2019. An overview of the stratospheric controlled perturbation experiment (SCoPEx) concept of operations. AGUFM 2019, GC33G–1408.

UNEP, 2018. The Emissions Gap Report 2018. United Nations Environment Programme, Nairobi.

Van Vuuren, D.P., Edmonds, J., Kainuma, M., et al., 2011. The representative concentration pathways: an overview. Clim. Change 109 (1), 5–31.

Van Vuuren, D.P., Stehfest, E., Gernaat, D.E., et al., 2018. Alternative pathways to the 1.5°C target reduce the need for negative emission technologies. Nat. Clim. Change 8 (5), 391–397.

Webb, M.J., Lambert, F.H., Gregory, J.M., 2013. Origins of differences in climate sensitivity, forcing and feedback in climate models. Clim. Dynam. 40 (3–4), 677–707.

Workman, M., Dooley, K., Lomax, G., Maltby, J., Darch, G., 2020. Decision making in contexts of deep uncertainty-An alternative approach for long-term climate policy. Environ. Sci. Pol. 103, 77–84.

Further reading

Anderson, T.R., Hawkins, E., Jones, P.D., 2016. CO_2, the greenhouse effect and global warming: from the pioneering work of Arrhenius and Callendar to today's Earth System Models. Endeavour 40 (3), 178–187.

Carbon Brief Q&A - How do climate models work? www.carbonbrief.org/qa-how-do-climate-models-work.

Carbon Pricing Leadership Coalition, 2017. Report of the High-Level Commission on Carbon Prices. The World Bank. Available at: www.carbonpricingleadership.org/report-of-the-highlevel-commission-on-carbon-prices.

Cramton, P., MacKay, D.J., Ockenfels, A., Stoft, S., 2017. Global Carbon Pricing: The Path to Climate Cooperation. The MIT Press.

Dictionary of the Climate Debate; An Informative Educational Tool Concerning The Cast of Characters and Terminology Of Climate Change: www.odlt.org/dcd.

EASAC, 2018. Negative Emission Technologies: What Role in Meeting Paris Agreement Targets? European Academies Science Advisory Council. EASAC policy report 35. Available at: easac.eu/publications/details/easac-net/.

Energy Transitions Commissions, 2018. Mission Possible: Reaching Net-Zero Carbon Emissions From Harder-To-Abate Sectors By Mid-Century. Available at: www.energy-transitions.org/mission-possible.

Farmer, J.D., Hepburn, C., Ives, M.C., et al., 2019. Sensitive intervention points in the post-carbon transition. Science 364 (6436), 132–134.

Global Commission on the Economy and Climate, 2018. Unlocking the Inclusive Growth Story of the 21st Century: Accelerating Climate Action in Urgent Times. Available at: www.newclimateeconomy.report.

Keeling, C.D., 1998. Rewards and penalties of monitoring the Earth. Annu. Rev. Energy Environ. 23 (1), 25–82.

Kellogg, W.W., 1987. Mankind's impact on climate: the evolution of an awareness. Clim. Change 10 (2), 113–136.

O'Neill, B.C., Kriegler, E., Riahi, K., et al., 2014. A new scenario framework for climate change research: the concept of shared socioeconomic pathways. Clim. Change 122 (3), 387–400.

Pacala, S., Socolow, R., 2004. Stabilization wedges: solving the climate problem for the next 50 years with current technologies. Science 305 (5686), 968–972.

Randall, D.A., Bitz, C.M., Danabasoglu, G., et al., 2019. 100 years of earth system model development. Meteorol. Monogr. 59, 12.1–12.66.

Riahi, K., Van Vuuren, D.P., Kriegler, E., et al., 2017. The Shared Socioeconomic Pathways and their energy, land use, and greenhouse gas emissions implications: an overview. Global Environ. Change 42, 153–168.

Salawitch, R.J., Canty, T.P., 2017. Paris Climate Agreement: Beacon of Hope. Springer Climate. Springer International Publishing, Cham, Switzerland (Open access).

Summerhayes, C.P., 2019. Archibald geikie and the ice age controversy. Geol. Soci., London, Spec. Publicat. 480 (1), 183–190.

Van Beek, L., Hajer, M., Pelzer, P., van Vuuren, D., Cassen, C., 2020. Anticipating futures through models: the rise of Integrated Assessment Modelling in the climate science-policy interface since 1970. Global Environ. Change 65, 102191.

Weart, S.R., 2010. The idea of anthropogenic global climate change in the 20th century. Wiley Interdiscip. Rev.: Clim. Change 1 (1), 67–81.

Weart, S.R., 2010. The development of general circulation models of climate. Stud. Hist. Philos. Sci. B Stud. Hist. Philos. Mod. Phys. 41 (3), 208–217.

Projects and organizations

Carbon180; Formerly the Center for Carbon Removal, NGO championing carbon removal through accelerated research, smart practices, and effective policies: www.carbon180.org.

Carbon Dioxide Removal Model Intercomparison Project (CDRMIP); Kiel Earth Institute project exploring CDR, using a suite of Earth System Models www.kiel-earth-institute.de/CDR_Model_Intercomparison_Project.html.

Carbon-Price.com; website devoted to the issue of global carbon pricing: www.carbon-price.com.

Carbon Brief; Website Covering the Latest Developments in Climate Science, Climate and Energy Policy: www.carbonbrief.org.

Energy Transitions Commission; Global Energy Industries Leadership Group Aiming to Accelerate the Low-Carbon Transition, Enable Robust Economic Development and Limit Global Temperature Rise To Well Below 2°C: www.energy-transitions.org.

Earth system models for the future (ESM2025); EU Funded Project to Develop Next Generation ESMs https://cordis.europa.eu/project/id/101003536.

Global Carbon Project; Tracking Global Carbon, Methane and N_2O Budgets: www.globalcarbonproject.org.

ICOS; Integrated Carbon Observation System portal for European and global GHG Data: www.icos-cp.eu.

IPCC; Assessment reports and other special reports available at: www.ipcc.ch/reports.

WITCH (World Induced Technical Change Hybrid) Integrated Assessment Model; www.witchmodel.org.

World Climate Research Programme; Facilitating Analysis and Prediction of Earth System Change, to Determine the Predictability of Climate and the Impact of Human Activities on Climate: www.wcrp-climate.org.

Chapter 2

Overview of negative emissions technologies

Steve Rackley

2.1 Early discussion of negative emissions

The possibility of intentional human intervention to control the Earth's climate was first discussed by Kellogg and Schneider (1974), almost 2 decades after Revelle and Suess (1957) had remarked on the inadvertent geoengineering experiment that had been in progress since the industrial revolution. At a time when there were many knowledge gaps concerning the processes and feedbacks in the climate system—as indeed there still are—and when climate modeling was still in its infancy, the technical challenge of predictability was naturally seen as the most important problem to be overcome if such an intervention was "deemed a worthy objective." Kellogg and Schneider also anticipated some of the ethical issues that such an intervention would raise.

At least in part due to these and other ethical concerns (discussed further in the next chapter), the debate was for a long time consigned to quiet conversations "in the corridors of scientific meetings rather than in print" (Jamieson, 1996). Negative emissions were, to a degree, implicit in the IPCC's First Assessment Report (FAR) and the 1992 FAR Supplement (IPCC, 1990, 1992), with references being made to afforestation and to increasing the use of wood as a replacement for energy-intensive materials as potential land-based mitigation options. This became explicit in the Second Assessment Report (SAR) where H_2 production from biomass was discussed as a fuel decarbonization option, resulting in negative emissions if the CO_2 would be captured and sequestered (Ishitani and Johansson, 1996, §19.2.6.4).

By 2001 and the publication of the Third Assessment Report (TAR; IPCC, 2001) such options were seen as more viable, the Working Group III report anticipated the start-up of pilot CCS projects by 2010, with the possibility of CCS making a major contribution to mitigation by 2020 (Metz et al., 2001). Inclusion of CCS was expected to result in a significant reduction in the overall cost of mitigation when compared to scenarios that excluded this option. Bioenergy with carbon capture and storage (BECCS, although denoted BECS in the early literature) began to be widely researched and discussed at around this time.

The IPCC's TAR was presented to the UNFCCC Seventh Conference of Parties (Marrakech) in 2001, and the Conference requested the IPCC to prepare a special report on CCS, which was duly completed and formally presented to COP11 (Montreal) in December 2005. The Special Report on Carbon Capture and Storage (SRCCS;IPCC, 2005) drew further attention to negative emissions by noting that, if available at sufficient scale, BECCS might provide an opportunity to reduce the CO_2 concentration in the atmosphere. A decade later, COP21 (Paris) called for a special report on the impact of global warming of 1.5°C which, on publication in 2018, concluded that pathways that limit warming to 1.5°C would require carbon dioxide removal—putting negative emissions firmly under the spotlight (SR15;IPCC, 2018). The rapid acceleration of work since the release of SR15 is amply demonstrated by the fact that cumulative scholarly publications under the search term "negative emissions" doubled from 2018 to 2021.

The potential for negative emissions technologies (NETs) to address legacy emissions as well as future "hard to abate" emissions has also contributed to rapid progress since 2018. Microsoft's pledge in January 2020, to remove by 2050 all the company's legacy emissions since its founding in 1975, was a significant milestone, marking and perhaps contributing to a

Negative Emissions Technologies for Climate Change Mitigation. https://doi.org/10.1016/B978-0-12-819663-2.00001-0

period of rapid growth in the voluntary carbon market (which tripled in volume from ~80 $MtCO_2$ in 2018 to ~250 $MtCO_2$ in 2021) as well as in the broader NET ecosystem, stimulating an explosion in entrepreneurial activity, both in technology development and in essential support functions such as verification, registration, and market making.

2.2 Negative emissions: geoengineering or mitigation?

Whether carbon dioxide removal (CDR) by the application of NETs should be classified as mitigation—a term originally reserved for actions aimed at emissions reduction—or as geoengineering has been the subject of much debate. The IPCC's SAR (IPCC, 1995) denoted geoengineering approaches such as stratospheric aerosol injection (SAI), and solar radiation management (SRM) more generally, as "nontraditional" mitigation options, and included the development of nonfossil energy sources in the same nontraditional category. Indeed, the first mention of BECCS in the SAR was in the context of energy sector decarbonation, and thus fell into this nontraditional mitigation category.

The IPCC introduced an explicit definition of mitigation in AR3 (IPCC, 2001), as "a human intervention to reduce the sources or enhance the sinks of greenhouse gases (GHG)," a sink being any process or mechanism that results in a net removal of GHG from the atmosphere. This definition had been implicit in earlier reports and, as noted above, afforestation and other practices to increase soil carbon stocks had been identified as mitigation options in the FAR (IPCC, 1990, 1992).

AR3 also offered a definition of geoengineering as "efforts to stabilize the climate system by directly managing the energy balance of the Earth, thereby overcoming the enhanced greenhouse effect," and this was later modified, with AR5 (Pachauri et al., 2014) redefined geoengineering as "a broad set of methods and technologies that aim to deliberately alter the climate system in order to alleviate the impacts of climate change." CDR and SRM were there described as two distinct clusters of geoengineering technologies, but it was noted that "[t]he boundary between CDR and mitigation is not clear," with CDR methods potentially being considered as geoengineering or mitigation, depending on the scale and impact of the activities. However, as Minx and colleagues observed, CDR via negative emissions technologies is similar to traditional mitigation measures (per the IPCC AR3 definition) in that it seeks to address rising atmospheric CO_2 concentration ([CO_2]) as the immediate cause of climate change, unlike SRM and similar geoengineering interventions which aim to manage the consequences of rising [CO_2] and may therefore be more akin to adaptation (Minx et al., 2018).

Why does it matter? At the time of AR3, "geoengineering" had become "a label for technologically overreaching proposals that are omitted from serious consideration in climate assessments" (Keith, 2001). The working definitions of mitigation and geoengineering have evolved over the decades, and technological progress (alongside political procrastination) has resulted in serious consideration being given to an increasingly broad range of possible interventions. In addressing the climate crisis, including NETs in the mitigation toolkit provides flexibility in the search for the least expensive, and most just and equitable solution (e.g., in terms of land-use and other impacts), and also provides flexibility to manage risks, for example from unexpected Earth system responses. In the coming decades, public acceptance will be critical to enable the large-scale deployment of NETs, and will be influenced by the way these methods and technologies are framed; emphasizing how they relate to and, in many cases, enhance natural processes may encourage acceptance, while classing them alongside proposals that are seen as technologically overreaching is more likely to create a barrier (Wright et al., 2014).

Recent scholarship has also pointed out that, within the portfolio of negative emissions approaches, any distinction between natural and artificial processes is itself problematic: that what is "unnatural" is not inherently bad, that "natural" climate solutions also need to be framed in technological terms, and that employing this false dichotomy risks a premature narrowing of options. These topics are discussed further in the following chapter.

NETs are an essential component of the portfolio of mitigation approaches and should be collectively framed, in the public debate and in policymaking, to support the development of policies that enable progress in the near future on a broad front toward the optimal development and deployment of this portfolio, while addressing the many challenges and uncertainties (Lomax et al., 2015a).

2.3 Introduction to negative emissions technologies

The term "negative emissions technologies" has become the default title for the range of practices, methods, and indeed technologies that can deliver negative carbon—or more broadly GHG—emissions, although it seems a somewhat inappropriate designation when discussing approaches such as planting trees or increasing the input of leaf litter and manures to agricultural soils. Nevertheless, for the most part, this tile is adhered to throughout this book, with the occasional relapse into "methods" or "techniques" when these terms seem more appropriate. This section briefly introduces the many and varied approaches to negative emissions to provide a context for later chapters, including the detailed treatment of individual NETs in Chapters 10–16. The following discussion focuses on CO_2 emissions and CDR, but non-CO_2 GHGs are also discussed under direct air capture (below) and in Chapter 14.

2.3.1 The negative emissions landscape

Following the IPCC definition of mitigation as any human intervention to reduce the sources or enhance the sinks of greenhouse gases, mitigation through negative carbon emissions may be defined as; any human intervention to enhance the natural processes removing CO_2 from the atmosphere to long-term carbon pools or to create addition flux to these pools through new engineered processes. A carbon pool is any system that can accumulate or release carbon; the quantity of carbon in the pool at any given time is termed the stock, or inventory, and the rate of transfer of carbon between pools is the flux.

Based on this definition, NETs may be classified in terms of the type of removal process and the nature of the storage sink. Removal processes may either enhance a natural flux or engineer an incremental one, while storage sinks may be terrestrial, oceanic, geological, or "functional," (i.e., aggregated CO_2 utilization in enduring products), the natural and engineered subclasses also applying to terrestrial and oceanic sinks.

As shown in Table 2.1, NETs may link a single CO_2 removal process with either a single carbon pool or with multiple pools. For example, ocean fertilization enhances the ocean's biological pump, removing CO_2 from the atmosphere and ultimately storing some portion of the removed carbon in seabed sediments, while afforestation links the natural flux through photosynthesis with a number of potential pools, both natural (e.g., standing biomass and soil carbon) and engineered (e.g., BECCS using forest wastes, building with biomass, biomass burial, etc.).

2.3.2 Afforestation and other land- and soil-based methods

The largest fluxes in the global carbon cycle (Fig. 4.1) are those between the atmosphere and the terrestrial biosphere with around 123 GtC/yr being removed by photosynthesis, of which around 120 GtC/yr is returned to the atmosphere as a result of plant and microbial respiration, and various biotic and abiotic decay processes.

The net flux into terrestrial carbon pools (the "land sink," denoted S_{LAND}) averaged 3.1 ± 0.6 GtC/yr during the decade to 2020, removing one-third of the 9.5 ± 0.5 GtC/yr of fossil fuel emissions during that period (Friedlingstein et al., 2021). Two-thirds of this flux is estimated to occur in temperate and boreal latitudes with the remainder in the tropics (Houghton et al., 2018). Enhancing this drawdown or reducing the back-flux by just a few percent can therefore result in significant CDR, and is the focus of a number of negative emissions methods.

TABLE 2.1 Classification of negative emissions techniques and technologies.

Pool		Flux: Enhanced natural flux	Engineered incremental flux
Terrestrial	Natural	Afforestation/reforestation, cultivation practices, habitat restoration (incl. coastal blue carbon), degraded land restoration, enhanced weathering; biochar	Enhanced photosynthesis
	Engineered	BECCS (including marine biomass) with geological storage (including in situ mineralization and subsea sediment storage); biomass storage (including terrestrial burial);	Direct air capture (DACCS), with geological storage (including in situ mineralization and subsea sediment storage)
Oceanic	Natural	Ocean alkalinity enhancement by mineral dispersal; ocean fertilization; enhanced upwelling	CO_2 capture by accelerated limestone weathering (AWL); ocean alkalinity enhancement by electrochemical methods
	Engineered	BECCS with direct ocean injection or contained seabed storage; marine biomass with deep ocean burial	DACCS with ocean storage (direct injection, contained seabed storage, etc.)
Functional	Engineered	Building with timber and other biomass; other long-lived products	Artificial photosynthesis; concrete product curing

TABLE 2.2 Bioenergy, carbon capture, and storage options.

BECCS system component	Main option(s)	Alternative options
Biomass input	Fast-growing grasses such as switchgrass, miscanthus, and bamboo	Fast rotation poplar, rice husks, sugar beet bagasse, marine biomass (kelp), water hyacinths (Dyson, 1977)
Conversion process	Combustion	Gasification, pyrolysis, hydrothermal conversion, enzymatic conversion
Capture process	Absorption (liquid solvent)	Adsorption, membrane separation
Carbon storage	Geological storage	CO_2 utilization, ocean storage

Afforestation/reforestation

Afforestation and reforestation (AR)—the establishment of new forests on land that historically has not been forested and the reinstatement of previously forested land—was the first approach to be widely investigated as a negative emissions method, and was included in the scope of the Clean Development Mechanism (CDM) established in 1997 under the Kyoto Protocol.

Forests occupy around 30% of the Earth's land surface, covering a total of some 4 billion hectares (40 million km^2), but contain 77% of the total carbon inventory in terrestrial vegetation and around 40% of soil organic carbon (SOC) stocks to 1 m depth. Roughly one-third of the Earth's forests (some 1.9 Gha or 19 million km^2) have been lost since preindustrial times and land-use change (including deforestation, which averaged 10 Mha/yr from 2015 to 2020—down from 16 Mha/yr in the 1990s; FAO and UNEP, 2020) continues to contribute around 10% to global CO_2 emissions. Stopping deforestation and allowing the undisturbed regrowth of secondary forests is therefore an essential precursor to avoid undermining AR as a negative emissions method, and could itself yield negative emissions averaging 1.5 GtC/yr over the remainder of this century (Houghton and Nassikas, 2018).

Large-scale planting of trees and other fast-growing plants was first proposed by Freeman Dyson, who estimated that 5 GtC/yr could be removed from the atmosphere by a growing 7 million km^2 of new American Sycamore forest, yielding 7 tC/ha/yr (Dyson, 1977). More recent estimates suggest a global range from 2.8 to 5.5 tC/ha/yr (Griscom et al., 2017).

Carbon stocks in forest biomass can be managed in a number of ways; for forests with large standing biomass, protecting the existing forest can be an effective approach, provided that the risk of loss as a result of fire, disease, and windthrow can be minimized. Alternatively, where high yields can be achieved and biomass markets are readily accessible, forests can be managed as renewable energy crops (Hall et al., 1991; Marland and Marland, 1992). The long-term carbon uptake of forested land can be increased by careful harvesting, as this avoids saturation—the gradual decline in growth of the pool as the crop matures. Harvested timber may then be "sequestered" as a construction material in buildings, in other long-lived wood products, or buried in environments where it would not decompose, such as deserts or peatlands (Scholz and Hasse, 2008; Freeman et al., 2012).

Bioenergy with carbon capture and storage

The photosynthetic "capture" of atmospheric CO_2 as biomass, its release through a variety of conversion processes producing bioenergy or biofuels, and the recapture and storage of CO_2 emitted in the conversion process is the negative emissions approach that, after afforestation and reforestation, has been most commonly deployed in integrated climate models. Biomass for bioenergy production with carbon capture and storage (BECCS) encompasses a broad range of options for biomass input (both terrestrial and marine), conversion and capture process, and storage, as summarized in Table 2.2.

Although bioenergy accounts for around 13% of primary energy consumption (Popp et al., 2021), only one pilot project integrating biomass-fueled power generation with carbon capture has been commissioned to date, at the Drax Power Station in Yorkshire, UK. The capture of CO_2 from power generation and other flue gases has been widely demonstrated and deployed as a means of reducing industrial emissions, and these methods, discussed in detail in Chapters 7–9, are directly applicable to CO_2 capture from biomass combustion or other conversion processes.

Due to its low cost and ready availability, the Drax project used biomass derived from existing forests (wood pellets from trees and forest waste material). However, reaching the 2–10 $GtCO_2$/yr scale anticipated by 2050 in the IPCC's AR5 will most likely require the dedication of very substantial fertile land areas to the cultivation of bioenergy crops, and competition with food production for both land-use and water availability may become significant constraints.

Soil carbon sequestration

Soils are the largest terrestrial carbon pool, containing an estimated 2050 GtC of organic and 1550 GtC of inorganic carbon in the upper 2 m, for a total of ~3600 GtC (Lal, 2020). Recent studies suggest that between 30 and 60 GtC of soil C has been lost due to land-use change since the 1850s, and up to 130 GtC over the last 12,000 years (Sanderman et al., 2017). These stocks can be replenished and sustained through a range of measures (collectively termed soil carbon sequestration—SCS) such as increasing the incorporation of plant litter and residues, thermochemically converted biomass (biochar/hydrochar), manure, etc., and reducing soil disturbance through no-till or conservation tilling practices, among others. The effectiveness of such measures will gradually decline due to saturation effects, as a new higher soil−carbon equilibrium is approached, which can take between 10 and 100 years for different SCS measures, depending on soil type and climate zone. Estimates of the potential to restore soil carbon levels range from as low as 8 GtC (Sanderman et al., 2017) to as high as 178 GtC (Lal, 2020), well in excess of the estimated 12,000 year global soil carbon debt.

Permanence is an important consideration for SCS since the implemented measures have to be continued to sustain the new equilibrium and avoid losses that would result from a reversion to previous practices, although even if these methods are seen as temporary carbon storage they may still contribute to reducing peak warming (Matthews et al., 2021). Other potential issues with large-scale deployment include possible increased CH_4 emissions as a result of greater use of manures and increased N_2O emissions and pollution from run-off if nitrogen-based fertilizers and other nutrients are applied to complement increased soil carbon availability (Mosier et al., 1991).

Habitat restoration—peatlands and blue carbon

Many habitats including peatlands, tidal marshes, mangroves, macroalgal "forests," and sea grass meadows have been degraded over the last century due to human activities such as agriculture and urban development, as well as due to rapid environmental changes linked to climate. These ecosystems provide a range of services that support human wellbeing and livelihoods, often linked to biodiversity, and degradation in this context implies a decline in the ecosystem's ability to sustain these services. Habitat degradation is invariably linked to the loss of carbon stocks, both in soils and in vegetation, and the restoration of carbon stocks to native conditions is generally one of the objectives of ecological restoration efforts.

Declining carbon stocks in coastal vegetated ecosystems—tidal marshes, mangroves, and sea grass meadows—collectively termed "blue carbon," may be releasing as much as $1GtCO_2$/yr to the atmosphere (Pendleton et al., 2012), which could be substantially reduced through restoration and sustainable management efforts, including the restoration of hydrological conditions and reduction of anthropogenic nutrient inputs.

Peatlands comprise just 3% of the Earth's land area but contain a vast soil carbon store, estimated at 390−450 GtC, or 260−300 tC/ha (Freeman et al., 2012), as a result of the anoxic conditions which prevent microbial decomposition of organic matter (Fenner and Freeman, 2011). The drainage of peatlands for forestry and agricultural uses restores the conditions for microbial decomposition and can result in the loss of 50—60 tC/ha. Rewetting drained peatlands can "jump-start" the restoration of these ecosystems, with the potential to restore carbon stocks to former levels (Kareksela et al., 2015), although this may be accompanied by an offsetting increase in CH_4 emissions (Hemes et al., 2018).

Enhanced weathering

If atmospheric [CO_2] is not returned to equilibrium with the geosphere via large-scale NET deployment, this will be achieved over a millennial timescale by the natural geochemical process of rock weathering, which currently removes around 0.9 $GtCO_2$/yr (0.22−0.29 GtC/yr) from the atmosphere and transfers it, as dissolved inorganic carbon, into the ocean (Hartmann et al., 2009). Enhanced weathering (EW) accelerates this process by the application of alkaline rock particles, typically containing silicate minerals, to inland or coastal land surface. Dissolution of this finely ground material results in the uptake of CO_2 through carbonation reactions (Chapter 10). Dunite, an ultramafic rock (see Glossary) containing >90% olivine has a sequestration potential of 1.1 tCO_2/t-dunite is an ideal EW feedstock that has been widely studied, and is being used in early trials. Other suitably prepared alkaline materials such as recycled concrete, mine, and other industrial wastes could also be used, although trace metals present in the later materials may constrain acceptable addition rates. The most suitable locations for EW would be tropical regions, where the warm humid conditions favor mineral dissolution, or in irrigated arid or semiarid regions.

As well as releasing alkalinity, EW would also release a variety of plant nutrients, including silicon, phosphorous, and potassium, which could be beneficial in areas where the natural availability of these elements limits plant growth (Hartmann et al., 2013). EW would also alter fluvial and marine biogeochemical processes downstream of the weathering site, as a result of increased pH and nutrient supply, with as yet poorly understood consequences.

2.3.3 Direct air capture

The technologies developed for flue gas capture, and being deployed in BECCS projects, can also be applied to the more challenging removal of CO_2 from the air at a concentration of around 400 ppm, several hundred times more dilute than for industrial capture applications. In direct air capture (DAC), wind-blown ambient air passes through a contacting structure where CO_2 is either absorbed by a solvent flowing over packing material or adsorbed in a sorbent bed. In the former case, the solvent from multiple contactors is regenerated in a central facility while in the latter case the sorbent is regenerated by sealing each individual contactor and applying a pressure, temperature, or humidity cycle. These absorption and adsorption methods are discussed in detail in Chapters 7 and 8. A wide range of different solvent and sorbent systems are being investigated for DAC, with the two leading developers employing absorption, using a hydroxide solvent (Carbon Engineering), and adsorption, using an amine-based sorbent (Climeworks).

Also similar to flue gas capture, the energy requirement and source for solvent or sorbent regeneration is a major factor in determining the cost and CDR effectiveness of the technology (Azarabadi and Lackner, 2019), and the availability of industrial waste heat is a significant enabler.

The "CS" component of DACCS and BECCS—carbon storage or sequestration—is expected to be largely realized through the injection of supercritical CO_2 into geological rock formations, such as saline aquifers or depleted oil and gas fields, or into porous basalts. Geological storage is one of the few components of a fully integrated NET system that has been implemented commercially at greater than 1 $MtCO_2$/yr—the minimum size that would be required for individual NET facilities if the technology is to achieve $GtCO_2$/yr globally. Other storage options, such as ocean storage (see below) could also be integrated into a BECCS or DACCS system, for example, if economically accessible geological storage capacity were inadequate to meet the needs of aggregated capture sources.

Deployment of the first DAC facility by Climeworks was enabled by the sale of the CO_2 product to enhance food production in local greenhouses, and other CO_2 utilization options, such as the manufacture of carbon-neutral fuels, will be important enablers for further commercial-scale deployment.

The direct removal of non-CO_2 greenhouse gases, particularly CH_4 but also N_2O and various halogenated gases, can make a significant contribution to reducing GSMT rise, and a range of options have been proposed, including the enhancement of natural sinks and (photo)catalytic methods, as discussed in Chapter 14.

2.3.4 Ocean based methods

As well as holding a carbon inventory in the order of 50 times that of the atmosphere, the ocean sink (denoted S_{OCEAN}) has taken up around one-quarter of total historical CO_2 emissions, at a rate that averaged 2.8 ± 0.4 GtC/yr during the decade to 2020, some 30% of the 9.5 ± 0.5 GtC/yr fossil fuel emissions during that period (Friedlingstein et al., 2021). This uptake has been accompanied by a significant increase in ocean acidity, one of the multiple climate change-related stressors that are adversely affecting marine ecosystems. Despite lagging behind the land-based approaches outlined above, the potential for the ocean to play a major role in achieving negative emissions targets is increasingly being recognized, and a number of ocean-based methods are under development, many of which could also mitigate the effects of ocean acidification if eventually deployed at scale.

Ocean fertilization

Ocean fertilization (OF) is achieved by the intentional introduction of micronutrients, such as iron, or macronutrients, such as nitrogen or phosphorus, to the well-lit (euphotic) zone in the upper ocean to enhance marine photosynthetic productivity and, through the biological pump (see Chapter 6), increase carbon deposition in marine sediments. This is expected to be particularly effective in ocean areas that are high in macronutrients yet low in chlorophyll (so-called HNLC regions) as a result of an iron deficiency. Some 13 mesoscale (50–100 km^2) OF experiments were conducted in the 1990s and the early 2000s, but the benefit in terms of organic carbon export to the ocean floor was highly uncertain, being demonstrated conclusively in only two cases.

In 2010, the London Convention on the prevention of marine pollution adopted an Assessment Framework for OF research, to ensure that further activities were conducted following environmental assessment and with appropriate oversight. In 2013, amendments to the Convention adopted in Resolution LP4(8) prohibited OF and other marine geo-engineering activities unless conducted for scientific research and in accordance with the assessment framework. Although not yet in force due to slow acceptance by contracting parties to the Convention, these governance mechanisms have nevertheless had a significant impact on further work on OF and are frequently referenced as exemplifying elements of the governance framework that is needed for ocean-based methods more generally (Hubert, 2021).

Ocean alkalinity enhancement

Ocean alkalinity enhancement (OAE), also referred to as artificial ocean alkalinization (AOA), encompasses a number of approaches that aim to increase the oceanic CO_2 inventory through the addition of alkalinity (Renforth and Henderson, 2017). This can be achieved either by the addition of reactive alkalinity, such as lime (CaO), the hydroxides of calcium, magnesium, and sodium ($Ca(OH)_2$, $Mg(OH)_2$, NaOH), or alkaline minerals such as olivine, or by various CO_2 mineralization processes such as the accelerated weathering of limestone (AWL) and terrestrial EW that add nonreactive alkalinity (Renforth and Kruger, 2013). In the former case, increased alkalinity results in the conversion of dissolved CO_2 to bicarbonate and carbonate ions, increasing the CO_2 partial pressure gradient between the atmosphere and ocean surface waters, and driving an air-to-sea CO_2 flux, effecting CDR. In the case of unreactive alkalinity addition, CO_2 captured in an upstream process is simply stored in the ocean in the form of bicarbonate and carbonate ions, adding to the ocean's already immense inventory of dissolved inorganic carbon.

Electrochemical processes can also be used to generate alkaline and acidic product streams, and a range of options have been proposed to combine renewable energy with alkalinity addition to the surface ocean (Rau and Baird, 2018; de Lannoy et al., 2018), including in combination with the injection of acidity into the deep ocean (Tyka et al., 2022).

Macroalgae cultivation

Macroalgae—including kelps—form the most productive vegetated ecosystems on the planet, accounting for an estimated net primary production of 1.5 GtC/yr, equivalent to a CO_2 uptake of around 16 tCO_2/ha/yr (Krause-Jensen et al., 2018). A significant fraction of that production—estimated to be in the region of 170 MtC/yr—is transported offshore and sequestered through deposition in shelf and deep ocean sediments.

The cultivation of macroalgae to achieve negative emissions couples this exceptional photosynthetic productivity with a number of utilization or storage options, including use as a BECCS feedstock or storage by deep ocean burial, and can also reduce the adverse effects of ocean acidification and deoxygenation by locally increasing seawater pH and oxygen supply. Additional benefits of creating new macroalgal habitats would include improved coastal protection and increased food supply from the fish communities supported by these habitats.

Ocean storage

Alongside techniques such as OF and OAE which enhance the natural uptake of CO_2 by the oceans, a number of options have also been proposed to engineer ocean-based storage of CO_2 captured by BECCS and DACCS. Disposal of CO_2 in sinking ocean currents was first envisaged by Marchetti (1977), and other options since proposed include direct injection of CO_2 as a gas, hydrate, or liquid into intermediate water depths from 1000 to 3000 m, and ocean floor disposal below the CO_2 buoyancy depth ($\sim$3000 m) either into topographically confined pools, which may be open or contained by geo-membranes (Capron et al., 2013) or beneath ocean floor sediments (House et al., 2006).

Natural pools of liquid CO_2 contained below a hydrate layer have been observed around ocean-floor geothermal vents and provide an analog for such storage options. Small-scale experiments on dissolution and on the environmental impact of seafloor CO_2 pools have also been conducted, notably by the Monterey Bay Aquarium Research Institute (Barry et al., 2004, 2014), but larger-scale trials ($\sim$5–40 tCO_2) proposed in the early 2000s in Hawaii and Norway were unable to secure the necessary environmental permits.

2.4 NETs as a portfolio of options

The diverse portfolio of negative emissions techniques and technologies outlined above broadens the range of possible climate change mitigation responses, and is likely to play a major part in delivering the Paris GMST goals. Concerns that the promise of these new options, mostly as yet unproven at scale, may encourage a risky deferment of essential emissions reduction measures (i.e., mitigation deterrence) are discussed in the next chapter.

Large-scale NET deployment is expected to reduce costs and add flexibility in the global efforts to limit warming, but will also face significant challenges (discussed further in Chapter 3), including competition for land and water availability versus other global needs. Internal competition between NETs for renewable energy, CO_2 storage capacity, and alkaline mineral resources may also become important when deployment reaches multigigatonne climate impact scale.

The global industrial and agricultural transformation that will be needed to achieve negative emissions at that scale will also provide opportunities and challenges in relation to other global concerns such as biodiversity loss and other sustainable development goals, including the possibility of synergistic cobenefits for food production as a result of co-deployment with land-based NETs (Amann and Hartmann, 2019).

The current high level of uncertainty in achievable scale and costs (discussed later)—a consequence of the low level of deployment to date—means that it is too early to predict what the eventual mix and scale of deployed NETs will be, although it is likely that employing many technologies in parallel will optimize both CDR scale and cost performance while hedging the risks that would arise from reliance on just one or two (Peters, 2016; Fuss et al., 2018). Although integrated assessment models (IAMs) have typically considered only BECCS, and more recently DAC, in mitigation scenarios, work is now underway on ways to reflect the full breadth of NET options in such scenarios, using multicriteria decision analysis to define optimum portfolios for a given set of performance criteria.

Fig. 2.1 summarizes the framework developed by Rueda and colleagues to define optimized NET portfolios to deliver GSMT targets under the set of performance criteria shown in Table 2.3 (Rueda et al., 2021). Each technology that is a candidate for inclusion in the portfolio is scored (1—10) against these criteria, and these scores are combined using a weighting scheme that reflects portfolio preferences, for example, toward the lowest cost or lowest risk (Table 2.4), to achieve a single ranking score for each technology. Based on this ranking, the negative emissions demand profile in a particular IAM scenario can then be filled by NET deployment in rank order, with individual NET profiles defined by the feasible deployment growth rate, peak carbon removal rate, and ultimate potential. In this work, negative emissions demand profiles were defined to be in line with the shared socioeconomic pathways (SSPs; Rogelj et al., 2018). The optimal portfolio that emerges from this analysis is of course strongly dependent on the choice of performance criteria and weightings, which therefore need in-depth discussion with and endorsement by those making decisions based on the results (Wang et al., 2009). For example, Table 2.4 shows an alternative and more extensive set of performance indicators that could be used.

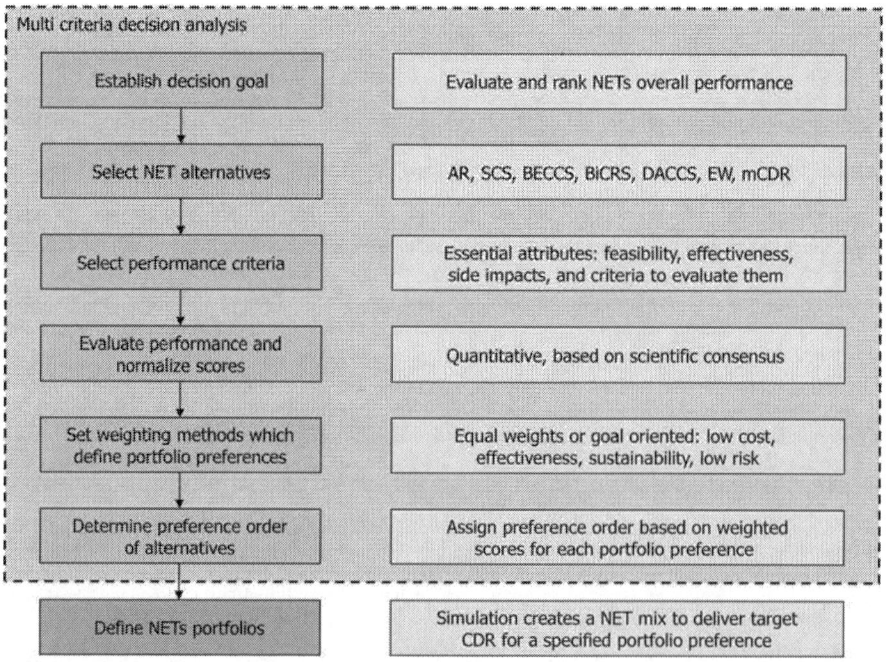

FIGURE 2.1 Decision framework to define NET portfolios. *From Rueda et al. (2021), with permission.*

TABLE 2.3 Assessment criteria and weighting schemes to define preferred NET portfolios.

		Weighting scheme for portfolio preferences		
Performance area	Assessment criteria	Economy	Effectiveness low risk	Sustainability
Feasibility	Technical feasibility	0.013	0.150	0.017
	Economic feasibility	0.900	0.150	0.017
	Governance feasibility	0.013	0.150	0.017
Effectiveness	Mitigation effect	0.013	0.150	0.017
	Timeliness	0.013	0.150	0.017
	Permanence	0.013	0.150	0.017
Side effects	Environmental impacts	0.013	0.033	0.300
	Economic impacts	0.013	0.033	0.300
	Social impacts	0.013	0.033	0.300

After Rueda et al. (2021).

TABLE 2.4 Alternative NET assessment criteria.

Performance area	Assessment criteria
Effectiveness	Global cooling potential, removal inertia, storage decay, storage maintenance requirements, albedo change, direct and indirect emissions.
Efficiency	Energy efficiency, resource intensity, cost
Scale	Technical potential, storage capacity, nonrivalrous versus deployment of other NETs, technical integration, juridical compatibility, market compatibility, acceptance
Risk	Diagnostic and prognostic uncertainty, investment risks, technical failure, tipping points
Synergies	Positive and negative synergies with other policy objectives, such as SDGs at the global level

After Fridahl et al. (2020).

Taking a global perspective, Rueda et al. found that optimized portfolios, primarily based on highly effective engineered NETs such as DAC and soil carbon sequestration, outperformed the default IAM reliance on BECCS. However, rather than being decided by a global level optimization, the NET portfolio that is implemented over this century will result from the interaction of policy at the state, national and regional level, with the innovation and investment from the growing ecosystem of start-ups and other companies pursuing these technologies. Different portfolios will emerge when different criteria, reflecting local constraints and preferences, are included in the ranking framework, and optimization studies at the appropriate level will be valuable aids to enable policy makers to identify and enable NET portfolios optimized for local conditions. National/regional resource constraints, current and future industrial and energy system capabilities (Creutzig et al., 2019), and synergies with other policies will come into play, and these assessments will need to be dynamic and flexible to avoid early lock-in and to accommodate future innovation and cost evolution. Effective international knowledge transfer mechanisms will help to ensure that local optimizations are based on the best available knowledge and experience (Peters, 2016).

Other tools originally developed in the financial sector are increasingly being applied to optimization problems in energy planning and climate response, including optimal (or modern) portfolio theory (Roques et al., 2010; deLlano-Paz et al., 2017), and real options theory (Fortin et al., 2008; Fuss et al., 2012; Gimbo et al., 2021), the later being primarily concerned with investment decisions under uncertainty which are particularly relevant for the emerging NET sector.

Among the performance criteria that are most likely to ensure an individual NET's role in an optimal state-level, national, or regional portfolio are its effectiveness, cost, and local potential, as discussed in the following sections.

2.4.1 Negative emissions effectiveness

A commonly used metric in comparing the relative effectiveness of different NETs is the CO_2e removal efficiency—the ratio of the net reduction in atmospheric CO_2 (i.e., CO_2 avoided) to the gross quantity of CO_2 removed, including the effect of any non-CO_2 GHG removals or emissions arising from the activity. Removal efficiency is determined by a variety of factors operating on global and local scales, including effects that are both generic and specific to individual technologies.

The latter, technology-specific, effects will be covered in Chapters 11–16; life cycle assessment here considers generic local factors, while global effects are discussed under carbon cycle and climate feedbacks.

Life cycle assessment

Life cycle assessment (LCA) is a tool to assess the impact of a system, product, or service on specified environmental or social issues, and originated during the 1960s in the energy industry. Concerns about the environmental and social impacts of large-scale biomass cultivation have driven growing interest in LCA in the context of bioenergy production, and it is particularly relevant for NETs in assessing CO_2e removal efficiency.

LCA commences with an inventory analysis stage in which the energy and resource requirements for the production of raw materials, use, and disposal of the subject system or product are determined, together with the emissions arising from each of these stages, within a predefined system boundary. Fig. 2.2 illustrates this schematically for a generic system or product. As shown in the figure, a system boundary may be drawn closely around a facility, considering only on-site emissions (gate-to-gate), it may extend back to include raw material production (cradle-to-gate), or it may also encompass the system end-of-life (cradle-to-grave). Fig. 2.3 shows a specific example for a biochar system with bioenergy production. Here, dashed arrows represent avoided processes and "T" indicated additional emissions from transportation.

Definition of the system boundary has important consequences for the conduct and conclusions of the analysis (Tillman et al., 1994). The most important boundary is that between the system and nature, because the analysis focuses on flows of energy and materials across this boundary, but other boundaries may also be relevant, such as an intersystem boundary between the production of a carbon-negative fuel and the production of the vehicles that would use it. A time boundary, or horizon, related to the requirement for storage permanence will also be relevant, particularly for those NETs where there is uncertainty regarding storage security. Geological storage for BECCS or DACCS is one such example (Alcalde et al., 2018), particularly if heavy reliance on these NETs leads to the availability of high-security storage sites becoming a constraint. In general, a range of both spatial and temporal boundary considerations apply to various NETs; for example, spatial mapping for AR, extended/international supply chains for biochar/BECCS and EW, rate of CO_2 capture, and sink saturation for AR/BECCS/Biochar and EW (Fajardy et al., 2019).

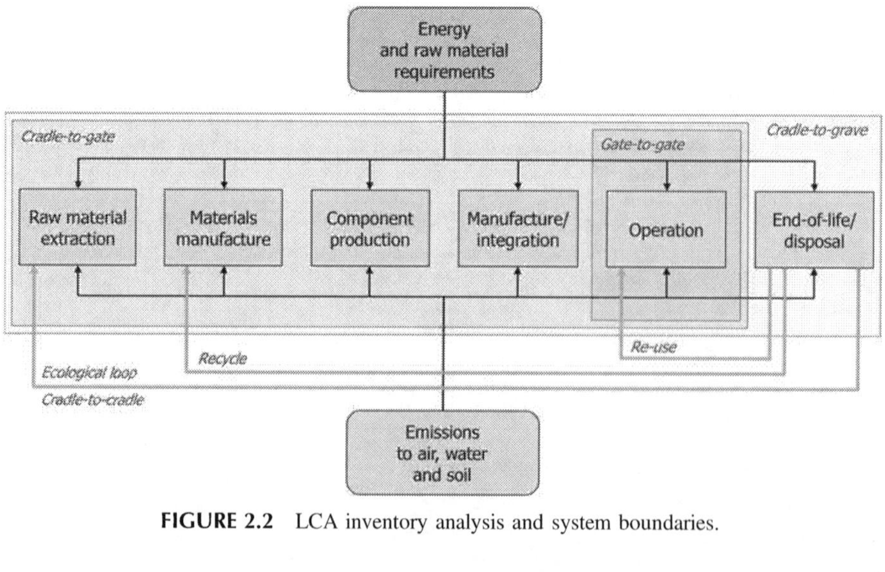

FIGURE 2.2 LCA inventory analysis and system boundaries.

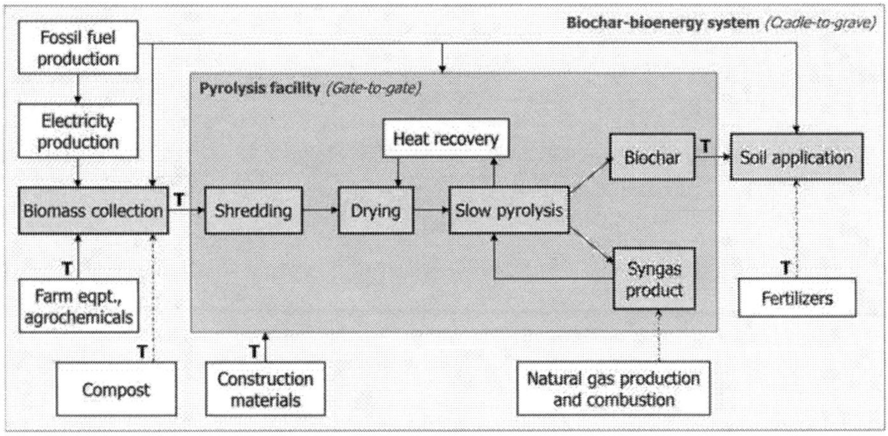

FIGURE 2.3 System boundary example for bioenergy-biochar system. *After Roberts et al. (2010).*

In applying LCA to assess CO_2e removal efficiency, the carbon—or strictly GHG—intensities of upstream and downstream activities will be the major focus of the data gathering and analysis, and these intensities will also be time dependent as the energy transition progresses and industrial processes are decarbonized.

While this initial inventory analysis stage will be sufficient for assessing the impact of the system's lifecycle on removal efficiency, a full LCA would then proceed to an impact assessment stage, in which the environmental, social, or other consequences of the input and output inventories are assessed. LCA is an effective tool to assess any given system configuration and can be used in combination with multiobjective optimization methods to perform life cycle optimization (LCO) with the aim of identifying alternative configurations that offer improved economic, environmental, and sustainability performance (Yue et al., 2016; Tan and Aviso, 2021).

The application of LCA to emerging technologies at the early development stages characteristic of many NETs is particularly challenging due to limited data availability, uncertainties around scale-up and deployment, and eventual functionality, and often an inability to make comparative checks due to the limited number of incumbents involved in a new technology (Terlouw et al., 2021). New frameworks are being developed to address these challenges although further research in this area is also needed (Bergerson et al., 2020; Sick et al., 2020; Thonemann et al., 2020).

Carbon cycle and climate feedbacks

The increase in atmospheric [CO_2] over the industrial era has been partially mitigated by the uptake of over half of all historic carbon emissions by the ocean and terrestrial ecosystems, through processes, discussed in Chapters 4–6, that have shifted these ecosystems toward equilibrium with rising atmospheric [CO_2]. If emissions reductions and NET deployments eventually reverse this [CO_2] trend, many of the processes that have buffered atmospheric [CO_2] increase will similarly buffer atmospheric [CO_2] decline through the release of CO_2 from systems that have reached or approached equilibrium with the higher [CO_2] atmosphere, reducing the effectiveness of NETs. This is illustrated schematically in Fig. 2.4, where the main carbon flows between the atmosphere, ocean, and terrestrial inventories are shown for conditions of positive and negative net emissions. Fig. 2.4B reflects global CO_2 removals primarily to geological storage via DACCS and BECCS, and the offsetting flows will depend on the deployed portfolio of NETs. Fluxes for some individual NETs are shown schematically in Fig. 2.5.

Although these feedbacks will not initially be noticeable, they will significantly reduce the overall CO_2 removal efficiency of NETs once these are deployed at impact scale (in the order of 10 GtC/yr). For example, Reith and colleagues found that when 70 GtC was removed by DAC over a period of 100 years, with storage by direct injection into the deep ocean, avoided emissions were 16%–30% lower than the gross removals as a result of these carbon cycle feedbacks (Reith et al., 2016). Similarly, in experiments designed to return [CO_2] from its 1998 value (348 ppm) to a pre-industrial 285 ppm by removing 135 GtC (i.e., 63 ppm reduction at 2.124 GtC/ppm) over 10 or 30 years, net removal was reduced by between 18 and 38 GtC (13%–28%), respectively (Vichi et al., 2013).

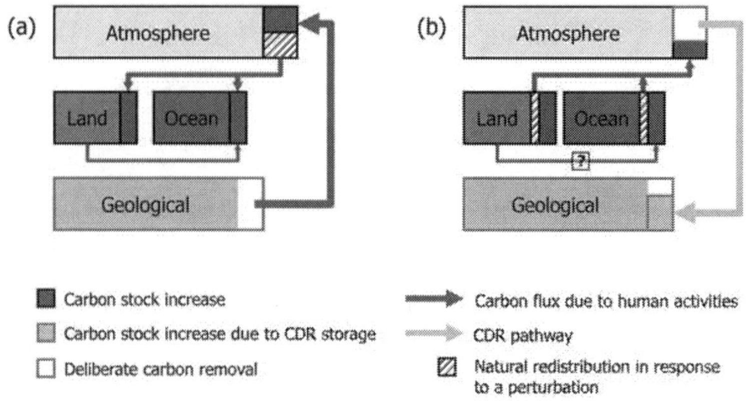

FIGURE 2.4 Main CO_2 fluxes under conditions of (A) positive and (B) negative net emissions. *After Keller et al. (2018).*

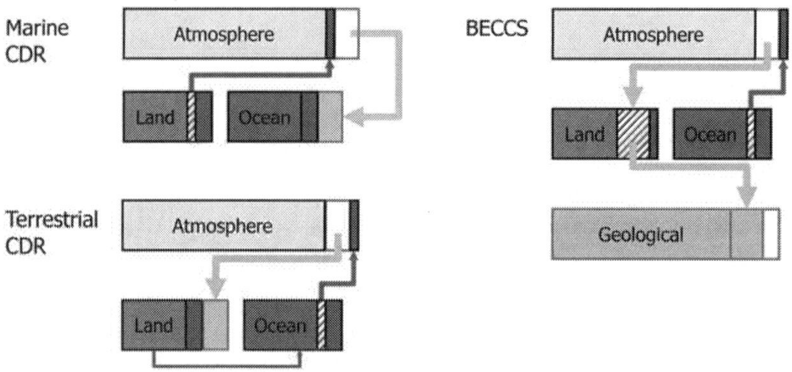

FIGURE 2.5 Schematic representation of the offsetting CO_2 fluxes from specific NETs. *After Keller et al. (2018).*

There are many uncertainties regarding how carbon cycle processes are responding to climate change and will in future respond to negative emissions, and improving our understanding of these effects and their inclusion in Earth system models are important research challenges toward an improved assessment of NET effectiveness (Rickels et al., 2018).

2.4.2 Negative emissions potentials and costs

An estimate of the scale of NET deployment required to achieve a given warming outcome can be provided by modeling studies although, given the many unknowns and uncertainties—Earth system, socioeconomic, and political—these estimates inevitably have very broad ranges. Pathways that limit warming to 1.5°C require CO_2 removals that range from 150 to 1180 $GtCO_2$ (40–320 GtC) during this century, reaching 5–15 $GtCO_2$/yr (1.4–4.1 GtC/yr) by 2050 (Fuss et al., 2018); for scale, recall that the average uptake of CO_2 by the global land sink over the decade to 2020 was 11 $GtCO_2$/yr (3.1 GtC/yr). Pathways can be constructed with a very low requirement for NETs in a 2°C scenario, and a substantially reduced requirement for 1.5°C (Van Vuuren et al., 2018), on the assumption of rapid decarbonization, reduced energy demand, and other societal and life-style changes, but most 2°C scenarios include CO_2 removals approaching those indicated for 1.5°C.

Many studies have addressed the challenge of estimating the potential for individual NETs to contribute to these immense volumes, the rate at which they could be deployed, and the costs of doing so, and recent comprehensive reviews have been reported by Fuss et al. (2018) and by the US National Academies (NASEM, 2019a, 2021). Table 2.5 and Fig. 2.6 summarize the likely ranges of achievable CO_2 removal rates and costs.

A rough benchmark for the magnitude of the biomass supply that would be required to achieve the 2050 CO_2 removal rate indicated for BECCS can be found in global grain production, which stood at 2.8 Gt/yr in 2021.

TABLE 2.5 CO_2 **removal rates and costs by 2050.**

NET	Gross CO_2 removal rate ($GtCO_2$/yr)	Cost range (\$/tCO_2)
Afforestation/Reforestation[a]	0.5–3.6	5–50
BECCS[a]	0.5–5	100–200
DACCS[a]	0.5–5	100–300
Enhanced weathering[a]	2–4	50–200
Biochar[a]	0.5–2	30–120
Soil carbon sequestration[a]	2–5	0–100
Ocean fertilization	0.1–1	<50
Seaweed cultivation	0.1–1	<100
Ocean alkalinity enhancement	0.1–1	100–150

[a]Data from Fuss et al., (2018) refer to rates and costs in 2050.
Data from Fuss et al. (2018) and NASEM (2021).

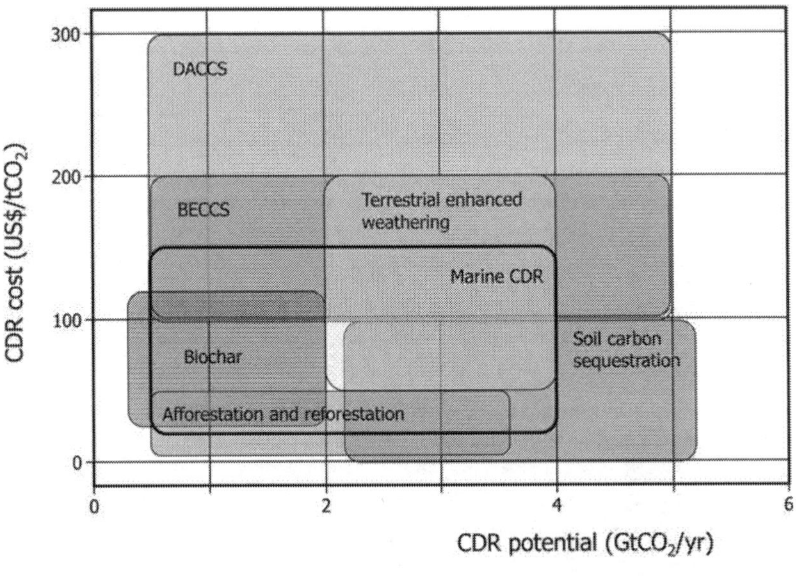

FIGURE 2.6 CO_2 removal rates and costs by 2050. *After Fuss et al. (2018).*

Current assessments of deployment potential and costs remain highly uncertain, in part because of the difficulty in estimating the impact of innovation and learning curve effects (Fuhrman et al., 2019), but it is clear that progressing the broadest possible range of NETs will increase the chance of delivering the 5–15 $GtCO_2$/yr removals expected to be needed by 2050 (Rau, 2019).

Growth rates of 15%–30% per year would be required over much of the 21st century to propel NETs to the multi-$GtCO_2$/yr scale, and sustained rates in this range have been observed historically in the growth of other energy technologies (Realmonte et al., 2019). New technologies typically generate economies of scale and grow either through an increase in unit numbers, for modular or commodity-type technologies, or an increase in unit capacities, for capital-intensive technologies involving large plant construction (Wilson, 2012). There is evidence that both these growth pathways will play a role in NET scale-up, particularly for DAC where two distinct sets of technologies—using either solvents or sorbents—are being progressed (see Chapter 13). For solvent-based technologies, economies of scale can be generated through the sharing of a central solvent regeneration plant, whereas for sorbent-based technologies, each sorbent bed needs to include stand-alone regeneration capability and economies will be generated through the commodity production of large numbers of such units. The scale-up plans of the two leading companies in this field provide some early evidence for this, with Carbon Engineering's solvent based system scaling in one step from a 1 tCO_2/d pilot to a 1$MtCO_2$/yr commercial plant—on the drawing board in 2021—while Climeworks' sorbent based system has scaled from a single 50 tCO_2/yr module to a 12-module (600 tCO_2/yr) demonstrator and to an 80-module (4000 tCO_2/yr) early commercial plant with the Orca project—starting operations in 2021.

2.5 The process of technology innovation

The maturation of any new technology toward commercial-scale deployment progresses through the stages of research, development, and demonstration (RD&D) stages, which can de characterized as follows;

Research—fundamental research and experimental proof of concept, led by academic or industrial research organizations; relatively low-cost exploration of a wide range of potential options; broad-brush estimate of eventual deployment costs and commercial viability.

Development—applied research focusing on process engineering/optimization and system integration; laboratory- and pilot-scale validation of the technology leading to an improved estimate of construction and operating cost, and economic assessment; additional fundamental research may be spawned as further implementation issues are identified.

Demonstration—near-commercial scale implementation to confirm expected performance, investigate process and design improvements, and test operating procedures, leading to a budget-level definition of construction and operating costs, and commercial assessment.

An example of specific RD&D tasks at each of these stages is shown in Table 2.6 for the CarbFix/CarbFix2 in situ CO_2 mineralization project (Gislason et al., 2018).

2.5.1 Technology readiness level classification

The progress of a new technology through the RD&D stages toward commercial deployment is commonly tracked using the Technology Readiness Level (TRL) scale, originally developed by NASA to classify the status of key mission components. Table 2.7 summarizes the TRL scale, using definitions based on those used by the EU Commission and the US Department of Energy (DoE), and including an indication of the current status (ca. 2023) of various NETs and NET system components, shown schematically in Fig. 2.7.

The TRL scale provides a common reference to assess development status that can be applied to all emerging technologies and is a useful tool for communication and risk management and to support decision-making in areas such as policy design and funding priorities.

TABLE 2.6 RD&D activities for CarbFix/CarbFix2 in situ CO_2 mineralization project.

Stage	Example RD&D tasks
Research (bench-scale testing)	Laboratory and field experiments to investigate basalt dissolution and reactivity
Development (pilot-scale testing)	Subsurface pilot injectivity test of CO_2-charged water to assess basalt dissolution and stable carbonate mineral precipitation
	Design and testing of dual feed (gas:water) injection system
	Design and execution of monitoring program (geochemical modeling, tracer testing, fluid sampling, and analysis)
	Recycle to applied research to address emerging issues (e.g., injection well biofouling)
Demonstration (field-scale testing)	Design and construction of field-scale plant (acid gas separation unit, injection facilities)
	Prepare field test site and install plant
	Perform field testing and technical evaluation at progressively increasing capacity
	Possible recycle to bench- or pilot-scale testing to resolve design, fabrication, or performance issues
	Test system integration options (e.g., with additional CO_2 supply from DAC)

TABLE 2.7 Technology readiness level definitions.

Technology readiness level	Definition and indication of scale	NET status (status ca. 2023)
TRL 1—concept	Basic principles observed and reported	Ocean storage (contained storage, subseabed sediments)
TRL 2—formulation	Technology concept and/or application formulated	MRV protocols for various land- and ocean-based methods; CO_2 marine transportation
TRL 3—proof of concept	Analytical or experimental proof of concept (e.g., bench scale pilot testing of key elements)	Ocean storage (direct injection)
TRL 4—lab prototype	Component and/or system validation in laboratory	Artificial photosynthesis
TRL 5—lab scale system	Lab-scale system validation in a relevant environment (e.g., continuous operation of a small scale pilot plant (<50 kWth, <100 tCO_2/yr))	Deep sea burial of macroalgae; OAE impact on various marine biota
TRL 6—pilot system	Pilot-scale system demonstrated in a relevant environment (e.g., continuous steady-state operation of (50 kW-1MW$_{th}$, 0.1–2 $ktCO_2$/yr) pilot at industrially relevant conditions)	BEC(CS) (Drax power plant, Yorkshire, UK. 0.4 $ktCO_2$/yr, capture without geological storage)
		DAC (liquid solvent; Carbon Engineering, 1 tCO_2/d pilot including fuel synthesis)
		Plot scale (10—100 m^2) assessment of biochar impact on crops
		OAE impact on coral biocalcification; coastal enhanced weathering
TRL 7—demonstration system	Prototype system demonstrated in an operational environment at precommercial scale (e.g., industrial pilot at >10 MW$_{th}$, > $ktCO_2$/yr)	DAC (enzyme enhanced solvent—CO_2 solutions, 30 tCO_2/d)
TRL 8—precommercial system	Complete system demonstrated and qualified in plant environment	DACCS (Climeworks, Hellisheiði, Iceland. 4000 tCO_2/yr capture with in-situ mineralization storage)
TRL 9—commercial system	Complete system operated over full range of expected conditions	Afforestation, reforestation; pipeline CO_2 transportation; CCS (Sleipner, Quest, Gorgon)

After Abanades et al. (2015).

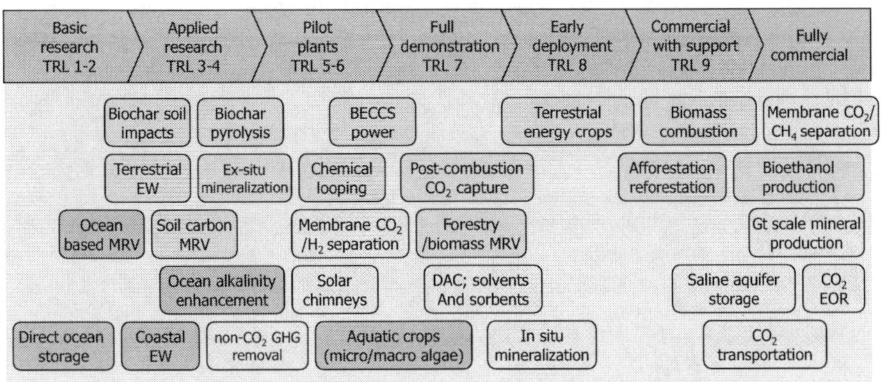

FIGURE 2.7 TRL status of various NETs and NET system components. *Adapted from Lomax et al. (2015b).*

2.5.2 The RD&D timescale

The timescale required to mature new technologies, illustrated in Fig. 2.8 for the CarbFix project and Fig. 2.9 for biomass carbon removal and storage (BiCRS), demonstrates the vision and sustained commitment required to bring such projects to commercial deployment.

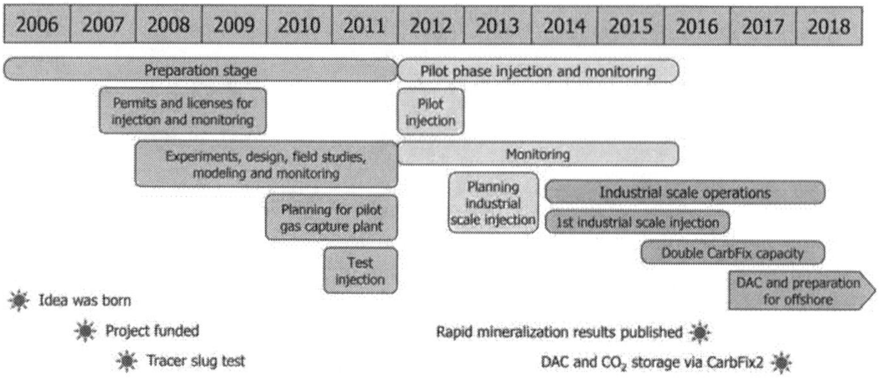

FIGURE 2.8 CarbFix technology RD&D timeline. *After Gislason et al. (2018).*

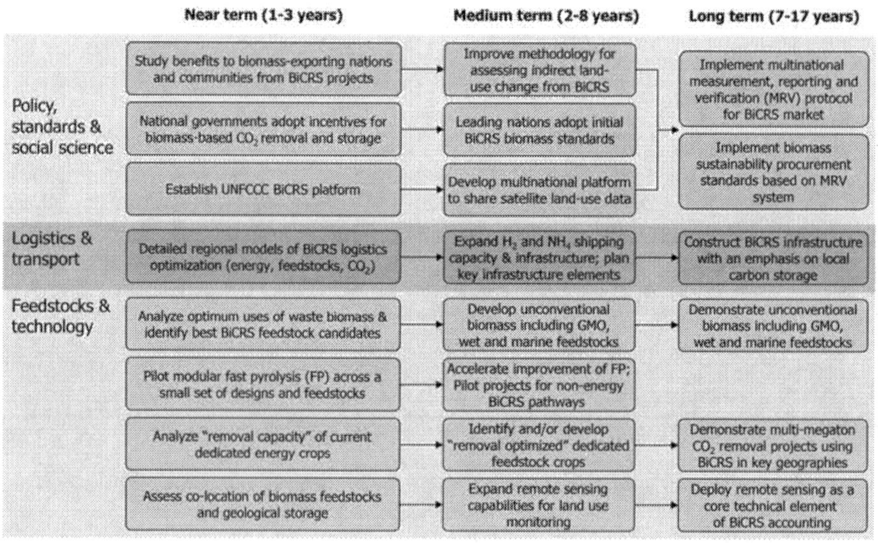

FIGURE 2.9 Innovation roadmap for BiCRS. *After Sandalow et al. (2021).*

The development of new technology is rarely as linear as that implied in these examples. In reality, recycle loops frequently occur, particularly from development back to research as additional investigations are needed to overcome unexpected obstacles; pilot projects may be shut-down prematurely for a variety of reasons; new regulations or policies may impact technical or commercial viability; economic conditions may lead to budget pressures and the curtailment or cancellation of work plans.

A technology may span several readiness stages at once, for example, being proven in one application or industry but not yet in another, and the integration of various elements that may individually be considered proven technology will often require an additional demonstration stage. While fundamental R&D work is generally directly transferable, the demonstration stage often has to be repeated to address location- or industry-specific aspects.

Typically, the costs involved increase as a technology advances through successive stages in the maturation process, and the availability of and competition for funding will be a major factor determining the rate of progress. Technical and commercial viability or, in the early stages, the prospect of that viability, are also key drivers of funding decisions. Other characteristics of the technology development process are shown in Table 2.8.

TABLE 2.8 Characteristics of the RD&D process.

Factor / Stage	Research	Development	Demonstration	Deployment
Diversity of options	Very wide	Wide	Narrow	Very narrow
Academic involvement	High	Moderate	Low	Low
Industry involvement	Variable	Moderate	High	High
Public engagement	Low	Moderate	High	High
Government involvement	General or targeted research funding	Innovation funding and other policy incentives	Project funding and other policy incentives	Market incentives
Costs	Low to moderate	Moderate	Moderate to high	High

Policy incentives, including innovation funding, play a major role in kick-starting the demonstration and early deployment phases of new energy-related technologies by enabling early projects that may be at best marginal from a commercial perspective. Continued deployment then invariably results in declining capital and operating costs, as operating improvements and economies of scale are realized.

2.6 Negative emissions RD&D agenda

It is clear from the foregoing discussion that deployment of a portfolio of NETs at a scale in the order of 10 $GtCO_2$/yr will be required in the second half of this century if the 1.5°C warming limit is not to be exceeded, and likely also for the 2°C limit. Rapid maturation of these technologies, from R&D through to demonstration projects at increasing scale, is essential to enable the large-scale deployment that needs to be well under way before 2050.

Research priorities for NETs have been discussed in a number of recent reviews and reports (Fuss et al., 2016; Minx et al., 2018; NASEM 2019a,b, 2021), addressing the main needs for specific technologies, as well as broader themes relevant to groups of NETs—such as land-based methods—or to the full NET portfolio. In this section, the current research priorities for the overall NET portfolio are summarized, while the research needs for specific NETs are reviewed in Chapters 11–16. The overarching knowledge gaps and research priorities fall into three broad categories: scale and impact, Earth system response, and governance and social license, as identified in Table 2.9.

NET research themes that address scale and impact have several parallels with the current research agenda in the agricultural sector, aimed at improving land-, water-, and nutrient-use efficiency, agricultural practices to maintain soil fertility and restore degraded soils, and harnessing the soil microbiome to improve productivity and resilience (NASEM, 2019b).

Policy frameworks, rooted in a deeper understanding of the necessary social, economic, and other changes, are an essential component of the enabling environment that will be needed to achieve rapid demonstration and deployment of NETs. Problems in this area, including discontinuities in government support, played a significant part in the under delivery to date of large-scale CCS projects—of the 77 projects that had been announced by 2010 barely one-third (26) were operating by 2021 (Martin-Roberts et al., 2021). NETs demonstration and deployment cannot afford another such "lost decade."

TABLE 2.9 NETs knowledge gaps and research priorities.

Research theme	Research priorities
Scale and impact	
Improved assessment of sustainable potentials	Further integration of biophysical, economic, and Earth system models together with regional bottom-up studies to improve estimates of sustainable potentials, particularly under resource constraints (land, water, energy, geological, and other storage capacity); improved understanding of future cost trajectories, including the impact of innovation
Reducing resource intensity and costs, and minimizing other negative impacts	Approaches to reducing resource utilization and monetary costs; improved understanding of the negative impacts of NETs (e.g., on water, biodiversity, SDGs); ways to minimize these; development of systemic indicators to assess both expected and unintended negative impacts of large-scale NET deployment (Fridahl et al., 2020).

Continued

TABLE 2.9 NETs knowledge gaps and research priorities.—cont'd

Research theme	Research priorities
Maximizing cobenefits	Potential for codeployment of land-based NETs to include synergies with and cobenefits for food production (Amann and Hartmann, 2019) and other SDGs; potential for ocean-based NETs cobenefits in countering ocean acidification.
Earth system response	
Global carbon cycle and its response to negative emissions	Improved understanding of the global carbon cycle; understanding and reducing modeling uncertainties and developing methodologies to evaluate and monitor the impact of NETs on the carbon cycle and climate system feedbacks (Schwinger and Tjiputra, 2018); incorporating the full portfolio of NETs into IAMs (Fuhrman et al., 2019); assessment of the differing responses to land- and ocean-based NET, to enable reliable accounting of NETs contribution to reducing atmospheric [CO_2] (Keller et al., 2018).
Carbon budget assessment	Further development and testing of Earth system models to narrow the range of uncertainty in estimating the remaining carbon budget linked to a specific GSMT goal.
Climate hysteresis and reversibility	Improved understanding of climate reversibility and hysteresis to guide longer-term global level decisions on mitigation (e.g., optimal postpeak [CO_2] trajectory to 2100 and beyond).
Governance and social license	
Governance frameworks	Governance systems to address deployment issues including cross-border accounting, monitoring and verification, legal and sustainability risks; sustainability standards for NET deployment; resolving trade-offs between NET impacts and climate impacts, and with SDGs.
MRV methods and protocols	Methods and protocols required including monitoring of impacts on food security, water resources, biodiversity, and other social and sustainability goals.
Policy frameworks	Policy mechanisms to foster rapid progress on R&D, demonstration, and deployment, including niche markets for early adoption; integration of ethical and social analysis, climate scenario modeling, and policy evaluation (Lenzi et al., 2018); deeper understanding of the social, economic, and other changes required to achieve the needed scale-up.
Social license	Improved understanding of factors influencing societal acceptance and preferences; governance approaches to address equity, justice, and human rights concerns; broader interdisciplinary and civil-society collaboration on mitigation assessments.

References and resources

References

Abanades, J.C., Arias, B., Lyngfelt, A., Mattisson, T., Wiley, D.E., Li, H., et al., 2015. Emerging CO_2 capture systems. Int. J. Greenhouse Gas Control 40, 126–166.

Alcalde, J., Flude, S., Wilkinson, M., Johnson, G., Edlmann, K., Bond, C.E., et al., 2018. Estimating geological CO_2 storage security to deliver on climate mitigation. Nat. commun. 9 (1), 2201.

Amann, T., Hartmann, J., 2019. Ideas and perspectives: synergies from co-deployment of negative emission technologies. Biogeosciences 16 (15), 2949–2960.

Azarabadi, H., Lackner, K.S., 2019. A sorbent-focused techno-economic analysis of direct air capture. Appl. Energy 250, 959–975.

Barry, J.P., Buck, K.R., Lovera, C.F., et al., 2004. Effects of direct ocean CO_2 injection on deep-sea meiofauna. J. Oceanogr. 60 (4), 759–766.

Barry, J.P., Lovera, C., Buck, K.R., et al., 2014. Use of a free ocean CO_2 enrichment (FOCE) system to evaluate the effects of ocean acidification on the foraging behavior of a deep-sea urchin. Environ. Sci. Technol. 48 (16), 9890–9897.

Bergerson, J.A., Brandt, A., Cresko, J., et al., 2020. Life cycle assessment of emerging technologies: evaluation techniques at different stages of market and technical maturity. J. Ind. Ecol. 24 (1), 11–25.

Capron, M.E., Stewart, J.R., Rowe, R.K., September 2013. Secure seafloor container CO_2 storage. In: 2013 OCEANS-San Diego. IEEE, pp. 1–8.

Creutzig, F., Breyer, C., Hilaire, J., Minx, J., Peters, G.P., Socolow, R., 2019. The mutual dependence of negative emission technologies and energy systems. Energy Environ. Sci. Energy Environ. Sci. 12, 1805–1817.

de Lannoy, C.F., Eisaman, M.D., Jose, A., et al., 2018. Indirect ocean capture of atmospheric CO_2: Part I. Prototype of a negative emissions technology. Int. J. Greenh. Gas Control 70, 243–253.

deLlano-Paz, F., Calvo-Silvosa, A., Antelo, S.I., Soares, I., 2017. Energy planning and modern portfolio theory: a review. Renew. Sustain. Energy Rev. 77, 636–651.

Dyson, F.J., 1977. Can we control the carbon dioxide in the atmosphere? Energy 2 (3), 287–291.

Fajardy, M., Patrizio, P., Daggash, H.A., Mac Dowell, N., 2019. Negative emissions: priorities for research and policy design. Front. Climate 1, 6.

FAO and UNEP, 2020. The State of the World's Forests 2020. Forests, Biodiversity and People. FAO, Rome, Italy.

Fenner, N., Freeman, C., 2011. Drought-induced carbon loss in peatlands. Nat. Geosci. 4 (12), 895–900.

Fortin, I., Fuss, S., Hlouskova, J., Khabarov, N., Obersteiner, M., Szolgayova, J., 2008. An integrated CVaR and real options approach to investments in the energy sector. J. Energy Mark. 1 (2), 61–86.

Freeman, C., Fenner, N., Shirsat, A.H., 2012. Peatland geoengineering: an alternative approach to terrestrial carbon sequestration. Phil. Trans. Math. Phys. Eng. Sci. 370 (1974), 4404–4421.

Friedlingstein, P., Jones, M.W., O'Sullivan, M., et al., 2021. Global carbon budget 2021. Earth Syst. Sci. Data Discuss. 1–191.

Fridahl, M., Hansson, A., Haikola, S., 2020. Towards indicators for a negative emissions climate stabilisation index: problems and prospects. Climate 8 (6), 75.

Fuhrman, J., McJeon, H., Doney, S.C., Shobe, W., Clarens, A.F., 2019. From zero to hero?: why integrated assessment modeling of negative emissions technologies is hard and how we can do better. Front. Climate 1, 11.

Fuss, S., Szolgayová, J., Khabarov, N., Obersteiner, M., 2012. Renewables and climate change mitigation: irreversible energy investment under uncertainty and portfolio effects. Energy Pol. 40, 59–68.

Fuss, S., Jones, C.D., Kraxner, F., et al., 2016. Research priorities for negative emissions. Environ. Res. Lett. 11 (11), 115007.

Fuss, S., Lamb, W.F., Callaghan, M.W., et al., 2018. Negative emissions—Part 2: costs, potentials and side effects. Environ. Res. Lett. 13 (6), 063002.

Ginbo, T., Di Corato, L., Hoffmann, R., 2021. Investing in climate change adaptation and mitigation: a methodological review of real-options studies. Ambio 50 (1), 229–241.

Gíslason, S.R., Sigurdardóttir, H., Aradóttir, E.S., Oelkers, E.H., 2018. A brief history of CarbFix: challenges and victories of the project's pilot phase. Energy Proc. 146, 103–114.

Griscom, B.W., Adams, J., Ellis, P.W., et al., 2017. Natural climate solutions. Proc. Natl. Acad. Sci. U. S. A. 114 (44), 11645–11650.

Hall, D.O., Mynick, H.E., Williams, R.H., 1991. Alternative roles for biomass in coping with greenhouse warming. Sci. Global Secur. 2 (2–3), 113–151.

Hartmann, J., Jansen, N., Dürr, H.H., Kempe, S., Köhler, P., 2009. Global CO_2-consumption by chemical weathering: what is the contribution of highly active weathering regions? Global Planet. Change 69 (4), 185–194.

Hartmann, J., West, A.J., Renforth, P., et al., 2013. Enhanced chemical weathering as a geoengineering strategy to reduce atmospheric carbon dioxide, supply nutrients, and mitigate ocean acidification. Rev. Geophys. 51 (2), 113–149.

Hemes, K.S., Chamberlain, S.D., Eichelmann, E., Knox, S.H., Baldocchi, D.D., 2018. A biogeochemical compromise: the high methane cost of sequestering carbon in restored wetlands. Geophys. Res. Lett. 45 (12), 6081–6091.

Houghton, R.A., Nassikas, A.A., 2018. Negative emissions from stopping deforestation and forest degradation, globally. Global Change Biol. 24 (1), 350–359.

Houghton, R.A., Baccini, A., Walker, W.S., 2018. Where is the residual terrestrial carbon sink? Global Change Biol. 24 (8), 3277–3279.

House, K.Z., Schrag, D.P., Harvey, C.F., Lackner, K.S., 2006. Permanent carbon dioxide storage in deep-sea sediments. Proc. Natl. Acad. Sci. U. S. A. 103 (33), 12291–12295.

Hubert, A.M., 2021. A code of conduct for responsible geoengineering research. Global Policy 12, 82–96.

IPCC, 1990. In: Houghton, J.T., Jenkins, G.J., Ephraums, J.J. (Eds.), Climate Change. The IPCC Scientific Assessment. Cambridge University Press, Cambridge, UK.

IPCC, 1992. In: Houghton, J.T., Callander, B.A., Varney, S.K. (Eds.), Climate Change 1992: The Supplementary Report to the IPCC Scientific Assessment. Cambridge University Press, Cambridge, UK.

IPCC, 2001. Climate Change 2001: The Scientific Basis. In: Houghton, J.T., Ding, Y., Griggs, D.J., et al. (Eds.), Contribution of Working Group I to the Third Assessment Report of the Intergovernmental Panel on Climate Change. Cambridge University Press, Cambridge.

IPCC, 1995. Climate Change 1995: The Science of Climate Change. Contribution of Working Group I to the Second Assessment Report of the Intergovernmental Panel on Climate Change. Cambridge University Press, Cambridge.

IPCC, 2005. In: Rubin, E., De Coninck, H. (Eds.), Special Report on Carbon Dioxide Capture and Storage. Cambridge University Press, Cambridge, UK.

IPCC, 2018. In: Masson-Delmotte, V.,P., Zhai, H.-O., Pörtner, D., et al. (Eds.), Special Report on the Impacts of Global Warming of 1.5°C. IPCC, Geneva, Switzerland.

Ishitani, H., Johansson, T.B., 1996. Energy supply mitigation options. In: Watson, R.T., Zinoyowera, M.C., Moss, R.H. (Eds.), Climate Change 1995: Impacts, Adaptations, and Mitigation of Climate Change: Scientific-Technical Analyses. Cambridge University Press, Cambridge, U.K.

Jamieson, D., 1996. Ethics and intentional climate change. Clim. Change 33, 323–336.

Kareksela, S., Haapalehto, T., Juutinen, R., Matilainen, R., Tahvanainen, T., Kotiaho, J.S., 2015. Fighting carbon loss of degraded peatlands by jump-starting ecosystem functioning with ecological restoration. Sci. Total Environ. 537, 268–276.

Keith, D.W., 2001. Geoengineering. Nature 409 (6818), 420.

Keller, D.P., Lenton, A., Littleton, E.W., Oschlies, A., Scott, V., Vaughan, N.E., 2018. The effects of carbon dioxide removal on the carbon cycle. Curr. Clim. Change Rep. 4 (3), 250–265.

Kellogg, W.W., Schneider, S.H., 1974. Climate stabilization: for better or for worse? Science 186 (4170), 1163–1172.

Krause-Jensen, D., Lavery, P., Serrano, O., Marbà, N., Masque, P., Duarte, C.M., 2018. Sequestration of macroalgal carbon: the elephant in the Blue Carbon room. Biol. Lett. 14 (6), 20180236.

Lal, R., 2020. Managing soils for negative feedback to climate change and positive impact on food and nutritional security. Soil Sci. Plant Nutr. 66 (1), 1–9.

Lenzi, D., Lamb, W.F., Hilaire, J., Kowarsch, M., Minx, J.C., 2018. Weigh the ethics of plans to mop up carbon dioxide. Nature 561 (7723), 303–305.

Lomax, G., Workman, M., Lenton, T., Shah, N., 2015a. Reframing the policy approach to greenhouse gas removal technologies. Energy. Pol. 78, 125–136.

Lomax, G., Lenton, T.M., Adeosun, A., Workman, M., 2015b. Investing in negative emissions. Nat. Clim. Change 5 (6), 498.

Marchetti, C., 1977. On geoengineering and the CO_2 problem. Clim. Change 1 (1), 59–68.

Marland, G., Marland, S., 1992. Should we store carbon in trees?. In: Natural Sinks of CO_2. Springer, Dordrecht, pp. 181–195.

Martin-Roberts, E., Scott, V., Flude, S., Johnson, G., Haszeldine, R.S., Gilfillan, S., 2021. Carbon capture and storage at the end of a lost decade. One Earth 4 (11), 1569–1584.

Matthews, H.D., Zickfeld, K., Dickau, M., et al., 2021. Temporary nature-based carbon removal can lower peak warming in a well-below 2°C scenario. Nat. Commun. Earth Environ. 3, 65.

Metz, B., Davidson, O., Swart, R., Pan, J. (Eds.), 2001. Climate Change 2001: Mitigation: Contribution of Working Group III to the Third Assessment Report of the Intergovernmental Panel on Climate Change, vol. 3. Cambridge University Press.

Minx, J.C., Lamb, W.F., Callaghan, M.W., et al., 2018. Negative emissions—Part 1: research landscape and synthesis. Environ. Res. Lett. 13 (6), 063001.

Mosier, A., Schimel, D., Valentine, D., Bronson, K., Parton, W., 1991. Methane and nitrous oxide fluxes in native, fertilized and cultivated grasslands. Nature 350 (6316), 330–332.

NASEM, 2019a. Negative Emissions Technologies and Reliable Sequestration: A Research Agenda. National Academies of Sciences, Engineering, and Medicine. The National Academies Press, Washington, DC.

NASEM, 2019b. Science Breakthroughs to Advance Food and Agricultural Research by 2030. National Academies of Sciences, Engineering, and Medicine. The National Academies Press, Washington, DC.

NASEM, 2021. A Research Strategy for Ocean-Based Carbon Dioxide Removal and Sequestration. National Academies of Sciences, Engineering, and Medicine. The National Academies Press, Washington, DC.

Pachauri, R.K., Allen, M.R., Barros, V.R., et al., 2014. Climate Change 2014: Synthesis Report. Contribution of Working Groups I, II and III to the Fifth Assessment Report of the Intergovernmental Panel on Climate Change. IPCC.

Pendleton, L., Donato, D.C., Murray, B.C., et al., 2012. Estimating global "blue carbon" emissions from conversion and degradation of vegetated coastal ecosystems. PLoS One 7 (9), e43542.

Peters, G.P., 2016. The 'best available science' to inform 1.5°C policy choices. Nat. Clim. Change 6 (7), 646–649.

Popp, J., Kovács, S., Oláh, J., Divéki, Z., Balázs, E., 2021. Bioeconomy: biomass and biomass-based energy supply and demand. N. Biotech. 60, 76–84.

Rau, G.H., Baird, J.R., 2018. Negative-CO_2-emissions ocean thermal energy conversion. Renew. Sustain. Energy Rev. 95, 265–272.

Rau, G.H., 2019. The race to remove CO_2 needs more contestants. Nat. Clim. Change 9 (4), 256-256.

Realmonte, G., Drouet, L., Gambhir, A., et al., 2019. An inter-model assessment of the role of direct air capture in deep mitigation pathways. Nat. Commun. 10 (1), 1–12.

Reith, F., Keller, D.P., Oschlies, A., 2016. Revisiting ocean carbon sequestration by direct injection: a global carbon budget perspective. Earth Syst. Dynam. 7, 797–812.

Renforth, P., Henderson, G., 2017. Assessing ocean alkalinity for carbon sequestration. Rev. Geophys. 55 (3), 636–674.

Renforth, P., Kruger, T., 2013. Coupling mineral carbonation and ocean liming. Energy Fuels 27 (8), 4199–4207.

Revelle, R., Suess, H.E., 1957. Carbon dioxide exchange between atmosphere and ocean and the question of an increase of atmospheric CO_2 during the past decades. Tellus 9 (1), 18–27.

Rickels, W., Reith, F., Keller, D., Oschlies, A., Quaas, M.F., 2018. Integrated assessment of carbon dioxide removal. Earth's Fut. 6 (3), 565–582.

Roberts, K.G., Gloy, B.A., Joseph, S., Scott, N.R., Lehmann, J., 2010. Life cycle assessment of biochar systems: estimating the energetic, economic, and climate change potential. Environ. Sci. Technol. 44 (2), 827–833.

Rogelj, J., Shindell, D., Jiang, K., et al., 2018. Mitigation pathways compatible with 1.5°C in the context of sustainable development. In: Masson-Delmotte, V., Zhai, P., Pörtner, H.-O., et al. (Eds.), Global Warming of 1.5°C. An IPCC Special Report on the Impacts of Global Warming of 1.5°C. Intergovernmental Panel on Climate Change.

Roques, F., Hiroux, C., Saguan, M., 2010. Optimal wind power deployment in Europe—a portfolio approach. Energy Pol. 38 (7), 3245–3256.

Rueda, O., Mogollón, J.M., Tukker, A., Scherer, L., 2021. Negative-emissions technology portfolios to meet the 1.5°C target. Global Environ. Change 67, 102238.

Sanderman, J., Hengl, T., Fiske, G.J., 2017. Soil carbon debt of 12,000 years of human land use. Proc. Natl. Acad. Sci. U. S. A. 114 (36), 9575–9580.

Scholz, F., Hasse, U., 2008. Permanent wood sequestration: the solution to the global carbon dioxide problem. Chem. Sus. Chem. 1 (5), 381–384.

Schwinger, J., Tjiputra, J., 2018. Ocean carbon cycle feedbacks under negative emissions. Geophys. Res. Lett 45 (10), 5062–5070.

Sick, V., Armstrong, K., Cooney, G., et al., 2020. The need for and path to harmonized life cycle assessment and techno-economic assessment for carbon dioxide capture and utilization. Energy Technol. 8 (11), 1901034.

Tan, R.R., Aviso, K.B., 2021. On life-cycle sustainability optimization of enhanced weathering systems. J. Clean. Prod. 125836.

Terlouw, T., Bauer, C., Rosa, L., Mazzotti, M., 2021. Life cycle assessment of carbon dioxide removal technologies: a critical review. Energy Environ. Sci. 14 (4), 1701–1721.

Thonemann, N., Schulte, A., Maga, D., 2020. How to conduct prospective life cycle assessment for emerging technologies? A systematic review and methodological guidance. Sustainability 12 (3), 1192.

Tillman, A.M., Ekvall, T., Baumann, H., Rydberg, T., 1994. Choice of system boundaries in life cycle assessment. J. Clean. Prod. 2 (1), 21–29.

Tyka, M.D., Van Arsdale, C., Platt, J.C., 2022. CO_2 capture by pumping surface acidity to the deep ocean. Energy Environ. Sci. 15, 786–798.

Van Vuuren, D.P., Stehfest, E., Gernaat, D.E., et al., 2018. Alternative pathways to the 1.5°C target reduce the need for negative emission technologies. Nat. Clim. Change 8 (5), 391–397.

Vichi, M., Navarra, A., Fogli, P.G., 2013. Adjustment of the natural ocean carbon cycle to negative emission rates. Clim. Change 118 (1), 105–118.

Wang, J.J., Jing, Y.Y., Zhang, C.F., Zhao, J.H., 2009. Review on multi-criteria decision analysis aid in sustainable energy decision-making. Renew. Sustain. Energy Rev. 13 (9), 2263–2278.

Wilson, C., 2012. Up-scaling, formative phases, and learning in the historical diffusion of energy technologies. Energy Pol. 50, 81–94.

Wright, M.J., Teagle, D.A., Feetham, P.M., 2014. A quantitative evaluation of the public response to climate engineering. Nat. Clim. Change 4 (2), 106–110.

Yue, D., Pandya, S., You, F., 2016. Integrating hybrid life cycle assessment with multiobjective optimization: a modeling framework. Environ. Sci. Technol. 50 (3), 1501–1509.

Further reading

Baumann, H., Tillman, A.-M., 2004. The Hitch Hiker's Guide to LCA - an Orientation in LCA Methodology and Application. Studentlitteratur, Lund.

Bullock, J.M., Aronson, J., Newton, A.C., Pywell, R.F., Rey-Benayas, J.M., 2011. Restoration of ecosystem services and biodiversity: conflicts and opportunities. Trends Ecol. Evol. 26 (10), 541–549.

Carton, W., Asiyanbi, A., Beck, S., Buck, H.J., Lund, J.F., 2020. Negative emissions and the long history of carbon removal. Wiley Interdis. Reviews: Clim. Change 11 (6), e671.

Faber, G., Mangin, C., Sick, V., 2021. Life Cycle and techno-economic assessment templates for emerging carbon management technologies. Front. Sustain. 2, 764057. https://doi.org/10.3389/frsus.

Hansen, J., Sato, M., Kharecha, P., Von Schuckmann, K., et al., 2017. Young people's burden: requirement of negative CO_2 emissions. Earth Syst. Dyn. 8 (3), 577–616.

IEAGHG, 2021. Assessing the Techno-Economic Performance, Opportunities, and Challenges of Mature and Nearly Mature Negative Emissions Technologies (NETs). IEAGHG. Technical Report 2021-04, December 2021.

Keller, D.P., Lenton, A., Scott, V., et al., 2018. The carbon dioxide removal model intercomparison project (CDRMIP): rationale and experimental protocol for CMIP6. Geosci. Model Dev. 11 (3), 1133–1160.

McDonald, T., Gann, G.D., Jonson, J., Dixon, K.W., 2016. International Standards for the Practice of Ecological Restoration–Including Principles and Key Concepts. Society for Ecological Restoration, Washington, DC, USA.

McManus, M.C., Taylor, C.M., 2018. Greenhouse gas balances of bioenergy systems: the role of life cycle assessment. In: Greenhouse Gas Balances of Bioenergy Systems. Academic Press, pp. 29–41.

Miller, J.R., Hobbs, R.J., 2007. Habitat restoration—do we know what we're doing? Restor. Ecol. 15 (3), 382–390.

Morrow, D.R., Buck, H.J., William, C.G., Nicholson, S., Turkaly, C., 2018. Why Talk about Carbon Removal? Carbon Removal Briefing Paper. Institute for Carbon Removal Law and Policy, American University, Washington, DC. (Available at). https://doi.org/10.17606/M6H66H.

Pörtner, H.O., Karl, D.M., Boyd, P.W., et al., 2014. Ocean systems. In: Climate Change 2014: Impacts, Adaptation, and Vulnerability. Part A: Global and Sectoral Aspects. Contribution of Working Group II to the Fifth Assessment Report of the Intergovernmental Panel on Climate Change. Cambridge University Press, Cambridge, UK, pp. 411–484.

Török, P., Helm, A., 2017. Ecological theory provides strong support for habitat restoration. Biol. Conserv. 206, 85–91.

Other resources

Climate Change Academy; on-Line Learning Resources on Climate Change cnd Cdr: climatechangeacademy.com/about.

Negative Emissions Platform; CDR Project Atlas: databasenegative-emissionsorg.webhosting.be.

World Climate Research Programme; Grand challenges: https://www.wcrp-climate.org/grand-challenges/grand-challenges-overview.

Chapter 3

Ethics, risks, and governance of NETs

Steve Rackley

3.1 Introduction

When the possibility of intentional human intervention to control the Earth's climate was first discussed by Kellogg and Schneider (1974), climate science was still in its infancy. Despite growing awareness, at least in scientific circles, of the real possibility of anthropogenic global warming and climatic change (Broecker, 1975), there was still a credible debate about whether the global average temperature decline of around 0.3°C observed from 1940 to the mid-1970s presaged a renewed period of glaciation. As well as highlighting the challenge of predictability, Kellogg and Schneider anticipated some of the ethical issues that would accompany any intentional climate intervention, such as the difficulty of arriving at a consensus "target" climate, and of equity, given that there would inevitably be regional winners and losers. In large part due to these concerns, the topic of the ethics of geoengineering was for two decades consigned to hushed break-time conversations in the corridors of scientific meetings, until the publication of Jamieson's seminal paper "Ethics and intentional climate change" (1996), the first attempt to take these issues head-on and to discuss under what conditions intentional climate intervention could be morally permissible.

This chapter introduces the major ethical issues that have been raised in relation to NETs, as well as the main uncertainties, risks, and appropriate risk-management strategies that arise both from NETs R&D and from their eventual large-scale deployment. Approaches to the governance of NETs, at the global, national, and lower levels, are then discussed. The perspective adopted here is what social scientists would call a "realist register" and assumes that (to some degree) uncertainties can be known and managed so that outcomes can be controlled by the decisions of rational agents and institutions, subject to economic, financial, and market forces. For some alternative perspectives the interested reader is referred to the work of Markusson and colleagues (2018).

3.2 Ethical concerns

Jamieson (1996) proposed four conditions that should be met for any intervention intended to mitigate climate change to be permissible:

1. The project is technically feasible
2. The consequences of implementation can be reliably predicted
3. Its implementation would result in an Earth-system state that is socioeconomically preferable to the alternatives (including do nothing or BAU)
4. Its implementing would not seriously and systematically violate any important, well-founded ethical principles or considerations.

This section addresses the last of these conditions, while the first three—technical and socioeconomic in nature—are dealt with below. Jamieson's hope that others would be stimulated to think through the ethics of geoengineering has been amply fulfilled, and considerable scholarly progress has been made over the last quarter century regarding the ethical aspects of intentional climate intervention.

The concerns summarized in Table 3.1 are applicable to a greater or lesser degree to all NETs, with some dependence on the specific technology under consideration and on the scale and stage of maturity of the activity (Preston, 2013; Cox et al., 2018). There are reasons to believe that some NETs may be less subject to ethical challenges, for example, those that

Negative Emissions Technologies for Climate Change Mitigation. https://doi.org/10.1016/B978-0-12-819663-2.00016-2

TABLE 3.1 Ethical concerns regarding negative emissions technologies.

Ethical concern	Description	Mitigation actions
Moral hazard (Preston, 2013; McLaren et al., 2019)	The promise of an unproven technological solution encourages risk-taking, specifically in reducing or delaying more politically difficult emissions reduction (mitigation deterrence), locking-in the need for the unproven solution and risking the exceedance of unpredictable climate tipping points.	Maintaining strictly separate targets for emissions reductions and removals. Lock-in risk is reduced by progressing a broad portfolio of CDR options.
Distributive justice (McLaren et al., 2019; Lawford-Smith and Currie, 2017)	How should the costs, risks, benefits, and potential harms be distributed? Technological solutions implemented by developed nations may perpetuate or deepen current injustices and inequities, including transferring risk both spatially and temporally.	Most concerns are land-based (e.g., land-grabbing, biodiversity loss); many ocean-based methods would impact, in most respects positively, on coastal communities.
Procedural justice (Lawford-Smith and Currie, 2017; Cox et al., 2018)	All those potentially affected should have a voice in decision-making, particularly regarding deployment, necessitating democratic, representative governance frameworks, spanning affected publics.	Increased transparency and inclusion in public consultations providing input to governance system development, at local/national and global levels.
Impact on human rights (Creutzig et al., 2013; Fuhrman et al., 2020)	Large-scale NET deployment would intensify land competition, with an impact on access to affordable food, water, and energy, particularly in the Global South.	NET-specific governance measures, depending on their resource intensity, e.g., AR (land), BECCS (land, water), DACCS (land, energy).
Hubris (Jamieson, 1996; Preston, 2013; Lenzi, 2018)	Unjustified confidence that technological solutions can solve sociobehavioral problems and will enable us to predictably control complex natural systems ("techno-fix"); concerns that starting research will inevitably lead to deployment ("slippery slope").	Mission-driven research to improve understanding of the complex natural processes impacted by NETs and strengthen confidence in eventual deployment decisions.
Tampering with nature (Gannon and Hulme, 2018; Corner et al., 2013)	For many stakeholders, notably indigenous groups, technological interference with natural processes is contrary to strongly held beliefs and values and is therefore morally unacceptable.	Some NETs can mitigate the ecological and environmental impacts of historical emissions, which have already tampered with nature.

enhance natural processes and may come with significant ecosystem cobenefits, while those requiring large areas of land (AR and BECCS) face particular concerns regarding their compatibility with sustainability, including issues of food security, land rights, and biodiversity (Shue, 2017).

3.2.1 Moral hazard and mitigation deterrence

The concept of a moral hazard originated in the insurance industry and denotes a situation where a person or company takes a risk that they might otherwise have avoided, because they will not bear the full consequences in the event of an adverse outcome—for example, because they are covered by insurance. In the climate context, the promise of future technological solutions presents a moral hazard if this encourages any delay or reduction in essential short-term efforts to reduce GHG emissions, as the full consequences of failure to deliver on these solutions would be borne by future generations, particularly those in more vulnerable locations. This possibility, termed "mitigation deterrence", has been at the heart of the ethical debate concerning negative-emissions technologies, and geoengineering more broadly, since its inception.

Against the background of the continued rise in global emissions, Anderson and Peters (2016) reiterated the concern that implementation delays in countries' mitigation plans were the result of undue faith that NETs would eventually solve the problem for them, describing NETs as "an unjust and high stakes gamble." This characterization drew a rapid response from a large group of climate scientists who contended that progressing NETs would strengthen the portfolio of options available to address climate change, and that "taking negative emissions technologies off the table would not make collective action any easier" (Lackner et al., 2016).

Contrary to Anderson and Peters' concern, Lawrence and Crutzen (2017) saw no clear evidence that mitigation deterrence or hubris were being manifested in the scientific literature and international climate policy sphere, and Merk and colleagues (2019) similarly found that expert proponents of geoengineering (specifically solar-radiation management) did not display moral hazard behavior when considering the allocation of research budgets across mitigation, adaptation, CDR, and geoengineering.

McLaren (2020), analyzing their role in IAMs, concludes that the failure of NETs to deliver, combined with possible indirect emissions arising from the pursuit of these technologies, could add 290–640 GtC to unabated emissions—potentially negating the benefits that any portfolio of NETs could deliver. The need for deep and rapid cuts in emissions is unquestioned in the climate science community and is the essential preamble to any discussion of NETs, which can only contribute to the achievement of climate targets if those reductions are also achieved. The possibility of mitigation deterrence is thus not only an ethical concern but is also a significant risk to the whole NET effort. Some approaches to managing this risk are touched on in Section 3.3.

3.2.2 Justice and equity

The difficulty of establishing consensus on the "ideal" climate was the first nontechnical issue raised by Kellogg and Schneider in their seminal 1974 paper, recognizing that the ideal climate for one region would likely have adverse consequences for another, and pointing to the challenges to procedural justice—the problem of participation and consent—posed by any intervention in the Earth system. Questions such as how to evaluate the impacts of NETs, whether or not to proceed with a particular technology, what entities should take charge of governance, and what targets or other criteria should be used to inform these decisions, all require stakeholder input (Preston, 2013). Kellogg and Schneider's challenge concerning the ideal climate also has a bearing on the eventual decision about when to stop CDR, and postimplementation issues, such as who bears responsibility for any continuing monitoring and maintenance requirement, also raise equity and justice concerns (Lawford-Smith and Currie, 2017). Just decision-making should include input from all those potentially affected and requires the establishment of new representative institutions tasked with developing the necessary governance frameworks at local, national, and international levels. The legitimacy of these frameworks will depend on the breadth and diversity of input and, as Lawford-Smith and Currie (2017) have noted, full legitimacy at the global level is probably unachievable as there is no institution for genuine global democracy at present, and no means of adequately representing the interests of future generations. Gardiner (2017) has proposed the establishment of a global constitutional convention to address this intergenerational justice challenge. Issues of equity and justice are also addressed in the UNESCO Declaration of Ethical Principles in relation to Climate Change (UNESCO, 2017), and related governance guidance such as the Oxford and Tollgate principles is discussed in Section 3.4.

Distributive justice and equity issues arise most clearly in the discussion of how the remaining emissions budget should be shared, as well as in discussing developed countries' carbon debt—the recognition that historical exploitation of the potential of the Earth system to absorb CO_2 has prevented developing and undeveloped countries from having a fair share of this common good. Gardiner (2010) provides an accessible introduction to the complex challenge of finding equitable answers to these questions. The possibility of large-scale CDR brings a new dimension to this discussion, as it affords countries the opportunity to address their historical carbon debts (in the same way that it has given companies such as Microsoft the opportunity to achieve life-of-venture net-zero by removing their legacy emissions). In IAM scenarios, CDR is typically distributed to minimize global costs and results in a significant burden falling on regions and countries, such as sub-Saharan Africa and India, that have contributed little to historical emissions. Under alternative, more equitable distribution mechanisms, based on cumulative per capita emissions (both past and future) or the ability to pay for CDR, this responsibility is shifted more to the United States in the former case, and predominantly to Europe and China in the latter (Fyson et al., 2020). While these countries and regional groups may choose to meet their obligations through financial transfers to other regions, appropriate governance would be required to ensure that risks and disbenefits are not also transferred. Equity can also be a political issue within developed nations, for example in New Zealand rural communities oppose measures that would expose them to the negative socioeconomic consequences of land-use change to balance continuing emissions arising primarily in urban areas (Schenuit et al., 2021).

At a project level, responsible management of issues related to justice and equity requires the active involvement of local stakeholders, and the establishment of a mechanism to ensure that local communities and others exposed to the risks and uncertainties inherent in the activity also share in the benefits generated. Such an approach is already being adopted in many pilot-scale NET projects and is well demonstrated in the social license activities of Project Vesta in the Dominican Republic (see *Projects and organizations* below).

3.2.3 Hubris and tampering with nature

Hubris—the excessive confidence in pursuing technological solutions to a sociobehavioral problem (fossil fuel addiction)—creates a moral hazard closely related to that arising from the promise of as yet unproved future solutions, in that the consequences of failure would largely fall on those who bore no responsibility for the pursuit of such a "techno-fix." Initial research into natural systems frequently—if not invariably—leads to a deeper appreciation of their complexity and to the identification of previously unknown unknowns so that, at the outset, there is a significant risk that interventions in these systems will prove not to be predictably controllable, with unknown consequences (thereby failing Jamieson's conditions 2 and 3—see above). A related concern, as Jamieson (1996) first pointed out, is that the momentum created by research work risks a slide into further development and deployment that may be inappropriate if downsides are not fully understood or assessed, or if there is excessive confidence that these will in turn yield to technological solutions—the so-called "slippery slope." Although these hazards can be controlled though the use of transparent, unbiased decision-making at each stage as a technology matures from research to implementation, governance systems that would impose this discipline are currently lacking.

All NETs modify natural processes and systems to some degree, and the perception that some amount to "tampering with nature," along with the belief that this is inherently a bad thing, are common features of public perception studies (Wolske et al., 2019). The fact that many "artificial" developments such as medical advances are generally considered "good" leads some ethicists to assert that this position is logically flawed (Lawford-Smith and Currie, 2017), and as Markussen (2022) has pointed out, "natural" climate solutions also need to be framed in technological terms—to be engineered, modellable within well-defined system boundaries, and tradable. Bellamy and Osaka (2020; see also Osaka et al., 2021) argue that the "natural" versus "artificial" dichotomy is a false one, simply reflecting the way solutions are framed rather than any substantive differences, and that the "natural" framing should be avoided as this risks a premature narrowing of options through policymaking decisions driven by perceived ease of public acceptance rather than effectiveness for climate restoration.

The "tampering with nature" concern has been taken further in relation to ocean fertilization (OF) by Hale and Dilling (2011), who ask "Is it permissible to remediate one pollutant by introducing a second pollutant into a system that has already been damaged, threatened, or altered?" They propose three additional moral claims that, taken together, would make a climate intervention impermissible, specifically:

- if it is **only** being undertaken to remediate the damage caused by a continuing action that is morally wrong (i.e., continued GHG emissions), and
- if it is not possible to secure the assent of **all** affected parties, which, they contend, is impossible for a global commons like the ocean, and
- if the intervention "moves the world to a **new, unknown state**," rather than returning the world to its original condition before the initial polluting act.

Hale and Dilling conclude that, as it fails all three of these tests, the implementation of OF is not morally permissible, although they argue (somewhat unconvincingly) that these rather stringent conditions would not exclude other NETs, such as DAC. A contrary position has been taken by Morrow and Svoboda (2016), who consider that, given our current nonideal situation with regard to atmospheric CO_2, any negative-emissions technique could be morally permissible if it is undertaken as a supplement to aggressive emissions reduction and provided that any adverse side effects are acceptably small when compared to the gains of justice and avoided injustice that would result.

Despite these professional perspectives, the perception of many NETs as "tampering with nature" is likely to remain a significant factor to be considered in public engagement.

3.3 Managing NET uncertainties and risks

At the heart of many of the ethical concerns discussed above is the potential for the adverse consequences of the many uncertainties to be inequitably borne by those who have contributed little to the causes of climate change and who have no opportunity to influence decisions on the development and deployment of mitigation options. This section briefly discusses the process of risk and uncertainty management and considers some aspects of risk mitigation related to NETs development and deployment.

3.3.1 Uncertainties and risks

Uncertainty is present when the occurrence or outcome of an event or process cannot be accurately predicted, and the research, development, and demonstration (RD&D) of NETs is subject to wide-ranging uncertainties, on every scale from the molecular to the global, some of which are listed in Table 3.2. The primary aim of RD&D is uncertainty reduction and learning from early deployment, although early research into complex systems invariably leads to an increased awareness of the knowledge gaps and an initial widening rather than a narrowing of uncertainty (Merk et al., 2019; Goldstein, 2021). Uncertainties, knowledge gaps, and research priorities for specific NETs will be discussed in subsequent chapters.

Uncertainties such as the optimal choice of feedstock for a specific project can be effectively reduced or eliminated during project development work while others, such as the pace of global emissions reductions or the carbon price in 2050, will be influenced by a wide range of factors and actors, and uncertainty will only be reduced with the passage of time.

Risk arises from uncertainty if undesirable consequences could follow from it and is defined as the "product" of the likelihood of occurrence of the uncertain event or outcome and the severity of its consequences. Likelihood and severity together determine the risk level or criticality and guide the effort to both reduce the uncertainty and manage the risk.

3.3.2 The process of risk management

Risk management has a significant role to play in addressing climate change, at all levels from individual projects up to multilateral governance. This section describes a typical risk-management process, as shown schematically in Fig. 3.1.

The identification of risks and the assessment of their potential consequences are the first steps in the process. The output of the risk-identification step is a clear risk statement that reflects both the uncertain outcome and the nature of the consequence, for example; "Uncertainty in the Earth system response to large-scale deployment of [*insert specific NET*] entails the risk of significant unexpected side effects emerging during scale-up, leading to a moratorium on further deployment."

As a preliminary to risk assessment and ranking, a scheme is required to classify risks along the two dimensions of likelihood and severity. Table 3.3 shows one such scheme, with a likelihood scale based on that used in the IPCC's Assessment Reports. Risk classification may be further refined by categorizing risks according to their main area of impact, for example, technical, economic/commercial, social/political, etc., and defining a severity scale specific to each category. For example, technical risks that could impact CO_2 injectivity in a BECCS or DACCS project might be classified as shown in Table 3.3.

Various tools such as quantitative risk assessment (QRA; Brown, 1985; Kobayashi et al., 2015), SWIFT; (Sollie et al., 2011), and others (Condor et al., 2011) are commonly used for systematic and numerical assessment of likelihood and

TABLE 3.2 Examples of NET-related uncertainties at various scales.

Level/Scale	Uncertainty examples
Global	• Pace of global GHG emissions reduction • Uncertainty in ESM/IAM model results due to input parameter values and theoretical models being poorly constrained by observations • Significant unexpected side effects (e.g., in Earth system response) from large-scale NET deployment • Evolution of the NET ecosystem (start-ups, energy majors, NGOs, etc.)
National	• Impact of other policy objectives (e.g., land-use, sustainability) on potential for NET deployment • Public acceptance of nationally relevant NETs and CO_2 storage options • Implementation of financial incentives and growth of carbon markets
Project	• Future carbon-price scenarios • Public acceptance of local project features (e.g., CO_2 storage) • Optimal choice and sourcing for biomass, mineral, or other feedstocks
R&D	• Minimum requirement for energy and other resources for individual NETs • Potential for genetic engineering to enhance biomass and other crop yields, particularly in arid and semiarid conditions • Progress on large-scale CO_2 reuse options

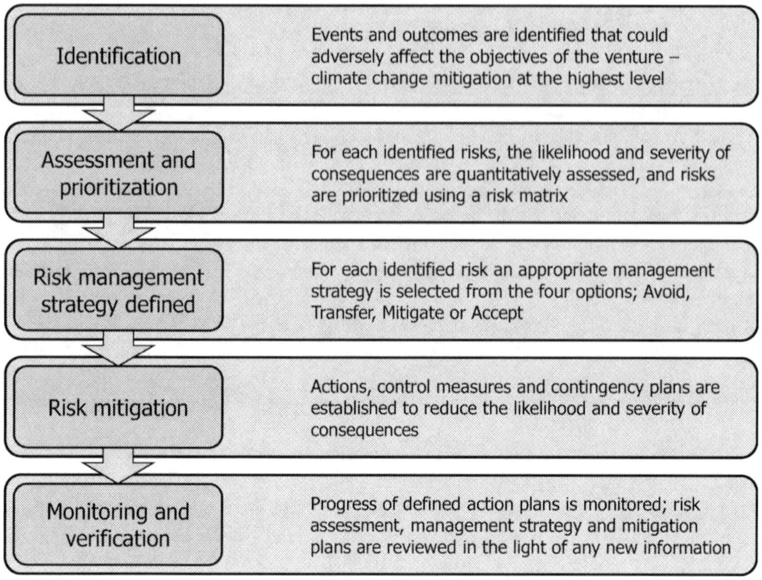

FIGURE 3.1 Schematic risk-management process.

severity to enable a quantitative ranking. This assessment is also, inevitably, subject to uncertainty; for example, if significant unexpected side effects were to emerge with the large-scale deployment of a particular NET, the consequences could be severe if few NETs were ready to scale, but potentially less so if a broad portfolio of methods was being pursued. While striving for a realistic evaluation, it is better to err on the side of caution, overestimating likelihood and severity where there is doubt, since this will drive more robust risk-mitigation planning.

The risk-identification and assessment outputs are frequently presented in the form of a risk register and a risk-assessment matrix (RAM). Table 3.4 shows a risk register, compiling the risk statements and classification for a number of risks that may impact on the large-scale deployment or effectiveness of NETs (see, for example, Fuss et al., 2014). The corresponding RAM is shown in Fig. 3.2. MacMartin and Kravitz (2019) have recently described the use of these tools to aid the prioritization of geoengineering research.

The RAM clusters risks of similar criticality—low, medium, high, and very high—as indicated by the shading in the figure. For example, a risk that is unlikely to occur but which would have severe consequences would have similar priority to a risk with less severe consequences but a higher likelihood.

Starting with the most critical risks, the next step is to decide which of four possible response strategies—terminate, transfer, treat, or take—to adopt for each risk:

Terminate:	Reduce the probability of occurrence to zero, for example by seeking another way to achieve the objectives that does not expose the endeavor to this risk
Transfer:	Transfer to another party that is better able to manage the risk (an example would be transferring long-term geological carbon-storage responsibility to a certified authority)
Treat:	Define and implement measures to reduce the likelihood (prevention) and/or the consequence severity (mitigation)
Take:	Once control measures for prevention and mitigation are in place, the residual risk must be taken, with appropriate monitoring

The high-level risks shown in Fig. 3.2 are all examples for which "Treat" would be the appropriate strategy. A risk-mitigation plan would then be prepared for each risk, identifying control measures to reduce the likelihood of occurrence, and contingency plans and other measures to reduce the severity of consequences. The aim of risk-mitigation planning is to reduce a risk to an acceptable residual level, by identifying a range of potential actions, evaluating and ranking them in

TABLE 3.3 Generic classification of risk likelihood and severity, and possible classification of risks to CO_2 storage injectivity in a BECCS or DACCS project.

Likelihood	Generic description	Injection project description	Severity	Generic description	Impact on CO_2 injectivity
Very likely (90%–100% probability)	An event or outcome that is almost certain to occur	Will occur several times during injection operations	Extreme	Extreme negative consequences that jeopardize the entire endeavor or severely impact the environmental and social conditions under which it operates	Significant permanent reduction Operation halted with no prospect of resumption
Likely (66%–90% probability)	A very common event or outcome	May occur several times during injection operations	Severe	Negative consequences that would seriously impact the endeavor or the environmental and social conditions under which it operates	Significant permanent reduction Operation halted until new injection wells are on-stream
As likely as not (33%–66% probability)	A commonly occurring event or outcome	Likely to occur during injection operations	Significant	Negative consequences that would pose a substantial threat to the achievement of the mission objectives	Significant temporary reduction Major injection volume shortfall with operation halted for days to weeks
Unlikely (10%–33% probability)	A relatively uncommon event or outcome	Unlikely to occur during injection operations	Minor	Small negative consequences that do not significantly impact the achievement of the mission objectives	Small reduction Reduced injected volume and interruption to operations for hours to days
Very unlikely (0%–10% probability)	A rare event or outcome	Unlikely to occur across multiple operations	Insignificant	No identifiable negative consequences for the endeavor	Small temporary reduction Reduced injected volume but no interruption to operations

After Sollie et al., 2011.

terms of their cost of implementation and effectiveness in reducing risk, and then implementing the most effective measures up to the point where the cost of implementing any additional measure outweighs the risk reduction benefit. In this way, the residual risk will be "as low as reasonably practicable." If mitigation plans are unable to reduce the risk to an acceptable level, an alternative risk strategy would have to be adopted.

Once mitigation plans are developed, the generic risk-management process (Fig. 3.1) would enter a monitoring and review phase, in which the risk assessment, analysis, and mitigation plans would be periodically reviewed in the light of new information. Communication and consultation take place in parallel with the other process steps, both internally and externally with the public and other stakeholders. The effectiveness of this communication and consultation effort will be an important factor in determining the quality and legitimacy of the resulting risk assessments and mitigation plans.

3.3.3 Managing critical NET-related risks and uncertainties

This section considers some risk management strategies applicable to the risks identified in Table 3.4.

Further delayed and inadequate global emissions reductions

Awareness of the climate crisis has undoubtedly grown in the years since the Paris Agreement, but the informed public has not yet reached the critical mass needed to force ambitious government action. In progressing negative emissions, the NET

TABLE 3.4 Register of risks to large-scale deployment or effectiveness of NETs.

Risk ID	Risk statement	Risk category	Likelihood	Severity	Risk classification
S1	Global emissions reductions efforts are inadequate, preventing the achievement of net-zero emissions	Social/Political	As likely as not	Extreme	Exceptional
S2	Lack of public acceptance constrains large-scale NET deployment leading to a shortfall in required removals to meet climate targets	Social/Political	Unlikely	Extreme	Very high
T1	Food production for a growing global population limits the achievable scale of CDR via land-based NETs (AR, BECCS, BiCRS)	Technical	As likely as not	Severe	Very high
T2	Unrealistic inclusion of NETs in IAM scenarios leads to unachievable national/regional CDR targets	Technical	Unlikely	Significant	Moderate
T3	Unexpected Earth system response reduces the effectiveness of some NETs, leading to a shortfall against target removals	Technical	Unlikely	Severe	High
T4	Growth of fossil-fuel CCS limits the storage resources available for BECCS and DACCS, constraining their scale-up	Technical	Very unlikely	Severe	Moderate
C1	Insurmountable accounting and permanence challenges prevent full market support for soil carbon methods, limiting their CDR impact	Commercial	As likely as not	Significant	High

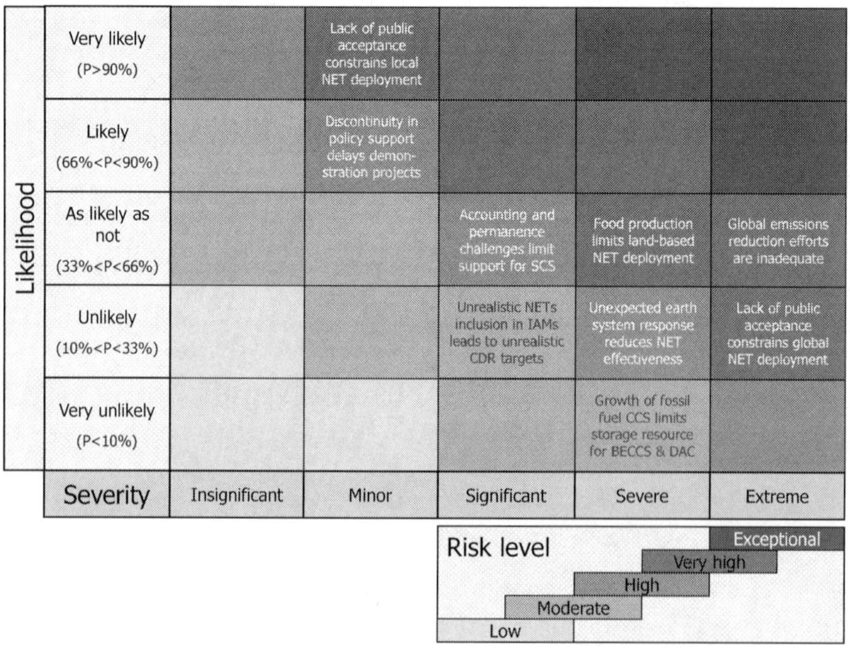

FIGURE 3.2 Example RAM for risks to NET delivery at scale.

community has an important role to play in countering any possibility that progress on NETs will lead to further delays and inadequate emissions cuts. This comes down to public engagement and will be aided by:

- prefacing every communication concerning NETs with a clear statement that the first priority remains rapid and deep emissions cuts
- avoiding undue enthusiasm that will quickly turn into the news of the latest "silver bullet"
- emphasizing the many uncertainties and challenges that need to be overcome before NETs can begin to have any climate impact.

Public acceptance of NETs

Acceptability of the social and environmental impacts is a significant uncertainty that could constrain NET deployment, and can only be resolved through open, transparent, and democratic public-engagement processes (Dooley and Kartha, 2018). Public awareness of CDR, whether as a necessity to meet an overall net-zero emissions climate goal, or as a portfolio of methods and technologies that will enable this, remains low, as one would expect given the immaturity of many of the technologies (Colvin et al., 2020).

Most studies of public acceptance of CDR have focused on BECCS, as the option most extensively evoked in IAMs (Bellamy et al., 2019), with solar radiation management (commonly via stratospheric aerosol injection) often taken as an alternative intervention at the "tampering with nature" end of the spectrum of options (Wolske et al., 2019). Unsurprisingly, more "natural" solutions such as large-scale afforestation are more readily accepted than progressively more "technological"—and therefore less intuitively understood—ones (Wright et al., 2014; Jobin and Siegrist, 2020), although using a naturalistic frame can lead to undue positivity (Corner and Pidgeon, 2015). Public acceptance is also inextricably linked with policy, as the type of policy instrument introduced to support a technology influences its acceptability (Bellamy et al., 2019), while policy decisions risk being driven by acceptability rather than effectiveness (Bellamy and Osaka, 2020). Public support for the incentivization of BECCS was found to be reduced if high costs would be borne by consumers and taxpayers, while policies that financially rewarded companies removing CO_2 from the atmosphere, as well as the use of standards and obligations, were well supported (Bellamy et al., 2019).

Continuing efforts to raise public awareness of the depth of the climate crisis and the range of possible responses, along with their respective risks and cobenefits, will help to reduce the likelihood that opposition will unduly constrain deployment. Further social science research into the impact of framing and other factors that influence public opinion on NETs will help in designing public-engagement and outreach efforts.

At the project level, the need for early engagement of potentially affected publics and other stakeholders is a common feature in proposed codes of conduct (discussed below). As well as responsible action in relation to social and environmental issues, open and transparent communication will be essential to build the foundation of trust needed to gain public acceptance and support for eventual NET deployment.

The consequences of opposition will be reduced by the pursuit of the widest possible portfolio of NETs (discussed further below) so that alternatives are available to accommodate national and regional preferences and capabilities, and by public engagement in deployment choices, including risk tradeoffs.

Resource competition and conflict with other policy objectives

Implementation of NETs at a large scale will have multiple implications for resources, including land, water, and energy, as well as financial resources (Smith et al., 2016). These requirements, which vary considerably between different technologies, need to be fully accounted for to arrive at robust assessments of sustainable NET potentials. Governance systems should therefore include an assessment of the threats posed by large-scale deployment to other national and regional policy objectives, such as agricultural production, sustainable development, and biodiversity (Dooley et al., 2021).

R&D to reduce energy requirements for DAC and various electrochemical technologies, and to increase biomass yield per hectare and biomass energy conversion efficiencies, are examples of current R&D focus areas for individual NETs which, while primarily driven by techno-economic considerations, will also mitigate the risks posed by resource competition. Further efficiency gains that are likely to arise as these technologies scale up will also contribute to risk mitigation. An important objective for biomass research in this regard is to minimize adverse impacts on biodiversity and ecosystem services, including the establishment of economically viable biomass-cultivation systems that are structurally complex and provide habitats for multiple native and other species (McCormack et al., 2016).

Realistic inclusion of NETs in IAM scenarios

Scenarios produced by IAMs are subject to wide-ranging parametric and model uncertainties, both in the underlying Earth system models and the socioeconomic frameworks, and typically rely on a fairly narrow set of assumptions about future pathways. With wide flexibility to make choices on model forms and input parameters, as well as on various ethical and political assumptions, almost any desired result can be demonstrated, giving a semblance of knowledge despite having limited scientific legitimacy (Pindyck, 2013; Beck and Krueger, 2016).

The difficulty of assessing loss arising from extreme events and potential climate tipping points, as well as the contentious question of discounting, are further challenges (Pindyck, 2013; Stern, 2016). Creutzig (2016) suggests that IAM studies would more effectively inform decision-makers if they shift emphasis from simply developing scenarios to also presenting and discussing the robustness of assumptions, and assessing uncertainties and risks, particularly those that become apparent when shifting from the usual global economic perspective of IAM's to look in more detail at ecological and biophysical processes at national and local scales.

Alternative approaches to dealing with uncertainties are also being applied, including stochastic dynamic programming and real-options analysis (Golub et al., 2014; Ginbo et al., 2021). The treatment of NET deployment as an "option" would be consistent with the standard approach to including unproven technologies in the evaluation of major industrial projects, where opportunities to cut costs, increase capacity, etc., are commonly included in sensitivity cases—to demonstrate the potential benefit if matured on the needed timescale—but are not included in the base-case evaluation until technically mature. If applied in the context of NETs, an options approach would encourage investment in R&D with the aim of maturing these options, but IAM outputs would not be based on this presumption until a portfolio of demonstration projects at increasing scale has confirmed technical feasibility and addressed the sociopolitical, institutional, economic, and other barriers to deployment. This approach will be facilitated by the construction of technological and institutional roadmaps (Fuss et al., 2014) that clearly specify the path and likely timescale for maturation of individual NETs to TRL8/9, at which point, once success at a climate-impact scale becomes "more likely than not," their inclusion in mitigation scenarios would be justified.

Earth system response to NET deployment at scale

The response of the Earth system to NET deployment at scale is currently poorly understood and quantified, although it will be an important factor in determining their effectiveness in mitigating climate change. Responses that will limit effectiveness include saturation effects affecting terrestrial carbon stocks, and carbon-cycle feedbacks such as ocean outgassing as atmospheric [CO_2] declines (Tokarska and Zickfeld, 2015; Keller et al., 2018). This an area where significant further research is needed, focusing on the following:

- improved understanding and quantification of the biological, chemical, and physical processes that underlie each NET
- reducing uncertainty in the sensitivity of carbon fluxes and storage to changing environmental conditions
- quantifying the extent to which negative emissions can reverse climate and carbon-cycle shifts caused by CO_2 emissions, reflecting the increased risk of irreversible climate change with further GMST increases
- developing methodologies to evaluate and monitor the impact of NETs on the carbon cycle and climate, to enable reliable negative-emissions accounting.

As Keller and colleagues (2018) noted, uptake of historical emissions by the land and ocean can be viewed as a "carbon debt" that will be paid back through reduced effectiveness as the deployment scale of NETs increases.

3.3.4 NET risk and benefit tradeoffs

Informed tradeoffs decisions—accepting one risk or benefit in preference to another—require quantitative, cross-disciplinary information on risks and benefits, and the close integration of scientific investigators with the affected public, decision-makers, and other stakeholders. Large-scale deployment of NETs will involve tradeoffs at many levels; weighing the desirability and scale of one method versus another, between benefits and disbenefits occurring at different times and locations (Kverndokk et al., 2014), or, at the highest level, between the residual risks from large-scale deployment (once these have been minimized through further RD&D) and the risk from foregoing this mitigation option.

The development of assessment methods to address tradeoffs has received considerable attention in the agricultural and conservation fields (Klapwijk et al., 2014; Kanter et al., 2018), notably in relation to land-use, biodiversity, and ecosystem services (McShane et al., 2011; Visseren-Hamakers et al., 2012; Dooley et al., 2018; Qiu et al., 2021). Tradeoff analysis (TOA) is a process designed to bring about integrated assessment and informed decision-making, as illustrated in Fig. 3.3.

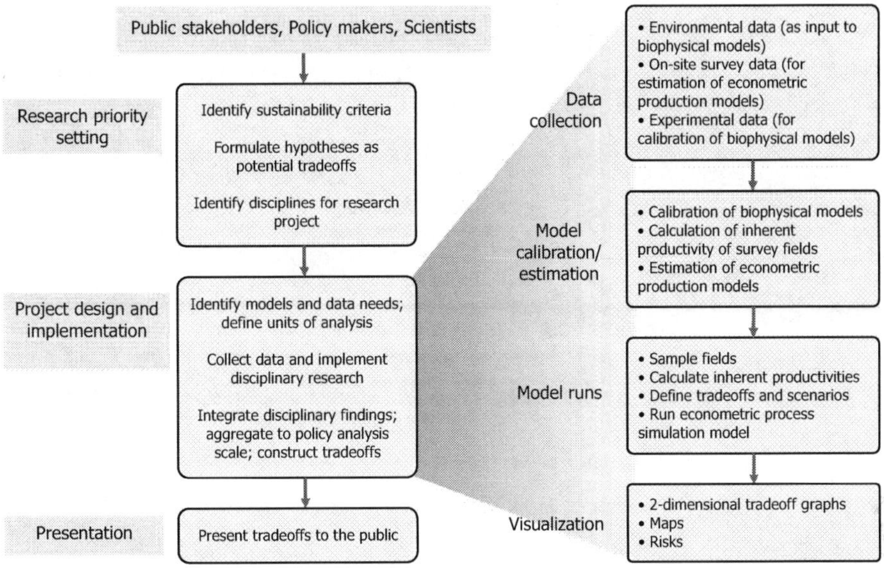

FIGURE 3.3 Tradeoff analysis process. *After Crissman et al. 1988; Stoorvogel, et al., 2004.*

An important feature of the TOA process is the involvement of all stakeholders (public, scientists, and decision-makers) from the outset, in defining the dimensions of the issue and the assessment criteria, which may be revised as further insights are gained. A full appreciation of uncertainty is also critical for informed tradeoff decisions, and sensitivity analysis is therefore an essential part of the assessment process. Stoorvogel and colleagues (2004) describe the application of the TOA process to agricultural production systems, while Giller and colleagues (2008) see tradeoffs as one aspect of a wider strategic societal negotiation, and describe a similar four-stage process that has been used to address stakeholders' competing claims by developing a shared understanding of problems and opportunities.

Certification schemes could also play a part in managing tradeoffs; for example, certification for rainfed biomass production or irrigated production covered by comprehensive water-management strategies that take into account other human and environmental needs would inform customer choice and create a market incentive to reduce water impact (Fehrenbach 2011; Pahl-Wostl et al., 2013), although further research is needed in this area (McCormack et al., 2016).

3.4 NET governance

3.4.1 Governance principles

In considering the pursuit of intentional climate interventions such as NETs, Jamieson (1996) called for ethical and other societal concerns to be included in decision-making about "what to research, at what level of funding, with what urgency." Five governance principles, aiming to ensure the societal legitimacy of such decisions, were proposed by Rayner and colleagues (2009) and quickly became known as the Oxford Principles. The principles aimed to address the governance of "climate change geoengineering," then defined by the Royal Society as "the deliberate large-scale manipulation of the planetary environment to counteract anthropogenic climate change," and therefore encompassing both solar radiation management and CDR (Rayner et al., 2013).

Rayner and colleagues recognized that these principles were only a starting point, and they have since been widely discussed, revised, and updated, notably at the 2010 Asilomar Conference and in the Tollgate principles (Gardiner and Fragnière, 2018). Table 3.5 summarizes and compares these three sets of principles, with the respective narratives modified slightly to focus on CDR rather than geoengineering more broadly.

These principles are reflected in a number of reports on the governance of CDR (and climate engineering more generally), recently reviewed by Morrow (2017), and are also broadly supported by the results of public-perception studies. Bellamy (2018) distilled five principles arising from such studies for the good governance of NET RD&D, which largely

TABLE 3.5 Oxford, Asilomar, and Tollgate principles applied to the governance of CDR.

Oxford principles	Asilomar principles	Tollgate principles
1: CDR to be regulated as a public good	1. Promoting collective benefit:	1: CDR should be administered by or on behalf of the global, intergenerational and ecological public, in light of their interests and other ethically relevant norms.
While the involvement of the private sector in the delivery of a CDR technique should not be prohibited, and may indeed be encouraged to ensure that deployment of a suitable technique can be effected in a timely and efficient manner, regulation of such techniques should be undertaken in the public interest by the appropriate bodies at the state and/or international levels.	Promoting the collective benefit of humankind and the environment must be the primary purpose of research conducted to develop and evaluate the potential for CDR technologies to moderate or reverse human-induced climate change.	CDR raises complex issues of global, intergenerational, and ecological ethics—including issues of justice, legitimacy, domination, rights, responsibility, and the human relationship to nature. It should be framed as such and administered by or on behalf of the global, intergenerational, and ecological public in light of relevant norms. This has implications for deployment, governance, and research.
2: Public participation in CDR decision-making.	5. Public involvement and consent	2: Authorization and 3: Consultation
Wherever possible, those conducting CDR research should be required to notify, consult, and ideally obtain the prior informed consent of, those affected by the research activities. The identity of affected parties will be dependent on the specific technique which is being researched.	Public participation and consultation in research planning and oversight, assessments, and development of decision-making mechanisms and processes must be provided to ensure consideration of the international and intergenerational implications of CDR	2: CDR decision-making should be done by bodies acting on behalf of the global, intergenerational, and ecological public, with appropriate authority and in accordance with suitably strong ethical norms, including justice and political legitimacy. 3: Decisions about CDR research activities should be made only after proper notification and consultation of those materially affected and their appropriate representatives, and after due consideration of their self-declared interests and values.
3: Disclosure of CDR research and open publication of results.	3. Open and cooperative research	4: Trust
There should be complete disclosure of research plans and open publication of results to facilitate a better understanding of the risks and to reassure the public as to the integrity of the process. It is essential that the results of all research, including negative results, be made publicly available.	CDR research should be conducted openly and cooperatively, preferably within a framework that has broad international support. Research activities with the potential to affect the environment in significant ways should be subject to risk assessment, taking into account the risks and their distribution associated with both the activity itself and the ongoing limits to understanding if the experiment is not conducted.	CDR policy should be organized to facilitate reliability, trust, and accountability. This suggests high levels of openness and disclosure in the publication of research plans and results, and in decision-making more generally. It also suggests that CDR policy should accept demanding norms of inclusion, across nations and other demographic groups, at all levels, including the basic design of policy and institutions.
4: Independent assessment of impacts	4. Iterative evaluation and assessment	4. Trust (continued)
An assessment of the impacts of CDR research should be conducted by a body independent of those undertaking the research; where techniques are likely to have a transboundary impact, such assessment should be carried out through the appropriate regional and/or international bodies. Assessments should address both the environmental and socioeconomic impacts of research, including mitigating the risks of lock-in to particular technologies or vested interests.	Iterative, independent technical assessments of research progress on climate-engineering approaches will be required. Assessing potential intended and unintended consequences, impacts, and risks will be critical to providing policy-makers and the public with the information needed to evaluate the potential for climate engineering to be implemented as a complement to mitigation and adaptation.	In addition, if any material is protected, it should be subject to a review process involving stakeholders or their representatives to maximize trust. Special attention should be paid to the concerns of those groups affected by decisions that cannot be effectively represented, such as future generations and nonhuman nature.

TABLE 3.5 Oxford, Asilomar, and Tollgate principles applied to the governance of CDR.—cont'd

Oxford principles	Asilomar principles	Tollgate principles
5: Governance before deployment	2. Establishing responsibility and liability:	5: Ethical accountability
Any decisions with respect to deployment should only be taken with robust governance structures already in place, using existing rules and institutions wherever possible.	Governments must clarify responsibilities for, and, when necessary, create new mechanisms for, the governance and oversight of large-scale climate engineering research activities that have the potential or intent to significantly modify the environment or affect society. These mechanisms should build upon and expand existing structures and norms for governing scientific research and, in the event of damaging outcomes, establish who would bear the cost and the degree of liability and proof that are required.	Robust governance systems (including authority, legitimacy, justification, and management) are needed and ethically necessary at each stage from advanced research to deployment. Robust structures of consultation, administration, and justification are ethically necessary before deployment, but a significant level should also be in place before commencing field testing (domestic and international), and some governance is desirable even before that, as advanced research programs emerge.

After Rayner et al., 2013; Asilomar Scientific Organizing Committee, 2010; Gardiner and Fragnière, 2018.

follow Oxford Principles 3–5, namely: programs should have transparent purposes, activities, and reporting; environmental impacts should be minimized and monitored; oversight of activities should be independent of private interests; the permissible scale of activities will depend on perceived controllability; technology and activity-specific governance protocols should be in place.

In developing the Tollgate Principles, Gardiner and Fragnière (2018) aimed to put greater emphasis on ethical issues, which they felt had been largely overlooked in the more procedurally oriented Oxford and Asilomar principles. They also noted that the Oxford Principles did not directly address what it would take for a CDR program to be justifiable and proposed additional, substantive principles (Table 3.6), building on those proposed by Jamieson (1996).

To support the application of the Oxford Principles, Rayner and colleagues (2013) also called for the development of specific governance protocols and stage-gate processes, and progress in the development of codes of conduct that embody and operationalize these principles is discussed below in Section 3.4.4.

TABLE 3.6 Comparison of Jamieson's permissibility conditions and Tollgate substantive principles.

Jamieson's permissibility conditions	Tollgate (substantive) principles
(1) The project is technically feasible	Technical availability: For a technique to be policy-relevant, ethically defensible forms of it must be technically feasible within the relevant timeframe.
(2) The consequences of implementation can be reliably predicted	Predictability: For a technique to be policy-relevant, ethically defensible forms of it must be reasonably predictable on the relevant timeframe and in relation to the threat being addressed.
(3) Its implementation would result in an Earth system state that is socioeconomically preferable to the alternatives (including do nothing or BAU)	Protection: Climate policies that include CDR should be socially and ecologically preferable to other available climate policies and focus on protecting basic ethical interests and concerns (e.g., human rights, capabilities, fundamental ecological values).
(4) Its implementing would not seriously and systematically violate any important, well-founded ethical principles or considerations	Respecting ethical and ecological norms: CDR policy should respect general ethical norms that are well founded and salient to global environmental policy (e.g., autonomy, justice), as well as well-founded ecological norms, including norms of environmental ethics and governance (e.g., sustainability, precaution, respect for nature, ecological accommodation).

After Jamieson, 1996; Gardiner and Fragnière, 2018.

3.4.2 Governance needs

Given the significant ethical concerns, uncertainties, and risks that arise in discussing the development and deployment of NETs, there is a clear need for robust governance systems to ensure that these activities take place in a way that is aligned with other societal needs and priorities (Lawford-Smith and Currie, 2017; Mace et al., 2021). Honegger and colleagues (2021) identify the following areas where concerted governance action is needed:

- gaining clarity on the intended role of NETs for limiting warming
- accelerating innovation
- ensuring participation
- ensuring robust MRV and accounting of removals
- managing side-implications
- transitioning to cost-effective long-term operations.

Robust public involvement is also essential to ensure the legitimacy of outcomes, particularly given the normative nature of many of these issues and the need to reconcile divergent views on appropriate responses. Governance mechanisms to address these areas would include the establishment and oversight of:

- reliable incentives to support multi-disciplinary R&D and encourage private-sector infrastructure development (e.g., for CO_2 transportation and storage hubs)
- systems to support synergy (mutual reinforcement) between multidisciplinary research programs, knowledge sharing, and capability development
- regulatory frameworks to address NET-specific deployment issues and concerns
- processes and systems to enable the equitable resolution of conflicting priorities (e.g., tradeoffs with sustainable development, biodiversity conservation, etc.)
- standards and guidelines to address the measurement, monitoring, life-cycle accounting, and verification of NET project performance, as well as the allocation of their benefits to the NDCs of participating nations
- safeguarding systems and appropriate redress/compensation mechanisms to address liability issues, including those arising from permanence uncertainties and long-term transfer of liabilities (e.g., for geological storage)
- carbon markets and other financial mechanisms to enable the cross-border flow of funding for NET projects, and to stimulate private-sector investment
- protocols for self-governance and independent oversight of activities, where these approaches are desirable
- deliberative, science-based processes for transparent and inclusive decision-making.

Governance systems will also need to reflect the specific requirements of individual NETs—such as storage permanence, conflicting policy priorities, etc.—and the appropriate governance level: international for those issues that have transborder dimensions; national for those more closely linked to other national or regional policies (Bellamy and Geden, 2019). Table 3.7 illustrates some international and national dimensions of differentiated NET governance needs.

Below the national level, some aspects may be amenable to governance through independent oversight by NGOs or, potentially, self-governance within the emerging NET sector. Effective governance of NETs will therefore require a multilevel and multilateral "regime complex."

3.4.3 Governance levels

In 2009, Nobel prize-winning economist Elinor Ostrom observed that "[t]he likelihood of developing an effective, efficient, and fair system to reduce greenhouse gas emissions that can be rapidly initiated at the global level appears to be very low. Given the severity of the threat, simply waiting for resolution of these issues at a global level, without trying out policies at multiple scales because they lack a global scale, is not a reasonable stance" (Ostrom, 2009). Notwithstanding the limited progress embodied in the Paris Agreement, the intervening years have not proved her wrong.

It is generally recognized that no single global institution can address all aspects of the governance of NETs. Public perception studies often reveal a desire for international governance—although tempered by a skepticism concerning the ability of international institutions to produce effective regulations—while self-governance or independent oversight at national and lower levels can be preferred for certain types of activities or for certain types of NETs (Bellamy et al., 2017; Bellamy, 2018). As Ostrom anticipated for the governance of emissions reduction, the effective governance of NETs (and climate engineering more broadly) will most likely require both top-down and bottom-up approaches, spanning international, national, and subnational levels, and different kinds of institutions and instruments at each level (Morrow, 2017). Aspects of governance at each level are discussed in the following sections.

TABLE 3.7 Aspects of differentiated NET governance.

NET	International	National
AR	Certification and accreditation frameworks; safeguards related to permanence; biodiversity and ecosystem services	Biodiversity and ecosystem services conservation
SCS	Consistent MRV protocols and accounting rules	Safeguards related to permanence; national agricultural policy integration
BECCS	Direct and indirect emissions accounting from biomass production and transborder supply chains; sustainability certification; international CO_2 transportation	Geological storage siting, planning, and regulation; integration with national energy, waste management, agricultural, and other land-use policies
DACCS	International CO_2 transportation	Geological storage siting; planning and regulation; integration into national energy systems
EW	MRV protocols and accounting rules; cross-border alkalinity transport in river basins	Integration with agricultural and other land-use policies; marine conservation and environmental permitting
Marine CDR (mCDR)	MRV protocols regarding ocean-based methods	MRV protocols for mCDR in national waters; marine conservation and environmental permitting

International governance

Current international conventions and legal instruments are generally restrictive rather than enabling for NET RD&D activities and, as Brent and colleagues (2018) have pointed out, there is an urgent need to reconcile this mismatch between the international legal position and the increasingly compelling need for their development. International policy instruments could bring many benefits in accelerating NET development, by addressing the scale of the financial incentives needed and the differentiation of national funding responsibilities, and by enabling the flexible allocation of financial and other resources, with due regard to economic efficiency (least-cost), sustainable development goals, and other international priorities (Honegger and Reiner, 2018).

International or multilateral governance will be required to address the transboundary issues that will accompany the implementation of specific NETs. Many of these issues are familiar from the recent experience of REDD + projects: international commercial interest in afforestation and biomass production creates the potential for local conflict due to the competing interests of other land users from the agricultural and sustainable forest communities, and equity concerns regarding benefit-sharing from large-scale projects (Kreuter and Lederer, 2021). Governance systems will be needed to ensure that these conflicting land-use issues are addressed, in line with the principles of good governance (Hunsberger et al., 2014; Creutzig, 2017). The need to estimate the total system-wide change in emissions and removals is a significant carbon accounting challenge (Brander et al., 2021), an important example being the assessment of emissions arising from direct and indirect land-use change when accounting for removals resulting from biomass production (Vaughan and Gough, 2016), so that MRV needs to extend beyond the supply chain. Given the process complexity, encompassing land-use change, cultivation, transportation, conversion, and storage, Torvanger (2019) argues for a standardized, internationally agreed MRV framework for BECCS, which should also clarify where the CDR benefits and offsetting emissions accrue.

Lawford-Smith and Currie (2017) propose that UNFCCC may be an appropriate governing body for NETs with lower risk, and extension of the REDD + framework would seem to be a logical and efficient approach to governing AR and biomass production, and related supply chains. REDD+ was initially envisaged to involve no tradeoffs or negative externalities and these only became clear once implementation had started (Kreuter and Lederer, 2021). Shortcomings in the REDD + framework—notably the failure to fully recognize and incorporate justice issues, the results-based payment scheme that undermines efforts to secure cobenefits, the overstatement of carbon storage outcomes, and inadequate management of storage permanence—have been recognized from project performance (Duchelle et al., 2018; Malkamäki et al., 2018). These learnings would be invaluable in establishing effective AR/biomass governance, although many of the tradeoffs have yet to be resolved.

Large-scale BECCS and DACCS deployment will have further international implications as a result of the need for access to geological storage—another unevenly distributed global resource (Kreuter and Lederer, 2021). While the governance of the subsurface may be seen largely as a national issue (comparable to the exploitation of fossil-fuel resources), issues relating to long-term storage liability will ideally be addressed through international governance mechanisms, to reduce the risk that measures to meet these liabilities fall victim to future national policy changes.

Implementation of ocean-based methods will also have transboundary impacts and therefore has international governance implications. Current ocean governance is a patchwork of international and sectoral regimes and principles of international law (Brent et al., 2019), and the lack of a single accountable body—a "coordinating node"—has been noted as a factor inhibiting governance efforts in relation to ocean acidification (Harrould-Kolieb and Hoegh-Guldberg, 2019). Ocean-based NET deployment at sea will fall within the scope of the London Convention and Protocol on the Prevention of Marine Pollution (LC/LP), which was amended between 2010 and 2013 to address marine geoengineering in response to the growing commercial interest in OF (Dixon et al., 2014), as well as the UN Convention on Biological Diversity (CBD) which similarly calls for a precautionary approach to geoengineering activities that may affect biodiversity (UNEP, 2010; 2012; Webb et al., 2021). The UN Intergovernmental Conference on Biodiversity beyond national jurisdiction (BBNJ) is currently negotiating an agreement under the UN Convention on the Law of the Sea, although it seems unlikely that this will help to stitch together the patchwork of "legal instruments and frameworks, and relevant global, regional and sectoral bodies" relevant to ocean-based NETs (De Santo et al., 2020).

These existing conventions have rightly aimed to minimize the risks to the marine environment, while still enabling well-defined and controlled research but, as R&D efforts progress and the risks (and cobenefits) become better understood, a comprehensive governance framework will be needed. Harrould-Kolieb and Hoegh-Guldberg (2019) consider it unlikely that a specific global agreement will be implemented to address ocean acidification (OA), and the same seems likely for ocean-based NETs.

Linking of national and regional financial mechanisms such as carbon markets can help to address challenges of leakage (companies emitting in low-cost and removing in high-cost markets), and address permanence uncertainty and response, and some degree of consistency—or at least of mutual confidence—in NET MRV procedures will be desirable to enable these linkages and to avoid the creation of nonequivalent removal credits (Bellassen et al., 2015; Oldfield, 2022).

National governance

Governance below the global level will be required to establish national[1] goals for CDR through NET deployment, as a contribution toward national commitments under the Paris Agreement. Financial mechanisms (e.g., cap and trade, carbon tax, subsidies, etc.) to encourage the development and demonstration of NETs will also need to be established at the national level (Bellamy, 2018).

Strategic environmental assessments and bottoms-up national reviews of CDR deployment potential, reflecting local constraints and conflicts, barriers and bottlenecks, are beginning to appear (Price et al., 2018); these will be essential to inform the choices that will determine the optimum (national or regional) portfolio of NETs to be supported (Fajardy et al., 2019; Thoni et al., 2020), as well as improving IAM scenario inputs (Fuss et al., 2020). A program of pilot projects with appropriate policy support will need to rapidly follow these assessments, scaling incrementally toward commercial deployment to test the technical feasibility and better understand the other barriers to deployment (Fuss et al., 2014).

Current policy recognition of and support for NETs in general remains limited, despite the highlighting of policy areas needing urgent attention by several authors (Lomax et al., 2015a, Peters and Geden, 2017). Lomax and colleagues anticipated the challenges faced by policymakers and proposed four principles to guide policies that will help to unlock the potential of NETs:

- support scientific research and technical demonstration to address technology-specific uncertainties and challenges, to constrain costs and risks, and to develop monitoring methodologies
- develop support for a range of early, low-risk and low-cost NET opportunities that can already provide economical carbon removal, ideally through existing policy mechanisms
- commit to integration of carbon removals into wider policy support, carbon-price mechanisms, and accounting, over the longer term, to stimulate investment, R&D, innovation, and planning for future opportunities
- develop and identify steps that address future barriers to NETs and lay the groundwork in relevant systems (e.g., in energy-system and infrastructure decisions), so that future opportunities are not locked-out and can be deployed rapidly when they become attractive.

Policymakers can either frame climate-change mitigation as a business opportunity or as a global crisis. In the first case, policies would primarily target the private sector, combining public innovation funding to support emerging NETs with

1. The term "national" is taken here to include regional and sub-national (e.g., state) governance, as may be appropriate in any particular jurisdiction.

market mechanisms to support deployment as they mature (Michaelowa et al., 2018). Framing as a national or global emergency seems less promising—at least for now—because public concern, in general, is low and unlikely to accept the strictures that this would involve.

Based on an analysis of the NET-related policies of nine OECD countries, Schenuit and colleagues (2021) proposed a spectrum of policy integration approaches, ranging from incremental modification to proactive entrepreneurship, as illustrated in Fig. 3.4. Whatever policy approach may be adopted, there is a need for this to be adaptive, given that further assessments could change the demand, feasibility, or acceptability of individual NETs (Fuss et al., 2020); or as increased awareness of the need to separate removals accounting from emissions reductions leads to a shift away from fungibility (which has been common when considering only LULUCF related removals) (Schenuit et al., 2021).

Although afforestation has been eligible for carbon credits issued under the Kyoto Clean Development Mechanism, there are currently few incentives for broader CDR development and demonstration work, which would provide input to improved technoeconomic analysis and kick-start the cost-reduction learning curve.

In the United States, two policy incentives are relevant for some NETs. The federal 45Q tax credit provides up to \$35/tCO_2 for CO_2 utilization and up to \$50/tCO_2 for CO_2 geological storage; in California, the low-carbon fuel standard (LCFS) gives credits, currently traded at around \$200/tCO_2, for transportation fuels with a carbon intensity below a specified cap. In addition, the Infrastructure Investment and Jobs Act (IIJA), enacted in 2021, will provide funding for a wide range of RD&D programs related to the energy transition, in areas from energy efficiency to grid resilience, including prize competitions for precommercial and commercial DAC technologies and other projects leading to the development of four carbon regional DAC hubs, as well as pilot and demonstration scale capture projects, and the establishment of a carbon utilization program.

The EU is considered a leader in international climate policy (Oberthür and Groen, 2017), yet progress to date on bringing CDR into EU climate policy has been limited (Geden et al., 2019). At a national level within the EU, the Dutch SDE++ scheme, introduced in 2021, extended renewable-energy subsidies to include other emissions-reduction measures such as CCS and H_2 production by electrolysis. In the United Kingdom, a 2020 open consultation on a carbon tax requested input on complementary policies that should be considered to include negative emissions in national climate policy (UK Government, 2020).

It is clear that NETs are not a substitute for aggressive emissions reduction, but instead might "crowd them out" if market mechanisms treat CDR as an alternative to emissions reduction (Markusson et al., 2018; Burke and Gambhir, 2022), pointing to the need for separate targets and incentives (McLaren et al., 2019). In developing policies to support NETs, greater scrutiny is needed of any decisions that would designate specific sectors, such as agriculture or aviation, as "hard to decarbonize," to avoid discouraging innovation (Scott and Geden, 2018). Care is also needed to ensure that private-sector removals supplying voluntary carbon markets are properly accounted as being additional to, and not a

	Incremental modification	**Early integration and fungibility**	**Proactive NET entrepreneurship**
NETs in mitigation targets	Strictly separate	Fungible	Fungible
Range of NETs addressed	Ecosystem-based only	Focus on ecosystem-based	Proactive technology support
Relation of NET policy to broader climate policies	Incremental opening	Full integration	Specific instruments
Support for niche NET development	Limited support	Limited support	Nurturing and empowering
View of NETs among incumbent regime actors	Restrained integration	Proactive integration	Proactive integration

FIGURE 3.4 Range of NET policy integration approaches. *After Schenuit et al., 2021.*

substitute for, existing NDC commitments (Honegger et al., 2021)—raising the question under what circumstances removals through NETs can be attributed to NDCs.

Existing policies such as carbon price (tax/markets) or tax credits at the ~\$50/tCO_2 level may do little to incentivize investment in developing or scaling-up NETs. Cost reduction through learning, dramatically achieved for renewable energy sources such as offshore wind, requires continued support for demonstration projects at progressively increasing scale and will require policy frameworks that provide targeted support for NETs alongside emissions reductions. For example, Bednar and colleagues (2021) propose a novel market approach to incentivize the repayment of legacy carbon debt by imposing "carbon removal obligations" on emitters, thereby explicitly linking emissions and removals.

Within individual jurisdictions, NET deployment will impinge on many other policies and regulations, including those governing agriculture, transport and other infrastructure, local development and planning, and environmental conservation. For example, the wide adoption of methods to increase soil carbon stocks will be largely driven by national incentives targeting the agricultural sector, and storage permanence will strongly depend on the continuity of related policies. Conflicts with other policy objectives will inevitably arise, risk tradeoff mechanisms will be needed, and local participation in decision-making will be essential to gain public acceptance in the face of these challenges—although processes to achieve this are rarely formalized in national regulations (Lezaun, 2021). In relation to ocean-based NETs, Boettcher and colleagues (2021) have noted that nations have both the right to use and the obligation to protect the ocean, and these rights and obligations could be constructively combined, for example, by implementing ocean alkalinity enhancement in line with international obligations while also aiming to protect coastal communities from the impacts of ocean acidification.

Subnational and self-governance

Private-sector involvement in NETs is in a rapid growth phase, propelled both by increasing awareness of the role NETs are expected to play in reaching net zero, and by a growing demand for carbon-removal credits on the voluntary carbon market, led by companies such as Microsoft. The expanding NET ecosystem includes technology developers, both energy-sector incumbents and a multitude of start-ups, aggregators/market-makers, standards and verification organizations, as well as advocacy and community-building organizations, and investors. The market for voluntary carbon-removal credits quadrupled from 16 ktCO_2 in 2020 to 60 ktCO_2 in 2021, a total dwarfed by the single purchase of 400 ktCO_2 by Airbus from Carbon Engineering announced in March 2022.

This explosion of commercial interest carries the risk that badly executed projects may prove ineffective or lead to adverse side effects, resulting in the erosion of public support and the undermining of future NET efforts (Honegger et al., 2021). Careful scrutiny and oversight of project activity, and of the growing markets for removal credits, are needed to ensure that projects are responsibly executed and generate removals of high quality, in terms of permanence, social and environmental responsibility, and sustainability. The importance being placed on these considerations by many project developers, market aggregators, and verifiers, such as Verro and Puro, and by buyers such as Microsoft, Swiss Re, Shopify, Stripe, and others are encouraging in this regard. For example, recognizing the controversy that had been generated by early BECCS proposals, Sandalow et al. (2021) were guided by three principles in formulating their approach to BiCRS: to do no harm; that social acceptability would be a key success factor for BiCRS; and that technology development should reflect social priorities.

Maher and Symonds (2022) have identified four areas where multilateral cooperation of motivated actors operating as "climate clubs" at the subnational level could contribute to support policy development, notably in: CDR accounting and reporting methods; the RD&D of the more "technological" NETs; the development of incentives for NET deployment; and defining governance and accountability mechanisms addressing issues of justice and social license. Such efforts will be most effective if they have a broad disciplinary base and are able to address not only techno-economics, but also the social, political, and cultural elements of NET development and deployment (Dowell et al., 2020).

The development of codes of practice or conduct, both generic (Hubert, 2020) and technology specific, the IUCN Global Standard for nature-based solutions (IUCN, 2020), and the espousal of corporate "planetary responsibility" statements explicitly aligned to the principles of responsible research and innovation (see e.g., Wren, 2022) also point to an emerging norm of responsible self-governance. The development of codes of conduct has received particular attention in relation to ocean-based methods, largely due to the current fragmented nature of ocean governance (discussed above) and the "sea of risks and uncertainties" that attend many of these methods. A framework for such a code of conduct has been proposed by Loomis and colleagues (2022), incorporating the principles of responsible research as well as provisions to ensure equity and fairness (Fig. 3.5), and these components are also reflected in the guidance for ocean-based CDR projects prepared by the Aspen Institute (2021).

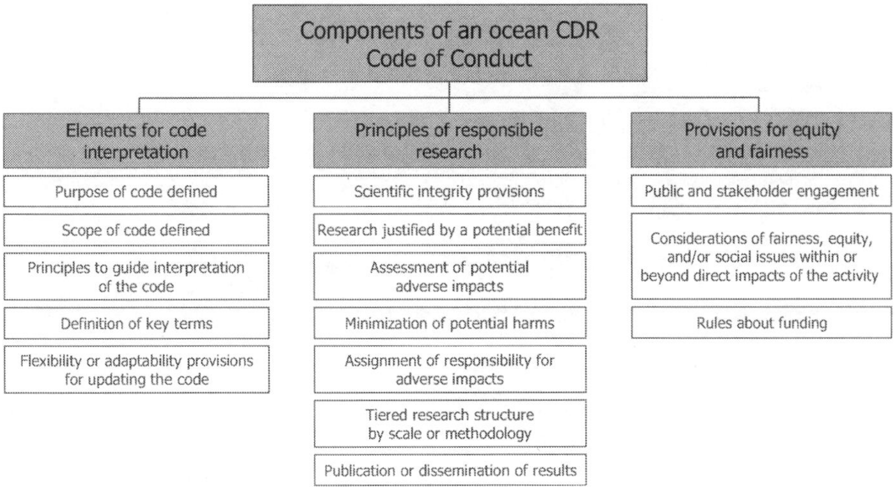

FIGURE 3.5 Components of an ocean CDR code of conduct. *After Loomis et al., 2022.*

3.4.4 Polycentric governance of NETs

It is widely recognized that a shift in climate governance away from a regime dominated by seemingly endless and ineffective international negotiations and national political hesitancy is essential if ambitious emissions reduction targets are to be achieved. Failure of the 2009 UN Climate Conference in Copenhagen (COP15) to deliver an effective follow-up to the Kyoto Protocol has been suggested as a turning point in this regard, although aspects of more distributed governance, such as the C40 Cities Climate Leadership Group, have earlier origins (2005 in this case; Fuhr et al., 2018). One reason for the failure of national and international governance of the "first-order" risk of climate change has been the preoccupation of political leaders and governments with "second-order" risks to their reputations and authority; in short, reputation management has impeded effective Earth system management (Howlett, 2014; Brown et al., 2019).

The IPCC Special Report on the impacts of global warming of 1.5°C (IPCC, 2018) called for broader governance approaches to reinforce global climate action, including strengthened, accountable, multilevel governance that includes nonstate actors partnering on technology development and transfer. Recognizing that the main impediments to early action are social factors, barriers and inertia, policy volatility, and governance uncertainties, Fuss and colleagues (2020) similarly called for "a broader alliance to move toward net-zero targets, which brings together the research and policy communities, industry, and the broader public to explore and find the common ground to push climate solutions forward."

In response to this gap between societal expectations and global (under)delivery, many new actors at all levels are increasingly taking up active roles (Bäckstrand et al., 2017). Terms such as "new climate governance" (Jordan et al., 2015), "meta-organization" (Chaudhury et al., 2016), and "hybrid-multilateral" (Bäckstrand et al., 2017) have been used to describe the new modes of distributed climate action and governance, which increasingly display polycentric characteristics.

Polycentric governance involves a diverse population of actors, including governments and government departments, subnational jurisdictions (states, provinces), private-sector companies, investors, scientific institutions, NGOs, and many other civil society groups, that form multiple overlapping centers of authority and decision-making and interact through various cooperative, competitive, and conflict-resolving mechanisms.

Fig. 3.6 illustrates four domains (governance levels, actors, cross-scale institutional and multiactor linkages, and knowledge production) that can be used to characterize different governance modes or regimes. The manner in which these domains and the attributes within them interact determines the nature of the regime, from international or nation-centric, to fully decentralized.

Although "polycentric" is often used simply as a metaphor to describe the diversity of actors involved in governance at different levels and scales, Dorsch and Flachsland (2017) go beyond this to identify the specific features and mechanisms through which a polycentric approach to "perceiving, designing, and implementing climate governance" can lead to more

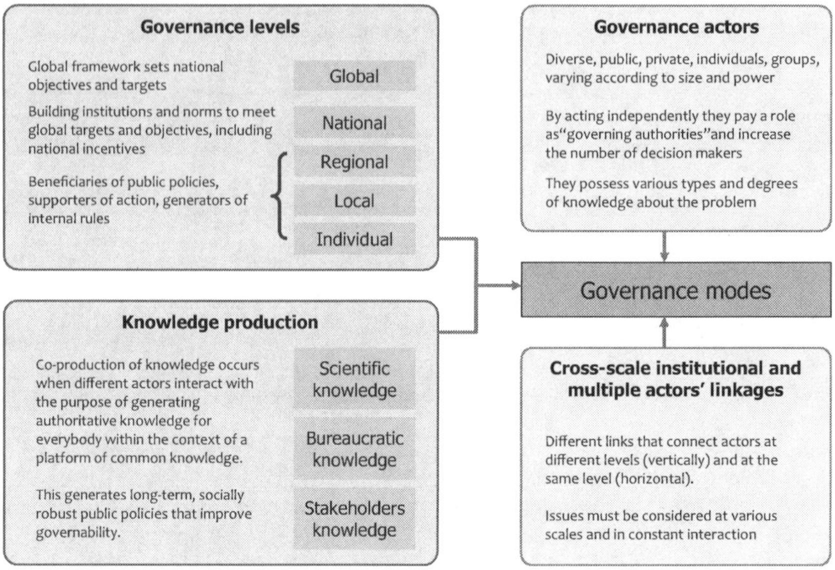

FIGURE 3.6 Domains and attributes for analyzing governance modes. *After Arriagada et al., 2018.*

ambitious and effective climate action. They identify the main challenge as being the coordination of these diverse actors and suggest that the vision for such a system is to create an adaptive consortium of multiple self-governing entities, at different scales and different levels, interacting with each other and deploying their site-specific capabilities toward a common goal. To ensure the effectiveness of polycentric governance, emphasis is therefore required in four areas:

1. Self-organization: with no groups or actors dominating and with actors having a high degree of autonomy in rule-setting
2. Understanding and consideration of site-specific conditions: of individual actors' preferences, competencies, constraints, and interactions
3. Enabling experimentation and learning: fostering innovation and adaptive institutional arrangements, norms, and knowledge diffusion
4. Trust to foster cooperation: enhanced through trust-building practices that can help overcome traditional cooperation dilemmas, such as free-riding.

The robust application of these four factors across the four domains illustrated in Fig. 3.6 provides a working definition of polycentric governance.

Nonstate actors can widen the range of issues gaining attention, for example, by broadening the narrow framing of climate justice and equity as predominantly international issues, and moving from a north versus south, developed versus developing view (Okereke, 2010) to look more closely at local inequities, vulnerabilities, and human rights (Shawoo and McDermott, 2020).

Polycentric modes of governance are well suited to manage issues such as climate change that are manifested at local up to global scales and also enable greater public involvement in decision-making and greater flexibility in institutional and policy design and experimentation. As Burns and Bryce (2022) have noted, state-level climate policy-making in the United States, such as the California LCFS and New York Carbon Dioxide Removal Leadership Act, provides good examples of experimentation in practice, with the states effectively acting as laboratories for climate policy.

Given the breadth of NET options, and the relative technological immaturity of many, a strategy of learning from diversity—rather than from duplication—would be most effective (Reiner, 2016), and the establishment of international mechanisms to coordinate project efforts and to ensure effective management and dissemination of knowledge will help to accelerate progress (Nemet et al., 2018). Coordinated project efforts and wide sharing of results will also contribute to building technical capability, as well as the organizational, marketing, and policy knowledge necessary to bring these technologies to scale (Bossink, 2017).

The need to emphasize trust and trust-building practices identified by Dorsch and Flachsland echoes the recognition by Ostrom (2009) of the conditions under which the collective action of individuals could lead to the sustainable management of common pool resources, such as village irrigation systems or grazing lands: reliable and high-frequency information about the issue of concern, frequent communication among at least subsets of participants, and an awareness of who else had agreed to change behavior and could be seen to have done so are all conditions and practices that build trust.

The varied perspectives and capabilities brought to bear by diverse actors are essential to address the complex interrelationships between social, environmental, financial, and other systems involved in climate governance, and can be expected to lead to better outcomes (Ostrom, 2010; Heikkila et al., 2018). Polycentric and other distributed modes of governance face a number of challenges, however, including:

1. Authority and accountability: the lack of a single (or small number of) recognized and responsible institutions means that potential problems of authority and accountability need to be addressed (Heikkila et al., 2018). However, while lacking formal authority, groups of nonstate actors, cooperating in communities, networks, or "clubs," can gain legitimacy by the adoption of overarching sets of rules (Bernstein, 2004).
2. Execution capability: the executive ability of nonstate actors can be constrained by limited capability and resources, as well as authority (Shawoo and McDermott, 2020) leading to isolated rather than coordinated action.
3. Communication and coordination: a deliberate strengthening of communication channels is needed, across the wide group of actors and levels (Arriagada et al., 2018); failures of coordination can lead to inefficiencies, such as duplicated or conflicting policies and actions, or to inequities in cost/benefit distribution (Dorsch and Flachsland, 2017; Heikkele et al., 2018).
4. Cost and impact assessment: aggregating costs and benefits across the diverse population of actors is problematic, and enhanced methods are needed to achieve this in a quantitative and transparent manner (Dorsch and Flachsland, 2017).

Although, as Dorsch and Flachsland observe, a recognition of its increasingly polycentric structure helps in understanding the dynamics of global climate governance, a robust demonstration of improved governance performance as a result of this shift is still a work in progress (Jordan et al., 2015).

References and resources

References

Anderson, K., Peters, G., 2016. The trouble with negative emissions. Science 354 (6309), 182−183.

Arriagada, R., Aldunce, P., Blanco, G., et al., 2018. Climate change governance in the anthropocene: emergence of polycentrism in Chile. Elementa Sci. Anthropocene 6, 68.

Aspen Institute, 2021. Guidance for Ocean-Based Carbon Dioxide Removal Projects: A Pathway to Developing a Code of Conduct. aspeninstitute.org/publications/ocean-carbon-dioxide-removal.

Asilomar Scientific Organizing Committee, 2010. The Asilomar Conference Recommendations on Principles for Research into Climate Engineering Techniques. Climate Institute, Washington DC. www.climateresponsefund.org/images/Conference/finalfinalreport.pdf.

Bäckstrand, K., Kuyper, J.W., Linnér, B.O., Lövbrand, E., 2017. Non-state actors in global climate governance: from Copenhagen to Paris and beyond. Environ. Polit. 26 (4), 561−579.

Beck, M., Krueger, T., 2016. The epistemic, ethical, and political dimensions of uncertainty in integrated assessment modeling. Wiley Interdiscip. Rev. Clim. Change 7 (5), 627−645.

Bednar, J., Obersteiner, M., Baklanov, A., et al., 2021. Operationalizing the net-negative carbon economy. Nature 596 (7872), 377−383.

Bellamy, R., Lezaun, J., Palmer, J., 2017. Public perceptions of geoengineering research governance: an experimental deliberative approach. Global Environ. Change 45, 194−202.

Bellamy, R., 2018. Incentivize negative emissions responsibly. Nat. Energy 3 (7), 532−534.

Bellamy, R., Geden, O., 2019. Govern CO_2 removal from the ground up. Nat. Geosci. 12 (11), 874−876.

Bellamy, R., Lezaun, J., Palmer, J., 2019. Perceptions of bioenergy with carbon capture and storage in different policy scenarios. Nat. Commun. 10 (1), 1−9.

Bellamy, R., Osaka, S., 2020. Unnatural climate solutions? Nat. Clim. Change 10 (2), 98−99.

Bellassen, V., Stephan, N., Afriat, M., et al., 2015. Monitoring, reporting and verifying emissions in the climate economy. Nat. Clim. Change 5 (4), 319−328.

Bernstein, S., 2004. Legitimacy in global environmental governance. J. Int. Law Int. Relat. 1, 139−166.

Boettcher, M., Brent, K., Buck, H.J., Low, S., McLaren, D., Mengis, N., 2021. Navigating potential hype and opportunity in governing marine carbon removal. Front. Climate 3, 664456.

Bossink, B.A., 2017. Demonstrating sustainable energy: a review based model of sustainable energy demonstration projects. Renew. Sustain. Energy Rev. 77, 1349–1362.

Brander, M., Ascui, F., Scott, V., Tett, S., 2021. Carbon accounting for negative emissions technologies. Clim. Pol. 21 (5), 699–717.

Brent, K., McGee, J., McDonald, J., Rohling, E.J., 2018. International law poses problems for negative emissions research. Nat. Clim. Change 8 (6), 451–453.

Brent, K., Burns, W., McGee, J., 2019. Governance of Marine Geoengineering. Centre for International Governance Innovation, Waterloo, ON, USA.

Broecker, W.S., 1975. Climatic change: are we on the brink of a pronounced global warming? Science 189 (4201), 460–463.

Brown, S.L., 1985. Quantitative risk assessment of environmental hazards. Annu. Rev. Publ. Health 6 (1), 247–267.

Brown, K., Adger, W.N., Cinner, J.E., 2019. Moving climate change beyond the tragedy of the commons. Global Environ. Change 54, 61–63.

Burke, J., Gambhir, A., 2022. Policy incentives for greenhouse gas removal techniques: the risks of premature inclusion in carbon markets and the need for a multi-pronged policy framework. Energy Climate Change 3, 100074.

Burns, W., Bryce, T., 2022. States Can Be Laboratories for Climate Policy. The Hill. www.thehill.com/opinion/energy-environment/598613-states-can-be-laboratories-for-climate-policy.

Chaudhury, A.S., Ventresca, M.J., Thornton, T.F., et al., 2016. Emerging meta-organisations and adaptation to global climate change: evidence from implementing adaptation in Nepal, Pakistan and Ghana. Global Environ. Change 38, 243–257.

Colvin, R.M., Kemp, L., Talberg, A., De Castella, C., Downie, C., Friel, S., et al., 2020. Learning from the climate change debate to avoid polarisation on negative emissions. Environmental Communication 14 (1), 23–35.

Condor, J., Unatrakarn, D., Wilson, M., Asghari, K., 2011. A comparative analysis of risk assessment methodologies for the geologic storage of carbon dioxide. Energy Proc. 4, 4036–4043.

Corner, A., Parkhill, K., Pidgeon, N., Vaughan, N.E., 2013. Messing with nature? Exploring public perceptions of geoengineering in the UK. Global Environ. Change 23 (5), 938–947.

Corner, A., Pidgeon, N., 2015. Like artificial trees? The effect of framing by natural analogy on public perceptions of geoengineering. Climatic Change 130 (3), 425–438.

Cox, E.M., Pidgeon, N., Spence, E., Thomas, G., 2018. Blurred lines: the ethics and policy of greenhouse gas removal at scale. Front. Environ. Sci. 38.

Creutzig, F., Corbera, E., Bolwig, S., Hunsberger, C., 2013. Integrating place-specific livelihood and equity outcomes into global assessments of bioenergy deployment. Environ. Res. Lett. 8 (3), 035047.

Creutzig, F., 2016. Economic and ecological views on climate change mitigation with bioenergy and negative emissions. Global Change Biol. Bioenergy 8 (1), 4–10.

Creutzig, F., 2017. Govern land as a global commons. Nature 546 (7656), 28–29.

Crissman, C.C., Antle, J.M., Capalbo, S.M. (Eds.), 1998. Economic, Environmental, and Health Tradeoffs in Agriculture: Pesticides and the Sustainability of Andean Potato Production. Kluwer Academic Publishers, Boston, USA, p. 281.

De Santo, E.M., Mendenhall, E., Nyman, E., Tiller, R., 2020. Stuck in the middle with you (and not much time left): the third intergovernmental conference on biodiversity beyond national jurisdiction. Marine Policy 117, 103957.

Dixon, T., Garrett, J., Kleverlaan, E., 2014. Update on the London protocol–developments on transboundary CCS and on geoengineering. Energy Proc. 63, 6623–6628.

Dooley, K., Kartha, S., 2018. Land-based negative emissions: risks for climate mitigation and impacts on sustainable development. Int. Environ. Agreements Polit. Law Econ. 18 (1), 79–98.

Dooley, K., Christoff, P., Nicholas, K.A., 2018. Co-producing climate policy and negative emissions: trade-offs for sustainable land-use. Global Sustain. 1 (E3), 1–10.

Dooley, K., Harrould-Kolieb, E., Talberg, A., 2021. Carbon-dioxide removal and biodiversity: a threat identification framework. Global Policy 12, 34–44.

Dorsch, M.J., Flachsland, C., 2017. A polycentric approach to global climate governance. Global Environ. Polit. 17 (2), 45–64.

Dowell, G., Niederdeppe, J., Vanucchi, J., et al., 2020. Rooting carbon dioxide removal research in the social sciences. Interface Focus 10 (5), 20190138.

Duchelle, A.E., Simonet, G., Sunderlin, W.D., Wunder, S., 2018. What is REDD+ achieving on the ground? Curr. Opin. Environ. Sustain. 32, 134–140.

Fajardy, M., Patrizio, P., Daggash, H.A., Mac Dowell, N., 2019. Negative emissions: priorities for research and policy design. Front. Climate 1, 6.

Fehrenbach, H., 2011. How bioenergy related water impacts are considered by certification schemes. Biofuels Bioproducts Biorefining 5 (4), 464–473.

Fuhr, H., Hickmann, T., Kern, K., 2018. The role of cities in multi-level climate governance: local climate policies and the 1.5°C target. Curr. Opin. Environ. Sustain. 30, 1–6.

Fuhrman, J., McJeon, H., Patel, P., Doney, S.C., Shobe, W.M., Clarens, A.F., 2020. Food–energy–water implications of negative emissions technologies in a +1.5°C future. Nat. Clim. Change 10 (10), 920–927.

Fuss, S., Canadell, J.G., Peters, G.P., et al., 2014. Betting on negative emissions. Nat. Clim. Change 4 (10), 850–853.

Fuss, S., Canadell, J.G., Ciais, P., et al., 2020. Moving toward net-zero emissions requires new alliances for carbon dioxide removal. One Earth 3 (2), 145–149.

Fyson, C.L., Baur, S., Gidden, M., Schleussner, C.F., 2020. Fair-share carbon dioxide removal increases major emitter responsibility. Nat. Clim. Change 10 (9), 836–841.

Gannon, K.E., Hulme, M., 2018. Geoengineering at the "edge of the world": exploring perceptions of ocean fertilisation through the Haida Salmon restoration corporation. Geo Geography Environ. 5 (1), e00054.

Gardiner, S.M., 2010. Ethics and climate change: an introduction. Wiley Interdiscip. Rev. Clim. Change 1 (1), 54–66.

Gardiner, S., 2017. Accepting collective responsibility for the future. J. Practical Ethics 5 (1).

Gardiner, S.M., Fragnière, A., 2018. The Tollgate Principles for the governance of geoengineering: moving beyond the Oxford Principles to an ethically more robust approach. Ethics Pol. Environ. 21 (2), 143–174.

Geden, O., Peters, G.P., Scott, V., 2019. Targeting carbon dioxide removal in the European Union. Climate Policy 19 (4), 487–494.

Giller, K.E., Leeuwis, C., Andersson, J.A., et al., 2008. Competing claims on natural resources: what role for science? Ecol. Soc. 13 (2), 34.

Ginbo, T., Di Corato, L., Hoffmann, R., 2021. Investing in climate change adaptation and mitigation: a methodological review of real-options studies. Ambio 50 (1), 229–241.

Goldstein, J.E., 2021. More data, more problems? Incompatible uncertainty in Indonesia's climate change mitigation projects, 132. Geoforum, pp. 195–204.

Golub, A., Narita, D., Schmidt, M.G., 2014. Uncertainty in integrated assessment models of climate change: alternative analytical approaches. Environ. Model. Assess. 19 (2), 99–109.

Hale, B., Dilling, L., 2011. Geoengineering, ocean fertilization, and the problem of permissible pollution. Sci. Technol. Hum. Val. 36 (2), 190–212.

Harrould-Kolieb, E.R., Hoegh-Guldberg, O., 2019. A governing framework for international ocean acidification policy. Mar. Pol. 102, 10–20.

Heikkila, T., Villamayor-Tomas, S., Garrick, D., 2018. Bringing polycentric systems into focus for environmental governance. Environ. Policy Governance 28 (4), 207–211.

Honegger, M., Reiner, D., 2018. The political economy of negative emissions technologies: consequences for international policy design. Clim. Pol. 18 (3), 306–321.

Honegger, M., Poralla, M., Michaelowa, A., Ahonen, H.M., 2021. Who is paying for carbon dioxide removal? Designing policy instruments for mobilizing negative emissions technologies. Front. Climate 3, 50.

Howlett, M., 2014. Why are policy innovations rare and so often negative? Blame avoidance and problem denial in climate change policy-making. Global Environ. Change 29, 395–403.

Hubert, A.M., 2020. A code of conduct for responsible geoengineering research. Global Policy 12, 82–96.

Hunsberger, C., Bolwig, S., Corbera, E., Creutzig, F., 2014. Livelihood impacts of biofuel crop production: implications for governance. Geoforum 54, 248–260.

IPCC, 2018. Special Report on the Impacts of Global Warming of 1.5°C. In: Masson-Delmotte, V.P., Zhai, H.-O., Pörtner, D., et al. (Eds.). IPCC, Geneva, Switzerland.

IUCN, 2020. Global Standard for Nature-Based Solutions. A User-Friendly Framework for the Verification, Design and Scaling up of NbS. IUCN, Gland, Switzerland.

Jamieson, D., 1996. Ethics and intentional climate change. Climatic Change 33, 323–336.

Jobin, M., Siegrist, M., 2020. Support for the deployment of climate engineering: a comparison of ten different technologies. Risk Analysis 40 (5), 1058–1078.

Jordan, A.J., Huitema, D., Hildén, M., et al., 2015. Emergence of polycentric climate governance and its future prospects. Nat. Clim. Change 5 (11), 977–982.

Kanter, D.R., Musumba, M., Wood, S.L., et al., 2018. Evaluating agricultural trade-offs in the age of sustainable development. Agric. Syst. 163, 73–88.

Keller, D.P., Lenton, A., Littleton, E.W., Oschlies, A., Scott, V., Vaughan, N.E., 2018. The effects of carbon dioxide removal on the carbon cycle. Curr. Clim. Change Rep. 4 (3), 250–265.

Kellogg, W.W., Schneider, S.H., 1974. Climate stabilization: for better or for worse? Science 186 (4170), 1163–1172.

Klapwijk, C.J., Van Wijk, M.T., Rosenstock, T.S., van Asten, P.J., Thornton, P.K., Giller, K.E., 2014. Analysis of trade-offs in agricultural systems: current status and way forward. Curr. Opin. Environ. Sustain. 6, 110–115.

Kobayashi, Y., Peters, G.M., Khan, S.J., 2015. Towards more holistic environmental impact assessment: hybridisation of life cycle assessment and quantitative risk assessment. Procedia CIRP 29, 378–383.

Kreuter, J., Lederer, M., 2021. The geopolitics of negative emissions technologies: learning lessons from REDD+ and renewable energy for afforestation, BECCS, and direct air capture. Global Sustain. 4.

Kverndokk, S., Nævdal, E., Nøstbakken, L., 2014. The trade-off between intra-and intergenerational equity in climate policy. Eur. Econ. Rev. 69, 40–58.

Lackner, K.S., Aines, R., Atkins, S., et al., 2016. The promise of negative emissions. Science 354 (6313), 714.

Lawford-Smith, H., Currie, A., 2017. Accelerating the carbon cycle: the ethics of enhanced weathering. Biol. Lett. 13 (4), 20160859.

Lawrence, M.G., Crutzen, P.J., 2017. Was breaking the taboo on research on climate engineering via albedo modification a moral hazard, or a moral imperative? Earth's Future 5 (2), 136–143.

Lenzi, D., 2018. The ethics of negative emissions. Global Sustain. 1 (E7), 1–8.

Lezaun, J., 2021. Hugging the shore: tackling marine carbon dioxide removal as a local governance problem. Front. Climate 3, 684063.

Lomax, G., Workman, M., Lenton, T., Shah, N., 2015a. Reframing the policy approach to greenhouse gas removal technologies. Energy Pol. 78, 125–136.

Loomis, R., Cooley, S.R., Collins, J.R., Engler, S., Suatoni, L., 2022. A code of conduct is imperative for ocean carbon dioxide removal research. Front. Mar. Sci. 9, 598.

Mace, M.J., Fyson, C.L., Schaeffer, M., Hare, W.L., 2021. Large-scale carbon dioxide removal to meet the 1.5°C limit: key governance gaps, challenges and priority responses. Global Policy 12, 67–81.

McCormack, C.G., Born, W., Irvine, P.J., et al., 2016. Key impacts of climate engineering on biodiversity and ecosystems, with priorities for future research. J. Integr. Environ. Sci. 13 (2–4), 103–128.

McLaren, D., 2020. Quantifying the potential scale of mitigation deterrence from greenhouse gas removal techniques. Climatic Change 162 (4), 2411–2428.

McLaren, D.P., Tyfield, D.P., Willis, R., Szerszynski, B., Markusson, N.O., 2019. Beyond 'Net-Zero': a case for separate targets for emissions reduction and negative emissions. Front. Climate 1, 4.

MacMartin, D.G., Kravitz, B., 2019. Mission-driven research for stratospheric aerosol geoengineering. Proc. Natl. Acad. Sci. USA 116 (4), 1089–1094.

McShane, T.O., Hirsch, P.D., Trung, T.C., et al., 2011. Hard choices: making trade-offs between biodiversity conservation and human well-being. Biol. Conserv. 144 (3), 966–972.

Maher, B., Symons, J., 2022. The international politics of carbon dioxide removal: pathways to cooperative global governance. Global Environ. Polit. 22 (1), 44–68.

Malkamäki, A., D'Amato, D., Hogarth, N.J., et al., 2018. A systematic review of the socio-economic impacts of large-scale tree plantations, worldwide. Global Environ. Change 53, 90–103.

Markusson, N., McLaren, D., Tyfield, D., 2018. Towards a cultural political economy of mitigation deterrence by negative emissions technologies (NETs). Global Sustain. 1 (E10), 1–9.

Markusson, N., 2022. Natural carbon removal as technology. Wiley Interdisciplinary Reviews: Climate Change 13 (2), e767.

Merk, C., Pönitzsch, G., Rehdanz, K., 2019. Do climate engineering experts display moral-hazard behaviour? Clim. Pol. 19 (2), 231–243.

Michaelowa, A., Allen, M., Sha, F., 2018. Policy instruments for limiting global temperature rise to 1.5°C can humanity rise to the challenge? Clim. Pol. 18 (3), 275–286.

Morrow, D., Svoboda, T., 2016. Geoengineering and non-ideal theory. Publ. Aff. Q. 30 (1), 83–102.

Morrow, D., 2017. International Governance of Climate Engineering: A Survey of Reports on Climate Engineering, 2009–2015.

Nemet, G.F., Zipperer, V., Kraus, M., 2018. The valley of death, the technology pork barrel, and public support for large demonstration projects. Energy Pol. 119, 154–167.

Oberthür, S., Groen, L., 2017. The European union and the Paris agreement: leader, mediator, or bystander? Wiley Interdisciplinary Reviews: Climate Change 8 (1), e445.

Okereke, C., 2010. Climate justice and the international regime. Wiley Interdiscip. Rev. Clim. Change 1 (3), 462–474.

Oldfield, E.D., 2022. Crediting agricultural soil carbon sequestration. Science 375, 1222–1225.

Osaka, S., Bellamy, R., Castree, N., 2021. Framing "nature-based" solutions to climate change. Wiley Interdiscip. Rev. Clim. Change 12 (5), e729.

Ostrom, E., 2009. A Polycentric Approach for Coping with Climate Change. World Bank Policy Research Working Paper 5095.

Ostrom, E., 2010. Polycentric systems for coping with collective action and global environmental change. Global Environ. Change 20 (4), 550–557.

Pahl-Wostl, C., Arthington, A., Bogardi, J., et al., 2013. Environmental flows and water governance: managing sustainable water uses. Curr. Opin. Environ. Sustain. 5 (3–4), 341–351.

Peters, G.P., Geden, O., 2017. Catalysing a political shift from low to negative carbon. Nature Climate Change 7 (9), 619–621.

Pindyck, R.S., 2013. Climate change policy: what do the models tell us? J. Econ. Lit. 51 (3), 860–872.

Preston, C.J., 2013. Ethics and geoengineering: reviewing the moral issues raised by solar radiation management and carbon dioxide removal. Wiley Interdiscip. Rev. Clim. Change 4 (1), 23–37.

Price, P., McGeever, A., Jones, M., McMullin, B., 2018. A Post-Paris Literature Review of Negative Emissions Technology, and Potential for Ireland. Dublin City University and Trinity College Dublin. Jan 2018. http://ienets.eeng.dcu.ie/documents/Lit-Review-2018-01.pdf.

Qiu, J., Queiroz, C., Bennett, E.M., et al., 2021. Land-use intensity mediates ecosystem service tradeoffs across regional social-ecological systems. Ecosystems People 17 (1), 264–278.

Rayner, S., Redgwell, C., Savulescu, J., Pidgeon, N., Kruger, T., 2009. Memorandum on Draft Principles for the Conduct of Geoengineering Research. House of Commons Science and Technology Committee, The Regulation of Geoengineering.

Rayner, S., Heyward, C., Kruger, T., Pidgeon, N., Redgwell, C., Savulescu, J., 2013. The Oxford principles. Climatic Change 121 (3), 499–512.

Reiner, D.M., 2016. Learning through a portfolio of carbon capture and storage demonstration projects. Nat. Energy 1 (1), 15011.

Sandalow, D., Aines, R., Friedmann, J., McCormick, C., Sanchez, D.L., 2021. Biomass Carbon Removal and Storage (BiRCS) Roadmap (No. LLNL-TR-815200). Lawrence Livermore National Lab.(LLNL), Livermore, CA (United States).

Scott, V., Geden, O., 2018. The challenge of carbon dioxide removal for EU policy-making. Nat. Energy 3 (5), 350.

Shawoo, Z., McDermott, C.L., 2020. Justice through polycentricity? A critical examination of climate justice framings in Pakistani climate policymaking. Clim. Pol. 20 (2), 199–216.

Schenuit, F., Colvin, R., Fridahl, M., et al., 2021. Carbon dioxide removal policy in the making: assessing developments in 9 OECD cases. Front. Climate 3, 7.

Shue, H., 2017. Climate dreaming: negative emissions, risk transfer, and irreversibility. J. Human Rights Environ. 8 (2), 203–216.

Smith, P., Davis, S.J., Creutzig, F., et al., 2016. Biophysical and economic limits to negative CO_2 emissions. Nat. Clim. Change 6 (1), 42–50.

Sollie, O.K., Bernstone, C., Carpenter, M.E., Selmer-Olsen, S., 2011. An early phase risk and uncertainty assessment method for CO_2 geological storage sites. Energy Proc. 4, 4132–4139.

Stern, N., 2016. Economics: current climate models are grossly misleading. Nat. News 530 (7591), 407.

Stoorvogel, J.J., Antle, J.M., Crissman, C.C., Bowen, W., 2004. The tradeoff analysis model: integrated bio-physical and economic modeling of agricultural production systems. Agric. Syst. 80 (1), 43–66.

Thoni, T., Beck, S., Borchers, M., et al., 2020. Deployment of negative emissions technologies at the national level: a need for holistic feasibility assessments. Front. Climate 2, 590305.

Tokarska, K.B., Zickfeld, K., 2015. The effectiveness of net negative carbon dioxide emissions in reversing anthropogenic climate change. Environ. Res. Lett. 10 (9), 094013.

Torvanger, A., 2019. Governance of bioenergy with carbon capture and storage (BECCS): accounting, rewarding, and the Paris agreement. Clim. Pol. 19 (3), 329–341.

UK Government, 2020. Carbon Emissions Tax – Open Consultation. HM Treasury. www.gov.uk/government/consultations/carbon-emissions-tax.

UNEP, 2010. Decision Adopted by the Conference of the Parties to the Convention on Biological Diversity at Its Tenth Meeting, X/33. Biodiversity and Climate Change; Convention on Biological Diversity: Rio de Janeiro, Brazil. UNEP/CBD/COP/DEC/X/33 8(w).

UNEP, 2012. Secretariat of the Convention on Biological Diversity. Geoengineering in Relation to the Convention on Biological Diversity: Technical and Regulatory Matters. UNEP Montreal, Technical Series, 66.

UNESCO, 2017. United Nations Educational, Scientific and Cultural Organization: Declaration of Ethical Principles in Relation to Climate Change. UNESCO, Paris. Available at: portal.unesco.org/en/ev.php-URL_ID=49457&URL_DO=DO_TOPIC& URL_SECTION=201.html.

Vaughan, N.E., Gough, C., 2016. Expert assessment concludes negative emissions scenarios may not deliver. Environ. Res. Lett. 11 (9), 095003.

Visseren-Hamakers, I.J., McDermott, C., Vijge, M.J., Cashore, B., 2012. Trade-offs, co-benefits and safeguards: current debates on the breadth of REDD+. Curr. Opin. Environ. Sustain. 4 (6), 646–653.

Webb, R., Silverman-Roati, K., Gerrard, M., 2021. Removing Carbon Dioxide through Ocean Alkalinity Enhancement and Seaweed Cultivation: Legal Challenges and Opportunities. Sabin Center for Climate Change Law, Columbia Law School, Columbia Public Law Research Paper, February 2021.

Wolske, K.S., Raimi, K.T., Campbell-Arvai, V., Hart, P., 2019. Public support for carbon dioxide removal strategies: the role of tampering with nature perceptions. Climatic Change 152 (3), 345–361.

Wren, 2022. The Wren Charter. Available at: www.notion.so/Wren-Charter-426a19cd3d414528bb5e95067e8a81be.

Wright, M., Teagle, D., Feetham, P., 2014. A quantitative evaluation of the public response to climate engineering. Nature Climate Change 4 (2), 106–110.

Further reading

Aven, T., 2016. Risk assessment and risk management: review of recent advances on their foundation. Eur. J. Oper. Res. 253 (1), 1–13.

Brent, K., Burns, W., McGee, J., 2019. Governance of marine geoengineering. Special report of the centre for international governance innovation. Available at: cigionline.org.

Bridge, D.J., 2021. The ethics of climate change: a systematic literature review. Account. Finance. https://doi.org/10.1111/acfi.12877.

Burch, S., Gupta, A., Inoue, C.Y., et al., 2019. New directions in earth system governance research. Earth System Governance 1, 100006.

Farhat, K., Benson, S.M., 2016. Translating risk assessment to contingency planning for CO_2 geologic storage: a methodological framework. Int. J. Greenh. Gas Control 52, 410–431.

Hale, B., 2012. Getting the bad out: remediation technologies and respect for others. In: Kabasenche, W.P., Rourke, M.O. (Eds.), The Environment: Philosophy, Science, and Ethics. MIT Press, pp. 223–243.

Hnottavange-Telleen, K., Chabors, E., Finley, R.J., Greenberg, S.E., Marsteller, S., 2011. Risk management in a large-scale CO_2 geosequestration pilot project, Illinois, USA. Energy Proc. 4, 4044–4051.

IMO (International Maritime Organization), 2010. Assessment Framework for Scientific Research Involving Ocean Fertilization. LC 32/15 Annex 6, 2.2.2.

IMO (International Maritime Organization), 2013. Resolution LP.4(8) on the Amendment to the London Protocol to Regulate the Placement of Matter for Ocean Fertilization and Other Marine Geoengineering Activities. LC 35/15 Annex 4.

Kartha, S., Dooley, K., 2016. The Risks of Relying on Tomorrow's 'negative Emissions' to Guide Today's Mitigation Action. Working paper 2016-08. Stockholm Environmental Institute, Somerville, MA, USA.

Lenton, A., Matear, R.J., Keller, D.P., Scott, V., Vaughan, N.E., 2018. Assessing carbon dioxide removal through global and regional ocean alkalinization under high and low emission pathways. Earth System Dynamics 9 (2), 339–357.

Li, Q., Liu, G., 2016. Risk assessment of the geological storage of CO_2: a review. In: Geologic Carbon Sequestration. Springer, Cham, Switzerland, pp. 249–284.

Lomax, G., Lenton, T.M., Adeosun, A., Workman, M., 2015b. Investing in negative emissions. Nat. Clim. Change 5 (6), 498.

UNEP, 2020. In: Olhoff, A., Christensen, J.M. (Eds.), Emissions Gap Report 2020. UNEP DTU Partnership.

Projects and organizations

Carbon Capture Coalition: US CCS and NETs advocacy group. www.carboncapturecoalition.org.

Carnegie Climate Geoengineering Governance Initiative (C2G2): Carnegie Council for Ethics in International Affairs initiative, aiming to catalyze the creation of effective governance for SRM and CDR. www.c2g2.net.

Centre for International Governance Innovation: an independent think tank focusing on governance of the global economy, global security and politics, and international law in collaboration. cigionline.org.

Earth System Governance: Social science research network working in the area of governance and global environmental change. www.earthsystemgovernance.org.

Future Earth: global collaboration of scientists, researchers, and innovators for a more sustainable planet. www.futureearth.org.

High Seas Alliance: Global NGO alliance working to support and strengthen ocean governance and conservation. www.highseasalliance.org.

Institute for Carbon Removal Law and Policy: research centre assessing social, legal, ethical, and political implications of CDR. www.american.edu/sis/centers/carbon-removal.

International Carbon Reduction and Offset Alliance (ICROA): non-profit organisation bringing together leading carbon reduction and offset providers in the voluntary carbon market. www.icroa.org.

Project Vesta: coastal enhanced weathering project with participatory governance approach to social license. www.vesta.earth/our-research#Social-License.

UN Convention on Biological Diversity: UN Convention addressing the conservation of biological diversity, its sustainable use, and the fair and equitable sharing of benefits arising therefrom. www.cbd.int.

US DOE Bipartisan Infrastructure Law Program: Funding support for US energy transition RD&D. https://www.energy.gov/bil/bipartisan-infrastructure-law-programs.

Chapter 4

The global carbon cycle

Steve Rackley

The carbon inventories of the atmosphere, ocean, and terrestrial biosphere are linked by fluxes driven by a complex set of biogeochemical processes, collectively known as the global carbon cycle. Fig. 4.1 illustrates the inventories (bold font, units of GtC) and fluxes (italic font, units of GtC/yr) that make up this cycle.

Estimates of the main carbon fluxes are published in the annually updated Global Carbon Budget (Friedlingstein et al., 2021, and references therein to previous years), which provides the basis for the nomenclature used throughout this book, as summarized in Table 4.1.

4.1 Carbon inventories

The main inventories relevant to the global carbon cycle are described in the following section.

4.1.1 Atmospheric carbon inventory

The atmospheric carbon inventory consists almost entirely of carbon dioxide, with a concentration [CO_2] reaching an annual mean value of 416.5 ppm, or 0.04% by volume, in 2021. This inventory has risen by 50% since preindustrial times (from ca. 277 ppm in 1750) as a net result of emissions from fossil fuel combustion and cement production (E_{FF}), and changes in land-use and land-management practices (E_{LUC}). The E_{FF} component has been predominant since the 1920s and accounts for around 80% of the overall historical increase in the atmospheric CO_2 inventory. The remaining atmospheric carbon inventory consists of methane (CH_4) at ~1.9 ppm, with traces of carbon monoxide (CO) and anthropogenic chlorofluorocarbons.

Detailed measurements of atmospheric [CO_2] were started by Charles Keeling at the National Oceanic and Atmospheric Administration Mauna Loa Observatory, Hawaii, in September 1957, with an average value of 315 ppm recorded for the first full year of measurements. The curve of increasing [CO_2] established since that time is known as the Keeling curve, and the data from the start of the Mauna Loa continuous high-precision record in March 1958 are shown in Fig. 4.2. The annual increase in [CO_2] has averaged 2.2 ppm/yr in the decade to 2021. The "milestone" of the first measured monthly average above 400 ppm was reached in April 2014 (or April 2015 when corrected for the seasonal trend) and a full year of monthly average [CO_2] >400 ppm was recorded in 2016.

The cyclical overprint on the rising trend is shown in Fig. 4.3 for the 5 years to 2020, and compared to an average for the period from 1960 to 1965. The cycle is synchronized with the northern hemisphere seasons, where [CO_2] is drawn down ~3.5 ppm below the annual average by photosynthetic production during spring and summer (May to September) and rebounds by a similar amount as a result of biomass decomposition in autumn and winter (October to April).

The amplitude of this annual [CO_2] cycle at Mauna Loa has increased by about 0.3% per year, from ~5.7 ppm in the late 1950s to 7.1 ppm over the 5 years to 2020, and at double that rate at high latitudes, the highest at Barrow, Alaska (13–18 ppm; Graven et al., 2013). This is attributed to an increase in summertime carbon uptake through increased photosynthetic production in northern terrestrial ecosystems, particularly boreal and temperate forests, as a result of increasing [CO_2] (i.e., CO_2 fertilization) and increasing temperatures. Observations confirm increased forest stand area and biomass, and the northward migration of evergreen shrubs and trees in response to climate change (Forkel et al., 2016).

The total inventory of carbon in the atmosphere with [CO_2] at 416 ppm is ~885 GtC (2.124 GtC/ppm [CO_2]; Ballantyne et al., 2012), with an annual growth (G_{ATM}) of ~5.1 GtC/yr in the decade to 2020 (Friedlingstein et al., 2021).

Negative Emissions Technologies for Climate Change Mitigation. https://doi.org/10.1016/B978-0-12-819663-2.00002-2

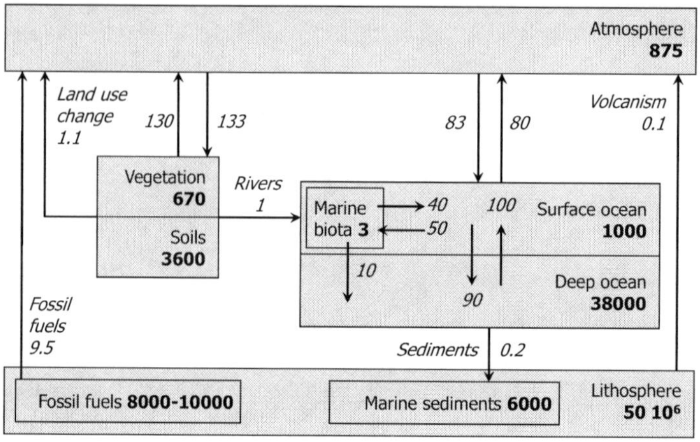

FIGURE 4.1 Inventories and fluxes in the carbon cycle (2011–20 estimates).

TABLE 4.1 Global carbon budget nomenclature.

Element	Description	Source
E_{FF} or E_{FFI}	Emissions from fossil fuel use and cement production	Energy statistics and industrial production data
E_{LUC}	Emissions from land use and land cover change	Land-cover change data, deforestation-related fire activity, and dynamic global vegetation models (DGVMs)
G_{ATM}	Annual growth in global atmospheric CO_2 inventory	Direct measurement of atmospheric CO_2 concentration [CO_2]
S_{OCEAN}	Uptake by the ocean carbon sink, including the impact of increasing atmospheric CO_2, and climate change	Observations and ocean models
S_{LAND}	Uptake by the terrestrial carbon sink due to increasing [CO_2], climate, and other environmental changes	Estimated from DGVMs

After Friedlingstein et al. 2021.

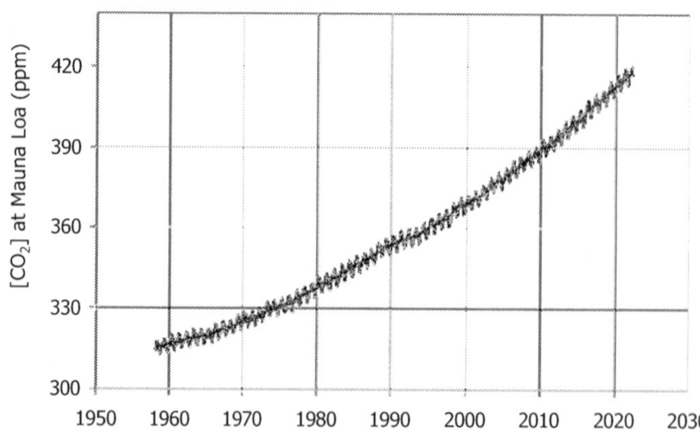

FIGURE 4.2 [CO_2] at Mauna Loa observatory. *Data courtesy NOAA, Earth System Research Laboratory, Global Monitoring Division (Scripps CO_2 Program).*

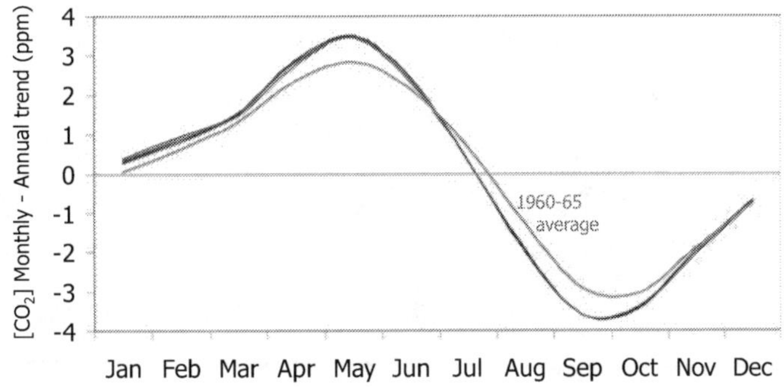

FIGURE 4.3 Annual $[CO_2]$ cycle, 2015–20. *Date source: as Fig. 4.2.*

TABLE 4.2 Soil carbon inventory.

Depth range (m)	Soil organic carbon (SOC; GtC)	Soil inorganic carbon (SIC; GtC))	Inventory (GtC)
0–0.3	750	173	923
0–1	1425	868	2293
0–2	2047	1558	3605

Data source: Lorentz and Lal 2018.

4.1.2 Terrestrial carbon inventory

The terrestrial carbon inventory is estimated to hold 4600 GtC, of which ~670 GtC is present as living biomass and ~3600 GtC as organic and inorganic carbon in soils to 2 m depth (Table 4.2). This inventory has declined by roughly 5% since preindustrial times, and predominantly since the mid-19th century, reaching ~3 GtC/yr losses in the period from 2001 to 2015, as a result of changes in land-use and land-management practices, with deforestation, conversion of grasslands to agricultural use, and intensive agricultural practices being the main contributors.

The soil carbon inventory can be classified according to the residence time of various components, ranging from plant and animal detritus, decomposing and partly emitted through respiration within a typical time scale of 1–10 years, to inert carbon, which is inaccessible to biological processes and remains in the soil until physically removed by water or airborne transport. These processes are discussed further in the next chapter.

4.1.3 Oceanic carbon inventory

The oceanic carbon inventory amounts to ~39,000 GtC, with more than 90% of this being present as bicarbonate ions (HCO_3^-), as shown in Table 4.3. In addition, some 2500 GtC is present in marine carbonate sediments, which are gradually transformed into sedimentary rock over geological time. Of the inorganic carbon inventory in the oceans, ~180 GtC is of anthropogenic origin, with an estimated uptake rate of 2.8 GtC/yr in the decade to 2020 (Friedlingstein et al., 2021).

Within the oceans, three key processes—the biological and solubility pumps, and the thermohaline circulation—drive the oceanic carbon inventory, its distribution between organic and inorganic fractions, and its transport and eventual deposition in sediments. These processes are discussed further in Chapter 6.

4.1.4 Carbon inventory of the lithosphere

The Earth's crust, which represents the upper part of the lithosphere, is the ultimate geological carbon sink and is estimated to hold 5×10^7 GtC in sedimentary rocks, ~20% of which is in the form of organic carbon and the remainder as limestone. Fossil fuels—coal, oil, and gas—together account for between 4000 and 6000 GtC, or ~0.05% of the total organic carbon present in sedimentary rocks.

TABLE 4.3 Oceanic carbon inventory.

Component	Inventory (GtC)
Bicarbonate ion	~36,000
Carbonate ion	~1300
Dissolved CO_2	~740
Dissolved organic carbon (DIC)	<700
Marine biomass	<10
Total	~39,000

4.2 Carbon fluxes

These carbon stocks are constantly changing as a result of both natural and anthropogenic fluxes, the net result of which has been that some 58% of historical anthropogenic CO_2 emissions ($E_{FF} + E_{LUC}$) to the atmosphere have been taken up in the terrestrial biosphere and the ocean.

4.2.1 Atmosphere ↔ terrestrial biosphere fluxes

Terrestrial photosynthesis removes an estimated 123 GtC/yr from the atmosphere as gross primary production, including ~28% of anthropogenic CO_2 emissions from fossil fuel combustion, industrial processes, and land-use changes. Roughly half of this is reemitted through plant (autotrophic) respiration, the remainder being retained as net primary production (NPP).

Soil respiration, primarily from the microbial communities that feed on plant detritus and root exudates (heterotrophic respiration), returns a further ~55 GtC/yr to the atmosphere. Under steady state, the balance is made up of emissions back to the atmosphere as a result of natural fires and dissolved organic carbon export by rainwater runoff into rivers.

The CO_2 flux from the atmosphere to the terrestrial carbon sink (S_{LAND}) is attributed to CO_2 fertilization, which increases NPP with increasing [CO_2], as a result of increased photosynthetic and water-use efficiency (the latter important in arid areas) and to warming which, as noted earlier, is resulting in the northward migration of forest ecosystems. Before 2016, S_{LAND} was inferred in the annual global carbon budget estimate as the balancing term in Eq. (4.1);

$$S_{LAND} = E_{FF} + E_{LUC} - S_{OCEAN} - G_{ATM} \tag{4.1}$$

Dynamic global vegetation models (DGVMs) provide an independent assessment of S_{LAND}, but up until 2015 the wide spread of DVGM results did not provide a narrower range for S_{LAND} than that derived from Eq. (4.1) (Le Quéré et al., 2016). From the 2016 estimate onwards, S_{LAND} has been determined from DGVMs, the overall budget being balanced by the introduction of an additional term (B_{IM}) to balance the now fully independent individual terms;

$$B_{IM} = E_{FF} + E_{LUC} - S_{LAND} - S_{OCEAN} - G_{ATM} \tag{4.2}$$

Over the decade to 2020, the net flux from the atmosphere to the terrestrial biosphere ($S_{LAND} - E_{LUC}$) is estimated to have been 2.0 GtC/yr (Fig. 1.2)—some 21% of fossil fuel emissions—the net result of an uptake of 3.1 GtC/yr offset by 1.1 GtC/yr of land-use change emissions (Friedlingstein et al., 2021).

4.2.2 Atmosphere ↔ ocean fluxes

The exchange of CO_2 between the atmosphere and the ocean occurs due to the difference in CO_2 partial pressure between the atmosphere and surface waters, with ~90 GtC being exchanged annually. This flux is controlled by two key processes: the global ocean circulation system, which exchanges surface and deep water on a 500- to 1000-year time scale, and the biogeochemistry of surface waters, in particular the removal of CO_2 by marine photosynthesis and of carbonate ions by ionic reactions and by precipitation.

An increase in dissolved CO_2 reduces the carbonate ion concentration ([CO_3^{2-}]) due to the following reactions:

$$CO_2 + H_2O \rightarrow H_2CO_3 \rightarrow H^+ + HCO_3^- \tag{4.3}$$

and

$$H^+ + CO_3^{2-} \rightarrow HCO_3^- \tag{4.4}$$

The conversion of carbonic acid (H_2CO_3) to bicarbonate ions by the consumption of carbonate ions mitigates the increase in acidity that would otherwise occur and is referred to as the carbonate buffer.

Ocean alkalinity, expressed in a simplified form as $[HCO_3^-] + 2 \times [CO_3^{2-}]$, is preserved in this reaction since one carbonate ion is replaced by two bicarbonate ions. However, the decline in $[CO_3^{2-}]$ means that the ion is less available to react with additional dissolved CO_2, resulting in a more pronounced increase in acidity with further CO_2 uptake (see Chapter 6).

The ocean provides a slow-acting buffer to stabilize atmospheric $[CO_2]$, and any atmospheric perturbation will be dissipated by absorption into the ocean over a time scale of centuries. The rise in atmospheric $[CO_2]$ shown in Fig. 4.2 has resulted in an increased rate of uptake of CO_2 by the oceans, reducing the atmospheric carbon inventory by ~2.3 GtC/yr, or roughly one-quarter of anthropogenic emissions, in the period from 1990 to 2020.

4.2.3 Atmosphere ↔ lithosphere fluxes

A carbon flux of ~0.5 GtC/yr from the atmosphere occurs as a result of the carbonate–silicate cycle in which carbonic acid, formed by the dissolution of CO_2 in rainwater (the first part of reaction 4.3), causes the weathering of exposed silicate rocks according to the overall reaction:

$$CaSiO_3 + 2CO_2 + 3H_2O \rightarrow Ca^{2+} + 2HCO_3^- + H_4SiO_4 \tag{4.5}$$

The dissolved minerals are transported by rivers into the sea, increasing ocean alkalinity and the DIC inventory, and supporting the carbonate buffer.

The carbonate–silicate cycle is eventually closed over geological time by the subduction of sedimentary rocks formed by the precipitation and sedimentation of the weathering products. Metamorphosis in subduction zones reforms the silicate minerals, while CO_2 is released through volcanoes. The CO_2 flux from volcanic venting is estimated to add on average ~0.1 GtC/yr to the atmospheric inventory.

The climatic impact of volcanic activity can however be far more significant, and more complex, that this relatively small CO_2 flux would suggest. One of the clearest deviations from the $[CO_2]$ growth trend shown in Fig. 4.2 occurred in the 2 years following the June 1991 eruption of Mount Pinatubo, on the Philippine island of Luzon. Rather than a continuous increase, the $[CO_2]$ trend shows a plateauing in 1992/93, which is attributed to the impact on the ocean and land carbon cycles of the release of some 100 Mt-SO_2.

4.2.4 Impact of eCO_2 and climate change on carbon fluxes

The impact of enhanced atmospheric CO_2 concentration (eCO_2) and climate change on terrestrial and oceanic carbon inventories and fluxes is one of the main sources of uncertainty affecting future climate predictions and is an active area of research (Ballantyne et al., 2012; Heinze et al., 2015; Pugh et al., 2016; IPCC, 2021). As discussed earlier, the terrestrial and oceanic sinks (S_{LAND} and S_{OCEAN}) have played a major role in mitigating the effect of historical CO_2 emissions, to the extent that only ~42% of anthropogenic CO_2 emissions ($E_{FF} + E_{LUC}$) remain in the atmosphere. Both of these sinks are expected to decline due to further increases in $[CO_2]$ and GMST, and a range of other climate change-related factors including; changing hydrology, increased fire and pest risk to standing biomass, and the impact of ocean heating and acidification on CO_2 solubility, marine food webs, and the ocean's biological carbon pump (discussed further in Chapters 5 and 6). These declines will have an important influence on the effectiveness of future climate change mitigation efforts, including NETs.

4.3 Carbon dioxide removal and the global carbon cycle

As emissions reductions and NET deployment reverse the rising $[CO_2]$ trend, adjustments in many of the terrestrial and oceanic processes that have mitigated the historical increase will influence the decline, including the release of CO_2 from systems that have reached or approached equilibrium with the higher $[CO_2]$ atmosphere. A reducing CO_2 fertilization effect on primary photosynthetic production will reduce S_{LAND} (Reith et al., 2016), while the eventual lowering of sea surface temperature would increase surface CO_2 solubility, potentially enhancing S_{OCEAN} (see Chapter 6; Ridgwell et al., 2011).

The magnitude and timing of these and other effects will depend on the pace and extent of [CO_2] decline as well as on whether removals are directly from the atmosphere (i.e., DAC), or via land- or ocean-based methods. Initial model comparison studies have shown that for a [CO_2] profile peaking at ~450 ppm around 2050 and declining to ~360 ppm by 2300 (RCP2.6 scenario) the land and ocean sinks will decline to a combined level of ~1.9 GtC/yr for the second half of this century (Jones et al., 2016), compared ~5.9 GtC/yr for the decade to 2020. At some point in the future, the land and ocean sinks may become negative, with net emissions from these carbon pools back into the atmosphere, although the timing and extent of these effects remain highly uncertain, particularly for the land sink (Keller et al., 2018).

Many Earth system adjustments to warming and increased [CO_2] are more or less reversible under future declining [CO_2], although some reversals, as well as some system adjustments, will be delayed or very slow (Hansen et al., 2022). For example, sea level rise due to thermal expansion from ocean heating is expected to continue for some decades after peak emissions, and take over 1000 years to return to preindustrial levels (Ehlert and Zickfeld 2018; Schwinger and Tjiputra 2018), neglecting the effects of ice sheet loss, and Earth system feedbacks may delay the onset of GSMT decline until some time after net-zero emissions are reached (Schleussner et al., 2022). The possibility of irreversible shifts in some elements of the Earth system—tipping points such as Atlantic deep-water formation, sea ice and ice sheet loss, and the shifting or instability of monsoons—leading to unknown future states of the climate system, also reminds us that returning [CO_2] to near its preindustrial level does not necessarily return the Earth system to the corresponding state (Lenton et al., 2008; Hansen et al., 2022).

4.4 Current carbon cycle research challenges

Considerable progress has been made in broadening and improving the range of "data, algorithms, statistics, model estimates, and interpretations" related to CO_2 emissions and the carbon cycle since the first global carbon budget estimate was published for the period 1959 to 2011 (Le Quéré et al., 2013), yet despite the equally considerable effort involved (evidenced by a near tripling of the number of contributing authors—from 34 to 95) the uncertainty ranges in the components of the budget (Eq. 4.1) have, with the exception of S_{LAND}, not decreased (Table 4.4).

As well as its importance in improving future climate predictions and informing climate policy, an important goal of this work is the reduction of uncertainties to the point where globally reported CO_2 emissions can be verified or challenged on the basis of Earth system observations. Further improvements in measurements, including high-accuracy atmospheric measurements and satellite data, in integrative modeling approaches, and in understanding the natural background variability of atmospheric [CO_2] are needed to enable a degree of independent verification of reported emissions, and to demonstrate the effectiveness of collective emissions reduction—and ultimately carbon removal—efforts (Peters et al., 2017).

Current research is progressing along three lines; bottoms-up studies focusing on improved understanding and modeling of atmospheric, terrestrial, and oceanic processes, top-down efforts to improve measurement and monitoring at a regional and global level, and inside-out studies to better understand the impact of climate change on the carbon cycle, including feedbacks, the potential for amplification of climate impacts, and overall predictability.

Improved process understanding and modeling capabilities, driven by observational and experimental studies, are at the heart of efforts to improve the predictability of global and regional carbon stocks and fluxes in both terrestrial (Sutherland et al., 2013; Luo et al., 2015) and oceanic environments (Jin et al., 2020), discussed further in the following two chapters.

TABLE 4.4 Uncertainty ranges for the annual carbon budget components; 2011 versus 2020 (Le Quéré et al., 2013; Friedlingstein et al., 2021).

Carbon budget component	Year 2011 estimate (GtC/yr)	Year 2020 estimate (GtC/yr)
E_{FF}	9.5 ± 0.5	9.5 ± 0.5
E_{LUC}	0.9 ± 0.5	0.9 ± 0.7
G_{ATM}	3.6 ± 0.2	5.0 ± 0.2
S_{OCEAN}	2.7 ± 0.5	3.0 ± 0.4
S_{LAND}	4.1 ± 0.9	2.9 ± 1.0

For terrestrial stocks and fluxes, models and measurements are well matched in Europe, less so elsewhere, and way out in South America, which may currently be a net source or a sink (Keenan and Williams 2018). Expansion of the global [CO_2] monitoring network to include regions where carbon uptake is sensitive to climate variability will also be essential to narrow carbon budget uncertainty (Ballantyne et al., 2012).

Identifying and assessing the major feedbacks and interactions between terrestrial and oceanic ecosystems and the atmosphere under a changing climate is among the most important current research questions (Sutherland et al., 2013). The World Climate Research Programme's grand challenge on carbon feedbacks in the climate system aims to improve our understanding of the land and ocean sinks—by addressing three guiding questions:

- What are the drivers of land and ocean carbon sinks?
- What is the potential for amplification of climate change over the 21st century via climate-carbon cycle feedbacks?
- How do greenhouse gases fluxes from highly vulnerable carbon reservoirs respond to changing climate, including climate extremes and abrupt changes (Ilyina and Friedlingstein 2016).

Improved understanding and reduced uncertainty in the many processes influencing carbon cycle fluxes and the correct incorporation of these processes into ESMs will improve the predictive value of these models and the effectiveness of their outputs in guiding climate policy decisions.

References and resources

References

Ballantyne, A.P., Alden, C.B., Miller, J.B., Tans, P.P., White, J.W.C., 2012. Increase in observed net carbon dioxide uptake by land and oceans during the past 50 years. Nature 488 (7409), 70.

Ehlert, D., Zickfeld, K., 2018. Irreversible ocean thermal expansion under carbon dioxide removal. Earth Syst. Dyn. 9 (1), 197–210.

Forkel, M., Carvalhais, N., Rödenbeck, C., et al., 2016. Enhanced seasonal CO_2 exchange caused by amplified plant productivity in northern ecosystems. Science 351 (6274), 696–699.

Friedlingstein, P., Jones, M.W., O'Sullivan, M., et al., 2021. Global carbon budget 2021. Earth Syst. Sci. Data Discuss. 1–191.

Graven, H.D., Keeling, R.F., Piper, S.C., et al., 2013. Enhanced seasonal exchange of CO_2 by northern ecosystems since 1960. Science 341 (6150), 1085–1089.

Hansen, J.E., Sato, M., Simons, L., Nazarenko, L.S., von Schuckmann, K., Loeb, N.G., et al., 2022. Global warming in the pipeline. *arXiv preprint arXiv*: 2212.04474.

Heinze, C., Meyer, S., Goris, N., et al., 2015. The ocean carbon sink–impacts, vulnerabilities and challenges. Earth Syst. Dyn. 6 (1), 327–358.

Ilyina, T., Friedlingstein, P., 2016. WCRP Grand Challenge–Carbon Feedbacks in the Climate System (Technical Report. WCRP).

IPCC, 2021. In: Masson-Delmotte, V.,P., Zhai, A., Pirani, S.L., et al. (Eds.), Climate Change 2021: The Physical Science Basis. Contribution of Working Group I to the Sixth Assessment Report of the Intergovernmental Panel on Climate Change. IPCC, Geneva, Switzerland.

Jin, D., Hoagland, P., Buesseler, K.O., 2020. The value of scientific research on the ocean's biological carbon pump. Sci. Total Environ. 749, 141357.

Jones, C.D., Ciais, P., Davis, S.J., Friedlingstein, P., Gasser, T., Peters, G.P., et al., 2016. Simulating the Earth system response to negative emissions. Environ. Res. Lett. 11 (9), 095012.

Keenan, T.F., Williams, C.A., 2018. The terrestrial carbon sink. Annu. Rev. Environ. Resour. 43, 219–243.

Keller, D.P., Lenton, A., Littleton, E.W., Oschlies, A., Scott, V., Vaughan, N.E., 2018. The effects of carbon dioxide removal on the carbon cycle. Curr. Clim. Change Rep. 4 (3), 250–265.

Le Quéré, C., Andres, R.J., Boden, T., et al., 2013. The global carbon budget 1959–2011. Earth Syst. Sci. Data 5 (1), 165–185.

Le Quéré, C., Andrew, R.M., Canadell, J.G., et al., 2016. Global carbon budget 2016. Earth Syst. Sci. Data 8 (2), 605–649.

Lenton, T.M., Held, H., Kriegler, E., et al., 2008. Tipping elements in the Earth's climate system. Proc. Natl. Acad. Sci. USA 105 (6), 1786–1793.

Lorenz, K., Lal, R., 2018. Soil carbon stock. In: Carbon Sequestration in Agricultural Ecosystems. Springer, Cham, Switzerland, pp. 39–136.

Luo, Y., Keenan, T.F., Smith, M., 2015. Predictability of the terrestrial carbon cycle. Global Change Biol. 21 (5), 1737–1751.

Peters, G.P., Le Quéré, C., Andrew, R.M., et al., 2017. Towards real-time verification of CO_2 emissions. Nat. Clim. Change 7 (12), 848–850.

Pugh, T.A.M., Müller, C., Arneth, A., Haverd, V., Smith, B., 2016. Key knowledge and data gaps in modelling the influence of CO_2 concentration on the terrestrial carbon sink. J. Plant Physiol. 203, 3–15.

Reith, F., Keller, D.P., Oschlies, A., 2016. Revisiting ocean carbon sequestration by direct injection: a global carbon budget perspective. Earth Syst. Dyn. 7, 797–812.

Ridgwell, A., Rodengen, T.J., Kohfeld, K.E., 2011. Geographical variations in the effectiveness and side effects of deep ocean carbon sequestration. Geophys. Res. Lett. 38 (17).

Schleussner, C.-F., Rogelj, J., Frölicher, T., MacDougall, A., Sanderson, B., Lejeune, Q., 2022. Carbon Dioxide Removal and warming reversal in the light of uncertain Earth System processes. In: EGU General Assembly 2022, Vienna, Austria, 23–27 May 2022, EGU22-9838.

Schwinger, J., Tjiputra, J., 2018. Ocean carbon cycle feedbacks under negative emissions. Geophys. Res. Lett. 45, 5062–5070.

Sutherland, W.J., Freckleton, R.P., Godfray, H.C.J., et al., 2013. Identification of 100 fundamental ecological questions. J. Ecol. 101 (1), 58–67.

Further reading

Arneth, A., Sitch, S., Pongratz, J., et al., 2017. Historical carbon dioxide emissions caused by land-use changes are possibly larger than assumed. Nat. Geosci. 10 (2), 79–84.

Cicerone, R., Orr, J., Brewer, P., et al., 2004. The ocean in a high CO_2 World. EOS Trans. Am. Geophys. Union 85, 351.

Friedlingstein, P., Meinshausen, M., Arora, V.K., et al., 2014. Uncertainties in CMIP5 climate projections due to carbon cycle feedbacks. J. Clim. 27 (2), 511–526.

Gruber, N., Landschützer, P., Lovenduski, N.S., 2019. The variable Southern Ocean carbon sink. Ann. Rev. Mar. Sci 11, 159–186.

Khatiwala, S., Tanhua, T., Mikaloff Fletcher, S., et al., 2013. Global ocean storage of anthropogenic carbon. Biogeosciences 10 (4), 2169–2191.

Lade, S.J., Donges, J.F., Fetzer, I., et al., 2018. Analytically tractable climate–carbon cycle feedbacks under 21st century anthropogenic forcing. Earth Syst. Dyn. 9 (2), 507–523.

Le Quéré, C., Raupach, M.R., Canadell, J.G., et al., 2009. Trends in the sources and sinks of carbon dioxide. Nat. Geosci. 2 (12), 831–836.

Sabine, C.L., Tanhua, T., 2010. Estimation of anthropogenic CO_2 inventories in the ocean. Ann. Rev. Mar. Sci 2, 175–198.

Wigley, T.M.L., Schimel, D.S. (Eds.), 2000. The Carbon Cycle. Cambridge University Press, Cambridge, UK.

Projects and organizations

Carboocean IP (EU-funded assessment of marine carbon sources and sinks, focusing on the Atlantic and Southern Oceans over the past 200 and next 200 years): www.carboocean.org.

Global Carbon Project, publishing the annual Global Carbon Budget: www.globalcarbonproject.org.

GLODAP (Global Ocean Data Analysis Project): http://cdiac.esd.ornl.gov/oceans/glodap/.

Monterey Bay Aquarium Research Institute (MBARI): www.mbari.org.

Ocean Chemistry of the Greenhouse Gases: www.mbari.org/ghgases.

National Maritime Research Institute (NMRI): www.nmri.go.jp/index_e.html North American Carbon Program (assessing the sources and sinks of CO_2, CH_4, and CO in North America and adjacent oceans): www.nacarbon.org.

Ocean Acidification Network: http://ocean-acidification.net/about/.

Ocean Carbon & Biogeochemistry: www.us-ocb.org.

State of the Carbon Cycle Report (SOCCR): The North American Carbon Budget and Implications for the Global Carbon Cycle: http://cdiac.ornl.gov/SOCCR/.

The Ocean in a High CO_2 world, UNESCO Intergovernmental Oceanographic Commission International Symposium Series: http://ocean-acidification.net/international-symposia/second-meeting-on-oa-in-a-high-co2-world/.

UNESCO Intergovernmental Oceanographic Commission; International Ocean Carbon Coordination Project: http://ioc-unesco.org/, www.ioccp.org.

U.S. Carbon Cycle Science Program: https://carboncyclescience.us/.

U.S. Department of Energy (DOE) Environmental Systems Science Data Infrastructure for a Virtual Ecosystem (ESS-DIVE); Global carbon budget data: http://ess-dive.lbl.gov/.

Woods Hole Oceanographic Institution: www.whoi.edu.

Café Thorium; WHOI marine chemistry and geochemistry group: http://cafethorium.whoi.edu.

World Climate Research Programme (WCRP): addressing frontier scientific questions related to the coupled climate system.

Chapter 5

Terrestrial carbon cycle processes

Steve Rackley

5.1 Introduction

Terrestrial ecosystems play an important part in the global carbon cycle, being both the repository of a carbon inventory some six times larger than the atmosphere, and the source and sink for closely balanced fluxes taking up CO_2 through photosynthesis and reemitting most of this uptake through plant and microbial respiration. The annual uptake effectively recycles the atmospheric carbon dioxide inventory through biomass roughly every 7 years, and the combined effect of these fluxes results in a net uptake of carbon from the atmosphere into the so-called land sink of terrestrial ecosystems that averaged 3.1 GtC/yr over the decade to 2020, 29% of total emissions (Friedlingstein et al., 2021), and has accounted for roughly one third of all historical anthropogenic carbon emissions.

The carbon inventory in terrestrial ecosystems is held in a variety of carbon pools (see Chapter 11), with widely different residence times, including litter and other plant products that are subject to rapid microbial decomposition, above- and belowground biomass such as trees and their root systems, and stabilized organic and inorganic carbon fractions in soils (Keenan and Williams, 2018).

The transfer of CO_2 from the atmosphere into terrestrial vegetation and soils begins with the process of carbon fixation through photosynthesis, in which plants convert CO_2, water, and energy in the form of absorbed photons into organic compounds.

5.2 Photosynthesis

Terrestrial net primary production (NPP), driven by photosynthesis, removes ~ 130 GtC from the atmosphere every year, and various methods such as afforestation and BECCS are being deployed to deliver negative emissions through this process. Marine photosynthesis occurs on a similar scale, removing ~ 80 GtC annually from euphotic waters, to be replaced by wind-driven CO_2 drawdown from the atmosphere. As discussed in Chapter 4, both of these immense fluxes are largely offset by respiration and by upwelling waters, reducing net fluxes to the terrestrial and oceanic sinks to around 3 GtC/yr. An informative review of our growing understanding of the mechanisms of photosynthesis, from the 18th century to the present day, can be found in Stirbet et al. (2020).

5.2.1 C3 photosynthesis

In plant chloroplasts, photosynthesis starts with the so-called light reactions, in which an electron is released from a chlorophyll molecule as a result of the adsorption of a photon. The electron is exchanged between a number of complex organic compounds present in the chloroplast and eventually results in the reduction of nicotinamide adenine dinucleotide phosphate (NADP; $C_{21}H_{29}N_7O_{17}P_3$) to NADPH, and the synthesis of adenosine-5′-triphosphate (ATP; $C_{10}H_{16}N_5O_{13}P_3$) from adenosine diphosphate (ADP; $C_{10}H_{15}N_5O_{10}P_2$).

A sequence of reactions known as the Calvin cycle (or Calvin−Benson−Bassham cycle) then takes place within the chloroplasts, in which the enzyme RuBisCO (ribulose-1,5-bisphosphate carboxylase/oxygenase) catalyzes the fixation of CO_2 to produce glyceraldehyde 3-phosphate (G3P), using the chemical energy stored in NADPH and ATP. The Calvin

Negative Emissions Technologies for Climate Change Mitigation. https://doi.org/10.1016/B978-0-12-819663-2.00013-7

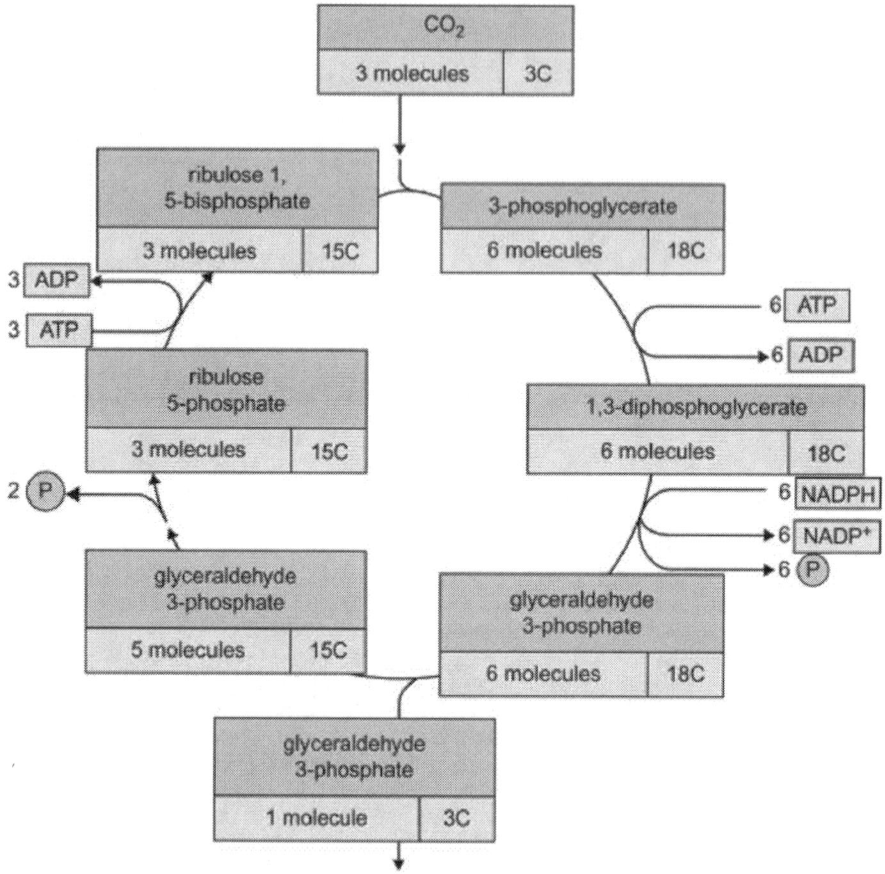

FIGURE 5.1 The Calvin–Benson–Bassham photosynthetic cycle.

cycle is shown schematically in Fig. 5.1, and the overall equations for the light reaction and the Calvin cycle can be written:

$$2H_2O + 2NADP{+}\ 2ADP + 2P_i + light \rightarrow 2NADPH + 2H^+ + 2ATP + O_2 \tag{5.1}$$

and

$$3CO_2 + 6NADPH + 9ATP + 6H^+ \rightarrow G3P + 6NADP + 9ADP + 3H_2O + 8P_i \tag{5.2}$$

where NADP+ is the oxidized form of NADPH, P_i is an inorganic phosphate, and G3P ($C_3H_7O_6P$), the end product of the cycle is a three-carbon compound—hence the name C3 photosynthesis. It is the precursor for the synthesis of glucose and other organic compounds, such as cellulose, and is thus the starting point for the growth of terrestrial biomass.

Some photosynthetically fixed carbon is lost from the chloroplast by photorespiration, a process that regenerates the CO_2 acceptor ribulose-1,5 bisphosphate that is lost in the oxygenase reaction pathway in which RuBisCO catalyses the fixation of O_2 rather than CO_2. The product of the oxygenation is recycled in the mitochondria to produce serine, an alternative precursor to G3P, with CO_2 being released. In total ~20%–40% of the carbon fixed by C3 photosynthesis is respired by this process of photorespiration.

In contrast to the oceans, where primary production is often limited by the availability of nutrients, terrestrial plants are generally able to increase gross primary production (GPP; see Glossary) with increasing CO_2 availability. Increasing atmospheric [CO_2] enhances leaf-scale photosynthesis, and evidence from ice cores and plant archives indicates a significant increase in global photosynthesis during the industrial era, of which around half is likely due to increasing [CO_2] (Walker et al., 2021). Increased water use efficiency has also been observed in a number of free air CO_2 enrichment (FACE) experiments (see later), as the stomata are able to remain closed more often. As a result, vegetation in arid or semiarid zones is particularly responsive to higher CO_2 concentrations, as the amount of water loss due to transpiration for a given CO_2 uptake is reduced as [CO_2] increases, enabling a longer growing season and greater biomass production for a given water input.

While increases in primary production can contribute to increasing carbon stocks, this will be mitigated in ecosystems with limited nitrogen availability—the progressive nitrogen limitation (PNL) theory. In this situation, relatively stable soil carbon will be decomposed to meet microbial and plant needs for nitrogen, resulting in higher soil respiration and offsetting the effect of increased GPP on soil carbon stocks. Overall, there is evidence that the global terrestrial carbon sink has increased historically, which is most likely attributed to increasing [CO_2] alongside a number of other changing environmental factors (Walker et al., 2021).

One feature of photosynthesis that has become an important research tool for tracking plant-derived carbon in the environment is its discrimination in favor of the lighter ^{12}C isotope. Although the heavier ^{13}C isotope has a natural abundance of ~1% in the atmosphere, when CO_2 availability is not the limiting factor for primary production photosynthesis discriminates against the heavier isotope with the result that carbon fixed in plant biomass is relatively depleted in ^{13}C. This discrimination is strongest in C3 plants, with C3 biomass being typically 2.5% depleted in ^{13}C (i.e., $\delta^{13}C = -2.2$ to -3.0% compared to the VPDB carbonate standard; see Glossary), while in C4 and crassulacean acid metabolism (CAM) plants (see later), the depletion is typically $\delta^{13}C \approx -0.8$ to -1.1%.

5.2.2 C4 and CAM photosynthesis

Two variants of the basic photosynthetic process have evolved that minimize water loss and the less efficient photorespiratory pathway and typically occur in plants adapted for arid or semiarid conditions. In so-called C4 plants, which include sugar cane, maize, and sorghum, CO_2 is captured in the inner mesophyll layer via the formation of oxaloacetate and malate. These compounds contain four carbon atoms, hence the name C4 photosynthesis. The malate is transported into bundle sheath cells (Fig. 5.2), where the CO_2 is released and the standard Calvin cycle proceeds within the chloroplasts. By this mechanism, RuBisCO is spatially isolated from oxygen present in the mesophyll, reducing photorespiration. High CO_2 concentrations are achieved at the site of the Calvin cycle without the need for the stomata to be wide open, thus reducing water loss.

CAM is another variant, occurring in plants such as pineapple, orchids, and cacti. In CAM plants, the stomata are opened at night, when the temperature is lower and humidity is generally higher, and water loss will therefore be minimized. As in C4 photosynthesis, the oxaloacetate and malate pathway is used to capture and store carbon dioxide. During the day the stomata are closed, reducing oxygen availability for photorespiration, and the CO_2 is released, enabling Calvin cycle (C3) photosynthesis.

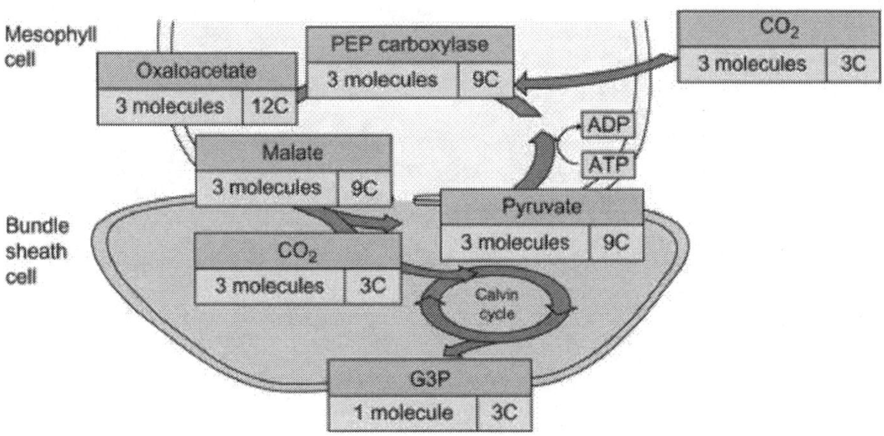

FIGURE 5.2 Photosynthetic mechanisms in C4 plants.

Water use efficiency, the amount of water needed by the plant to produce a unit of dry biomass, is an order of magnitude higher in CAM plants than in common C3 and C4 crops (Mason et al., 2015), allowing these plants to flourish in semiarid deserts and other seasonally dry environments. This opens up a potentially important role for these plants as bioenergy crops, since the growth of CAM plants for biomass on semiarid lands that are not viable for arable farming could help to mitigate the land use conflict (food vs. fuel) that may arise with large scale BECCS deployment. The possibility of engineering the CAM pathway into existing C3 and C4 crops has also been highlighted (Borland et al., 2014; Yang et al., 2015) and is opening up a wide field of bioengineering R&D.

5.2.3 Aboveground and belowground carbon allocation

Up to 50% of the mono- and disaccharides produced by plants (e.g., glucose and sucrose, respectively) are delivered to the root system, where they are used to build root biomass, or exuded and accessed by soil microbes, particularly the mycorrhizal fungi that form a symbiotic relationship with the plant—with 5%–20% of a plant's total carbon budget being allocated to these fungal associations. The exuded carbohydrates are partly consumed by heterotrophic respiration (see Glossary), releasing some fixed CO_2 back to the atmosphere, and partly used to build microbial biomass. In return, the plant gains access to the filamentary hyphae of the fungal mycelium, effectively increasing the surface area and fine-scale penetrative capacity of the root system for the uptake of water and nutrients.

As a plant's primary reproductive task—the production of flowers, fruits, and seeds—takes place mainly above ground, the belowground allocation of GPP will be optimized to maximize reproductive success under specific ecological and environmental conditions. Some of the factors affecting belowground allocation are described in Table 5.1.

5.3 Biomineralization

The production of minerals by living organisms is widespread in nature, with end products ranging from coral and mollusk shells to bones and teeth. These processes are collectively called biomineralization, and two general types occur. In biologically controlled mineralization (BCM), mineral synthesis takes place either within or on the surface of the cell, at a specific location under the control of the organism, while in biologically induced mineralization (BIM) mineral deposition takes place outside the cell as a result of the organism's metabolic activity. Simplistically, in BCM mineral synthesis is an objective of the process while in BIM the mineral deposit is a by-product.

5.3.1 Calcite precipitation by cyanobacteria

Cyanobacteria (blue-green algae) are among the oldest lifeforms on Earth, and their photosynthetic activity played a central role in the oxygenation of the ocean and atmosphere that was a precursor to the Cambrian explosion, some 540 million years ago. Calcite precipitation by cyanobacteria and other microorganisms—in the form of stromatolites—is common in the geological record stretching back at least 3.5 billion years. Whether there is a physiological reasons for this induced mineralization—for example, related to the need to control calcium levels inside the cell—or whether it is simply a result of alkalinity increase in their environment as a result of CO_2 removal is still under debate. Recent work does however suggest that functional groups on the cell surface provide a template for calcite crystallization—an indication that extracellular mineralization benefits the organism, and that at least some cyanobacterial strains also produce intracellular calcite—possibly indicative of a controlled (BCM) rather than an induced (BIM) process.

TABLE 5.1 Factors influencing belowground carbon allocation in plants.

Factor	Description
Nutrient availability	Increased nutrient availability, for example, through fertilization, reduces below-ground allocation since the plant is more easily able to meet its nutrient requirements.
Water availability	Similar to nutrients, less NPP needs to be invested below ground when water availability is high.
Soil disturbance	Intensive cultivation may result in a shift in soil microbial community composition toward bacteria rather than fungi, which can result in a compensating increase in belowground allocation.

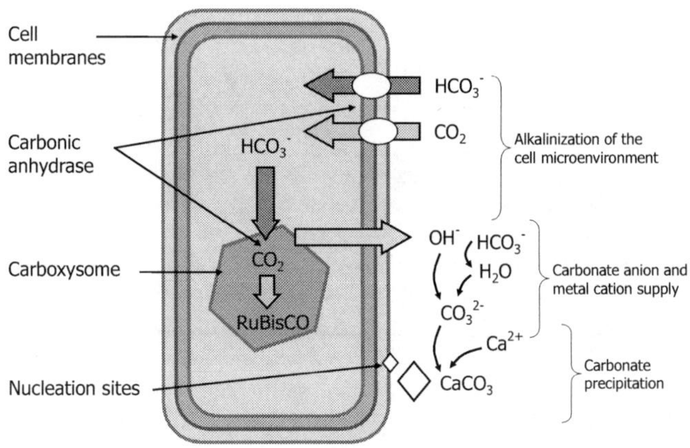

FIGURE 5.3 Extracellular biomineralization in cyanobacteria. *After Power et al. (2013).*

The extracellular mineralization process, shown schematically in Fig. 5.3, involves a number of inorganic chemical reactions, starting with the dissolution of CO_2 in the aqueous environment to form carbonic acid and its dissociation to bicarbonate ions and protons, with carbonic anhydrase (CA; see next section) playing an important part in this initial step of bicarbonate ion formation. Bicarbonate ions are then transported into the cell and provide the CO_2 needed for photosynthesis, CA again playing an important role in concentrating CO_2 within the cell. Hydroxyl ions are formed within the cell as a byproduct of the dehydration of bicarbonate ions by CA in the carboxysome (where photosynthesis takes place). These are transported out of the cell and react with bicarbonate ions in the aqueous environment:

$$OH^- + HCO_3^- \rightarrow H_2O + CO_3^{2-} \tag{5.3}$$

Finally, the carbonate anion combines with a calcium or magnesium cation from the aqueous medium, transported from within the cell or desorbed from the cell wall, and calcite is precipitated:

$$Ca^{2+} + CO_3^{2-} \rightarrow CaCO_3\downarrow \tag{5.4}$$

The last step is facilitated by carboxyl and amine functional groups on the cell surface which provide nucleation sites for crystallization. Biologically mediated carbonate precipitation thus combines a renewable energy source (sunlight), with the generation of alkalinity (OH^- ions), while fixing CO_2 as biomass (Power et al., 2013).

The carbon mineralization ability of cyanobacteria is being studied as a potential capture technology that could either be combined with microalgal biomass production for biofuel synthesis, providing two parallel capture routes as shown in Fig. 5.4, or deployed as an engineered biomimetic capture process (Kamennaya et al., 2012). The scope of current research includes the following:

- The physiological functions achieved by mineralization
- The stages and controls on the mineralization process (e.g., nucleation at functional sites, crystallization, and crystal morphology and aggregation)
- The interaction of photosynthetic production and mineralization, and

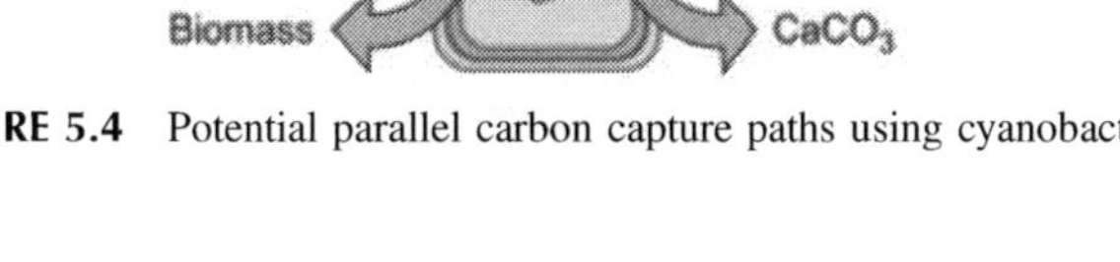

FIGURE 5.4 Potential parallel carbon capture paths using cyanobacteria.

- The genetic profiles of bacterial strains affecting mineralization capacity under varying conditions, particularly in low-nutrient and alkaline environments,
  The overall aim being to provide the fundamental understanding necessary to engineer viable industrial processes, possibly including the genetic engineering or directed evolution of bacterial strains.

Apart from the photosynthetic-related BIM seen in cyanobacteria, calcite is also deposited by other bacteria, such as ureolytic bacteria commonly found in soils, which metabolize urea ($CO(NH_2)_2$) as a source of nitrogen, producing calcite as a byproduct. This process, in which urea is hydrolyzed to ammonia (NH_4) and carbonate ions (CO_3^{2+}), is the basis of the microbially induced calcium carbonate precipitation (MICP) process (Achal and Mukherjee, 2015). Such induced mineralization processes have been widely studied, with potential applications from wastewater clean-up and bioremediation of environmental pollution (phycoremediation) to the fabrication and strengthening of building materials and ground consolidation (Terzis and Lalui, 2019), as well as in the remediation of leakage paths in well cement, microfractures, and permeable caprock in geological storage reservoirs (see Chapter 15).

5.3.2 Biocatalytic carbon mineralization

When the calcite precipitation process described earlier takes place in seawater, without biological mediation, the formation of bicarbonate ions is the rate-limiting step. A family of enzymes has evolved in various organisms that catalyze the hydration of CO_2 to HCO_3^- and can be applied to accelerate the process of biomineral formation. These enzymes, collectively known as CAs, are metalloenzymes—meaning that the active site contains a metal ion (commonly zinc). The catalyzed hydration mechanism is illustrated in Fig. 5.5. The figure shows the active site—the zinc atom—which is held in place by bonds to three amino acids (histidines—His94, etc.). The complex organic form of the anhydrase enzyme, which includes a pocket or cleft that specifically accommodates a CO_2 molecule, bringing it close to the active site, is omitted for clarity.

The chemical bonding geometry of zinc in CA is tetrahedral (known as tetrahedral coordination), and in the first step of the hydration process (1 in the figure), the fourth coordination position is occupied by a water molecule. Polarization of the water molecule weakens the H—O bonds and a fourth histidine (His64), located close to the water, accepts a proton (H+) and transfers it to the bulk solution, leaving a hydroxide attached to the zinc (2). A CO_2 molecule can then enter the active space and the excess electron from the OH^- ion transfers to the CO_2 (3) causing a realignment of the C—O bonds and the formation of a zinc-bound bicarbonate (4). After isomerization (5—the shift of the H to the nonzinc-bound oxygen), the scene is set for the release of the bicarbonate ion and its replacement by a new water molecule (1). For several types of carbonic anhydrase, including the human variety hCAII, each CA molecule can execute this elegant mechanism more than 10^6 times per second, many orders of magnitude faster than in the absence of a catalyst.

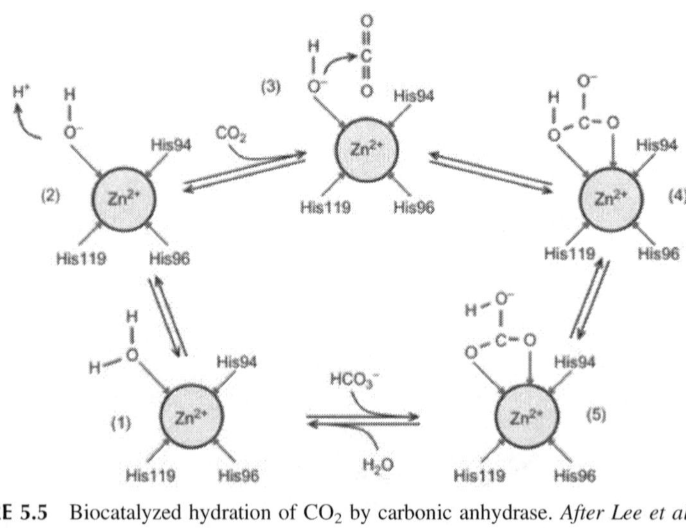

FIGURE 5.5 Biocatalyzed hydration of CO_2 by carbonic anhydrase. *After Lee et al. (2010).*

Once the bicarbonate ion is released into a bulk medium containing calcium ions, calcite precipitation can proceed according to Reactions 5.3 and 5.4. Experimental investigations into the use of carbonic anhydrase to catalyze carbon mineralization in artificial seawater began in the 1990s, and intensive research since then has focused on sustaining catalytic activity at industry-relevant temperatures, including the identification and extraction of suitable CA from microorganisms and the chemical or physical immobilization of the catalyst to increase stability and longevity while sustaining performance by ensuring that the active site remains accessible. The commercial potential of this technology has driven rapid R&D progress and has led to a number of proprietary systems being developed and demonstrated at laboratory scale (Lee et al., 2010). Ongoing research, much of it aimed at the application of CA to enhance CO_2 absorption systems (see Chapter 8), continues to focus on (Di Fiore et al., 2015);

- Characterization of thermostable CAs, typically isolated from thermophilic microorganisms
- Improvement of immobilization techniques
- Protein engineering of wild-type CAs to improve thermal stability
- Preparation and characterization of novel catalytic molecules to mimic enzyme activity
- Investigation of biological processes such as rapid repair of damaged shell by mollusks.

5.3.3 Carbon sequestration in phytoliths

One biological process that has recently been identified as a significant contributor to long-term carbon sequestration in soils is the production by many plants of phytoliths—literally "plant stones"—which are microscopic silica bodies deposited in cell walls, cell lumina, and intracellular spaces (Fig. 5.6), most likely as a way for the plant to get rid of excess silica adsorbed by plant roots as monosilicic acid (Piperno, 2006).

Organic carbon trapped within these biomineralizations, know as phytolith occluded carbon or PhytOC, and incorporated into soils via leaf litter, has been estimated to account for the sequestration of 156 ± 92 $MtCO_2$/yr (Song et al., 2017). PhytOC is more highly resistant to decomposition than other soil organic carbon (SOC) components, discussed

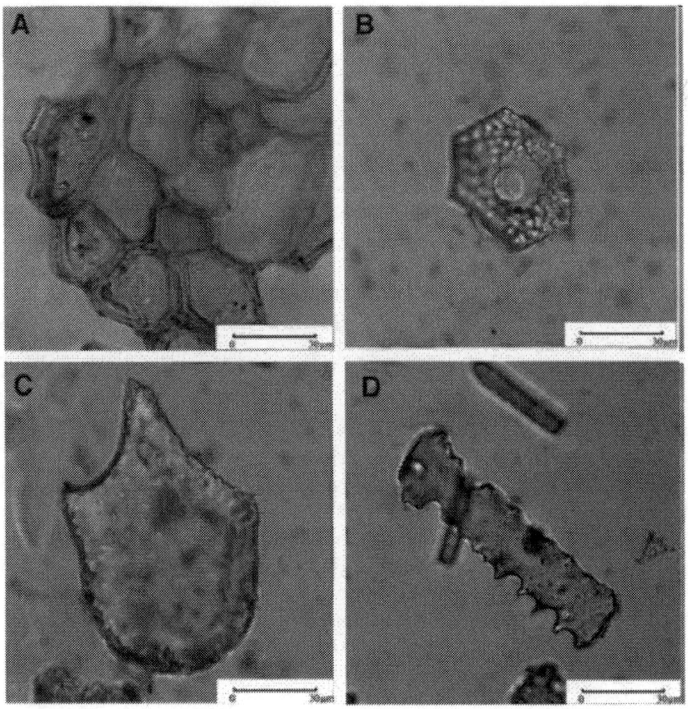

FIGURE 5.6 Fossil phytoliths from Byron Bay, New South Wales, Australia dated c. 8710 ± 50 BP. *From Parr and Sullivan (2005), with permission.*

later, and accounts for up to 82% of the total carbon in some well-drained soils after 1000 years (Parr and Sullivan, 2005), with fossil evidence indicating the potential for sequestration for millions of years.

A range of practices could substantially increase the carbon sink resulting from phytolith deposition, including crop and grassland species selection, silicate mineral soil amendments, and silicon fertilization, for example from organic mulching (Song et al., 2017), while genetic engineering to enhance PhytOC in cereal crops and bamboo afforestation has the potential to increase this sink to $GtCO_2$ scale (Parr et al., 2010; Jansson et al., 2010).

5.4 Biogeochemical features and processes in soils

The complex symbiotic and competitive processes taking place in the immediate vicinity of root systems, the rhizosphere, have a major impact on the fate of soil organic matter (SOM) and are therefore a key to understanding the carbon inventory and the potential for enhancing carbon storage in soils. Fixed carbon accumulates as above- and belowground biomass and is incorporated into soils through litterfall from leaves and branches, from root growth and mortality, and from microbial biomass produced from root exudates.

5.4.1 Soil microbiota

Microbial communities play an important and only partially understood role in the biogeochemical processes taking place within soils. The soil microbiome comprises two dominant microbial groups—bacteria and fungi—each having a different influence on soil processes (Six et al., 2006). Bacteria are generally seen to respire a greater proportion of the carbon they metabolize as CO_2 rather than using it to build biomass, and dead bacterial biomass is more rapidly broken down and metabolized by other organisms. Fungi respire a lower proportion of metabolized carbon, their cellular material is itself more resistant to decomposition (more recalcitrant), they produce extracellular enzymes that aid in the humification of lignitic material (see below), and they help to increase the residence time of carbon in soil through the part their fine hyphae play in maintaining fine-scale soil structure (see also below). In addition, mycorrhizal fungi have a symbiotic relationship with most plants, providing nutrients to the plant "in exchange" for root exudates, and therefore contribute directly to primary productivity.

The influence of microbes on soil processes, including those that impact enhanced carbon storage in soils, and the way in which these processes will be affected by climate change—increasing temperature and $[CO_2]$—are active research areas, as discussed in Section 5.5.

5.4.2 Humification

Carbon typically represents ~57% by weight of organic matter incorporated into soils. This material is progressively broken down and decomposed by detritus-feeding soil animals, as well as by the action of microbes with ~70% being remineralized to CO_2 within 1 year as a result of root and microbial respiration. Note that in the context of soil processes, "mineralization" is commonly used to denote the conversion of organic carbon to CO_2, in contrast to the usage elsewhere, where mineralization refers to conversion of CO_2 to carbonate materials. The residues from plant biomass and the byproducts from microbial activity (microbial organic matter; MOM) are a complex mixture of organic compounds including sugars, amino acids, proteins, cellulose, lignin, lipids, tannins, and fragments of these molecules. Humic material is the end product of the process of decomposition and comprises the remnants of plant-, microbe-, and animal-derived compounds, broken down under partially oxidizing conditions into smaller organic molecules that are held together in supramolecular structures by weak (nonchemical) binding forces. Lignin is the least susceptible of biomass residues to decomposition and is therefore a major component of humus and of soil organic carbon stocks.

These residues make up the majority of the slow and passive carbon pools in soils, as shown in Fig. 5.7. The figure shows a high-level model of soil carbon pools, fluxes, and controls, from gross primary production to the long-lived passive pool of soil carbon, and indicates a number of points at which respiratory activity contributes to the flux of CO_2 back to the atmosphere. The light/dark shading denotes respectively above- and belowground components of the model. For simplicity, only two process linkages are shown in the figure: tillage, incorporating surface litter into the subsurface litter pool, and the supply of nutrient from subsurface microbial activity into GPP.

Complete decomposition of these organic fragments would eventually result in full mineralization to CO_2 and is hindered or prevented either by biochemical recalcitrance or physicochemical protection, or by low activity or abundance of decomposer organisms. Biochemical recalcitrance results from the blocking of microbial catabolic pathways due to the lack of essential enzymes, while the physicochemical protection of SOM occurs as a result of chemical adsorption onto

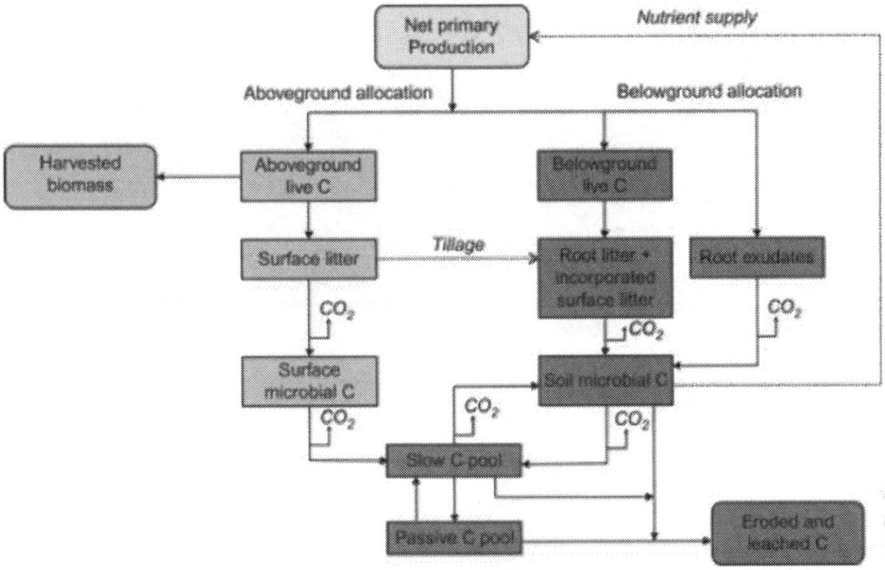

FIGURE 5.7 Partial model of soil carbon pools, fluxes, and controls.

mineral surfaces, facilitated by the presence of polyvalent cations such as calcium, iron, and magnesium, or the complete envelopment of SOM particles within aggregated soil particles. The ability of soils to physically protect organic material is an important determinant of soil carbon retention and is influenced by a range of factors described in Table 5.2.

Various studies have shown that root-derived carbon is preferentially stabilized in soil compared to aboveground-derived carbon, due to the physical protection of root hairs within soil aggregate, the biochemical recalcitrance of root components, and the protection of root exudates or decomposition products by adsorption onto clay particles. The resulting SOM provides a source of nutrients for plant growth, as well as helping to retain water in the soil, further contributing to productivity. Complete mineralization of SOM will occur if it is exposed to fully oxidative conditions, promoting aerobic microbial activity, for example as a result of disturbance of the aggregated soil structure by tilling. The addition of readily available carbon compounds to soil microorganisms may also promote their decomposer activity on SOM (so-called microbial priming, discussed below), an effect that has been observed to cause a reduction of soil carbon content under changing environmental conditions. Conversely, the stability and activity of the enzymes and oxidants needed for

TABLE 5.2 Factors influencing the physical protection of SOM.

Factor	Description
Clay content and mineralogy	Clays promote the growth of microbial biomass by providing protection from predation and desiccation, by regulating the soil pH, and by absorbing contaminants detrimental to microbial growth.
Pore-size distribution in aggregate particles	Small pores (<10 μm) in soil micro- and macroaggregates provide refuge to bacteria and fungal hyphae from predation by larger predators such as protozoa and nematodes.
Soil microclimate, including wetting and drying cycles	Repeated wetting and drying leads to soil erosion with SOM export, and to breakdown of soil microaggregates, exposing SOM particles to further decomposition.
Soil aggregate stability	Soil disturbance, for example, as a result of tillage, results in the breakdown of soil macroaggregates leading to SOM erosion/deposition, and increased soil aeration leading to SOM decomposition.

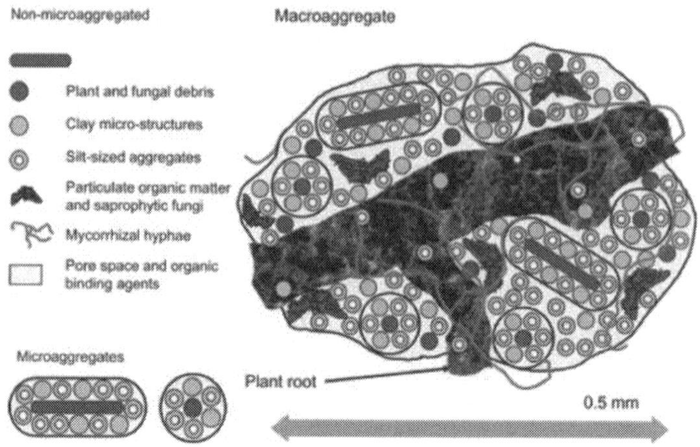

FIGURE 5.8 Macro- and microaggregate particles in soil. *After Jastrow and Miller (1998).*

microbial decomposition of organic material are reduced under acidic conditions and low temperatures, such as those historically occurring in northern peatlands, resulting in reduced rates of decomposition.

5.4.3 Soil structure

The structure of soils is an important factor in determining the residence time of SOC, playing a role in protecting SOM from microbial access and in the balance between bacterial and fungal pathways by which this material is cycled through the food web. Particulate matter in soils can be aggregated at a microaggregate level (aggregate particle diameter <0.25 mm) or at a larger macroaggregate level, as illustrated in Fig. 5.8.

The hyphal networks of mycorrhizal fungi play an important role in stabilizing macroaggregates, and loss of these networks can result in a breakdown of macroaggregates and a loss of SOC through increased bacterial activity. As a result of the clay-binding process described earlier, microaggregate particles provide a degree of physical protection to SOM, which becomes less susceptible to microbial action when incorporated into these particles (Jastrow and Miller, 1998). Microaggregates are thus an essential repository of recalcitrant carbon in soils, alongside unprotected black (pyrogenic) carbon.

When soil structure is disturbed by changes in land use, land-management practices, or environmental conditions, soil carbon stocks will tend toward a new quasi-equilibrium. In particular, practices or conditions that affect the protection of SOC within aggregates or the rate of decomposition in the soil will lead to reduced carbon stocks (Blanco-Canqui and Lal, 2004). These and other factors affecting carbon stocks are summarized in Table 5.3.

5.5 Modeling plant productivity and soil carbon processes

The incorporation of photosynthesis, plant and microbial respiration, and other soil carbon processes into Earth system models, and particularly the time-dependent response of these processes to environmental changes, is essential if such

TABLE 5.3 Factors increasing soil carbon stocks and related practices.

Factor	Related practices
Reduced soil disturbance to sustain soil structure	Tillage, grazing, tree harvesting, deforestation
Reduced soil aeration to reduce rate of decomposition	Tillage
Increased return of biomass and retention of belowground inputs to the soil	Conservation tillage, pasture crop selection for root mass and litter, crop rotation
Nutrient management, including addition of exogenous organic matter	Fertilization, composts/manures, cover crops, crop rotation

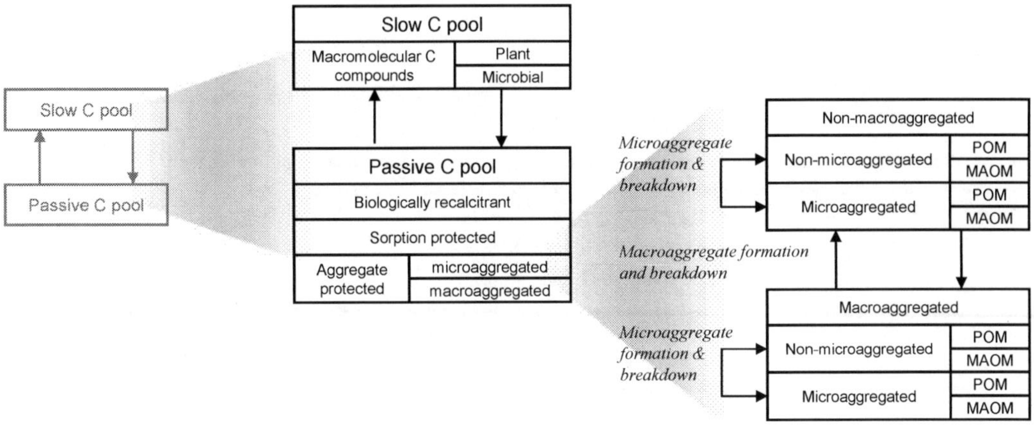

FIGURE 5.9 A simple conceptual model of soil carbon pools.

models are to realistically simulate climate-driven feedbacks from terrestrial ecosystems to the atmosphere, and considerable progress has been made in this area since the 1990s (Parton et al., 1993; Campbell and Paustian, 2015; Abramoff et al., 2018).

Considering soil carbon, the high-level model shown in Fig. 5.7 provides a framework to integrate more detailed models of soil carbon pools and processes. For example, Figs. 5.9 and 5.10 show further levels of detail of the soil carbon pools and the interactions between soil microbial carbon and the progressively longer-lived slow carbon and passive carbon pools (Segoli et al., 2013). The figures indicate some of the factors that influence these pools and processes.

Many of the interactions occurring within soils are highly complex and, in some cases, counterintuitive, as is well illustrated by considering microbial priming—one of the factors indicated in Fig. 5.10 as controlling the degree to which "passive" carbon may become available for further biological processing. Some soil carbon may be stabilized due to the absence of microbial activity, for example, in deeper soils where a lack of readily available substrates limits microbial activity. A change to a deeper rooting pattern may be induced by a reduction in local precipitation, increasing [CO_2], or other environmental changes, leading to higher NPP and increased belowground allocation. Cultivating plants with higher belowground carbon inputs and deeper roots may also cause increased belowground allocation. Deeper rooting in turn will cause the addition of labile carbon—that proportion of soil C that is most rapidly mineralized back to CO_2—in the form of root exudates into deeper soils, stimulating microbial activity, which may then result in the decomposition of previously stable soil carbon. Similar priming effects have been invoked to explain numerous experimental studies where soil carbon has reduced as a result of labile carbon addition or where an expected increase in SOM has not occurred, for example, in

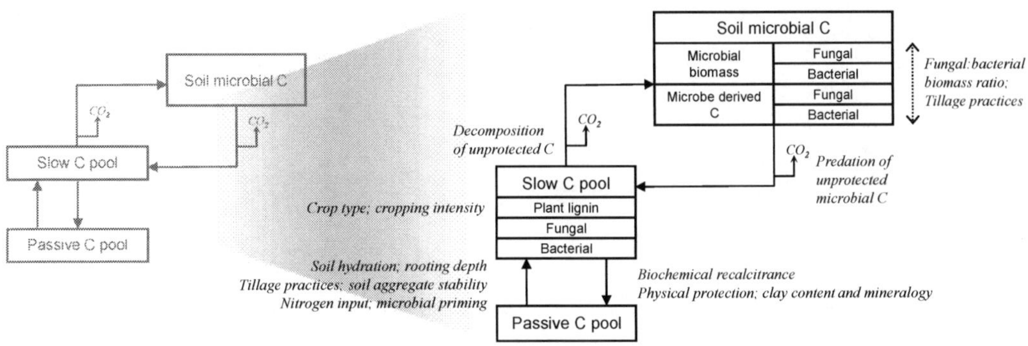

FIGURE 5.10 Second-level model of soil microbial, slow and passive carbon pools.

free air CO_2 enrichment experiments (see below and Heimann and Reichstein, 2008). Increased microbial activity can also contribute to an increase in soil carbon storage, as has been observed where this increased activity is associated with increased plant diversity (Lange et al., 2015).

Recent Earth system models have generally incorporated two main feedback mechanisms between the climate and the terrestrial carbon cycle;

- Increasing [CO_2] stimulates photosynthetic production (CO_2 fertilization), increasing the flux of CO_2 from the atmosphere into the terrestrial ecosystem and providing a negative feedback to rising [CO_2]
- Rising temperatures stimulate respiratory activity, resulting in increased respiratory flux of CO_2 from terrestrial ecosystems to the atmosphere and a positive feedback on mean temperatures.

A more complete representation of the coupling between terrestrial carbon pools and climate change requires many more factors than simply [CO_2] and mean temperature to adequately reflect these complex interactions, as discussed below. Many of these factors require fine spatial resolution in the models, as well as location and ecosystem-specific field data to calibrate input parameters and validate model results. Models also need to incorporate the time-dependent nature of ecosystem responses, from instantaneous response, through acclimation to longer-term adaptation (Smith and Dukes, 2013).

Such models, incorporating the many complex processes occurring within terrestrial ecosystems and their interaction with climate on multiple spatial and temporal scales, will be needed to accurately assess the nature of climate–ecosystem coupling, the impact of climate change on the current terrestrial carbon inventory, and the carbon storage potential of the land- and soil-based NET options described in later chapters. Long-term ecosystem-scale observations and experiments will also be required to improve understanding and validate the simulation models, with a particular focus on ecosystems with major contributions to carbon inventories and fluxes.

5.6 Impact of eCO_2 and climate change on photosynthesis and soil carbon processes

Although the response of vegetation and soil carbon to climate change remains a major uncertainty in assessing the overall impact of climate change on the global carbon cycle (Friedlingstein et al., 2003), as noted above, recent Earth system model studies predict a continued global uptake of carbon into terrestrial ecosystems with rising [CO_2] and mean global temperatures in the coming few decades. Depending on the assumed parameters in the various models, a gradual slowing occurs as a result of the increasing respiratory release of CO_2, in some cases resulting in the terrestrial ecosystem becoming a net CO_2 source rather than a sink. Fig. 5.11 shows schematically the range of predictions of net carbon uptake from a number of such climate models from the Coupled Carbon Cycle Climate Model Intercomparison Project (CMIP4). This comparison was based on the IPCC SRES-A2 carbon emissions scenario in which global CO_2 emission from energy production rises to ~16 GtC/yr in 2050, without significant CCS deployment, and a very similar picture emerges in the CMIP5 comparison, which uses the IPCC AR5 RCP8.5 scenario (Heimann and Reichstein, 2008; Stocker, 2013; Friedlingstein et al., 2014).

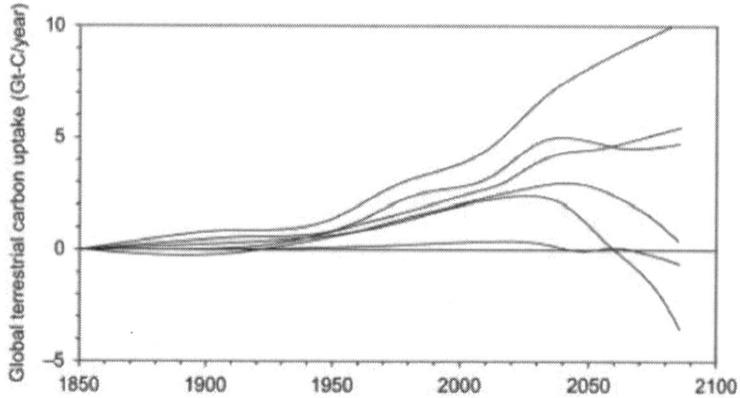

FIGURE 5.11 Range of predicted terrestrial carbon uptake from 11 carbon cycle—climate models. *After Heimann and Reichstein (2008).*

TABLE 5.4 Major factors influencing terrestrial carbon uptake.

Factor	Description
Atmospheric composition effects	CO_2 fertilization and interaction with water and nutrient (N, P) availability (Coskun et al., 2016; Terrer et al., 2019); damage by pollution
Direct and indirect climate effects	Changes in mean precipitation, temperature, radiation regime, stress from atmospheric aridity (water vapor deficit); climate impact on microbial community dynamics and SOC decomposition, including priming effects; climate impact on the accessibility of slow and passive carbon pools (e.g., impact on physico-chemical protection mechanisms); impact of climate on soil hydrological conditions (e.g. (dis)aggregation, drying, erosion, and SOC leaching; Frank et al., 2015a);
Extreme disturbance effects	Changing rates of occurrence of hurricanes and high winds, wildfires, pests and pathogens, and extreme precipitation events.

After Keenan and Williams (2018).

The range of model predictions of cumulative terrestrial carbon uptake to 2100 differ by ~500 GtC and different models predict that terrestrial ecosystems could then be either a net source or sink of ~5 GtC/yr, or anything in between. Clearly, this difference represents a significant uncertainty for global climate change prediction, necessitating an improvement in the fundamental understanding of the processes and interactions involved.

Alongside the impact of land-use changes (deforestation, afforestation, and agricultural practices) and the legacy of historical land-use change, the main climate change-related influences on terrestrial carbon uptake are summarized in Table 5.4.

Ecosystem responses to these factors have been the subject of a wide range of experimental and modeling studies, summarized in the following sections. Some additional areas of uncertainty and research regarding the impact of climate change on terrestrial processes include the availability of previously protected soil carbon as a result of permafrost melting, microbial priming effects, and C–N cycle interactions (Heimann and Reichstein, 2008), increased rate of N mineralization at higher temperature, enhancing nutrient supply for plant growth, the impact of extreme events (fire risk, wind-throw, pests, and pathogens) on standing carbon stocks (Reichstein et al., 2013; Arneth et al., 2017), the regional distribution of the land sink, particularly in tropical regions, and the impact of localized climatic effects such as El Niño (Keenan and Williams, 2018).

5.6.1 Impact of changing atmospheric composition

Among the influences of atmospheric composition on photosynthesis and soil carbon processes, that of enhanced [CO_2] (eCO_2) is the most important and most widely studied. Investigations over the last 20 years have used data acquired both from purpose-built facilities and also from FLUXNET, the established global network of bio-meteorological monitoring stations. Some of the key results of this work, which have a particular bearing on land-based approaches to achieving negative emissions, are described below.

The impact of eCO_2—FACE experiments

FACE experiments are an important class of midscale experiments that investigate the carbon cycle in a variety of terrestrial ecosystems, with emphasis on the impact of eCO_2 and in some cases also elevated ozone concentration (eO_3). A typical FACE facility consists of a number of plots within which the relevant gas concentration is maintained at an elevated level during the hours of photosynthetic activity.

The Aspen (Rhinelander) FACE, operated by the Brookhaven National Laboratory and the Michigan Technological University between 1997 and 2009, was a typical forest ecosystem example. The layout of this facility, shown in Fig. 5.12, consisted of 12 circular experimental plots, each 30 m in diameter, within which [CO_2] and [O_3] could be controlled. The study gases were introduced from a set of vertical vent pipes positioned on the circumference of each plot, with gas being vented only from those pipes upwind of the plot. The effect of eCO_2 was investigated by introducing an excess [CO_2] of ~200 ppm or ~50% above ambient concentration. Sensors mounted within the canopy provided feedback to a computer control system that maintained the gas concentration at the desired level, and the system was typically able to maintain the annual average [CO_2] within $\pm 10\%$ of the target level for $>90\%$ of the canopy volume within each experimental plot, although short-term excursions from the target level were larger.

FIGURE 5.12 Aspen (Rhinelander) FACE experimental configuration. *Courtesy; Michigan Technological University. Photo Credit David F. Karnosky.*

As the plots were unconfined, other aspects of the natural environment remained unchanged. The design allowed assessment of the effects of these gases, either alone or in combination, on many attributes of the ecosystem, including above- and belowground growth and soil carbon.

FACE experiments have been performed in a range of ecosystems, from forests to desert scrub, and the main features of a number of these experiments are summarized in Tables 5.5 and 5.6. Despite the importance of understanding climate feedback effects in these ecosystems and their potential significance for terrestrial carbon storage, many of the experiments listed have been closed, often as a result of competing funding priorities rather than the completion of their respective scientific programs, while others continue on a monitoring only basis, without CO_2 enrichment. Largely for logistical reasons, the first generation of forest FACE experiments focused on young trees in temperate environments (including some species from boreal forests), while second-generation experiments expanded this coverage to mature (hemi)boreal and tropical forests, which is important since the latter ecosystem accounts for over 2/3 of live terrestrial plant biomass. Full details of individual projects can be found in the US DOE database, referenced at the end of this chapter.

Plot scale experiments of this type improve our understanding of a wide range of processes relating to terrestrial carbon storage, including (Kuzyakov et al., 2019):

- Carbon and nitrogen cycling,
- Microbial growth and function, and extracellular enzyme activity,
- Turnover of SOM, including induced priming effects,
- N mobilization/immobilization through microbial activity,
- Processes affecting nutrient mobilization from organic sources and from accelerated biochemical weathering of soil minerals,
- Water use efficiency and impact on soil moisture,
- The formation and leaching of soil carbonates, including by organic acids.

The extension of these experiments to mature forests would help to clarify whether increased growth observed in young trees under eCO_2 represents simply an acceleration of growth (potentially with no impact on final C storage) or a change to a higher equilibrium biomass at maturity, as illustrated in Fig. 5.13. The curves show schematically the accumulation of biomass with time since establishment, with two FACE experimental periods shown by the gray shaded areas. In the first FACE period, representing a young forest experiment, an increase in biomass accumulation rate is observed, but no conclusion can be drawn as to whether the final equilibrium will be affected by eCO_2 or determined by other preexisting constraints such as light or nutrient availability. In the second FACE period, representing a mature forest experiment, any increase in the observed accumulation rate must reflect a change to a higher equilibrium under the influence of eCO_2.

Results from FACE experiments have been compiled in a number of review articles (Jones et al., 2014, and references therein) and have improved our understanding of many of the processes listed above. These experiments have consistently demonstrated that increased atmospheric $[CO_2]$ leads to an increase in NPP. FACE results have also been validated by investigations at the sites of natural CO_2 springs, where similar increases in photosynthesis and other plant physiological responses have been found (Saban et al., 2019). These findings also support the use of FACE results to predict the longer term impact of eCO_2 on their respective ecosystems.

TABLE 5.5 Free air CO_2 enrichment (FACE) experiments in forest ecosystems.

FACE facility	Organization	Location	Ecosystem	Species	Key parameters
First generation experiments					
Aspen (Rhinelander) FACE	Brookhaven National Laboratory, Michigan Technological University	Harshaw Experimental Forest, Rhinelander, Wisconsin, USA	Temperate deciduous forest	Aspen (*Populus tremuloides*) Birch (*Betula pendula*) Sugar Maple (*Acer Saccharum*)	9 FACE + 3 control, $[CO_2]$ = 560 ppm, $[O_3]$ = 1.5 × ambient
Bangor FACE	University of Wales	Henfaes Experimental Farm, Bangor, UK	Temperate deciduous forest	Birch (*Betula pendula*) Alder (*Alnus glutinosa*) Beech (*Fagus sylvatica*)	4 FACE + 4 control, 8 m diameter rings, $[CO_2]$ = +200 ppm
Duke Forest FACE	Brookhaven National Laboratory	Duke Forest, Orange County, North Carolina, USA	Temperate pine forest	Loblolly Pine (*Pinua taeda*)	4 FACE + 8 control, 30 m diameter rings, $[CO_2]$ = ambient + 200 ppm
Euro FACE	University of Tuscia	Tuscania, Viterbo, Italy	Temperate deciduous forest	White Poplar (*Populus alba*)	3 FACE + 3 control,
				Black Poplar (*Populus nigra*)	22 m diameter rings
				Eastern Cottonwood (*Populus deltoides*), and hybrids	$[CO_2]$ = 380, 200 ppm
ORNL-FACE	Oak Ridge National Laboratory	Oak Ridge National Environmental Research Park, Oak Ridge, Tennessee, USA	Temperate deciduous forest	American Sweetgum (*Liquidambar styraciflua*)	2 FACE + 3 control, 25 m diameter rings $[CO_2]$ = 565 ppm
Second-generation experiments—mature forests					
AmazonFACE	Universidade Estadual Paulista, Brazil and Oak Ridge National Laboratory, USA	Manaus, Brazil	Old growth tropical rain forest	Diverse (52 tree species across the 2 pilot plots)	>200 yr, 30 m canopy, pilot phase: 1 FACE + 1 control, 30 m plots, later 3+3 additional plots, $[CO_2]$ = +200 ppm, eCO_2 start Dec 2016
BIFor-FACE	The Birmingham Institute of Forest Research, University of Birmingham	Staffordshire, United Kingdom	Temperate deciduous forest	Oak (Quercus robur), Sycamore (*Acer pseudoplatanus*), Silver Birch (*Betula pendula*), with Hazel coppice (*Corylus avellana*) understorey	~150 yr, 25 m canopy, 9 × 28 m plots, $[CO_2]$ = +150 ppm, eCO_2 start April 2016
EucFACE	Hawkesbury Institute for the Environment, Western Sydney University	Richmond, NSW, Australia	Subtropical dry forest	Eucalyptus (*Eucalyptus tereticornus*)	>100 yr, 22 m canopy; 3 FACE + 3 control, 25 m plots, $[CO_2]$ = +150 ppm, eCO_2 start Sept 2012
SwedFACE	Lund University, Linnaeus University, and Gothenburg University	Asa Research Park, Småland, Sweden	Boreal pine forest	Scots Pine (*Pinus sylvestris*)	34 yr, 12–15 m canopy, 6 × 25 m plots, $[CO_2]$ = +150 ppm coupled with nutrient (P, N) treatment, eCO_2 planned Apr 2017

TABLE 5.6 Free air CO_2 enrichment (FACE) experiments in nonforest ecosystems.

FACE facility	Organization	Location	Ecosystem	Species	Key parameters
AG FACE	Grains R&D Corporation and the University of Melbourne	First site; Horsham, Victoria, Australia Second site; Walpeup, Victoria, Australia	Agroecosystem	Wheat (*Triticum* spp.)	8 FACE + 8 control, $[CO_2] = 350$, 550 ppm, site 1: 12 m diameter rings, site 2: 4 m diameter rings
Arizona FACE	University of Arizona	Maricopa Agriculture Centre, Maricopa, Arizona, USA	Agroecosystem	Cotton (*Gossypium hirsutum*) Wheat (*Triticum* spp.) Sorghum (*Sorghum* spp.)	4 FACE + 4 control, $[CO_2] = +200$ ppm
BioCON FACE	University of Minnesota	Cedar Creek Ecoscience Reserve, Minneapolis, Minnesota, USA	Grassland	16 species in 4 groups; warm-season grasses (C4), cool-season grasses (C3), forbs, and nitrogen-fixing legumes	3 FACE + 3 control, 20 m diameter rings, each with ~60 2 m × 2 m plots, $[CO_2] = 368$, 56 ppm
GiFACE	University of Gießen	Gießen, Germany	Temperate grassland	Diverse grass species (incl. Oat grass *Arrhena-theretum elatioris*), non-leguminous herbs and legumes	3 FACE +3 control, 8 m diameter rings, $[CO_2] = +20\%$
Nevada Desert FACE	Brookhaven National Laboratory, University of Nevada	Mojave Desert, Mercury, Nevada, USA	Desert scrub	Creosote Bush (*Larrea tridentate*), Bursage (*Ambrosia dumosa*)	3 FACE + 6 control, 23 m diameter rings, $[CO_2]$ = ambient + 50%
Oz FACE	CSIRO and James Cook university	Yabulu Refinery, Townsville, Queensland, Australia	Coastal tropical savannah	Kangaroo Grass (*Themeda triandra*), Eucalyptus, and Acacia	4 FACE + 2 control, 15 m diameter rings, $[CO_2] = 370$, 460, 550 ppm
Soy FACE	University of Illinois	Urbana-Champaign, Illinois, USA	Agroecosystem	Soy (*Glycine max*): 16 rings under elevated $[CO_2]$ and $[O_3]$; Maize (*Zea mays*): 8 rings under elevated $[CO_2]$	16 FACE + 8 control, 15 m diameter rings, $[CO_2] = 550$ ppm, $[O_3] = 1.2 \times$ ambient

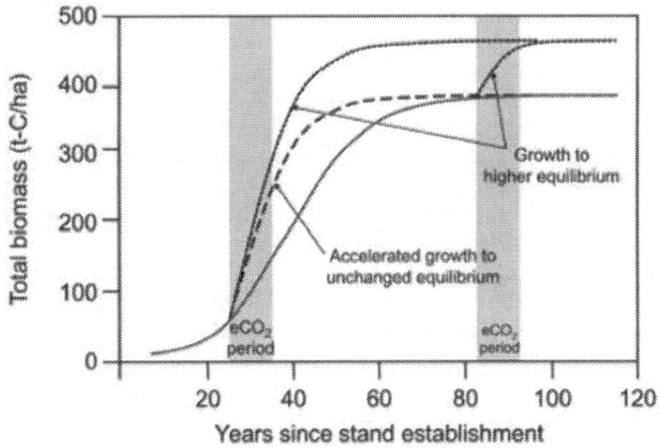

FIGURE 5.13 Forest biomass accumulation curves, showing potential impacts of eCO_2.

However, FACE experiments have been geographically biased toward temperate regions, giving less insight into the responses in the important tropical and boreal forests. The impact on SOC has also been less widely investigated, with inconsistent or unexpected results demonstrating that the processes and feedbacks affecting SOC are even more complex than anticipated, and are influenced by a wide range of site specific properties and histories (Jones et al., 2014).

eCO_2 driven increase in water use efficiency

Increasing [CO_2] has also been shown to result in an increase in biomass growth per unit of water transpired (water use efficiency or W_e) in temperate and boreal forests. Using regional and site-specific data from the global FLUXNET network of bio-meteorological stations, Keenan and colleagues (2013) identified increasing water use efficiency linked to CO_2 fertilization, a trend that has been confirmed in a number of other studies (Frank et al., 2015b; van der Sleen et al., 2015).

Although this bodes well for timber yields and forest carbon uptake, precipitation in some regions relies on water transpired from these forests, and reduced humidity can lead to increased local air temperatures and contribute to droughts in other parts of the world (Keenan et al., 2013).

5.6.2 Impacts of climate change

Alongside the direct impact of rising [CO_2], carbon stocks and fluxes in all terrestrial ecosystems are also affected by the resulting changes in climatic conditions, as biological processes respond to rising temperatures, changing patterns of precipitation or atmospheric aridity, etc.

As well as supporting CO_2 fertilization effects, rising temperatures will enhance nutrient supply by increasing mineralization rates, leading to an increase in carbon uptake in biomass, but will also stimulate microbial decomposition of soil carbon, leading to an increase in CO_2 respiration. There remains considerable uncertainty concerning the overall impact on the terrestrial carbon budget (Arneth et al., 2017), with different ecosystems likely to respond differently. Atmospheric monitoring and dynamic vegetation modeling studies for high northern latitudes (above 40°N) have shown that the increased seasonal [CO_2] cycle since 1960 is driven by higher photosynthetic production in temperate and boreal forests, rather than higher ecosystem respiration (Forkel et al., 2016). On the other hand, CO_2 release from Arctic tundra has been increasing as a result of higher temperatures, particularly in the early winter period, with October through December emissions increasing almost 75% since 1975 (Commane et al., 2017).

Large uncertainties remain about future carbon fluxes from tropical terrestrial ecosystems. In tropical forests, notably the Amazon, increasing temperature and a decreasing rainfall have led to a decline in GPP, despite CO_2 fertilization, and an increase in ecosystem respiration (Raddatz et al., 2007; Seiler, 2014). This trend is expected to continue and lead to a significant weakening or potentially a reversal of this important carbon sink (Baccini et al., 2017; Mitchard, 2018), exacerbated by extreme events such as droughts and fires. Similar outcomes are indicated in other tropical forests (Qie et al., 2017).

The response of SOC stocks to climate change remains highly uncertain. Sulman and colleagues recently reviewed current SOC models and experimental results and showed that while models uniformly predicted an increase in CO_2 efflux and a decline in SOC stocks, many experiments showed the opposite effect, with CO_2 flux decreasing in one-third of experiments and stocks increasing in almost one half (Sulman et al., 2018). As discussed above, soil microbes have an important influence on SOC, and an improved understanding of the impact of climate change on these communities and their activities is also needed (Bardgett et al., 2008). Areas of uncertainty include the impact of climate-related mechanisms such as permafrost melting (Commane et al., 2017) and changes in wetlands, and of microbial priming and carbon-nitrogen cycle interactions on the availability for microbial decomposition of previously inaccessible SOC (Heimann and Reichstein, 2008; Bond-Lamberty et al., 2018). The current and future role of nutrients as factors that might limit carbon uptake by land plants is a further uncertainty in relation to the future of soil carbon stocks. Since the global carbon inventory of soils is several times that of the atmosphere, these uncertainties represents a fundamental limitation in our ability to predict feedback from terrestrial ecosystems in response to climate change (Bradford et al., 2016; Bond-Lamberty et al., 2018).

Improved understanding of the magnitude, regional distribution, and future evolution of the impact of climate extremes on the carbon cycle, is being progressed through improved observations, targeted regional and rapid response assessments, and manipulation experiments (Reichstein et al., 2013; Frank et al., 2015a) alongside higher resolution Earth system modeling, particularly of droughts and related processes (convection, cloud formation, soil hydrology, etc.) and of processes that modulate the impact of climate extremes, such as nutrient cycling and biological acclimation/adaptation (Reichstein et al., 2013; Smith and Dukes, 2013).

While the emphasis here has been on direct climate effects on carbon stocks and fluxes, the impact of climate change on methane fluxes in the biosphere is also of crucial importance, in view of methane's high warming potential (Dean et al., 2018). In particular, wetlands and melting permafrost soils may become net methane sources in the near term, creating an additional positive climate feedback (Kayranli et al., 2010).

5.6.3 Extreme disturbance effects

Carbon uptake and residence in terrestrial ecosystems can be affected both directly and indirectly as a result of extreme meteorological conditions, as shown schematically in Fig. 5.14. Direct impacts are shown here as solid arrow, with the relative importance being indicated by the width of the arrow. Indirect impacts are shown as dashed arrows.

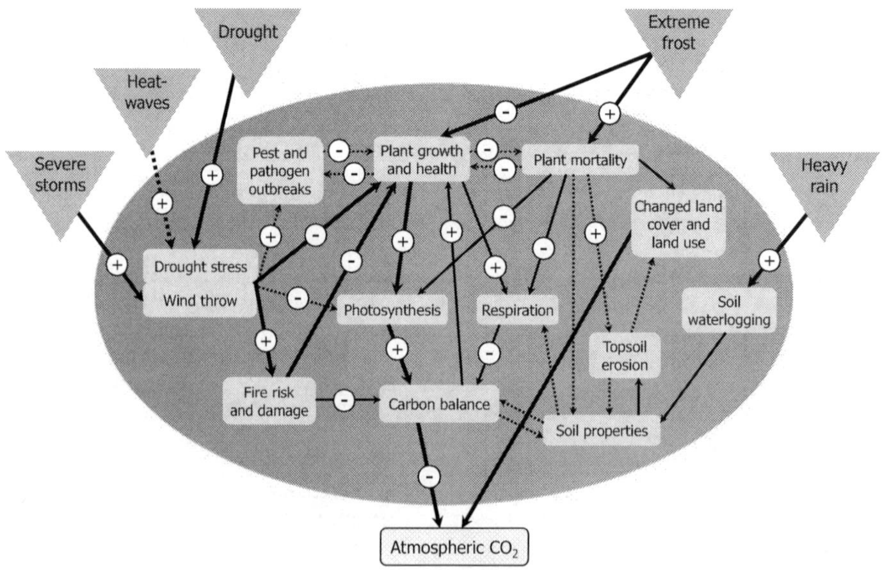

FIGURE 5.14 Processes and feedbacks triggered by extreme climate events. *After Reichstein et al. (2013).*

In forests, for example, wind throw from increasingly frequent and intense storms transforms the standing biomass stock to dry, dead wood, increasing the risk of fires and pathogen outbreaks. During extended hot, dry periods, reduced water availability impacts plant physiology, seasonal patterns (phenology), and above/belowground carbon allocation, and can lead to increased tree mortality, with the consequent risk of fire and susceptibility to pathogens.

As an example, the Amazon rainforest, which contains about half of all tropical forest carbon stocks, is experiencing frequent severe droughts due to climate variability linked to El Nino and to tropical North Atlantic warming anomalies. The drought of 2005 switched affected areas of the forest from net sinks to sources and is estimated to have resulted in the loss of 1 GtC from carbon stocks over the 2005 to 2008 period (Yang et al., 2018).

The immediate impact and recovery from disturbance events, as well as the longer-term effect on ecosystems of climate-driven change in disturbance regimes, currently suffer from low predictability, which is the subject of ongoing research (Luo et al., 2015). Extreme disturbance effects will not only affect the fate of carbon stocks in existing terrestrial ecosystems, but will also be important in assessing the effectiveness of various NET approaches, including afforestation and large-scale biomass cultivation. Present climate and vegetation models are designed to optimally represent long-term responses, and do not adequately model extreme climatic events and their ecosystem impacts. As climate change-related extreme events are expected to increase in future, improved models, underpinned by extensive field observations, will be needed to enable reliable predictions of the future impact of climate extremes on the terrestrial carbon inventories and fluxes (Frank et al., 2015a).

References and resources

References

Abramoff, R., Xu, X., Hartman, M., et al., 2018. The Millennial model: in search of measurable pools and transformations for modeling soil carbon in the new century. Biogeochemistry 137 (1), 51–71.

Achal, V., Mukherjee, A., 2015. A review of microbial precipitation for sustainable construction. Construct. Build. Mater. 93, 1224–1235.

Arneth, A., Sitch, S., Pongratz, J., et al., 2017. Historical carbon dioxide emissions caused by land-use changes are possibly larger than assumed. Nat. Geosci. 10 (2), 79–84.

Baccini, A., Walker, W., Carvalho, L., et al., 2017. Tropical forests are a net carbon source based on aboveground measurements of gain and loss. Science 358 (6360), 230–234.

Bardgett, R.D., Freeman, C., Ostle, N.J., 2008. Microbial contributions to climate change through carbon cycle feedbacks. ISME J. 2 (8), 805–814.

Blanco-Canqui, H., Lal, R., 2004. Mechanisms of carbon sequestration in soil aggregates. Crit. Rev. Plant Sci. 23 (6), 481–504.

Bond-Lamberty, B., Bailey, V.L., Chen, M., Gough, C.M., Vargas, R., 2018. Globally rising soil heterotrophic respiration over recent decades. Nature 560 (7716), 80.

Borland, A.M., Hartwell, J., Weston, D.J., et al., 2014. Engineering crassulacean acid metabolism to improve water-use efficiency. Trends Plant Sci. 19 (5), 327–338.

Bradford, M.A., Wieder, W.R., Bonan, G.B., et al., 2016. Managing uncertainty in soil carbon feedbacks to climate change. Nat. Clim. Change 6 (8), 751.

Campbell, E.E., Paustian, K., 2015. Current developments in soil organic matter modeling and the expansion of model applications: a review. Environ. Res. Lett. 10 (12), 123004.

Commane, R., Lindaas, J., Benmergui, J., et al., 2017. Carbon dioxide sources from Alaska driven by increasing early winter respiration from Arctic tundra. Proc. Natl. Acad. Sci. USA 114 (21), 5361–5366.

Coskun, D., Britto, D.T., Kronzucker, H.J., 2016. Nutrient constraints on terrestrial carbon fixation: the role of nitrogen. J. Plant Physiol. 203, 95–109.

Dean, J.F., Middelburg, J.J., Röckmann, T., et al., 2018. Methane feedbacks to the global climate system in a warmer world. Rev. Geophys. 56 (1), 207–250.

Di Fiore, A., Alterio, V., Monti, S.M., De Simone, G., D'Ambrosio, K., 2015. Thermostable carbonic anhydrases in biotechnological applications. Int. J. Mol. Sci. 16 (7), 15456–15480.

Forkel, M., Carvalhais, N., Rödenbeck, C., et al., 2016. Enhanced seasonal CO_2 exchange caused by amplified plant productivity in northern ecosystems. Science 351 (6274), 696–699.

Frank, D., Reichstein, M., Bahn, M., et al., 2015a. Effects of climate extremes on the terrestrial carbon cycle: concepts, processes and potential future impacts. Global Change Biol. 21 (8), 2861–2880.

Frank, D., Poulter, B., Saurer, M., et al., 2015b. Water-use efficiency and transpiration across European forests during the Anthropocene. Nat. Clim. Change 5 (6), 579–583.

Friedlingstein, P., Dufresne, J.L., Cox, P.M., Rayner, P., 2003. How positive is the feedback between climate change and the carbon cycle? Tellus B: Chem. Phys. Meteorol. 55 (2), 692–700.

Friedlingstein, P., Meinshausen, M., Arora, V.K., et al., 2014. Uncertainties in CMIP5 climate projections due to carbon cycle feedbacks. J. Clim. 27 (2), 511–526.

Friedlingstein, P., Jones, M.W., O'Sullivan, M., et al., 2021. Global carbon budget 2021. Earth Syst. Sci. Data Discuss. 1–191.

Heimann, M., Reichstein, M., 2008. Terrestrial ecosystem carbon dynamics and climate feedbacks. Nature 451 (7176), 289.

Jansson, C., Wullschleger, S.D., Kalluri, U.C., Tuskan, G.A., 2010. Phytosequestration: carbon biosequestration by plants and the prospects of genetic engineering. Bioscience 60 (9), 685–696.

Jastrow, J.D., Miller, R.M., 1998. Soil aggregate stabilization and carbon sequestration: feedbacks through organomineral associations. In: Lal, R., Kimble, J.M., Follett, R.F., Stewart, B.A. (Eds.), Soil Processes and the Carbon Cycle. CRC Press, Boca Raton, FL.

Jones, A.G., Scullion, J., Ostle, N., Levy, P.E., Gwynn-Jones, D., 2014. Completing the FACE of elevated CO_2 research. Environ. Int. 73, 252–258.

Kamennaya, N.A., Ajo-Franklin, C.M., Northen, T., Jansson, C., 2012. Cyanobacteria as biocatalysts for carbonate mineralization. Minerals 2, 338–364.

Kayranli, B., Scholz, M., Mustafa, A., Hedmark, Å., 2010. Carbon storage and fluxes within freshwater wetlands: a critical review. Wetlands 30 (1), 111–124.

Keenan, T.F., Hollinger, D.Y., Bohrer, G., et al., 2013. Increase in forest water-use efficiency as atmospheric carbon dioxide concentrations rise. Nature 499 (7458), 324–327.

Keenan, T.F., Williams, C.A., 2018. The terrestrial carbon sink. Annu. Rev. Environ. Resour. 43, 219–243.

Kuzyakov, Y., Horwath, W.R., Dorodnikov, M., Blagodatskaya, E., 2019. Review and synthesis of the effects of elevated atmospheric CO_2 on soil processes: No changes in pools, but increased fluxes and accelerated cycles. Soil Biol. Biochem. 128, 66–78.

Lange, M., Eisenhauer, N., Sierra, C.A., et al., 2015. Plant diversity increases soil microbial activity and soil carbon storage. Nat. Commun. 6 (1), 1–8.

Lee, S.W., Park, S.B., Jeong, S.K., Lim, K.S., Lee, S.H., Trachtenberg, M.C., 2010. On carbon dioxide storage based on biomineralization strategies. Micron 41 (4), 273–282.

Luo, Y., Keenan, T.F., Smith, M., 2015. Predictability of the terrestrial carbon cycle. Global Change Biol. 21 (5), 1737–1751.

Mason, P.M., Glover, K., Smith, J.A.C., Willis, K.J., Woods, J., Thompson, I.P., 2015. The potential of CAM crops as a globally significant bioenergy resource: moving from 'fuel or food' to 'fuel and more food. Energy Environ. Sci. 8 (8), 2320–2329.

Mitchard, E.T., 2018. The tropical forest carbon cycle and climate change. Nature 559 (7715), 527.

Parr, J.F., Sullivan, L.A., 2005. Soil carbon sequestration in phytoliths. Soil Biol. Biochem. 37 (1), 117–124.

Parr, J., Sullivan, L., Chen, B., Ye, G., Zheng, W., 2010. Carbon bio-sequestration within the phytoliths of economic bamboo species. Glob. Chang. Biol. 16 (10), 2661–2667.

Parton, W.J., Scurlock, J.M.O., Ojima, D.S., et al., 1993. Observations and modeling of biomass and soil organic matter dynamics for the grassland biome worldwide. Global Biogeochem. Cycles 7 (4), 785–809.

Piperno, D.R., 2006. Phytoliths: A Comprehensive Guide for Archaeologists and Paleoecologists. Rowman Altamira.

Power, I.M., Harrison, A.L., Dipple, G.M., et al., 2013. Carbon mineralization: from natural analogues to engineered systems. Rev. Mineral. Geochem. 77 (1), 305–360.

Qie, L., Lewis, S.L., Sullivan, M.J., et al., 2017. Long-term carbon sink in Borneo's forests halted by drought and vulnerable to edge effects. Nat. Commun. 8 (1), 1–11.

Raddatz, T.J., Reick, C.H., Knorr, W., et al., 2007. Will the tropical land biosphere dominate the climate–carbon cycle feedback during the twenty-first century? Clim. Dynam. 29 (6), 565–574.

Reichstein, M., Bahn, M., Ciais, P., et al., 2013. Climate extremes and the carbon cycle. Nature 500 (7462), 287–295.

Saban, J.M., Chapman, M.A., Taylor, G., 2019. FACE facts hold for multiple generations; Evidence from natural CO_2 springs. Global Change Biol. 25 (1), 1–11.

Segoli, M., De Gryze, S., Dou, F., et al., 2013. AggModel: a soil organic matter model with measurable pools for use in incubation studies. Ecol. Model. 263, 1–9.

Seiler, C., 2014. The Sensitivity of Tropical Forests to Climate Variability and Change in Bolivia. Wageningen University.

Six, J., Frey, S.D., Thiet, R.K., Batten, K.M., 2006. Bacterial and fungal contributions to carbon sequestration in agroecosystems. Soil Sci. Soc. Am. J. 70 (2), 555–569.

Smith, N.G., Dukes, J.S., 2013. Plant respiration and photosynthesis in global-scale models: incorporating acclimation to temperature and CO_2. Global Change Biol. 19 (1), 45–63.

Song, Z., Liu, H., Strömberg, C.A., Yang, X., Zhang, X., 2017. Phytolith carbon sequestration in global terrestrial biomes. Sci. Total Environ. 603, 502–509.

Stirbet, A., Lazár, D., Guo, Y., Govindjee, G., 2020. Photosynthesis: basics, history and modelling. Ann. Bot. 126 (4), 511–537.

Stocker, T.F. (Ed.), 2013. Climate Change 2013: The Physical Science Basis, Working Group 1. Contribution to the IPCC 5th Assessment Report (AR5). Cambridge University Press, Cambridge, UK.

Sulman, B.N., Moore, J.A., Abramoff, R., et al., 2018. Multiple models and experiments underscore large uncertainty in soil carbon dynamics. Biogeochemistry 141 (2), 109–123.

Terrer, C., Jackson, R.B., Prentice, I.C., et al., 2019. Nitrogen and phosphorus constrain the CO_2 fertilization of global plant biomass. Nat. Clim. Change 9 (9), 684–689.

Terzis, D., Laloui, L., 2019. A decade of progress and turning points in the understanding of bio-improved soils: a review. Geomech. Energy Environ. 19, 100116.

van der Sleen, P., Groenendijk, P., Vlam, M., et al., 2015. No growth stimulation of tropical trees by 150 years of CO_2 fertilization but water-use efficiency increased. Nat. Geosci. 8 (1), 24–28.

Walker, A.P., De Kauwe, M.G., Bastos, A., et al., 2021. Integrating the evidence for a terrestrial carbon sink caused by increasing atmospheric CO_2. New Phytol. 229 (5), 2413–2445.

Yang, X., Cushman, J.C., Borland, A.M., et al., 2015. A roadmap for research on crassulacean acid metabolism (CAM) to enhance sustainable food and bioenergy production in a hotter, drier world. New Phytol. 207 (3), 491–504.
Yang, Y., Saatchi, S.S., Xu, L., et al., 2018. Post-drought decline of the Amazon carbon sink. Nat. Commun. 9 (1), 1–9.

Further reading

Ainsworth, E.A., Long, S.P., 2021. 30 years of free-air carbon dioxide enrichment (FACE): what have we learned about future crop productivity and its potential for adaptation? Global Change Biol. 27 (1), 27–49.
Anbu, P., Kang, C.H., Shin, Y.J., So, J.S., 2016. Formations of calcium carbonate minerals by bacteria and its multiple applications. SpringerPlus 5 (1), 250.
Dean, J.F., Middelburg, J.J., Röckmann, T., et al., 2018. Methane feedbacks to the global climate system in a warmer world. Rev. Geophys. 56 (1), 207–250.
Fatichi, S., Pappas, C., Zscheischler, J., Leuzinger, S., 2019. Modelling carbon sources and sinks in terrestrial vegetation. New Phytol. 221 (2), 652–668.
Heveran, C.M., Williams, S.L., Qiu, J., Artier, J., Hubler, M.H., Cook, S.M., Cameron, J.C., Srubar III, W.V., 2020. Biomineralization and successive regeneration of engineered living building materials. Matter 2 (2), 481–494.
Horwath, W.R., Kuzyakov, Y., 2017. Climate change impacts on soil processes and ecosystem properties. Dev. Soil Sci. 35 (Elsevier).
Jackson, R.B., Lajtha, K., Crow, S.E., Hugelius, G., Kramer, M.G., Piñeiro, G., 2017. The ecology of soil carbon: pools, vulnerabilities, and biotic and abiotic controls. Annu. Rev. Ecol. Evol. Syst. 48, 419–445.
Lal, R., Negassa, W., Lorenz, K., 2015. Carbon sequestration in soil. Curr. Opin. Environ. Sustain. 15, 79–86.
Paul, E.A., 2016. The nature and dynamics of soil organic matter: plant inputs, microbial transformations, and organic matter stabilization. Soil Biol. Biochem. 98, 109–126.
Six, J., Bossuyt, H., Degryze, S., Denef, K., 2004. A history of research on the link between (micro) aggregates, soil biota, and soil organic matter dynamics. Soil Tillage Res. 79 (1), 7–31.
Zhang, X., Yan, S.Y., Tyagi, R.D., Surampalli, R.Y., Zhang, T.C., 2015. Enzymatic sequestration of carbon dioxide. In: Surampalli, R.Y., et al. (Eds.), Carbon Capture and Storage: Physical, Chemical, and Biological Methods. American Society of Civil Engineers, Reston, Virginia, USA.

Projects and organizations

BioMASON; using MICP technology to “grow” construction materials: www.biomason.com.
FLUXNET; portal for micrometeorological measurement of trace gas fluxes between the land and atmosphere: fluxnet.org.
US DoE Global List of FACE Experiments: http://facedata.ornl.gov/global_face.html.

Chapter 6

Ocean carbon cycle processes

Steve Rackley

In describing the processes that control the interchange of CO_2 between the atmosphere and the ocean, this chapter links the discussion of the global carbon cycle (Chapter 4) with the detailed review of ocean-based NETs to follow in Chapter 16. Current knowledge of the impact of climate change on these physical and biological processes is summarized, as well as the emerging understanding of the potential impact of negative emissions.

6.1 Introduction

The world's oceans contain an estimated 39,000 GtC (143,000 $GtCO_2$), around 10 times greater than the terrestrial carbon inventory and over 40 times that of the atmosphere, and are already playing a critical role in mitigating climate change, having taken up a quarter of total CO_2 emissions during the industrial era (Khatiwala et al., 2013; Friedlingstein et al., 2021). This uptake of CO_2 from the atmosphere to the ocean was first noted by Revelle and Suess (1957), who concluded, based on carbon isotope measurements, that the ocean was a major sink of anthropogenic emissions.

Deep ocean waters provide a CO_2 sink that is coupled to the atmosphere via the solubility and biological pumps, discussed later, and changes in atmospheric [CO_2] are damped on a time scale determined by the slow turnover of deep ocean waters. This is illustrated in Fig. 6.1, which shows schematically the evolution of atmospheric [CO_2] resulting from a 100-year "pulse" of ~6 $GtCO_2$/yr emissions to the atmosphere, starting at year zero. In this model, which considers only the partitioning of the emitted volume between atmosphere and ocean and excludes other sinks and feedbacks, the ocean takes up 50% of the emitted CO_2 after 150 years and almost 75% after 1000 years. From an initial level of 350 ppm, atmospheric [CO_2] peaks at 550 ppm at the end of the emission pulse and declines to ~430 ppm over the final 100 years. Using more complete models, the total uptake of emitted CO_2 into terrestrial and oceanic sinks is estimated to be 85%–90% over a millennial time scale (Sabine et al., 2004).

CO_2 is a minor component of the Earth's atmosphere because of its transformation into carbonate rocks and ocean alkalinity as a result of surface mineral weathering on a geological timescale (Schuiling and Krijgsman, 2006). Similarly, the ocean will be the natural long-term sink of all anthropogenic CO_2 emissions, on a ~100 kyr timescale, as a result of the action of mineral weathering alongside two ocean processes—the solubility pump and the biological pump—which drive the ocean carbon cycle.

6.2 The solubility pump

The solubility pump, or physical pump, is the term used to describe the physical and chemical process by which CO_2 is transported from the atmosphere to deep ocean waters. The process is illustrated in Fig. 6.2, together with the interlinked biological pump described later.

The two main elements of the solubility pump are the dissolution of CO_2 into surface waters and the successive downwelling of surface water and upwelling of deep water via the global thermohaline circulation system. The surface layer of the ocean is well mixed by wind and wave action (hence also known as the mixed layer) and CO_2 in the atmosphere reaches thermodynamic equilibrium with dissolved CO_2 in this layer on a time scale of weeks to months,

Negative Emissions Technologies for Climate Change Mitigation. https://doi.org/10.1016/B978-0-12-819663-2.00004-6

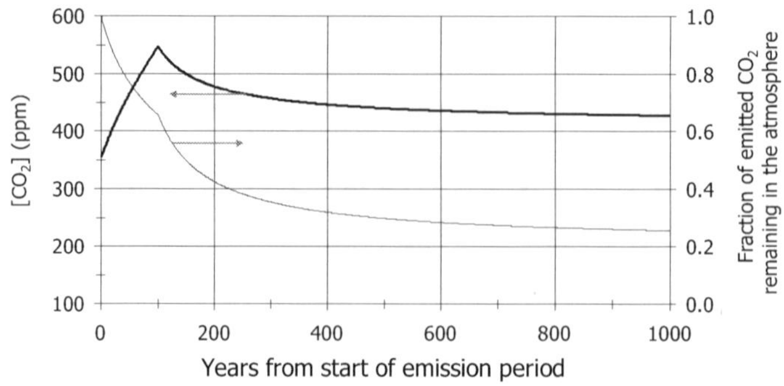

FIGURE 6.1 Schematic long-term partitioning of an emitted CO_2 pulse.

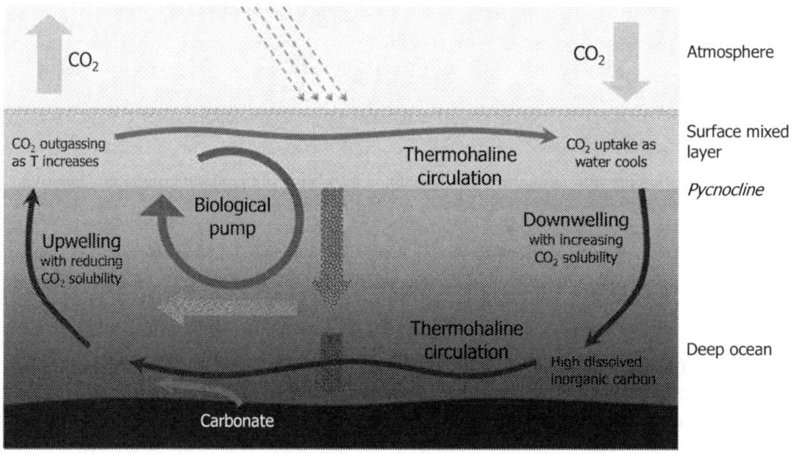

FIGURE 6.2 The ocean carbon cycle: solubility and biological CO_2 pumps.

depending on mixed layer depth and wind conditions. The equilibrium solubility of a gas is directly proportional to its partial pressure (Henry's law):

$$[CO_{2aq}] = pCO_2/K_{H.CO2} = pCO_2 \cdot K_{CO2} \tag{6.1}$$

where $[CO_{2aq}]$ is the dissolved CO_2 concentration, pCO_2 is the partial pressure of CO_2 in the atmosphere, $K_{H.CO2}$ is the Henry's law constant, and K_{CO2} is the solubility of CO_2. The solubility K_{CO2} is a function of pressure, salinity, and acidity (pH) and, most importantly, of temperature, with solubility increasing from ~0.03 molCO_2/kg at 20°C to ~0.05 molCO_2/kg at 10°C for current atmospheric and oceanic conditions.

The dissolution of CO_2 in water initially forms carbonic acid (H_2CO_3), which will dissociate into bicarbonate and then carbonate ions, the partitioning depending on the water temperature, salinity, and pH. Taken together, these species—dissolved free CO_2, H_2CO_3, bicarbonate ions (HCO_3^-), and carbonate ions (CO_3^{2-})—make up the total dissolved inorganic carbon (DIC). The reactions that determine the contributions of these species to total DIC are as follows:

$$CO_2 + H_2O \leftrightarrow H_2CO_3 \leftrightarrow HCO_{3^-} + H^+ \leftrightarrow CO_{3^{2-}} + 2H^+ \tag{6.2}$$

The equilibrium for this set of reactions moves to the left as pH decreases, and the partitioning is commonly represented using a Bjerrum plot, shown in Fig. 6.3.

At current average mixed layer conditions (pH ~ 8.0), DIC is ~91% bicarbonate and 8% carbonate ions, with the remaining ~1% being dissolved CO_2 plus carbonic acid. The DIC content of surface waters varies from a low of ~1850 μmolC/kg in the Bay of Bengal to a high of ~2200 μmolC/kg in the Weddell Sea.

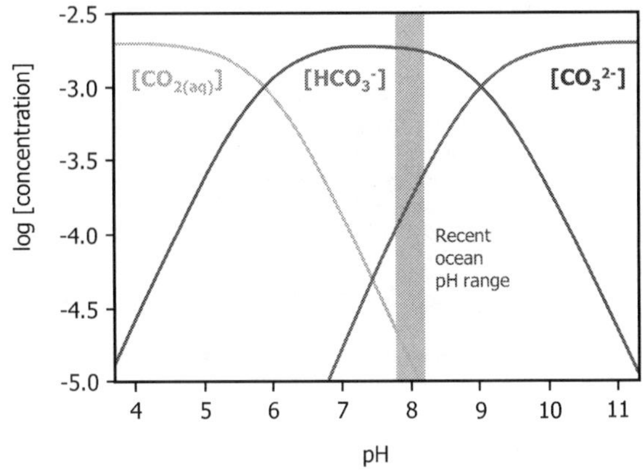

FIGURE 6.3 DIC partitioning versus pH (Bjerrum plot).

The decrease in pH resulting from the increased CO_2 dissolution in the ocean during the industrial era has been mitigated due to the significant carbonate ion presence, in a process known as carbonate buffering. The shift in the equilibrium point of reactions 6.2 consumes carbonate ions and mitigates the acidification that would otherwise arise by converting carbonic acid to alkaline bicarbonate ions, according to the overall reaction:

$$CO_{2(aq)} + H_2O + CO_3{}^{2-} \rightarrow 2HCO_3{}^{-} \tag{6.3}$$

However, the reduction in carbonate ion concentration has consequences both for future pH buffering and for calcifying marine organisms, as discussed later.

The second component of the solubility pump is the formation of deep waters through meridional overturning that occurs in the North and South Atlantic Oceans (Fig. 6.4). A parcel of water moving northward in the Gulf Stream experiences strong evaporative cooling due to wind action, resulting in increasing uptake of CO_2 from the atmosphere as a result of its temperature-dependent solubility, as well as an increase in salinity and density as a consequence of evaporation. By the time a parcel of water reaches the Norwegian Sea, the density becomes high enough to cause it to sink—the Atlantic meridional overturning circulation (AMOC)—forming North Atlantic Deep Water. This water flows back in a southerly direction into the abyssal Atlantic basin (Kuhlbrodt et al., 2007). A similar process occurs in the sub-Antarctic

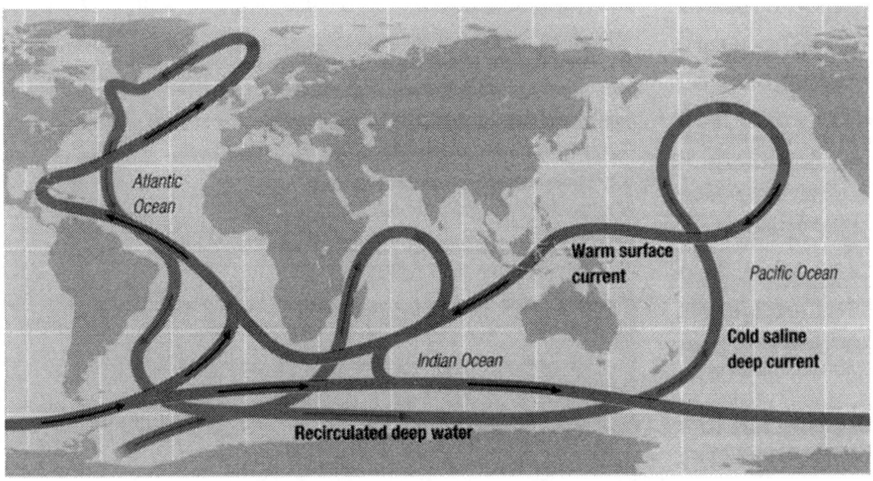

FIGURE 6.4 Global thermohaline circulation.

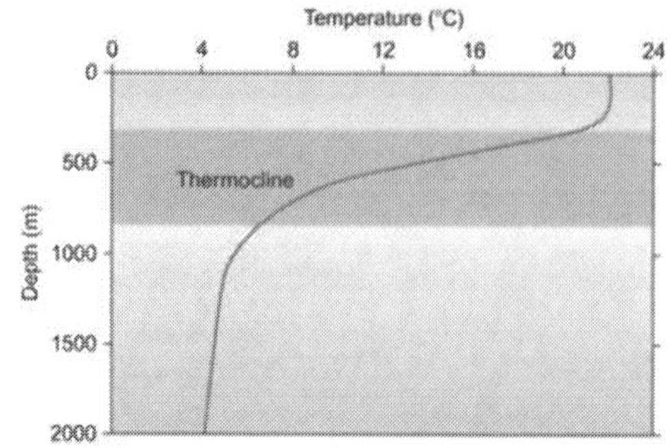

FIGURE 6.5 Temperature versus depth in a tropical ocean.

South Atlantic, forming Antarctic Bottom Water. The total rate of downwelling is estimated at 30–40 Sv (1 Sv (Sverdrup) = 10^6 m^3/day), and it is roughly evenly distributed between the North and South Atlantic locations (Weijer et al., 2019).

Other than at these two downwelling locations, a layering or stratification occurs in the ocean as a result of temperature and density gradients. This stratification can be seen in the tropical ocean temperature profile shown in Fig. 6.5. A warmer, less dense upper layer is separated from the colder, denser deep water by an interval known as the thermocline, where the temperature drops rapidly with depth. The thermocline marks the approximate limit of the mixed layer, within which fairly rapid thermodynamic equilibrium is reached between the surface water and the atmosphere.

Deep ocean waters, with temperatures down to 2°C and pressures >10 MPa, are highly CO_2 undersaturated, and it is this capacity for additional dissolution that enables the ocean to provide a sink for increased atmospheric CO_2 over millennial time scales.

The deep water from the Atlantic basin circulates eastward around the Antarctic Ocean basin, branching off and subsequently upwelling in the Indian and Pacific Oceans as well as in the Southern Ocean. The time scale of this circulation system is millennial. Early estimates of the time required for downwelling surface water to fully displace the deep ocean volume based on ^{14}C radioactive tracer measurements put this "ventilation time" of the deep ocean at ~1400 years. Subsequent improved understanding of the processes involved has reduced this time scale to ~250 years for the Atlantic and ~550 years for the Pacific (England, 1995).

When deep water upwells, CO_2 outgases to the atmosphere at the air-sea interface since the partial pressure of CO_2 in the seawater is higher than the partial pressure in the atmosphere. It is thought that in preindustrial times this cycle was at equilibrium, with average surface and deep waters containing ~2000 and 2100 μmolC/kg of DIC, respectively (Goodwin et al., 2007). The anthropogenic increase in CO_2 partial pressure in the atmosphere, from 277 ppm in preindustrial times to the current ~420 ppm, has resulted in an increase in the average DIC of surface water to ~2100 μmolC/kg. As a result, the ocean has already sequestered an estimated 650 Gt of anthropogenic CO_2 emissions during the industrial era. This has been accompanied by a decline in pH of ~0.05 and, as noted above, a consequent reduction in the carbonate ion concentration. Under these conditions, the production of shells, corals, etc., by calcifying organisms becomes progressively more difficult, as observed in many locations where mariculture has been adversely affected by ocean acidification. If further CO_2 addition reduces carbonate ion concentration to the point that surface waters become undersaturated, carbonate minerals such as shells, coral reefs, and limestone sediments will start to dissolve.

6.3 The biological pump

The biological pump is the term used to describe the biological mechanisms that control the export of particulate organic carbon (POC) from the surface waters to the deep ocean. Fig. 6.6 illustrates the biological pump and its close interconnection with the solubility pump.

The starting point for the biological pump is the process of photosynthesis, through which the production of organic carbon takes place in well-lit (euphotic) shallow ocean waters. Photosynthesis starts with the absorption of a photon of light by a chlorophyll molecule, contained in the chloroplast of marine algae or within cyanobacteria. As described in Chapter 5, the energy from the photon is used in a complex sequence of biochemical reactions known as the Calvin cycle (or Calvin-Benson-Bassham or CBB cycle), which eventually results in the production of simple sugars that are used to synthesize

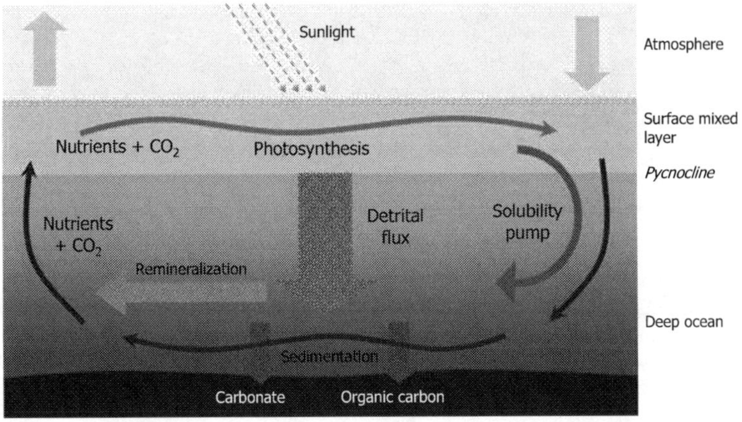

FIGURE 6.6 The ocean carbon cycle: biological pump.

other organic compounds and to support the metabolism of the organism. This set of reactions can be expressed in the highly simplified form:

$$6CO_2 + 6H_2O + \text{photons} \rightarrow C_6H_{12}O_6 + 6O_2 \tag{6.4}$$

which illustrates that CO_2 is consumed and O_2 is produced in the overall chain of reactions. CO_2 may be either absorbed directly from the air, as is the case for terrestrial plants, or obtained from CO_2 dissolved in the surface seawater. This process of generating organic carbon via photosynthesis is termed "primary production" and is the base of the marine food web.

Phytoplankton also requires macronutrients such as nitrate, phosphate, and silicic acid, which are delivered to the euphotic zone by upwelling deep ocean waters or by terrestrial runoff, as well as micronutrients, such as iron, which are typically deposited as windborne (aeolian) dust from arid, upwind land masses. The molar ratio of carbon to the macronutrients nitrogen and phosphorus in phytoplankton is a relatively stable number and has been found to be equal to the ratio of these elements in deep ocean water, known as the Redfield ratio (C:N:P = 106:16:1). This congruence demonstrates the closeness of the interaction between the biochemistry of phytoplankton and the chemistry of the deep ocean water body that regulates their environment, and is an example of ecological stoichiometry—the interconnection between the chemistry of the environment and the biochemistry of the biome, mediated by the supply, consumption, and recycling of nutrients. This ratio, named after American oceanographer Alfred Redfield (1890–1983), is the reason that the marine photosynthetic reaction is sometimes written as:

$$106CO_2 + 16HNO_3 + H_3PO_4 + 122H_2O \rightarrow C_{106}H_{263}O_{110}N_{16}P + 138O_2 \tag{6.5}$$

where $C_{106}H_{263}O_{110}N_{16}P$ represents the average plankton composition.

The biomass of phytoplankton is successively consumed by small, then larger marine organisms, which release organic carbon as fecal pellets. This biological activity results in the production of dissolved organic carbon (DOC), much of which is recycled by these heterotrophic organisms, as well as particulate inorganic carbonate (PIC) from the shells or tests of calcifying marine organisms such as foraminifera and free-swimming mollusks. As noted above, these organisms are able to produce carbonate shells because surface waters are supersaturated with carbonate ions. The PIC element of the biological pump is referred to as the hard tissues pump, with the remainder being the soft tissues pump.

Dead phytoplankton and other biomass are partly remineralized into inorganic carbon, nitrate, and ammonia by bacterial action, contributing to the total DIC. Remaining dead biomass and decay products, comprising 20%–25% of primary production, are exported from the euphotic zone as particulate organic matter (POM), the falling material being known as "marine snow" (Le Moigne, 2019). Microbial processes in shallow waters also transform bioavailable organic carbon into recalcitrant dissolved organic carbon (RDOC) via a number of poorly understood processes collectively termed the microbial carbon pump (Jiao et al., 2010; Robinson et al., 2018). The RDOC component of the oceanic carbon stock is estimated to be ~660 GtC, equivalent to ~75% of the current atmospheric carbon stock, and has a permanence in excess of 1000 years (Zhang et al., 2018).

In contrast to surface waters, deep waters are undersaturated in carbonate, and falling POM can dissolve as it reaches these undersaturated waters, further increasing DOC in deep waters. Material reaching the ocean floor is a nutrient source for abyssal ecosystems, and recent studies have shown that, as well as cycling organic material, bacterial communities on the abyssal seafloor can also convert DIC into biomass, creating another link between the solubility and biological pumps (Herndl and Reinthaler, 2013; Sweetman et al., 2019). Carbon cycling in these ecosystems is the final process that determines the amount of carbon sequestered into ocean floor sediments, currently estimated at 0.2 GtC/yr.

6.3.1 Factors controlling ocean productivity

Phytoplankton are responsible for the annual net primary production (NPP) of biomass containing around 50 GtC in the ocean, despite comprising less than 0.2% of the planet's photosynthetically active biomass (ca. 1 GtC vs. 600 GtC in terrestrial ecosystems). Oceanic NPP is estimated from satellite data, by sensing the absorption of blue and blue-green wavelengths to estimate the concentration of photosynthetic pigments including chlorophyll. The estimated concentration ranges from $\sim 0.2 \text{ mg/m}^3$ in mid-ocean to $>5 \text{ mg/m}^3$ in coastal regions, driven primarily by nutrient supply (Falkowski et al., 1998).

NPP limitations often occur due to nutrient deficiency, for example, productivity in low latitude, (sub)tropical oceans is limited as a result of nitrogen deficiency, while in the eastern Mediterranean and subtropical north Atlantic, both phosphorus and nitrogen are limiting (Moore et al., 2013). Nutrient supply is largely the result of the upwelling of deeper nutrient-rich water into the euphotic mixed layer, and as a result of storms, coastal upwelling and other mixing processes driven by convection and winds. These processes are all highly variable, with the result that oceanic production is also highly variable, and is characterized by intermittent phytoplankton "blooms" initiated by nutrient availability and briefly sustained as a result of the slower reproductive cycle of the zooplankton that graze on them (Falkowski et al., 1998).

Supply of nutrients from subsurface waters is not always sufficient to sustain primary production and three ocean areas, the Southern Ocean, and the equatorial and subarctic regions of the eastern Pacific, are unproductive, despite having high nutrient content. It was first suggested in the 1930s, by English biologist Joseph Hart, that there might be an iron deficiency in these high-nutrient, low-chlorophyll (HNLC) regions (Hart, 1934). Iron is required as a micronutrient by phytoplankton, and this suggestion was confirmed in the 1980s by oceanographer John Martin and colleagues, based on experiments demonstrating increased phytoplankton growth in iron-enriched water samples from an HNLC area in the Southern Ocean (Martin et al., 1990).

In contrast to macronutrients, iron is primarily delivered to the ocean as a component of wind-borne dust, and the low level of primary production in the Southern Ocean and eastern Pacific can then be understood as a result of the lack of upwind land masses. The limited iron supply in these regions is largely derived from sediments on the continental margins, while hydrothermal vents on the oceanic ridges are a further source of dissolved iron, transported to the surface by upwelling deep waters (Tagliabue et al., 2017). Recent analysis of sea-bed cores from the Southern Ocean has confirmed that an increase in dust-borne iron from Patagonia to the sub-antarctic Atlantic resulted in a drop in atmospheric [CO_2] of some 40 ppm during the most recent glacial maximum (Martinez-Garcia et al., 2014), and modeling studies have shown that given sufficient iron availability to deplete other nutrients in the Southern Ocean, this [CO_2] drawdown effect could reach 60−100 ppm (Joos et al., 1991).

6.3.2 Calvin cycle alternatives in the ocean

Alongside the CBB cycle, five other biochemical pathways for fixing carbon have been identified in different families of bacteria and archaea, and three of these play a significant role in primary production in specific oceanic environments (Hügler and Sievert, 2011). The organisms that use these pathways are known as chemoautotrophic or chemolithoautotrophic, as they manufacture their food using chemical energy derived from inorganic chemical and mineral inputs, in contrast to the photoautotrophs that use solar energy to drive the CBB cycle.

At deep sea hydrothermal vents, chemoautotrophic bacteria have been found to use both the CBB cycle and the more energy-efficient reductive tricarboxylic acid (rTCA) cycle, with the rTCA cycle dominating at temperatures above 20°C (Nakagawa and Takai, 2008). In the deep ocean, archaea using the 3-hydroxypropionate/4-hydroxybutyrate (3-HP/4-HB) cycle are estimated to fix around 0.4 GtC per year, and play a role in transforming relatively labile carbon from decomposing organic material into more recalcitrant (sometimes termed refractory) forms, with very long residence times in ocean bottom sediments (as observed in the geological record).

Oxygen-deficient zones in the oceans, which are increasing as a result of climate change (Keeling et al., 2010), are traditionally thought to be a source of CO_2 as a result of heterotrophic respiration. However, a variety of bacteria known as planctomycetes have also been found to fix carbon under these conditions, using the reductive acetyl-CoA, or Wood-Ljungdahl pathway (Nakagawa and Takai, 2008).

The overall contribution of chemoautotrophic organisms to carbon fixation in the ocean remains a subject of ongoing investigation (Hügler and Sievert, 2011).

6.4 Climate change and ocean carbon cycle processes

The impact of anthropogenic CO_2 emissions and climate change on oceanic processes is important for negative emissions technologies on a number of fronts. Firstly, the continued effectiveness of the ocean as a sink for a large fraction of our CO_2 emissions is a major factor in determining the future trajectory of atmospheric [CO_2], and therefore, in part, determining the size of the negative emissions challenge. Secondly, insofar as the oceans have a potentially significant part to play in the implementation of NETs (Chapter 16) the response of oceanic processes to climate change and to the eventual drawdown in [CO_2] will impact on the overall effectiveness of NETs. Thirdly, if implemented on a sufficient scale, some NETs, including enhanced weathering and ocean alkalinity enhancement, could help to address another oceanic consequence of rising [CO_2], namely ocean acidification.

These physicochemical changes, and the biogeochemical consequences, are briefly described in this section, and a number of general references are included in the Resources section as a gateway to the enormous body of research that is building on the impact of a high [CO_2] on the oceans.

6.4.1 Climate change impact on marine physicochemical processes

Increasing sea surface temperature resulting from climate change will have a number of physicochemical effects on the ocean, including reduced CO_2 solubility, reduced surface water density, and greater stratification. As a result, the uptake of CO_2 into surface waters will decrease, the export of surface waters to the deep ocean will decrease, and the outgassing of CO_2 from upwelling deep water will increase, all of which point to a future decline in the net uptake of anthropogenic CO_2 by the ocean.

Thermohaline circulation

Changes in the global hydrological cycle (Oki and Kanae, 2006), including increased precipitation over the north Atlantic, and increased meltwater runoff from the Greenland ice sheet, reinforce the reduction in surface water density, weakening the AMOC and the formation of North Atlantic Deep Water (Joos et al., 1999; IPCC, 2014). A permanent AMOC monitoring network became operational in 2004, and results for the first decade confirm a weakening of the circulation by about 3% per year (Srokosz and Bryden, 2015). Modeling studies supported by these observations suggest a further weakening of about 30% by the end of the century, significantly reducing heat transport to NW Europe.

Upwelling and carbon uptake in the Southern Ocean have also been shown to be affected by decadal-scale climate-driven variability in the configuration of the southern polar jet stream (Gruber et al., 2019), although it is still unclear whether this is linked to anthropogenic climate change.

Ocean acidification

The uptake of anthropogenic CO_2 emissions by the ocean is causing its gradual acidification and has resulted in a decline in the average pH of ocean surface water from ~8.2 to ~8.1 since preindustrial times, with the potential for this to drop further to 7.7–7.9 depending on future emissions scenarios. As noted above, pH decline shifts the equilibrium of reactions 6.2 toward the bicarbonate ion, at the expense of a decreasing carbonate ion concentration, buffering the pH decline (Doney et al., 2009). However, the decline in carbonate ion concentration has important consequences for calcifying organisms, as discussed in the next section.

Although ocean acidification to date has been driven by the increase in atmospheric [CO_2], aided and abetted by increased freshwater input due to ice melt (Woosley and Millero, 2020; Terhaar et al., 2021), climate-driven feedbacks that reduce the terrestrial carbon sink could become additional drivers of future acidification (Matear and Lenton, 2018).

Ocean deoxygenation

The dissolved oxygen content in the ocean, currently averaging ~178 μmol/kg, is essential for aerobic respiration and the biogeochemical cycling of carbon, nitrogen, iron, and phosphorus that sustain marine life. Similar to CO_2, the solubility of oxygen declines at increasing temperature, so that overall warming of the oceans results in reduced uptake. Most oceanic deep water is formed at high latitudes, and oxygen uptake during the high latitude winter—where solubility increases by 6% per °C of cooling—is therefore important in determining the initial oxygen content of subsurface waters globally.

Dissolved oxygen is then utilized primarily in the bacterial decomposition of organic matter—part of the biological pump discussed above. As a parcel of water is transported by the thermohaline circulation (Fig. 6.4), oxygen depletion will continue when such biological activity is present, and reoxygenation will only occur when the parcel once again enters the surface mixed layer. As a result, waters with long ventilation time that have been exposed to high biological activity can become severely oxygen depleted, resulting in oxygen minimum zones (OMZ).

An oxygen level of 60 μmol/L is commonly taken as the threshold for hypoxia—the point at which organisms will begin to feel stress as a result of oxygen deficiency. As well as being the largest oxygen-depleted (suboxic) region, the North Pacific is also the most extensively studied, and 50 years of data confirm a declining [O_2] trend of about 7 μmol/L per decade for waters at mid-depth (Keeling et al., 2010). This trend is due in part to reduced upwelling and ventilation in the northeast Pacific.

Due to the long timescale of oceanic circulation, the response of the ocean to climate change lags behind atmospheric and terrestrial impacts. This means that the oceans will continue to warm for centuries after CO_2 emissions start declining, with attendant declines in the oceanic O_2 inventory, growth in OMZs, and as yet unknown consequences for marine productivity and food webs.

6.4.2 Climate change impact on marine biological processes

As described in the last section, marine ecosystems are experiencing the impact of three related stressors as a consequence of anthropogenic CO_2 emissions: ocean warming, acidification, and deoxygenation. Other aspects of climate change, such as changes in the hydrological cycle, are also having an effect on nutrient availability, through reduced overturning and changes in micronutrient delivery by aeolian dust.

These physicochemical changes will have a range of impacts on marine biological processes and ecosystems, as briefly discussed in this section.

Net production and the food web

How climate change and ocean acidification will influence primary production is one of the most important current questions to improve our understanding of marine ecosystems (Sutherland et al., 2013). The mixing of surface waters and the upwelling of deep ocean waters as part of the thermohaline circulation provide the main sources of nutrients for autotrophic organisms, and both of these mechanisms are being impacted by climate change.

Modeled reduction in the AMOC, for example, could lead to a 10%–20% global decline in carbon export via the biological pump as a result of nutrient limitations. Within the North Atlantic, this could lead to a 50% decline in phytoplankton populations, with implications for the food web all the way up to fisheries (Brierley and Kingsford, 2009). More widely, this would also be expected to result in the growth of the areas in the Atlantic, Pacific, and Indian Ocean subtropical gyres where productivity is limited by nitrogen availability (Moore et al., 2013).

In both the Atlantic and the Pacific, the eastern boundary currents of the subtropical gyres cause coastal upwelling of nutrient-rich deep waters that support the majority of the world's fisheries. Studies of the impact of interdecadal climate variability on the California Current ecosystems have demonstrated that increasing sea surface temperature (SST) leads to increased stratification, reducing upwelling, but this effect can be countered by increased wind stress. Upwelling waters here are also affected by the suboxic conditions in the northeast Pacific OMZ, as discussed earlier.

In ocean regions that are strongly stratified and do not benefit from nutrient supply from upwelling deep waters, a decline in NPP has also been shown to be linked to strengthened stratification caused by increasing SST (Behrenfeld et al., 2006). Most of the ocean surface between 40°N and 40°S exhibits such permanently stratified conditions, and continued SST rise is therefore expected to have a significant impact on productivity in these regions.

Changes have also been observed in the supply of important micronutrients such as iron, with evidence of an increase in dust supply from desert regions over the last century (Moore et al., 2013). The supply of iron from fossil fuel combustion has also been recently identified as an important source for oceans adjacent to highly populated areas, notably the North Atlantic (Sholkovitz et al., 2009), but also to the Southern Ocean as a result of coal combustion in South Africa, Australia, and Indonesia (Matsui et al., 2018).

As well as nutrient availability, many other environmental factors including increasing SST and changes in inorganic carbon availability and speciation due to declining ocean pH will have an impact on carbon export via the

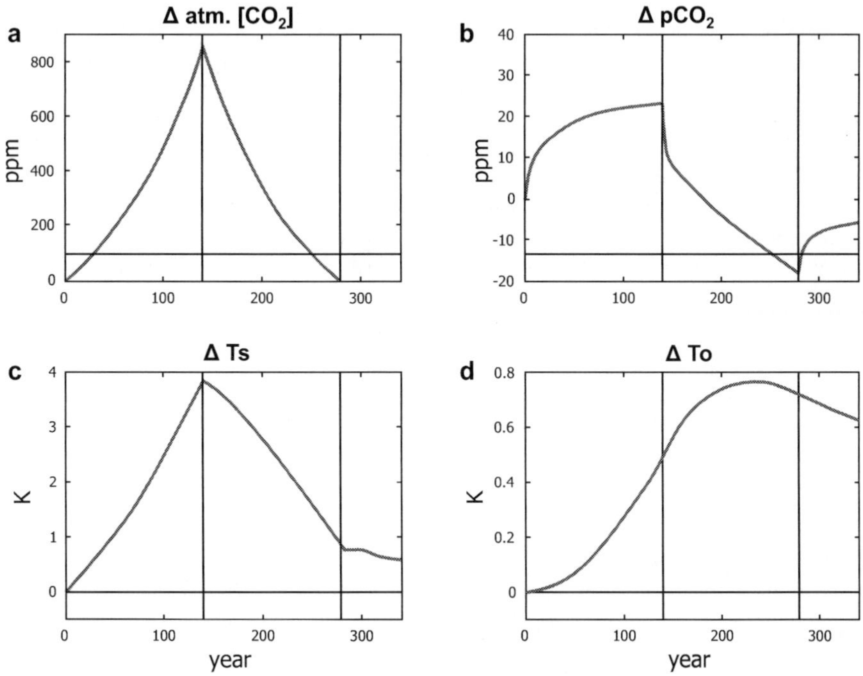

FIGURE 6.7 Changes in (A) prescribed atmospheric $[CO_2]$, (B) air-sea pCO_2 disequilibrium, (C) sea surface temperature, and (D) potential ocean temperature, relative to a preindustrial control simulation. (*after* Schwinger and Tjiputra, 2018).

biological pump, and the overall regional and integrated global impact remains uncertain (Passow and Carlson, 2012). For example, most models identify the Southern Ocean as a key place for oceanic carbon uptake but the current variability of that sink and its response to future changes remain very poorly assessed and represented in models (Sutherland et al., 2013).

Biomineralization

Calcium carbonate ($CaCO_3$) exists in three crystalline forms, calcite, aragonite, and vaterite, in order of increasing solubility, with calcite and aragonite being the forms most commonly produced in biological processes (Kamennaya et al., 2012). The solubility product (K_{sp}) for a mineral at a given temperature, salinity, and pressure is defined as the product of calcium and carbonate ion concentrations in a solution saturated with respect to the mineral;

$$K_{sp} = \left[Ca^{2+}\right] \times \left[CO_3^{2-}\right] \tag{6.6}$$

and the saturation state (Ω) of a solution is defined as follows:

$$\Omega = \left[Ca^{2+}\right] \times \left[CO_3^{2-}\right]/K_{sp} \tag{6.7}$$

An environment with a high calcite or aragonite Ω is conducive to biocalcification, and any reduction in the saturation state as a result of pH decline and decreasing carbonate ion concentration will impede and ultimately prevent the production of shells and skeletons in marine organisms, including plankton, corals and coralline algae, and molluscs (see also Chapter 5 and Doney et al., 2009). Some organisms may be able to compensate by diverting more of their metabolic energy to the calcification process, but this will put them at a competitive disadvantage compared with noncalcifying organisms (Turley and Gattuso, 2012).

There are also cases where elevated CO_2 (eCO_2) has been observed to have a positive impact on biomineralization. For example, the internal biomineralization process in some strains of coccolithophores has been observed to increase under eCO_2—suggesting that the process is limited by CO_2 availability (Doney et al., 2009). However, the overall growth of the algae was slower under eCO_2, indicating a diversion of metabolic resources that could have negative consequences for the organism within its ecosystem.

6.4.3 Ocean response to negative emissions

Various studies have been undertaken using Earth system models to look at the response of the ocean to the eventual drawdown of atmospheric [CO_2] as a result of emissions reductions and, ultimately, the large-scale deployment of NETs. In general, these investigations have concluded that the major climate-induced changes in the ocean, such as increased surface temperature, reduced primary production and meridional overturning are largely reversible, reverting more or less in line with declining [CO_2].

The overall average temperature of the ocean (expressed as the average potential temperature, T_o—see Glossary) behaves differently as a result of the long ventilation time scale. This continues to rise for several decades after the peak of emissions and declines toward its preindustrial level over a millennial timescale (Fig. 6.7; Schwinger and Tjiputra, 2018). CO_2 uptake by the ocean also continues for several decades after the start of CO_2 drawdown, until the partial pressure of CO_2 (P_{CO_2}) drops below the product [C_{CO_2}] × K_{HCO_2} (Eq. 6.1), and even after [CO_2] returns to preindustrial levels, the ocean would still hold some 400 GtC that would be returned to the atmosphere on a millennial timescale.

Sea level rise due to thermal expansion is expected to show a similar trend to average ocean temperature, continuing to rise for several decades after peak emissions and declining on a millennial timescale (Ehlert and Zickfeld, 2018).

Other model studies indicate that during [CO_2] drawdown, the imbalance between declining SST compared to sustained average T_o reduces the upper ocean density gradient, leading to deeper mixing and increased nutrient supply to surface waters. Coupled with an SST that remains somewhat elevated compared with the preindustrial baseline, this leads to an increase in NPP in some ocean regions (John et al., 2015).

As discussed in Chapter 1, modeling studies of the impact and efficacy of NETs, including the reversibility of climate-induced changes in the oceans, are still in their infancy, and the protocols for comparison of model results have only recently been established by the carbon dioxide removal model intercomparison project (CDRMIP; Keller et al., 2018). Further work in this area will allow a more robust assessment of the ocean response to negative emissions, including the wide range of ocean-based NETs discussed in Chapter 16.

References and resources

References

Behrenfeld, M.J., O'Malley, R.T., Siegel, D.A., et al., 2006. Climate-driven trends in contemporary ocean productivity. Nature 444 (7120), 752–755.

Brierley, A.S., Kingsford, M.J., 2009. Impacts of climate change on marine organisms and ecosystems. Curr. Biol. 19 (14), R602–R614.

Doney, S.C., Fabry, V.J., Feely, R.A., Kleypas, J.A., 2009. Ocean acidification: the other CO_2 problem. Ann. Rev. Mar. Sci 1, 169–192.

Ehlert, D., Zickfeld, K., 2018. Irreversible ocean thermal expansion under carbon dioxide removal. Earth System Dynamics 9 (1), 197–210.

England, M.H., 1995. The age of water and ventilation timescales in a global ocean model. J. Phys. Oceanogr. 25 (11), 2756–2777.

Falkowski, P.G., Barber, R.T., Smetacek, V., 1998. Biogeochemical controls and feedbacks on ocean primary production. Science 281 (5374), 200–206.

Friedlingstein, P., Jones, M.W., O'Sullivan, M., et al., 2021. Global carbon budget 2021. Earth Syst. Sci. Data Discuss. 1–191.

Goodwin, P., Williams, R.G., Follows, M.J., Dutkiewicz, S., 2007. Ocean-atmosphere partitioning of anthropogenic carbon dioxide on centennial timescales. Global Biogeochem. Cycles 21 (1), GB1014.

Gruber, N., Landschützer, P., Lovenduski, N.S., 2019. The variable Southern Ocean carbon sink. Ann. Rev. Mar. Sci 11, 16.1–16.28.

Hart, T.J., 1934. On the phytoplankton of the southeast Atlantic and the Bellingshausen sea, 1929–1931. Discov. Rep. 8, 1–268.

Herndl, G.J., Reinthaler, T., 2013. Microbial control of the dark end of the biological pump. Nat. Geosci. 6 (9), 718–724.

Hügler, M., Sievert, S.M., 2011. Beyond the Calvin cycle: autotrophic carbon fixation in the ocean. Ann. Rev. Mar. Sci 3, 261–289.

IPCC, 2014. Long-term climate change: projections, commitments and irreversibility. In: Climate Change 2013 – the Physical Science Basis: Working Group I Contribution to the Fifth Assessment Report of the Intergovernmental Panel on Climate Change (1029–1076). Cambridge University Press, UK.

Jiao, N., Herndl, G.J., Hansell, D.A., et al., 2010. Microbial production of recalcitrant dissolved organic matter: long-term carbon storage in the global ocean. Nat. Rev. Microbiol. 8 (8), 593–599.

John, J.G., Stock, C.A., Dunne, J.P., 2015. A more productive, but different, ocean after mitigation. Geophys. Res. Lett. 42 (22), 9836–9845.

Joos, F., Siegenthaler, U., Sarmiento, J.L., 1991. Possible effects of iron fertilization in the Southern Ocean on atmospheric CO_2 concentration. Global Biogeochem. Cycles 5 (2), 135–150.

Joos, F., Plattner, G.K., Stocker, T.F., Marchal, O., Schmittner, A., 1999. Global warming and marine carbon cycle feedbacks on future atmospheric CO_2. Science 284, 464–467.

Kamennaya, N.A., Ajo-Franklin, C.M., Northen, T., Jansson, C., 2012. Cyanobacteria as biocatalysts for carbonate mineralization. Minerals 2 (4), 338–364.

Keeling, R.F., Körtzinger, A., Gruber, N., 2010. Ocean deoxygenation in a warming world. Ann. Rev. Mar. Sci 2, 199–229.

Keller, D.P., Lenton, A., Scott, V., et al., 2018. The carbon dioxide removal model intercomparison project (CDRMIP): rationale and experimental protocol for CMIP6. Geosci. Model Dev. (GMD) 11 (3), 1133–1160.

Khatiwala, S., Tanhua, T., Mikaloff Fletcher, S., et al., 2013. Global ocean storage of anthropogenic carbon. Biogeosciences 10 (4), 2169–2191.

Kuhlbrodt, T., Griesel, A., Montoya, M., Levermann, A., Hofmann, M., Rahmstorf, S., 2007. On the driving processes of the Atlantic meridional overturning circulation. Rev. Geophys. 45 (2), RG2001.

Le Moigne, F.A., 2019. Pathways of organic carbon downward transport by the oceanic biological carbon pump. Front. Mar. Sci. 6, 634.

Martin, J.H., Fitzwater, S.E., Gordon, R.M., 1990. Iron deficiency limits phytoplankton growth in Antarctic waters. Global Biogeochem. Cycles 4 (1), 5–12.

Martínez-García, A., Sigman, D.M., Ren, H., et al., 2014. Iron fertilization of the Subantarctic Ocean during the last ice age. Science 343 (6177), 1347–1350.

Matear, R.J., Lenton, A., 2018. Carbon–climate feedbacks accelerate ocean acidification. Biogeosciences 15 (6), 1721–1732.

Matsui, H., Mahowald, N.M., Moteki, N., et al., 2018. Anthropogenic combustion iron as a complex climate forcer. Nat. Commun. 9 (1), 1–10.

Moore, C.M., Mills, M.M., Arrigo, K.R., et al., 2013. Processes and patterns of oceanic nutrient limitation. Nat. Geosci. 6 (9), 701–710.

Nakagawa, S., Takai, K., 2008. Deep-sea vent chemoautotrophs: diversity, biochemistry and ecological significance. FEMS Microbiol. Ecol. 65 (1), 1–14.

Oki, T., Kanae, S., 2006. Global hydrological cycles and world water resources. Science 313 (5790), 1068–1072.

Passow, U., Carlson, C.A., 2012. The biological pump in a high CO_2 world. Mar. Ecol. Prog. Ser. 470, 249–271.

Revelle, R., Suess, H.E., 1957. Carbon dioxide exchange between atmosphere and ocean and the question of an increase of atmospheric CO_2 during the past decades. Tellus 9 (1), 18–27.

Robinson, C., Wallace, D., Hyun, J.H., et al., 2018. An implementation strategy to quantify the marine microbial carbon pump and its sensitivity to global change. Natl. Sci. Rev. 5 (4), 474–480.

Sabine, C.L., Feely, R.A., Gruber, N., et al., 2004. The oceanic sink for anthropogenic CO_2. Science 305 (5682), 367–371.

Schuiling, R.D., Krijgsman, P., 2006. Enhanced weathering: an effective and cheap tool to sequester CO_2. Climatic Change 74 (1–3), 349–354.

Schwinger, J., Tjiputra, J., 2018. Ocean carbon cycle feedbacks under negative emissions. Geophys. Res. Lett. 45, 5062–5070.

Sholkovitz, E.R., Sedwick, P.N., Church, T.M., 2009. Influence of anthropogenic combustion emissions on the deposition of soluble aerosol iron to the ocean: empirical estimates for island sites in the North Atlantic. Geochem. Cosmochim. Acta 73 (14), 3981–4003.

Srokosz, M.A., Bryden, H.L., 2015. Observing the Atlantic Meridional Overturning Circulation yields a decade of inevitable surprises. Science 348 (6241), 1255575.

Sutherland, W.J., Freckleton, R.P., Godfray, H.C.J., et al., 2013. Identification of 100 fundamental ecological questions. J. Ecol. 101 (1), 58–67.

Sweetman, A.K., Smith, C.R., Shulse, C.N., et al., 2019. Key role of bacteria in the short-term cycling of carbon at the abyssal seafloor in a low particulate organic carbon flux region of the eastern Pacific Ocean. Limnol. Oceanogr. 64 (2), 694–713.

Tagliabue, A., Bowie, A.R., Boyd, P.W., Buck, K.N., Johnson, K.S., Saito, M.A., 2017. The integral role of iron in ocean biogeochemistry. Nature 543, 51–59.

Terhaar, J., Torres, O., Bourgeois, T., Kwiatkowski, L., 2021. Arctic Ocean acidification over the 21st century co-driven by anthropogenic carbon increases and freshening in the CMIP6 model ensemble. Biogeosciences 18 (6), 2221–2240.

Turley, C., Gattuso, J.P., 2012. Future biological and ecosystem impacts of ocean acidification and their socioeconomic-policy implications. Curr. Opin. Environ. Sustain. 4 (3), 278–286.

Weijer, W., Cheng, W., Drijfhout, S.S., et al., 2019. Stability of the Atlantic meridional overturning circulation: a review and synthesis. J. Geophys. Res.: Oceans 124 (8), 5336–5375.

Woosley, R.J., Millero, F.J., 2020. Freshening of the western Arctic negates anthropogenic carbon uptake potential. Limnol. Oceanogr. 65 (8), 1834–1846.

Zhang, C., Dang, H., Azam, F., et al., 2018. Evolving paradigms in biological carbon cycling in the ocean. Natl. Sci. Rev. 5 (4), 481–499.

Further reading

Cicerone, R., Orr, J., Brewer, P., et al., 2004. The ocean in a high CO_2 World. EOS Transact. Am. Geophysical Union 85, 351.

Heinze, C., Meyer, S., Goris, N., et al., 2015. The ocean carbon sink–impacts, vulnerabilities and challenges. Earth System Dynamics 6 (1), 327–358.

Sarmiento, J.L., Slater, R., Barber, R., et al., 2004. Response of ocean ecosystems to climate warming. Global Biogeochem. Cycles 18 (3), GB3003.

Stanford University Ocean Biogeochemistry Lab: Lecture notes for the Marine Chemistry course, http://ocean.stanford.edu/courses/bomc/lnotes.html.

Tanhua, T., Bates, N.R., Körtzinger, A., 2013. The marine carbon cycle and ocean carbon inventories. Intern. Geophys. 103, 787–815.

Williamson, P., Turley, C.M., Ostle, C., 2017. Ocean acidification. Marine Climate Change Impacts Partnership MCCIP Science Review.

Zhang, C., Dang, H., Azam, F., et al., 2018. Evolving paradigms in biological carbon cycling in the ocean. Natl. Sci. Rev. 5 (4), 481–499.

Projects and organizations

Carboocean IP (EU-funded assessment of marine carbon sources and sinks, focusing on the Atlantic and Southern Oceans over the past 200 and next 200 years), www.carboocean.org.

GEOTRACES International study of marine biogeochemical cycles of trace elements, www.geotraces.org.

GLODAP (Global Ocean Data Analysis Project), http://cdiac.esd.ornl.gov/oceans/glodap/.

Monterey Bay Aquarium Research Institute (MBARI), www.mbari.org.

National Maritime Research Institute (NMRI), www.nmri.go.jp/index_e.html North American Carbon Program (assessing the sources and sinks of CO_2, CH_4, and CO in North America and adjacent oceans): www.nacarbon.org.

Ocean Acidification Network, http://ocean-acidification.net/about/.

Ocean Carbon & Biogeochemistry, www.us-ocb.org.

Ocean Chemistry of the Greenhouse Gases, www.mbari.org/ghgases.

The Ocean in a High CO_2 world, UNESCO Intergovernmental Oceanographic Commission International Symposium Series, http://ocean-acidification.net/international-symposia/second-meeting-on-oa-in-a-high-co2-world/.

UNESCO Intergovernmental Oceanographic Commission; International Ocean Carbon Coordination Project, http://ioc-unesco.org/, www.ioccp.org.

Woods Hole Oceanographic Institution, www.whoi.edu.

Café Thorium; WHOI marine chemistry and geochemistry group, http://cafethorium.whoi.edu.

Chapter 7

CO_2 absorption

Steve Rackley

Separation of CO_2 from gas streams based on chemical or physical absorption has been applied on an industrial scale for more than 70 years, driven by the requirement to treat natural gas to a sales gas specification with low content of acid gases, such as CO_2 and H_2S, or by the need to condition syngas from coal gasification as a feed for other chemical processes such as Fischer–Tropsch liquids production.

These processes have generated an extensive body of experience that can be applied to CO_2 recovery at the lower partial pressures that will occur in negative emissions applications such as biomass conversion (Chapter 12) or direct air capture (Chapter 13).

7.1 Chemical and physical fundamentals

Processes to capture CO_2 based on absorption are distinguished depending on whether the solvent reacts chemically with the sorbate (CO_2) to form chemical compounds from which the CO_2 is subsequently recovered, or is chemically inert and absorbs the sorbate without a chemical reaction. These distinct processes are termed chemical and physical absorption respectively. The most important requirements for absorption processes to achieve cost-efficient CO_2 separation include the following:

- Fast absorption and desorption rates to minimize equipment size
- High loading of CO_2 per unit solvent volume, coupled with low solvent cost
- Low heat of desorption to reduce the energy penalty for solvent regeneration,
- Low solvent loss through thermal and oxidative degradation, and
- Solvent tolerance to contaminants, to reduce degradation and unwanted by-product formation.

7.1.1 Chemical CO_2 absorption

Chemical absorption for CO_2 capture is based on the exothermic reaction of a sorbent with the CO_2 present in a gas stream, preferably at low temperature. The reaction is then reversed in a so-called stripping, or regeneration, process at a higher temperature. Chemical absorption is particularly suitable for CO_2 capture at low partial pressure, with amine or carbonate solutions being the predominant solvents.

Amine-based CO_2 absorption

Amines are organic compounds derived from ammonia (NH_3) in which one or more of the hydrogen atoms are replaced by organic components or substituents. Depending on the number of substituents, these compounds are termed primary, secondary, or tertiary amines. The simplest primary amine is methylamine (CH_3NH_2 or CH_5N), with one hydrogen replaced by a methane (CH_3) group. To simplify the nomenclature, the organic group is commonly replaced by R, giving R^1-NH_2, R^1R^2-NH, and $R^1R^2R^3$-N for primary, secondary, and tertiary amines, respectively.

The most commonly used amine for CO_2 capture is ethanolamine (or monoethanolamine—MEA), a primary amine with R = CH_2CH_2OH. In an aqueous solution, MEA acts as a weak base, which can neutralize an acidic molecule such as CO_2. In this reaction, a weakly bonded compound called a carbamate is formed (Fig. 7.1):

$$2\text{R-NH}_2 + CO_2 \rightarrow \text{R-NH}_3^+ + \text{R-NHCOO}^- + \text{heat} \tag{7.1}$$

Negative Emissions Technologies for Climate Change Mitigation. https://doi.org/10.1016/B978-0-12-819663-2.00009-5

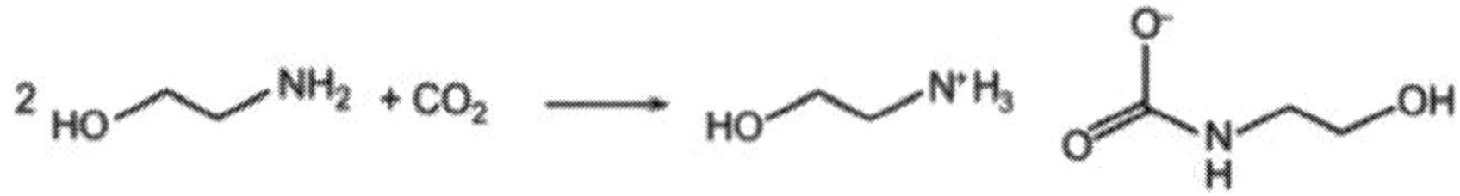

FIGURE 7.1 Carbamate ion structure.

the heat of absorption for CO_2 in MEA being 2.0 GJ/tCO_2.

Two other reactions also occur that result in CO_2 dissolution, base-catalyzed hydration of CO_2:

$$\text{R-NH}_2 + CO_2 + H_2O \rightarrow \text{R-NH}_3^+ + HCO_3^- + \text{heat} \tag{7.2}$$

and the formation of carbonic acid:

$$CO_2 + H_2O \rightarrow H_2CO_3 \tag{7.3}$$

However, compared to the rate of reaction 7.1 for MEA, the other two reactions make a minimal contribution to the overall CO_2 absorption rate.

For a secondary amine, the equivalent carbamation reaction would be as follows:

$$2R^1R^2\text{-NH} + CO_2 \rightarrow R^1R^2\text{-NH}_2^+ + R^1R^2\text{-NCOO}^- + \text{heat} \tag{7.4}$$

While for tertiary amines, the reaction is the base-catalyzed hydration of CO_2:

$$R^1R^2R^3\text{-N} + CO_2 + H_2O \rightarrow R^1R^2R^3\text{-NH}^+ + HCO_3^- \tag{7.5}$$

The carbamate ion nitrogen—carbon bond is easily broken down by the application of heat, leading to the reverse reaction in which the original solvent is regenerated. For example, for MEA regeneration, typically at ~100–120°C:

$$\text{R-NH}_3^+ + \text{R-NHCOO}^- + \text{heat} \rightarrow 2\text{R-NH}_2 + CO_2 \tag{7.6}$$

Regeneration of the solvent in reaction 7.2 similarly proceeds by the reverse reaction:

$$HCO_3^- + \text{R-NH}_3^+ + \text{less heat} \rightarrow \text{R-NH}_2 + CO_2 + H_2O \tag{7.7}$$

The maximum CO_2 loading of the MEA solvent that can be theoretically achieved in reaction 7.1 is 0.5 (i.e., 1 mole CO_2 in 2 moles of solvent), while the bicarbonate forming reactions reactions 7.2 and 7.5 achieve twice this loading (1 mole of CO_2 per mole of solvent). This higher loading can be achieved in practice by steric hindrance of the carbamation reaction—the use of a larger amine molecule in which the carbamation reaction is hindered because of the physical structure of the amine molecule. Although the energy required for regeneration is lower for these hindered amines, this is compensated by the slower overall reaction rates, and a rate promoter such as piperazine (see later) is commonly added to the solvent. Proprietary formulations of sterically hindered amine solvents with rate-promoting additives are used in the KEPCO/MHI's Advanced KM-CDR Process, and strong bicarbonate forming systems were developed as part of the (2014–2017) EU HiPerCap research project (Kim et al., 2018).

When considering the three key requirements for effective CO_2 solvents—fast reaction rate, low regeneration energy, and high loading capacity—the choice between primary, secondary, and tertiary amines is not straightforward. In general, the relative merits are as follows: for reaction rate, primary, followed by secondary, then tertiary, while for regeneration energy and loading capacity, the order is reversed. Mixing primary, secondary, and tertiary amines in blended amine solvents is one approach to combine the advantages of the different types—for example, increasing the reaction rate of a tertiary amine such as methyldiethanolamine (MDEA) by adding a secondary amine activator to produce activated MDEA (aMDEA).

The absorption reaction rate is important because this affects the size of the physical process equipment required to treat a given gas flow rate; a faster reaction allows equipment to be smaller, thus reducing the capital cost of the capture plant.

Oxidative and thermal degradation of amines results in the formation of a range of organic reaction products. SO_2 and NO_2—common flue gas contaminants—also react with amines to form heat-stable salts (sulfates and nitrates), which lead to a loss of solvent absorption capacity and additional requirements for solvent make-up and waste stream disposal. Amine degradation products are a hazardous waste that requires treatment before disposal. Conventional aerobic or anaerobic wastewater treatment can be used for diluted streams, while concentrated waste (such as amine reclaimer waste; ARW—see Section 7.2.1) can be combusted, for example in cement kilns, on account of its high calorific value (LHV ~16 GJ/t). A further operational disadvantage of amines is their corrosivity, necessitating the use of corrosion-resistant alloys or corrosion inhibitors in equipment experiencing high temperatures or high CO_2 loading.

Amine-based solvents have been used in two CO_2 capture pilots at the UK Drax Group's BECCS project in North Yorkshire. Blended amine solvents developed by C-Capture were used in the first 1 tCO_2/d capture pilot undertaken in 2019, while the KEPCO/MHI's Advanced KM-CDR Process, using sterically hindered amine solvents with rate-promoting additives, was used in a second 300 $kgCO_2/d$ pilot starting in mid-2020.

Aqueous carbonate-based CO_2 absorption

The dissolution of carbonate rocks by carbonic acid—formed from CO_2 and rainwater—is a process that removes CO_2 from the atmosphere as part of the global carbon cycle (Chapter 4):

$$H_2O + CO_2 \rightarrow H_2CO_3 \tag{7.8}$$

$$CaCO_3 + H_2CO_3 \rightarrow Ca^{2+} + 2HCO_3^- \tag{7.9}$$

Acceleration of this natural process has been proposed as a method of removing CO_2 from power plant or cement plant flue gases, using a reactor in which the gas is flowed through a bed of crushed limestone, wetted by a continuous flow of water (Rau et al., 2007). Rather than regenerating the limestone sorbent, the concept is that the bicarbonate-rich effluent stream would be disposed of into the ocean. The accelerated weathering of limestone process is described further in Chapter 16.

Regenerable aqueous carbonate solutions have also been applied for CO_2 capture by chemical absorption in a range of industrial applications. A potassium carbonate (K_2CO_3) absorption system, known as the Benfield process, is based on the following reaction:

$$K_2CO_3 + CO_2 + H_2O \leftrightarrow 2KHCO_3 \tag{7.10}$$

the rate-determining step being the formation of the bicarbonate ion:

$$CO_2 + OH^- \leftrightarrow HCO_3^- \tag{7.11}$$

This process has the advantage of lower desorption energy requirements compared to amine-based systems (0.9−1.6 GJ/tCO_2 vs. 2.0 GJ/tCO_2 for MEA) balanced by the disadvantage of lower rates of reaction at the low CO_2 pressures typical for flue gases, controlled by reaction 7.11. However, the rate of absorption and solvent absorption capacity can be increased by adding reaction rate promoters such as amines or amino acids, boric acid, carbonic anhydrase (see later), or piperazine to the reaction.

Piperazine (PZ) is a cyclohexane ring with two opposing carbon atoms replaced by amine functional groups (Fig. 7.2). The piperazine enhanced reaction scheme can be simplified as follows, following reactions 7.10 and 7.11:

$$KHCO_3 \rightarrow K^+ + HCO_3^- \tag{7.12}$$

$$PZ + HCO_3^- + H^+ \leftrightarrow PZCOO^- + H_3O^+ \tag{7.13}$$

$$PZCOO^- + HCO_3^- + H^+ \leftrightarrow PZ(COO^-)_2 + H_3O^+ \tag{7.14}$$

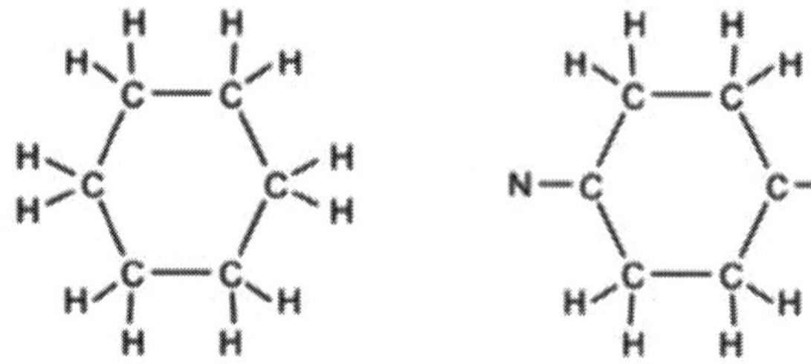

Cyclohexane Piperazine

FIGURE 7.2 Cyclohexane (C_6H_{12}) and piperazine ($C_4H_{10}N_2$).

The carbamate reactions 7.13 and 7.14 dominate the CO_2 absorption process and, as for the straight K_2CO_3 system, require significantly lower heat input than MEA for regeneration.

Enzyme-catalyzed chemical CO_2 absorption

In aqueous absorption systems, the hydration of dissolved CO_2 to form bicarbonate ions (HCO_3^-) according to;

$$CO_2 + H_2O \leftrightarrow HCO_3^- + H^+ \tag{7.15}$$

is commonly either a rate-limiting step (e.g., in the case of K_2CO_3) or a secondary absorption mechanism that makes minimal contribution to the overall CO_2 loading of the solvent solution (e.g., for amine systems). The use of the enzyme carbonic anhydrase (CA) as a catalyst to address these limitations has been widely investigated since it was first proposed in 1961.

Carbonic anhydrases are a broad class of enzymes found in all living organisms—from microbes to man—that catalyze the hydration of CO_2. The catalyst has the largest effect if it is dissolved in the solvent solution, typically at <1 g/L, with a 12-fold rate improvement observed in lab tests with K_2CO_3 solvent. As the enzyme is susceptible to process conditions such as high temperatures and pH, it can be protected from these extremes by a range of immobilization methods, including microparticles (see Section 7.3.5), as a coating on porous materials, and in composite membranes (Molina-Fernández and Luis, 2021).

Aqueous ammonia-based CO_2 absorption

The reaction of ammonia and its derivatives with CO_2 also has the advantage of lower heat of reaction than the equivalent amine-based reactions, opening up the possibility of significant energy-efficiency improvement and cost reduction compared to an amine-based system.

Ammonia may be used directly, with a lean solvent containing 10%–15% NH_3 by weight in water giving the lowest overall energy requirements. Aqueous reactions with CO_2 and water produce a range of reaction products—ammonium carbonate ($(NH_4)_2CO_3 \cdot H_2O$), ammonium bicarbonate (NH_4HCO_3), ammonium sesqui-carbonate ($(NH_4)_2CO_3 \cdot 2NH_4HCO_3$), and ammonium carbamate (NH_2COONH_4)—depending on the operating temperature and CO_2 loading. The various reaction products may reach their respective solubility limits, in which case the rich solvent will form a slurry.

An aqueous solution of ammonium carbonate ($(NH_4)_2CO_3$) may also be used as a solvent, reacting with CO_2 and water to form ammonium bicarbonate (called the AC/ABC reaction):

$$(NH_4)_2CO_3 + CO_2 + H_2O \leftrightarrow 2NH_4HCO_3 \tag{7.16}$$

As ammonia-based scrubbing is also applied to flue gas SO_2 and NO_2 removal, the use of ammonia for CO_2 capture potentially allows a single solvent system to be used for multicomponent flue gas emissions control.

Sodium hydroxide-based CO_2 absorption

The principal reactions taking place in the absorption of CO_2 by sodium hydroxide are firstly carbonic acid formation from CO_2 and water, and secondly sodium bicarbonate or carbonate formation according to:

$$H_2CO_3 + NaOH \rightarrow NaHCO_3 + H_2O \tag{7.17}$$

$$NaHCO_3 + NaOH \rightarrow Na_2CO_3 + H_2O \tag{7.18}$$

The relative proportions of the carbonate and bicarbonate end products can be controlled by adjusting the pH in the absorption stage, for example, by adding hydrochloric acid (HCl).

Unlike other solvents, regeneration of sodium hydroxide cannot be achieved by a mild temperature swing and requires a further chemical step. For example, the addition of lime (CaO):

$$CaO + H_2O \rightarrow Ca(OH)_2 \tag{7.19}$$

$$Ca(OH)_2 + Na_2CO_3 \rightarrow 2NaOH + CaCO_3 \tag{7.20}$$

Calcium carbonate is separated from the resulting slurry and is in turn regenerated by heating to drive off water, followed by calcination at ~900°C:

$$CaCO_3 + \text{heat} \rightarrow CaO + CO_2 \tag{7.21}$$

This process has the advantage of using chemicals that are inexpensive and abundantly available, but a significant disadvantage is the high energy requirement for $CaCO_3$ recovery in the calciner. An alternative solvent recovery route has been investigated using the solid-state reaction between sodium carbonate and sodium trititanate ($Na_2O \cdot 3TiO_2$), the so-called titanate cycle. The solid carbonate and trititanate are heated to ~860°C, at which point the pentatitanate is formed and CO_2 is released:

$$5(Na_2O \cdot 3TiO_2)_{(s)} + 7Na_2CO_{3(s)} \rightarrow 3(4Na_2O \cdot 5TiO_2)_{(s)} + 7CO_{2(g)} \tag{7.22}$$

The sodium pentatitanate is then cooled and leached with water to regenerate the caustic solvent:

$$3(4Na_2O \cdot 5TiO_2)_{(s)} + 7H_2O \rightarrow 14NaOH_{(aq)} + 5(Na_2O \cdot 3TiO_2)_{(s)} \tag{7.23}$$

Overall, including heating and heat recovery, the titanate cycle has approximately half the energy requirement of the calcination route at ~2.9 GJ/tCO_2 vs. ~5.7 GJ/tCO_2 for calcination.

An alternative to the regeneration of the caustic solvent is to continuously produce sodium hydroxide by electrolysis of NaCl brine:

$$2NaCl + 2H_2O \rightarrow Na^+ + Cl^- + H^+ + OH^- \rightarrow 2NaOH + Cl_2 + H_2 \tag{7.24}$$

This has been proposed for a capture system that would produce solid sodium bicarbonate as a product stream (Section 7.3.6).

Phase-change solvents

Phase change solvents have been developed through R&D work aimed at improving amine solvent performance and offer several advantages in terms of solvent loading, regeneration energy requirements, stability, and corrosivity when compared to conventional amine solvents (Zhang et al., 2019). Two types of phase change have been exploited in this way, biphasic or phase-separation liquid systems and precipitating systems.

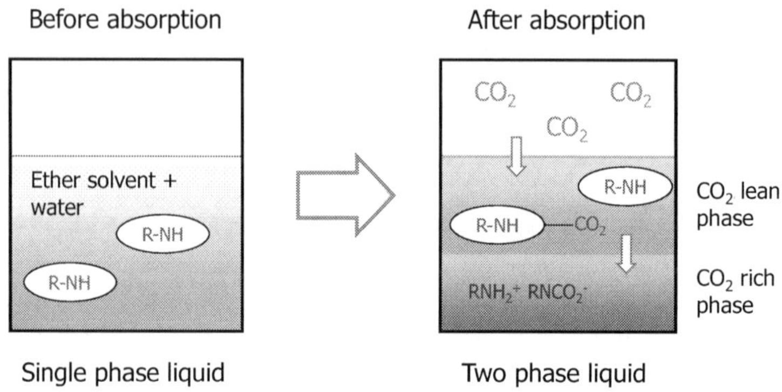

FIGURE 7.3 Phase-separation solvent mechanism. *After Machida et al. (2018).*

Biphasic solvents are a class of mixed solvents that form two immiscible liquid phases under certain conditions of temperature and CO_2 loading—one phase being CO_2 rich and the other CO_2 lean. These solvents operate as a single homogeneous phase during gas absorption, typically at ~40°C, and separation is induced either directly as a result of CO_2 loading or after heating to a regeneration temperature above the liquid phase separation temperature. Systems that become immiscible on heating are termed thermomorphic biphasic solvents. Fig. 7.3 illustrates the mechanism of the phase-separation system developed by Machida and coworkers.

Initially, the amine and water form a single liquid phase. Absorbed CO_2 reacts with the amine to form a hydrophilic carbamate that becomes immiscible with the hydrophobic organic solvent. Different phase-separation behaviors can be obtained by combining amine and ether groups that differ in their relative hydrophobicities.

After the separation of the two phases in a liquid–liquid separator, only the rich phase then needs to be regenerated, typically at ~80–90°C, the CO_2 lean light phase being mixed with the regenerated solvent and returned to the absorber. The regeneration energy requirement is significantly reduced when compared to MEA, due to the lower operating temperature and the reduced volume of solvent that needs to be stripped, with values as low as 1.6 GJ/tCO_2 being reported compared to 3.7 GJ/tCO_2 for a reference 30%wt MEA solvent (Machida et al., 2018).

In precipitating systems, a high CO_2 loading is achieved without the solvent becoming saturated by removing a precipitated reaction product containing CO_2 (e.g., as a carbamate, carbonate, or bicarbonate), while recirculating the CO_2 lean supernatant liquid to the absorber. The precipitate is removed as a concentrated slurry from which CO_2 is released on heating, regenerating the solvent. A variety of solvents have been tested for this application, including potassium carbonate (the UNO MK3 process developed by CO_2CRC), hydrous ammonium carbonate, precipitating ammonium bicarbonate, gamma amino propyl tetramethyldisiloxane (GAP-0), a solvent developed by GE Global Research, precipitating GAP-0 carbamate, and amino acid salt solutions (e.g., salts of taurine ($C_2H_7NO_3S$), proline ($C_5H_9NO_2$) and sarcosine ($C_3H_7NO_2$)), which produce various precipitates depending on amino acid structure and solubility.

The main advantages of precipitating systems are the high solvent loading, achieved by rapid removal of the carbonated reaction product, and the possibility to regenerate the solvent at high pressure, reducing subsequent energy required for CO_2 compression. Challenges with these systems include slow reaction kinetics, addressed for K_2CO_3 solvent by the addition of a rate promoter such as glycine, the coprecipitation of salts containing contaminants such as SO_2 and NO_x, which cannot be easily separated from CO_2-derived precipitates, and the potential for equipment fouling due to precipitate build-up.

"Solvent-free" chemical absorption

Deionized water has also been investigated as a CO_2 solvent, using an electrochemical method for regeneration. Membrane capacitative deionization (MCDI) is commonly used to remove ions from water in desalination and purification systems and involved the application of a potential difference across an aqueous solution flowing between electrodes located behind ion-exchange membranes. In this application, carbonic acid (H_2CO_3) is separated into H^+ and HCO_3^- ions, which pass through cation and anion exchange membranes and are held in the porous electrodes (Fig. 7.4). On reversal of the potential difference, the ions recombine and a pure CO_2 stream can be recovered.

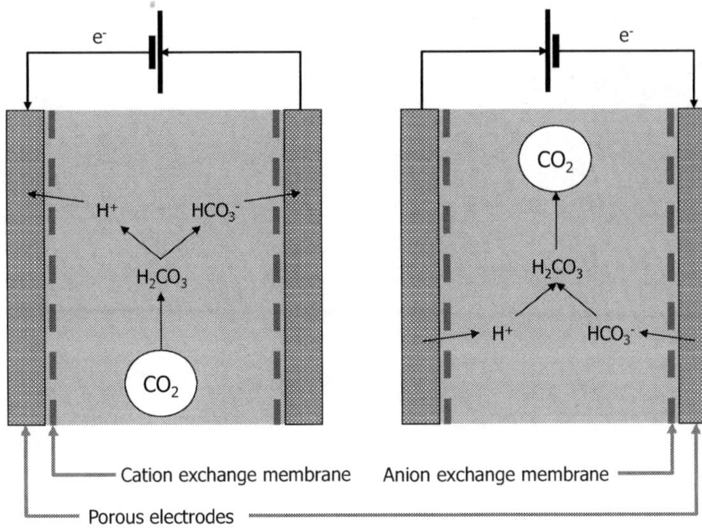

FIGURE 7.4 CO_2 capture and release in an MCDI cell. *Adapted from Legrand et al. (2018).*

This technology is being developed by UK start-up Mission Zero Technologies as part of a water-based direct air capture system (Chapter 13).

7.1.2 Physical CO_2 absorption

Physical absorption processes for gas separation use organic or inorganic physical solvents to absorb gas components rather than reacting with them chemically. CO_2 absorption by a physical solvent is determined by the vapor–liquid equilibrium of the mixture via Henry's law, which states that, at a given temperature, the concentration of a gas dissolved in unit volume of a solvent is proportional to the partial pressure of the gas in equilibrium with the solvent.

Thus the solubility of CO_2 (K_{CO2}) is expressed as follows:

$$K_{CO2} = [CO_2]/pCO_2 = 1/K_{HCO2} \quad (7.25)$$

where $[C_{CO2}]$ is the concentration of dissolved CO_2, pCO_2 is the CO_2 partial pressure in equilibrium with the solvent, and K_{HCO2} is the Henry's law constant for CO_2. Since the concentration of the CO_2 solute is proportional to its partial pressure, high solvent loading will be achieved for feed gas streams at high CO_2 partial pressure.

Physical absorption has the advantage over chemical absorption in that the heat required for desorption is significantly lower; however, this also necessitates low operating temperatures at the absorption stage to achieve high solvent loading. Methanol (CH_3OH) is used in the Rectisol process (see below) as a physical solvent for CO_2 separation from synthesis gas, produced, for example, by biomass gasification. The dependence of CO_2 solubility in methanol on temperature and partial pressure is illustrated in Figs. 7.5 and 7.6. Solubility rises steeply at reducing temperatures below −40°C, enabling very effective separation of CO_2 at operating temperatures in the range from −60°C to −70°C.

Figs. 7.5 and 7.6 also illustrate that starting from an absorption operating point at low temperature, high pressure, or both, a change in operating conditions to higher temperature or lower pressure will result in the release of the solute as a result of a decrease in solubility. Regeneration of a physical solvent can thus be achieved by an increase in temperature (temperature swing) or a reduction in pressure (pressure swing).

Other physical solvents that have been applied to CO_2 capture include dimethyl ethers of polyethylene glycol ($CH_3(CH_2CH_2O)_nCH_3$) in the Selexol process (see later). Unlike chemical solvents, physical solvents are chemically inert to the treated gas stream, avoiding the formation of heat-stable salts that can be problematic in some chemical absorption systems.

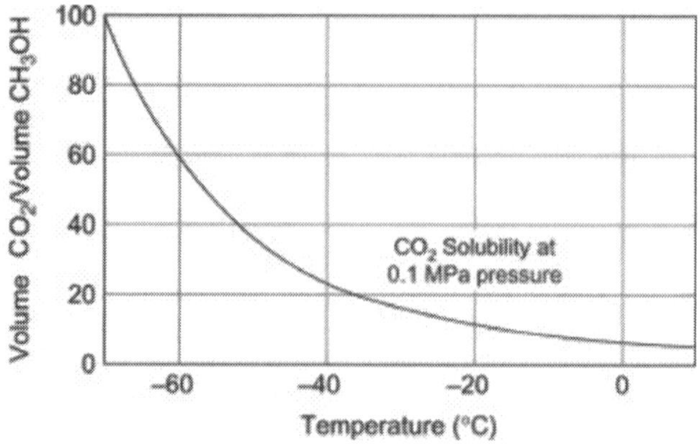

FIGURE 7.5 Temperature dependence of CO_2 solubility in methanol.

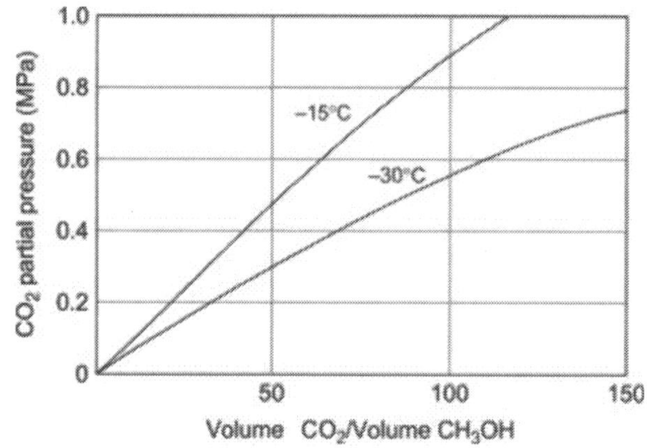

FIGURE 7.6 Partial-pressure dependence of CO_2 solubility in methanol.

7.2 CO_2 absorption applications

Due to the relative maturity of the technology, absorption-based CO_2 separation systems—both chemical and physical—dominate the landscape of large-scale operating and planned CO_2 capture projects.

7.2.1 Chemical absorption applications

Amine-based CO_2 chemical absorption

Chemical absorption processes using a variety of amine-based solvents have been deployed for postcombustion CO_2 capture, with installed single-train capacities reaching ~4800 tCO_2/day, for example, in the Petra Nova project (see later). Fig. 7.7 illustrates the process flow scheme of a typical amine-based CO_2 capture system.

Flue gas entering the process at close to atmospheric pressure is cooled to the required operating temperature in the region of 40–60°C. Cooling by direct water contact also beneficially removes fine particulate matter from the gas stream. The lean solvent (low content of CO_2 reaction products) is brought into contact with cooled flue gas in a packed absorber tower (amine scrubber). Flue gas exiting the top of the absorber is water washed, to reduce the entrainment and carryover of solvent droplets and vapor, and is then vented to the atmosphere.

Rich solvent (high content of CO_2 reaction product) exits the base of the scrubber and is pumped to the top of the amine stripping tower. A heat exchanger heats the rich solvent, recovering heat from the regenerated solvent cycling back to the absorber. The stripping tower typically operates at 100–140°C and at marginally higher pressure than the absorber. The

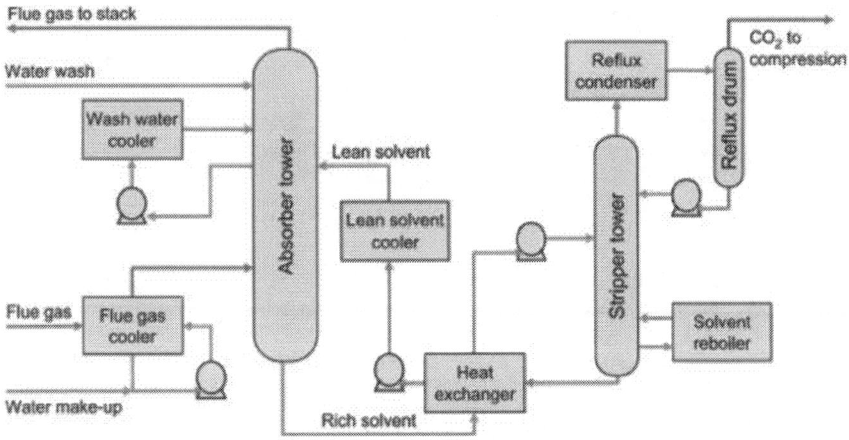

FIGURE 7.7 Process flow scheme for amine-based CO_2 capture with thermal solvent reclaiming.

heat required to reverse the absorption reaction, releasing pure CO_2 and regenerating the lean solvent, is supplied by a reboiler, which would typically be integrated into the steam cycle of the host plant. The reboiler energy requirement for a 30%wt MEA solvent (3.7 GJ/tCO_2) is often used as a benchmark against which the energy penalty for other CO_2 capture processes is compared.

Steam and released CO_2 exit the top of the stripping tower, where the steam is condensed from the CO_2 product stream. Lean solvent from the base of the tower passes through the reclaimer unit where degradation products settle out, and is then cooled and cycled back to the absorber.

Alternatives to the traditional packed tower configuration have also been investigated for both absorption and stripping, including hollow-fiber membrane contactors (Chapter 9) and solvent microencapsulation systems, discussed below.

A number of amine-based CO_2 separation processes have been commercialized, the key features of which are summarized in Table 7.1. The two largest commercial CO_2 capture projects to date in the power generation sector use amine-

TABLE 7.1 Relative solubilities of sorbates at 25°C (CO_2 = 1).

Component		(a) Selexol solvent	(b) Rectisol solvent (methanol)	(b) Fluor solvent (propylene carbonate)
Hydrogen	H_2	0.013	0.0054	0.0078
Methane	C_1	0.066	0.051	0.038
Ethane	C_2	0.42	0.42	0.17
Carbon dioxide	CO_2	1.00	1.0	1.00
Propane	C_3	1.01	2.35	0.51
nButane	nC_4	2.37	–	1.75
Carbonyl sulfide	COS	2.30	7.06	3.29
nPentane	nC_5	5.46	–	5.0
Hydrogen sulfide	H_2S	8.82	3.29	7.06
nHexane	nC_6	11.0	–	13.5
nHeptane	nC_7	23.7	–	29.2
Water	H_2O	730	–	300

based capture systems, at Saskpower's Boundary Dam power station near Estevan in Saskatchewan, Canada (Shell's Cansolv process), and NRG Energy's Petra Nova project located in Thompsons, Texas (KEPCO/MHI Advanced KM CDR process process). The KM CDR process is also planned to be used by the UK Drax Group in moving its BECCS project from pilot scale to full commercial deployment.

Outside the power generation sector, activated MDEA has been in use since 1996 and 2008 respectively to capture 0.9 and 0.7 $MtCO_2$/yr from natural gas produced at Statoil's Slepiner and Snøhvit fields in Norway, since 2015 at Shell's Quest Project (Shell ADIP-X process), where up to 1.2 $MtCO_2$/yr is captured during hydrogen production from steam methane reforming (SMR) units at the Scotford bitumen upgrading plant in Alberta, Canada, and since 2016 to capture up to 4 $MtCO_2$/yr from natural gas at Chevron's Gorgon Project, in Western Australia. For all these projects, CO_2 is initially destined for geological storage.

Amine-based systems therefore offer a mature technology to achieve negative emissions through CO_2 capture from flue gases resulting from biomass combustion or conversion processes, for example in BECCS projects, as discussed in Chapter 12.

Ammonia-based CO_2 chemical absorption

A flue gas CO_2 capture process using a chilled slurry of dissolved and suspended ammonium carbonate and ammonium bicarbonate in ammonia has been developed by Alstom Power Systems (now GE) and the Electric Power Research Institute.

The system uses a typical absorber tower configuration, operating at near-freezing temperatures (0−10°C), in which cooled flue gas flows upward in countercurrent to the absorbent slurry. The low operating temperature allows high CO_2 loading of the solvent slurry and reduces "ammonia slip"—the carryover of entrained ammonia droplets and suspended solids with the clean flue gas exiting the tower. Ammonia slip is further reduced by a cold-water wash of the cleaned flue gas, which consists mainly of nitrogen, excess oxygen, and a low residual concentration of CO_2.

The solvent slurry regenerator operates at temperatures >120°C and pressures >2 MPa. Ammonia slip on regeneration is also controlled by water washing, yielding a high-pressure CO_2 stream with low moisture and ammonia content. This high-pressure regeneration has the advantage of reducing the energy requirement for subsequent compression and delivery of the CO_2 product stream for storage. The flow scheme of the chilled ammonia process (CAP) is shown in Fig. 7.8.

As part of an extensive development program, proof of concept pilot testing at a 15 $ktCO_2$/yr scale demonstrated the capabilities of the technology and was followed by two larger scale pilot projects: a 30 MW_e slip-stream test at AEP's Mountaineer power plant in West Virginia, capturing up to 110 $ktCO_2$/yr with storage on-site at the plant in the Mount Simon Sandstone saline aquifer, and a test at Technology Center Mongstad in Norway, capturing up to 82 $ktCO_2$/yr (Augustsson et al., 2017).

7.2.2 Physical absorption applications

Fig. 7.9 illustrates the relative solvent loading of chemical and physical solvents as a function of the partial pressure of the sorbate gas. Although chemical solvents can achieve high loading at low partial pressure, physical solvents are favored at high pressure. Physical absorption is therefore applicable to CO_2 separation in higher pressure processes, such as from a biomass gasification synthesis gas stream (see Chapter 12), while chemical absorption is preferred for lowerpressure applications, such as capture from biomass combustion flue gases. Use of physical absorption for CO_2

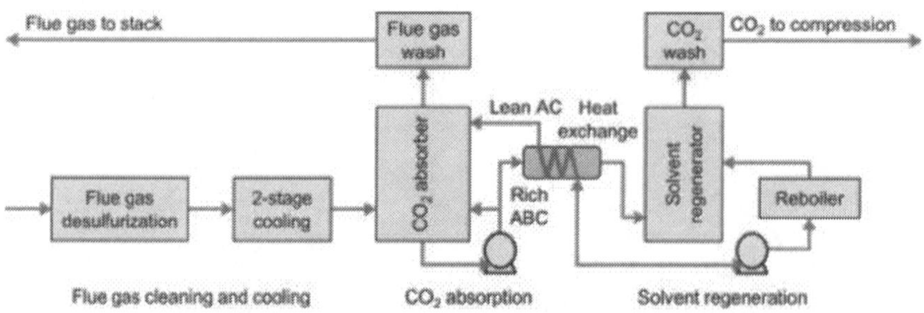

FIGURE 7.8 Chilled ammonia process flow scheme.

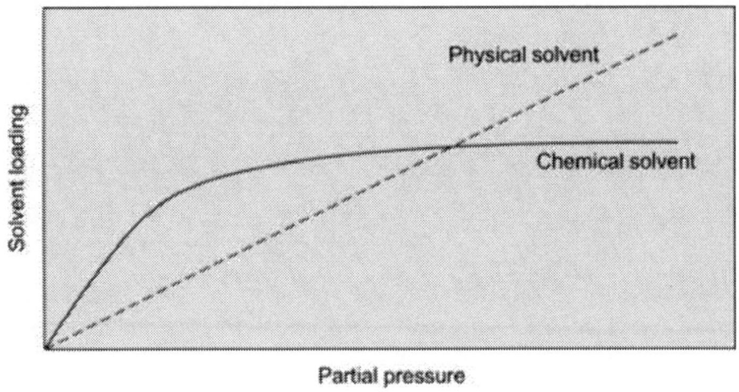

FIGURE 7.9 Relative solvent loading of chemical and physical solvents.

capture from flue gas would require compression of large volumes of gas, the major component being nitrogen, which would subsequently be blown down for release to the atmosphere. The attendant energy penalty would make this process unattractive.

Selexol process

The Selexol process uses a liquid physical solvent to remove acid gas from synthetic or natural gas streams and has been in commercial application since the early 1970s, with well over 50 units currently operating worldwide. The proprietary Selexol solvent was originally developed by Allied-Signal (now Honeywell) in the 1950s and is a mixture of the dimethyl ethers of polyethylene glycol with the formula $[CH_3(CH_2CH_2O)_nCH_3]$, where n is 3–9.

Table 7.2 shows the relative solubilities of typical gas stream components in a number of solvents. Because of the high solubility of water, the partial pressure of water vapor in the feed gas stream to a Selexol plant must be kept low to avoid impairing the CO_2 loading capacity of the solvent.

Fig. 7.10 illustrates the basic Selexol process for CO_2 removal from a gas stream. The solvent is regenerated in two or more flash stages, with offgas from the first (HP) flash being returned to the absorber tower. CO_2 is recovered as offgas from the second and later flash drums, total CO_2 recovery and product purity increasing with additional flash stages.

The Selexol process has been deployed in some of the world's largest operating CO_2 capture plants in both natural gas processing and power generation sectors. Occidental Petroleum's Century Plant in Pecos County, Texas, in operation since 2010, is the world's largest capture plant in terms of capture rate, with a capacity to remove 8.4 $MtCO_2$/yr from high CO_2 natural gas. At Mississippi Power's Kemper County Energy facility, in operation since October 2016, the Selexol process is used to capture 3 $MtCO_2$/yr from syngas at an IGCC power plant, reducing the CO_2 emission from the plant by ~65%.

Rectisol process

The Rectisol process, using refrigerated methanol as a physical solvent, was initially developed by Lurgi GmbH (now part of Air Liquide SA) and found its first commercial application in 1955 in the purification of coal gasification syngas for ammonia and Fischer–Tropsch liquids production. It has become the dominant process applied worldwide for syngas purification and now purifies 75% of syngas produced globally.

The Rectisol process can be configured in many different ways, depending on the types of contaminants to be removed and the required product streams—for example, whether food-grade CO_2 is required. When applied to CO_2 capture from a syngas production process, raw syngas is first treated to remove trace components such as H_2S, HCN, and NH.

Pre-treated syngas is water–gas shifted (see Section 8.4.2), producing a stream with ~33% CO_2 content (Fig. 7.11). This is cooled and enters the CO_2 absorber, where CO_2, typically at a partial pressure >1 MPa, is removed by contact with refrigerated methanol at −10°C to −70°C.

The largest operating Rectisol application is at the Great Plains synfuel plant, which produces natural gas by gasification of lignite. The Dakota Gasification Company plant has been operating since 1984, and the retrofitted capture process has been capturing 3 $MtCO_2$/yr since 2000.

TABLE 7.2 Amine-based CO_2 capture systems in commercial use and under development.

Process	Description	Application limits/operational conditions	Further development focus areas
Kerr-McGee/ABB Lummus Crest technology process	MEA solvent, using low amine concentration (15%–20%) to limit corrosion. Single train capacity up to 400 tCO_2/d.	+ SO_2 tolerant to 100 ppm − High solvent flow rate and regeneration heat requirement − Lower oxygen tolerance leading to solvent degradation	Higher solvent concentration Process improvements to reduce oxygen degradation of solvent
Fluor Econamine FG Plus[SM] process	Inhibited aqueous MEA solvent (30% MEA by weight). Single train capacity up to ~5000 tCO_2/d.	+ Improved oxygen tolerance with low corrosivity − Higher solvent consumption (1.5–2.0 kg/tCO_2) Low SO_2 tolerance (<10 ppm)	Improved solvent formulation to increase reaction rate and reduce absorber tower height Higher solvent capacity leading to lower solvent flow rate and reduced regeneration energy cost Process improvements in demonstration scale plants
Kansai/Mitsubishi KM-CDR process	Sterically hindered amine solvents (KS1, KS2, KS3) plus a rate promoter. Single train capacity up to 4776 tCO_2/d at the Petra Nova site. Pilot application at the Drax BECCS power plant.	+ Low regeneration heat requirement/steam consumption + Low corrosion tendency + Low solvent flow rate, degradation, and losses − Higher solvent cost compared to MEA − Solvent loss through thermal degradation − Low SO_2 tolerance (<10 ppm)	Further development of hindered amines with lower heat of desorption Reducing other process heat requirements Reducing solvent volatility and losses Process improvement in demonstration scale plants
Shell Cansolv® process	World's first commercial scale postcombustion CCS deployment at SaskPower's Boundary Dam power plant in Canada	+ Low regeneration energy and faster kinetics compared to conventional amines + High loading capacity + Improved oxidative and thermal degradation resistance	Process optimization in commercial scale plant e.g., reducing postregeneration compression costs at higher pressure/temperature.

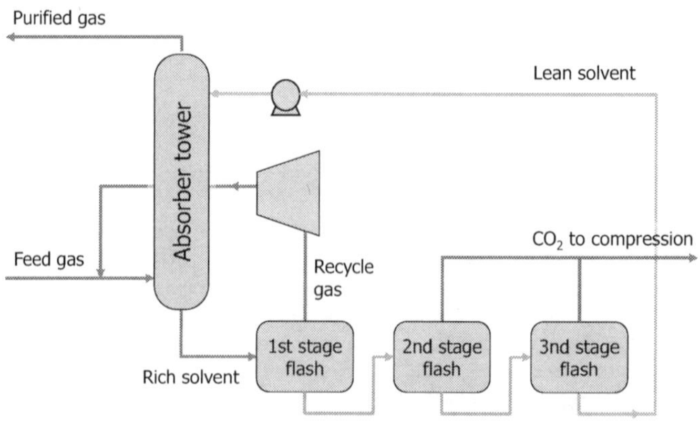

FIGURE 7.10 Selexol process scheme.

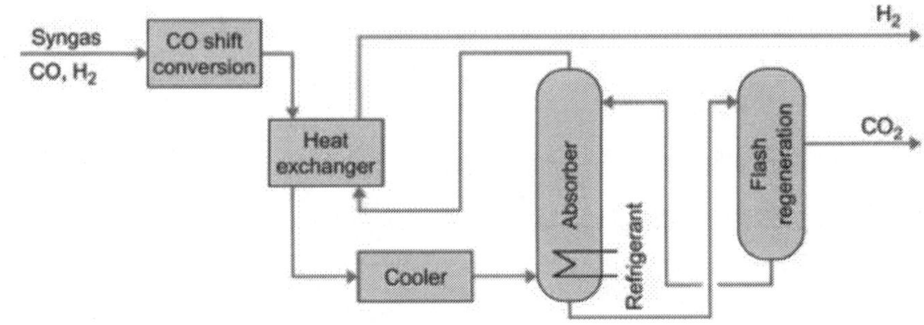

FIGURE 7.11 Rectisol process for CO_2 removal from water−gas-shifted syngas.

7.3 CO_2 absorption technology RD&D status

Although absorption processes for CO_2 capture have over half a century of industrial application, it remains an area of significant ongoing RD&D effort, applied both to the continuous improvement of existing processes and also to the development of novel systems aiming for a step-change improvement in cost-effectiveness (Table 7.3). Many solvents are also being investigated for use in hybrid systems, including both absorption + adsorption systems and membranes; these developments are discussed in Chapters 8 and 9.

The fundamental challenge, on which most R&D effort is focused, is to reduce the overall process energy requirement, particularly for solvent regeneration. Early (first generation) amine systems required 3.7 GJ/tCO_2 captured for solvent regeneration, and the current achievable level for amine systems is ~2.6 GJ/tCO_2. Research on new mixed solvents, such as biphasic amine solvents, has reduced this to ~1.6 GJ/tCO_2 (2019), and results on catalytic regeneration promise further reduction to ~1.1 GJ/tCO_2, the latter results both currently at lab scale only. As the thermodynamic minimum work of separation of CO_2 from flue gas is <0.2 GJ/tCO_2, the potential for further breakthrough improvements remains, and motivates the extensive global R&D effort in this area.

7.3.1 Improved amine-based systems

All of the major suppliers of amine-based systems, summarized in Table 7.2, have ongoing development programs aimed at improving the effectiveness and market competitiveness of their proprietary processes.

Areas of research and development to improve the performance of these systems include the following:

- Improving solvent formulations and regeneration processes to reduce energy requirement and solvent degradation
- Sterically hindered amines (such as 2-amino-2-methyl-1-propanol) to increase loading via strong bicarbonate formation
- Use of catalysts to improve absorption kinetics (e.g., piperazine, carbonic anhydrase) and to reduce solvent regeneration energy requirements (e.g., γ-Al_2O_3)
- Genetic engineering, for example, by directed evolution (see Glossary), to produce temperature and alkalinity tolerant carbonic anhydrase variants
- Increasing contact area and reducing pressure drop by modified tower packing
- Reducing energy requirements by increasing heat integration with the host plant
- Producing higher amine concentration with reduced corrosion by using new additives, such as phenoxide salts
- Treatment of amine degradation products

Several of these topics are focused on reducing the energy requirement for solvent regeneration, and novel approaches to this problem are also under investigation including CO_2 desorption via crystalization (Garrabrant et al., 2019) or by electrochemical means (Stern et al., 2013). Fig. 7.12 illustrates the method proposed by Stern and coworkers.

In this process, the CO_2-rich solvent stream is pumped from the absorber into the anode compartment of an electrochemical cell where copper cations are generated from the copper anode. This drives the dissociation of the loaded amine toward CO_2 desorption, according to

$$Cu^{2+} + R\text{-}N\text{-}CO_2 \rightarrow Cu\text{-}R\text{-}N^{2+} + CO_2 \quad (7.26)$$

TABLE 7.3 CO_2 absorption technologies and RD&D summary.

Absorption technology	Advantages	Disadvantages	RD&D focus areas
Physical absorption			
Fluor process (propylene carbonate)	Well-proven technology with 50+ years of commercial application	Low H_2S tolerance in feed gas stream	
	High selectivity for CO_2 relative to methane	Feed gas must be dehydrated due to high water solubility	
	Noncorrosive solvent	Irreversible reaction with CO_2 and water at 90°C precludes temperature swing regeneration	
Rectisol process (chilled methanol)	High selectivity for H_2S and CO_2	High refrigeration energy cost	
	Able to remove many contaminants in a single process	Solvent volatility and emissions concerns	
	Low solvent cost	Higher selectivity for H_2S over CO_2	
	Long-proven commercial application in gasification projects	Feed gas must be dehydrated due to high water solubility	
Selexol process (dimethyl ethers of polyethylene glycol [DMEPG])	Well-proven technology with 30+ years of commercial application	Requires high partial pressure of CO_2	Additives to reduce solvent costs
	Chemically inert solvent not subject to degradation; very low vapor pressure	High solvent viscosity	Process optimization to reduce energy requirements and operating costs
	Nonaqueous solvent allows low-cost carbon steel plant construction	High solvent costs	
	Dual-stage process can capture CO_2 and H_2S, e.g., from syngas		
Chemical absorption			
Amine systems (e.g., MEA, DMEA)	Well-proven technology with extensive commercial application, leading to low capital costs	Low amine concentrations to resist corrosion also limit loading capacity Hazardous degradation products	Evaluation of new amine-based solvents, including amine-blends and aromatic amines with lower heat of adsorption, higher loading, and reduced degradation
	Range of solvents have been developed	Low tolerance to SO_2, NO_x, and O_2 in flue gas	Catalyzed absorption and regeneration/desorption (e.g., solid acid catalysts, carbonic anhydrases)
	Technical and economic benchmark for other solvent systems	High energy requirement for sorbent regeneration Solvent loss due to mist formation	Additives to improve system performance (higher amine concentration and loading, faster reaction rate, inhibited solvent degradation)
			Process improvements to reduce energy requirements, improve heat integration, increase solvent capacity (e.g., absorber

TABLE 7.3 CO_2 **absorption technologies and RD&D summary.—cont'd**

Absorption technology	Advantages	Disadvantages	RD&D focus areas
			tower inter-cooling), reduce solvent loss
			Quantum-chemical modeling to predict solvent properties based on molecular structure
			Reduced energy solvent regeneration, e.g., electro-chemically mediated amine regeneration (EMAR), application of photoacids
			Performance of precipitating amine + amino acid solvents
Bi-phasic liquid solvents	Higher solvent loading capacity Regeneration at higher pressure and lower temperature reduces energy requirement Lower corrosivity in some cases than standard amine solvents	Typically higher loaded solvent viscosity than amines, inducing higher process pressure drops and slow phase separation	Regulation of solvent phase change Alternative desorption techniques (e.g., ultrasonic, direct gas stripping) Identification of degradation products Development and validation of liquid:liquid phase change thermodynamic models
Potassium carbonate systems	Higher solvent loading capacity Lower regeneration energy requirement	Slow reaction rate without promoters such as piperazine High cost of additives such as piperazine	Energy saving through improved tower packing and multistage stripping
	Lower corrosivity Low solvent cost	System fouling due to reaction product precipitation	Alternative additives (e.g., CA, amino acids) to improve CO_2 absorption rate and reduce stripping heat requirement
			Precipitating systems to reduce regeneration energy
Aqueous ammonia	Low regeneration energy requirement	High solvent volatility requires reduced operating temperature	Increasing solvent loading and reaction rate, inter alia through process optimization
	Lack of solvent degradation during the absorption and regeneration cycle	High solvent loss during regeneration at high temperature	Reducing solvent losses, for example using ionic liquids as additives
	Low solvent cost High-pressure regeneration		Integration with industrial production of ammonium bicarbonate fertilizer
	Tolerant to O_2 and other contaminants		
	Sellable ammonium sulfate and ammonium nitrate fertilizer by-products		
Chilled ammonia	Low regeneration energy requirement Lack of solvent degradation during the absorption and regeneration cycle Low solvent cost	Near-freezing operating conditions required to limit solvent losses	Process optimization in demonstration scale plant Reducing solvent emissions by improving water wash Improved heat integration and control systems
		Potential for fouling due to ammonium bicarbonate solids	
	Sellable ammonium sulfate fertilizer by-product		

Continued

TABLE 7.3 CO_2 **absorption technologies and RD&D summary.—cont'd**

Absorption technology	Advantages	Disadvantages	RD&D focus areas
Task-specific and reversible ionic liquids	Very low volatility and solvent loss	High production cost of ionic liquids	Synthesis and performance testing of new, lower cost ILs e.g., using amino acids
	Very low heat of adsorption (10%–15% of MEA) High chemical and thermal stability	High viscosity, reducing absorption rate and increasing parasitic energy load for solvent transport Absorption efficiency reduced if water is present	Increased absorption capacity and rate; reduced visosity Energy efficient removal of water from anhydrous ILs
	Dual mode (physical plus chemical solvents) with high loading capacity Less corrosive than amines		Use of TSILs in membrane contactors Coproduction of saleable byproducts
Sodium hydroxide	Low cost and abundance of required chemicals	High energy requirement in calcining intermediate regeneration product ($CaCO_3$)	Alternative methods of solvent recovery (titanate cycle)
	All process steps are currently proven technologies, applied in other industries	High water and solvent loss in spray tower contactor	Laboratory-scale spray tower optimization and commercial-scale concept development
			Process simulation and optimization
"Solvent free": Water + MCDI	Solvent free; system components are individually commercialized	Low "solvent" loading	Enzymatic "solvent" enhancement (e.g., CA); reduction of cell resistance to reduce ohmic losses

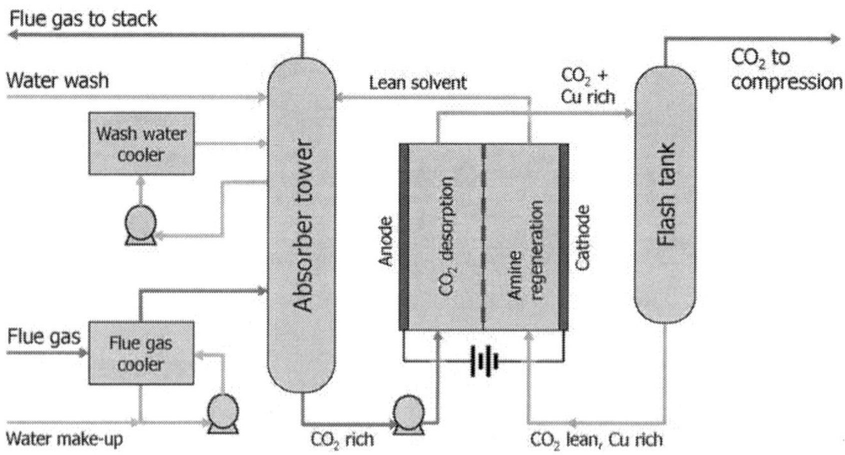

FIGURE 7.12 Electrochemically mediated amine regeneration. *After Stern et al. (2013).*

CO_2 is removed in a flash tank and the solvent returns to the cathode compartment where regeneration is accompanied by the electrochemical plating of copper onto the cathode;

$$\text{Cu-R-N}^{2+} + 2e \rightarrow \text{Cu} + \text{R-N} \tag{7.27}$$

In addition to the potential for lower energy demand, the electrochemically mediated amine regeneration process enables CO_2 desorption at lower temperature and higher pressure, resulting in reduced amine degradation and evaporation losses, and lower CO_2 compression costs. Similar electrochemical approaches have been investigated for solvent regeneration of K_2CO_3 (Rahimi et al., 2020) and of ammonia (Wang et al., 2022), the latter achieving a remarkably low regeneration energy requirement of 334 MJ/tCO_2, an order of magnitude below the MEA benchmark.

Other novel electric and electromagnetic approaches to reduce the energy demand for amine regeneration include microwave heating (McGurk et al., 2017) and dielectric heating (Tsubaki et al., 2020).

7.3.2 Enzyme-catalyzed aqueous carbonate solvents

Building on the extensive body of industrial experience in the application of aqueous carbonate systems for CO_2 and H_2S removal from industrial gases, there has been rapid commercial uptake in the application of carbonic anhydrase to improve the performance of these systems.

Canadian company CO_2 Solutions Inc. (CSI) completed 1 tCO_2/d pilot testing removing CO_2 from coal and natural gas flue gases at the University of North Dakota, followed in 2015 by a small (10 tCO_2/d, 2500 h) demonstration test in association with Husky Oil,. A proprietary carbonic anhydrase, enhanced through directed evolution, was dissolved in the K_2CO_3 solvent, with regeneration using hot water at 75–85°C rather than steam. This solvent has the advantage that it uses low-cost environmentally-friendly commodity chemicals that do not form undesirable byproducts or generate aerosol emissions from the absorber (Madore and Fradette, 2017).

The technology reached TRL8 with the first commercial-scale deployment in 2018/19, capturing 30 tCO_2/d at Resolute's Kraft pulp mill in Saint-Félicien, Québec, and CSI's technology at the Saint-Félicien plant was acquired by Saipem in 2020. The solvent reboiler energy requirement of 2.4 GJ/tCO_2 was provided entirely by low-grade waste heat from the pulp mill, with captured CO_2 being reused at the adjacent Serres Toundra greenhouse facility. These projects have included the first carbon capture application of rotating packed bed (RPB) technology, replacing a 370 m^3 absorber tower with two 6 m^3 RPBs, significantly reducing both capital cost and deployment space requirements. The use of RPBs has also been reported using MEA as the solvent (Lee et al., 2017).

Other focus areas for enzyme-related R&D include genetic engineering and synthesis of CAs with higher thermal stability and prolonged activity, low-cost immobilization methods that maintain high enzyme activity and stability, and development of novel multiphase contacting systems.

7.3.3 Phase change solvents

Bi-phasic liquid solvent R&D

A mixed solvent system comprising the tertiary alkanolamine diethylethanolamine and *N*-methyl-1,3-diaminopropane, which has one primary and one secondary amine functional group, was investigated as part of the EU iCap project. In a lab-scale pilot, this solvent achieved a one-third reduction in regeneration energy requirement (2.4 GJ/tCO_2) compared to the MEA benchmark. The process flow scheme is similar to the basic amine process with thermal regeneration (see Fig. 7.7), with the addition of a liquid–liquid separator between two towers. This can be located either before or after the heat exchanger, depending on whether additional heat is required to initiate phase separation. Fig. 7.13 illustrates the process for the latter case.

An industry scale, 3.5 MW_e equivalent pilot of a biphasic solvent developed by IFP Energies Nouvelles (the DMX process) was planned as part of the EU-funded OCTAVIUS project. This process, using a dipropil-methyl-xanthine-based solvent has a low reported solvent regeneration requirement, at <2.3 GJ/tCO_2. Unfortunately, the retrofit of the process at ENEL's 48 MW Brindisi power plant was too expensive to be funded within the project budget. A pilot unit to demonstrate the process is planned to be built at ArcelorMittal's steel plant in Dunkirk, France capturing 12 tCO_2/d from blast furnace gases, as part of a wider EU-funded project to develop a North Sea Capture and Storage Cluster at Dunkirk.

Precipitating solvent R&D

Precipitating solvents were investigated as part of the EU FP7-funded HiPerCap project (Kim et al., 2018). As part of its scope on absorption capture, this project focused on inorganic solvents that can achieve high CO_2 loading by precipitating reaction products, while avoiding the problems of emissions and degradation associated with organic solvents.

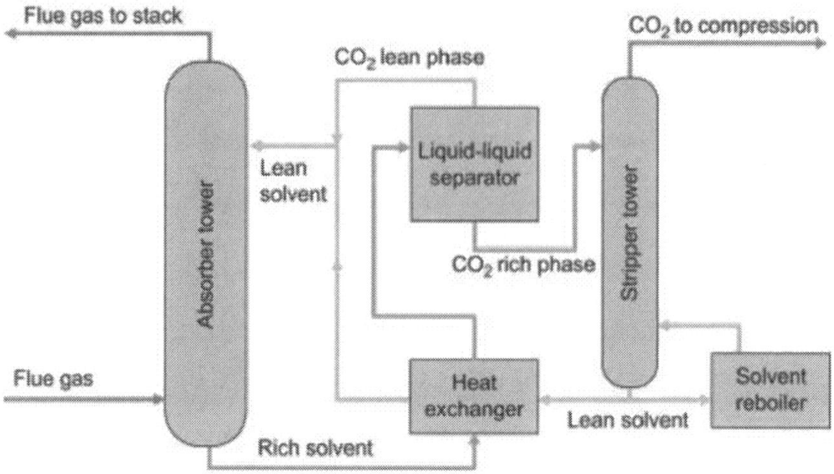

FIGURE 7.13 Simplified flow scheme for biphasic solvent capture process.

An industrial pilot capturing 1 tCO_2/d has been conducted using a concentrated potassium carbonate precipitating system—the UNO MK3 process developed by CO2CRC (Anderson et al. 2014)—at the GDF-SUEZ Australia Energy operated Hazelwood Power Station in Victoria, Australia, and confirmed low regeneration energy, good environmental performance and low overall costs (Smith et al., 2017). Further R&D work toward large-scale demonstration has included testing the impact of rate promoters such as diethylenetriamine and the amino acid glycine.

7.3.4 Ionic liquid solvents

One of the disadvantages of liquid solvents for gas separation is the loss to the gas stream of solvent, and water in the case of aqueous solutions, due to the vapor pressure of the solvent solution. Ionic liquids (ILs), formed by the melting of certain organic and inorganic salts, have negligible vapor pressure as a result of the strong Coulombic attraction between the constituent ions, and could therefore avoid solvent loss.

ILs typically comprise a large organic cation and a smaller organic or inorganic anion, and it is this asymmetry between the cation and anion sizes that reduce the strength of the binding force and results in a low melting point. Commercial applications include their use in supercapacitors and for nanoparticle synthesis, while other uses as lubricants, fluid additives, and for metal extraction are under development. Gas separation applications, first proposed by Blanchard and coworkers (Blanchard et al., 2001), are still at the R&D stage, and the potential for IL properties such as CO_2 absorption capacity, viscosity, and thermal stability to be fine tuned by varying the IL structure offer advantages over other solvents, although high costs (up to 100 times the cost of benchmark organic solvents) present a significant hurdle for commercial applications.

Task-specific ionic liquids

By preparing ionic liquids that include ions with specific functional groups, so-called task-specific ionic liquids (TSILs) can be designed for particular separation applications. Amine functional groups have been successfully used, capturing CO_2 as a carbamate ion in chemical absorption reactions equivalent to reaction 7.1.

R&D work on TSILs has focused on the engineering of new low-cost liquids with high absorption capacity. TSIL molecular structures are readily amenable to modification, so that structure-dependent physical and thermodynamic properties can be optimized. TSILs were investigated in the EU FP7-funded IoLiCAP project (2011–2016), which synthesized and tested the properties (absorption capacity, corrosivity, etc.) of a wide range of ionic liquids. The project also developed improved molecular simulation models to better understand the structural, thermodynamic, and transport properties of ILs, and equation of state models for IL phase behavior in the presence of CO_2. A pilot plant was constructed in Megalopolis, Greece, capturing CO_2 from a coal-fired power plant flue gas slipstream using a solvent based on a mixture of Ils. Further development of Ils for a variety of applications is being progressed by IoLiTec GmbH.

Although the low regeneration cost and tunable nature of ILs offer considerable potential for CO_2 capture, the low absorption capacity and high viscosity have limited further work on their use as liquid phase solvents. However, these disadvantages can be turned to advantages by applying ILs in other structures such as supported liquid membranes (see Chapter 10), microencapsulation (see below), doped metal-organic frameworks (Cota and Martinez, 2017), or phase-separation gels (Kataoka et al., 2022).

Reversible ionic liquids

Reversible (also known as switchable or smart) ionic liquids are a class of IL solvents in which a key property, for example, CO_2 solubility, undergoes a step change in response to an external stimulus, such as heat, light, or pH. This abrupt property change allows the regeneration method to be engineered into the structure of the solvent, resulting in substantially lower regeneration energy requirements. These novel solvents also offer higher CO_2 loading capacity compared to benchmark solvents, as they can operate as "dual-mode" solvents, absorbing CO_2 by both chemical and physical absorption (Jessop et al., 2005).

Ongoing research focuses on the design of reversible IL solvents using molecular modeling to optimize the solvent properties for CO_2 capture and regeneration, followed by synthesis, characterization, and testing of the chemical and physical properties of the new solvents.

7.3.5 Solvent microencapsulation

Many of the limitations of current solvent systems—such as corrosivity, solvent loss, degradation, or absorption efficiency loss due to impurities—are the result of direct contact between the solvent and the feed gas or system hardware. One way to address these challenges is to isolate the solvent by encapsulation within a permeable shell, and two approaches have been proposed to achieve this, as illustrated in Fig. 7.14.

In the dry alkanolamine approach, patented by Yang et al. (2015), the solvent is vigorously stirred together with silica (SiO_2) nanoparticles, ~0.3 μm in diameter, that have been fluorinated to make the surface oleophobic. Encapsulated so-called "liquid marbles" are formed (Fig. 7.14A), ~80 μm in diameter, which flow and behave like a dry powder. The resulting dry solvent, using either MEA or DEA as the encapsulated solvent, showed improved adsorption capacity compared to the free solvent. The low thermal conductance and high infrared albedo of the silica shell made conventional conductive/radiative heating inefficient to achieve the ~120°C required for solvent regeneration, and this was achieved by microwave heating.

Microencapsulated carbon sorbents (MECSs) take a different approach to encapsulation, using a solid rather than a powder aggregate shell. MECSs were first demonstrated in 2015 by Vericella and coworkers, who fabricated permeable silicone microcapsules containing K_2CO_3 solvent together with a pH indicator dye (Vericella et al., 2015). On absorption, CO_2 diffuses through the ~30 μm thick high permeability capsule wall and reacts with the solvent to form a bicarbonate precipitate (the Benfield process—see Section 7.1), the drop in pH being indicated by a change in dye color from blue to yellow. Heating to ~60°C then redissolves the precipitate releasing CO_2 that diffuses out of the microcapsule. Compared

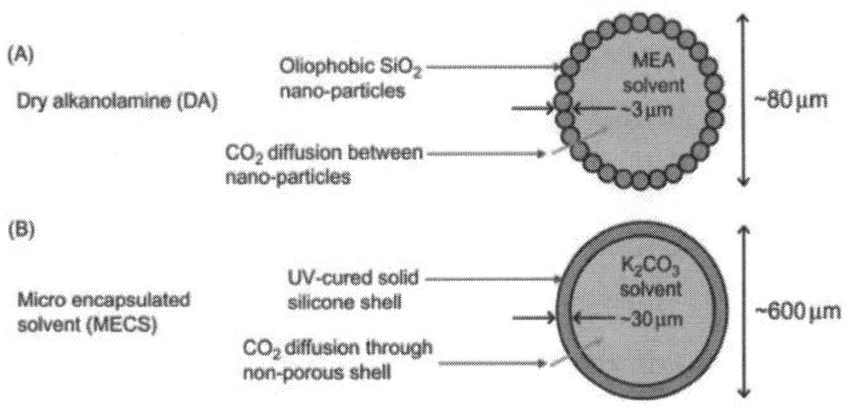

FIGURE 7.14 Schematic structure of dry alkanolamine (DA) and microencapsulated carbon (MECS) sorbents.

to a standard packed tower absorber geometry, the fabricated microcapsules exhibited a 10–100-fold increase in surface area per unit solvent volume, and a further enhancement in absorption rate was demonstrated by including Zn-cyclen (Zn-1,4,7,10-tetraazacyclododecane—a carbonic anhydrase mimic) with the encapsulated solvent. The microcapsules also proved sufficiently robust to use in a fluidized bed-type system.

Current R&D themes in solvent microencapsulation include techniques to encapsulate a range of solvents (Stolaroff et al., 2016), capsule and process design aspects, including novel absorber configurations (Hornbostel et al., 2019), and heat recovery systems. Larger scale and longer term pilot testing is also required to demonstrate sustainable performance, for example in the presence of flue gas impurities, but the potential to tailor the capsule material and to encapsulate a wide range of different solvent types opens up a wide field of opportunity to develop improved capture processes.

7.3.6 Sodium and potassium hydroxide-based systems

Capture of CO_2 using chemical absorption by sodium or potassium hydroxide is at the stage of initial deployment, both for flue gas and direct air capture.

Flue gas CO_2 capture using sodium hydroxide

A flue gas capture process, applying carbonation reactions in 7.17 and 7.18 to cooled flue gas at 30°C, has been developed by CarbonFree Chemicals (formerly Skyonic Inc.) in the Skymine process. Few technical details of the process have been released, but the concept shown in Fig. 7.15 indicates that hydrogen and chlorine, together with carbon as sodium bicarbonate, in solid form or as a concentrated solution, are the process products.

The process also generates other intermediate components; sodium hydroxide is generated in the process by the electrolysis of NaCl brine, and HCl is generated from hydrogen and chlorine products of the brine electrolysis and used to acidify the brine ahead of electrolysis as well as to control pH in the absorption columns. As noted earlier, pH in the absorption column determines the ratio of carbonate to bicarbonate reaction products.

A first deployment of this technology began operations at a Capitol Aggregates cement plant in San Antonio, Texas in 2015. The plant was designed to capture 75 $ktCO_2$/yr and avoid a further 225 $ktCO_2$/yr by producing carbon negative Na_2CO_3. While the process benefits economically from the value of the byproducts (H_2, Cl, HCl, Na_2CO_3), the limited markets for these products would be a constraint on large-scale deployment.

Direct air CO_2 capture using sodium or potassium hydroxide solvents

The overall process for direct air capture using NaOH as a solvent is illustrated in Fig. 7.16. Air is drawn through a contactor where CO_2 is captured by the solvent. Physically the contactor could be a packed tower configuration, in which the solvent drips down through packing material designed to maximize contact area, an empty tower similar to the familiar evaporative cooling tower of a power plant, with a solvent spray dropping in a cocurrent (downward) air flow, or a cross-flow contactor in which a horizontal (wind-assisted) air flow passes through solvent wetted packing.

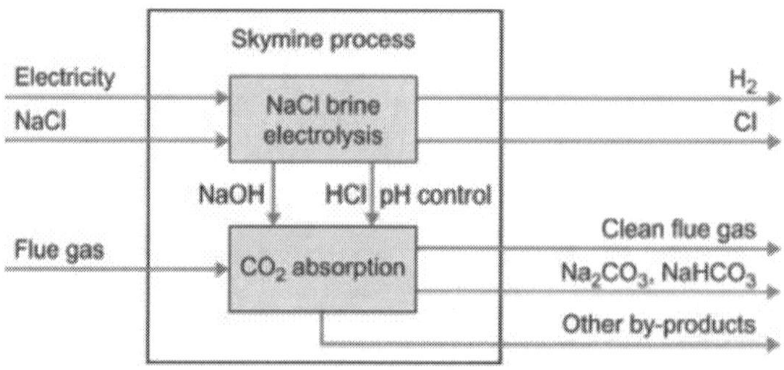

FIGURE 7.15 Skymine process schematic.

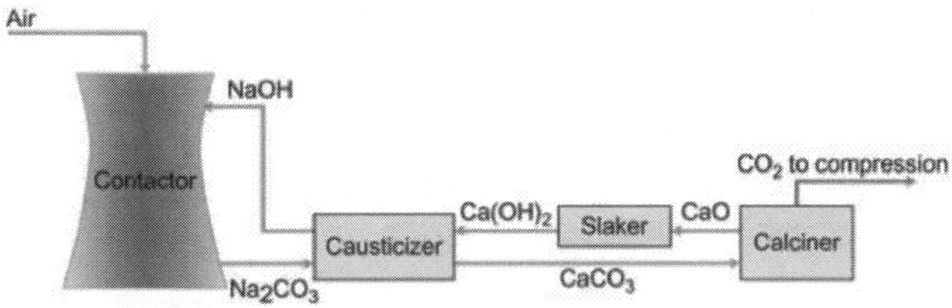

FIGURE 7.16 Process concept for air capture using sodium hydroxide.

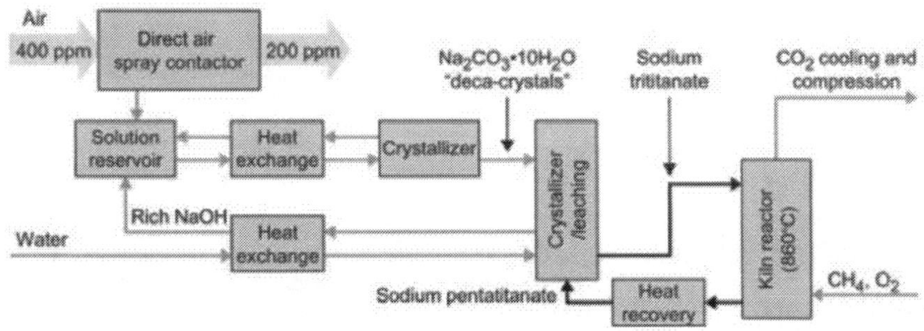

FIGURE 7.17 Direct air capture process using sodium titanate cycle.

In the causticizer, NaOH is recovered by the addition of slaked lime ($Ca(OH)_2$) produced by the hydration of lime (CaO) according to reaction 7.18. The calcium carbonate mud produced in the causticizer is heated in the calciner, driving off water as steam and yielding lime and CO_2. In the subsequent development of this process by Carbon Engineering (discussed in Chapter 14), the causticizer takes the form of a pellet reactor in which $CaCO_3$ pellets are formed by crystalization (Giacomin et al., 2020). Calcining is used on an industrial scale in the production of cement, and a conventional high-efficiency calciner comprises a fluidized bed fired by the combustion of natural gas. The CO_2 content of the exhaust gas stream is enhanced as a result of the offgas from the calcining reaction, and CO_2 capture from the exhaust gas stream can then be achieved by any absorption process.

The energy efficiency of the process can be enhanced by integrating heat recovery from steam produced in the calciner, as well as recovering low-level heat from the exothermic slaking reaction ($\Delta H = -82$ kJ/mol).

Fig. 7.17 illustrates an alternative process scheme that uses the titanate cycle (Section 7.1.1) rather than calcination to regenerate the caustic solvent. Following the air contactor, a first crystallization stage precipitates $Na_2CO_3 \cdot 10H_2O$ ("deca-crystals") by cooling the rich solvent to $\sim 10°C$. Lean solvent is recycled to the spray tower reservoir, while the deca-crystals are conveyed to a second crystallizer/leaching unit.

In this unit, operating at 100°C, Na_2CO_3 is recrystallized in an anhydrous form as a result of the higher temperature and, together with sodium trititanate, is conveyed to an oxycombustion gas-fired reactor, where the trititanate is converted to sodium pentatitanate at 860°C (reaction 7.22) and CO_2 is released. By using oxycombustion in this reactor, the exhaust gas comprises CO_2 plus steam; cooling to 25°C condenses the steam to produce make-up water, while the pure CO_2 is compressed for transportation and storage. After cooling, the pentatitanate is leached in the crystallizer and leaching unit, regenerating NaOH and sodium trititanate and completing the titanate cycle. The NaOH is recycled into the spray tower reservoir.

A 50% capture efficiency is estimated to be achievable for such a system with a 1.5 m high packed bed or a 100 m high spray tower. A commercial DAC system derived from this approach but now using a potassium-based solvent is under development by the Canadian company Carbon Engineering and partners, the engineering design of a 1 $MtCO_2$/year capture unit having commenced in 2019. This is further discussed, alongside other DAC methods, in Chapter 13.

References and resources

References

Anderson, C., Hooper, B., Qader, A., et al., 2014. Recent developments in the UNO MK 3 process—a low cost, environmentally benign precipitating process for CO_2 capture. Energy Proc. 63, 1773–1780.

Augustsson, O., Baburao, B., Dube, S., et al., 2017. Chilled ammonia process scale-up and lessons learned. Energy Proc. 114, 5593–5615.

Blanchard, L.A., Gu, Z., Brennecke, J.F., 2001. High-pressure phase behavior of ionic liquid/CO_2 systems. J. Phys. Chem. B 105 (12), 2437–2444.

Cota, I., Martinez, F.F., 2017. Recent advances in the synthesis and applications of metal organic frameworks doped with ionic liquids for CO_2 adsorption. Coord. Chem. Rev. 351, 189–204.

Garrabrant, K.A., Williams, N.J., Holguin, E., Brethomé, F.M., Tsouris, C., Custelcean, R., 2019. Energy-efficient CO_2 capture from flue gas by absorption with amino acids and crystallization with a bis-iminoguanidine. Ind. Eng. Chem. Res. 58 (24), 10510–10515.

Giacomin, C.E., Holm, T., Mérida, W., 2020. $CaCO_3$ growth in conditions used for direct air capture. Powder Technol. 370, 39–47.

Hornbostel, K., Nguyen, D., Bourcier, W., Knipe, J., Worthington, M., McCoy, S., Stolaroff, J., 2019. Packed and fluidized bed absorber modeling for carbon capture with micro-encapsulated sodium carbonate solution. Appl. Energy 235, 1192–1204.

Jessop, P.G., Heldebrant, D.J., Li, X., Eckert, C.A., Liotta, C.L., 2005. Reversible nonpolar-to-polar solvent. Nature 436 (7054), 1102-1102.

Kataoka, T., Orita, Y., Shimoyama, Y., 2022. CO_2 absorption and desorption using phase-separation gel. Chem. Eng. J. 428, 131126.

Kim, I., Vevelstad, S.J., Knuutila, H., et al., 2018. Solvent technology development in the EU project HiPerCap: from molecular simulation to technological roadmap for large scale demonstration. In: 14th Greenhouse Gas Control Technologies Conference Melbourne, pp. 21–26.

Lee, J., Kolawole, T., Attidekou, P., 2017. Carbon capture from a simulated flue gas using a rotating packed bed adsorber and mono ethanol amine (MEA). Energy Proc. 114, 1834–1840.

Legrand, L., Schaetzle, O., De Kler, R.C.F., Hamelers, H.V.M., 2018. Solvent-free CO_2 capture using membrane capacitive deionization. Environ. Sci. Technol. 52 (16), 9478–9485.

Madore, E., Fradette, S., 2017. CO_2 solutions: a biomimetic approach to mitigate CO_2 emissions—the use of carbonic anhydrase in an "industrial lung". In: De Gonzalo, G., de María, P.D. (Eds.), Biocatalysis: An Industrial Perspective. Royal Society of Chemistry, pp. 477–496.

McGurk, S.J., Martín, C.F., Brandani, S., Sweatman, M.B., Fan, X., 2017. Microwave swing regeneration of aqueous monoethanolamine for post-combustion CO_2 capture. Appl. Energy 192, 126–133.

Machida, H., Ando, R., Esaki, T., et al., 2018. Low temperature swing process for CO_2 absorption-desorption using phase separation CO_2 capture solvent. Int. J. Greenh. Gas Control 75, 1–7.

Molina-Fernández, C., Luis, P., 2021. Immobilization of carbonic anhydrase for CO_2 capture and its industrial implementation: a review. J. CO2 Util. 47, 101475.

Rahimi, M., Catalini, G., Hariharan, S., Wang, M., Puccini, M., Hatton, T.A., 2020. Carbon dioxide capture using an electrochemically driven proton concentration process. Cell Rep. Phys. Sci. 1 (4), 100033.

Rau, G.H., Knauss, K.G., Langer, W.H., Caldeira, K., 2007. Reducing energy-related CO_2 emissions using accelerated weathering of limestone. Energy 32 (8), 1471–1477.

Smith, K.H., Harkin, T., Mumford, K., et al., 2017. Outcomes from pilot plant trials of precipitating potassium carbonate solvent absorption for CO_2 capture from a brown coal fired power station in Australia. Fuel Process. Technol. 155, 252–260.

Stern, M.C., Simeon, F., Herzog, H., Hatton, T.A., 2013. Post-combustion carbon dioxide capture using electrochemically mediated amine regeneration. Energy Environ. Sci. 6 (8), 2505–2517.

Stolaroff, J.K., Ye, C., Oakdale, J.S., Baker, S.E., Smith, W.L., Nguyen, D.T., Spadaccini, C.M., Aines, R.D., 2016. Microencapsulation of advanced solvents for carbon capture. Faraday Discuss. 192, 271–281.

Tsubaki, S., Furusawa, K., Yamada, H., Kato, T., Higashii, T., Fujii, S., Wada, Y., 2020. Insights into the dielectric-heating-enhanced regeneration of CO2-rich aqueous amine solutions. ACS Sustain. Chem. Eng. 8 (36), 13593–13599.

Vericella, J.J., Baker, S.E., Stolaroff, J.K., et al., 2015. Encapsulated liquid sorbents for carbon dioxide capture. Nat. Commun. 6 (1), 1–7.

Wang, C., Jiang, K., Jones, T.W., Yang, S., Yu, H., Feron, P., Li, K., 2022. Electrowinning-coupled CO_2 capture with energy-efficient absorbent regeneration: towards practical application. Chem. Eng. J. 427, 131981.

Yang, J., Tan, H.Y., Low, Q.X., Binks, B.P., Chin, J.M., 2015. CO_2 capture by dry alkanolamines and an efficient microwave regeneration process. J. Mater. Chem. 3, 6440.

Zhang, S., Shen, Y., Wang, L., Chen, J., Lu, Y., 2019. Phase change solvents for post-combustion CO_2 capture: principle, advances, and challenges. Appl. Energy 239, 876–897.

Further reading

Dai, Z., Noble, R.D., Gin, D.L., Zhang, X., Deng, L., 2016. Combination of ionic liquids with membrane technology: a new approach for CO_2 separation. J. Membr. Sci. 497, 1–20.

Greer, A.J., Jacquemin, J., Hardacre, C., 2020. Industrial applications of ionic liquids. Molecules 25 (21), 5207.

Nessi, E., Papadopoulos, A.I., Seferlis, P., 2021. A review of research facilities, pilot and commercial plants for solvent-based post-combustion CO_2 capture: packed bed, phase-change and rotating processes. Int. J. Greenh. Gas Control 111, 103474.

Ochedi, F.O., Yu, J., Yu, H., Liu, Y., Hussain, A., 2020. Carbon dioxide capture using liquid absorption methods: a review. Environ. Chem. Lett. 1–33.

Pellegrini, L.A., Gilardi, M., Giudici, F., Spatolisano, E., 2021. New solvents for CO_2 and H_2S removal from gaseous streams. Energies 14 (20), 6687.

Wu, Y., Xu, J., Mumford, K., Stevens, G.W., Fei, W., Wang, Y., 2020. Recent advances in carbon dioxide capture and utilization with amines and ionic liquids. Green Chem. Eng. 1 (1), 16–32.

Projects and organizations

Baker Hughes; compact modular capture systems: www.bakerhughes.com/process-solutions/compact-carbon-capture.

Carbon Clean Solvents Ltd.; post-combustion capture solvent development: www.carboncleansolutions.com.

Carbon Engineering Ltd.; commercializing direct air capture): carbonengineering.com.

CarbonFree Chemicals; Skymine® developer: carbonfree.cc/our-technologies/#skymine.

C-Capture Ltd.; developing solvents for CO_2 capture, including the Drax BECCS project: www.c-capture.co.uk.

CO_2 Solutions Inc.; enzyme accelerated amine solvents: www.co2solutions.com.

HiPerCap; EU research project including enzyme catalyzed absorption, precipitating solvents and strong bicarbonate forming hindered amine solvents: www.sintef.no/hipercap.

iCAP bi-phasic solvent research project: http://cordis.europa.eu/result/rcn/156228_en.html.

IoLiTec Ionic Liquids Technologies GmbH; commercializing ionic liquid technologies: www.iolitec.de/en.

ION Engineering; developing absorption based capture technology: ioncleanenergy.com/our-technology/.

Mission Zero Technologies; DAC developer using water as solvent and MCDI regeneration: www.missionzero.tech/our-technology.

Mitsubishi Heavy Industries; KM CDR process: www.mhi.com/products/environment/carbon_dioxide_recovery_process_outline.html.

Shell Cansolv® process: www.shell.com/business-customers/global-solutions/shell-cansolv-gas-absorption-solutions.html.

Chapter 8

CO_2 adsorption

Steve Rackley

In contrast to absorption, in which the absorbed component (the sorbate) enters into the bulk of the solvent, adsorbed atoms or molecules (known as adparticles) remain on the surface of the sorbent. Similar to absorption, however, the bonding of the sorbate to the surface may be through either a chemical bond (chemical adsorption or chemisorption) or a weaker physical attractive force (physical adsorption or physisorption).

Gas separation or purification based on adsorption has a history of industrial application as long as that for absorption-based technologies, having been initially driven primarily by air purification applications. Adsorption processes using solid sorbents have a number of potential advantages when compared to absorption into liquid solvents, including a very wide range of operating temperatures, a lack of liquid waste streams, and in many cases solid wastes that are environmentally benign and pose few problems for disposal.

A rapid expansion in the application of adsorption as a gas separation technology has occurred in the past decades, due in part to the development of new sorbent materials. These sorbents can be combined with a broad spectrum of process options, including steady-state and cyclic processes, yielding a very fertile field for performance optimization and innovation in negative emissions applications.

8.1 Physical and chemical fundamentals

The most important characteristic of a sorbent is the quantity of sorbate that a given quantity of sorbent can hold, at the relevant operating temperatures and pressures. For applications such as CO_2 capture that require separation of one specific component from a mixed feed stream, the selectivity of the sorbent for that component is equally important and is determined by the relative adsorptive rates and capacities of the sorbent for the various components in the feed.

8.1.1 Physical adsorption thermodynamics

The process of physical adsorption is described by the Langmuir adsorption equation, or Langmuir isotherm, which can be derived from a simple model of the equilibrium between adsorption and desorption. For an adparticle of mass m (kg), at pressure P (Pa), the kinetic theory of gases gives the frequency of collisions, ν (s^{-1}), of the particle with the containing surface as follows:

$$\nu = P/(2\pi mkT)^{1/2} \tag{8.1}$$

where k is the Boltzmann constant (J/K) and T is the absolute temperature (K).

If θ is the fraction of adsorption sites that are occupied, and α is the probability of a particle being retained on the surface during any single collision, then the rate of adsorption (A) per unit surface area will be:

$$A = \alpha\ \nu(1 - \theta) \tag{8.2}$$

The rate of desorption, B, will be given by:

$$B = \beta\ \theta\ e^{-Q/RT} \tag{8.3}$$

Negative Emissions Technologies for Climate Change Mitigation. https://doi.org/10.1016/B978-0-12-819663-2.00012-5

where β is the rate constant for desorption, Q is the activation energy for desorption, equal to the heat of adsorption ($-\Delta H_{abs}$), and R is the gas constant.

At equilibrium, the rates of adsorption and desorption are equal, giving:

$$\alpha\, \nu(1 - \theta) = \beta\, \theta\, e^{-Q/RT} \tag{8.4}$$

which can be solved to give the equilibrium coverage θ_{eq}:

$$\theta_{eq} = KP/(1 + KP) \tag{8.5}$$

where

$$K = \alpha\, e^{Q/RT} / \beta\, (2\pi mkT)^{1/2} \tag{8.6}$$

For an n-component system, the equilibrium surface coverage of component i is given by

$$\theta_{eq,i} = K_iP_i \Big/ \left(1 + \sum_{j=1}^{n} K_jP_j\right) \tag{8.7}$$

Adsorption from a multicomponent feed is thus a competitive process, since the equilibrium surface coverage of one component will be reduced as the summation in the denominator of Eq. (8.7) is increased by the adsorption of other components.

The general form of the Langmuir isotherm is illustrated in Fig. 8.1 for two different temperatures. The temperature dependence of the Langmuir constant is $K \sim e^{Q/RT}\, T^{-1/2}$. As physical adsorption is always exothermic (i.e., ΔH_{abs} is negative), Q in Eq. (8.3) is always positive and, as a result, K decreases rapidly with increasing temperature, as shown by the two adsorption isotherms in the figure.

The temperature and pressure dependence of θ_{eq} are exploited as the basis of the two main adsorption-based gas separation processes: temperature swing adsorption (TSA) and pressure swing adsorption (PSA). In their simplest form, these processes rely on the preferential adsorption of one gas component from a mixed feed stream at a certain operating pressure and temperature (P_H, T_L), followed by desorption of the adsorbed component at either a reduced pressure or increased temperature (P_L, T_H) (or both in the case of a pressure and temperature swing). The differential adsorptive capacity of the sorbent between P_H and P_L, or T_L and T_H, is called the working capacity (Δ) of the sorbent, expressed as mol/mol or mol/kg of sorbent (Fig. 8.2).

Recent work on metal-organic frameworks (MOFs) has revealed strikingly different forms of adsorption isotherms compared to the classical Langmuir type illustrated, when CO_2 adsorption induces dynamic changes in the crystal structure of the MOF or "cooperative" adsorption behavior occurs between adjacent functionalized sites. These developments are further discussed in Section 8.4.

Fig. 8.2 shows the ideal case in which adsorption and desorption follow exactly the same isotherm. In practice, as shown in Fig. 8.3, many sorbent/sorbate combinations exhibit hysteresis, with the desorption isotherm falling above the adsorption isotherm as a result of capillary condensation. As sorbate loading increases, multiple layers of adparticles may

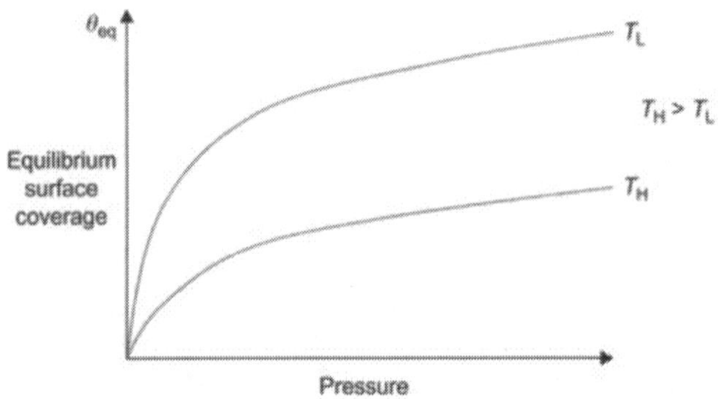

FIGURE 8.1 Langmuir adsorption isotherms versus temperature and pressure.

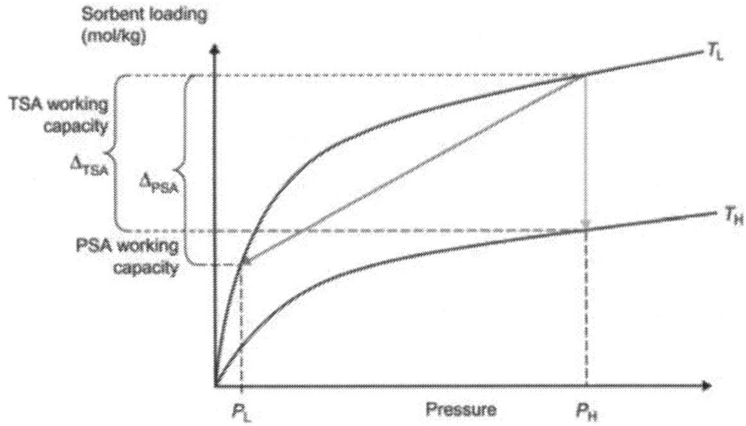

FIGURE 8.2 Sorbent working capacity for temperature swing and pressure swing adsorption.

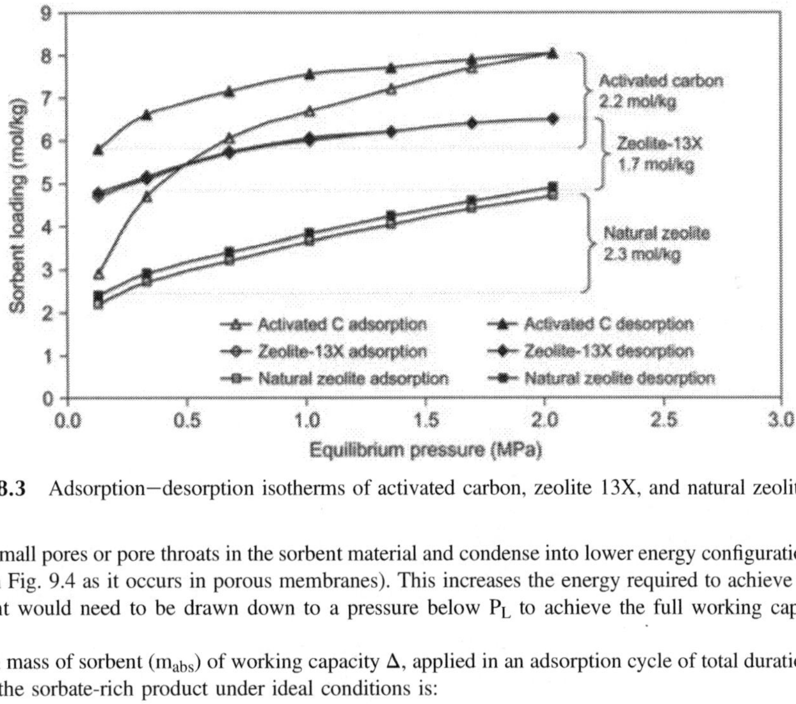

FIGURE 8.3 Adsorption–desorption isotherms of activated carbon, zeolite 13X, and natural zeolite at 25°C.

bridge across small pores or pore throats in the sorbent material and condense into lower energy configurations (the process is illustrated in Fig. 9.4 as it occurs in porous membranes). This increases the energy required to achieve full desorption, and the sorbent would need to be drawn down to a pressure below P_L to achieve the full working capacity shown in Fig. 8.2.

For a given mass of sorbent (m_{abs}) of working capacity Δ, applied in an adsorption cycle of total duration t_c, the rate of production of the sorbate-rich product under ideal conditions is:

$$q = m_{abs}\ \Delta/t_c \tag{8.8}$$

For a given process throughput rate, the sorbent mass and related plant size and cost are therefore minimized by shortening the duration of the cycle (Azarabadi and Lackner, 2019). Rapid cycling is also desirable since the release of the heat of adsorption results in the heating of the sorbent bed during adsorption. Excessive heating will reduce the capacity of the sorbent bed and conversely, if the bed is allowed to cool before pressure swing desorption occurs, the increase in adsorptive capacity of the sorbent as the temperature falls will reduce the working capacity of the bed. One approach to mitigate temperature fluctuations—incorporating phase change materials into the sorbent bed—is discussed in Section 8.4.

In general, TSA cycles require relatively longer cycle times for heating and cooling, leading to larger sorbent beds for a given throughput, while PSA cycles can be faster and employ smaller beds, as pressurization and depressurization can be achieved more rapidly.

Sorption—desorption characteristics

The desired properties of sorbents for CO_2 separation are summarized in Table 8.1, and the adsorption—desorption isotherms for three low-temperature CO_2 sorbents (activated carbon, natural zeolite, and zeolite 13X) at 25°C are shown in Fig. 8.3, which illustrate the desired properties listed in the table.

Activated carbon has the highest CO_2 adsorption across most of the pressure range, reaching 8 mol/kg at 2 MPa, as well as the steepest adsorption isotherm. However, the desorption isotherm shows a high degree of hysteresis, reducing the working capacity over the pressure range shown to ~2.2 mol/kg. Zeolite 13X shows an intermediate adsorption capacity, with very low hysteresis, but the shallow gradient of the isotherm results in a lower working capacity of 1.7 mol/kg. Natural zeolite has the lowest adsorption capacity, little more than half that of activated carbon at 2 MPa, but the steeper isotherm and relatively low hysteresis result in a working capacity over this pressure range of 2.3 mol/kg, marginally higher than activated carbon.

Chemical and physical CO_2 adsorbents

Due to the weaker binding forces and resulting lower heat of adsorption, physical sorbents tend to find application at low temperatures, while chemical sorbents, with higher heats of adsorption, are able to sustain high capacity at higher temperatures. The division between physical and chemical sorbents is therefore also in general a division between low and high operating temperatures, although there are exceptions such as surface-modified porous media—a class of low-temperature chemical sorbents. A wide variety of sorbents can be used for low-temperature CO_2 capture, including those summarized in Table 8.2. These low-temperature sorbents are typically applied at temperatures from ambient up to ~100°C, in a few cases up to a few hundred degrees, since their capacity drops to very low levels at higher temperatures.

A range of sorbents has also been developed with high capacities at the operating temperatures typical of hot flue gas applications (400—600°C). Table 8.3 summarizes a number of these high-temperature sorbents, while further novel materials under development are described in Section 8.4.

8.1.2 Chemical sorbents

Metal oxide sorbents

Several of the absorbents described in this Chapter have also been applied in the solid state as chemical adsorbents, the most common examples being the carbonation of metal oxides, such as calcium, sodium, and potassium oxide, as well as the further carbonation to produce bicarbonates:

TABLE 8.1 Desired properties of CO_2 sorbents.

Property	Description
Preferential CO_2 adsorption	High selectivity for CO_2 relative to other relevant gases (N_2, CO, CH_4) is essential for an efficient adsorption/desorption cycle.
Low heat of adsorption	Low heat of adsorption ensures that the drop of pressure or increase in temperature required in the desorption stage does not result in a high energy penalty.
High sorbent working capacity	Increased working capacity reduces the volume of sorbent required for a given throughput, reducing sorbent bed and related equipment size, capital costs, and energy requirements.
Low adsorption hysteresis	Hysteresis of the adsorption isotherm results in an increased pressure or temperature swing to achieve the same sorbent working capacity, increasing the energy penalty.
Steep adsorption isotherm	A sorbent with a steep adsorption isotherm delivers a given working capacity with the lowest pressure or temperature swing.

TABLE 8.2 Low temperature CO_2 sorbents.

Sorbent	Description
Activated alumina	Nonreactive synthetic amorphous sorbents with low heats of adsorption produced from aluminum trihydrate. The sorbent can be a beaded material or powder, and surface areas can be controlled during manufacturing to tailor the sorbent to various applications.
Activated carbon	Carbonaceous crystalline sorbents are available with a wide range of properties, depending on the raw material and activation method. Activated carbons are used in a wide range of purification processes, including the purification of drinking water.
Ion-exchange resins	Ionic polymers typically prepared as macroporous beads, in which either the anion or the cation is bound to the resin structure while the counter-ion is free. Ions (e.g., CO_3^{2-}) with the same charge as the free counter-ion can be exchanged from a solute onto an anion exchange resin.
Metal-organic frameworks (MOFs)	Networks of metal–ions or metal–ion clusters connected into an extended nano-porous structure by organic linkers. Selection of different combinations of metal ion + linker allows control of the MOF pore size and shape, enabling tuning of sorption capacity and selectivity.
Surface-modified porous media	Silica, carbon- or polymer-based meso-, micro-, or nanoporous media with high specific areas, modified by the incorporation of functional groups such as amines and related organic compounds.
Zeolites	Naturally occurring and synthetic microporous minerals composed of hydrated sodium, potassium, calcium, or magnesium aluminosilicate that are able to trap adparticles within their open crystal structure.

TABLE 8.3 High temperature CO_2 sorbents.

Sorbent	Description
Metal oxides (e.g., CaO)	High-temperature (600–1000°C) chemical sorbents, transformed to carbonates and bicarbonates and regenerated by calcination.
Hydrotalcites	Naturally occurring anionic clays that become CO_2 sorbents in the temperature range ~200–400°C, displaying superior stability than CaO but lower capacity and slower reaction kinetics.
Lithium zirconate (Li_2ZrO_3)	Mid- to high-temperature (600–750°C) chemical sorbent. Capacity can be significantly increased by adding potassium carbonate (K_2CO_3).

$$CaO_{(s)} + CO_{2\,(g)} \leftrightarrow CaCO_{3(s)} \tag{8.9}$$

$$Na_2O_{(s)} + CO_{2(g)} \leftrightarrow Na_2CO_{3(s)} \tag{8.10}$$

The high temperatures required for the calcination step make these sorbents particularly suited to flue gas or syngas processes, typically taking place at 400–600°C. However, at temperatures above 533°C (the Tammann temperature for $CaCO_3$) sintering occurs, leading to loss of adsorption capacity over repeated cycles. To address this problem, novel CaO-based sorbents are being developed, the most promising being certain calcium aluminates—solid solutions of CaO and Al_2O_3.

Alkali metal carbonate sorbents

The carbonates of alkali metals such as Al, K, and Na, react reversibly with CO_2 and water to form bicarbonates, the carbonate being regenerated by heating.

The reaction of sodium carbonate (Na_2CO_3) to form sodium bicarbonate (Na_2HCO_3) or Wegscheider's salt ($Na_2CO_3 \cdot 3NaHCO_3$) has been investigated for CO_2 capture from combustion flue gases. Carbonation and regeneration occur as a result of the following reactions:

$$Na_2CO_3 + CO_2 + H_2O \leftrightarrow 2NaHCO_3 \tag{8.11}$$

$$Na_2CO_3 + 0.6\ CO_2 + 0.6\ H_2O \leftrightarrow 0.4\ (Na_2CO_3 \cdot 3NaHCO_3) \tag{8.12}$$

The carbonation reactions occur in the temperature range 60–70°C and are exothermic, with a heat of absorption of approximately $\Delta H_{abs} = -136$ kJ/molCO_2 (3.1 GJ/tCO_2). Regeneration of the sorbent is then achieved in the reverse reactions in the temperature range 120–200°C. The disadvantages of these sorbents are the relatively low loading capacity (3–11 wt%) and the high desorption energy requirement.

Supported amine sorbents

Amines and related organic absorbents have also been applied as adsorbents, by attaching the functional groups onto the surface of high-specific-area solids such as meso- or microporous silica, alumina, or polymer-based materials to form supported amine sorbents or amine-solids. The rate and efficiency of adsorption are increased as a result of the high surface area of the supporting material, which allows easy access of CO_2 to the amine or other functional groups.

The adsorption isotherms for amine-solids (Fig. 8.4) also differ from the typical Langmuir form shown in Fig. 8.1 because sorbent loading is the result of both chemical adsorption onto the amine groups and physical adsorption onto the surface of the supporting material (Gelles et al., 2020). Under optimal conditions, synergy between the support and amine can exceed the additive contributions indicated in the figure as a result of the increased density of adsorption sites that are exposed to the sorbate (Xu et al., 2002). The choice of support material is therefore an important factor in determining the overall sorbent performance, and the optimization of these synergistic effects is an active research area.

Sorbent loading significantly higher than aqueous amine solutions and a regeneration energy roughly 1/3 of the 30% MEA benchmark have also been reported for supported amine sorbents. Oxidative degradation and loss of the sorbent by evaporation are minimized as a result of the surface binding, while the operational problems typical of aqueous solvent systems, notably corrosion, foaming, and contaminated wastewater, are eliminated in the dry gas–solid system.

A supported amine sorbent known as HSC$^+$, using polyethyleneimine bonded to acrylic-based polymer beads with regeneration by vacuum swing desorption, has been used since the early 1990s for CO_2 removal in space shuttle life-support systems, and has also been investigated for wider CO_2 capture application, including in direct air capture (Satyapal et al., 2001). A variety of supported amine sorbents have also been developed for direct air capture, including nanofibrillated cellulose, produced from biogenic raw material (Fig. 8.5), which has been commercialized by Climeworks in combination with a TSA regeneration cycle (Gebald et al., 2011, 2014), and a polypropylene resin supported sorbent with a moisture swing regeneration, initially developed by Klaus Lackner at Arizona State University and currently being commercialized by Carbon Collect (Wang et al., 2011; see also Chapter 14).

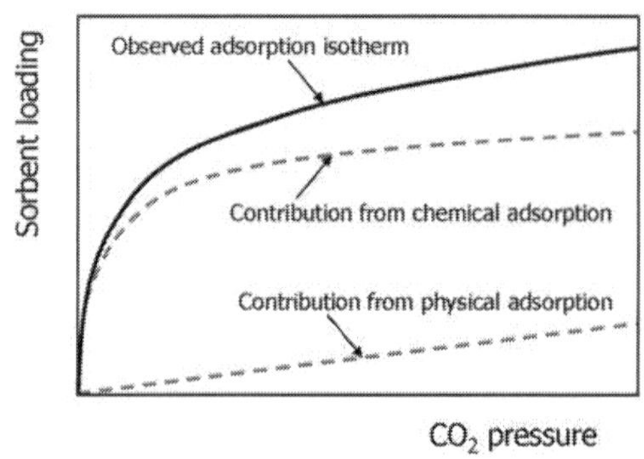

FIGURE 8.4 Physical and chemical adsorption uptake on a supported amine sorbent.

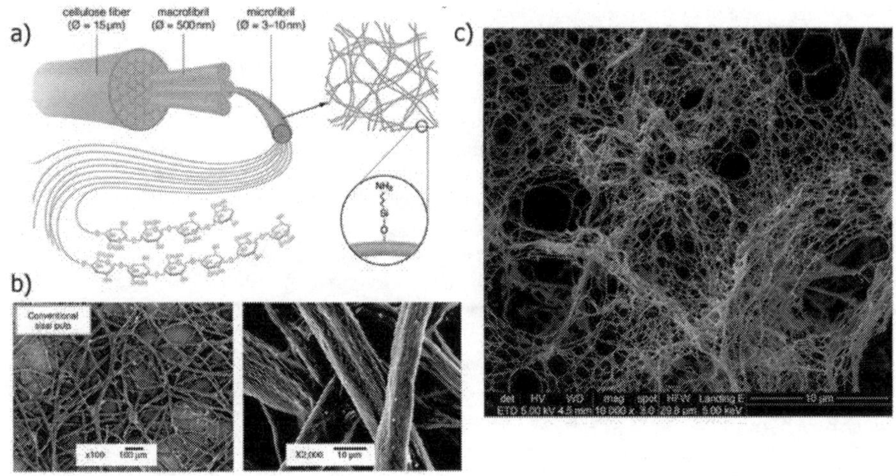

FIGURE 8.5 Amine functionalized nanofibrillated cellulose sorbent; (A) physical configuration within macro- and microfibrils, (B) SEM images of cellulose fibers, (C) SEM image of nanofibrillated cellulose. *(A) From Gebald et al. (2014), with permission; (B) From Ardanuy Raso et al. (2012), with permission; (C) Courtesy; Nanografi.*

Hydrotalcites

Hydrotalcites (HTCs) are a class of high-temperature chemical sorbents that have been widely investigated for application in sorption-enhanced reactions (Section 8.4.2). Also known as layered double hydroxides or Feitknecht compounds, hydrotalcites are a family of naturally occurring anionic clays with the general formula:

$$M^{2+}{}_{1-x}\, M^{3+}{}_{x}\, (OH)_2(A^{n-})_{x/n} \cdot mH_2O$$

Here M^{2+} and M^{3+} are divalent (e.g., Mg, Ni, Zn, Cu, Mn) and trivalent (e.g., Al, Fe, Cr) metal ions, most commonly magnesium and aluminum respectively, A^{n-} is an anion (most commonly $CO_3{}^{2-}$, but also $SO_4{}^{2-}$, $NO_3{}^{2-}$, Cl^-, OH^-), and the value of x is 0.1–0.5. The metal ions are held in an octahedrally coordinated hydroxide layer, similar to the mineral brucite, while the anions and water of hydration occur in an anionic interlayer (see Fig. 8.6). The net positive charge of the metal ion layer is compensated by the net negative charge of the anion–water interlayer.

The mineral hydrotalcite is the most common of the class of HTCs and has the formula:

$$Mg_{0.75}Al_{0.25}(OH)_2(CO_3{}^{2-})_{0.125} \cdot 0.5H_2O$$

The thermal decomposition (calcination) of HTC is a complex three-stage process, starting in the temperature range from ambient up to ~300°C, with the dehydration of water molecules held in the anion–water interlayer. At temperatures up to 700°C, this is followed by dehydroxilation, in which OH groups plus oxygen from the metal ion layer are released as H_2O, while two oxygen atoms from a carbonate ion are included in the metal ion layer. This leads to the start of

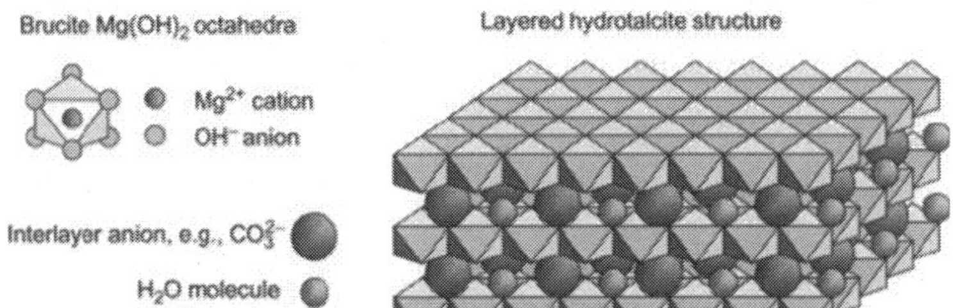

FIGURE 8.6 Layered structure of hydrotalcite.

decarbonation and the release of CO_2. Up to about 500°C, the so-called hydrotalcite memory effect ensures that the layered structure is restored on rehydration, either in a carbonate solution or by steam in the presence of CO_2. Above 700°C, decarbonation of the interlayer is completed, resulting in the final decomposition products MgO and $MgAl_2O_4$ in the case of mineral hydrotalcite.

The calcination process is accompanied by a rapid increase in the specific surface area as the progressive loss of the anion–water causes the clay layers to separate. The resulting high density of strong basic sites on the layer surfaces make HTCs effective as high-temperature CO_2 sorbents, with the optimum calcination temperature for a given HTC being determined by the balance between surface area and active site density.

The CO_2 loading capacity of HTC can be significantly increased by impregnating (or promoting) the mineral with 10–20 wt% of potassium carbonate (K_2CO_3) and also by the presence of steam, although loading capacity and carbonation and decarbonation rates remain inferior to CaO. It is mainly in the area of sorbent stability that HTCs performance exceeds that of the simpler metal oxides. Ongoing R&D efforts focus on the development of new HTCs combining higher sorption capacities and kinetics with good stability through multiple sorption/desorption cycles under operation conditions encountered in the sorption-enhanced steam reforming process.

8.1.3 Physical sorbents

Two classes of physical sorbents that have received considerable attention for postcombustion CO_2 capture are zeolites and MOFs.

Zeolites

Zeolites (from the Greek for "boiling stone" due to their release of water on heating) are naturally occurring and synthetic microporous minerals composed of hydrated sodium, potassium, calcium, or magnesium aluminosilicate. The open crystal structure of zeolites incorporates cages of various sizes within which adparticles can be trapped. This is illustrated in Fig. 8.7, which shows the structure and structural elements of zeolites Sodalite, Linde A and Faujasite.

Zeolites have high adsorption capacity at low temperatures (<100°C) and partial pressures (<1 bar CO_2) and exhibit very rapid adsorption, in many cases reaching equilibrium in a few minutes, their main disadvantage being structural instability in the presence of water.

Although zeolite 13X and 5A are the most commonly investigated, a range of different structures, and trap sizes, can be synthesized at low cost, and their CO_2 selectivity and adsorption capacity can be improved by surface treatment, for example by amine impregnation. Computational screening methods, including state-of-the-art molecular simulation combined with process modeling tools, can be used to rapidly screen sorbent materials for CO_2 capture applications.

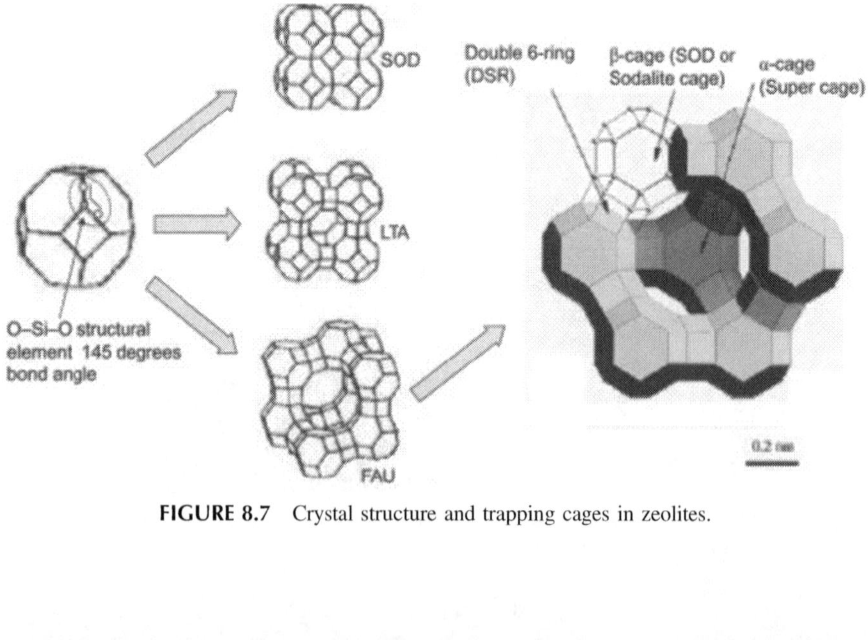

FIGURE 8.7 Crystal structure and trapping cages in zeolites.

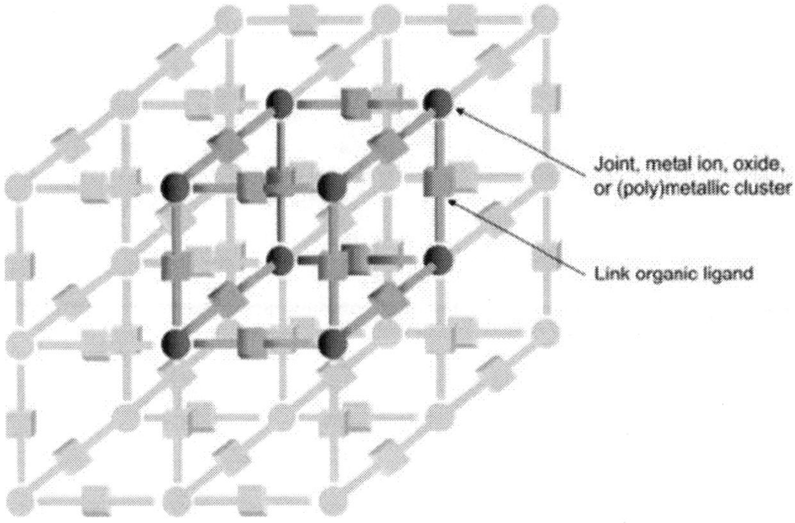

FIGURE 8.8 Example of a simple cubic MOF structure.

Subramanian Balashankar and Rajendran (2019) screened 120,848 existing and theoretical zeolites, the 24 top performers in terms of energy consumption having CO_2 heats of adsorption in the ranges of 32–42 kJ/mol, equivalent to 0.73–0.96 GJ/tCO_2. This is a factor of 3–4 lower than the regeneration energy requirement of benchmark solvents, and demonstrates the significant advantage of physical over chemical methods.

Metal-organic frameworks

MOFs, are crystalline nanoporous materials composed of networks of metal–ions or clusters of metal–ions connected into extended nanoporous structures by organic linkers—known as ligands. Some typical MOF structures are illustrated in Fig. 8.8. Selection of different metal ion + ligand combinations allows the pore size and shape of the framework to be controlled, enabling tuning of sorption capacity and selectivity, so that this class of materials has considerable potential for efficient CO_2 capture.

Zeolitic imidazolate frameworks (ZIFs), are a type of MOF composed of transition metal ions (e.g., Fe, Co, Cu, Zn) connected by imidazolate ($C_3H_4N_2$) linkers. Because the metal-imidazole-metal angle is very close to the Si-O-Si angle of 145 degrees, as shown in Fig. 8.7, ZIFs form lattice structures that are topologically isomorphic to zeolites. Fig. 8.9 shows

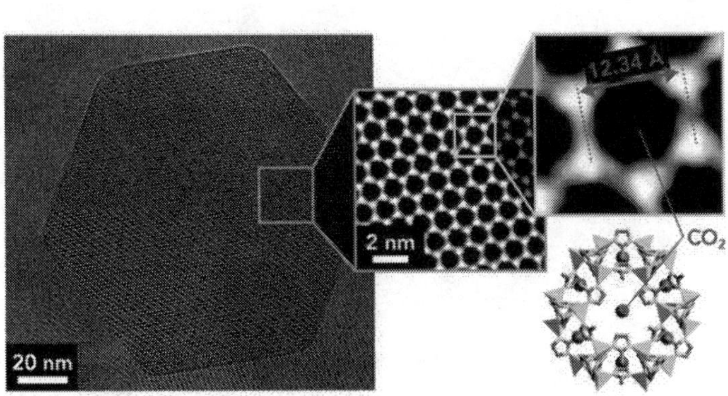

FIGURE 8.9 Host–guest structures within ZIF-8. (A) CO_2-filled ZIF-8 particle, (B) denoised image of boxed region, (C) magnified image of a single ZIF-8 unit cell, (D) simulated structure of ZIF-8 with predicted binding site of CO_2. *From Li et al. (2019) with permission.*

a near-atomic-resolution micrograph of the host-guest interaction between ZIF-8 and CO_2, obtained by Li and coworkers using cryogenic electron microscopy (Li et al., 2019).

As well as being studied as selective gas sorbents, ZIFs have also been investigated for applications as reaction catalysts, and in a hybrid absorption plus adsorption system combining a glycol solvent with a ZIF sorbent (Liu et al., 2014).

As CO_2 sorbents, MOFs have high sorption capacity at low temperatures and high CO_2 partial pressure, but generally perform less well at high temperature, low partial pressure, and in the presence of impurities. In particular, water can disrupt the MOF lattice structure by displacing the ligands, although the incorporation of MOFs into a hydrophobic polymer has been shown to slow down degradation under humid conditions (Sadiq et al., 2020).

8.2 Adsorption process configurations

A variety of different physical configurations can be used in the sorption–desorption process, the main distinctions being between sorbent bed structures which include fixed, moving, rotating, and fluidized systems (Dhoke et al., 2021).

8.2.1 Fixed adsorption bed systems

In simple fixed bed systems the sorbent is held in a vessel with feed gas entering at one end and light product, from which the sorbate has been removed, produced at the other. Fig. 8.10 illustrates the fixed bed sorbent system employed in the first Climeworks-CarbFix DAC-mineralization pilot in Iceland, discussed further in Chapter 13.

When a bed is first put into feed mode, transfer of the sorbate from the feed gas to the sorbent starts at the feed end of the bed. Fig. 8.11 illustrates the progressive movement of this mass transfer zone through a fixed bed.

The figure shows that throughout the adsorption period, only a small part of the sorbent inventory is actively adsorbing, while the majority of the sorbent is either fully loaded, and therefore awaiting regeneration, or not yet contacted by the sorbate. The release of the heat of adsorption within the localized mass transfer zone also means that heat transfer requires careful attention in fixed bed design. This is needed to avoid overheating of the sorbent, leading to a reduction in carrying capacity, or to a waste of the liberated heat, which would increase the energy penalty for regeneration.

8.2.2 Moving adsorption bed systems

Moving bed processes overcome these difficulties, as each parcel of sorbent can be more promptly exported from the adsorption bed for regeneration once it has been fully loaded, while mass transfer between sorption and desorption zones provides an extra degree of freedom for overall heat management within the process. The moving bed concept was initially patented in the 1920s as a method of separating components from coal gasification syngas and was renamed hypersorption when renewed interest in the late 1940s led to the construction of a number of commercial moving bed units. As illustrated in Fig. 8.12, the moving bed or hypersorption process is functionally equivalent to a distillation tower, and one of the feature of the process is that the flexible operating conditions—temperature, pressure, and sorbent circulation rate—can be tuned to perform sharp separations.

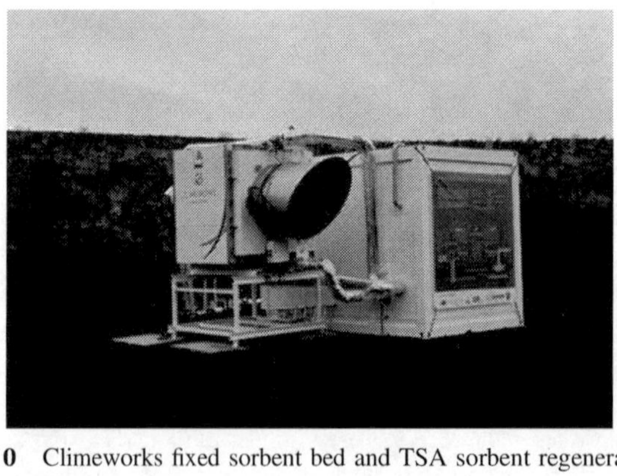

FIGURE 8.10 Climeworks fixed sorbent bed and TSA sorbent regeneration system.

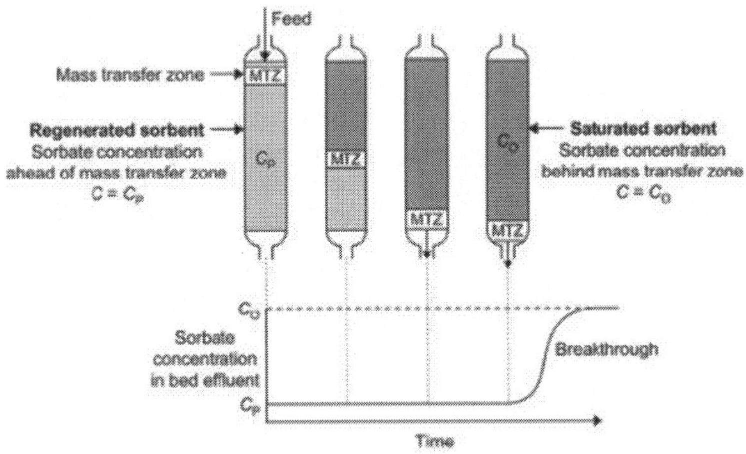

FIGURE 8.11 Movement of the mass transfer zone through a fixed sorbent bed.

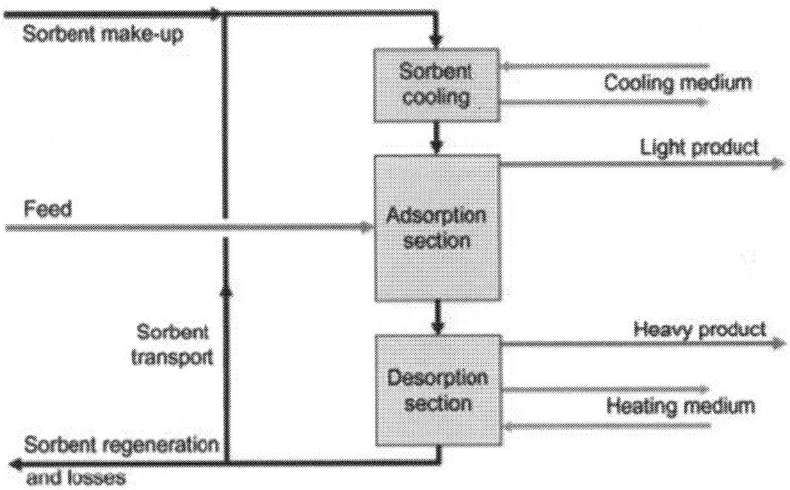

FIGURE 8.12 Schematic of hypersorption gas separation unit.

Feed is introduced into the middle of the column and contacts the sorbent, which is moving down under gravity. More-adsorbing components in the feed move down with the sorbent, while less-adsorbing components move up and exit the top of the column. The decreasing temperature toward the top of the column, controlled by cooling the circulating regenerated sorbent, increases the adsorptive capacity of the sorbent and improves separation.

Heat is provided to the bottom of the column by a heated reflux of the heavy product, equivalent to a distillation column reboiler. The increasing temperature toward the bottom of the column progressively releases the sorbate, with the potential for recovery of intermediate as well as heavy product streams. Reflux of product side streams back into the hypersorption columns can be used to achieve increased purity of these products, while multistage hypersorption, using two or more such columns in series, can also be used for separating multiple feed components.

A pilot scale (5 tCO_2/d) moving bed postcombustion CO_2 capture system, using amine impregnated porous sorbent particles, has been demonstrated by Kawasaki Heavy Industries (Okumura et al., 2017). This system has the advantage of

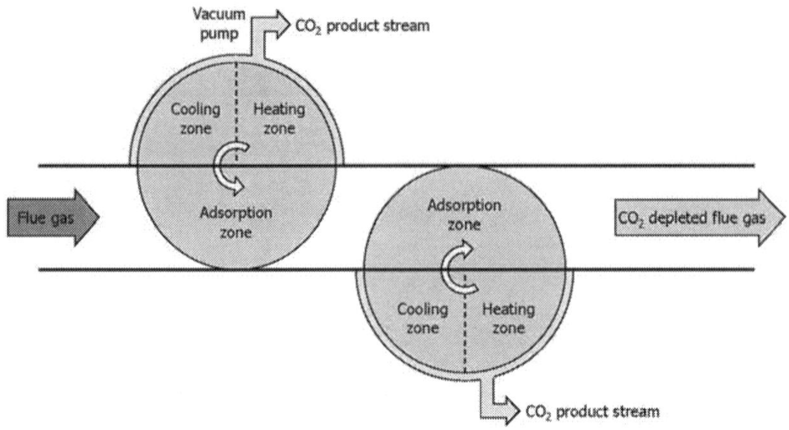

FIGURE 8.13 Rotating packed bed adsorption and desorption scheme. *After Gupta and Ghosh (2015).*

using low temperature (60°C), waste steam for regeneration, and pilot results indicate a low regeneration energy penalty of 1.3 GJ/tCO_2.

An alternative form of moving bed that has been proposed for CO_2 adsorption applications is the rotating packed, which has also been studied for absorption using liquid solvents. In this configuration, shown schematically in Fig. 8.13, circular sorbent disks rotate between an adsorption chamber where they are exposed to flue gases and a desorption chamber where the sorbent is regenerated under a pressure-temperature swing. Conceptual studies of this system, have been reported using activated carbon and zeolite 13X as sorbents and configured for postcombustion CO_2 capture from a 500 MW power plant (Gupta and Ghosh, 2015).

8.2.3 Simulated moving beds

A simulated moving bed is an intermediate step between fixed and moving beds in which sorbent in a fixed bed is cycled through the flows, temperatures, and pressures as experienced by a parcel of sorbent circulating in a moving bed process. This is illustrated in Fig. 8.14, where a moving bed system is broken into four process steps. The fixed bed here undergoes the same adsorption–regeneration process experienced by a parcel of sorbent moving through the hypersorption column:

Zone	Process step
A	Sorbent cooling by chilled light product reflux
B	Feed and light product recovery
C	Heavy product recovery
D	Further heavy product recovery by temperature swing

This scheme is conceptually identical to the cyclic schemes described in Section 8.4.1, where each individual fully loaded bed is taken through a sequence of regeneration steps and then rejoins the cascade of adsorbing beds.

A moving bed can also be simulated using a long fixed bed by sequentially moving the inlet and outlet feed and regeneration points along the bed. This technique is applied to small-scale fractionation for chemical analysis and is known as moving-port chromatography.

8.2.4 Fluidized beds

Fluidized beds are the most common form of moving bed adsorber, and have been developed to a high state of refinement due to their widespread application in many industrial processes, including power generation combustion and gasification. Fig. 8.15 illustrates a fluidized bed absorber.

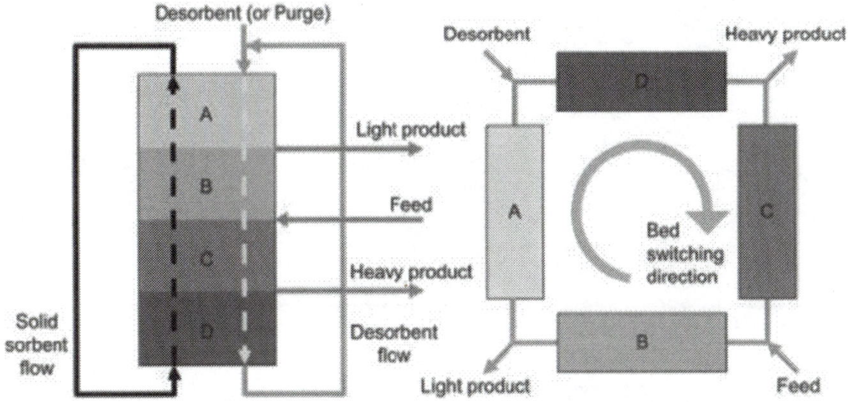

FIGURE 8.14 Simulated moving bed using cascaded fixed beds.

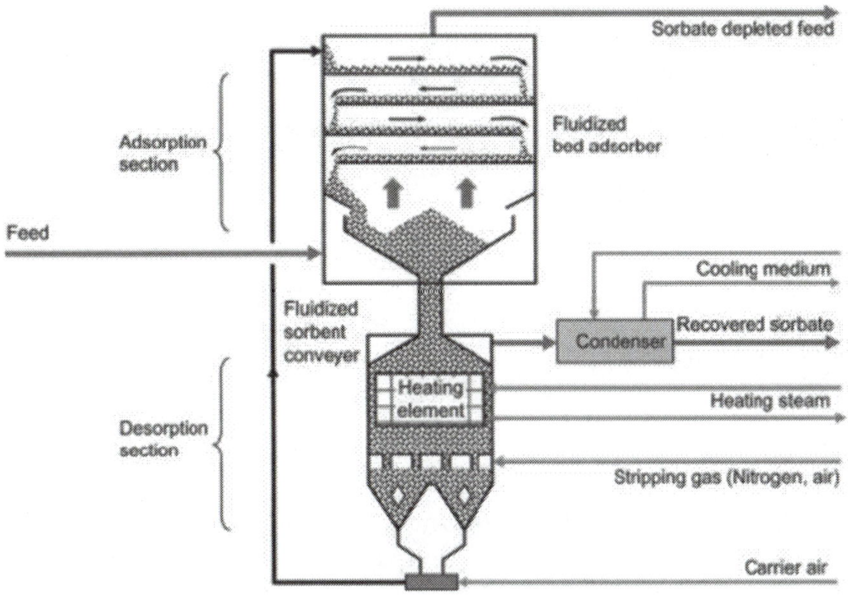

FIGURE 8.15 Fluidized adsorption bed configuration.

Fluidized beds introduce additional technical requirements, including the need for the sorbent particles to resist attrition, and for equipment to circulate sorbent and to remove attrition products.

Chemical looping

Chemical looping is one example of a fluidized bed system in which the sorbent is circulated between interconnected fluidized bed reactors where adsorption and desorption take place. The concept is illustrated in Fig. 8.16, which shows schematically a chemical looping system for postcombustion CO_2 capture, first proposed by Ishida and Jin (1994).

The chemical looping concept has also been extensively investigated and pilot tested for use at the combustion stage, as a variant of oxyfueling, and chemical looping combustion (CLC) has emerged over the last decade as a leading clean combustion technology. In CLC, a metal oxide sorbent carries oxygen to a fuel reactor where combustion takes place; the reduced oxygen carrier is then conveyed to an air reactor where it is regenerated (oxidized in air), closing the loop.

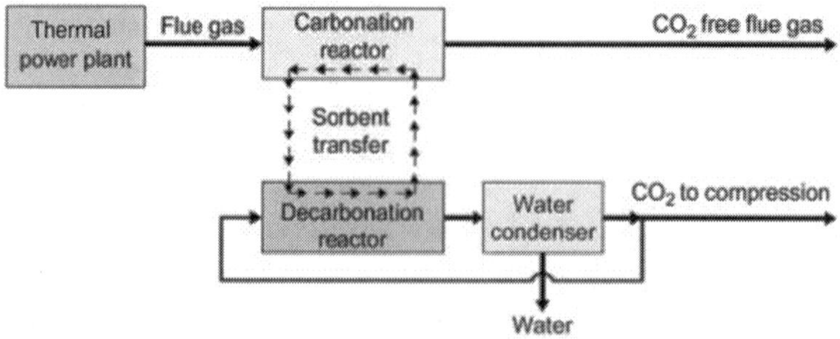

FIGURE 8.16 Schematic chemical looping system for postcombustion CO_2 capture.

Chemical looping combustion with oxygen uncoupling (CLOU), first proposed by Mattisson et al. (2009), is a related process in which fuel is combusted in oxygen released as a gas phase from the oxygen carrier rather than by directly reducing the carrier. CLOU requires an oxygen carrier that reacts with oxygen in the air reactor and then decomposes to a reduced metal oxide and gas-phase oxygen in the fuel reactor.

CuO (copper (II) or cupric oxide) is a common carrier that releases oxygen at ca. 900°C to form Cu_2O (copper (I) or cuprous oxide). Oxygen uncoupling is beneficial in the case of solid fuel combustion (e.g., biomass), as fuel components such as char and volatile compounds are more efficiently oxidized in gaseous oxygen than by direct reaction with the metal oxide carrier.

Current development of chemical looping systems is further discussed in Section 8.4.5.

8.3 Sorbent regeneration processes

Sorbent regeneration can be achieved through a variety of different processes which, depending on the type of sorbent being used, will involve a change in one or more physical conditions, with temperature, pressure, or humidity being the most commonly applied.

8.3.1 Temperature swing adsorption

The use of CaO as a regenerable sorbent for CO_2 is an example of a TSA process that was first proposed in the 19th century. In the temperature range 600–800°C, the carbonation reaction proceeds rapidly and high sorbent capacity can be achieved, while the familiar calcining reaction, releasing CO_2, is favored at >900°C. A pilot-scale postcombustion separation plant has been demonstrated, using interconnected fluidized bed carbonation and calcining reactors, as described in Section 8.4.5.

TSA processes being investigated for CO_2 capture typically operate at much lower temperatures (as low as 80°C), and are favored in situations where waste heat can be utilized for the desorption step. The Climeworks capture unit shown in Fig. 8.10 is one such example, using waste heat from a geothermal power station for desorption in a fixed bed TSA configuration, and the use of low-grade heat from solar heated water has also been reported (Zhao et al., 2018).

One disadvantage of TSA is the relatively long time needed to heat the sorbent bed for desorption and then cool the bed for the next adsorption step; extended cycle times lead to larger adsorption beds for a given capture rate and to higher capital costs, and the additional energy required to heat the support material and other system components is a further inefficiency. TSA requires efficient heat transfer within the sorbent medium, which can be problematic for low thermal conductivity materials such as MOFs. A number of approaches have been investigated to overcome this conductivity challenge, including the use of a hot CO_2 product as a purge gas stream (Ntiamoah et al., 2016), the use of light (Lyndon et al., 2015), and magnetic induction heating (Sadiq et al., 2020).

The performance of a TSA cycle can be enhanced by applying a partial vacuum to the sorbent in a temperature–vacuum swing adsorption (TVSA) cycle and/or by using a steam purge in a steam-assisted TSA or S-TSA cycle. In TVSA, reducing the gas pressure in the desorption cycle will increase CO_2 recovery, but at the cost of higher electrical energy requirements. Steam-assisted TSA was first proposed by Li and coworkers (Li et al., 2010) as a way to enhance desorption kinetics although at an additional thermal energy cost. In combination with TVSA a steam purge enables higher CO_2 production at milder vacuum pressures, resulting in a reduction in specific energy requirement. S-TSA systems have been widely investigated for direct air capture (see e.g., Zhu et al., 2021).

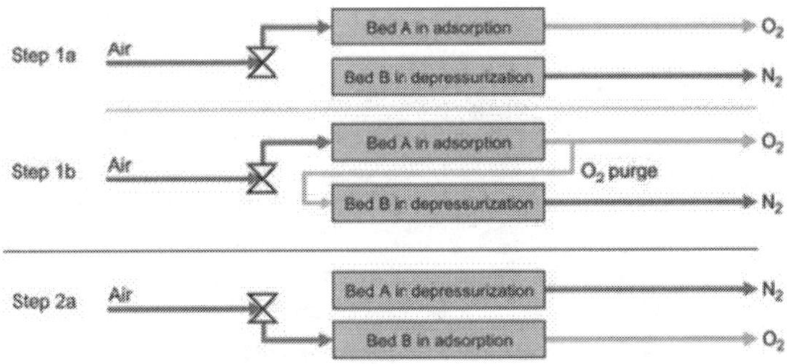

FIGURE 8.17 Skarstrom PSA cycle.

TSA using MOF-based sorbents has also been identified as a promising technique to carbon capture from internal combustion engine-driven vehicle exhaust, with power for the system recovered from the ~65% of fuel energy present in cooling system and exhaust waste heat. In combination with an exhaust heat-driven organic Rankine cycle to generate mechanical work for compression, the envisaged system could capture 90% of the CO_2 from exhaust gases without an energy penalty (Sharma and Maréchal, 2019) and could find applications in the marine and road freight transport sectors (García-Mariaca and Llera-Sastresa, 2021).

8.3.2 Pressure swing adsorption

The pressure swing adsorption concept was developed by the Esso Research and Engineering Company and published in a 1960 patent that disclosed a dual bed, two-step cycle known as the Skarstrom cycle. The process is illustrated in Fig. 8.17 as applied to air separation.

In this air separation example, feed air is pretreated by compression, drying, and filtering and then flows into one of two parallel adsorption beds (A). Here nitrogen is adsorbed and oxygen passes through the bed and is recovered as the light product stream. While the first bed is operating in this mode, the sorbent material in the second bed (B) is being regenerated by releasing the pressure, recovering nitrogen as the heavy product stream. After depressurization, nitrogen recovery is increased by purging bed B with a fraction of the bed A product stream (light product reflux), the cycle ending before the purge stream breaks through the bed.

In the second part of the cycle, the functions of the two beds are reversed, with repressurization and adsorption taking place in bed B and regeneration (N_2 release and purge) in bed A. The process can be represented by the scheme shown in Fig. 8.18, in which the upper schematic shows (from left to right) the successive cycle steps for bed A, while the table shows how these steps are phased over the two beds. Arrows indicate increasing or decreasing pressure in the bed. This format will be useful in describing more complex PSA cycles in the following sections.

The nomenclature of this schematic, including terms that will appear in the later discussion, is summarized in Table 8.4.

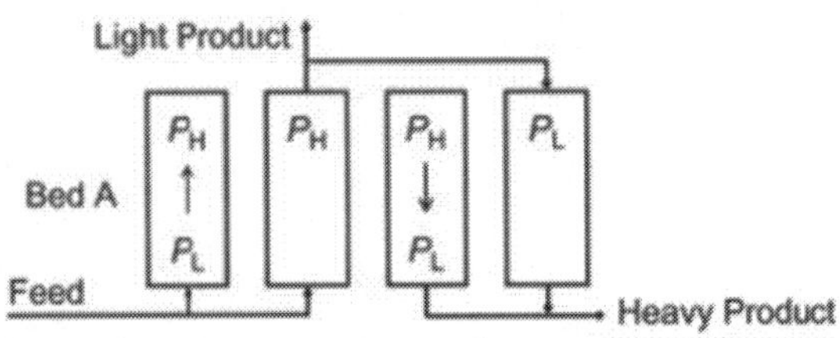

Step		1a	1b	2a	2b
Bed	A	FP	F	CnD	LR
	B	CnD	LR	FP	F

FIGURE 8.18 Schematic and cycle sequence chart of the two bed Skarstrom PSA cycle.

TABLE 8.4 PSA process schematic nomenclature.

Symbol	Description
CnD	Counter-current depressurization
CoD	Cocurrent depressurization
E	Evacuation (VSA)
$EQ_{1,2,3}$	Pressure equalization step 1, 2, 3
F	Feed
FP	Feed pressurization
HP	Heavy product (sorbate-rich)
HR	Heavy product reflux
LP	Light product (sorbate-lean)
LPP	Light product pressurization
LR	Light product reflux
P	Purge
P_H	High operating pressure
P_I	Intermediate pressure
$P_{1,2,3}$	Intermediate pressures 1, 2, 3
P_L	Low operating pressure

A disadvantage of this simple cycle is that the light product reflux step (the O_2 purge in this case) causes a dilution of the heavy (sorbate rich) component relative to the initial feed, and as a result, it is difficult to achieve high purity for the heavy product. Several improvements have been made to the Skarstrom cycle, including initial pressurization using light product, cocurrent depressurization after the initial feed step, one or more pressure-equalization steps, and a heavy product reflux step, as described in Table 8.5. While the addition of reflux steps increases product purity and recovery, either the flow rate or duration of the reflux step must be controlled to avoid the breakthrough of the reflux into the other product stream, reducing the throughput of the overall system.

TABLE 8.5 Skarstrom cycle improvement steps.

Cycle step	Description
Light product pressurization	Light product purity can be improved by using a stream of light product rather than raw feed to bring the sorption bed up to high pressure before the main feed step. This necessitates a shift from two bed to four bed configuration for continuous operation, because with LPP the feed is active in only one of the four steps.
Cocurrent depressurization	Cocurrent depressurization reduces the mass of feed plus LP reflux in the pore space of the sorption bed and allows the heavy product to start desorption while the light product is being produced, resulting in a higher-purity heavy product.
Pressure equalization	Interconnection of two beds to achieve pressure equalization improves energy efficiency by transferring momentum from a bed that is in the pressure-reducing stage after feed to another bed in the pressure-increasing stage leading up to the feed step.
Heavy reflux	Addition of a heavy product reflux to the cocurrent depressurization step increases heavy product purity in the later counter-current depressurization and light product reflux steps but requires recompression of the heavy product up to the full feed pressure.

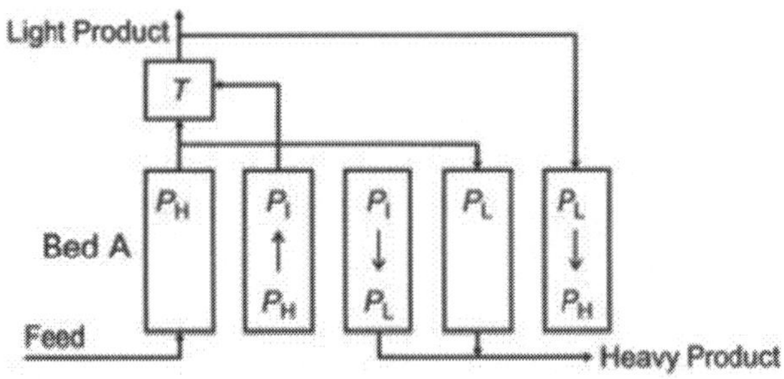

Step		1	2	3	4	5
Bed	A	F	CoD	CnD	LR	LPP
	B	CoD	CnD	LR	LPP	F
	C	CnD	LR	LPP	F	CoD
	etc.					

FIGURE 8.19 Five-step Skarstrom PSA cycle with LPP and CoD.

To illustrate these improvements, Fig. 8.19 shows the Skarstrom cycle with light product pressurization (LPP) and cocurrent depressurization. Note that "T" in the figure represents a tank holding light product for the LPP and LR steps. Further improvements to these adsorption cycles are being developed to optimize the process for CO_2 capture from flue gases, and are described in Section 8.4.1.

Vacuum swing adsorption

If the feed gas to a separation process is at or near ambient pressure, the energy penalty associated with compression to achieve an elevated feed pressure typical of a PSA cycle can be minimized or avoided by operating the cycle with P_L below ambient pressure. Vacuum swing adsorption (VSA) is such a variant of PSA in which a partial vacuum is applied to the downstream end of the sorbent bed during the feed stage, to draw in a low-pressure feed stream, and also during desorption to improve product purity. The product purity and separation efficiency of a PSA system may also be increased by applying a vacuum to the bed under regeneration, resulting in a hybrid vacuum and pressure swing adsorption (VPSA) cycle.

A VSA cycle incorporating a depressurization step to an intermediate pressure followed by evacuation (E) at a pressure below ambient was patented in 1957 by French engineers P. Guerin de Montgareuil and D. Domine. This cycle can be implemented with a wide range of operating schemes and bed numbers, and a schematic of two bed version is shown in Fig. 8.20.

The Guerin–Domine cycle performs significantly better in air separation than the Skarstrom cycle and has been commercially applied for mid-sized air separation units (up to ~340 tO_2/d) since the 1970s.

Pressure swing adsorption has been used since the 1980s for the purification of hydrogen produced by steam methane reforming (SMR) followed by the water–gas shift (WGS) reaction. The first commercial-scale SMR-WGS CCS demonstration project at the Valero refinery, Port Arthur, Texas has been in operation since 2012. The process uses a zeolite sorbent in the VPSA cycle, and completed its demonstration period in 2017, with over 4 $MtCO_2$ captured (Baade et al., 2012; Preston, 2018).

The SMR and WGS product stream is typically delivered at a pressure of ~2 MPa and temperature of 850–950°C. CO_2 removal from the hydrogen product stream by PSA can recover 90% of the hydrogen in the feed stream with a purity >99.99%. Systems can simultaneously deliver H_2 and CO_2 at high purity and typically operate with a 3–10 min cycle time. Continuous development has resulted in commercial systems that deliver high reliability at low capital and operating costs.

High-frequency pressure cycling

In both TSA and PSA, the amount of sorbent required to achieve a given sorbate capture rate is directly related to the frequency of the pressure cycle (Eq. 8.8), with a high frequency resulting in a higher throughput rate. Raising the cycle

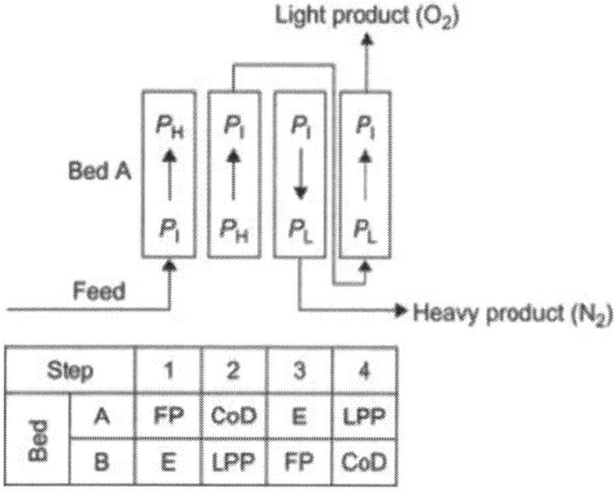

FIGURE 8.20 Two bed air separation unit using the Guerin–Domine cycle.

frequency therefore has the potential to reduce the capital cost of adsorption systems by reducing sorbate volume and the related physical system size. Conventional sorbent beds using beads or pellets exhibit high-pressure drops at the high gas velocities that occur in rapid cycle systems and monolithic structured sorbents are favored to overcome this limitation (Sharma et al., 2020). Since mechanical switching of valves can result in early mechanical failure, alternative methods of rapidly cycling the system pressure are under investigation. Possible approaches include the use of rotary valves, reciprocating (piston) or pressure wave-based systems.

8.3.3 Moisture swing adsorption

Humidity swing for sorbent regeneration was first investigated by Wang and coworkers using an amine-based ion exchange resin sorbent (Wang et al., 2011). The heat of sorption was measured at 32 kJ/molCO_2 (0.73 GJ/tCO_2), while the CO_2 partial pressure above the resin increased by two orders of magnitude for humidity swing from 15.8% to 100%, an effect which is too pronounced to be entirely due to the preferential adsorption of water molecules.

Moisture swing has since been investigated for a range of sorbents, and the desorption mechanism has been shown to be the result of a change in the free energy of the bicarbonate ion deprotonation reaction (8.13). Under humid conditions, an increase in the number of water molecules in the hydration shell around the bicarbonate ions drives this reaction to the right;

$$2HCO_3 \rightarrow CO_3^{2-} + CO_2 + H_2O \tag{8.13}$$

$$CO_3^{2-} + H_2O \rightarrow HCO_3^- + OH^- \tag{8.14}$$

while drying then drives the reverse reaction (8.14) producing hydroxyl ions that then bind with CO_2 in the adsorption stage (Shi et al., 2020).

8.3.4 Electromagnetic swing processes

Electric swing adsorption (ESA) and magnetic induction swing adsorption (MISA) are TSA variants that can achieve fast heating of the sorbent bed, overcoming the long desorption period inherent in traditional TSA. In an ESA cycle, the sorbent is heated by Joule heating, passing a current through the resistive sorbent material (Grande and Rodrigues, 2007), while MISA achieves a similar result using magnetic framework composites (MFCs) in which magnetic nanoparticles, such as $MgFe_2O_4$, embedded within the sorbent medium are heated by induction. Sadiq and coworkers have reported a remarkably low regeneration energy requirement of 1.29 GJ/tCO_2 using this technique (Sadiq et al., 2016).

Microwave swing adsorption has also been proposed as an approach to rapid desorption (Cherbański and Molga, 2009; van Schagen et al., 2022), with improved efficiency over traditional TSA since it can deliver selective volumetric heating, although achieving a uniform temperature profile and avoiding local overheating (hot spots) are challenges. Recent investigations using zeolite 13X show a 50% reduction in desorption time and a similar reduction in energy requirement (Ellison et al., 2021).

As well as the use of electromagnetic fields for sorbent heating, the possibility of using an electrostatic field to directly interfere with the sorption sites has also been investigated as an alternative to TSA or PSA. Farooq and coworkers demonstrated the effectiveness of this approach for the desorption of H_2S from a granular activated carbon sorbent in a biogas upgrading system (Farooq et al., 2018).

When integrated into a larger process, TSA can usually benefit from the cheap availability of low-grade waste whereas electrical and magnetic swing methods require high-grade electrical energy. This dictates the use of renewable energy to reduce both the operating costs and the carbon penalty of the process.

Charge modulated and supercapacitive swing adsorption

Electric fields have also been applied to the sorption/desorption cycle for nonconductive sorbent materials, using the technique of charge-modulated adsorption. This possibility was first investigated by Sun and coworkers who found that CO_2 strongly adsorbed onto boron nitride nanomaterials (nanosheets and nanotubes) when these materials were charged with excess electrons, and desorbed spontaneously when the negative charge was removed, with very high selectivity for CO_2 over CH_4 and H_2 (Sun et al., 2013).

A wide range of materials has subsequently been found to display similar sorption behavior under charge modulation, including nanosheets of borophene, C_3N, graphitic carbon nitride (g-C_4N_3), *Me*—N—C (*Me* = Fe, Cu, and Co), and N-doped penta-graphene, with selectivity for CO_2 over CH_4, H_2 and various short chain hydrocarbons being demonstrated. Sorbate loadings of up to 42 wt% and energy requirements as low as 1.2 GJ/tCO_2 have been observed in these investigations.

Charge-modulated selectivity for CO_2 over N_2 has also been demonstrated using a nanoporous graphene membrane and using calcite, the latter material having the advantage of natural abundance and low cost (Tao et al., 2020).

In supercapacitive swing adsorption, CO_2 is adsorbed onto the cathode of a supercapacitor during charging and desorbed during discharging. First demonstration by Landskron and coworkers (Kokoszka et al., 2014) is that a single module using an activated carbon cathode and NaCl electrolyte is able to concentrate a simulated flue gas (15% CO_2, 85% N_2) to a 46% CO_2 discharge stream with an energy requirement of 57 kJ/molCO_2 (1.2 GJ/tCO_2) (Liu and Landskron, 2017).

8.4 CO_2 adsorption technology RD&D status

8.4.1 Advanced PSA/VPSA cycles

Optimization of the PSA cycle configuration is an active area of research, with the aim of achieving high-purity product streams with minimal energy consumption (Webley et al., 2017). Early improvements described above included the introduction of refluxes of either light (LR) or heavy product (HR) to improve the purity of the product streams. Combining LR and HR into a single dual-reflux cycle has been shown in process simulations to outperform other cycle configurations in maximizing the recovery of a high-purity CO_2 heavy product stream from high-temperature flue gases.

The dual-reflux cycle is shown schematically in Fig. 8.21. The heavy reflux can be delivered either from the counter-current depressurization step, as shown here, or from the subsequent light reflux step, the former having the advantage of requiring the reflux product to be compressed only from an intermediate pressure to P_H rather than from P_L to P_H.

With the use of dual refluxes, the PSA cycle begins to mimic the operation of a distillation column. Optimization of PSA/VPSA cycles is ongoing, building on this analogy to develop even more flexible cycles, the current state of the art being represented by the 8 bed 14 step VPSA system developed by Sharma and coworkers for CO_2 capture from small-scale SMR processes (Sharma et al., 2020).

Adsorption heat storage

As noted earlier, temperature management is an important aspect of sorption process optimization, since heating during adsorption and cooling before or during desorption reduces working capacity, and various heat exchange systems have been employed to mitigate temperature change in adsorption beds. One option is the incorporation of phase change materials (PCMs) within the sorption bed, either as part of the bed structure, as separate elements interspersed among the sorbent particles, or in the form of hybrid sorbent-PCM particles (Horstmeier et al., 2016), or fibers (DeWitt et al., 2018).

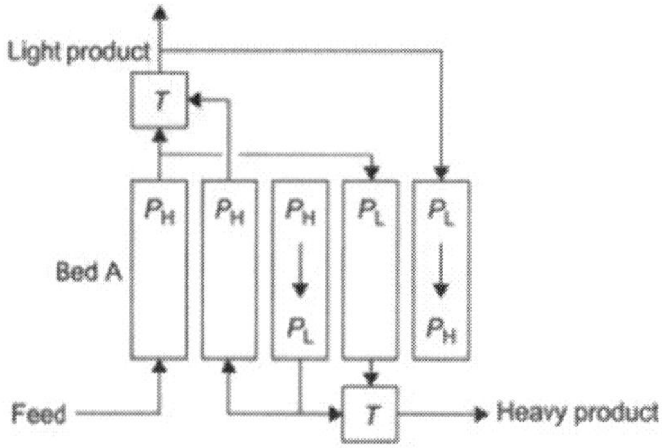

Step		1	2	3	4	5
Bed	A	F	HR	CnD	LR	LPP
	B	HR	CnD	LR	LPP	F
	C	CnD	LR	LPP	F	HR
	D	LR	LPP	F	HR	CnD
	E	LPP	F	HR	CnD	LR

FIGURE 8.21 Five bed dual-reflux PSA cycle with heavy reflux from countercurrent depressurization.

A phase change material, such as paraffin wax encapsulated in a copper shell, with a phase change temperature just above the bed temperature at the start of adsorption, will store the released heat of adsorption as latent heat of fusion and make this heat available to maintain bed temperature on desorption, limiting temperature fluctuations in the bed, increasing working capacity and improving product purity.

8.4.2 Sorption-enhanced reactions

High-temperature adsorption has potential applications in CO_2 capture from biomass gasification, and is the subject of a number of R&D projects. The production of hydrogen through SMR and WGS reactions (8.15 and 8.16) is an important example. This process, traditionally applied to a natural gas feedstock, can also be applied to produce hydrogen from biomethane generated, for example, from biomass pyrolysis or anaerobic digestion.

$$\text{SMR: } CH_4 + H_2O \leftrightarrow CO + 3H_2 \tag{8.15}$$

$$\text{WGS: } CO + H_2O \leftrightarrow CO_2 + H_2 \tag{8.16}$$

$$CO_2 \text{ removal: } CO_2 + CaO \leftrightarrow CaCO_3 \tag{8.17}$$

$$\text{Overall reaction: } CH_4 + 2H_2O + CaO \rightarrow CaCO_3 + 2H_2 \tag{8.18}$$

Reversible chemical reactions, such as the WGS reaction, can reach an equilibrium state before the full consumption of the reactants as a result of the buildup of reaction products and the increasing rate of the reverse reaction. Removal of one of the reaction products from the gas phase in the reaction zone, for example by adsorption, enables the reaction to reach completion.

Pressure swing adsorption has been commercially applied to the purification of the hydrogen product stream from these reactions for many years. One aim of current research is the development of elevated temperature pressure swing adsorption (ET-PSA) to enable application at the higher temperatures and pressures required for the SMR and WGS reactions.

The conditions in SMR and WGS reactors put severe requirements on sorbents, including the need for:

- High CO_2 loading capacity and sufficiently fast adsorption and desorption kinetics

TABLE 8.6 SEWGS feed and product stream compositions.

Component	Steam-reformed feed gas	SEWGS product gas
H_2	57%	89%
H_2O	16%	8%
CO	16%	0.5%
CO_2	10%	2%
CH_4	0.5%	0.5%

- Mechanical and chemical stability for long periods, at temperatures up to 900°C and pressures up to 4.0 MPa
- Tolerance to high stream partial pressures, with p_{steam}/p_{CO2} typically >20 in steam reforming

Fig. 8.22 shows a process schematic for a sorption-enhanced reactor-based hydrogen production system, using a simple dual bed reactor, feeding a zero-emissions power plant. A methane plus steam mixture is fed into reactor bed 1, containing a high-temperature CO_2 sorbent and an SMR catalyst. The sorption-enhanced reaction results in a hydrogen plus steam product stream, while CO_2 is adsorbed within the reactor. The reaction step continues until the H_2 effluent purity drops to a preset value. Concurrently, reactor 2 is fed with steam as a purge gas to remove CO_2, with desorption enhanced either by a pressure or temperature swing, or both. Water is condensed from the desorbed stream, yielding a CO_2 product stream suitable for compression. Using a biomethane feedstock, and with storage of the CO_2 stream, results in negative-emissions power generation.

Laboratory-scale proof-of-concept experiments have confirmed the high H_2 and very low CO and CO_2 concentrations in the light product stream compared to conventional SMR (Antzara et al., 2016). Further development is ongoing, focusing on:

- Identifying sorbent–catalyst combinations with improved properties, including dual-function materials combining steam reforming and in situ CO_2 removal
- Optimization of reaction conditions (temperature, pressure, water vapor presence) and reactor configuration (fixed and circulating systems)
- Natural sorbent selection criteria and pretreatment for low attrition
- For circulating systems (CFB), optimizing operating conditions to reduce sorbent attrition.

8.4.3 Novel sorbent materials

The search for novel sorbent materials which address the shortcomings of current systems is an active research area, and some ongoing lines of investigation are summarized here.

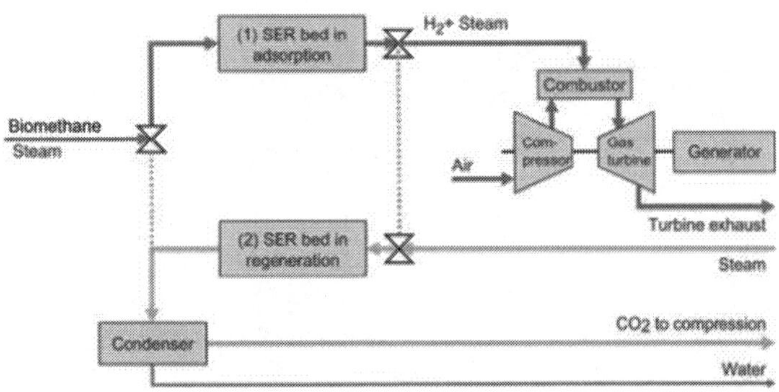

FIGURE 8.22 Hydrogen fueled negative-emissions power using sorption-enhanced steam reforming of biomethane.

High temperature sorbents

Natural calcium-based minerals, such as limestone and dolomite, have a number of advantages as high-temperature CO_2 sorbents, including fast carbonation and calcination reaction kinetics, natural abundance, and low cost. However, as noted earlier, they suffer from a gradual loss of sorption capacity with repeated cycling and degradation in the presence of sulfur due to the formation of nonregenerable calcium sulfate.

In applications where a catalyst is added, such as in SMR, any mismatch between catalyst and sorbent lifetimes is problematic, requiring either the premature disposal of still-active catalyst or the separation of catalyst and sorbent. Considerable research effort has been aimed at developing more reliable high-temperature sorbents, based on lithium and sodium, as well as calcium. Among the candidate sorbents that have been tested are:

- Lithium zirconate (Li_2ZrO_3)
- Lithium orthosilicate (Li_4SiO_4)
- Sodium zirconate (Na_2ZrO_3)
- Calcium aluminate ($CaAl_2O_4$)

These sorbents can be carbonated in SMR reactions, for example:

$$Li_4SiO_4 + CH_4 + 2H_2O \rightarrow Li_2SiO_3 + Li_2CO_3 + 4H_2 \tag{8.19}$$

and regenerated through the analogous calcining reaction:

$$Li_2SiO_3 + Li_2CO_3 + \text{heat} \rightarrow Li_4SiO_4 + CO_2 \tag{8.20}$$

The zirconate sorbents are found to give high hydrogen yield in SMR experiments but suffer from slow reaction kinetics (lithium zirconate) and catalyst poisoning (sodium zirconate). Further investigation of carbon-based sorbents also continues, investigating synthetic CaO materials and treatment methods to address the performance issues of natural calcium-based sorbents. One such area is the chemical modification of high-surface-area carbon-based materials by the introduction of CaO and $CaO \cdot MgO$, resulting in significantly enhanced high-temperature sorption capacity for CO_2.

Composite sorbents

Composite sorbents, in which the active sorbent material is incorporated into a supportive or protective structure, can overcome some of the limitations of bare sorbents, such as reducing degradation to extend lifetime, as well as exhibiting other beneficial characteristics.

Examples of such composite sorbents include the incorporation of functionalized amine sorbents into mesocellular siliceous foams or of solid sorbent powder into a substrate comprising a network of hydrophobic polymer filaments (Berger et al., 2017). The later example, under development by the EPRI and W. L. Gore, enables TSA regeneration by direct steam injection which would normally not be possible due to the adverse effect of moisture on sorbent performance (capacity, degradation, etc.). This has numerous advantages including improved energy efficiency and reduced cycle time leading to increased system throughput.

Novel methods for composite sorbent preparation, including 3D printing, are also under development with the aim of reducing sorbent cost and enabling large-scale production of sorbents with tunable porosity, sorbent loading, and mechanical properties (Thakkar et al., 2016; Lawson et al., 2021). Fig. 8.23 shows a zeolite 5A sorbent monolith fabricated by

FIGURE 8.23 3D printed zeolite 5A monolith. *From Thakker et al. (2016).*

3D printing. The monolith, with a high zeolite loading (90 wt%), contained a network of interconnected micro-, meso-, and macropores, and showed improved adsorption capacity and mechanical stability, potentially avoiding the attrition commonly encountered in packed beds using beads or pellets.

8.4.4 Metal-organic frameworks

The identification and assessment of new MOF sorbents to achieve high CO_2 selectivity and capacity have been a very active and fruitful area of research in recent years. Novel adsorption phenomena observed in MOFs, such as gate opening and breathing, and non-Langmuir adsorption behavior, have the potential to achieve a step change reduction in energy penalty for sorbent regeneration. These significant developments are discussed below. Other areas of current MOF research include (Petit, 2018):

- Reducing the impact of impurities, including water, on sorbent capacity
- MOF modification (e.g., functionalization, impregnation, doping, decoration) with a variety of materials, either post-synthetic or during synthesis as MOF structural elements
- Design of TSA and PVSA operating cycles to optimize sorbent performance
- MOF sorbents to improve PVSA energy efficiency by removing the vacuum requirement
- Impact of synthesis methods on MOF morphology and properties
- Scale-up of MOF synthesis methods, for low-cost large-scale production
- Development of MOF monolithic structures through additive manufacturing (3D printing)

Much of this work is being progresses in the private sector, with many start-up companies active in this area (Petit, 2018).

Gate opening and breathing phenomena in flexible MOFs

Reversible structural transitions in flexible MOFs as a result of external stimuli, including light, heat, and the adsorption of guest molecules, were first reported in the early 2000s and have been extensively investigated since then. Two of the most widely studied such transitions are "gate opening" that refers to the transition from a closed, nonporous phase to an open, porous one, and "breathing" that refers to an expansion and contraction of the MOF unit cell. These transitions, illustrated schematically in Fig. 8.24, occur as a result of the movement (rotation, bending, etc.) of the MOF organic linkers and in many cases occur exclusively in response to the presence of CO_2. This is because the energy of adsorption of the guest molecule in the MOF is sufficient to activate the transition for CO_2 but not for other gases. As also shown in the figure, the adsorption isotherms differ significantly from the classical Langmuir type discussed above.

Such flexible MOFs exhibit an intrinsic capability to address the problem of thermal management in sorbent beds (discussed in Section 8.1) since the exothermic heat of adsorption is partially offset by the endothermic energy required to expand the framework during gate-opening, with the opposite effect occurring during desorption and gate-closing (Mason et al., 2015).

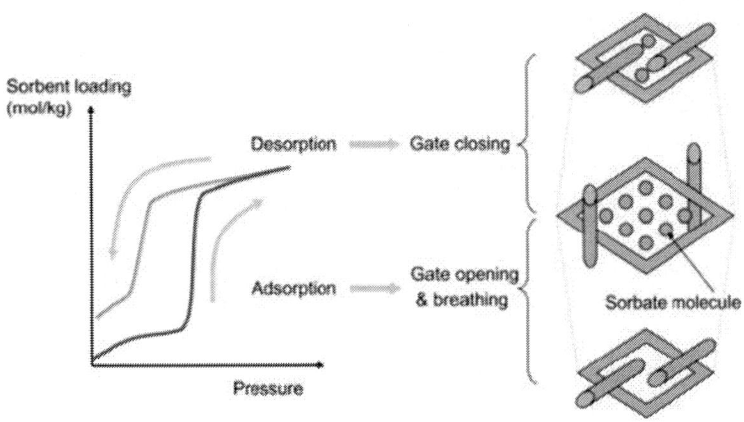

FIGURE 8.24 Schematic of gate opening and unit cell breathing in flexible MOFs. *After Hyun et al. (2016).*

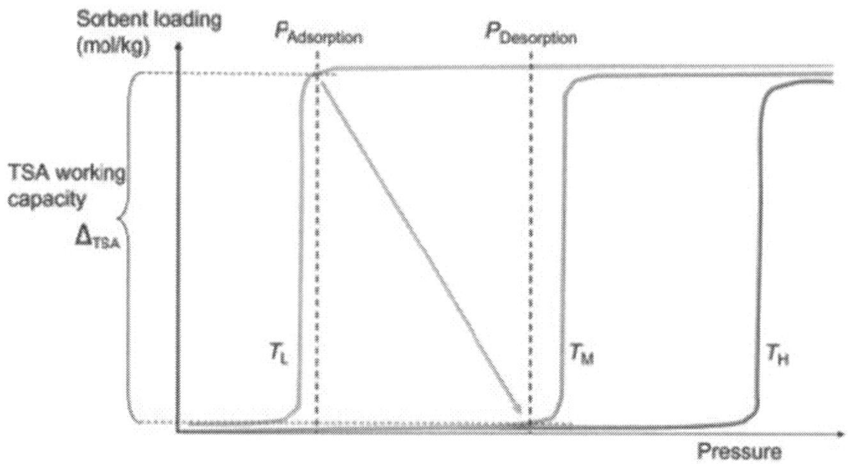

FIGURE 8.25 Step-shaped adsorption isotherms of a "phase change" sorbent.

Phase change functionalized MOF sorbents

A new class of "phase change" functionalized MOF sorbents has recently been reported by McDonald and coworkers (McDonald et al., 2015), which also exhibit non-Langmuir adsorption behavior. The step-shaped adsorption isotherms of these sorbents (Fig. 8.25) show that, at a given temperature, sorbate loading increases from near zero to full capacity over a very narrow pressure range (compare to Figs. 8.2 and 8.3) and that the full working capacity can be achieved with a much narrower temperature swing than for sorbents exhibiting typical Langmuir isotherms.

This behavior was demonstrated in diamine-appended MOFs and arises because the adsorption of a CO_2 molecule by the amine at one metal site (forming ammonium carbamate) destabilizes the amines at adjacent metal sites resulting in a chain reaction that drives the adsorption of CO_2 and the reorganization of the amines into an ordered chain of ammonium carbamates.

Unlike classical sorbents, where a low adsorption temperature is needed to maximize working capacity, the efficiency of this type of phase-change sorbent increases at higher temperatures, widening the potential operating temperature range for adsorption processes.

8.4.5 Chemical looping

Chemical looping combustion

A large number of CLC pilots have been executed by research teams around the world, the two largest being a 3 MW_{th} plant using CaO as a carrier at Alstom Windsor Labs, CT, USA, and a 4MW_{th} CLC circulating fluidized bed boiler at Chalmers University of Technology. These and many other pilot plants have been recently reviewed by Adánez et al. (2018). Further work to close identified technology gaps in the Alstom CaO carrier pilot, was undertaken in a 2013–2017 project, supported by the US DoE NETL's Advanced Combustion Systems Program, with the aim of achieving readiness for an industrial scale demonstration project (10–50 MW_{th}) beyond 2017 (Braun et al., 2017). The design for a 10 MW_{th} CLC-based steam generation plant, planned to be constructed by Cenovus Energy Inc., was also reported in 2013 (Sit et al., 2013), but subsequent progress on the project was halted due to budget constraints driven by low oil prices in 2015.

Chemical looping combustion with oxygen uncoupling has been proof-of-concept tested at laboratory-scale (1.5 kW_{th}), using a variety of biomass and coal fuels and with CuO as the oxygen carrier, and demonstrated a carbon capture efficiency of 93%–97% (Adánez-Rubio et al., 2021).

Alongside these demonstration efforts, continuing areas of CLC/CLOU research include the screening of alternative oxygen carrier materials for long lifetime and high reactivity toward specific fuels, and the optimization of oxygen carrier particles, including support materials and production processes, for long-term mechanical stability under fluidized bed conditions. In addition, for solid fuels such as biomass, challenges related to the solid–solid interactions include sustaining carrier reactivity in the presence of ash components, and processes to recover carrier particles from the ash stream, such as the use of magnetic supports (Coppola and Scala, 2021).

Calcium looping postcombustion capture

Calcium or CaO looping (CaL) is being investigated for postcombustion CO_2 capture, using interconnected fluidized bed adsorption and regeneration reactors. Fig. 8.26 illustrates the physical configuration, in which hot flue gases enter the base of the carbonation reactor (A), along with a feed of hot CaO. The fine sorbent particles rise in the column along with the flue gas stream and are separated in a cyclone separator (C1).

CO_2-depleted flue gases are released while the carbonated sorbent ($CaCO_3$) is conveyed to the base of the regenerating bed (B). Calcination at 900°C results in the release of CO_2, which is separated from regenerated sorbent by a second cyclone separator (C2). The loop is completed by conveying the regenerated sorbent to the base of the carbonation reactor. An equivalent CaL concept can also be used to remove CO_2 from the $H_2 + CO_2$ syngas product stream from the WGS reaction, a separation more commonly achieved using fixed bed systems.

Ongoing research into synthetic calcium-based sorbents for chemical looping aims to sustain sorbent capacity through repeated regeneration cycles. Mixed sorbents such as CaO(75%)/$Ca_{12}Al_{14}O_{33}$(25%) or CaO(90%)/Al_2O_3(10%) have been shown to sustain higher carrying capacity over repeated cycles than pure CaO, attributed to the higher sintering temperature.

Alternatives to CaO, such as lithium silicate (Li_4SiO_4), have also been investigated for high-temperature chemical looping reactors, with the aim of identifying sorbents with lower regeneration temperatures. Lithium silicate can be regenerated at a temperature of ~800°C, significantly lower than the temperature required for calcining $CaCO_3$.

Several pilot scale CaL capture systems have been reported, including a 1.9 MW_{th} plant capturing 1 tCO_2/hr from flue gas at the Taiwan Cement Company's Ho-Ping cement plant in Hualien, Taiwan, and a 1.7 MW_{th} plant capturing up to 95% of CO_2 from a flue gas slip stream at the Hunosa operated 50 MW_e CFBC power plant at La Pereda in Spain (Arias et al., 2013), the latter being part of the EU-funded "CaOling" project. Further process evaluation has also been undertaken using a 1 MW_{th} pilot plant constructed at Technische Universität Darmstadt a part of the EU's SCARLET project, with the aim of providing a basis for scale up to a 20 MW_{th} demonstration plant (Helbig et al., 2017). CaL is also seen as a viable decarbonization route for the pulp and paper industry, particularly since $CaCO_3$ in lime mud is the main biowaste from the Kraft process, which could be economically retrofitted to existing plants given a negative emissions credit of ~42 €/tCO_2 (Santos et al., 2021).

Hybrid combustion–gasification by chemical looping

The versatility of chemical looping is well illustrated by the proposed combination of two calcium-based chemical loops into a hybrid combustion–gasification process. The first loop uses $CaSO_4$ as an oxygen carrier to gasify the fuel (e.g., biomass) in a partial oxidization step, completing the loop by oxidation of CaS in air:

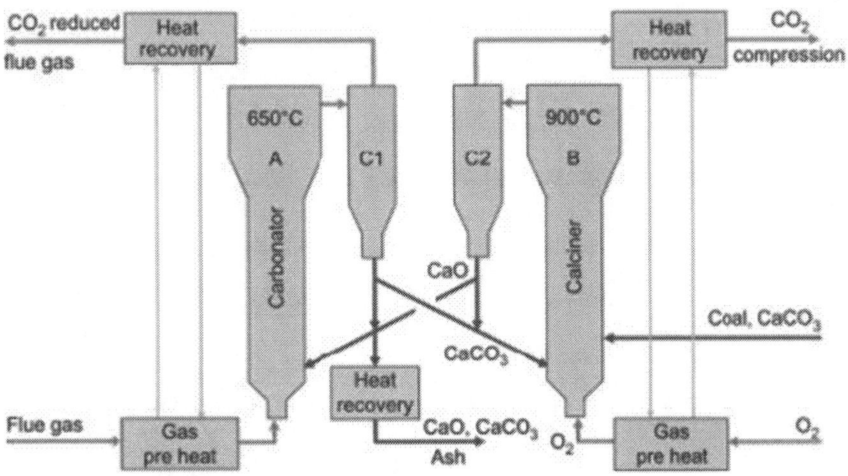

FIGURE 8.26 Postcombustion CO_2 capture using CaO looping.

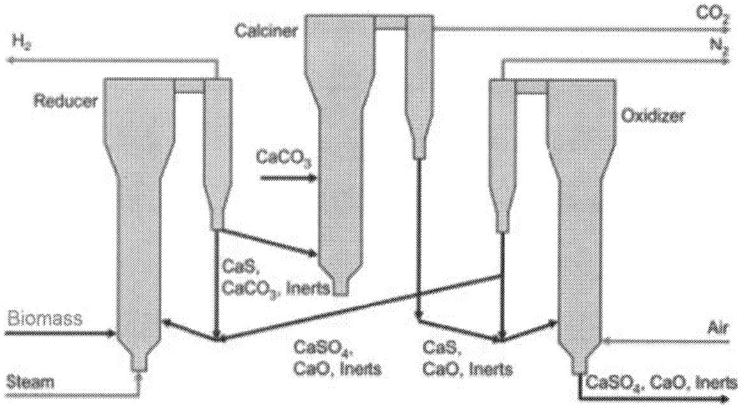

FIGURE 8.27 Hybrid biomass combustion–gasification chemical looping process.

$$4C + CaSO_4 \rightarrow 4CO + 4CaS \tag{8.21}$$

$$CaS + 2O_2 \rightarrow CaSO_4 \tag{8.22}$$

In the presence of steam, the partial oxidation product is water-shifted to produce H_2 and CO_2, and a second CaO loop is then used to remove CO_2 from the syngas stream.

A schematic of this hybrid system is shown in Fig. 8.27. Three interconnected fluidized bed reactors are shown: a reducer, calciner, and oxidizer. The reducer is fed with a carbonaceous fuel such as biomass, steam, and the chemical sorbents $CaSO_4$ and CaO. The fuel gasification, WGS, and CaO carbonation reactions take place in this reactor, producing a product gas stream of pure hydrogen. The solid reaction products are carried out of the top of the reactor with the gas product stream and separated in the first of three cyclone separators. The solids are then conveyed to the second reactor, where the $CaCO_3$ is calcined and CO_2 released. Heat for the calcination reaction is provided by a heat exchange loop from the combustion reactor, using bauxite as the heat transfer medium (not shown in the figure). Solid products are then conveyed to the final reactor, where CaS is oxidized in a heated air stream.

Attrition products from the $CaSO_4$ and CaO particles, as well as other inert combustion products, are removed from circulation, while make-up carbonate is added to the calciner as required. Reflux of the solid reaction products from the reducing and oxidation reactors can be used for additional process control.

Development of a 3 MW_{th} prototype chemical looping-gasification facility was progressed by Alstom Power, in parallel with the CLC development work described earlier. An industrial-scale demonstration project was initially planned for implementation from 2017 to 2020, but no further progress has been reported since the 2015 takeover of Alstom by General Electric.

References and resources

Adánez, J., Abad, A., Mendiara, T., Gayán, P., De Diego, L.F., García-Labiano, F., 2018. Chemical looping combustion of solid fuels. Prog. Energy Combust. Sci. 65, 6–66.

Adánez-Rubio, I., Samprón, I., Izquierdo, M.T., Abad, A., Gayán, P., Adánez, J., 2021. Coal and Biomass Combustion With CO_2 Capture by CLOU Process Using a Magnetic Fe-Mn-Supported CuO Oxygen Carrier, p. 122742. Fuel.

Antzara, A., Heracleous, E., Lemonidou, A.A., 2016. Energy efficient sorption enhanced-chemical looping methane reforming process for high-purity H_2 production: experimental proof-of-concept. Appl. Energy 180, 457–471.

Ardanuy Raso, M., Claramunt Blanes, J., Arévalo Peces, R., Parés Sabatés, F., Aracri, E., Vidal Lluciá, T., 2012. Nanofibrillated cellulose (NFC) as a potential reinforcement for high performance cement mortar composites. Bioresources 7 (3), 3883–3894.

Arias, B., Diego, M.E., Abanades, J.C., et al., 2013. Demonstration of steady state CO_2 capture in a 1.7 MWth calcium looping pilot. Int. J. Greenh. Gas Control 18, 237–245.

Azarabadi, H., Lackner, K.S., 2019. A sorbent-focused techno-economic analysis of direct air capture. Appl. Energy 250, 959–975.

Baade, W., Farnand, S., Hutchinson, R., Welch, K., 2012. CO_2 capture from SMRs: a demonstration project. Hydrocarb. Process. 91 (9), 63–68.

Berger, A.H., Horowitz, J.A., Machalek, T., Wang, A., Bhown, A.S., 2017. A novel rapid temperature swing adsorption post-combustion CO_2 capture process using a sorbent polymer composite. Energy Proc. 114, 2193–2202.

Braun, T.J., Sloan, D.G., Turek, D.G., Unker, S.A., Vitse, F., 2017. Improvement of Alstom's Limestone-Based Chemical Looping Combustion Process for Higher Purity Flue Gas Production (NETL Report DOE-API-0025073-02a). https://doi.org/10.2172/1440031.

Cherbański, R., Molga, E., 2009. Intensification of desorption processes by use of microwaves—an overview of possible applications and industrial perspectives. Chem. Eng. Process 48 (1), 48–58.

Coppola, A., Scala, F., 2021. Chemical looping for combustion of solid biomass: a review. Energy Fuels 35 (23), 19248–19265.

DeWitt, S.J., Rubiera Landa, H.O., Kawajiri, Y., Realff, M., Lively, R.P., 2018. Development of phase-change-based thermally modulated fiber sorbents. Ind. Eng. Chem. Res. 58 (15), 5768–5776.

Dhoke, C., Zaabout, A., Cloete, S., Amini, S., 2021. Review on reactor configurations for adsorption-based CO_2 capture. Ind. Eng. Chem. Res. 60 (10), 3779–3798.

Ellison, C., Hoffman, J., Shekhawat, D., 2021. Comparison of microwave and conventional heating for CO_2 desorption from zeolite 13X. Int. J. Greenh. Gas Control 107, 103311.

Farooq, M., Almustapha, M.N., Imran, M., Saeed, M.A., Andresen, J.M., 2018. In-situ regeneration of activated carbon with electric potential swing desorption (EPSD) for the H_2S removal from biogas. Bioresour. Technol. 249, 125–131.

García-Mariaca, A., Llera-Sastresa, E., 2021. Review on carbon capture in ICE driven transport. Energies 14 (21), 6865.

Gebald, C., Wurzbacher, J.A., Tingaut, P., Zimmermann, T., Steinfeld, A., 2011. Amine-based nanofibrillated cellulose as adsorbent for CO_2 capture from air. Environ. Sci. Technol. 45 (20), 9101–9108.

Gebald, C., Wurzbacher, J.A., Borgschulte, A., Zimmermann, T., Steinfeld, A., 2014. Single-component and binary CO_2 and H_2O adsorption of amine-functionalized cellulose. Environ. Sci. Technol. 48 (4), 2497–2504.

Gelles, T., Lawson, S., Rownaghi, A.A., Rezaei, F., 2020. Recent advances in development of amine functionalized adsorbents for CO_2 capture. Adsorption 26 (1), 5–50.

Grande, C.A., Rodrigues, A.E., 2007. Electric swing adsorption for CO_2 removal from flue gases. Int. J. Greenh. Gas Control 2, 194–202.

Gupta, T., Ghosh, R., 2015. Rotating bed adsorber system for carbon dioxide capture from flue gas. Int. J. Greenh. Gas Control 32, 172–188.

Helbig, M., Hilz, J., Haaf, M., Daikeler, A., Ströhle, J., Epple, B., 2017. Long-term carbonate looping testing in a 1 MWth pilot plant with hard coal and lignite. Energy Proc. 114, 179–190.

Horstmeier, J.F., Gomez Lopez, A., Agar, D.W., 2016. Performance improvement of vacuum swing adsorption processes for CO_2 removal with integrated phase change material. Int. J. Greenh. Gas Control 47, 364–375.

Hyun, S.M., Lee, J.H., Jung, G.Y., et al., 2016. Exploration of gate-opening and breathing phenomena in a tailored flexible metal–organic framework. Inorg. Chem. 55 (4), 1920–1925.

Ishida, M., Jin, H., 1994. A new advanced power-generation system using chemical-looping combustion. Energy 19 (4), 415–422.

Kokoszka, B., Jarrah, N.K., Liu, C., Moore, D.T., Landskron, K., 2014. Supercapacitive swing adsorption of carbon dioxide. Angew. Chem. Int. Ed. 53 (14), 3698–3701.

Lawson, S., Li, X., Thakkar, H., Rownaghi, A.A., Rezaei, F., 2021. Recent advances in 3D printing of structured materials for adsorption and catalysis applications. Chem. Rev. 121 (10), 6246–6291.

Li, W., Choi, S., Drese, J.H., et al., 2010. Steam-stripping for regeneration of supported amine-based CO_2 adsorbents. ChemSusChem 3 (8), 899–903.

Li, Y., Wang, K., Zhou, W., et al., 2019. Cryo-EM structures of atomic surfaces and host-guest chemistry in metal-organic frameworks. Matter 1 (2), 428–438.

Liu, H., Liu, B., Lin, L.C., et al., 2014. A hybrid absorption–adsorption method to efficiently capture carbon. Nat. Commun. 5 (1), 1–7.

Liu, C., Landskron, K., 2017. Design, construction, and testing of a supercapacitive swing adsorption module for CO_2 separation. Chem. Commun. 53 (26), 3661–3664.

Lyndon, R., Konstas, K., Thornton, A.W., Seeber, A.J., Ladewig, B.P., Hill, M.R., 2015. Visible light-triggered capture and release of CO_2 from stable metal organic frameworks. Chem. Mater. 27 (23), 7882–7888.

Mattisson, T., Lyngfelt, A., Leion, H., 2009. Chemical-looping with oxygen uncoupling for combustion of solid fuels. Int. J. Greenh. Gas Control 3 (1), 11–19.

McDonald, T.M., et al., 2015. Cooperative insertion of CO_2 in diamine appended metal-organic frameworks. Nature 519, 303–308.

Mason, J.A., Oktawiec, J., Taylor, M.K., et al., 2015. Methane storage in flexible metal–organic frameworks with intrinsic thermal management. Nature 527 (7578), 357–361.

Manzolini, G., Giuffrida, A., Cobden, P.D., van Dijk, H.A.J., Ruggeri, F., Consonni, F., 2020. Techno-economic assessment of SEWGS technology when applied to integrated steel-plant for CO_2 emission mitigation. Int. J. Greenh. Gas Control 94, 102935.

Ntiamoah, A., Ling, J., Xiao, P., Webley, P.A., Zhai, Y., 2016. CO_2 capture by temperature swing adsorption: use of hot CO_2-rich gas for regeneration. Ind. Eng. Chem. Res. 55 (3), 703–713.

Okumura, T., Yoshizawa, K., Nishibe, S., Iwasaki, H., Kazari, M., Hori, T., 2017. Parametric testing of a pilot-scale design for a moving-bed CO_2 capture system using low-temperature steam. Energy Proc. 114, 2322–2329.

Petit, C., 2018. Present and future of MOF research in the field of adsorption and molecular separation. Curr. Opin. Chem. Eng. 20, 132–142.

Preston, C., 2018. The carbon capture project at Air Products' Port Arthur hydrogen production facility. In: 14th Greenhouse Gas Control Technologies Conference Melbourne, pp. 21–26.

Sadiq, M.M., Li, H., Hill, A.J., Falcaro, P., Hill, M.R., Suzuki, K., 2016. Magnetic induction swing adsorption: an energy efficient route to porous adsorbent regeneration. Chem. Mater. 28 (17), 6219–6226.

Sadiq, M.M., Konstas, K., Falcaro, P., Hill, A.J., Suzuki, K., Hill, M.R., 2020. Engineered porous nanocomposites that deliver remarkably low carbon capture energy costs. Cell Rep. Phys. Sci. 1 (6), 100070.

Santos, M.P., Manovic, V., Hanak, D.P., 2021. Unlocking the potential of pulp and paper industry to achieve carbon-negative emissions via calcium looping retrofit. J. Clean. Prod. 280, 124431.

Satyapal, S., Filburn, T., Trela, J., Strange, J., 2001. Performance and properties of a solid amine sorbent for carbon dioxide removal in space life support applications. Energy Fuels 15 (2), 250–255.

Sharma, S., Maréchal, F., 2019. Carbon dioxide capture from internal combustion engine exhaust using temperature swing adsorption. Front. Energy Res. 7, 143.

Sharma, I., Friedrich, D., Golden, T., Brandani, S., 2020. Monolithic adsorbent-based rapid-cycle vacuum pressure swing adsorption process for carbon capture from small-scale steam methane reforming. Ind. Eng. Chem. Res. 59 (15), 7109–7120.

Shi, X., Xiao, H., Kanamori, K., Yonezu, A., Lackner, K.S., Chen, X., 2020. Moisture-driven CO_2 sorbents. Joule 4 (8), 1823–1837.

Sit, S.P., Reed, A., Hohenwarter, U., Horn, V., Marx, K., Proell, T., 2013. Cenovus 10 MW CLC field pilot. Energy Proc. 37, 671–676.

Subramanian Balashankar, V., Rajendran, A., 2019. Process optimization-based screening of zeolites for post-combustion CO_2 capture by vacuum swing adsorption. ACS Sustain. Chem. Eng. 7 (21), 17747–17755.

Sun, Q., Li, Z., Searles, D.J., Chen, Y., Lu, G., Du, A., 2013. Charge-controlled switchable CO_2 capture on boron nitride nanomaterials. J. Am. Chem. Soc. 135 (22), 8246–8253.

Tao, L., Huang, J., Dastan, D., et al., 2020. CO2 capture and separation on charge-modulated calcite. Appl. Surf. Sci. 530, 147265.

Thakkar, H., Eastman, S., Hajari, A., Rownaghi, A.A., Knox, J.C., Rezaei, F., 2016. 3D-printed zeolite monoliths for CO_2 removal from enclosed environments. ACS Appl. Mater. Interfaces 8 (41), 27753–27761.

van Schagen, T.N., van der Wal, P.J., Brilman, D.W.F., 2022. Development of a novel, through-flow microwave-based regenerator for sorbent-based direct air capture. Chem. Eng. J. Adv. 100187.

Wang, T., Lackner, K.S., Wright, A., 2011. Moisture swing sorbent for carbon dioxide capture from ambient air. Environ. Sci. Technol. 45 (15), 6670–6675.

Webley, P.A., Qader, A., Ntiamoah, A., Ling, J., Xiao, P., Zhai, Y., 2017. A new multi-bed vacuum swing adsorption cycle for CO_2 capture from flue gas streams. Energy Proc. 114, 2467–2480.

Xu, X., Song, C., Andresen, J.M., Miller, B.G., Scaroni, A.W., 2002. Novel polyethylenimine-modified mesoporous molecular sieve of MCM-41 type as high-capacity adsorbent for CO_2 capture. Energy Fuels 16 (6), 1463–1469.

Zhao, R., Zhao, L., Wang, S., Deng, S., Li, H., Yu, Z., 2018. Solar-assisted pressure-temperature swing adsorption for CO_2 capture: effect of adsorbent materials. Sol. Energy Mater. Sol. Cell. 185, 494–504.

Zhu, X., Li, S., Shi, Y., Cai, N., 2019. Recent advances in elevated-temperature pressure swing adsorption for carbon capture and hydrogen production. Prog. Energy Combust. Sci. 75, 100784.

Zhu, X., Ge, T., Yang, F., Wang, R., 2021. Design of steam-assisted temperature vacuum-swing adsorption processes for efficient CO_2 capture from ambient air. Renew. Sustain. Energy Rev. 137, 110651.

Further reading

Ben-Mansour, R., Habib, M.A., Bamidele, O.E., et al., 2016. Carbon capture by physical adsorption: materials, experimental investigations and numerical modeling and simulations–a review. Appl. Energy 161, 225–255.

Fennell, P., Anthony, B. (Eds.), 2015. Calcium and Chemical Looping Technology for Power Generation and Carbon Dioxide (CO_2) Capture. Elsevier, Amsterdam.

Hanak, D.P., Michalski, S., Manovic, V., 2018. From post-combustion carbon capture to sorption-enhanced hydrogen production: a state-of-the-art review of carbonate looping process feasibility. Energy Convers. Manag. 177, 428–452.

Hussin, F., Aroua, M.K., 2020. Recent trends in the development of adsorption technologies for carbon dioxide capture: a brief literature and patent reviews (2014–2018). J. Clean. Prod. 253, 119707.

Kang, J.S., Kim, S., Hatton, T.A., 2021. Redox-responsive sorbents and mediators for electrochemically based CO_2 capture. Curr. Opin. Green Sustain. Chem. 100504.

Lee, S.-Y., Park, S.-J., 2015. A review on solid adsorbents for carbon dioxide capture. J. Ind. Eng. Chem. 23, 1–11.

Lemonidou, A., 2017. Sorption Enhancement of Chemical Processes. Academic Press.

Liu, R.S., Shi, X.D., Wang, C.T., et al., 2021. Advances in post-combustion CO_2 capture by physical adsorption: from materials innovation to separation practice. ChemSusChem 14 (6), 1428–1471.

Lyngfelt, A., 2020. Chemical looping combustion: status and development challenges. Energy Fuels 34 (8), 9077–9093.

Mendiara, T., García-Labiano, F., Abad, A., et al., 2018. Negative CO_2 emissions through the use of biofuels in chemical looping technology: a review. Appl. Energy 232, 657–684.

Ruthven, D.M., Farooq, S., Knaebel, K.S., 1994. Pressure Swing Adsorption. John Wiley & Sons, Inc., New York.

Trickett, C.A., Helal, A., Al-Maythalony, B.A., Yamani, Z.H., Cordova, K.E., Yaghi, O.M., 2017. The chemistry of metal–organic frameworks for CO_2 capture, regeneration and conversion. Nat. Rev. Mater. 2 (8), 1–16.

Wilcox, J., Haghpanah, R., Rupp, E.C., He, J., Lee, K., 2014. Advancing adsorption and membrane-separation processes for the Gigaton carbon capture challenge. Annu. Rev. Chem. Biomol. Eng. 5, 479–505.

Yang, R.T., 1987. Gas Separation by Adsorption Processes. Butterworth Publishers, London, UK.

Zhou, W., Meng, X., Gao, J., Zhao, H., Zhao, G., Ma, J., 2021. Electrochemical regeneration of carbon-based adsorbents: a review of regeneration mechanisms, reactors, and future prospects. Chem. Eng. J. Adv. 5, 100083.

Projects and organizations

Adsorption.org; Adsorption knowledge sharing site: www.adsorption.org.

Alstom Power; Chemical looping combustion/gasification development: www.ge-alstom.com.

CACHET/CESAR; Early development of SEWGS technology: https://www.ecn.nl/publicaties/PdfFetch.aspx?nr=ECN-L–09-157.

ECN (Energy Research Centre of the Netherlands; Novel sorbents and sorption-enhanced reactions (SEWGS)): www.ecn.nl.

IFE (Institute for Energy Technology; High-temperature sorbents for sorption-enhanced reactions): www.ife.no/en/ife/main_subjects_new/energy_environment/klima.

International Conference on Chemical Looping: http://chemical-looping2016.com.

International Zeolite Association: Promoting the development of zeolite science and technology: www.iza-online.org/.

STEPWISE project (SEWGS industrial pilot using blast furnace gas): www.stepwise.eu/home/.

Chapter 9

Membrane CO_2 separation

Steve Rackley

The first application of membranes for gas separation was in the 1980s, when techniques were developed for hydrogen separation in ammonia production, oxygen/nitrogen air separation, and the separation of CO_2 from light hydrocarbons. These early applications are summarized in Table 9.1. The technological breakthrough that brought membranes into large-scale industrial use was the development of composite polymer membranes, in which a thin selective layer is deposited on a thicker, nonselective, and inexpensive layer that provides mechanical support.

Membranes have a number of potential applications for negative-emissions CO_2 capture, particularly in BECCS projects, including hydrogen separation in biomass gasification and CO_2 separation from biomass combustion flue gases (Wilcox et al., 2014). A wide variety of different membrane types and materials are potentially applicable, using proven technologies that are already deployed on an industrial scale as well as novel technologies, based on advanced materials, that are under research and development.

9.1 Physical and chemical fundamentals

Conceptually, a membrane acts as a filter, separating one specific component (the permeate) from a mixture of gases in a feed stream. The two most important parameters defining a membrane's separation performance are its gas permeance (or permeability, which takes into account the thickness of the active layer), indicating how fast gas molecules are transported though the membrane, and its selectivity, indicating the relative speeds through the membrane of different gas molecules from a mixture. This "filtration" can involve a number of physical and chemical processes, depending on the membrane materials and design, and these processes are discussed in this section.

The two common membrane separation flow schemes are illustrated schematically in Fig. 9.1. A feed gas stream enters a module housing the membrane, and the permeate exits on the downstream side, having passed through the membrane. The feed stream minus permeate is termed the retentate and may either exit the module having been depleted of the permeate (as in the cross-flow example [A]) or be continuously replenished (as in the dead-end flow case [B]).

The key characteristic of membranes that determines the mechanism for transporting permeate molecules is porosity, and membranes are classified as either porous or nonporous. For porous membranes, the pore size determines the size of particles (molecules at the smallest scale) that can pass through the membrane, as shown in Table 9.2. Porous membranes with pore sizes >1 nm (micro- to macroporous) are mostly employed as supports for a thin dense gas separation layer.

At the macropore scale, gas flow is dominated by viscous forces, giving way to diffusion as the pore size reduces below the mean free path of gas molecules. Below 10 Å, as the pore size approaches molecular dimensions, flow is increasingly dominated by surface interaction on the pore walls, and is termed molecular sieving. Once the pore size drops to the 1 Å scale, a solution−diffusion transport mechanism occurs rather than molecular sieving, and the membrane is considered to be nonporous or "dense." These transport mechanisms are described in the following sections.

9.1.1 Porous membrane transport processes

The physical nature of the gas transport process occurring in porous membranes depends on the size of pores in the membrane and on the temperature and pressure of the feed gas. The different transport regimes are characterized by the Knudsen number N_{Kn}, which is equal to the ratio of the mean free path of gas molecules for component i (λ_i) to the average pore diameter in the membrane (d_0):

$$N_{Kn\ i} = \lambda_i / d_0 \quad (9.1)$$

Negative Emissions Technologies for Climate Change Mitigation. https://doi.org/10.1016/B978-0-12-819663-2.00007-1

TABLE 9.1 Early membrane separation applications and technologies.

Application	Company	Technology
Hydrogen recovery from ammonia production	Monsanto	Prism hollow-fiber polymeric membranes
Nitrogen separation from air	Dow, Air Products	Generon hollow-fiber polymeric membranes Prism nitrogen membrane systems
CO_2 separation from methane and ethane	Cynara/NATCO	Acetyl activated nonporous cellulose acetate membrane
	UOP-Honeywell	Separex spiral-wound and hollow-fiber cellulose acetate membranes

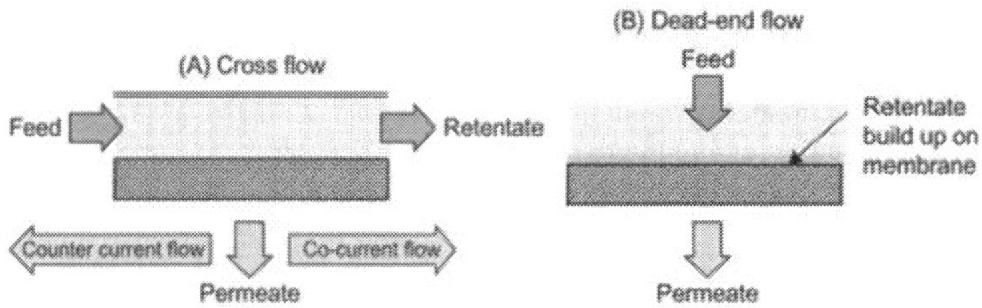

FIGURE 9.1 Membrane configurations: (A) cross flow and (B) dead-end flow.

TABLE 9.2 Porous membrane classification.

Pore classification	Pore size range	Filtration classification
Macroporous	>50 nm (500 Å)	Microfiltration
Mesoporous	2–50 nm (20–500 Å)	Ultrafiltration
Microporous	1–2 nm (10–20 Å)	Nanofiltration
Nanoporous	<1 nm (10 Å)	Molecular sieving

Viscous capillary flow

If the membrane pore diameter is very large compared to the mean free path of the gas molecules ($N_{Kn} < 0.001$), the gas flux is governed by Poiseuille's equation for viscous flow through capillaries:

$$J_i = r^2 \Delta P \, P_{av}/(8 \, \mu RTL) \tag{9.2}$$

where J_i is the flux of gas component i, ΔP is the differential pressure across the membrane, P_{av} is the average pressure within a capillary, r is the representative pore radius, μ is the gas viscosity, R is the gas constant, T is the temperature (K), and L is the length of the capillary (which, ignoring pore tortuosity, equals the membrane thickness). The flux can also be expressed in terms of the porosity (ε) of the membrane as follows:

$$J_i = r^2 \, \Delta P \, \varepsilon/(8 \, \mu \, \mathrm{L}) \tag{9.3}$$

A mixture of two gases in capillary flow will be characterized by a single average pressure and a single viscosity that will depend on the ratio of gases in the mixture. The ratio of fluxes of the two components (J_i/J_k) will depend on their partial pressure in the mixture, and the transport will exhibit no selectivity between the components because there is no dependence on their individual characteristics.

Knudsen diffusion

If the mean free path is very much larger than the membrane pore radius ($N_{Kn} > 100$), gas molecules will collide more frequently with the pore walls than they will with other gas molecules. This situation is termed free-molecule flow, or Knudsen flow (Fig. 9.2), and the flux rate is determined by the Knudsen equation:

$$J_i = D_i\ \Delta P/RTL = 2r\ \Delta P\ (8RT/\pi M_i)^{1/2}/3RTL \tag{9.4}$$

where for component i, $D_i = 2/3\ r\ (8RT/\pi M_i)^{1/2}$ is the Knudsen diffusion coefficient and M_i is the molecular weight.

Eq. (9.4) shows that if two gas species are being transported through a membrane under Knudsen flow, the flux ratio (J_i/J_k) varies inversely with the ratio of the molecular weights:

$$(J_i/J_k) = (M_k/M_i)^{1/2} \tag{9.5}$$

In the transition region between Knudsen flow and Poiseuille flow ($0.001 < N_{Kn} < 100$) correction factors are required to account for the varying effect of molecular and boundary collisions. Eqs. (9.4) and (9.5) apply to pure gases moving in straight cylindrical pores, and a correction factor is also needed to account for pore size distributions and for tortuous flow paths through actual membrane pore networks. Eq. (9.4) can be corrected with a porosity–tortuosity factor e/T by replacing D_i for a single pore by D_i for a porous medium:

$$D_{i\ \text{Porous medium}} = D_{i\ \text{Single pore}} \cdot \varepsilon/\tau \tag{9.6}$$

The molecular weight dependence introduces the possibility of selectivity through materials exhibiting Knudsen flow. However, as shown in Table 9.3, the selectivity that can be achieved for common carbon capture-related gas separations is limited due to the relatively small molecular weight differences. Materials exhibiting Knudsen flow are therefore not attractive for direct gas separation applications, although this can be a consideration in choosing the support material for a composite membrane, because Knudsen flow in the support can have an impact on overall selectivity.

Membranes exhibiting Knudsen flow have been investigated for use in gas–liquid membrane contactors being developed as an alternative to absorption towers for CO_2 recovery from flue gases. In this application flue gas flows through the bores of a hollow-fiber bundle membrane modules (Section 9.2), with a CO_2 absorbent, such as an amine solvent, flowing in counter-current through the shell side of the module.

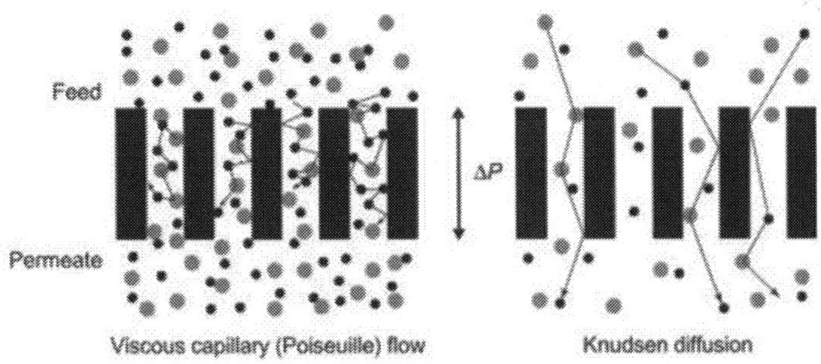

FIGURE 9.2 Viscous capillary flow and Knudsen diffusion.

TABLE 9.3 Porous membrane selectivity under Knudsen flow.

Gas components	Molecular weight ratio	Knudsen flux ratio
CO_2/H_2	21.83	1/4.67
CO_2/CH_4	2.75	1/1.66
CO_2/N_2	1.57	1/1.25
O_2/N_2	1.14	1/1.07

Fig. 9.3 illustrates this membrane gas absorption process. In this application the membrane itself offers no selectivity, having a nominal pore diameter typically in the 10–200 nm range. However, it serves to provide an interface that offers a very high surface-to-volume ratio for mass exchange between the gas and solvent.

An important assumption underlying the calculation of flux in Knudsen flow is that the interaction between permeate molecules and the pore surface is simply one of momentum exchange in collisions. Adsorption and desorption are assumed to be in equilibrium and surface diffusion effects are ignored. As the membrane pore diameter approaches the dimensions of the permeate molecules, these assumptions are no longer valid and the interactions between permeate molecules and the pore surfaces become significant factors in determining flux.

Surface diffusion and capillary condensation

In Knudsen flow of a gas mixture, a component with a higher molecular weight and with higher polarity and polarizability will tend to be preferentially adsorbed onto the pore surface (Fig. 9.4). This will increase the permeation rate of the adsorbed component by adding a surface diffusion flux to the Knudsen diffusion flux. Selectivity can also be increased, because the adsorbed species can substantially reduce the pore space available for the diffusion of other components, particularly if the pore size is in the range of two to three times the kinetic diameter of the adsorbed species. Surface effects can be promoted by modifying the pore surfaces, for example, by the addition of functional groups that promote adsorption.

If adsorption continues beyond a monolayer, capillary condensation can take place (Fig. 9.4). Ultimately the pores may become blocked with the condensate and diffusion of other gas components will be prevented, further increasing selectivity.

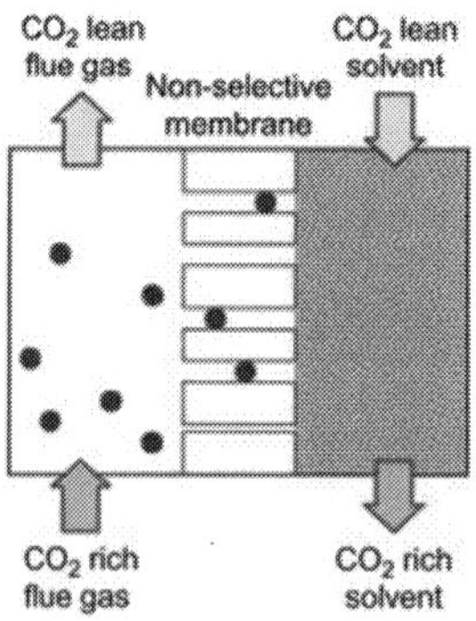

FIGURE 9.3 Gas–liquid membrane contactor for CO_2 removal from flue gas.

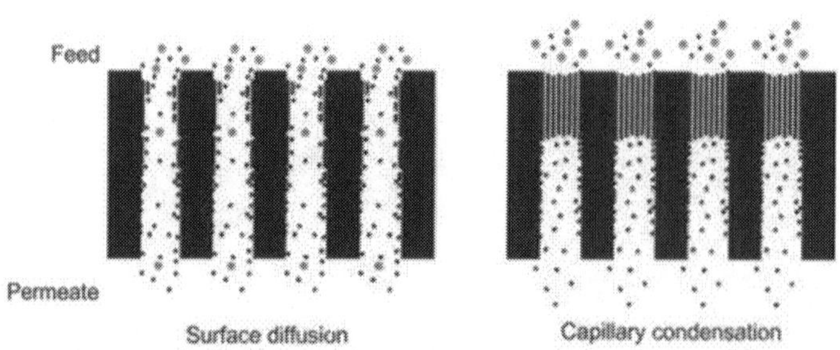

FIGURE 9.4 Surface diffusion and capillary condensation effects.

Molecular sieving

Molecular sieving is the process that takes over from Knudsen flow when the pore diameter approaches the diameter of the molecules attempting to pass through a porous membrane (Fig. 9.5).

Molecular sieves are materials with pore sizes of molecular dimensions within which, ignoring surface effects, selectivity is a function of molecular size, with smaller molecular species having higher diffusion rates.

In molecular sieve studies, permeate molecules are commonly known as "guest" molecules. At this scale, neither the pore walls nor the guest molecules can be considered rigid; thermal vibration of the lattice structure of the membrane material and of the guest molecule will give both a degree of flexibility. The effective size of a hydrogen, nitrogen, or CO_2 molecule is determined by a combination of the bond length between the constituent atoms plus the van der Waals's radius, which characterizes the range of the electron cloud surrounding each atomic nucleus (Table 9.4). The size of a guest molecule is characterized by its minimum diameter when in an equilibrium state and is known as the minimum kinetic diameter.

Some important examples of molecular sieve materials, zeolite, activated carbon (carbon molecular sieves), and microporous silica, are compared in Table 9.5.

Transport of gas molecules through the tortuous pore network of a molecular sieve takes place through a combination of diffusion within the pore space and adsorption and surface diffusion on the pore surfaces. Adsorbed gas atoms or molecules (adatoms or adparticles) diffuse on the pore surfaces down the concentration gradient between the upstream (feed) and downstream (permeate) sides of the membrane. These surface diffusion effects will be important only at lower temperatures because adsorption will no longer occur once the thermal energy of the gas molecule exceeds the heat of adsorption.

For a binary gas mixture, diffusivity within the pores will be dependent on the molecular sizes. If the adsorption strengths of the two species are similar, the membrane will be selective for the smaller molecule. However, if molecular sizes are similar, selectivity will be driven by adsorption strength, because the component with the higher adsorption strength will dominate the available adsorption sites and the surface diffusion contribution to permeate flux.

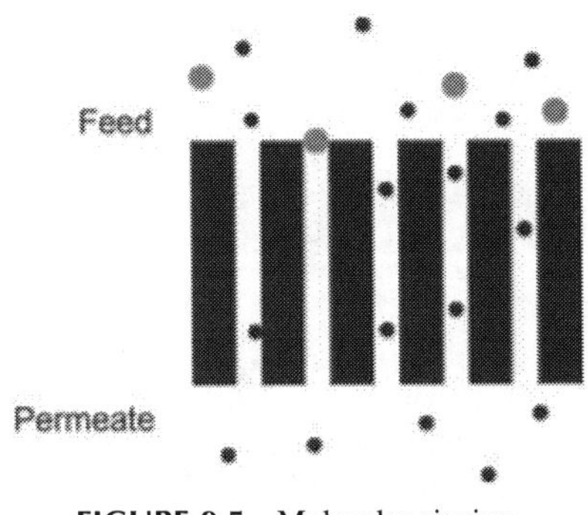

FIGURE 9.5 Molecular sieving.

TABLE 9.4 Common molecular dimensions.

Element	Bond length (Å = 10^{-10} m)	Van der Waal's radius (Å)	Kinetic diameter (d_m Å)
Hydrogen	1.97	1.20	2.89
Nitrogen	1.04	1.55	3.64
Oxygen	1.21	1.52	3.46
CO_2	1.42 (single bond)	1.70 (carbon)	3.35
Methane	1.09 (single bond)	2.00	3.80
H_2O	0.96 (single bond)	3.12	2.65

TABLE 9.5 Molecular sieve materials compared.

Material	Description	Example applications
Zeolite	Microporous aluminosilicate minerals with well-defined pore sizes in the range of 3–13 Å, determined by the specific chemical composition and crystal structure. Selectivity can be manipulated over a wide range by modifying the structure and chemistry of synthetic zeolites	Hydrogen separation for 3 Å and CO_2/N_2 or CO_2/CH_4 separation for 4 Å sieves
Activated carbon	Activated carbon is an amorphous, disordered network of graphitic carbon platelets with interstitial spaces due to the presence of cross-linking chains of carbon atoms and foreign atoms between the layers. Carbon molecular sieves (CMSs) use activated carbon with a molecular scale pore network	H_2/CO_2 separation from water–gas shift (WGS) reaction products in gasification processes. Oxygen separation from air using CMS and pressure-swing adsorption
Composite silica–alumina or mixed metal oxide (e.g., Mg–Al–O)	Microporous silica (pore diameter ~3Å) deposited on a mesoporous alumina or mixed metal oxide support	H_2/CO_2 separation from WGS reaction products in gasification process

Among carbon capture-relevant gases (CO_2, N_2, CH_4, H_2), CO_2 has the highest heat of adsorption onto zeolites and other minerals with polar surfaces, and membranes made of these materials are therefore selective for CO_2 at low temperatures. If adjacent adsorption sites are separated by an energy barrier, as is the case for many adsorbates on zeolite, an activation energy is required to enable an adparticle to hop from one adsorption site to another. This variant of surface diffusion is known as activated diffusion. Another example of activated diffusion is the passage of a guest molecule through a pore aperture that is the same size as or slightly smaller than the diameter of the diffusing molecule.

Multicomponent, multiphase transport through a molecular sieve, including surface diffusion, is described mathematically using the Maxwell–Stefan formulation, which accounts for both membrane–guest and $guest_1$–$guest_2$ interactions such as competitive adsorption. This occurs when the available adsorption sites become saturated by the most strongly adsorbing component, blocking the transport of the other component either due to the lack of sites or because the adsorbed layers reduce the free pore size below the kinetic diameter of the second component.

When simplified for a single component, the Maxwell–Stefan expression for surface diffusivity is also known as the Darken equation, and gives the flux due to surface diffusion in the form:

$$J_s(T) \sim \rho q_{sat} D_s^0(0)(1-\theta)-^1\exp(-E_{D,S}/RT)d\theta/dx \quad (9.7)$$

where J_S is the flux due to surface diffusion at temperature T, ρ is the density of the membrane material, q_{sat} is the saturated molar volume of gas adsorbed by the membrane, $D_S^0(0)$ is the limiting surface diffusivity, θ is the fraction of available adsorption sites that are occupied, $E_{D,S}$ is the surface diffusion activation energy, R is the gas constant, and x is the linear dimension parallel to the flow through the membrane.

The dependence of permeance and selectivity of a molecular sieve on the surface interactions between the guest molecule and the pore surfaces opens up the opportunity to improve membrane separation performance by "nano-engineering" the pore surfaces to increase the surface diffusion rate of the desired permeate. Synthetic zeolites have been widely investigated in this respect because the ionic properties of the surface can be controlled by varying the composition of the zeolite. Some examples of the CO_2/N_2 selectivity and CO_2 permeance for these ion-exchange zeolite molecular sieve membranes in the temperature range 35–40°C are shown in Table 9.6 and illustrate the generally inverse relationship between permeance and selectivity for a particular material type. Application of a surface charge has also been investigated as a means of altering the surface interactions to influence selectivity (Sun et al., 2019).

A molecular sieve with a pore diameter of a few Angstroms is the finest form of microporous membrane. As pore sizes reduce further, and finally disappear, the transport process that takes over in a nonporous membrane—solution–diffusion—is a far simpler process than the complex, surface-mediated adsorption–surface diffusion occurring in the finest-scale porous membranes.

TABLE 9.6 Zeolite ion-exchange membranes: CO_2–N_2 molecular sieving performance.

Material	CO_2/N_2 selectivity	CO_2 permeance (mol $m^{-2}s^{-1}Pa^{-1}$)
Li(20%)Y	10	7×10^{-7}
LiY	4	2×10^{-6}
K(62%)Y	39	5×10^{-7}
KY	30	1.4×10^{-6}

9.1.2 Solution–diffusion transport process

Gas separation using nonporous membranes occurs in a two-step process in which the permeate first dissolves into the membrane and then diffuses through it (Fig. 9.6). Hydrogen separation using a dense metal membrane is one such example, with adsorption of hydrogen on the feed side of the metal membrane followed by dissociation of hydrogen molecules and diffusion of protons in the membrane lattice. Hydrogen molecules are then regenerated on the permeate side of the membrane and desorbed into the permeate stream.

The effectiveness of the solution–diffusion process is determined by how well the permeate is absorbed into the membrane and then how fast it diffuses through it. Permeate solubility is controlled by Henry's law, which states that, at a given temperature, the amount of a gas dissolved in a unit volume of a medium is proportional to the partial pressure of the gas in equilibrium with the medium. This is expressed for permeate component i as follows:

$$C_i = P_i/K_{Hi} = P_i\ K_i \tag{9.8}$$

where C_i is the dissolved concentration, P_i is the partial pressure, K_{Hi} is Henry's law constant, and K_i is the solubility of component i. Considering CO_2 separation, the gas solubility (K_{CO_2}) is expressed as follows:

$$K_{CO_2} = C_{CO_2}/P_{CO_2} = 1/K_{HCO_2} \tag{9.9}$$

Diffusion through a nonporous membrane is governed by Fick's first law of diffusion, which states, again considering CO_2 separation, that:

$$J_{CO_2} = D_{CO_2}\Delta C_{CO_2}/L \tag{9.10}$$

where J_{CO_2} is the flux of CO_2 through unit area of the membrane, D_{CO_2} is the diffusion coefficient or diffusivity of CO_2 in the membrane, and ΔC_{CO_2} is the difference in CO_2 concentration in the membrane between the upstream and downstream surfaces. Replacing ΔC_{CO_2} using Eq. (9.9) gives:

$$J_{CO_2} = K_{CO_2}D_{CO_2}\Delta P_{CO_2}/L \tag{9.11}$$

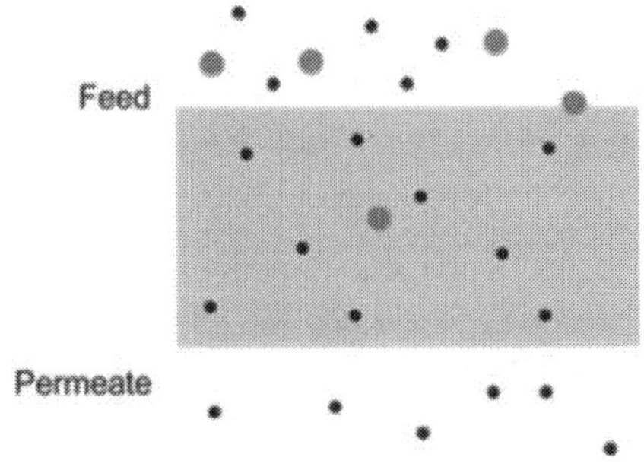

FIGURE 9.6 Solution–diffusion transport process.

where ΔP_{CO_2} is the difference in CO_2 partial pressure across the membrane. The product of solubility and diffusivity is the permeability of the membrane to the molecular component under consideration.

Other gases in the feed stream will also dissolve in and diffuse through the membrane, depending on their specific solubility and diffusivity. The ratio of the flux rates for two components i and k determines the selectivity of the membrane:

$$\alpha_{i/k} = J_i/J_k \tag{9.12}$$

Polymeric membranes are an important class of nonporous membranes operating through the solution–diffusion mechanism, and typically consist of a thin dense selective layer supported by a less dense, porous, and therefore nonselective supporting layer (Section 9.2.1). The thinness of the selective layer is dictated by the need to maximize permeance, and therefore flux through the membrane, which will reduce the needed membrane area and minimize cost, the selective layer being high cost due to the customized materials and preparation requirements.

The performance of polymeric membranes can be enhanced by increasing the solubility and diffusion rate of the desired permeate. Solubility is controlled by changing the chemical composition of the polymeric material, while diffusion rate is controlled by changing the physical packing of polymer within the membrane—a characteristic that is strongly dependent on the method of membrane fabrication. Very high selectivity can be achieved by these means but can be sustained only at relatively low temperatures, typically up to ~100°C.

The physical arrangement of polymers within the membrane depends on whether the operating temperature of the membrane is above or below the glass transition temperature of the membrane material (T_g, see Glossary). Polymeric membranes operating above T_g are termed rubbery, and are flexible and generally of higher permeability and lower selectivity. Although hybrid (polymer + inorganic) membranes have been reported operating at up to 400°C, the rapid drop in selectivity with increasing temperature above T_g imposes a limit on the applicability of many polymeric membranes.

Below T_g, polymeric membranes are glassy, hard, and less permeable, but also have microscopic voids between the polymer chains as a result of imperfect packing. This excess free volume has an important effect on the performance of glassy membranes as a result of increased solubility due to adsorption of the permeate onto the void surfaces. This adsorption is described by Langmuir's adsorption equation:

$$C_{Li} = C_{max.i}\, a_i\, P_i/(1 + a_i\, P_i) \tag{9.13}$$

where C_{Li} is the adsorbed concentration on the microvoid surfaces, $C_{max.i}$ is the maximum adsorbed concentration, a_i is the Langmuir adsorption constant, and P_i is the partial pressure of component i. The total bulk concentration, from Eqs. (9.8) and (9.13), is therefore:

$$C_i = P_i\, K_i + C_{max.i}\, a_i\, P_i/(1 + a_i\, P_i) \tag{9.14}$$

Glassy and rubbery materials have also been combined in composite polymeric (copolymer) membranes, which aim to exploit the selectivity of glassy polymers and the permeability of rubbery polymers to optimize overall membrane performance.

Nonporous ceramic metal oxide membranes based on solution–diffusion have also been investigated for CO_2 separation at high temperatures. Oxides such as Li_2ZrO_3 are particularly suitable, having high CO_2 adsorption capacity (0.1 kg/kg at 600°C and $P_{CO_2} = 70$ kPa) and negligible absorption of other gases such as hydrogen.

9.1.3 Mixed matrix membranes

In mixed matrix membranes (MMMs) two or more materials with dissimilar properties are combined with the aim of creating new materials which display the best features of the components. Polymers, either glassy or rubbery, have typically been used as the continuous phase with a variety of different embedded particles, including zeolites, carbon molecular sieves and nanotubes, mesoporous and dense silica, mixed-metal oxides, zeolitic imidazolate frameworks (ZIFs) and other metal–organic frameworks (MOFs) (Fig. 9.7).

The pioneering work on MMMs was performed in the 1970s and involved molecular sieve particles, such as zeolite 5A, incorporated into rubbery polymers. Transport through the membrane involves a combination of Knudsen diffusion in the molecular sieve component and solution-diffusion in the polymer component. The main fabrication challenge is to ensure that the sieving particles are well bonded with the polymer to avoid microcavities within the membrane, which can result in a loss of selectivity. Surface modification of the embedded particles, for example using organic coatings, is one approach to

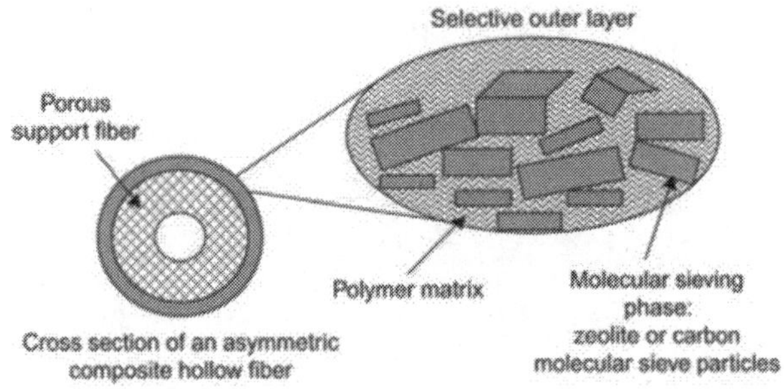

FIGURE 9.7 Mixed matrix hollow-fiber membrane structure.

improve membrane cohesion while also improving permeance and selectivity through surface functionalization (Zhu et al., 2019).

As well as the wide range of individual filler particles noted earlier, binary combinations of particles have also been investigated, including mesoporous silica + mixed-metal oxides, and zeolite + ZIFs. This approach has a number of advantages, including improved dispersion of particles within the matrix during fabrication, higher permeability and selectivity, and the potential to engineer multifunctional MMMs for separation of multiple components from gas streams. Going one step further, ternary MMMs incorporate a third component, typically a low molecular weight organic additive, that improves the bonding between matrix and filler particles and fills any void space, with the aim of further improving membrane performance.

MMM filler particles inspired by the catalytic effect of carbonic anhydrase enzyme (CA—see Section 7.1.1) have also been investigated. The high cost and poor stability of CA have been addressed by the preparation of more durable complexes that mimic the CA active site, the first such example being Co-BBP, synthesized by Sahoo and coworkers, in which cobalt (II) is coordinated with 2, 6-bis(2-benzimidazolyl) (Sahoo et al., 2013). Such nanomaterial-based artificial enzymes, or nanozymes, have since been investigated for a very wide range of applications, from cancer therapy to CO_2 removal (Wu et al. 2019). Significant increases in the CO_2 permeance and selectivity of polymer membranes have been demonstrated by incorporating Co-BBP activated nanoparticles (porous organic polymers or POPs) into a polyether block amide (Pebax) polymer matrix (Wang et al., 2020).

9.1.4 Facilitated transport membranes

The solution–diffusion process described earlier is an example of a passive membrane transport mechanism, in which the permeate travels by diffusion down a concentration gradient. The application of membranes based on this process is often constrained by low permeate flux rates due to low solubility and/or low diffusivity. In contrast, facilitated, or carrier-assisted, transport is an active transport mechanism that increases flux rate by transporting the permeate across the membrane attached to a carrier (Ward and Robb, 1976). Many biological processes rely on facilitated transport across cell walls, and in principle facilitated transport can enable the permeate to move against a concentration gradient (from a region of low concentration to one of high concentration).

In facilitated transport, the permeate reacts reversibly with a component present in the membrane (the carrier) to form complexes or reversible reaction products at the feed side of the membrane. These complexes then diffuse across the membrane driven by a concentration gradient in the complex rather than a gradient in the permeate. At the downstream surface of the membrane, the complexing process or reaction is reversed, liberating the transported component into the permeate stream.

Carriers can be either fixed within the membrane structure or mobile, or a combination of fixed and mobile may be used. Fig. 9.8 illustrates the facilitated transport of CO_2 across a cellulose acetate membrane containing an aqueous carbonate solution. At the feed side of the membrane, bicarbonate anions are formed by the dissolution of CO_2 in water (reaction 9.15), which is energetically aided by the presence of the mobile carbonate anion carrier:

$$CO_2 + H_2O + CO_3^{2-} \leftrightarrow 2HCO_3^- + \text{heat} \tag{9.15}$$

while, at the permeate side of the membrane, the reverse reaction occurs and CO_2 is liberated.

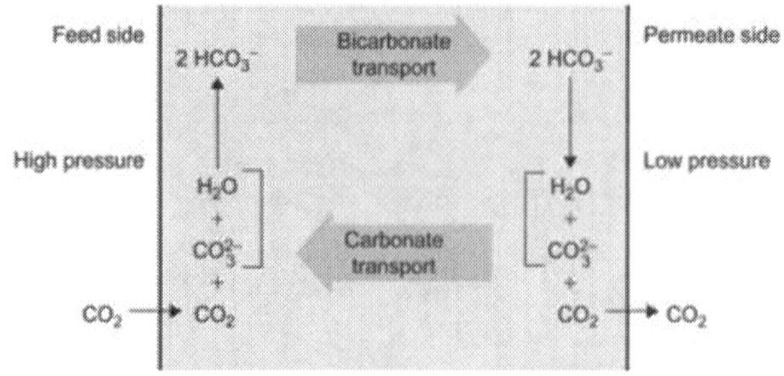

FIGURE 9.8 Facilitated transport of CO_2 in a supported liquid membrane (SLM) with carbonate carrier.

An example of a fixed carrier facilitated transport membrane for CO_2 separation is shown in Fig. 9.9. Here, CO_2 diffuses into the membrane and reacts with water and fixed amine groups, attached to the polymer backbone, to form bicarbonate ions:

$$CO_2 + H_2O + R{-}NH_2 \leftrightarrow HCO_3^- + R{-}NH_3^+ \tag{9.16}$$

Bicarbonate ions diffuse across the membrane between carrier sites, while protons are transported along the polymer backbone, with CO_2 being released and diffusing out of the membrane at the permeate side.

Secondary amines have also been investigated as carriers, the advantage being the lower binding energy and therefore faster desorption compared to primary amines. The reaction of a secondary amine (R_1R_2NH) with dissolved CO_2 is described by a two-stage process, starting with the formation of a zwitterion reaction product ($R_1R_2NH^+CO_2^-$), a polar molecule that is electrically neutral but carries formal charges on the amine and oxygen:

$$CO_2 + R_1R_2NH \rightarrow R_1R_2NH^+CO_2^- \tag{9.17}$$

In the second step, the zwitterion donates its excess proton to a second amine, forming a carbamate ion that transports CO_2 across the membrane:

$$R_1R_2NH^+CO_2^- + R_1R_2NH \rightarrow R_1R_2NCO_2^- + R_1R_2NH_2^+ \tag{9.18}$$

the reverse reactions taking place on desorption at the permeate side of the membrane. Amine-based compounds, including ethylenediamine and amino acid salts, such as potassium, lithium, and piperazine glycinate, have also been investigated as mobile carriers, either alone or in combination with fixed amine carriers.

Several examples of facilitated transport are summarized in Table 9.7. The permeability and selectivity values shown in the table illustrate the range of experimental results that have been achieved by varying the carrier chemistry for a given polymer type. The "Barrer" is a commonly used unit of membrane permeability and is equal (in SI) to 3.4×10^{-16} mol $m^{-1}s^{-1}Pa^{-1}$. Note that for a membrane with a 1 μm active layer thickness, the permeance measured in gas permeance units is equal to the permeability in Barrer.

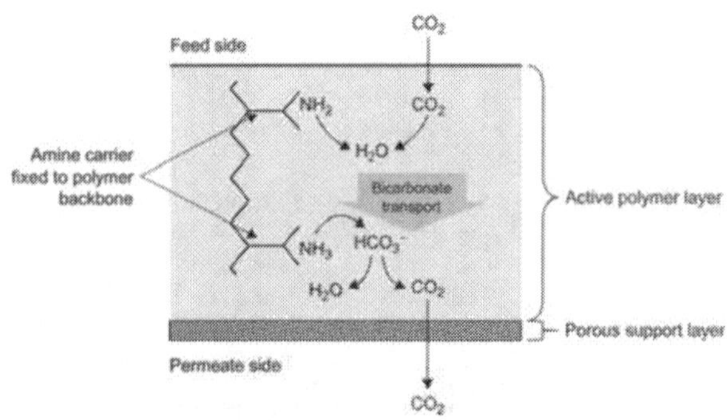

FIGURE 9.9 Amine-facilitated transport membrane for CO_2 separation.

TABLE 9.7 Experimental facilitated transport membrane performance.

Membrane type: Facilitator	Experimental conditions	CO_2 permeability (barrer)	CO_2/X selectivity
Cellulose acetate: Pure water	$CO_2 + O_2$, 20°C, 4 kPa	400	22 (CO_2/O_2)
Cellulose acetate: 2N $KHCO_3$ + 0.5N $NaAsO_2$	$CO_2 + O_2$, 20°C, 4 kPa	2000	600–800 (CO_2/O_2)
Vinyl alcohol–acrylate copolymer: 2N K_2CO_3	90%N_2, 10% CO_2, 20°C, 98 kPa	6100	317 (CO_2/N_2)
Vinyl alcohol–acrylate copolymer: 2N K_2CO_3 + 0.5N EDTA	90%N_2, 10% CO_2, 20°C, 98 kPa	2400	1417 (CO_2/N_2)
Polyvinylalcohol: Glycine + polyethylenimine	40/20/40 $H_2/CO_2/N_2$, 20°C, 200 kPa	194	28 (CO_2/H_2)
Polyvinylalcohol (cross-linked formaldehyde): Dimethylglycine + polyethylenimine + KOH	40/20/40 $H_2/CO_2/N_2$, 20°C, 200 kPa	338	1782 (CO_2/H_2)
Polyethersulphone: Polyvinylamine + aminoacid salts	80%N_2, 20% CO_2, 57°C, 10 kPa	110	210 (CO_2/N_2)

A recent development in facilitated transport is the use of a molecularly thin amine layer prepared on a commercial supported polymer membrane to form a so-called hybrid-integrated membrane (Sandru et al., 2022). The combination of the high CO_2 selectivity of the amine nano-layer with the high permeability of the polymer resulted in a membrane displaying ultra-selectivity (CO_2/N_2 >1000) and high permeability (CO_2 permeability > 1000 Barrer; see Fig. 9.17). Both elastomeric polydimethylsiloxane and glassy polytetrafluoroethylene (PTFE AF) polymer membranes were investigated in this proof-of-concept study.

9.1.5 Ion transport membranes

Ion transport membranes, also known as ion exchange membranes or mixed conducting membranes, are a class of facilitated transport membranes in which selective transport of ionic species is achieved as a result of the structure and composition of the membrane. Fig. 9.10 shows an example of an ion exchange membrane in which oxygen ions are transported through a mixed-metal oxide lattice by jumping between oxygen vacancies in the lattice structure. The vacancies are in effect fixed carrier sites facilitating oxygen transport. Electrons liberated at the permeate side as the ions recombine into oxygen molecules are conducted back through the matrix to maintain electrical neutrality. A high electrical conductivity is therefore a desired characteristic for such a membrane.

Minerals such as perovskite (see Glossary) and synthetic perovskite-like ceramics are suitable for this type of membrane and can be engineered for selectivity and permeability by changing the stoichiometric ratios of the metal components and oxygen vacancies. These materials have the advantage of very high-temperature operation (in the range 700–900°C) and are under development for use in syngas generation and oxygen separation for oxycombustion (Section 9.4). In the

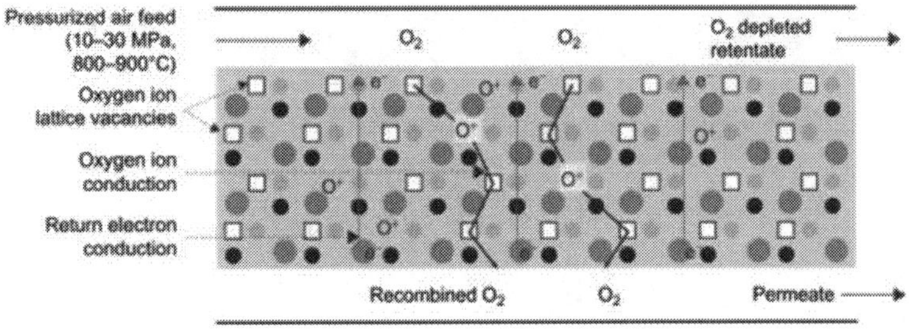

FIGURE 9.10 Schematic oxygen ion transport in a metal oxide lattice.

latter application, the oxygen separator may be either a low-temperature unit or a high-temperature reactor closely integrated with the combustion and heat cycle.

Dual ion exchange membranes have also been investigated for CO_2 separation, with CO_2 transported as a carbonate ion (CO_3^{2-}) within a mixed Li/Na/K carbonate phase which is molten at the high operating temperature of the membrane. This class of membranes is also known as dual-phase membranes, because the molten carbonate is supported in a solid metal or ceramic framework. Formation of the carbonate ion requires the presence of oxygen in the feed gas stream. In the case of a mixed electron carbonate conductor (MECC), the second ion exchange process is the return of the excess electrons, typically through a stainless steel or silver framework, or in a mixed oxide carbonate conductor (MOCC), an oxygen ion (O^{2-}) is returned in the second exchange process through a ceramic mixed metal oxide (e.g., $Bi_{1.5}Y_{0.3}Sm_{0.2}O_3$, or BYS). A molten carbonate MOCC dual ion exchange membrane being developed to enable CO_2 separation from high-temperature and high-pressure streams in applications such as biomass integrated gasification combined cycle (BIGCC) is described further in Section 9.4.

At lower temperature, bipolar membrane electrodialysis (BPMED) has been investigated for the regeneration of caustic solvents such as sodium or potassium hydroxide, and for CO_2 capture from seawater (Nagasawa et al., 2009; Eisaman et al., 2012). A BPMED unit comprises a stack of alternating anion exchange and bipolar exchange membranes, the latter having both anion and cation exchange layers, at which protons and hydroxide ions are generated in by the dissociation of water. Fig. 9.11 illustrates the BPMED configuration investigated for KOH regeneration as part of a direct air capture system (see Chapter 14).

Ion exchange membranes for this type of application can also be fabricated as mixed matrix membranes (see earlier discussion) by using filler particles that enable ion transport. This is one approach to addressing the high cost of ion exchange membranes, which is a significant economic barrier to deployment (Sabatino et al., 2020).

9.1.6 Supported liquid membranes

Immobilized or supported liquid membranes (SLMs) are membranes in which a liquid is immobilized by capillary pressure within the pore space of a support material, and in which gas permeation occurs typically by the solution-diffusion mechanism. A wide variety of support materials have been tested, including alumina and titanium oxides, PTFE, mesoporous carbon, ZIFs and MOFs, while liquids most commonly employed in SLMs have been amine-based solvents, such as diethylenetriamine, diaminoethane, and diglycolamine.

More recently ionic liquids (see Chapter 7) have been investigated as sorbents in supported ionic liquid membranes (SILMs), having the dual advantages of low vapor pressure and high viscosity, which can reduce liquid losses from the membrane through evaporation or expulsion due to a high trans-membrane differential pressure (Dai et al., 2016).

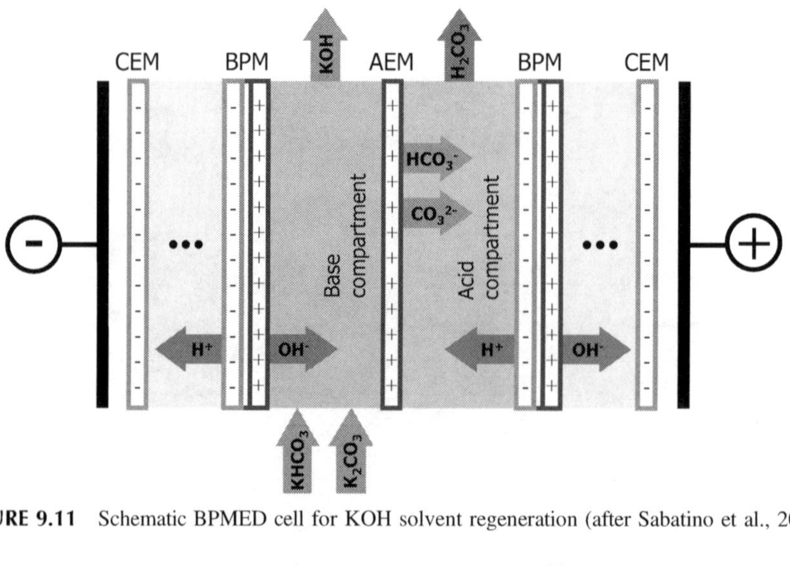

FIGURE 9.11 Schematic BPMED cell for KOH solvent regeneration (after Sabatino et al., 2020).

In the case of SILMs using functionalized or task-specific ILs, the amine functional groups introduce a facilitated transport mechanism of the type illustrated in Fig. 9.9, alongside the solution-diffusion mechanism.

The combination of ILs, which can be tailored by the selection of a variety of cations and anions, with different membrane types (thin film composite, mixed matrix, etc.) provides a fertile field for generating novel membranes with enhanced properties. Current areas of research in SLMs include (Dai et al., 2016);

- alternative functional groups that can enhance gas separation (high capacity, fast kinetics)
- methods to prepare composite membranes with an ultra-thin SIL layer on a highly permeable support, to improve gas permeability while retaining high selectivity, and
- improved mechanical strength and stability at high temperature and in the presence of contaminants from industry-relevant gas mixtures.

9.1.7 Molecular gate membranes

A novel combination of supported liquid and facilitated transport mechanisms has been exploited in the development of molecular gate membranes which show promise for CO_2/H_2 separation in gasification applications (Kazama et al., 2005; Kai et al., 2013). These are a type of polymeric membrane in which a dendrimer (see Glossary) such as poly(amidoamine), which is a room temperature liquid, is immobilized in a polymeric network. Typically in polymer membranes, the solution-diffusion process gives little or no CO_2/H_2 selectivity—CO_2 being more soluble but H_2 having higher diffusivity due to its smaller size. However, in a molecular gate membrane, CO_2 reacts with the amine groups forming bicarbonate or carbamate ions which facilitate the transport of CO_2 across the membrane while at the same time reducing its permeability to H_2 by 10–100-fold, depending on the dendrimer content (20–50 wt%).

9.2 Membrane configuration and preparation, and module construction

Although widely used for natural gas processing and air separation, membrane technology for negative emissions-related gas separation applications is still confined to the laboratory or small pilot scale. As well as the physical and chemical investigation and optimization of membrane materials and structures, the industrial-scale use of membranes in NET applications will also benefit from further development of technologies for fabricating reliable and economical membrane modules.

The main challenges here include:

- reducing the thickness of the membrane selective layer to increase flux and reduce cost, while preserving selectivity
- large-scale production of defect-free membranes with the desired physical and chemical properties
- automation of membrane and module fabrication processes to cost-effectively package the membrane into reliable modules, achieving maximum membrane surface area within a given module size
- chemical and mechanical compatibility and stability within the industrial process environment, to give a commercially viable membrane lifetime
- plant-level analysis to optimize system configuration and operating parameters, and cost versus performance trade-offs.

Methods for industrial-scale preparation of membranes and fabrication of modules have been developed since the 1960s, driven by applications in reverse osmosis for water purification and desalination. Many of these techniques are directly applicable to gas separation problems but novel membrane materials and applications in CO_2 separation will also require new developments in this area.

In this section, membrane preparation and module fabrication methods are described, with reference to specific applications and to current areas of technology development.

9.2.1 Membrane types

Membrane preparation methods and suitable module configurations are strongly tied to the structure of the membrane material. Several types of membrane structure can be identified, as shown in Fig. 9.12, the main features of these structures being summarized in Table 9.8. A fascinating account of the origin of the Loeb-Sourirajan membrane, which was central to the development of desalination by reverse osmosis, can be found in (Loeb, 1981).

Most nonliquid membranes, whether polymeric, ceramic, or metallic, can be prepared as either symmetric or asymmetric structures. However, optimization for high flux rate and low cost generally drives toward asymmetric structures with

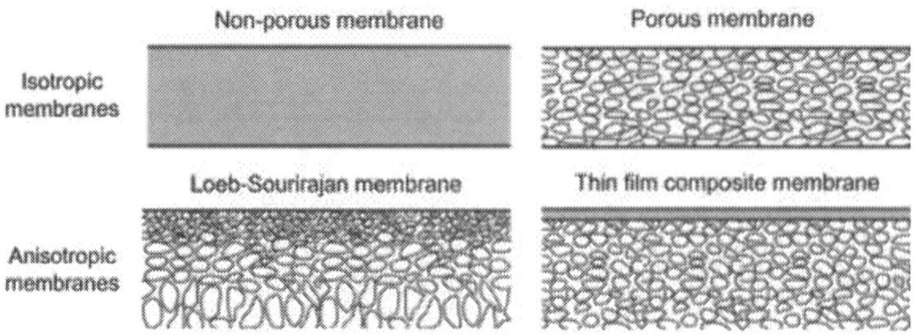

FIGURE 9.12 Physical structure of membranes.

TABLE 9.8 Structural characteristics of membranes.

Structure	Characteristics
Symmetric (or isotropic) membranes	Porous or dense nonporous membranes with a uniform structure throughout the membrane thickness
Asymmetric (or anisotropic) membranes	Porous or dense nonporous membranes with a thin selective layer either bonded to a thicker porous nonselective layer that provides mechanical support (thin film composite membrane) or resulting from the specific membrane preparation process (Loeb-Sourirajan membrane)
Liquid membranes	A liquid selective phase, supported, contained, or immobilized within the pore space of a polymeric or ceramic membrane

a thinner selective surface combined with a thicker, nonselective supporting layer. In composite membranes, the influence of the substrate material extends beyond mechanical support, and the properties of the substrate and the substrate-active layer interface can have an important influence on overall membrane performance (Ramon et al., 2012).

Zeolite membranes are one example of asymmetric membranes and are typically prepared by in situ hydrothermal synthesis of a thin zeolite layer onto support tubes or disks of porous stainless steel or alumina.

9.2.2 Membrane module configurations

Symmetric or asymmetric membranes can be prepared either as laminar membranes, as hollow tubes or fibers, or as wafers, and packaged in a variety of different module configurations. The earliest module designs consisted of flat sheets of membrane material held in a frame (plate-and-frame modules) or bundles of relatively large diameter (1−3 cm) tubes (tubular modules). This module design does not in general provide the high surface area to bulk volume ratio required for gas separation applications (Table 9.9), and current commercially available gas separation modules use either spiral-wound or hollow-fiber modules.

Spiral-wound modules

Spiral-wound membrane modules consist of a number of membrane envelopes wound onto a central perforated collecting tube. Each membrane envelope, or leaf, consists of two membranes separated by a permeate spacer and sealed on three

TABLE 9.9 Area-to-volume ratios for various membrane configurations.

Module configuration	Typical area-to-volume ratios (m^2/m^3)
Plate and frame	200
Spiral wound	500−1000
Hollow fiber	1500−10,000

sides, leaving one side open for permeate removal. Early designs used a single membrane envelope, but this requires a long spiral path to achieve a high surface area-to-volume ratio, resulting in a high pressure drop along the spiral permeate flow path. Fig. 9.13A illustrates the configuration of a spiral-wound membrane element.

The feed gas mixture flows axially along the module through feed spacers held between successive membrane leaves. The permeate passes through the membranes and then flows spirally through the space held open by the permeate spacer, exiting the open edge of the membrane leaf into the central collecting tube.

The construction of spiral-wound modules in commercial production for CO_2 separation is illustrated in Fig. 9.13B. Optimization for a specific application requires selection of the number of membrane leaves, module length, and diameter, which determine the pressure drop, weight, and cost of the module. Weight and ease of handling are important practical considerations in view of the need to replace modules after performance degradation or failure.

Typically four to six modules are mounted inside a tubular pressure vessel, and many of these tubular elements are mounted in series or parallel to form the complete membrane separation unit. Fig. 9.14 shows an example of a natural gas sweetening unit (i.e., CO_2 or H_2S removal) comprising two banks of six tubular membrane modules.

Hollow-fiber modules

Hollow-fiber modules are also extensively used in natural gas treatment because of the very high surface-to-bulk volume ratio noted earlier. Modules consist of bundles of fibers enclosed in a pressure vessel, with the permeate passing either out of the fibers (bore-side feed) or into them (shell-side feed), as illustrated in Fig. 9.15. Fibers are mechanically able to withstand much higher differential pressures applied externally than internally, so for high-pressure applications the shell-side feed configuration is used. Optimization of hollow-fiber modules for specific applications involves the selection of the fiber diameter, and hence the packing density of fibers in the module, the flow direction (shell- or bore-side), and whether the permeate flow is cocurrent or counter-current.

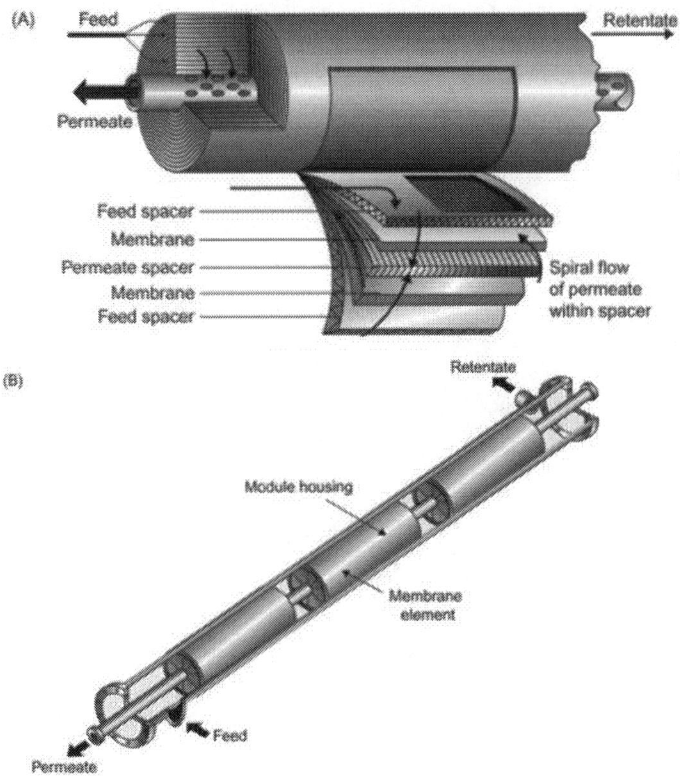

FIGURE 9.13 Spiral-wound membranes: (A) element configuration and (B) module construction.

FIGURE 9.14 Natural gas sweetening membrane unit. *(Courtesy: Membrane Technology and Research Inc.).*

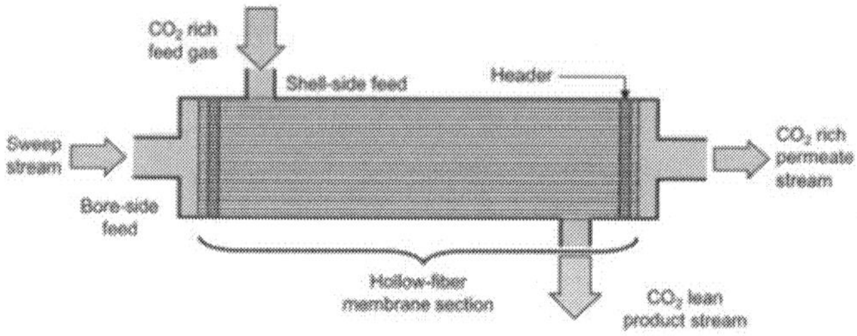

FIGURE 9.15 Hollow-fiber membrane module structure.

Hollow-fiber membrane separation units for gas separation are constructed in a similar manner to the spiral-wound units. Fiber modules are commonly 1 m in length and 0.2–0.25 m in diameter, again for physical handling reasons, and contain $\sim 10^6$ individual fibers with internal and external diameters of $\sim$50 and 100 μm. Hollow-fiber modules for gas–liquid contactors use polymer fibers, which are typically an order of magnitude larger in cross-section, with inner and outer diameters of $\sim$600 and 1000 μm.

Hollow-fiber-based separation units are typically smaller than those based on spiral-wound modules, although spiral-wound modules are able to handle higher operating pressures and are more resistant to fouling caused by particles in the feed gas stream that can block fine fibers.

Ceramic wafer stack modules

Advanced membrane materials such as metal oxide ion transport membranes require new module structures to cope with the high operating temperatures and pressures. One example is the ceramic wafer stack configuration shown in Fig. 9.16, a derivative of the early plate-and-frame type module. This module has been developed for oxygen separation for integrated gasification combined cycle (IGCC) and oxyfueling applications by a U.S. Department of Energy National Energy Technology Laboratory-funded consortium led by Ceramatec and Air Products.

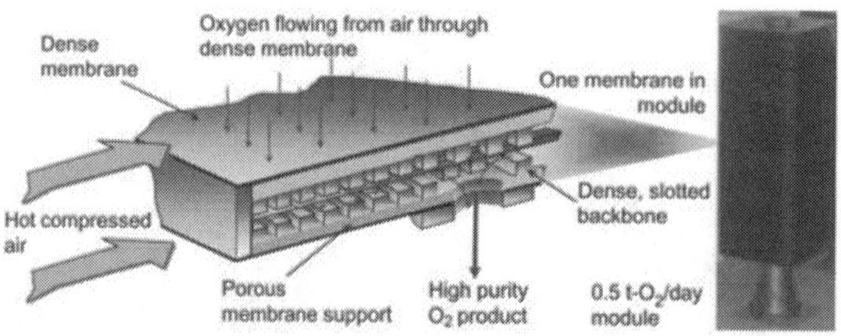

FIGURE 9.16 Ceramic membrane wafer stack module. *(Courtesy; Air Products and Chemicals Inc.).*

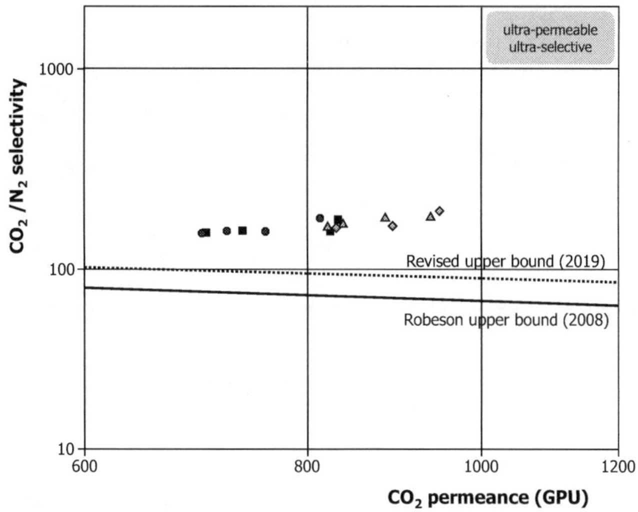

FIGURE 9.17 Robeson limit plot and recent experimental data for CO_2/N_2 separation using facilitated transport membranes (after Zhang et al., 2021).

The module consists of a stack of wafers, each comprising a back-to-back pair of perovskite-based oxygen ion transport membranes. Air is fed into the module at 800−900°C and a pressure of 0.7−2.0 MPa and oxygen is transported through the wafers and collected from the central tube.

9.3 Membrane technology R&D status

The previous sections illustrate the diverse range of membrane technologies that are available for CO_2 separation in negative emissions applications. Tables 9.10 and 9.11 summarize the main features of these membrane types and indicate the current areas of research, development, and demonstration work. The R&D focus areas listed are not intended to be exhaustive but to give a flavor of the range of topics being investigated for each membrane type.

A major R&D focus for many membrane types is on the improvement of permeance and selectivity, and membrane performance is often compared using a crossplot of these two characteristics (Robeson, 1991). The Robeson upper bound, reflecting the limit of experimentally achieved permeance and selectivity has been periodically updated since its introduction, as R&D has delivered progressively improved performance. Fig. 9.17 illustrates the concept, including a range of recently reported experimental results for facilitated transport membranes comprising polyvinylamine fixed carriers with different amino acid salt mobile carriers, synthesized for postcombustion CO_2/N_2 separation. It should be noted that the active layer thickness in novel materials experiments is often one to two orders of magnitude greater than for membranes that have been optimized for commercial production (10s to 100s of μm compared to a few hundred nm), so that a comparison based on permeance does not directly reflect the inherent performance of the different materials.

The negative slope of the upper bound means that, in any particular separation application, a trade-off is required between permeance and selectivity, which essentially balances the cost of membrane area against the cost of compression.

Although membrane R&D for negative emissions is primarily focused on flue gas separation, which would be applicable for example in a BECCS power plant, the potential for membrane-based direct air capture (mDAC) is also under investigation. mDAC has generally been discounted, based on a very low flux rate that could be achieved without excessive energy consumption, but the simplicity of a membrane-based system continues to attract attention as new ultrapermeable membranes with high CO_2 selectivity are developed (Fujikawa et al., 2020; Castro-Muñoz et al., 2022).

9.4 Membrane separation applications

Membrane separation processes have potential in a range of negative-emissions-related applications, where the continuous simple operating process, low physical footprint with the potential for modular scale-up, lack of secondary or waste products, and high energy efficiency are important advantages (Castel and Favre 2018).

TABLE 9.10 Membrane technologies and RD&D summary: microporous membranes.

Membrane technology	Description	Advantages/Disadvantages	RD&D focus areas
Carbon molecular sieve (CMS) membranes	Activated CMS composed of an amorphous, disordered network of graphitic carbon platelets	+ High CO_2 selectivity against N_2 and CH_4	Selection and carbonization of precursor material to control pore structure; membrane doping to improve mechanical strength and reduce cracking; techniques to improve regeneration and eliminate aging effects (e.g., due to chemical degradation)
		+ Durable and high-temperature operation	
		− Low CO_2 permeability	
		− Optimal precursor materials and membrane fabrication are expensive	
		− Brittle nature and thermal expansion make membrane module construction complex	
Zeolite molecular sieve membranes	Microporous lattice with well-controlled pore sizes in the range 3−13Å	+ Perfect selectivity and high permeability in specific molecular sieving applications (e.g., H_2 separation) + Resistant to CO_2 induced plasticization	Optimizing adsorption−diffusion by substituting various cations (Li^+, Na^+, Ca^{2+}, K^+) into the zeolite to produce ion-exchanged zeolite membranes and reduce H_2S sensitivity; modify surface properties to overcome water sensitivity
		− Limited selectivity for molecules with similar kinetic diameter (e.g., CO_2, O_2, N_2) − Water and H_2S cause degradation in performance	
Silica membranes	Microporous membrane operating either by Knudsen diffusion or by molecular sieving and surface diffusion, depending on pore size	+ Chemical, thermal, and structural stability + Structure (pore size and distribution) easily tailored by preparation methods and conditions − Generally low selectivity	Silica deposition methods and conditions to achieve desired, defect-free pore structure; fine-tuning pore size (e.g., by deposition of alkoxysilane monolayers)
Hybrid membranes (e.g., amine− or polyimide−silica membranes)	Composite membranes with organic groups incorporated into a microporous silica matrix or chemically bonded to the pore surfaces to create so-called functional membrane pores	+ Increased selectivity compared to a nonfunctional ceramic membrane as a result of the enhanced surface diffusion of the permeate + Thermal and mechanical stability of ceramic membranes combined with the selectivity of polymers	Selection of polymers and alternative techniques to uniformly incorporate and activate the functional groups in the ceramic matrix; optimized pore structure of composite membranes; integration of membranes with absorption or adsorption systems to reduce regeneration energy requirement

9.4.1 Membranes in IGCC applications

Palladium (Pd) or palladium and silver (Pd/Ag) alloy metallic membranes, which are 100% selective for hydrogen, have an important potential application in IGCC power plants (Bredesen et al., 2004). A Pd-based water−gas shift membrane reactor (WGSMR) can be used to separate hydrogen directly from the WGS reaction zone (Fig. 9.18), resulting in improved efficiency of the reaction. The membrane typically consists of a thin Pd or Pd/Ag film of a few microns thickness, on a macroporous support such as a stainless steel tube. Removing hydrogen as it is formed results in a more complete conversion of CO to CO_2, by preventing the shift from becoming limited by the reverse reaction. Alternatively, the reaction temperature to achieve a certain conversion rate can be reduced, thereby reducing the energy requirements.

Water is removed from the retentate by condensation, to leave a near-pure CO_2 stream at a pressure that is close to the high operating pressure of the reactor (>10 MPa), reducing the additional energy required for compression before transport and storage.

TABLE 9.11 Membrane technologies and RD&D summary: dense (nonporous) membranes.

Membrane technology	Description	Advantages/Disadvantages	RD&D focus areas
Polymeric membranes	Predominantly glassy polymer materials, of which polyimides are the most extensively investigated; also copolymer (glassy + rubbery) varieties, combining selectivity and permeability, and hybrid (rubbery + inorganic backbone) to improve mechanical properties	+ Thermal stability, chemical resistance, and mechanical strength of polyimide membranes + High degree of control over permeability and selectivity through control of polymer preparation, and chemical composition, including functional groups – Swelling and plasticization tendency with CO_2 sorption – Limited to relatively low-temperature applications	Polymer selection and preparation methods to improve mechanical strength, durability, and chemical stability and reduce plasticization tendency and manufacturing cost; polymers for higher-temperature applications; increasing selectivity in rubbery membranes by functional group substitution and in glassy membranes using high CO_2 solubility pore-filling materials; integration with adsorption (PSA/VSA) systems to improve energy efficiency
Mixed matrix membranes (MMMs)	Glassy or rubbery polymer membranes with molecular sieving particles (zeolite, activated carbon) or other micro- or nanoparticles (silica, MOFs) embedded in the polymer matrix	+ Opportunity to combine the high permeability of rubbery polymers with the selectivity of molecular sieves – Formation of microcavities due to poor bonding between polymer and included particles, particularly for glassy polymers – Poor dispersion of nanoparticles in the polymer matrix, reducing performance and mechanical strength	New polymers, preparation and posttreatment methods to improve particle bonding, reduce swelling and plasticization; filler particle selection (incl. computational approaches), preparation and pretreatment; combination of multiple filler particle types; use of hollow fiber membranes
Fixed carrier facilitated transport membranes	Solid polymeric membranes incorporating fixed carriers that react reversibly with the permeate and increase the selectivity and permeability of the solution–diffusion transport mechanism	+ High selectivity and permeate flux + Less carrier loss than mobile carrier membranes – Lack of mobility of the functional group	Identification and testing of polymer-carriers combinations; membrane fabrication methods; incorporation of multiple fixed and mobile carriers to improve transport properties
Immobilized (supported) liquid facilitated transport membranes (SILMs)	A mobile carrier facilitated transport membrane, typically a polymeric or microporous support containing an aqueous solution immobilized in the pore space	+ Very high CO_2 permeability and selectivity to N_2 + Low solvent inventory required – Loss of liquid through evaporation or leakage – Aging or degradation of the carrier due to chemical instability – Carrier saturation at high permeate pressure	Reaction catalysts and other solution additives to improve separation performance; improved carriers to reduce degradation at realistic operating conditions; fabrication methods and support structures; low-volatility liquid selection to limit liquid loss through drying
Metal membranes	Hydrogen-selective solution–diffusion membranes formed by depositing a micron-scale metallic layer, such as palladium, onto a porous support	+100% hydrogen selectivity – Low permeability – Poor mechanical properties and chemical degradation due to sulfur and chlorine sensitivity	Metal alloys to improve mechanical properties and resistance to chemical degrading; membrane treatment to improve permeability; membrane and module fabrication methods
Metal oxide ion-exchange membranes	Dense metal oxide membranes, typically perovskite-like minerals, conducting oxygen ions by facilitated transport between oxygen vacancies in the lattice	+ High-temperature operation + High O_2 or CO_2 absorption capacity + Low susceptibility to contaminant gases (SO_x, H_2S)	Scale-up of pilot scale (5 t-O_2/day) oxygen separation plant for oxycombustion

Continued

TABLE 9.11 Membrane technologies and RD&D summary: dense (nonporous) membranes.—cont'd

Membrane technology	Description	Advantages/Disadvantages	RD&D focus areas
Dual-phase ion-exchange membranes	Ion exchange in a molten carbonate electrolyte, held in the pore space of a metal (MECC) or ceramic metal oxide (MOCC) structure	+ Very-high-temperature application — close to combustion point in power generation + Absolute selectivity for CO_2 − Support material performance in high-temperature oxidizing environment − MOCC CO_2 permeance typically limited by oxygen ion transport in the ceramic phase	Robustness of metal framework under high-temperature oxidizing conditions (e.g., use of silver vs. stainless steel); membrane composition and construction to achieve high permeance

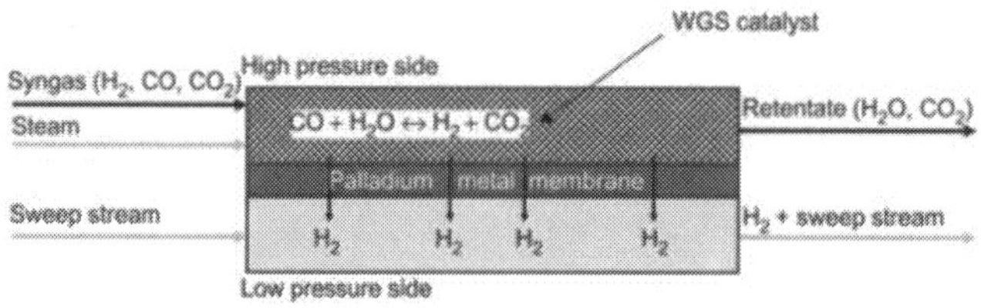

FIGURE 9.18 Water−gas shift membrane reactor.

Laboratory-scale WGS membrane reactors have been demonstrated using Pd or Pd/Ag alloy membranes operating in the temperature range 300−600°C, and have achieved 100% hydrogen selectivity and almost complete CO conversion. However, Pd membranes have been found to be very sensitive to H_2S, a common flue gas contaminant, and alternative membrane materials are also being evaluated for IGCC applications (Chi et al., 2020).

Microporous membranes (silica, alumina, carbon, and zeolite), typically asymmetric membranes on mesoporous alumina supports, have also been investigated for WGSMR applications (Bredesen et al., 2004). In this case, hydrogen selectivity is the result of molecular sieving and surface diffusion through the selective layer, with pore diameters in the region of 7−20 Å (3−7 Å for zeolite) although these membranes have lower H_2 versus CO_2 selectivity than Pd-based metallic membranes.

9.4.2 Oxygen ion transport membranes for syngas production

The formation of syngas (H_2 + CO) in gasification processes requires a supply of oxygen for the partial oxidation step. In 1988, Standard Oil Co. (subsequently Amoco and now BP) patented the Electropox (electrochemical partial oxidation) process that integrated oxygen separation, methane oxidation and methane steam reforming into a single step. The process is based on a hollow tube-configured ceramic oxygen transport membrane that contains the methane-plus-steam mixture and allows the passage of oxygen ions and electrons. The membrane is composed of a perovskite-like solid metal oxide material, in which the oxidation state of cations in the membrane has been altered to introduce oxygen vacancies into the structure. This allows oxygen ion transport through the membrane by a hopping mechanism between vacancies, as described in Section 9.1.4.

The Electropox technology has been the subject of an intensive, collaborative research and development effort since 1997, with the aim of bringing the technology to the stage of a demonstration-scale plant. Table 9.12 summarizes some of the key issues that have been investigated and illustrates the breadth of the R&D effort required to bring such a technology to the demonstration stage.

As an example, optimization of the oxygen conductivity and stability of the membrane material involves investigating the properties of various types of mixed-ion electronic conductors of the general form $A_xA'_{1-x}B_yB'_{1-y}O_{3-\delta}$ where A, A′, B and B′ are metal ions and δ indicates the oxygen nonstoichiometry (i.e., the proportion of vacancies in the lattice). Here

TABLE 9.12 Electropox R&D focus areas.

Key issue	Research tasks
Oxygen diffusion kinetics	Measurement of kinetics of oxygen uptake and transport in ceramic membrane materials under commercially relevant operating conditions
	Measurement of surface activation and reaction rates in ion transport membranes
Grain structure and atomic segregation	Evaluate the effect of defect configuration on ceramic membrane conductivity and long-term chemical and structural stability
	Assess the microstructure of membrane materials to evaluate the effects of vacancy–impurity association, defect clusters, and vacancy–dopant association on the membrane performance and stability
Phase stability and stress development	Evaluate phase stability and thermal expansion of candidate perovskite membranes and develop techniques to support these materials on porous metal structures
Mechanical property evaluation in thermal and chemical stress fields	Determine material mechanical properties under conditions of high temperature and reactive atmospheres
	Preparation and characterization of dense ceramic oxygen-permeable membranes
Graded ceramic and metal seals	Design, fabrication, and evaluation of ceramic-to-metal seals based on graded ceramic powder and metal braze joints

replacement of a proportion of the A-site ions (e.g., trivalent La^{3+}) by a lower-valency A′-ion (such as divalent Sr^{2+}) increases the oxygen vacancies in the lattice. One such example is $La_{0.7}Sr_{0.3}Co_{0.2}Fe_{0.8}O_{3-\delta}$, which exhibited high oxygen permeance combined with good chemical and mechanical stability in laboratory investigations.

A similar application for oxygen ion transport membranes is in the partial oxidation of methane as an initial step in a gas-to-liquids process, for example using bio-methane. As shown in Fig. 9.19, this is essentially the Electropox process but with the steam reforming reaction left out. The partial oxidation reaction:

$$2CH_4 + O_2 = 2CO + 4H_2 \tag{9.19}$$

produces a syngas product stream with a stoichiometric ratio close to the ideal for the Fischer–Tropsch gas-to-liquids process, which is used to convert syngas to liquid fuel through catalyzed chemical reactions of the form:

$$(2n + 1)H_2 + nCO \rightarrow C_nH_{2n+2} + nH_2O \tag{9.20}$$

As an alternative to oxygen supply from cryogenic air separation, which is capital- and energy-intensive, further developments in oxygen ion transport membranes could simplify the gas-to-liquids process and significantly reduce production costs.

9.4.3 Membrane and molecular sieve applications in oxyfuel combustion

A 500 MW power plant using oxyfuel combustion, operating at currently achievable plant efficiency (ca. 45%), requires an oxygen supply of 8 ktO_2/d. Cryogenic air separation is the only currently commercial process for delivering oxygen at this

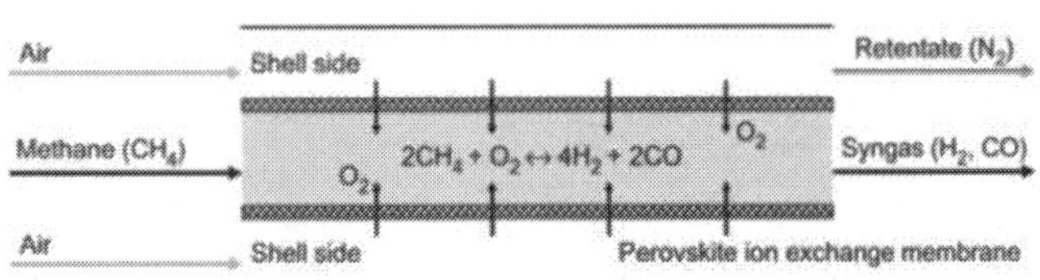

FIGURE 9.19 Oxygen transport membrane application to methane partial oxidation.

rate, and cryogenic air separation units (ASUs) with capacities up to ~4 ktO_2/day are installed worldwide. However, with a typical energy requirement of ~200 kWh/tO_2 to deliver oxygen at 170 kPa for oxyfuel combustion, oxygen supply alone would consume >10% of the output of an oxyfueled power plant.

The IPCC threshold of 0.1 $MtCO_2$/yr for defining a large stationary CO_2 source corresponds to a power plant of ~12 MW for current efficiencies. A plant of this size would require ~200 tO_2/d for oxyfuel combustion. This is close to the maximum rate currently achievable using molecular sieve-based ASUs and is in the range of oxygen production rates targeted by technology development work for membrane-based oxygen production.

Molecular sieves for oxygen production

The industrial separation of air into oxygen and nitrogen by pressure swing adsorption (PSA) or vacuum and pressure swing adsorption (VPSA), discussed in Chapter 8, is an established application of carbon and zeolite molecular sieves. Carbon molecular sieves have very similar adsorption capacities for oxygen and nitrogen, but the adsorption rates differ significantly, oxygen adsorbing considerably faster due to its smaller kinetic diameter (3.46 Å) compared to that of nitrogen (3.64 Å). PSA air separation exploits this difference; when a carbon molecular sieve comes into contact with air, the adsorbed phase becomes oxygen-rich while the gaseous phase becomes nitrogen-rich.

Carbon molecular sieve PSA (CMS-PSA) is widely used for nitrogen generation with, in this case, the adsorbed oxygen being vented back to the atmosphere during sorbent regeneration. Alternatively, the released oxygen can be captured and the process used for oxygen generation. PSA units can yield an oxygen stream with 95% purity, with plant capacities up to 200 tO_2/d.

Ion transport membranes for oxygen production

Ion transport membranes have been the subject of an extensive RD&D project since 1999, partly funded by the U.S. Department of Energy National Energy Technology Laboratory, and a 5 tO_2/d prototype unit (shown in Fig. 9.20) started operation in 2005. The perovskite-based metal oxide ion transport membrane modules have a wafer stack configuration (Section 9.2), and multiple modules are housed within the high-pressure membrane vessel. The plant delivers a >99% pure oxygen permeate stream and has successfully demonstrated that this technology can be applied on a commercial scale.

This technology reduces the energy cost of oxygen production by a third compared to cryogenic methods. An intermediate scale pilot capable of producing 100 tO_2/d has been constructed and tested but future development work, which aimed to scale capacity to in excess of 1000 tO_2/d, was terminated in 2015.

9.4.4 Membrane applications in postcombustion CO_2 separation

Membrane separation of CO_2 from flue gases is another very active research field, which aims to reduce the energy penalty and cost of membranes to enable large-scale demonstration and deployment (Favre 2007; Khalilpour et al., 2015). The key characteristics that need to be combined in a commercial membrane are;

- Sufficient permeate flux rate to achieve desired operating throughput
- Low space requirements and capital cost, including membrane material cost
- Low permeation for nonselected gases (typically N_2, CH_4, and H_2)
- Chemical resistance to contaminants (SO_x, NO_x, H_2O, H_2S)
- Thermal stability at required operating temperature
- Inexpensive fabrication methods for a variety of module configurations.

Achieving a combination of sufficient selectivity, to yield high CO_2 purity in the permeate stream, with high permeability, to give a high flow rate without excessive pressure drop, is the key to unlocking the potential of membranes for CO_2 separation from high-volume flue gases.

A 1 tCO_2/d membrane-based pilot for postcombustion capture was tested at the US DoE's National Carbon Capture Center, processing a flue gas slipstream from Alabama Power's E. C. Gaston power plant in Wilsonville, Alabama, USA. The system used a proprietary polymer membrane developed by Membrane Technology Research (MTR Polaris) in spiral wound modules (White et al., 2015). The tested process configuration includes two membrane separation stages, as illustrated in Fig. 9.21.

In the first stage, CO_2 is drawn through the membrane by a vacuum pump, yielding a 70% CO_2 permeate stream which would be further processed by dehydration, compression, and chilling to yield CO_2 for sequestration. In the second stage,

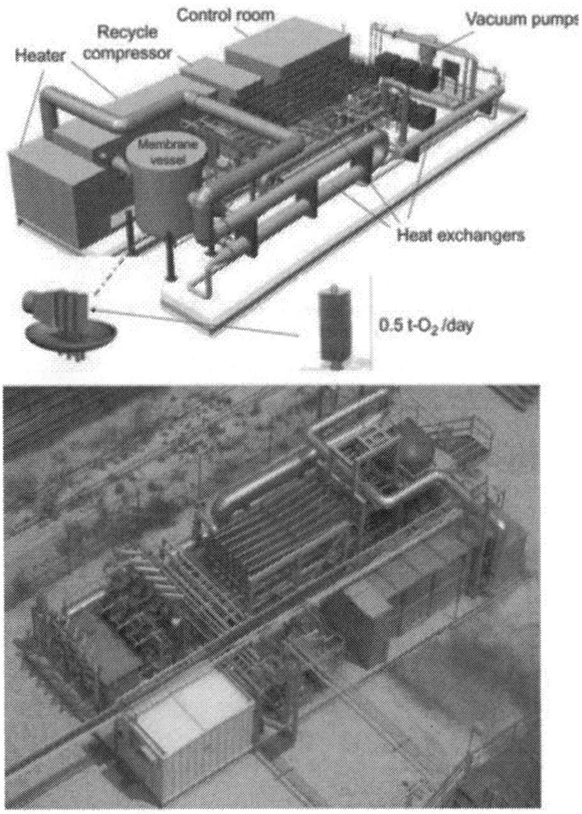

FIGURE 9.20 Prototype 5 tO_2/d ITM oxygen separation plant. *(Courtesy: Air Products).*

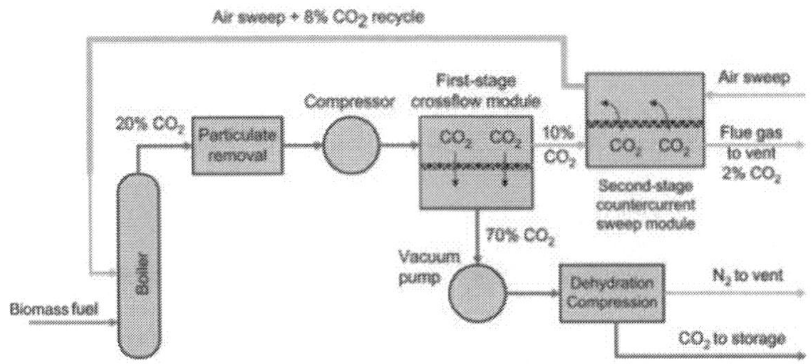

FIGURE 9.21 Multistage membrane separation system for postcombustion capture.

CO_2 permeate is swept by a counter-current air flow, and the resulting 8% CO_2 permeate stream is recycled to the boiler, boosting the flue gas CO_2 content from the 13%−14% typical for air-drafted boilers to 20%.

Following extensive pilot testing from 2012 to 2014, a 20 tCO_2/d demonstration-scale plant began operation at the same facility in 2015 (Kniep et al., 2018). The engineering design for a commercial-scale deployment of the MTR Polaris membrane system at the Duke Energy's 600 MW_e coal-fired East Bend Station in Kentucky has recently been reported (Dillon et al., 2021). The proposed installation would result in a net capture of 52% of the site emissions (including a natural gas-fired turbine to power the capture unit) at a cost of \$83/tCO_2 captured and \$124/tCO_2 avoided.

A recent comparative assessment suggests that, given current performance, membrane-based postcombustion capture has a 10%–15% higher energy penalty than the benchmark monoethanolamine (MEA) absorption system and, coupled with higher capital and periodic replacement costs for membrane modules, a ca. 20% higher levelized cost of capture (van der Spek et al., 2019), although a number of promising polymeric membrane materials have the potential to achieve cost competitiveness with the MEA benchmark after further development (Roussanaly et al., 2016). Other membrane technologies being investigated for postcombustion capture are described in the following sections.

High-temperature molten carbonate membrane

A dual ion conduction membrane using a molten carbonate is under development for high-temperature and high-pressure CO_2 separation from flue gas streams close to the point of combustion, or to remove CO_2 from a syngas stream after the production of hydrogen in the WGS reaction (Xu et al., 2012). The membrane, illustrated in Fig. 9.22, is based on a composite material that combines oxygen ion exchange through a solid metal oxide with CO_2 transport as a carbonate anion (CO_3^{2-}) through the molten carbonate phase.

At the feed surface of the membrane, an oxygen anion from the metal oxide combines with CO_2 to form a carbonate anion:

$$CO_2 + O^{2-} \rightarrow CO_3^{2-} \tag{9.21}$$

This ion is transported through the molten carbonate electrolyte, which is immobilized in the pores of a metal or mixed-metal oxide matrix (e.g., silver oxide and perovskite). At the permeate side, the carbonate anion releases CO_2 into the gas phase while the oxygen ion enters a lattice vacancy in the metal oxide and is transported back to the feed side, driven by the gradient in vacancies within the metal oxide lattice.

Current R&D efforts focus on the identification and testing of alternative metal oxide compositions to optimize CO_2 permeance and selectivity, surface modification of metal oxide matrix (e.g., alumina coating) via chemical or atomic deposition to improve performance, and on varying fabrication methods to reduce cost and improve stability and efficiency, the latter for example through control of the volume fractions and pore structure of the composite material (Zhang et al., 2021).

Facilitated transport membranes

Many different membrane material and functional group combinations have been tested for application in postcombustion separation, a typical example being amine carriers in a cross-linked polyvinyl alcohol (PVA) polymer membrane. The membrane showed good CO_2 permeability and CO_2 and N_2 selectivity at operating temperatures up to 170°C. Mathematical modeling studies indicated that a 0.6 m long hollow-fiber module containing $\sim 10^6$ fibers could recover >95% of the CO_2 from a 0.5 m^3/s flue gas stream into a 98% CO_2 permeate stream, yielding ~ 12 tCO_2/d per module (Huang et al., 2008). At this rate around 800 modules would be required to process the $\sim 1.4\ 10^6$ m^3/h of flue gas from a 500 MW power plant.

A polyvinylamine (PVAm) fixed carrier membrane has also been tested in a small-scale pilot as part of the EU-funded Nanoglowa project. A plate and frame membrane module containing 1.5 m^2 of the PVAm membrane was tested at the

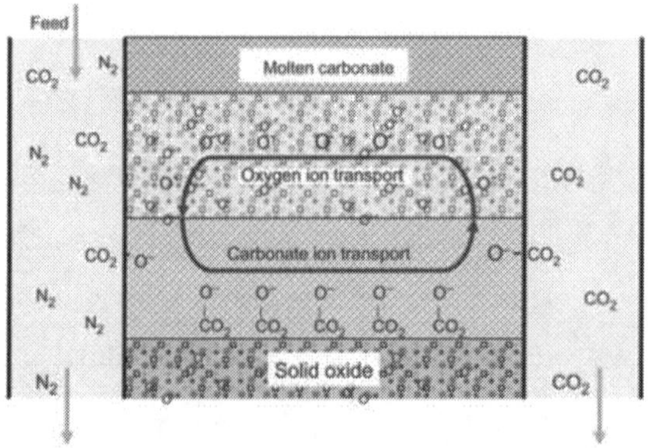

FIGURE 9.22 Molten carbonate dual ion exchange membrane.

EDP's 314 MW_e coal-fired power plant in Sines, Portugal, processing a 24 m^3/h flue gas slip-stream and producing a 75% CO_2 permeate stream (Sandru et al., 2013). The permeability (88–266 Barrer), and CO_2/N_2 selectivity (80–300) are similar to other facilitated transport membranes summarized in Table 9.7. Other pilot scale tests have recently been reviewed by Han et al. (2020).

Carbon molecular sieve membranes

CMS membranes are promising for flue gas CO_2 separation applications because of their high thermal and chemical stability in nonoxidizing environments, essential characteristics for this application (Huang and Dai 2020). Although high selectivity for CO_2/N_2 (>100) and for CO_2/CH_4 (>250) have been demonstrated in laboratory experiments, these have been achieved only at very low permeabilities in the range from 1 to 10 Barrer ($3-30 \times 10^{-16}$ mol $m^{-1}s^{-1}Pa^{-1}$).

Current research efforts for this type of membrane are focusing on:

- Selection, preparation, and pretreatment of the precursor material to reduce the cost of membrane production
- Parameters of the carbonization process (temperature, heating rate, pyrolysis atmosphere) to control the membrane pore structure
- Postcarbonization treatment of the membrane, for example additional carbon deposition by chemical vapor deposition (CVD)
- Membrane doping to improve mechanical strength and reduce cracking
- Techniques to improve regeneration and eliminate aging effects, for example due to chemical degradation.

Gas–liquid membrane contactors

Gas–liquid membrane contactors make use of low-cost macroporous membranes, typically hollow-fiber modules, to provide an interface between a gas and a liquid absorbent and are being developed for a number of applications including CO_2 removal from flue gases (Qi and Cussler 1985; Koutsonikolas et al., 2016) and from shifted syngas in an IGCC process (Usman et al., 2018). The relatively large pore diameter of the membrane, typically in the range of 10–200 nm, means that the membrane itself is nonselective and the selectivity of the contactor results from the physical or chemical absorption process at the permeate side of the membrane.

This hybrid membrane–absorbent configuration offers significant advantages compared to the traditional column-type adsorption contactor, notably very high surface area, as well as some disadvantages, summarized in Table 9.13.

An important characteristic of the membrane–absorbent combination is that the membrane should not be wetted by the solvent solution, so that the pore space remains fully gas-filled to maximize mass transfer through the membrane. The critical differential pressure at which pores will become wetted is determined by Laplace–Young equation;

$$\Delta P_c = -4\ \gamma \cos \theta / d_{max} \tag{9.22}$$

TABLE 9.13 Advantages and disadvantages of gas–liquid membrane contactors.

Advantages	**Description**
High area-to-volume ratio	Currently available hollow-fiber membrane modules offer packing densities up to 7000 m^2/m^3, compared to <350 m^2/m^3 for traditional packed and plate column contactors, resulting in mass transfer rates 100–1000 times higher, as well as significant weight and space reductions
No interpenetration of gas and liquid phases	Ability to independently control gas and liquid flow rates to optimize the absorption process, avoiding operational problems encountered in distillation processes such as liquid entrainment in the gas phase or column flooding
Disadvantages	
Additional mass transfer resistance through the membrane	Resistance will be exacerbated if the membrane becomes partially wetted
Membrane wetting	Precise pressure control required to prevent membrane wetting and solvent flow across the membrane

where ΔP_c, is the critical breakthrough (i.e., wetting) pressure, γ is the surface tension of the solvent, θ is the contact angle in degrees for the gas solvent membrane system, and d_{max} is the maximum diameter of the membrane pores. The need to avoid pore wetting, and the precise pressure control across the membrane that is needed, both to ensure this and to avoid liquid breakthrough, are significant operational challenges.

Other solvent requirements include the ability to achieve a high CO_2 loading at the relatively low partial pressures achievable on the permeate side of the membrane, low viscosity to reduce energy requirements for solvent circulation, and low volatility to reduce solvent losses.

Traditional amine-based solvents have been used in experimental work but require the use of more expensive membrane materials, such as polytetrafluoroethylene (PTFE—Teflon) or polyvinylidene fluoride, to avoid the problem of membrane wetting. Recent research has focused on the use of commercially available hollow-fiber membrane modules, using inexpensive membrane materials such as polypropylene, requiring either alternative nonwetting solvents (higher γ and θ) or treatment of the membrane material to make it less hydrophilic (increase θ), such as plasma treatment or the application of a thin superhydrophobic coating (Lv et al., 2012).

A number of pilot-scale examples of membrane contactors for CO_2 capture have been reported, including a demonstration unit using Teflon (PTFE) membranes and a conventional amine solvent developed by Aker Kvaerner. This unit was able to recover 85% of the CO_2 from the exhaust gas of a 520 kW gas turbine engine, at a rate of 195 kgCO_2/h from 2610 kg/h of exhaust gas, with a solvent flow rate of 4.5 m^3/h (Falk-Pedersen et al., 2005).

Significant recent effort has also been devoted to the development of mathematical modeling techniques for membrane contactors, to enable the optimization of solvent and contactor characteristics, operating conditions, and integration into the overall capture process (Al-Marzouqi et al., 2008). Recent technoeconomic analysis shows that membrane contactors require further development before they can be cost-competitive with traditionally packed tower absorbers (Kimball et al., 2014), and current R&D focus areas include the development of novel membrane materials, including nonporous membranes, to minimize wetting and improve thermal and chemical stability, the potential for use in solvent regeneration, and the search for novel solvents with low viscosity and fast absorption kinetics (Zhao et al., 2016). The chemical stability of polymeric membranes in amine-based solvents is a particular challenge, and the incorporation of immobilized carbonic anhydrase as biocatalytic nanoparticles into polymer fibers has also been investigated to address the slow absorption kinetics of more benign solvents such as K_2CO_3 (Hou et al., 2016; Xu et al., 2019).

References and resources

References

Al-Marzouqi, M.H., El-Naas, M.H., Marzouk, S.A., Al-Zarooni, M.A., Abdullatif, N., Faiz, R., 2008. Modeling of CO_2 absorption in membrane contactors. Separ. Purif. Technol. 59 (3), 286–293.

Bredesen, R., Jordal, K., Bolland, O., 2004. High-temperature membranes in power generation with CO_2 capture. Chem. Eng. Process 43, 1129–1158.

Castel, C., Favre, E., 2018. Membrane separations and energy efficiency. J. Membr. Sci. 548, 345–357.

Castro-Muñoz, R., Ahmad, M.Z., Malankowska, M., Coronas, J., 2022. A new relevant membrane application: CO_2 direct air capture (DAC). Chem. Eng. J. 446 (2022), 137047.

Chi, J., Li, K., Zhang, S., Zhu, X., Zhao, L., Wang, B., Xiao, Y., 2020. Process simulation and integration of IGCC systems with novel mixed ionic and electronic conducting membrane-based water gas shift membrane reactors for CO_2 capture. Int. J. Hydrogen Energy 45 (27), 13884–13898.

Dai, Z., Noble, R.D., Gin, D.L., Zhang, X., Deng, L., 2016. Combination of ionic liquids with membrane technology: a new approach for CO_2 separation. J. Membr. Sci. 497, 1–20.

Dillon, D., Swisher, J., Freeman, B., et al., 2021. Engineering design of a full scale membrane post-combustion CO_2 capture system retrofit for Duke Energy's East Bend station. In: Proceedings of the 15th Greenhouse Gas Control Technologies Conference 15-18 March 2021.

Eisaman, M.D., Parajuly, K., Tuganov, A., Eldershaw, C., Chang, N., Littau, K.A., 2012. CO_2 extraction from seawater using bipolar membrane electrodialysis. Energy Environ. Sci. 5 (6), 7346–7352.

Falk-Pedersen, O., Grønvold, M.S., Nøkleby, P., Bjerve, F., Svendsen, H.F., 2005. CO_2 capture with membrane contactors. Int. J. Green Energy 2 (2), 157–165.

Favre, E., 2007. Carbon dioxide recovery from post-combustion processes: can gas permeation membranes compete with absorption? Membr. Sci. 294 (50–59).

Fujikawa, S., Selyanchyn, R., Kunitake, T., 2020. A new strategy for membrane-based direct air capture. Polym. J. 1–9.

Han, Y., Yang, Y., Ho, W.S., 2020. Recent progress in the engineering of polymeric membranes for CO_2 capture from flue gas. Membranes 10 (11), 365.

Hou, J., Zulkifli, M.Y., Mohammad, M., Zhang, Y., Razmjou, A., Chen, V., 2016. Biocatalytic gas-liquid membrane contactors for CO_2 hydration with immobilized carbonic anhydrase. J. Membr. Sci. 520, 303–313.

Huang, J., Zou, J., Ho, W.W., 2008. Carbon dioxide capture using a CO_2-selective facilitated transport membrane. Ind. Eng. Chem. Res. 47 (4), 1261–1267.

Huang, K., Dai, S., 2020. Carbon membranes for CO_2 separation. In: Materials for Carbon Capture. John Wiley and Sons Ltd, pp. 215–236.

Kai, T., Taniguchi, I., Duan, S., et al., 2013. Molecular gate membrane: poly (amidoamine) dendrimer/polymer hybrid membrane modules for CO_2 capture. Energy Proc. 37, 961–968.

Kazama, S., Duan, S., Ohno, C., et al., 2005. Molecular gate membrane for CO_2 capture. Greenhouse Gas Control Technol. 7, 1947–1950.

Khalilpour, R., Mumford, K., Zhai, H., Abbas, A., Stevens, G., Rubin, E.S., 2015. Membrane-based carbon capture from flue gas: a review. J. Clean. Prod. 103, 286–300.

Kimball, E., Al-Azki, A., Gomez, A., et al., 2014. Hollow fiber membrane contactors for CO_2 capture: modeling and up-scaling to CO2 capture for an 800 MWe coal power station. Oil Gas Sci. Technol.–Revue d'IFP Energies Nouvelles 69 (6), 1047–1058.

Kniep, J., Baker, R., Casillas, C., et al., 2018. Small pilot test results from a Polaris membrane 1 MWe CO_2 capture system. Proc. 14th Greenhouse Gas Control Technol. Conf. Melbourne 21–26.

Koutsonikolas, D., Pantoleontos, G., Mavroudi, M., et al., 2016. Pilot tests of CO_2 capture in brick production industry using gas–liquid contact membranes. Int. J. Energy Environ. Eng. 7 (1), 61–68.

Loeb, S., 1981. The Loeb-Sourirajan membrane: how it came about. In: Turbak, A. (Ed.), Synthetic Membranes, ACS Symposium Series. American Chemical Society, Washington, DC, USA.

Lv, Y., Yu, X., Jia, J., Tu, S.T., Yan, J., Dahlquist, E., 2012. Fabrication and characterization of superhydrophobic polypropylene hollow fiber membranes for carbon dioxide absorption. Appl. Energy 90 (1), 167–174.

Nagasawa, H., Yamasaki, A., Iizuka, A., Kumagai, K., Yanagisawa, Y., 2009. A new recovery process of carbon dioxide from alkaline carbonate solution via electrodialysis. AIChE J. 55 (12), 3286–3293.

Qi, Z., Cussler, E.L., 1985. Microporous hollow fibers for gas absorption: I. Mass transfer in the liquid. J. Membr. Sci. 23 (3), 321–332.

Ramon, G.Z., Wong, M.C., Hoek, E.M., 2012. Transport through composite membrane, part 1: is there an optimal support membrane? J. Membr. Sci. 415, 298–305.

Robeson, L.M., 1991. Correlation of separation factor versus permeability for polymeric membranes. J. Membr. Sci. 62 (2), 165–185.

Roussanaly, S., Anantharaman, R., Lindqvist, K., Zhai, H., Rubin, E., 2016. Membrane properties required for post-combustion CO_2 capture at coal-fired power plants. J. Membr. Sci. 511, 250–264.

Sabatino, F., Mehta, M., Grimm, A., et al., 2020. Evaluation of a direct air capture process combining wet scrubbing and bipolar membrane electrodialysis. Ind. Eng. Chem. Res. 59 (15), 7007–7020.

Sahoo, P.C., Jang, Y.N., Lee, S.W., 2013. Enhanced biomimetic CO_2 sequestration and $CaCO_3$ crystallization using complex encapsulated metal organic framework. J. Cryst. Growth 373, 96–101.

Sandru, M., Kim, T.-J., Capala, W., Huijber, M., Hägg, M.-B., 2013. Pilot scale testing of polymeric membranes for CO_2 capture from coal fired power plants. Energy Proc. 37, 6473–6480.

Sandru, M., Sandru, E.M., Ingram, W.F., et al., 2022. An integrated materials approach to ultrapermeable and ultraselective CO_2 polymer membranes. Science 376 (6588), 90–94.

Sun, C., Zhu, S., Liu, M., Shen, S., Bai, B., 2019. Selective molecular sieving through a large graphene nanopore with surface charges. J. Phys. Chem. Lett. 10 (22), 7188–7194.

Usman, M., Hillestad, M., Deng, L., 2018. Assessment of a membrane contactor process for pre-combustion CO_2 capture by modelling and integrated process simulation. Int. J. Greenh. Gas Control 71, 95–103.

van der Spek, M., Bonalumi, D., Manzolini, G., Andrea, R., André, F., 2019. Part-load techno-economic performance analysis of CCGT with advanced membrane concept. In: Proceedings of the 14th International Conference on Greenhouse Gas Control Technologies, GHGT-14, 1–5.

Wang, Z., Zhang, Y., Wang, J., Zhang, Y., 2020. Bioinspired porous organic polymer-functionalized membranes for efficient CO_2 capture. Sustain. Energy Fuels 4 (3), 1191–1198.

Ward, W.J., Robb, W.L., 1976. Carbon dioxide–oxygen separation: facilitated transport of carbon dioxide across a liquid film. Science 156, 1481–1484.

White, L.S., Wei, X., Pande, S., Wu, T., Merkel, T.C., 2015. Extended flue gas trials with a membrane-based pilot plant at a one-ton-per-day carbon capture rate. J. Membr. Sci. 496, 48–57.

Wilcox, J., Haghpanah, R., Rupp, E.C., He, J., Lee, K., 2014. Advancing adsorption and membrane-separation processes for the Gigaton carbon capture challenge. Annu. Rev. Chem. Biomol. Eng. 5, 479–505.

Wu, J., Wang, X., Wang, Q., Lou, Z., Li, S., Zhu, Y., Qin, L., Wei, H., 2019. Nanomaterials with enzyme-like characteristics (nanozymes): next-generation artificial enzymes (II). Chem. Soc. Rev. 48 (4), 1004–1076.

Xu, N., Li, X., Franks, M.A., Zhao, H., Huang, K., 2012. Silver-molten carbonate composite as a new high-flux membrane for electrochemical separation of CO_2 from flue gas. J. Membr. Sci. 401, 190–194.

Xu, Y., Lin, Y., Chew, N.G.P., Malde, C., Wang, R., 2019. Biocatalytic PVDF composite hollow fiber membranes for CO_2 removal in gas-liquid membrane contactor. J. Membr. Sci. 572, 532–544.

Zhang, P., Tong, J., Huang, K., Zhu, X., Yang, W., 2021a. The current status of high temperature electrochemistry-based CO_2 transport membranes and reactors for direct CO_2 capture and conversion. Prog. Energy Combust. Sci. 82, 100888.

Zhang, Z., Rao, S., Han, Y., Pang, R., Ho, W.W., 2021b. CO_2-selective membranes containing amino acid salts for CO_2/N_2 separation. J. Membr. Sci. 638, 119696.

Zhao, S., Feron, P.H., Deng, L., et al., 2016. Status and progress of membrane contactors in post-combustion carbon capture: a state-of-the-art review of new developments. J. Membr. Sci. 511, 180–206.

Zhu, B., Liu, J., Wang, S., et al., 2019. Mixed matrix membranes containing well-designed composite microcapsules for CO_2 separation. J. Membr. Sci. 572, 650–657.

Further reading

Baker, R., 2004. Membrane Technology and Applications. John Wiley and Sons, Chichester, UK.

Cejka, J., van Bekkum, H., Corma, A., Schueth, F., 2007. Introduction to Zeolite Molecular Sieves. Elsevier, Oxford, UK.

Chang, P.T., Ng, Q.H., Ahmad, A.L., Low, S.C., 2021. A critical review on the techno-economic analysis of membrane gas absorption for CO_2 capture. Chem. Eng. Commun. 1–17.

Chen, G., Feldhoff, A., Weidenkaff, A., et al., 2021. Roadmap on sustainable mixed ionic-electronic conducting membranes. Adv. Funct. Mater. 2105702.

Chuah, C.Y., Kim, K., Lee, J., Koh, D.Y., Bae, T.H., 2019. CO_2 absorption using membrane contactors: recent progress and future perspective. Ind. Eng. Chem. Res. 59 (15), 6773–6794.

IUPAC Terminology for Membranes and Membrane Processes. Available at: pac.iupac.org/publications/pac/pdf/1996/pdf/6807x1479.pdf.

Koros, W.J., Ma, Y.H., Shimidzu, T., 1996. Terminology for Membranes and Membrane Processes. IUPAC Recommendations.

Varcoe, J.R., Atanassov, P., Dekel, D.R., et al., 2014. Anion-exchange membranes in electrochemical energy systems. Energy Environ. Sci. 7 (10), 3135–3191.

Wu, Y., Xu, J., Mumford, K., Stevens, G.W., Fei, W., Wang, Y., 2020. Recent advances in carbon dioxide capture and utilization with amines and ionic liquids. Green Chem. Eng. 1 (1), 16–32.

Chapter 10

Carbon mineralization

Adrienne Sewel, Pol Knops, and Steve Rackley

10.1 Introduction

The capture methods described in Chapters 7–9 aim to produce a pure or near-pure CO_2 stream that can then be utilized or stored using one or other of the approaches described in Chapters 15 to 17. Storing CO_2 in its molecular unmineralized form has numerous disadvantages, for example, the need for continued monitoring to confirm storage integrity and the potential adverse environmental impacts in the case of ocean storage or leakage from geological storage sites. In contrast, the mineralization methods discussed in this chapter seek to store carbon in the form of solid, chemically stable, and relatively benign carbonates, with the potential added advantage of providing useful mineral and mineral-based products.

The weathering of rocks and transportation of the products of weathering into the oceans are key steps in the global carbon cycle, described in Chapter 4. Mineral carbonation, or CO_2 mineralization, first proposed by Seifritz (1990) and Lackner et al. (1995), is a carbon storage technology that accelerates this natural geological process by reacting CO_2 with minerals containing oxides or silicates of magnesium, iron, and calcium to form stable carbonates. This accelerated mineralization can either occur in situ, with CO_2 being transported to storage sites and injected into subsurface formations rich in reactive minerals, or ex situ, using mined and pretreated minerals as the source of cations for carbonation in engineered reactors or other artificial settings.

Carbon mineralization can also exploit alkalinity from a wide variety of industrial waste-derived feedstocks containing similar oxides or silicates, ranging from combustion ash to mine tailings, and wastes from aluminium production, cement kilns, and paper mills. CO_2 mineralization also makes a generally small contribution to the trapping of CO_2 injected for geological storage into sedimentary formations, as a result of the reaction of CO_2 with alkaline mineral components in the sedimentary rock fabric (Chapter 15).

Another useful division is between passive mineralization techniques, such as allowing reactive olivine to change slowly under natural ambient surface conditions, and active mineralization techniques where the passive process is accelerated by adjusting one or more of the environmental conditions such as temperature and pressure. Passive and active techniques can be used both in situ and ex situ. There is usually a financial and energy cost associated with active techniques in return for faster carbonation. A further division is between carbon capture and storage (CCS) and carbon capture storage and utilization (CCSU) applications. An example of CCS is the in situ Carbfix project in Iceland, discussed later, that simply traps and stores carbon underground in the form of carbonate, while, in CCSU, the mineral products that incorporate carbon into their structure are put to further use, for example as building materials.

10.2 Mineral carbonation chemistry

The aim of carbon mineralization is to bind CO_2 into a stable mineral structure, where it will remain secure over geological time. Although there are a wide variety of intermediate steps and mineral states, the ideal end result is that carbon becomes incorporated into a mineral carbonate which is either stored or put to further use (Dunsmore, 1992).

The simplest carbonation reaction occurs when a metal oxide, such as magnesium oxide, is reacted with CO_2 to yield magnesium carbonate:

$$MgO + CO_2 \rightarrow MgCO_3 + \text{heat} \tag{10.1}$$

Negative Emissions Technologies for Climate Change Mitigation. https://doi.org/10.1016/B978-0-12-819663-2.00010-1

The reaction product is stable over geological time because a carbonate is the lowest energy state for a carbon atom (Fig. 10.1) and, depending on the mineral reactant, the carbonation reaction releases 60–180 kJ/mol. This gives mineral carbonation an inherent energetic advantage over options such as geological storage, where CO_2 is stored as a gas, because the reaction heat can in principle be recovered to power other process steps.

Calcium or magnesium oxides would be ideal candidates as mineral feedstocks for carbonation, because their carbonates ($CaCO_3$, $MgCO_3$) have low solubility in water. This is an important consideration if the carbonation products would ultimately be stored on land (e.g., back-filling an open cast mine) to prevent ground water contamination through leaching. However, as described below, the carbonation reaction typically requires some 2–3 t-mineral/tCO_2 sequestered. Natural abundance is therefore an important characteristic for a suitable feed mineral, and these oxides are rare in nature as a result of their reactivity.

Silicate rocks are an alternative source of metal oxides and make up the majority of the earth's mantle. In particular, mafic and ultramafic rocks are igneous silicate rocks that consist predominantly of magnesium, calcium, and iron minerals. Silicate minerals occurring in these rocks include olivine ($(Fe_x, Mg_{1-x})_2SiO_4$), wollastonite ($CaSiO_3$), and serpentine ($Mg_3Si_2O_5(OH)_4$). In the case of olivine, (Fe_x, Mg_{1-x}) in the chemical formula indicates that the mineral is a mixture (known as a solid solution) of iron and magnesium silicates, with forsterite (Mg_2SiO_4) and fayalite (Fe_2SiO_4) being the pure end-members of this range.

The three minerals noted above are representatives of different classes of silicates, which are defined according to the way the silicate ion (SiO_4^{4-}) is bonded in the crystal lattice. Orthosilicate minerals, of which olivine is an example, have the simplest structure and consist of a single SiO_4^{4-} anion bonded to two divalent metal cations (Mg^{2+} or Fe^{2+}). In contrast, pyroxenes form single chains as a result of the sharing of an oxygen atom between two SiO_4 units. Wollastonite is an example of a single-chain silicate. Serpentine is an example of the more complex class of phyllo- (or leaf-like) silicates, with sheet-like structures in which three of the four oxygen atoms from each SiO_4 unit are shared with adjacent units to form hexagonal rings that extend into sheets (Fig. 10.2).

In minerals containing hydroxyl ions (OH^-), including serpentine, the OH^- ions bond at the center of the hexagonal ring. Metal cations form bonds with oxygen and hydroxyl to form a second, parallel sheet in which each metal ion is surrounded by eight oxygen or hydroxyl ions. Fig. 10.3 shows the resulting lamellar structure for the lizardite form of serpentine. In some cases, including the antigorite and chrysotile forms of serpentine, the combined two-sheet structure takes up a curved form as a result of the longer spacing of the metal ion layer compared to the Si_2O_5 layer. Chrysotile (asbestos) is an extreme example of this curvature in which the structure is rolled up to form fibers. The different structures of these minerals have an impact on the relative ease with which they react with CO_2, necessitating various pretreatment options described below.

The carbonation reactions for a number of minerals that have been considered as possible candidates for CO_2 milerализation are shown in reactions 10.2–10.5:

Olivine (Forsterite)

$$Mg_2SiO_4 + 2CO_2 \rightarrow 2MgCO_3 + SiO_2 \quad (10.2)$$

Olivine (Fayalite)

$$Fe_2SiO_4 + 2CO_2 \rightarrow 2FeCO_3 + SiO_2 \quad (10.3)$$

Wollastonite

$$CaSiO_3 + CO_2 \rightarrow CaCO_3 + SiO_2 \quad (10.4)$$

Serpentine

$$Mg_3Si_2O_5(OH)_4 + 3CO_2 \rightarrow 3MgCO_3 + 2SiO_2 + 2H_2O \quad (10.5)$$

In an aqueous reaction, the second end product will generally be orthosilicic acid (H_4SiO_4) rather than silicon dioxide (SiO_2).

The enthalpy of reaction from carbonation of these minerals is compared in Table 10.1 to the heat initially released in the combustion of carbon to produce a mole of CO_2 ($C + O_2 \rightarrow CO_2$). Clearly, the enthalpy of reaction is a significant fraction of the initial heat from combustion, and the effective use of this heat can reduce the energy and carbon penalty of an engineered carbonation process.

Carbonates are salts that include the CO_3^{2-} anion, in which the carbon atom has a covalent double bond to one oxygen atom and single bonds to the others. Mineral carbonates always have this CO_3^{2-} anion in their structure, often combining

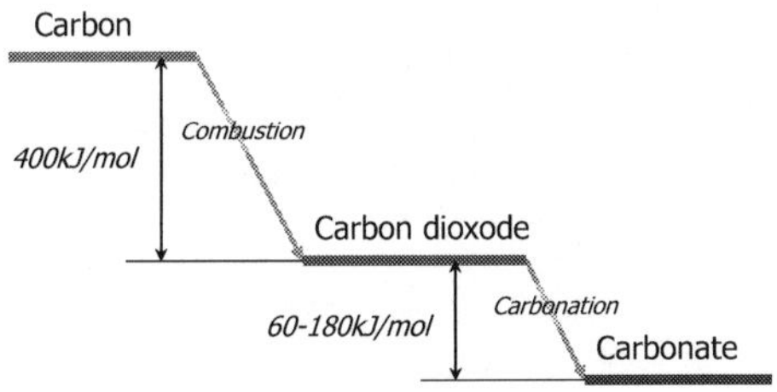

FIGURE 10.1 Energy states of carbon.

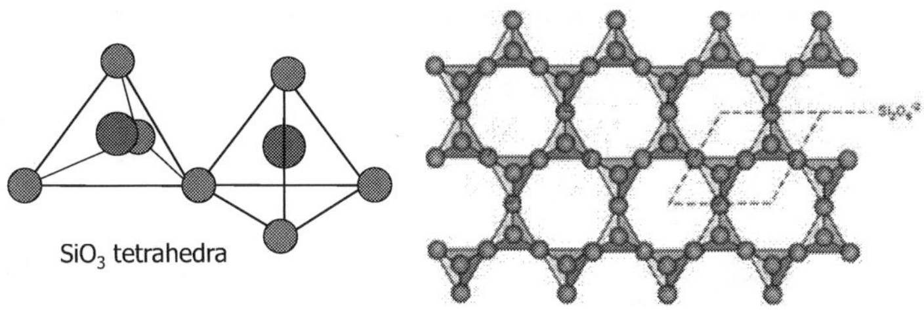

FIGURE 10.2 Si_2O_5 sheet structure of phyllosilicate minerals.

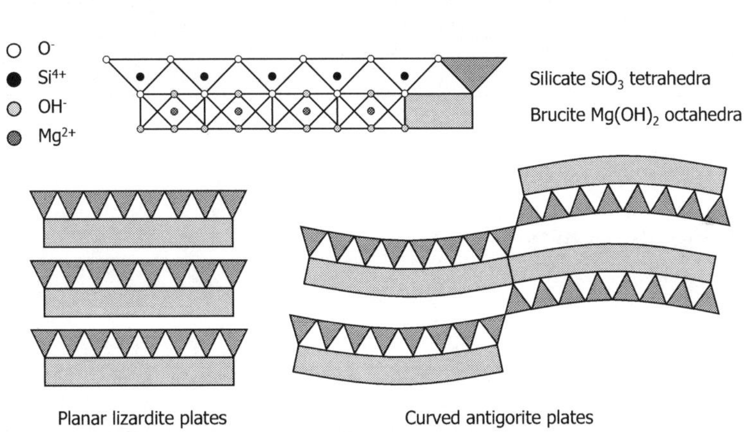

FIGURE 10.3 Structure of the sheet silicate serpentine: (A) lizardite and (B) antigorite.

with metal cations such as Fe, Mg, K, Na, Ca, and a wide variety of less common elements. Carbonates comprise classes 13–17 in the new Dana mineralogy index (Gaines et al., 1997), being further subdivided into acid, anhydrous, hydrated, hydroxyl, halogen, and compound classes of carbonate. The types of carbonate most relevant to carbon mineralization are the anhydrous carbonates (class 14), such as calcite, dolomite, magnesite, and siderite.

Although the carbonation reactions are thermodynamically favored, they proceed at very slow rates at room temperature and pressure, due to the slow dissolution of the metal from the mineral lattice. Chemical process optimization to achieve reaction rates that are viable for industrial applications is therefore a major focus of ongoing mineral carbonation R&D work. Many process options have been considered, including methods where CO_2 is reacted directly with the mineral feedstock, and indirect methods where the reactive mineral component is extracted from the feedstock in an initial step to improve the reaction rate. Following the initial extraction step, the reactive component can be carbonated using any of the direct route options, as summarized in Table 10.2.

The wide range of mineralization approaches—direct/indirect, passive/active, in situ/ex situ—are discussed in further detail below.

TABLE 10.1 Heat released in carbonation reactions.

Mineral	Enthalpy of reaction $-\Delta H^0_r$ (kJ/mol)	% of heat from carbon combustion ($-\Delta H_r$/393.8)%
Calcium oxide (CaO)	179	45%
Magnesium oxide (MgO)	118	30%
Olivine (forsterite)	89	23%
Wollastonite	87	22%
Serpentine	64	16%

TABLE 10.2 Potential mineral carbonation reaction routes.

Reaction route	Options	Suboptions
Direct carbonation	Dry gas–solid reaction	Gaseous CO_2; Supercritical CO_2
	Aqueous reaction	With or without additives
	Electrochemical	pH swing—two-step aqueous route Seawater and brines
Indirect carbonation	Acid extraction	Hydrochloric acid; Sulfuric acid; Acetic acid
	Molten salt extraction Ammonium sulfate extraction	
	Sodium hydroxide Microbially accelerated	

10.3 Direct carbonation routes

10.3.1 Direct gas–solid carbonation

The direct carbonation of a solid mineral exposed to gaseous CO_2 is the simplest carbonation reaction in terms of the process design and offers the best prospect of effective utilization of the heat released in the reaction. However, using silicate minerals the reaction rate is too slow for industrial-scale application, even at elevated temperatures and pressures, for example, using supercritical CO_2 (sCO_2) at 10–15 MPa.

The direct gas–solid reaction is, however, a viable carbonation route if the metal is first extracted from the feedstock as an oxide or hydroxide, because the carbonation of these compounds is rapid. For example:

$$Mg(OH)_2 + CO_2 \rightarrow MgCO_3 + H_2O \quad (10.6)$$

This route has also been proposed for the carbonation of bag house dust—a byproduct of steelmaking—which comprises >85% oxides of Fe, Ca, and Mg and typically has a median particle size of $\sim 100\ \mu m$. In this case, the gas–solid route has the advantage that it avoids the leaching of heavy metals such as lead, chromium, and cadmium that cause this material to be classified as a hazardous waste. Carbonation can also lead to the encapsulation and immobilization of these metals, making the end product less hazardous.

10.3.2 Direct aqueous carbonation

Water accelerates the carbonation reaction and direct carbonation in an aqueous medium, widely considered to be the most promising reaction route, has been the subject of extensive research. When CO_2 is dissolved into water, whether naturally or via industrial process, it forms carbonate ions while also slightly acidifying the water via the release of H^+ (reactions 10.7–10.9):

$$CO_{2(aq)} + H_2O \leftrightarrow H_2CO_3 \text{ (carbonic acid)} \quad (10.7)$$

$$H_2CO_3 \leftrightarrow HCO_3^- \text{ (bicarbonate ion)} + H^+ \quad (10.8)$$

$$HCO_3^- \leftrightarrow CO_3^{2-} \text{ (carbonate ion)} + H^+ \quad (10.9)$$

This reduced pH can become a limiting factor for the formation and precipitation of carbonate minerals, especially as further H^+ can be released into the fluid when the products of reactions 10.7 to 10.9 react with silicate rocks, causing them to release metal cations such as calcium, which combine with the carbonate ions in the water to produce carbonate minerals (reactions 10.10, 10.11):

$$(Ca, Mg, Fe)^{2+} + HCO_3^- \rightarrow (Ca, Mg, Fe)CO_3 + H^+ \quad (10.10)$$

or a neutral combination:

$$(Ca, Mg, Fe)^{2+} + CO_3^{2-} \rightarrow (Ca, Mg, Fe)CO_3 \quad (10.11)$$

The carbonate products formed by reaction 10.11 can then precipitate out of the fluid, commonly crystalizing in rock fissures and pore spaces of the host rock (Matter and Keleman, 2009). The mineral products will vary according to the partial pressure of CO_2 (pCO_2) in the fluid. Low pCO_2 produces secondary calcites, clays, and zeolites (Neuhoff et al., 1999) whereas high pCO_2 produces secondary iron, magnesium, and calcium carbonates. High or low pCO_2 can produce quartz/chalcedony (Rogers et al., 2006). For carbonate minerals to form and precipitate the excess H^+ needs to be removed, and this is possible via Ca-plagioclase dissolution and clay formation:

$$CaAl_2Si_2O_8 \text{ (Ca-plagioclase)} + 2H^+ + H_2O \rightarrow Ca^{2+} + Al_2Si_2O_5(OH)_4 \text{ (kaolinite)} \quad (10.12)$$

$$2Mg_2SiO_4 \text{ (forsterite)} + Mg_2Si_2O_6 \text{ (Mg-pyroxene)} + 4H_2O \rightarrow 2Mg_3Si_2O_5(OH)_4 \text{ (serpentine)} \quad (10.13)$$

The inclusion of H_2O into secondary minerals during serpentinization and carbonate formation requires a volumetric expansion that can often exceed 44% (Keleman et al., 2011). When this expansion occurs in situ it will regularly cause the reaction-driven fracturing of surrounding rocks (Keleman and Hirth, 2012). This can increase permeability and expose fresh mineral surfaces for continued reaction, although a negative consequence can be the occurrence of microseismic events (Keleman and Matter, 2008).

More important for carbon mineralization than serpentinization and clay formation is the precipitation of CO_2 as carbonate, for example the carbonation of olivine (reaction 10.14) and of olivine and pyroxene (reaction 10.15) (Keleman and Hirth, 2012):

$$Mg_2SiO_4 \text{ (forsterite)} + 2CO_2 \rightarrow 2MgCO_3 \text{ (magnesite)} + SiO_2 \text{ (quartz)} \quad (10.14)$$

$$Mg_2SiO_4 \text{ (forsterite)} + CaMgSi_2O_6 \text{ (CaMg-pyroxene)} + 2CO_2 + 2H_2O \rightarrow$$
$$Mg_3Si_2O_5(OH)_4 \text{ (serpentine)} + CaCO_3 \text{ (calcite)} + MgCO_3 \text{ (magnesite)} \quad (10.15)$$

The volumetric expansion due to water incorporation can also lead to reduced pore and fracture network permeability, which may be exacerbated by carbonate deposition (reactions 10.14 and 10.15). Reactivity may also be reduced by the deposition of a passivating layer on grain surfaces; for example carbonation of olivine can be slowed by up to 2 orders of magnitude by a $\leq$40 nm layer of silica forming on the grain surface (Daval et al., 2011). All of these effects—clogging, cracking, and passivation—have been observed in experimental investigations (Keleman et al., 2018), and assessing these competing feedback mechanisms is a major objective of ongoing research and pilot testing. Other important areas of ongoing carbon mineralization research include in situ CO_2/brine multiphase flow (Wang et al., 2015) on a pore scale (Chen and Zhang, 2010; Bandara et al., 2011), core sample scale (Perrin et al., 2009, 2011; Gunde et al., 2010), and large project scale (Ghomian et al., 2008; Yamamoto et al., 2009; Doughty, 2010; Chasset et al., 2011).

In natural environments, the variation of chemistry is virtually endless and the reactions described earlier are intended to highlight some of the more common and simple examples. Considerable research is underway on the full complexity of the geochemical reactions that occur. The type of secondary precipitated minerals will depend to a large degree on the initial fluid acidity and solute content, as well as the fluid throughflow rate, and the water-to-rock ratio during the reaction.

Where the above reactions occur in natural subsurface basaltic settings over geological time frames, neutralization of water acidity by reaction 10.11 commonly proceeds beyond neutral pH, with hyperalkaline waters ($pH > 11$) reaching the Earth's surface (Pentecost, 1995; Keleman and Matter, 2008; Blank et al., 2009; Arcilla et al., 2011). These alkaline springs are Ca rich, Mg poor (Chavagnac et al., 2013), and are sometimes called travertines. Travertine springs are common in ophiolite regions, where oceanic crust has been obducted onto the continent, examples of which are the Gros Morne park in Canada (Szponar et al., 2013), Othrys in Greece (Tsikouras et al., 2013), Del Puerto in California (Blank et al., 2009), and Samail in Oman (Olsson et al., 2014).

Due to the diversity of secondary mineral possibilities, combined with the problem of identifying a conservative reference element unaffected by the weathering reactions, it is extremely difficult to assess the rate of chemical alteration (and carbon drawdown) in ophiolites and other basalt regions simply by correlation with the volumetric quantity of secondary carbonate products observable (White and Buss, 2013). This difficulty of quantifying carbonation in environments outside the laboratory is also true in the heterogeneous mineralogy of mining wastes, where it has been found a lizardite to forsterite weight ratio of 0.3–0.4 is optimal for carbonation, with other minerals such as brucite having little effect on carbonate production (Li et al., 2022a,b).

10.3.3 Carbonation in seawater and other brines

The rate-limiting step in direct aqueous carbonation—metal ion liberation from the mineral matrix—can be overcome by using seawater, which already contains sufficient quantities of cations. Seawater contains 0.01 mol-Ca/kg and 0.05 mol-Mg/kg, and precipitation of the carbonates can therefore capture 1 tCO_2 in 370 t-seawater, producing 2.1 t of carbonates, comprising 81% $MgCO_3$ and 19% $CaCO_3$. Precipitation requires a pH of ~ 10 so that addition of an alkali such as NaOH is necessary for a seawater feed with a pH of ~ 8 (Wang et al., 2010).

An advantage of this technique is the low cost of utilizing seawater compared to other chemical solvents. Alkaline industrial waste products such as cement kiln dust (CKD) have been found to store 185 $kgCO_2$/tonne-CKD and produce 419 kg-$CaCO_3$/tonne-CKD when added to CO_2-saturated seawater (Jeon and Kim, 2019). Despite natural impurities in seawater the $CaCO_3$ product was 99.4% pure.

A demonstration project to capture CO_2 from power plant flue gas by absorption into seawater, with precipitation of calcium and magnesium carbonates for use as cement additives, was opened in August 2008 by Calera Corp. The pilot project, located adjacent to the Dynegy Inc. power plant at Moss Landing, Monterey County, CA, initially used a synthetic flue gas stream. With a 0.2 Mt/d seawater supply system on site, the plant had the capacity to eventually capture ~ 600 tCO_2/d from the power plant's flue gas stream but was not scaled up beyond ~ 1 tCO_2/d due to the high cost of the seawater pumping and limited availability of the alkaline waste needed to raise the pH to enable precipitation (Sanna et al., 2014).

Other potential sources of metal cations for carbon mineralization include aqueous waste streams such as produced water associated with oil and gas operations (Zhu et al., 2022), reject brine from seawater desalination plants (Bang et al., 2021), and synthetic seawater (Ma and Yoon, 2015).

The desalination of seawater is a vital process in many parts of the world to produce water for human consumption or irrigation. The residual concentrated brines can be used for direct aqueous mineralization and experiments, for example using brucite ($Mg(OH)_2$) as a carbonation feedstock, show that ions present in these brines reduce the amount of CO_2 required to achieve a given Mg conversion ratio, with 99% of the CO_2 supplied for reaction being precipitated as carbonate (Bang et al., 2021).

10.3.4 Direct electrochemical carbonation

Electrochemical methods have been investigated for CO_2 mineralization, and are mentioned here for completeness, while being more fully discussed in other chapters. Electrochemical mineralization in NaCl brine has been developed and deployed as a CO_2 capture technique, discussed in Chapter 7. Electrochemical enhancement of mineral weathering has been investigated as a means of increasing CO_2 uptake by the ocean (Chapter 16), and recently the potential for electrochemical methods to capture and mineralize CO_2 while also generating electrical power and H_2 has been reported (Chapter 7).

Numerous electrochemical methods use pH swing to capture CO_2, including electrolysis, bipolar membrane electrodialysis, reversible redox reactions, and capacitive deionization (Sharifian et al., 2021), although one downside to electrochemical methods is the high cost of membrane materials and of high-grade energy requirements.

10.4 Indirect carbonation routes

10.4.1 Microbially accelerated carbonation

Microorganisms have been demonstrated to accelerate carbon mineralization via extracellular biologically induced mineralization (BIM; see also Chapter 5) in a number of settings. Increases of 12%–16% in carbonation reactions in steel slag have been demonstrated using *Bacillus mucilaginosus* (Zhang et al., 2021), while also increasing the compressive strength of the carbonated product. Ureolytic bacteria have also been widely studied to precipitate calcium carbonate through the hydrolysis of urea to ammonium, with the carbonate used in stabilizing concrete and recycled aggregate (Sonmez et al., 2021).

Cyanobacteria, discussed further in Chapter 5, can also be used. Scanning electron microscope images of mine tailings that had been inoculated with a cyanobacteria culture after an acid leaching pretreatment reveal the deposition of magnesium carbonate minerals directly on the cell surface (Fig. 10.4), as a result of the concentration of carbonate ions and the presence of nucleation sites (McCutcheon et al., 2016).

A small-scale field trial on two 1.0 m^2 experimental plots at the abandoned Woodsreef asbestos mine in New South Wales, Australia, partially reproduced the lab results, although the degree of carbonation was reduced due to dry meteorological conditions and limited CO_2 availability at depth (McCutcheon et al., 2017). Experiments have also been performed using a small-scale open flow-through bioreactor, and there are clear synergies here with bioreactors and similar systems being developed for algal biomass cultivation (see Chapter 12).

10.4.2 Acid extraction

Indirect carbonation via acid extraction involves using acidic solutions to increase the rate of cation release from the alkaline feedstock into the solution. Strong acids are more effective than weaker ones yet can corrode metal process equipment, increasing equipment costs (Woodall et al., 2019). The use of hydrochloric acid (HCl) with a serpentine feedstock proceeds according to the following steps:

1. HCl is used to extract magnesium as magnesium chloride:

$$Mg_3Si_2O_5(OH)_4 + 6HCl + H_2O \rightarrow 3MgCl_2 \cdot 6H_2O + 2SiO_2 \quad (10.16)$$

2. HCl is recovered by heating the solution to ~250°C:

$$MgCl_2 \cdot 6H_2O \rightarrow MgCl(OH) + HCl + 5H_2O \quad (10.17)$$

3. Water is introduced to reform MgCl(OH) to $Mg(OH)_2$ which is precipitated:

$$2MgCl(OH) \rightarrow Mg(OH)_2 + MgCl_2 \quad (10.18)$$

The precipitated hydroxide is then carbonated via the direct carbonation step in reaction 10.6. The energy cost of Step 2 in this extraction route is very high and has been estimated as four times the energy released by carbon combustion for the quantity of CO_2 eventually sequestered. Clearly, without a substantial reduction in energy requirements, this is not a feasible sequestration process.

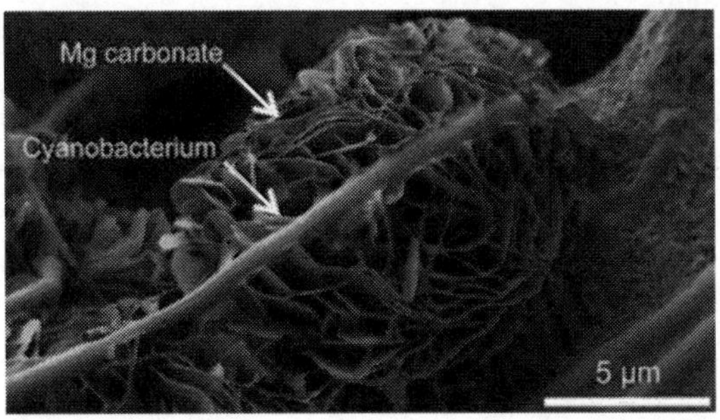

FIGURE 10.4 Magnesium carbonate mineral precipitation on filamentous cyanobacteria. *(From McCutcheon et al., 2016).*

A similar route has been proposed using sulfuric acid, either forming magnesium hydroxide for carbonation as in the HCl route or carbonating the magnesium sulfate directly in solution according to:

$$Mg_3Si_2O_5(OH)_4 + 3H_2SO_4 \rightarrow 3MgSO_4 + 5H_2O + 2SiO_2 \tag{10.19}$$

followed by:

$$3MgSO_4 + 3H_2O + 2SiO_2 + 3CO_2 \rightarrow 3MgCO_3 + 2SiO_2 + 3H_2SO_4 \tag{10.20}$$

The sulfuric acid process does not require the energy-intensive dehydration step that was necessary for HCl extraction. Although in principle these reaction routes result in recovery of the acid, some make-up volume is likely to be needed in practice. This would be required, for example, if the feedstock contains alkali metals that would react with HCl to form soluble chlorides, consuming chlorine.

Acetic acid extraction has also been proposed, using wollastonite as a feedstock to produce calcium carbonate. This has the advantage of being a less corrosive medium and therefore puts less stringent constraints on reactor material selection, although laboratory trials indicate that recovery of the acetic acid may be problematic.

Citric, ascorbic, ethylenediaminetetraacetic, and nitric acids have also been investigated for use in mineral carbonation via acid extraction (Woodall et al., 2019).

10.4.3 Ammonium sulfate extraction

A promising indirect carbonation route has been developed by Zevenhoven and colleagues (2013) at the Åbo Akademi University in Finland, using ammonium sulfate in a multistep process to produce magnesium hydroxide. Milled silicate rock is first heated at 450°C with ammonium sulfate $(NH_4)_2SO_4$ to produce magnesium sulfate;

$$Mg_3Si_2O_5(OH)_4 + 3(NH_4)_2SO_4 \leftrightarrow 3MgSO_4 + 2SiO_2 + 5H_2O + 6NH_3 \tag{10.21}$$

The solids are mixed with water and the insoluble rock residue is removed. Ammonia released in this step is used to raise the solution pH, first to ~7–8 to precipitate iron oxyhydroxide (FeOOH - goethite) from sulfates that will have formed from iron silicate components in the rock. This by-product can be used for iron making. The pH is further raised by additional ammonia dissolution to ~10–11 and magnesium precipitates as $Mg(OH)_2$;

$$MgSO_4 + 2NH_4OH \leftrightarrow (NH_4)_2SO_4 + Mg(OH)_2 \tag{10.22}$$

Finally, ammonium sulfate is recovered from the solution either by complete evaporation using heat recovery from flue gases or, with greater energy efficiency, using mechanical vapor recompression (MVR—see Glossary). The $Mg(OH)_2$ can then be carbonated in a fluidized bed to capture CO_2 from industrial flue gas, or used in other capture processes such as direct air capture (DAC; Chapter 13) or ocean alkalinity enhancement (OAE; Chapter 16). The overall process is shown schematically in Fig. 10.5.

This route was used for a pilot scale application at a 200 t/d lime kiln at Parainen, Finland (Slotte et al., 2013). Heat recovery from the kiln was sufficient to process 13.2 t/d of serpentine rock; with 80% Mg extraction and 90% $Mg(OH)_2$ carbonation this would result in the capture of ~4.5 tCO_2/d. External electrical energy input of 0.8 GJ/tCO_2 captured would also be required for rock crushing/grinding and gas compression.

10.4.4 Molten salt extraction

This indirect carbonation technique, using a molten salt such as magnesium chloride as the extraction medium, can partially reduce the energy cost of the dehydration step in the HCl extraction process. In its simplest form, the carbonation reaction takes place in a single direct step in a molten salt medium ($MgCl_2 \cdot 3.5H_2O$) at ~200°C. Alternatively, a multistep process similar to HCl extraction can be used, with the intermediate precipitation, separation, and carbonation of magnesium hydroxide. Once again, the energy cost and corrosive nature of the extraction medium are challenges for this option.

10.4.5 Sodium hydroxide extraction

The use of sodium hydroxide for indirect carbonation has also been proposed as an alternative to acid extraction. In this option calcium or magnesium is released from the mineral matrix and precipitated as hydroxide, with sodium taking the

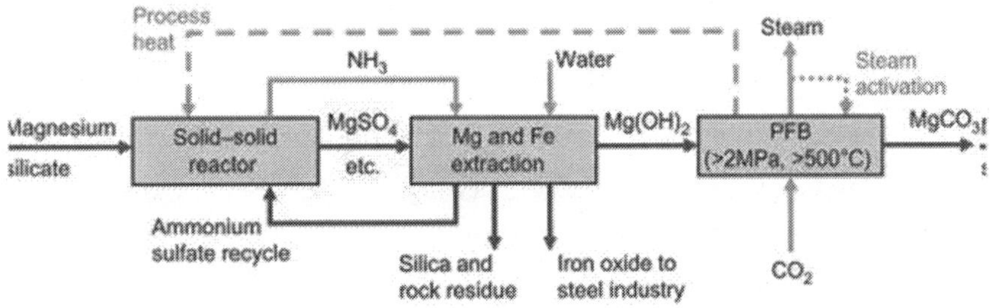

FIGURE 10.5 Mineral carbonation by ammonium sulfate extraction.

place of the extracted metal ion in the mineral. For example, with plagioclase (calcium feldspar) as a feedstock, the overall reaction including carbonation of the intermediate $Ca(OH)_2$ is:

$$3CaAl_2Si_2O_8 + 8NaOH + 4CO_2 \rightarrow 3CaCO_3 + Na_8(AlSiO_4)_6CO_3 \cdot 2H_2O + 2H_2O \quad (10.23)$$

Because sodium hydroxide is consumed in the reaction, substantial quantities (1.8 t-NaOH/tCO_2) would be required if this reaction route were to be applied on an industrial scale.

10.5 Passive ex situ carbon mineralization

Extensive research supports adding olivine or olivine-rich basalt to agricultural soils, which weather and sequester CO_2 in mineral carbonate form (Schuiling and Krijgsman, 2006; Edwards et al., 2017). Adding olivine to coastal marine environments has also been proposed (Montserrat et al., 2017), as well as the use of olivine in other passive industrial roles such as aggregate beneath railway tracks and in road construction. These methods, including the current field trial status, are further discussed in Chapter 11.

10.5.1 Terrestrial enhanced weathering

Adding olivine in the form of fine sand to agricultural soils can passively sequester CO_2 directly from the atmosphere and is referred to as enhanced weathering (EW). Olivine addition can act as a pH buffer in acidic soils, replacing traditional liming (Renforth et al., 2015), and in agricultural soils can also (ten Berge et al., 2012):

- Increase plant growth by up to 15.6%.
- Increase plant potassium content by up to 16.5%.
- Increase the bioavailability of magnesium and nickel.
- Increase plant uptake of Mg, Si, and Ni.
- Increase CO_2 sequestration.
- Suppress plant Ca uptake.

There is the danger of potentially harmful amounts of trace elements such as Ni and Cr being released into the soils, and olivine rich in these elements may be less applicable for terrestrial EW (Amann et al., 2020). It can be difficult to monitor sequestration rates due to the sheer complexity and site-specific variability of natural soils (Zinke, 2021). However, monitored field-scale experiments are in progress to assess the potential of this method, and its effect on crop yield and water runoff. There is also a need to build public trust in a technique that affects the food supply (Beerling et al., 2018). The potential carbon removal on cropland areas could be as large as 95 GtCO_2/yr for dunite addition, and 4.9 GtCO_2/yr for basalt (Strefler et al., 2018), with warm and wet countries such as India, Brazil, South-East Asia, and China, possessing 75% of the global potential.

10.5.2 Coastal-enhanced weathering

Coastal-enhanced weathering involves artificially sanding beaches and shallow coastal regions with olivine sand and allowing passive carbon mineralization to proceed over decadal periods. As noted earlier, olivine dissolution and alteration can be slowed by up to two orders of magnitude by the formation of a silica passivating layer on the grain surface (Béarat et al., 2006; Daval et al., 2011), but wave action is expected to remove this passivating layer so that grains can carbonate to completion (Schuiling and de Boer, 2011). In nonmarine experimental settings, grinding the silicates prior to carbonation has been found to yield 600% more carbonate, with dunite shown to be more reactive compared to olivine and lizardite (Rashid et al., 2021).

The timescale for particle dissolution is a further challenge that has led to skepticism about the viability of this approach. Assuming a mean seawater temperature of 15–25°C, for carbon sequestration to complete within 15–20 years requires grain sizes <10 μm, which necessitates significant preparation with associated financial and carbon emissions penalties, as well as the difficulties of transportation and dispersion (Hangx and Spiers, 2009). The use of larger grain sizes would avoid these difficulties, but the 700–2100 year dissolution time estimated by Hangx and Spiers (2009) for a 300-μm grain size would need to be significantly reduced by wave action for this to contribute on a timescale relevant to climate mitigation.

10.6 Active ex situ industrial carbon mineralization

New CO_2 removal policies and the potential for beneficial use of mineralized products are spurring many companies, including MCi International, SCW Systems, Green Minerals, and others discussed later, to develop mineralization processes using a range of direct and indirect reaction routes and feedstocks (olivine, serpentine, etc.) for application in various carbon-intensive industries.

10.6.1 Cement and aggregate manufacture

Since 1995, there has been rapidly growing interest in the utilization of captured CO_2 in the cement industry, where mineral carbonation can enhance the binding quality of industrial recycled waste aggregate (Lange et al., 1995). Carbon 8 Aggregates Ltd. is one of those leading the commercialization of these techniques.

CO_2 can be both a waste product and an ingredient in the cement-making process. Manufacturing a metric tonne of average cement produces ca. 0.73–0.99 tCO_2 as a waste product (Hasanbeigi et al., 2012). However, CO_2 that has been mineralized as carbonate can then be used to produce cement via a number of pathways. One example is magnesium carbonate trihydrate ($MgCO_3 \cdot 3H_2O$; the mineral nesquehonite), which can be obtained via staged precipitation from desalination brines (Galvez-Martos et al., 2018), or alternatively from MgO hydrating into brucite (Vandeperre and Al-Tabbaa, 2007). Nesquehonite, or alternatively dypingite ($Mg_5(CO_3)_4(OH)_2 \cdot 5H_2O$) and hydromagnesite ($Mg_5(CO_3)_4(OH)_2 \cdot 4H_2O$), can then be processed into a cement through a variety of pathways such as combining with pulverized fly ash (Zhang et al., 2018). This cement has a CO_2 content of ca. 30% and comes under the umbrella term "eco bricks". Formation of nesquehonite from brucite can occur in as little as 30 min in a laboratory, using diluted HCl as a catalyst (Zhao et al., 2010).

Carbonated reactive magnesia cement (CRMC) incorporates reactive magnesia with various waste powders, creating an end product that can adsorb CO_2 from the environment into the cementitious matrix (Soares et al., 2022). Calera Corporation (under the name Fortera) is at the forefront of this technology, while other companies active in cement CO_2 sequestration include Solidia Technologies Inc., Novacem Ltd., and CarbonCure Technologies Inc.

An ongoing project in this field is CO_2MIN led by HeidelbergCement, the Institute for Advanced Sustainability Studies (IASS), and Rheinisch-Westfälische Technische Hochschule (RWTH) at the University of Aachen, which has been supported by the German Federal Ministry of Education and Research (BMBF) with a €3 million investment. Its remit is to explore the absorption of CO_2 from industrial flue gas by olivine and basalt, and evaluate the potential use of the secondary products as aggregates or in cement production. Results from the project include evidence for up to 60% carbonation of olivine feedstocks and CO_2 uptake increases of 83.7%, corresponding to 305 $kgCO_2$/t of olivine (Kremer et al., 2022). Further, the project identified that supplementary cementitious materials (SCM) produced via CO_2 carbonation could be produced at scale with a levelized cost of 120 €/tSCM, which lies in the range of current cement prices (Strunge, 2021).

10.6.2 Paper manufacture

Precipitated calcium carbonate (PCC) is used in the paper-making process, where calcium carbonate ($CaCO_3$) is heated to create lime (CaO), which is then used in several ways (Miner and Upton, 2002; National Lime Association, 2022):

- As a causticizing agent in sulfate (Kraft process) plants.
- As a reactant to produce calcium bisulfite used to dissolve noncellulosic constituents of wood chips.

- Reacted with chlorine to produce calcium hypochlorite, which is used to bleach and whiten paper.
- In various other roles such as a coagulant in color removal, a filtration conditioner, as a pH neutralizer, or to recover by-products from process waste such as alcohol, calcium lignosulfonate, and yeast.

Carbonate created via carbon mineralization can replace other sources of PCC in paper processing, as well as in other industries.

10.6.3 Steel production

To produce an average tonne of steel results in ca. 0.2–0.4 t of by-products such as slags, dust, sludges, and other materials (Kumar et al., 2019). The global iron and steel industry generates 400–500 Mt/yr of slag, with the potential to capture 160–220 $MtCO_2$/yr (Renforth et al., 2011; Kumar et al., 2019), reducing the carbon intensity of the product by up to 20%. In addition, the global stockpile of slag from historical production is estimated to amount to 6–8 Gt, with the potential to sequester 2–4 $GtCO_2$, depending on the degree to which this has already been naturally weathered. A number of synergies would result from the integration of mineral carbonation with steel production, including:

- Similar requirements for mining and preprocessing of iron ore and carbonation feedstock.
- Potential use of peridotite (a magnesium- and iron-rich mineral—see Glossary) as a common feedstock for iron ore and magnesium silicate.
- Sequestration of CO_2 produced both from iron ore reduction and from the rest of the steelmaking process, totaling ~1.7 tCO_2/t-steel produced.
- Carbonation of steel slag with added value as a sellable product, including the potential for use in creating artificial reefs (Li et al., 2022a,b).

Fig. 10.6 shows the concept of an integrated steel mill with CO_2 sequestration by mineral carbonation that has been investigated by the American Iron and Steel Institute together with Colombia University. Although not shown explicitly in this schematic, slag from the steelmaking process would also be fed back into the carbonation plant.

Pilot-scale results on carbonation of steel slag under elevated pCO_2 (0.40–0.46 atm) reported CO_2 uptake of ca. 5% by weight after a 30 min reaction time in a rotating drum reactor, increasing to ca. 10% by weight after curing for 12 days at 25°C and 100% relative humidity (Librandi et al., 2017). Ongoing research is aimed at optimizing operating conditions to sustain this high CO_2 uptake while the slag particle size is increased to maximize its value in construction applications.

The carbonation of stockpiled iron and steel slag for direct air capture is a possible early-stage demonstrator project that is being pursued by several companies. This relatively simple, low-tech option, illustrated in Fig. 10.7, consists of a number of large basins (each ~10,000 m^2) into which the prepared waste material is loaded. Water is sprayed onto the basin as a fine mist, absorbing CO_2 as the drops fall through the air. The solution reacts with calcium hydroxide ($Ca(OH)_2$) as it moves through the bed and is recycled, dissolving more CO_2 and $Ca(OH)_2$, and precipitating $CaCO_3$ on each cycle.

10.6.4 Polymer production

Reaction products of carbon mineralization can also be used in industrial polymers. Wollastonite, talc, etc., are carbon mineralization reaction products that can be used as functional fillers to manage properties such as shrinkage, strength, and thermal conductivity in the final polymer products (Hadal et al., 2004; Lee et al., 2006).

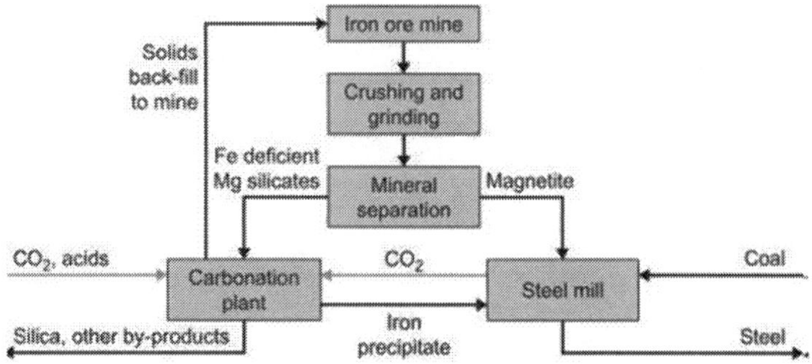

FIGURE 10.6 Integrated steel mill with CO_2 sequestration by mineral carbonation.

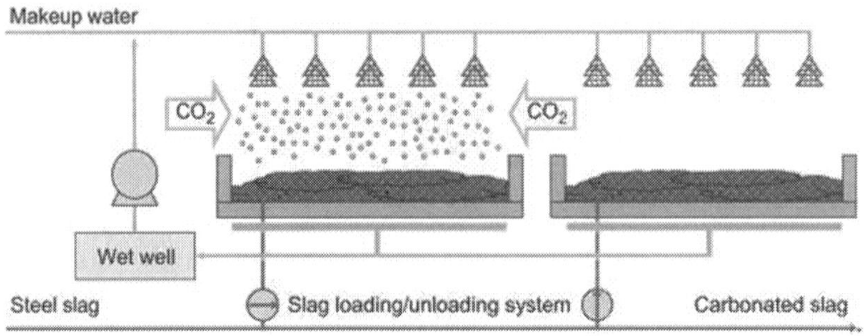

FIGURE 10.7 Direct air capture of CO_2 by steel slag carbonation.

10.6.5 Mining industries

Exploiting the obvious synergies between carbon mineralization and the mining industry requires that projects reflect site-specific conditions, and a variety of approaches are therefore possible, as illustrated in Fig. 10.8.

Using this approach, Power et al. (2014) have outlined a number of projects that could be implemented at the Mount Keith nickel mine in Western Australia, starting with enhanced passive carbonation, which could sequester >150 $ktCO_2$/yr (44% of the mine's total emissions). Microbially assisted carbonation could add >50 $ktCO_2$/yr, while larger scale mineralization of the 2.5% brucite in the tailings could render the mine carbon neutral at a cost of ~\$10/tCO_2. The latter proposal envisaged the mineralization of CO_2 captured from additional external point sources, and the use of bioleaching (the generation of acidic conditions by iron and sulfur-oxidizing microbes) to enhance the release of Mg ions from the tailings.

Reaction of CO_2-rich gases with the ultramafic mine tailings from the Baptiste nickel deposit in British Columbia, Canada, for example, has been assessed to have the capacity to offset 42%–53% of the mine's emissions (Power et al., 2020). The rate of sequestration was estimated at 3.5 $kgCO_2/m^2$/yr and a site total of 17 $ktCO_2$/yr based on an area of 5 km^2. Interestingly, this rate of sequestration can be increased fivefold when aerating the tailings (Power et al., 2020).

Nickel mine tailings offer significant global potential. Recent analysis, which includes tailings associated with future production, indicated a worldwide carbonation capacity of around 7 $GtCO_2$ (Siegrist et al., 2017), which could be

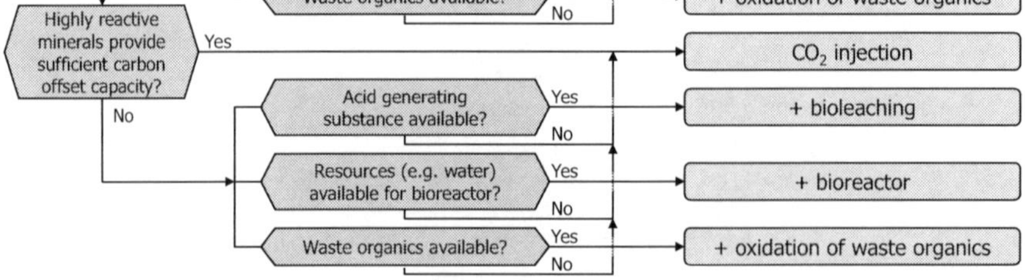

FIGURE 10.8 Decision tree for mineral carbonation pilot project design (after Power et al., 2014).

economically exploited given a carbon price of ~\$50/tCO_2. Over 60% of this estimated capacity is located in Australia, Canada, and South Africa, and some mines have the potential to capture many times their emissions, resulting in significant CO_2 drawdown. The main obstacle preventing progress toward the implementation of these projects is the lack of a financial incentive to the industry in the form of a carbon price.

Prefeasibility studies on acid and metalliferous drainage mitigation combined with carbon sequestration in the nickel mine waste in Finland have also been conducted (Zevenhoven et al., 2017; Savage et al., 2019). These studies apply the indirect "Åbo Akademi" carbonation routes discussed above, including the potential for process integration with iron and steel or cement industries.

Waste materials from these industries contain a significant proportion of primarily calcium oxides and hydroxides, 10%−15% for concrete waste, 40%−50% for steel slag (Stolaroff et al., 2005), and, when stockpiled as waste or used for landfill, are subject to natural carbonation at a slow rate (see, for example, Mayes et al., 2018). This process can be accelerated using any of the reaction routes applied to mineral feedstock, direct aqueous carbonation being the front-runner. The use of these materials in a direct air capture system has also been considered and is described in Chapter 13.

Many of these wastes have a reuse value in the US\$5−10/t range that is comparable to or exceeds the expected mined cost of mineral feedstocks. However, carbonation can in many cases be effectively integrated with the reuse process, adding value to the end product. Examples are the production of precipitated calcium carbonate or the reuse of steel slag as a cement additive. This process requires the slag to be crushed and finely ground, and these are also necessary preparation steps for carbonation, so that integration of a carbonation step in the recycling process would not incur additional preparation costs.

A key advantage of utilizing mine tailings such as asbestos (chrysotile serpentine) is that the hazardous waste is effectively remediated by the carbonation reactions. The quantity of asbestos mine tailings available globally is estimated to be in the range from 2 to 6 Gt, with a potential to sequester up to 2.5 GtCO_2. Much of this is located in the United States and Canada, the Black Lake mine in Quebec, Canada, alone having a potential storage capacity of up to 0.7 GtCO_2. In many cases, carbonation of tailings could fully offset emissions from mine operations as well as deliver other environmental benefits.

10.6.6 Biomass gasification power plant with integrated mineral carbonation

Gasification of biomass or other fuels is a highly efficient process that is applied in power generation and the production of synthetic liquid fuels or hydrogen. The gasification product − synthesis gas or syngas − is a mixture of carbon monoxide and hydrogen and is generated by the combustion of the feedstock in an oxygen-poor environment, leading to partial oxidation. Depending on the application, some or all of the carbon monoxide in the syngas can be converted to CO_2 using a water−gas shift (WGS) reaction, typically carried out at a temperature of 300−500°C:

$$CO + H_2O \rightarrow CO_2 + H_2 \tag{10.24}$$

For example, if the syngas is to be used for liquid fuel synthesis using the Fischer−Tropsch process, ~35% of the syngas needs to be shifted to achieve the required CO to H_2 ratio. For a typical syngas composition, this results in a post-WGS syngas containing ~60% CO_2 by weight.

Syngas leaves the partial oxidation step at a temperature of ~1500°C, and the heat available in cooling the gas to the desired temperature for the WGS reaction (~1.47 MJ/kg-syngas) can be used for heat pretreatment of mineral carbonation feedstock. Each kg of syngas cooled can heat-activate ~1.1 kg-serpentine (1.32 MJ/kg) and yields 0.75 kgCO_2 if 35% of the CO is water-shifted. Because the quantity of material required for CO_2 capture (R_{CO2}; see below) is 2.1 kg/kgCO_2 for heat-activated serpentine, 0.52 kgCO_2 could be sequestered via carbonation, which is 70% of the CO_2 generated in the WGS reaction.

Fig. 10.9 shows a schematic diagram of a biomass gasification, liquid fuel synthesis plant with integrated CO_2 sequestration by mineral carbonation. Additional process heat integration is shown, with the cooled syngas exiting the gasification reactor being used to preheat the incoming crushed and coarsely ground serpentine feedstock. Fine grinding of the mineral is not required because larger particles, up to a few millimeters, will fragment due to the expansion of steam released during dehydroxylation.

10.7 In situ carbon mineralization

In situ carbon mineralization is a direct aqueous method that injects hot water supersaturated in CO_2 into a suitable geological formation using the geothermal gradient combined with the exothermic nature of the mineral reactions to sustain

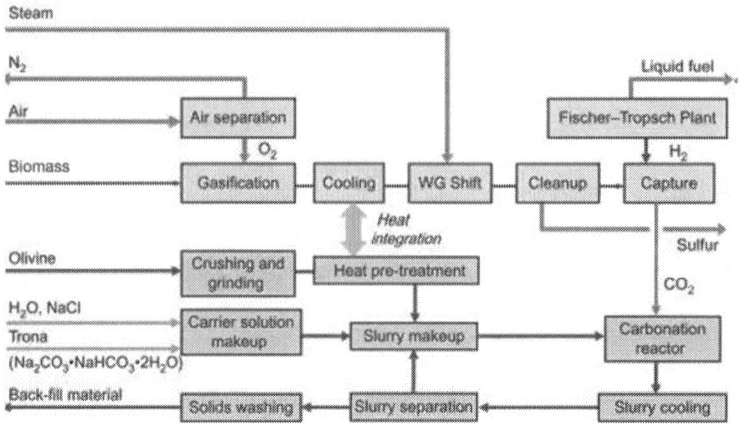

FIGURE 10.9 Biomass gasification, liquid fuel synthesis plant with integrated mineral carbonation.

the temperature. High temperature high pressure reactions proceed faster than low temperature low pressure reactions, with temperature exerting a greater influence over reaction rate than pressure, which primarily affects gas solubility. Mineral carbonation progresses at an optimal rate at temperatures ~185°C (Keleman and Matter, 2008).

Experiments have demonstrated that mineral carbonation occurs up to 40 times more efficiently in olivine-rich basalt compared with pure olivine (Sissmann et al., 2014). Such results imply that in situ mineralization in basalt targets may be far more efficient for sequestering CO_2 than ex situ techniques that use pure olivines.

In situ carbon mineralization is therefore most suitable in regions where the target mineralogy contains high amounts of olivine. Flood basalts and other volcanic settings are one example, as are ocean seafloors or ophiolites (where the seabed has been obducted onto the land). Such rock types are readily available and account for a significant amount of global surface area. It is estimated that in situ carbon mineralization could sequester up to 60,000,000 $GtCO_2$, which far exceeds the required CO_2 sequestration requirement for climate change mitigation (Kelemen et al., 2019).

10.7.1 CarbFix project

The CarbFix project, a collaboration between Reykjavik Energy, the University of Iceland, Columbia University, and the Centre Nationnal de la Recherche Scientifique (France), began in 2007 and remains the most high-profile in situ carbon mineralization project. Situated in the Western Hengill, Iceland, the project drilled numerous holes into an aquifer located at 550 m beneath a cap rock of impermeable glassy basalt at 100–300 m depth (Alfredsson et al., 2013).

The host olivine tholeiite basalt comprises olivine, plagioclase, pyroxene (augite), magnetite, ilmenite, chromium spinel, and glass (Alfredsson et al., 2013). Basalts are naturally rich in divalent cations, notably Mg^{2+}, Fe^{2+}, Ca^{2+}, and the project site possesses an estimated 6 mol/kg of cations potentially available for reaction (Alfredsson et al., 2013).

Geothermally heated water at ~25°C was super saturated with 0.8–0.42 mol/kg of pure CO_2 equilibrated to 25 bar and CO_2–H_2S–H_2 was equilibrated to 14 bar at a ratio 75:24.2:0.8. This fluid mix was then injected into the basalt subsurface at a rate of 0.5–0.7 kg/s (Alfredsson et al., 2013; Matter et al., 2009). Generally, a high water-to-rock ratio and high through flow rate will increase aqueous cation removal from a system, reducing carbonate growth potential. Fluid injection rate is therefore an important factor to consider for in situ carbon mineralization.

Mineral olivine dissolution increases markedly as pH lowers in a process called proton-promoted dissolution (Pokrovsky and Schott, 2000; Rosso and Rimstidt, 2000; Liu et al., 2006). This dissolution occurs at the grain boundaries, reducing grain size, and within grains, where it widens veins and fractures. This can be achieved by the addition of hydrogen sulfide (H_2S) or hydrogen, reducing pH to 3.7–4.0 to accelerate the liberation of cations from the basalt (Alfredsson et al., 2013). CarbFix now includes the SulFix project which disposes of H_2S (produced along with CO_2 from geothermal wells) by injection at a depth of ~1350 m (Aradóttir et al., 2015), where the pressure and temperature are more conducive to sulfide mineral stability and thus long term storage (Juliusson et al., 2015; Ragnarsson, 2015).

The in situ ambient water was found to be ~30–60°C at 300–800 m depth and 180°C at 1500 m depth, with an alkaline pH of 8.4–9.8 (Alfredsson et al., 2013). Tracer fluids of fluorescent Na dye and sulfur hexafluoride SF_6 formed part of the injection fluid for the purpose of observing any leaks in the cap rock, and monitoring the transport fate of the fluids (Matter et al., 2009). The site has a natural groundwater through flow rate of 25 m/yr (Alfredsson et al., 2013), and no leakage of tracer fluid has been detected.

The thermodynamic database initially used by CarbFix (EQ3/6 V7.2b) consisted of just 36 mineral reactions, and the researchers discovered that other databases also lacked the range of mineral reaction data required to model CarbFix injection conditions. Al-hydroxy complexes were particularly lacking and the thermodynamic database SUPCRT92 was used to compute solubility constants where experimental data were lacking. 110 minerals are now included in the CarbFix kinetic database, although there was a lack of kinetic experimental data and PHREEQC was used to fill empirical data gaps. Swift progress has since been made in developing kinetic databases for minerals associated with carbon mineralization (Heřmanská et al., 2022).

Currently (2023), CarbFix has an injection capacity of 12,000 tCO_2/yr and has injected 78,000 tCO_2 since 2014.

CarbFix has partnered with Rio Tinto on a further carbon mineralization project at the ISAL aluminum smelter in Straumsvík, Hafnarfjarðarkaupstaður, Iceland. This is the first smelter where carbon capture and storage is implemented in the aluminum industry. Carbfix drilled the first injection wells at the Coda Terminal in 2022, with CO_2 injection planned for 2025. The Coda Terminal is a cross-border carbon transport and storage hub, accepting CO_2 from multiple industrial sites across Northern Europe.

10.7.2 Wallula project, Columbia river basalts

The potential for carbon storage in continental flood basalts was pointed out by McGrail and colleagues (2006), and a pilot CCS project—the Wallula Basalt Sequestration Pilot Project—was planned by this group as part of the US Department of Energy (DoE) Big Sky Regional Carbon Sequestration Partnership. The initial pilot scope included the capture of up to 0.8 $MtCO_2$/yr from the power plant of the Boise White Paper mill, located on the banks of the Columbia River in Washington state, and the injection of $scCO_2$ into permeable flow units in the Columbia River Basalt Group (CRBG) that underlies $>200,000\ km^2$ of Washington, Oregon, and Idaho.

Following site assessment using 2D seismic, a pilot well was drilled in 2009 for site characterization and identified a suitable injection zone at ca. 850 m depth, with 28 m net thickness and an average permeability of 44−90 milliDarcy (mD) estimated from constant-rate pumping tests. Injection tests also confirmed the presence of three intervals that would act as flow barriers above the injection zone, with permeabilities of 0.01−0.1 microDarcy (μD). Unfortunately, the integrated CCS pilot was shelved in 2011 when DoE funding was not secured for the next phase. Funds were however available for an injection pilot and, in 2013, 977 tCO_2 was injected over 25 days in the world's first $scCO_2$ injection test into flood basalt.

Postinjection fluid sampling showed a 100−1000-fold increase in Ca, Mg, Fe, and Mn cation concentrations compared to preinjection samples, as well as ^{13}C and ^{18}O isotopic shifts consistent with the onset of geochemical reactions. Geochemical analysis of sidewall core samples recovered before well abandonment in 2015 confirmed the rapid conversion of injected $scCO_2$ to carbonate minerals, which were visible as nodules of ankerite ($Ca(Fe,Mg,Mn)(CO_3)_2$) distributed within the basalt matrix (McGrail et al., 2017).

Further results based on 50 sidewall core samples indicated that ~60% of the injected CO_2 was mineralized within 24 months. Iron-rich ankerite nodules were the prevalent carbonate mineralogy, with secondary silicates such as clays and zeolites (Holliman et al., 2021).

10.7.3 Subsea in situ mineralization

Building on the results of the Wallula and CarbFix projects, prefeasibility studies for a sub-seabed in situ mineralization in the eastern flank of the Juan de Fuca plate, offshore Washington state, and British Columbia started in 2018 (Goldberg et al., 2018). The Juan de Fuca plate as a whole has a carbon mineralization potential estimated at 762 $GtCO_2$ (Kazemifar, 2022).

The project is envisaged to capture up to 2 $MtCO_2$/yr from industrial sources such as Shell's Puget Sound refinery and Alcoa's Italco aluminum smelter and transport this to an injection site some 250 km offshore. Both pipeline and ship are being considered as transportation options, although the work of Geske and colleagues (2015) suggests that pipeline interconnectors between coastal sources, with ship transport to the injection site, would be an optimal solution, given the long offshore distance and (relatively) small volume.

The target reservoir is fairly well appraised as a result of three scientific drilling expeditions in the region, with permeability of 0.1−1 Darcy and porosity of 10%−20% expected in the top 300 m of basalt, which here lies beneath 300−400 m of sediments in (somewhat challenging) water depths of around 2600 m.

10.7.4 Flood basalt and ophiolite storage capacity

There are over 60 ophiolites around the world in easily accessible areas where large-scale carbon mineralization projects could be implemented. Extensively studied examples of in situ carbon mineralization are the Samail Ophiolite, in the Sultanate of Oman, the Columbia River, and Juan de Fuca plate basalts, in the United States, and the entirety of Iceland (Kelemen and Matter, 2008; McGrail et al., 2006; Gislason et al., 2010). Younger silicate rocks weather faster than older ones, particularly when <8 Ma old (Gislason et al., 1994), therefore rock age is an important factor when deciding upon sequestration sites.

Estimates suggest that the Colombia River and Juan de Fuca basalts alone may have a capacity >1000 $GtCO_2$ (Power et al., 2013), sufficient to sequester North America's combined emissions for over 200 years; while for deep-sea basalts worldwide total sequestration capacity could range from 30,000 to 150,000 $GtCO_2$ (Goldberg and Slagle, 2009), many times greater than the emissions that would result from the combustion of all producible fossil fuels on Earth.

Research suggests that subaerial exposure of the ocean floor in the New Caledonian terrain of the Pacific created a carbon sink that reduced atmospheric CO_2 concentration by ~ 100 μatm during the Eocene (~ 35 Ma ago), which may have triggered the Antarctic glaciation (Reusch, 2011). Transforming whole planetary environments through the power of mineral weathering is clearly feasible.

10.8 Technoeconomic considerations

Even if a technique has been demonstrated to be technically successful in mineralizing carbon, a number of further factors need to be considered when determining whether the technique is viable for deployment at scale. These factors include the financial and CO_2 cost in extracting and transporting material feedstocks, and any reaction optimization required such as mineral pretreatment and activation. In some cases, the CO_2 emitted in the end-to-end process may be more than that eventually sequestered, or high costs may simply make a project economically unviable; in other cases, both the economics and net CO_2 sequestered can be positive. Such assessments are highly case specific, reflecting unique feedstock characteristics, locations, and transportation routes.

10.8.1 Identification of mineral feedstock

Important characteristics that would need to be fulfilled by an optimal feedstock for mineral carbonation include:

- Low cost of supply and preparation
- High reaction rate
- Global distribution and availability close to CO_2 sources
- High alkalinity content and conversion efficiency
- Low process energy requirements

Cost of supply

Low cost of supply and preparation is essential because typically 2–3 tonnes of mineral feedstock are required to capture 1 tCO_2. Depending on the active mineral concentration, costs therefore cannot exceed a few US$/t, which compares with current hard rock mining costs of ~$10/t. Supply and preparation steps include mining, transportation, grinding, and removal of any impurities or components that would impede the carbonation reaction or cause problems in storing the reaction products (e.g., components that could leach out of mine back-filling). Many potential feedstocks contain 5%–20% by weight of iron oxides, as well as lower concentrations of aluminum, chromium, and nickel, which can yield valuable by-products, reducing the net cost of supply.

Reaction rate

A high reaction rate, either natural or through reaction optimization, is essential for carbonation on an industrial rather than a geological time scale. Carbonation proceeds at a greater rate for alkali metals such as sodium and potassium than for the alkaline–earth metals calcium and magnesium. The higher reactivities of sodium and potassium result in their carbonates being soluble in water, which poses significant issues for the storage of reaction products.

Of the alkaline–earth metals, the carbonation reaction proceeds faster for calcium (e.g., wollastonite) than for magnesium (e.g., forsterite olivine), although wollastonite is far less abundant. Quantitative empirical knowledge of carbon mineralization kinetics is currently limited, and further investigations of mineral surface alteration processes, precipitation controls, and reaction pathways are also needed (Wang et al., 2015).

Availability

Global distribution and availability close to major CO_2 capture sites are clearly very important for a feedstock if mineral carbonation is to be applied on a climate impact scale. Of the alkaline–earth silicates, olivine and serpentine occur in nature in vast quantities. The igneous rocks that contain these minerals originated from upwelling magma on ancient ocean floors and, as noted above, occur abundantly in a number of countries. CCS studies of deposits within the United States have shown a good match on a regional scale between the distribution of these resources and the regional CO_2 emissions from current power plants. However, cost-effective exploitation of the mineral resources to achieve negative emissions may require the development of CO_2 transportation infrastructure (for example, between BECCS power plants and carbonation plants at mineral locations), the transport of extracted reactive products to capture sites, or the location of CO_2 sources (DAC hubs or new BECCS plants) adjacent to mineral resources.

Alkalinity content and conversion efficiency

A high molar fraction of reactive minerals in the feedstock will minimize the quantity of rock that needs to be mined per tCO_2 captured. Two ratios commonly used to compare feedstocks are R_{CO2}—the quantity of mineral required to capture $1 tCO_2$, and R_C—the quantity of mineral required to capture the CO_2 produced by combustion of 1 tC (equal to R_{CO2} times 44/12, the ratio of the molar weights of CO_2 and C).

For a given carbonation feedstock, the conversion efficiency of the carbonation reaction is expressed by the parameter R_x, which measures the percentage conversion of available metal cations into carbonate. The conversion efficiency depends on the specific reaction conditions (temperature, pressure, and time) and mineral pretreatment (see below). These three ratios are shown in Table 10.3 for some potential mineral feedstocks.

R_C and R_{CO2} are determined purely by the mineral feedstock composition and, in the case of a pure mineral, by the ratios of molecular weights. Table 10.3 shows two R_{CO2} values for each mineral, first the theoretical value for a pure mineral and secondly a typical value for an actual prepared mineral sample. The R_{CO2} values show that the magnesium minerals (forsterite, lizardite) are more efficient in terms of quantities required, due to the lower molar weight of magnesium (24) versus calcium (40). The R_x values shown for each mineral illustrate the range of values achieved in various experiments, the lower figure without and the higher figure with activation, either by ultrafine grinding or by heat or other pretreatment.

Process energy requirement

Low process energy requirements will help to reduce costs and will also contribute to the net capture efficiency by reducing CO_2 emissions attributed to process energy consumption. Olivine and wollastonite are slightly favored over serpentine due to the higher exothermic heat of the reaction, although the practicality of usefully applying this reaction heat in an integrated carbonation process is the subject for ongoing investigation. If an indirect acid extraction method is used the carbonation reaction heat could be directly used in acid recovery.

10.8.2 Mineral feedstock ranking

Table 10.4 compares a number of potential feedstocks for mineral carbonation using the characteristics discussed earlier. Based on these considerations, most early development work has been focused on either olivine or serpentine as the preferred feedstock for large-scale application, although alkaline industrial wastes, particularly mine tailings, are increasingly attracting attention due to their low supply cost.

TABLE 10.3 Relative mineral feedstock quantities and conversion efficiencies.

Mineral	R_{CO2} (kg/kg)	R_C (kg/kg)	R_x (%)
Olivine (forsterite)	1.6–1.8	5.9–6.6	60–80
Olivine (fayalite)	2.3–2.8	8.4–10.3	60–65
Wollastonite	2.6–2.8	9.5–10.3	40–80
Serpentine (lizardite)	2.1–2.1	7.7–7.7	9–40
Serpentine (antigorite)	2.1–2.5	7.7–9.2	60–90

TABLE 10.4 Ranking of potential mineral carbonation feedstocks.

Characteristic	Olivine	Serpentine	Wollastonite	Alkaline industrial waste
Cost of supply and preparation	=	=	=	+ +
Reaction rate	=	–	+	+ +
Availability	+ +	+ +	–	– –
Global distribution	+	+	–	+
High molar fraction of alkaline–earth metal	+	+	–	=
Low reaction energy requirements	=	–	=	+

Note: = represents the basis for comparison. + and + + imply better or much better performance against a particular characteristic; – and – – imply poorer performance.

10.8.3 Reaction optimization, including mineral pretreatment and activation

Early investigations of aqueous carbonation identified an optimal carrier solution for the reaction of both olivine and serpentine as a sodium bicarbonate plus sodium chloride solution with strength 0.64 M for $NaHCO_3$ and 1 M for NaCl (Gerdemann et al., 2007). The addition of sodium bicarbonate increases the concentration of HCO_3^-, accelerating the carbonation reaction. CO_2 solubility was also increased by a factor of 20 in this carrier solution (20 g/L vs. <1 g/L in distilled water), allowing the reaction to take place at lower CO_2 pressure. Finally, the bicarbonate–salt mixture buffered the pH of the carrier solution to a slightly alkaline 7.7–8.0, aiding the precipitation of the carbonate. The optimal reaction temperature and CO_2 pressure, established in the same studies were, for olivine, T = 185°C and P_{CO2} = 15 MPa, and for heat-treated serpentine (see below) 155°C and 10.5 MPa (Gerdemann et al., 2007).

In aqueous carbonation, the liquid to solid ratio (L/S) is a further optimization parameter; a lower L/S improves the heat efficiency of the process while a higher L/S makes stirring and pumping easier and improves the conversion efficiency. The optimal L/S will also depend on the type of carbonation reactors, such as stirred tank or pipe configuration.

As noted above, serpentine is an attractive feedstock for industrial-scale carbonation, despite the slow reaction rate which arises as a result of the relative inaccessibility of the magnesium ion within the layered phyllosilicate structure. Various pretreatment options have been investigated with the aim of increasing the reaction rate, as summarized in Table 10.5. Combined pretreatments, such as mechanical plus chemical activation have also been found to be effective.

Investigations aimed at optimizing the carbonation reaction have spawned additional basic research into the molecular-level controls on the reaction process. Examples of this work are X-ray diffraction (XRD) and scanning electron microscopy (SEM) studies of the dehydroxylation process in heat treatment of serpentine and of the surface morphology of mineral particles during the carbonation reaction. This is an example of the nonsequential nature of the technology development process. The latter work has led to insights into the development of a surface passivating layer and the accelerated exfoliation of such layers by the addition of abrasive, nonreactive particles such as quartz or by ultrasonic agitation, further improving the reaction rate. Any additives used would need to be efficiently recovered to avoid an economic penalty.

Although heat pretreatment of serpentine is energy-intensive, a possible route to mitigate this high energy cost is through integrating mineral carbonation into an industrial process such as biomass gasification (Section 10.6.6). The considerable process heat from the initial partial combustion step in the gasification process could be used to heat-treat the serpentine feedstock, with CO_2 produced from the WGS reaction captured in a carbonation reaction with the heat-treated serpentine.

10.8.4 Disposal and reuse options for carbonation end products

Mineral carbonation generates in excess of 2.5 t-reaction product per tCO_2 sequestered. For carbon mineralization to be applied on a climate impact scale, many gigatonnes of reaction products, including magnesium carbonate, silica, and other compounds, would be created annually. The eventual fate of these products, including potential reuse options, is therefore a significant issue warranting further research. In addition to mine back-filling and land reclamation, which could consume large quantities, other options that have been proposed include reuse as an additive in cement or asphalt (Nejad et al., 2020), for soil enhancement (Chapter 11), or ocean liming (Chapter 16).

TABLE 10.5 Mineral pretreatment options.

Pretreatment option	Description
Dry and wet grinding	Grinding to increase the specific surface area is the first pretreatment step. Early investigations revealed that a reduction of average particle size increased the extent of conversion (R_X) for a given residence time in line with the surface area increase. Attrition grinding (vigorously stirring the mineral particles together with hard balls) can further increase reactivity by creating dislocations in the mineral lattice structure, increasing the accessibility of the metal ion. The feedstock is typically reduced to a mean particle diameter of <75 μm, a further stage reduction to <38 μm being a trade-off between additional energy costs (~80 kWh per tonne) and increased mineral conversion.
Heat treatment	Heating serpentine results in progressive removal of the hydroxyl ion (for lizardite, 45% residual OH at 580°C, 10% at 650°C) and eventually, at ~1100°C, to the crystallization of forsterite and enstatite ($MgSiO_3$). Above 580°C the dehydroxylation results in the breakup of the layered structure into an amorphous phase, yielding a substantially increased carbonation rate compared to the untreated mineral. However, the energy required is substantial (specific heat to 580°C plus enthalpy of dehydroxylation ~330 kWh/t for lizardite feedstock) and heat treatment is likely to be viable only when waste process heat is available.
Chemical activation	A wide range of organic and inorganic acids, bases, and salts have been investigated as options to increase feedstock reactivity, with sulfuric acid being a particularly effective agent. These treatments can increase specific surface area by a factor of 40, equivalent to reducing mean particle diameter by a factor of 6 through additional grinding. Disadvantages are the consumption of chemicals and the reduction in alkaline–earth metal content of the feedstock as a result of leaching. Organic ligands have also been shown to aid the leaching of Mg.
Ultrasonic and microwave activation	Early work on mineral activation by ultrasonic or microwave pretreatment failed to demonstrate increased reactivity; subsequently microwave heating has been shown to result in dehydroxylation of serpentine to olivine, increasing reactivity and improving grinding and slurry rheology. Exfoliation of the passivating layer by ultrasonic agitation is also beneficial during the carbonation reaction.

10.8.5 Business case for carbon mineralization

Although CDR is now considered essential for the achievement of climate goals, on an individual business level carbon removal options remain costly and are generally uneconomic (Roussanaly et al., 2017; Voldsund et al., 2018). However, CO_2 mineralization has the potential not only to sequester an estimated 3 $GtCO_2$ per year (Woodall et al., 2019) but also to produce a range of secondary products that can support a positive business case.

The cost of mineralizing a tonne of CO_2 is estimated between \$75 to \$510 (€65 to €443; Strunge et al., 2022), although this varies depending on the type and particle size of the feedstock, and the technique employed. For enhanced weathering the cost is estimated at \$60/tCO_2 using dunite (>90% olivine), and \$200/tCO_2 for basalt (Strefler et al., 2018), while in applications producing useable secondary products a positive revenue of €14-€700/tCO_2 could be generated (Ostovari et al., 2020).

10.9 Conclusions

Carbon mineralization has the potential to contribute to net-zero targets and deliver negative emissions in a number of industries, but has for a long time been perceived as a fringe and relatively untested experimental technology, and one unlikely to be deployed on a large scale. This is no longer the case, with a wealth of experimental data and studies now in place, as well as numerous small to medium-sized demonstration projects. The cement industry is just one example, where CO_2 emissions could be reduced by 8%–33% (Strunge et al., 2022), and mineralization could potentially shift the industry from carbon positive to carbon negative (Ostovari et al., 2021).

A lack of financial incentives and a general policy vacuum are the factors currently limiting the development and deployment of carbon mineralization at an increasing scale. This chapter demonstrates that carbon mineralization is a proven technology that can be financially viable in the right circumstances. With several potential target sites around the globe available for further in situ carbon mineralization, and with passive ex situ techniques, such as enhanced weathering, ready for deployment at increasing scale, a supportive policy framework could enable rapid expansion in the field of carbon mineralization over the coming decades.

References and resources

Alfredsson, H.A., Oelkers, E.H., Hardarsson, B.S., Franzson, H., Gunnlaugsson, E., Gislason, S.R., 2013. The geology and water chemistry of the Hellisheidi, SW-Iceland carbon storage site. Int. J. Greenh. Gas Control 12, 399–418.

Amann, T., Hartmann, J., Struyf, E., Garcia, W.O., Fischer, E.K., Janssens, I., Meire, P., Schoelynck, J., 2020. Biogeosciences 17, 103–119.

Aradóttir, E.S.P., Gunnarsson, I., Sigfússon, B., Gunnarsson, G., Júliusson, B.M., Gunnlaugsson, E., Sigurdardóttir, H., Arnarson, M.T., Sonnenthal, E., 2015. Toward cleaner geothermal energy utilization: capturing and sequestering CO_2 and H_2S emissions from geothermal power plants. Transp. Porous Media 108 (1), 61–84.

Arcilla, C.A., Pascua, C.S., Alexander, W.R., 2011. Hyperalkaline groundwaters and tectonism in the Philippines: significance to natural carbon capture and sequestration. Energy Proc. 4, 5093–5101.

Bandara, U.C., Tartakovsky, A.M., Palmer, B.J., 2011. Pore-scale study of capillary trapping mechanism during CO_2 injection in geological formations. Int. J. Green Gas Control 5, 1566–1577.

Bang, J., Chae, S., Song, K., Lee, S., 2021. Optimizing experimental parameters in sequential CO_2 mineralization using seawater desalination brine. Desalination 519, 115309.

Blank, J.G., Green, S.J., Blake, D., Valley, J.W., Kita, N.T., Treiman, A., Dobson, P.F., 2009. An alkaline spring system within the Del Puerto Ophiolite (California, USA): a Mars analog site. Planet. Space Sci. 57 (5–6), 533–540.

Béarat, H., McKelvy, M.J., Chizmeshya, A.V.G., Gormley, D., Nunez, R., Carpenter, R.W., Squires, K., Wolf, G.H., 2006. Carbon sequestration via aqueous olivine mineral carbonation: role of passivating layer formation. Environ. Sci. Technol. 40 (15), 4802–4808.

Beerling, D.J., Leake, J.R., Long, S.P., Scholes, J.D., Ton, J., Nelson, P.N., Bird, M., Kantzas, E., Taylor, L.L., Sarkar, B., Kelland, M., DeLucia, E., Kantola, I., Müller, C., Rau, G., Hansen, J., 2018. Farming with crops and rocks to address global climate, food and soil security. Nat. Plants 4, 138–147.

Chasset, C., Jarsjö, J., Erlström, M., Cvetkovic, V., Destouni, G., 2011. Scenario simulations of CO_2 injection feasibility, plume migration and storage in a saline aquifer, Scania, Sweden. Int. J. Green Gas Control 5, 1303–1318.

Chavagnac, V., Monnin, C., Ceuleneer, G., Boulart, C., Hoareau, G., 2013. Characterization of hyperalkaline fluids produced by low-temperature serpentinization of mantle peridotites in the Oman and Ligurian ophiolites. G-cubed 14 (7), 2496–2522.

Chen, C., Zhang, D., 2010. Pore-scale simulation of density-driven convection in fractured porous media during geological CO_2 sequestration. Water Resour. Res. 46, W11527.

Daval, D., Sissmann, O., Menguy, N., Saldi, G.D., Guyot, F., Martinez, I., Corvisier, J., Garcia, B., Machouk, I., Knauss, K.G., Hellmann, R., 2011. Influence of amorphous silica layer formation on the dissolution rate of olivine at 90°C and elevated pCO_2. Chem. Geol. 284 (1–2), 193–209.

Doughty, C., 2010. Investigation of CO_2 plume behavior for a large scale pilot test of geologic carbon storage in a saline formation. Transp. Porous Media 82, 49–76.

Dunsmore, H.E., 1992. A geological perspective on global warming and the possibility of carbon dioxide removal as calcium carbonate mineral. Energy Convers. Manag. 33 (5–8), 565–572.

Edwards, D.P., Lim, F., James, R.H., Pearce, C.R., Scholes, J., Freckleton, R.P., Beerling, D.J., 2017. Climate change mitigation: potential benefits and pitfalls of enhanced rock weathering in tropical agriculture. Biol. Lett. 13, 4.

Gaines, R.V., Skinner, H.C.W., Foord, E.E., Mason, B., Rosenzweig, A., King, V.T., Dowty, E., 1997. Dana's New Mineralogy, eighth ed. Wiley-Interscience, New York.

Galvez-Martos, J.L., Elhoweris, A., Morrison, J., Al-horr, Y., 2018. Conceptual design of a CO_2 capture and utilisation process based on calcium and magnesium rich brines. J. CO_2 Util. 27, 161–169.

Gerdemann, S.J., O'Connor, W.K., Dahlin, D.C., Penner, L.R., Rush, H., 2007. Ex situ aqueous mineral carbonation. Environ. Sci. Technol. 41, 2587–2593.

Geske, J., Berghout, N., van den Broek, M., 2015. Cost-effective balance between CO_2 vessel and pipeline transport: Part I – impact of optimally sized vessels and fleets. Int. J. Greenh. Gas Control 36, 175–188.

Ghomian, Y., Pope, G.A., Sepehrnoori, K., 2008. Reservoir simulation of CO_2 sequestration pilot in frio brine formation, USA Gulf Coast. Energy 33, 1055–1067.

Gislason, S.R., Wolff-Boenisch, D., Stefansson, A., Oelkers, E.H., Gunnlaugsson, E., Sigurdardottir, H., Sigfusson, B., Broecker, W.S., Matter, J.M., Stute, M., Axelsson, G., 2010. Mineral sequestration of carbon dioxide in basalt: a pre-injection overview of the CarbFix project. Int. J. Greenh. Gas Control 4, 537–545.

Gislason, S.R., Arnorsson, S., Armannsson, H., 1994. Present chemical weathering of basalt in Iceland. Mineral. Mag. 58 (A), 333.

Goldberg, D., Slagle, A.L., 2009. A global assessment of deep-sea basalt sites for carbon sequestration. Energy Proc. 1, 3675–3682.

Goldberg, D., Aston, L., Bonneville, A., Demirkanli, I., Evans, C., Fisher, A., Garcia, H., Gerrard, M., Heesemann, M., Hnottavange-Telleen, K., Hsu, E., 2018. Geological storage of CO_2 in sub-seafloor basalt: the CarbonSAFE pre-feasibility study offshore Washington State and British Columbia. Energy Proc. 146, 158–165.

Gunde, A.C., Bera, B., Mitra, S.K., 2010. Investigation of water and CO_2 (carbon dioxide) flooding using micro-CT (microcomputed tomography) images of Berea sandstone core using finite element simulations. Energy 35, 5209–5216.

Hadal, R.S., Dasari, A., Rohrmann, J., Misra, R.D.K., 2004. Effect of wollastonite and talc on the micromechanisms of tensile deformation in polypropylene composites. Mater. Sci. Eng. A 372 (1–2), 296–315.

Hangx, S.J.T., Spiers, C.J., 2009. Coastal spreading of olivine to control atmospheric CO_2 concentrations: a critical analysis of viability. Int. J. Greenh. Gas Control 3 (6), 757–767.

Hasanbeigi, A., Price, L., Lin, E., 2012. Emerging energy-efficiency and CO_2 emission-reduction technologies for cement and concrete production: a technical review. Renew. Sustain. Energy Rev. 16 (8), 6220–6238.

Heřmanská, M., Voigt, M.J., Marieni, C., Declercq, J., Oelkers, E.H., 2022. A comprehensive and internally consistent mineral dissolution rate database: Part I: Primary silicate minerals and glasses. Chem. Geol. 597, 120807.

Holliman, J., Schaef, H., Miller, Q., Horner, J., McGrail, B., Polites, E., 2021. Carbon Sequestration in Basalts: Sidewall Core Characterization Data from the Wallula Basalt Pilot Project. AGU 2021 Fall Meeting, New Orleans, LA and virtual, 13–17 December 2021.

Jeon, J., Kim, M., 2019. CO_2 storage and $CaCO_3$ production using seawater and an alkali industrial by-product. Chem. Eng. J. 378, 122180.

Juliusson, B.M., Gunnarsson, I., Matthiasdottir, K.V., Markusson, S.H., Bjarnason, B., Sveinsson, O.G., Gislason, T., Thorsteinsson, H.H., 2015. Tackling the challenge of H_2S emissions. In: Proceedings World Geothermal Congress 2015.

Kazemifar, F., 2022. A review of technologies for carbon capture, sequestration, and utilization: cost, capacity, and technology readiness. Greenh. Gas Sci. Technol. 12 (1), 200–230.

Kelemen, P.B., Matter, J., 2008. In situ carbonation of peridotite for CO_2 storage. Proc. Natl. Acad. Sci. USA 105 (45), 17295–17300.

Kelemen, P., Matter, J., Streit, E., Rudge, J., Curry, W., Blusztajn, J., 2011. Rates and mechanisms of mineral carbonation in peridotite: natural processes and recipes for enhanced, in situ CO_2 capture and storage. Annu. Rev. Earth Planet Sci. 39, 545–576.

Kelemen, P.B., Hirth, G., 2012. Reaction-driven cracking during retrograde metamorphism: olivine hydration and carbonation. Earth Planet Sci. Lett. 345–348, 81–89.

Kelemen, P.B., Aines, R., Bennett, E., Benson, S.M., Carter, E., Coggon, J.A., et al., 2018. In situ carbon mineralization in ultramafic rocks: natural processes and possible engineered methods. Energy Proc. 146, 92–102.

Kelemen, P., Benson, S.M., Pilorgé, H., Psarras, P., Wilcox, J., 2019. An overview of the status and challenges of CO_2 storage in minerals and geological formations. Front. Clim. 9. https://doi.org/10.3389/fclim.2019.00009.

Kremer, D., Dertmann, C., Etzold, S., Telle, R., Friedrich, B., Wotruba, H., 2022. Ex-situ mineral carbonation – a parameter study on carbon mineralization in an autoclave as part of a large-scale utilisation process. J. CO_2 Util. 58, 101928.

Kumar, S., Dhara, S., Kumar, V., Gupta, A., Prasad, A., Keshari, K., Mishra, B., 2019. Recent trends in slag management & utilization in the steel industry. Miner. Met. Rev. 94–109.

Lackner, K.S., Wendt, C.H., Butt, D.P., Joyce Jr., E.L., Sharp, D.H., 1995. Carbon dioxide disposal in carbonate minerals. Energy 20, 1153–1170.

Lange, L.C., Hills, C.D., Poole, A.B., 1995. Preliminary investigation into the effects of carbonation on cement-solidified hazardous wastes. Environ. Sci. Technol. 30 (1), 25–30.

Lee, G.W., Park, M., Kim, J., Lee, J.I., Yoon, H.G., 2006. Enhanced thermal conductivity of polymer composites filled with hybrid filler. Compos. Appl. Sci. Manuf. 37 (5), 727–734.

Li, J., Jacobs, A.D., Hitch, M., 2022a. The effect of mineral composition on direct aqueous carbonation of ultramafic mine waste rock for CO_2 sequestration, a case study of Turnagain ultramafic complex in British Columbia, Canada. Int. J. Min. Reclamat. Environ. 36 (4), 267–286.

Li, J., Zhao, S., Song, X., Ni, W., Mao, S., Du, H., Zhu, S., Jiang, F., Zeng, H., Deng, X., Hitch, M., 2022b. Carbonation curing on magnetically separated steel slag for the preparation of artificial reefs. Materials 15 (6), 2055.

Librandi, P., Costa, G., Souza A., A., Stendardo, S., Luna, A.S., Baciocchi, R., 2017. Carbonation of steel slag: testing of the wet route in a pilot-scale reactor. Energy Proc. 114, 5381–5392.

Liu, Y., Olsen, A.A., Rimstidt, J.D., 2006. Mechanism for the dissolution of olivine series minerals in acidic dissolution. Am. Mineral. 91, 455–458.

Ma, J., Yoon, R.H., 2015. In: Morreale, B., Shi, F. (Eds.), CO_2 Mineralization in Artificial Seawater, Chapter 12 in Novel Materials for Carbon Dioxide Mitigation Technology. Elsevier, Amsterdam, the Netherlands.

Matter, J.M., Broecker, W.S., Stute, M., Gislason, S.R., Oelkers, E.H., Stefánsson, A., Wolff-Boenisch, D., Gunnlaugsson, E., Axelsson, G., Björnsson, G., 2009. Permanent carbon dioxide storage into basalt: the CarbFix pilot project, Iceland. Energy Proc. 1, 3641–3646.

Matter, J.M., Keleman, P., 2009. Permanent storage of carbon dioxide in geological reservoirs by mineral carbonation. Nat. Geosci. 2, 837–841.

Mayes, W.M., Riley, A.L., Gomes, H.I., Brabham, P., Hamlyn, J., Pullin, H., Renforth, P., 2018. Atmospheric CO_2 sequestration in iron and steel slag: Consett, county Durham, United Kingdom. Environ. Sci. Technol. 52 (14), 7892–7900.

McGrail, B.P., Schaef, H.T., Ho, A.M., Chien, Y.J., Dooley, J.J., Davidson, C.L., 2006. Potential for carbon dioxide sequestration in flood basalts. J. Geophys. Res. 111, B12201.

McGrail, B.P., Schaef, H.T., Spane, F.A., Horner, J.A., Owen, A.T., Cliff, J.B., Qafoku, O., Thompson, C.J., Sullivan, E.C., 2017. Wallula basalt pilot demonstration project: post-injection results and conclusions. Energy Proc. 114, 5783–5790.

McCutcheon, J., Wilson, S.A., Southam, G., 2016. Microbially accelerated carbonate mineral precipitation as a strategy for in situ carbon sequestration and rehabilitation of asbestos mine sites. Environ. Sci. Technol. 50 (3), 1419–1427.

McCutcheon, J., Turvey, C.C., Wilson, S.A., Hamilton, J.L., Southam, G., 2017. Experimental deployment of microbial mineral carbonation at an asbestos mine: potential applications to carbon storage and tailings stabilization. Minerals 7 (10), 191.

Miner, R., Upton, B., 2002. Methods for estimating greenhouse gas emissions from lime kilns at kraft pulp mills. Energy 27 (8), 729–738.

Montserrat, F., Renforth, P., Hartmann, J., Leermakers, M., Knops, P., Meysman, F.J.R., 2017. Olivine dissolution in seawater: implications for CO_2 sequestration through enhanced weathering in coastal environments. Environ. Sci. Technol. Lett. 51 (7), 3960–3972.

National Lime Association, 2022. Pulp and Paper at. https://www.lime.org/lime-basics/uses-of-lime/other-uses-of-lime/pulp-and-paper/. (Accessed 15 April 2022).

Nejad, F.M., Geraee, E., Azarhoosh, A.R., 2020. The effect of nano calcium carbonate on the dynamic behaviour of asphalt concrete mixture. Eur. J. Environ. Civil Eng. 24 (8), 1219–1228.

Neuhoff, P.S., Fridriksson, T., Arnorsson, S., Bird, D.K., 1999. Porosity evolution and mineral paragenisis of basaltic lavas at Teigarhorn, Eastern Iceland. Am. J. Sci. 299, 467–501.

Olsson, J., Stipp, S.L.S., Gislason, S.R., 2014. Element scavenging by recently formed travertine deposits in the alkaline springs from the Oman Semail Ophiolite. Mineral. Mag. 78 (6).

Ostovari, H., Sternberg, A., Bardow, A., 2020. Rock 'n' use of CO_2: carbon footprint of carbon capture and utilization by mineralization. Sustain. Energy Fuels 4, 4482–4496.

Ostovari, H., Müller, L., Skocek, J., Bardow, A., 2021. From Unavoidable CO_2 source to CO_2 sink? A cement industry based on CO_2 mineralization. Environ. Sci. Technol. 55, 5212–5223.

Pentecost, A., 1995. The quaternary travertine deposits of Europe and Asia Minor. Quat. Sci. Rev. 14 (10), 1005–1028.

Perrin, J.C., Krause, M., Kuo, C.W., Miljkovic, L., Charoba, E., Benson, S.M., 2009. Core-scale experimental study of relative permeability properties of CO_2 and brine in reservoir rocks. Energy Proc. 1, 3515–3522.

Perrin, J.C., Falta, R.W., Krevor, S., Zuo, L., Ellison, K., Benson, S.M., 2011. Laboratory experiments on core-scale behaviour of CO_2 exolved from CO_2-saturated brine. Energy Proc. 4, 3210–3215.

Power, I.M., Harrison, A.L., Dipple, G.M., Wilson, S.A., Kelemen, P.B., Hitch, M., Southam, G., 2013. Carbon mineralization: from natural analogues to engineered systems. Rev. Mineral. Geochem. 77 (1), 305–360.

Power, I.M., McCutcheon, J., Harrison, A.L., Wilson, S.A., Dipple, G.M., Kelly, S., Southam, C., Southam, G., 2014. Strategizing carbon-neutral mines: a case for pilot projects. Minerals 4 (2), 399–436.

Power, I.M., Dipple, G.M., Bradshaw, P.M.D., Harrison, A.L., 2020. Prospects for CO_2 mineralization and enhanced weathering of ultramafic mine tailings from the Baptiste nickel deposit in British Columbia, Canada. Int. J. Greenh. Gas Control 94, 102895.

Povrovsky, O.S., Schott, J., 2000. Forsterite surface composition in aqueous solutions: a combined potentiometric, electrokinetic, and spectroscopic approach. Geochem. Cosmochim. Acta 64, 3299–3312.

Ragnarsson, Á., 2015. Geothermal development in Iceland 2010–2014. In: Proceedings World Geothermal Congress 2015.

Rashid, M.I., Benhelal, E., Farhang, F., Oliver, T.K., Stockenhuber, M., Kennedy, E.M., 2021. Application of concurrent grinding in direct aqueous carbonation of magnesium silicates. J. CO_2 Util. 48, 101516.

Renforth, P., Washbourne, C.L., Taylder, J., Manning, D.A.C., 2011. Silicate production and availability for mineral carbonation. Environ. Sci. Technol. 45 (6), 2035–2041.

Renforth, P., Pogge von Strandmann, P.A.E., Henderson, G.M., 2015. The dissolution of olivine added to soil: implications for enhanced weathering. Appl. Geochem. 61, 109–118.

Reusch, D.N., 2011. New Caledonian carbon sinks at the onset of Antarctic glaciation. Geology 39 (9), 807–810.

Rogers, K.L., Neuhoff, P.S., Pedersen, A.K., Bird, D.K., 2006. CO_2 metasomatism in a basalt-hosted petroleum reservoir, Nuussuaq, West Greenland. Lithos 92 (1–2), 55–82.

Rosso, J.J., Rimstidt, J.D., 2000. A high resolution study of forsterite dissolution rates. Geochem. Cosmochim. Acta 64, 797–811.

Roussanaly, S., Fu, C., Voldsund, M., Anantharaman, R., Spinelli, M., Romano, M., 2017. Techno-economic analysis of MEA CO_2 capture from a cement kiln – impact of steam supply Scenario. Energy Proc. 114, 6229–6239.

Sanna, A., Uibu, M., Caramanna, G., Kuusik, R., Maroto-Valer, M.M., 2014. A review of mineral carbonation technologies to sequester CO_2. Chem. Soc. Rev. 43 (23), 8049–8080.

Savage, R.J., Pearce, S., Mueller, S., Barnes, A., Renforth, P., Sapsford, D., 2019. Methods for assessing acid and metalliferous drainage mitigation and carbon sequestration in mine waste: a case study from Kevitsa mine, Finland. In: Fourie, A.B., Tibbett, M. (Eds.), Mine Closure 2019: Proceedings of the 13th International Conference on Mine Closure. Australian Centre for Geomechanics, Perth, pp. 1073–1086.

Schuiling, R.D., Krijgsman, P., 2006. Enhanced weathering: an effective and cheap tool to sequester CO_2. Clim. Change 74, 349–354.

Schuiling, R.D., de Boer, P.L., 2011. Rolling stones; fast weathering olivine in shallow seas for cost-effective CO_2 capture and mitigation of global warming and ocean acidification. Earth Syst. Dyn. Discuss. 2, 551–568.

Seifritz, W., 1990. CO_2 disposal by means of silicates. Nature 345, 486.

Sharifian, R., Wagterveld, R.M., Digdaya, I.A., Xiang, C., Vermaas, D.A., 2021. Electrochemical carbon dioxide capture to close the carbon cycle. Energy Environ. Sci. 14, 781–814.

Siegrist, M., Southam, C., Southam, G., 2017. An economic analysis of the worldwide potential for CO_2 sequestration through bacteria-mediated carbon mineralization at nickel mine sites. 30th Australasian Finance and Banking Conference. 10.2139/ssrn.3020584.

Sissmann, O., Brunet, F., Martinez, I., Guyot, F., Verlaguet, A., Pinquier, Y., Daval, D., 2014. Enhanced olivine carbonation within a basalt as compared to single-phase experiments: reevaluating the potential of CO_2 mineral sequestration. Environ. Sci. Technol. Lett. 48 (10), 5512–5519.

Slotte, M., Romão, I., Zevenhoven, R., 2013. Integration of a pilot-scale serpentinite carbonation process with an industrial lime kiln. Energy 62, 142–149.

Soares, E.G., Castro-Gomes, J., Sitarz, M., Zdeb, T., Hager, I., Hassan, K., Al-Kuwarid, M.S., 2022. Feasibility for co-utilisation of Carbonated Reactive Magnesia Cement (CRMC) and industrial wastes in circular economy and CO_2 mineralization. Construct. Build. Mater. 323, 126488.

Sonmez, M., Ilcan, H., Dundar, B., Yildirim, G., Ersan, Y.C., Sahmaran, M., 2021. The effect of chemical- versus microbial-induced calcium carbonate mineralization on the enhancement of fine recycled concrete aggregate: a comparative study. J. Build. Eng. 44, 103316.

Stolaroff, J.K., Lowry, G.V., Keith, D.W., 2005. Using CaO- and MgO-rich industrial waste streams for carbon sequestration. Energy Convers. Manag. 46, 687–699.

Strefler, J., Amann, T., Bauer, N., Kriegler, E., Hartmann, J., 2018. Potential and costs of carbon dioxide removal by enhanced weathering of rocks. Environ. Res. Lett. 13, 034010.

Strunge, T., 2021. The Costs of CO_2 Carbonation in the Cement Industry. SINTEF Proceedings, vol 7. SINTEF Academic Press.

Strunge, T., Renforth, P., Van der Spek, M., 2022. Towards a business case for CO_2 mineralization in the cement industry. Commun. Earth Environ. 3, 59.

Szponar, N., Brazelton, W.J., Schrenk, M.O., Bower, D.M., Steele, A., Morrill, P.L., 2013. Geochemistry of a continental site of serpentinization, the tablelands ophiolite, Gros Morne National park: a mars analogue. Icarus 224 (2), 286–296.

Ten Berge, H.F.M., Van der Meer, H.G., Steenhuizen, J.W., Goedhart, P.W., Knops, P., Verhagen, J., 2012. Olivine weathering in soil, and its effect on growth and nutrient uptake in ryegrass (Lolium perenne L.): a pot experiment. PLoS One 7 (8), e42098.

Tsikouras, B., Etiope, G., Ifandi, E., Kordella, S., Papatheodorou, G., Hatzipanagiotou, K., 2013. Petrological implications for the production of methane and hydrogen in hyperalkaline springs from the Othrys ophiolite, Greece. In: Proceedings of the 13th International Congress, Chania, XLVII. Bulletin of the Geological Society of Greece, 1–449.

Vandeperre, L.J., Al-Tabbaa, A., 2007. Accelerated carbonation of reactive MgO cements. Adv. Cement Res. 19 (2), 67–79.

Voldsund, M., Anantharaman, R., Berstad, D., De Lena, E., Fu, C., Gardarsdottir, S.O., Jamali, A., Pérez-Calvo, J.F., Romano, M., Roussanaly, S., Ruppert, J., Stallmann, O., Sutter, D., 2018. D4. 6: CEMCAP Comparative Techno-Economic Analysis of CO_2 Capture in Cement Plants. H2020 Project: CO_2 Capture from Cement Production.

Wang, W., Hu, M., Ma, C., 2010. Possibility for CO_2 sequestration using sea water. In: 4th International Conference on Bioinformatics and Biomedical Engineering (iCBBE). IEEE.

Wang, D., Dong, B., Breen, S., Zhao, M., Qiao, J., Liu, Y., Zhang, Y., Song, Y., 2015. Review: approaches to research on CO_2/brine two-phase migration in saline aquifers. Hydrogeol. J. 23, 1–18.

White, A.F., Buss, H.L., 2013. Natural weathering rates of silicate minerals. In: Holland, H.D., Turekian, K.K. (Eds.), Treatise on Geochemistry. Elsevier Science Ltd, pp. 115–155.

Woodall, C.M., McQueen, N., Pilorgé, H., Wilcox, J., 2019. Utilization of mineral carbonation products: current state and potential. Greenh. Gas. Sci. Technol. 9 (6), 1096–1113.

Yamamoto, H., Zhang, K., Karasaki, K., Marui, A., Uehara, H., Nishikawa, N., 2009. Large-scale numerical simulation of CO_2 geologic storage and its impact on regional groundwater flow: a hypothetical case study at Tokyo Bay, Japan. Energy Proc. 1, 1871–1878.

Zevenhoven, R., Fagerlund, J., Nduagu, E., Romão, I., Jie, B., Highfield, J., 2013. Carbon storage by mineralization (CSM): serpentinite rock carbonation via $Mg(OH)_2$ reaction intermediate without CO_2 pre-separation. Energy Proc. 37, 5945–5954.

Zevenhoven, R., Slotte, M., Koivisto, E., Erlund, R., 2017. Serpentinite carbonation process routes using ammonium sulfate and integration in industry. Energy Technol. 5 (6), 945–954.

Zhang, H., Shen, C., Xi, P., Chen, K., Zhang, F., 2018. Study on effect of the activated magnesia carbonized building blocks based on the content of fly ash. Construct. Build. Mater. 185 (10), 609–616.

Zhang, X., Qian, C., Yi, H., Ma, Z., 2021. Study on carbonation reactivity of silicates in steel slag accelerated by Bacillus mucilaginosus. Construct. Build. Mater. 292, 123433.

Zhao, L., Sang, L., Chen, J., Ji, J., Teng, H.H., 2010. Aqueous carbonation of natural brucite: Relevance to CO_2 sequestration. Environ. Sci. Technol. 44 (1), 406–411.

Zhu, B., Wilson, S.A., Zeyen, N., et al., 2022. Unlocking the potential of hydraulic fracturing flowback and produced water for CO_2 removal via mineral carbonation. Appl. Geochem. 105345.

Zinke, L., 2021. Wearing down olivine. Nat. Rev. Earth Environ. 2, 8.

Further reading

Bobicki, E., Liu, Q., Xu, Z., Zeng, H., 2012. Carbon capture and storage using alkaline industrial wastes. Prog. Energy Combust. Sci. 38, 302–320.

IEA Greenhouse Gas Programme, 2005. Carbon Dioxide Storage by Mineral Carbonation. IEA Greenhouse Gas Programme. Report 2005/11.

IPCC, 2005. Mineral carbonation and industrial Uses of carbon dioxide. In: Special Report on Carbon Dioxide Capture and Storage. Cambridge University Press, Cambridge, UK (Chapter 7)).

Li, J., Hitch, M., Power, I.M., Pan, Y., 2018. Integrated mineral carbonation of ultramafic mine deposits—a review. Minerals 8, 147.

National Academy of Sciences, 2018. Carbon mineralization of CO_2. In: Negative Emissions Technologies and Reliable Sequestration: A Research Agenda. The National Academies Press, Washington, DC.

Projects and organizations

44.01 Project; Mineralizing captured CO_2 in the Samail ophiolite, Oman: 4401.earth.

Åbo Akademi University, Finland. http://web.abo.fi/fak/tkf/vt/Eng/research_CarbonDioxineCapture.htm.

CarbFix Project. Icelandic in situ carbonation project. www.or.is/english/carbfix-project/about-carbfix.
Carbon8. Company who treats industrial residues to produce carbonate products. https://c8s.co.uk.
Carbonought. Developing EW for the agricultural industry. www.carbonaught.io.
ClimeWorks. DAC system developer and CO_2 supplier to the CarbFix project. www.climeworks.com.
De Beers Group. Carbonation of diamond mine tailings. www.debeersgroup.com/building-forever/protecting-the-natural-world/mineral-carbonation.
Eion Carbon. Developing EW for the agricultural industry. https://eioncarbon.com/.
Fortera. Reactive Calcium Carbonate supplier. https://forterausa.com/.
Green Minerals. Mineral carbonation carbon capture technology developer. https://www.green-minerals.nl/.
Heirloom Carbon. DAC by passive mineralization. www.heirloomcarbon.com.
Mineral Carbonation International (MCI). Mineral carbonation carbon capture technology developer. www.mineralcarbonation.com.
Neustark. Swiss start-up mineralizing CO_2 in waste concrete. https://www.neustark.com/.
OCO Technologies. waste carbonation and construction materiala. https://oco.co.uk/.
Oman Drilling Project. Investigation of the Oman ophiolite formation and CO_2 uptake in the weathering processes. www.omandrilling.ac.uk.
Solidia Technologies. eco cement production. https://www.solidiatech.com/.
VITO. Carbstone slag carbonation technology. https://vito.be/en/carbstone.
Wallula Basalt Pilot Project. In situ mineral carbonation project. www.bigskyco2.org/research/geologic/basaltproject.

Chapter 11

Afforestation and other land- and soil-based methods

Steve Rackley

11.1 Introduction

Alongside decarbonization of the global economy and the more technological negative emissions methods such as BECCS and DACCS, nature-based solutions (NBS) have the potential to play a major role in climate change mitigation. These solutions comprise "conservation, restoration, and improved land management actions that increase carbon storage and/or avoid greenhouse gas emissions across global forests, wetlands, grasslands, and agricultural lands," while also including safeguards for food security and biodiversity conservation (Griscom et al., 2017).

The total estimated carbon inventory in terrestrial plant and microbial biomass (ca. 670 GtC) and soils (ca. 3600 GtC to 2 m depth) is almost four times the atmospheric inventory. Fig. 11.1 summarizes the distribution of terrestrial carbon stocks among biomes. The annual interchange of carbon between the atmosphere and land stocks through photosynthesis, plant respiration, and decay amounts to ~15% of the atmospheric inventory, with a net uptake to the land sink averaging 3.1 GtC/yr in the decade to 2020 (Friedlingstein et al., 2022).

Carbon emissions from land-use, land-use change, and forestry (LULUCF), termed E_{LUC} in the global carbon budget, contributed a net 1.1 ± 0.7 GtC/yr to total anthropogenic emissions in the decade to 2020, and overall emissions of 235 GtC from LULUCF account for roughly one-third of total anthropogenic emissions during the industrial era (Friedlingstein et al., 2022). These fluxes are the net result of a range of human activities, including deforestation, forest harvesting and degradation, and shifting cultivation, balanced by forest regrowth and afforestation. Losses affect both vegetated biomass, which is estimated to have declined by 155 GtC between 1901 and 2012 (Li et al., 2017) and soils, where stocks in the top 2 m are estimated to have been historically reduced by a similar amount (133 GtC) (Sanderman et al., 2017).

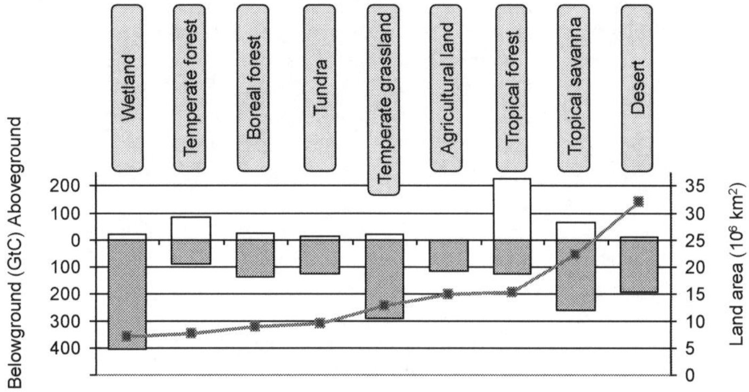

FIGURE 11.1 Global biome areas and aboveground/belowground carbon stocks. *After Villa and Bernal (2018).*

Negative Emissions Technologies for Climate Change Mitigation. https://doi.org/10.1016/B978-0-12-819663-2.00018-6

Reversing these losses, through a variety of methods, can therefore have a significant negative emissions impact (Jansson et al., 2010). These methods include:

- Increasing forest cover and bioenergy crop cultivation,
- Increased use of biomass in durable products, or sequestration by burial,
- Increased soil carbon stocks through a range of approaches including improved land management practices, biochar and mineral addition, and
- Genetic improvement of plants to enhance photosynthesis, improve stress tolerance, and increased soil carbon through direct allocation or more recalcitrant decay products.

Although an increased allocation of land for forestry and biomass crop cultivation can cause competition with food production, many of these methods contribute to improving soil fertility and therefore have cobenefits for sustainable agriculture, and the challenge of feeding the planet's estimated 9 billion population by 2050 (Rockström et al., 2017).

11.2 Afforestation, reforestation, and forest management

Forests occupy around 31% of the earth's land surface, covering a total of some 4 billion hectares (4000 Mha or 40 million km^2) (FAO and UNEP, 2020), some 93% being naturally regenerating forests and 7% (290 Mha) planted, the latter area having expanded by 120 Mha since 1990 (FAO, 2020). Around 60% of terrestrial carbon stocks are contained in forest ecosystems (Fig. 11.1), with live and dead trees accounting for around 60% of this (McKinley et al., 2011), or around 77% of the total global carbon stock in vegetation (IPCC, 2001).

Roughly one-third (ca. 1900 Mha) of the earth's forests have been lost since preindustrial times and deforestation, averaging 14 Mha/yr since 1990 (FAO and UNEP, 2020), largely due to agricultural expansion, is a major contributor to LULUCF emissions. In tropical forests, 200–300 GtC of carbon stocks have been lost historically, and deforestation and forest degradation are currently releasing 1–2 GtC/yr, while secondary forests recovering from harvest and shifting agriculture are accumulating as much as 3 GtC/yr (Houghton et al., 2015).

In addition to storing carbon, forests also provide a variety of other ecosystem services such as controlling soil erosion, supporting biodiversity, mitigating floods and droughts, and water filtration, and these benefits can also accrue to well-planned afforestation and reforestation and other forest management projects.

Afforestation and reforestation (AR)—the establishment of new forests on land that historically has not been forested, or the reinstatement of previously forested land—was the first negative emissions option to be widely investigated, and was included in the scope of the clean development mechanism (CDM) established under the Kyoto Protocol in 1997, which allowed emission reduction or removal projects in developing countries to earn Certified Emission Reductions (CERs) that could be sold in international markets to companies or individuals wishing to offset their unavoidable emissions. Unfortunately, uptake was extremely limited, with only 15 AR projects among the 2500 registered with the UNFCCC by 2010, due in part to the onerous administrative and governance requirements, including complex project development and measurement, reporting, and verification (MRV) procedures (Corbera and Friedli, 2012).

As well as afforestation and reforestation projects, carbon storage in forests can also be enhanced through conservation activities that protect old-growth forests from deforestation, and from improved forest management practices, including sustainable harvesting and the increased substitution of forest products for traditional materials with higher carbon intensity (Canadell and Raupach, 2008). Management practices to sustain and enhance carbon stocks were not included in the CDM due to expected difficulties in establishing methodologies to ensure additionality (e.g., the extent to which project impacts can be considered additional to the impact of pre-existing national policies) and to prevent leakage (i.e., detrimental effects resulting from a project's activities but occurring outside the project area). UNFCCC parties did agree at the 2007 COP13 to include these activities as part of the "Bali Action Plan," and methodological guidance was adopted at the 2009 Copenhagen Summit (COP13) although concerns about the effectiveness and side effects of this guidance have been raised and are discussed below.

Large-scale planting of trees and other fast-growing plants to achieve negative emissions was first proposed by Freeman Dyson, who argued that, since annual photosynthetic production is many times greater than the annual increase in atmospheric carbon stock "it should be possible in case of a world-wide emergency to plant enough trees and other fast-growing plants to absorb the excess CO_2 and bring the annual increase to a halt." (Dyson, 1977). Dyson saw this as a stop-gap measure to buy time for the decarbonization of global energy supply and estimated that 5 GtC/yr—his mid-case 1976 estimate of emissions in 2000—could be removed by planting 7 million km^2 of American Sycamore forest (7.5 tC/ha/yr). More recent estimates of the achievable rate of carbon storage in forests are lower than Dyson's, ranging from 2.8 to 5.5 tC/ha/yr globally, with an additional 0.2–1.2 tC/ha/yr achievable through improved forest management practices such as more rapid restocking after harvest, optimizing stand density and rotation period between harvests, and restoring degraded forests (Griscom et al., 2017). Houghton and colleagues (2015) estimated that planting 500 Mha (5 million km^2) could sequester more than 1 GtC/yr for decades and suggested that reduction of emissions and increase in drawdown in tropical forests totaling ca. 5 GtC/yr could be implemented relatively quickly, noting, as had Dyson, that this could provide a bridge to a fossil-free future. The potential for 1500–2600 Mha of land to be under AR by 2100, representing a 40%–65% increase in the current global forest area, has been suggested by some studies. However, deployment of AR to deliver $GtCO_2$ scale negative emissions faces a number of significant challenges, including potential conflict with local development goals (livelihood, access/exclusion, etc.), land and water competition with food and bioenergy crops, climate feedbacks, and reduced albedo at higher latitudes (Yosef et al., 2018).

The potential for new forests to increase terrestrial carbon stocks depends on many factors, an important one being the latitude of the site, with tropical locations such as south America and south-east Asia having the greatest potential (Fig. 11.2). New forests in tropical and temperate locations show a rapid initial growth, slowing after around 50 years as saturation is approached, while boreal forests show a more linear accumulation. Fig. 11.2 assumes no forest management after establishment, but where markets for timber or biomass are accessible, periodic harvest and regrowth can substantially increase cumulative carbon storage over a century timescale. Fig. 11.3 shows the cumulative carbon stock resulting from a hypothetical forestry project located in the south east USA or central Europe, including afforestation and harvest/reforestation with the harvest being put to a variety of uses, including biofuel. The carbon balance also includes emissions savings with respect to fossil-fueled and more energy-intensive alternative products and services.

While mature unharvested forests are expected to reach an equilibrium in which the carbon absorbed by growth is matched by that released through decomposition, recent observations show that mature forests are net sinks as a result of the CO_2 fertilization effect (Chapter 5) as well as the fertilization of forest soils by nitrogen deposited from industrial and agricultural activity (Bellassen and Luyssaert, 2014). How the global forest carbon sink will develop by 2100 will depend

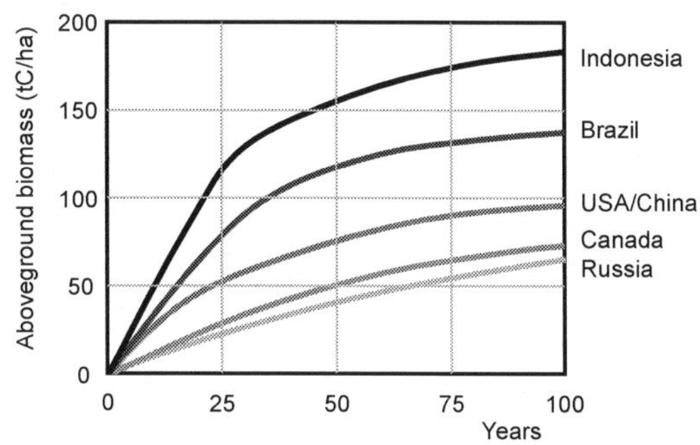

FIGURE 11.2 Accumulation of above-ground carbon stock in newly forested agricultural land. *After Doelman et al. (2020).*

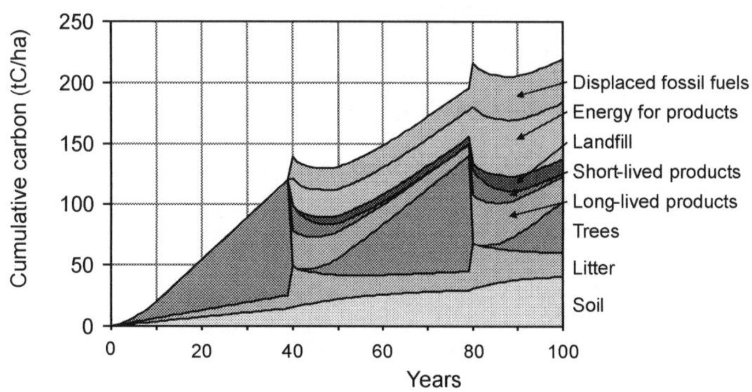

FIGURE 11.3 Illustrative carbon balance for a managed temperate afforestation project. *After IPCC (2001), AR3 WGIII.*

on a variety of natural and anthropogenic factors, and remains a significant unknown. Improved photosynthesis due to higher temperatures and further increase in [CO_2] may increase the global forest sink by a factor of 5 from the current level to 10 GtC/yr or, if other factors lead to respiration becoming dominant, this sink could become a source of 6 GtC/yr, a factor of 5 increase in LULUCF emissions over recent decades (Friedlingstein et al., 2022).

This uncertainty is a crucial factor in determining optimal strategies for forest management; conservation of mature forests will be preferred over-harvesting if they remain net sinks, while harvesting would be preferred if they become net sources. Further studies are required to improve our understanding of this complex problem but, in the meantime, some no-regret management strategies can be identified, such as;

- increasing wood production for carbon efficient uses such as bioenergy fuels or construction, ideally harvested from dying or low productivity stands (Amiri et al., 2020),
- maximizing carbon stock retained and recovery after harvesting, for example, by planting locally appropriate tree species mixes, including a nitrogen-fixing understorey (Bellassen and Luyssaert, 2014).

11.2.1 Climate feedbacks from afforestation and reforestation

The drawdown of atmospheric CO_2 and increase in terrestrial carbon stocks through AR has a direct climate change mitigation benefit, but increased forest cover also results in biophysical feedbacks with warming consequences that are more complex, such as the impact on land surface albedo in the boreal zone and on evaporation, cloud cover, and precipitation in the tropics.

Albedo impact of boreal zone forest cover

Boreal forests, primarily in Canada, Scandinavia, and northern Russia, have lower cumulative storage potential than temperate or tropical forests, and AR in this region can also induce a positive climate feedback due to the impact of forest cover on albedo, which is reinforced by subsequent loss of sea ice. The cooling effect resulting from the conversion of forest to cropland has long been recognized (Bonan, 1997; Brovkin et al., 1999; Lee et al., 2011), and the implications for afforestation as a climate mitigation strategy were first investigated by Betts (2000) who found that the positive radiative forcing resulting from reduced albedo could offset the benefits from carbon storage in boreal forests, calling into question the viability of AR as a climate change mitigation strategy in these regions. Forest cover in mid-latitudes (30—50°N, essentially temperate forests) also has a net warming effect (Claussen et al., 2001), while the reverse is true for tropical forests, where cooling due to increased evapotranspiration exceeds the warming impact of reduced albedo (Gibbard et al., 2005).

Boreal and temperate forests can potentially play an important part in national climate policy for many countries, and various approaches including species selection and forest management practices have been investigated with the aim of reducing this adverse effect. As the feedback from reduced albedo occurs largely during the winter, the planting of deciduous species would mitigate this effect, and this has been proposed for the reforestation of open woodland areas (i.e., crown cover $<25\%$) in Canada that have failed to recover from repeated natural disturbances. Similarly in the Russian far east, Shuman and colleagues found a positive feedback from increasing temperatures due to the replacement of Siberia's deciduous larch forests through out-competition by evergreen coniferous species (Shuman et al., 2011). This undesirable feedback could therefore be countered by using deciduous species in reforestation projects and by other measures to sustain forest diversity.

Biophysical impacts of semiarid zone afforestation

Land-use competition, particularly for food production, is a significant challenge for large-scale afforestation as a climate change mitigation strategy, although this can be avoided by the establishment of forests on otherwise unproductive lands such as occur in arid and semiarid zones. The biophysical feedbacks arising from large-scale arid or semiarid zone afforestation make this a particularly interesting option.

In semiarid desert margins such as the Sahel, south of the Sahara, any increase in albedo and heat loss as a result of reduced vegetation causes additional drying through a reduction and (in this case) southward shift in rainfall, exacerbating the vegetation loss. This feedback mechanism was first recognized by Charney (1975) and the implications of the reverse mechanism—an increase in rainfall as a consequence of increased evapotranspiration after afforestation—led Ornstein and colleagues to investigate the potential for irrigated afforestation of the Sahara and Australian Outback as a mitigation strategy (Ornstein et al., 2009). This study concluded that desert afforestation could be initiated with carefully controlled irrigation, sufficient only to replace local evapo-transpiraion and applied directly to the root zone, accompanied by the interplanting of nitrogen-fixing species to counter the lack of soluble nitrogen in desert sands and soils, and by measures to promote the reestablishment of a functioning soil rhizospheric microbial community (Ayangbenro and Babalola, 2020). Once established, the induced increase in seasonal precipitation (April to November) would substantially reduce the need for irrigation, although recent studies using improved coupled models suggest more modest increases in precipitation (Kemena et al., 2017). Several investigators have extended these results to consider other locations and plant types (Becker et al., 2013; Branch and Wulfmeyer, 2019), as well as possible afforestation of semiarid regions without artificial irrigation, based on the characteristics of a small-scale semiarid forest on the edge of the Negev desert (Yosef et al., 2018).

11.2.2 Addressing deforestation and forest degradation

Efforts to mitigate climate change through AR will be undermined unless ongoing deforestation and forest degradation are halted. The UNFCCC definition of a forest, as expressed in the Kyoto Protocol CDM, is any area larger than 0.5–1.0 ha with a minimum tree crown cover of 10%–30%, where "tree" is defined as a plant reaching a height greater than 2–5 m at maturity. Individual national definitions, which should be consistent with these ranges, span the full spectrum from [1 Ha, 30%, 5 m] in Brazil, Colombia, and Uganda among others, to [0.05 Ha, 15%, 2 m] in India. Deforestation occurs when a previously forested area drops below the national thresholds and is largely driven by clearance for agriculture, while logging for timber and, predominantly in Africa, charcoal fuelwood gathering and livestock grazing are the main drivers of forest degradation. Houghton and Nassikas (2018) have estimated that stopping deforestation and degradation, and allowing the undisturbed regrowth of secondary forests could add up to 120 $GtCO_2$ to the cumulative CO_2 uptake to the terrestrial sink by 2100.

The 2008 launch of the UN REDD program, a cooperative effort between the UN Food and Agriculture Organization (FAO), the United Nations Development Programme (UNDP), and the United Nations Environment Programme (UNEP) with the aim of reducing emissions from deforestation and forest degradation in developing countries, and its extension to REDD+ at COP19 in Warsaw, 2013, to include the sustainable management of forests and the conservation and enhancement of forest carbon stocks, demonstrated the political will and importance of managing forests for climate change mitigation (Houghton et al., 2015). These programs provide results-based incentives to developing countries, with largely beneficial—although neither ubiquitous nor uncontested—impacts on carbon stocks as well as other outcomes affecting community well-being, tenure/land rights, and biodiversity (Duchelle et al., 2018; West et al., 2020).

Various other international efforts have also been directed at solving the problems of deforestation and forest degradation, such as the UN Convention for Biological Diversity which, in its 2010 Strategic Plan for Biodiversity (2011–20) stated the goal that, by 2020 "the rate of loss of all natural habitats, including forests, [would be] at least halved and where feasible brought close to zero, and degradation and fragmentation [would be] significantly reduced." The 2014 New York Declaration on Forests (NYDF), a voluntary, nonbinding pledge reiterated the goal of halving deforestation by 2020, and sought to end it by 2030, but, in common with other forest-related targets for 2020, such as the Bonn Challenge (see below), this was not achieved. International efforts did however result in a 1/3 reduction in deforestation (from 16 to 10 Mha/yr) between the 1990s and the period from 2015 to 2020 (FAO, 2020).

The current UN Strategic Plan for Forests takes as its first goal to "reverse the loss of forest cover worldwide through sustainable forest management, including protection, restoration, afforestation and reforestation, and increase efforts to prevent forest degradation and contribute to the global effort of addressing climate change," with a target to increase forest area by 3% (120 Mha) worldwide by 2030. A commitment to halt and reverse forest loss and land degradation by 2030 was also endorsed by leaders from 141 countries as part of the Glasgow COP26 Declaration on Forest and Land Use.

Tackling deforestation is not just a challenge for those countries in which it happens, since in most cases the economic driver originates in the demand for agricultural commodities in consumer countries. The concept of "embodied deforestation" or "imported deforestation" explicitly recognizes the deforestation associated with such imports, and its incorporation into economic and environmental policies is under discussion in a number of countries and economic blocs, including the EU. The 2021 update to the NYDF goals recognized the impact of economic supply chains in driving deforestation, proposing a systemic, multisectoral approach to end this pressure, supporting other international efforts to ensure the sustainability of imported products.

11.2.3 Forest management for carbon storage

No single forest management strategy can maximize carbon uptake and long-term storage across all conditions. For forests with large standing biomass protection can be an effective strategy, while for land with little standing biomass and low yields, the most effective strategy will be to reforest the land for carbon storage. However, where high yields are expected and markets for biomass are readily accessible, managing forests as harvestable energy crops for fossil fuel substitution can be a highly effective strategy, particularly since harvesting—especially when done sustainably—increases long-run yields by avoiding saturation (Hall et al., 1991; Marland and Marland, 1992).

Forests, and the carbon stocks they contain, face a number of climate-related stressors, including increased incidence of droughts, fires, insect pests, and pathogens, and a range of practices are being developed and applied to aid forest adaptation to these changing conditions while maintaining or enhancing carbon stocks. Ontl and colleagues (2020) developed a suite of management strategies and practices to enhance forest carbon storage, summarized in Table 11.1.

Sustainable harvesting, also known as a continuous cover regime, is one important practice that aims to follow a forest's natural regeneration process by harvesting mature trees as they approach mortality, providing space and light for new growth with minimal disturbance to the forest habitat and other ecosystem services (Assmuth and Tahvonen, 2015). Techniques for identifying trees to be "evicted" include adopting methods from political science to combine expert

TABLE 11.1 Forest carbon management strategies.

Strategy 1: Maintain or increase extent of forest ecosystems
1.1 Avoid forest conversion to nonforest land uses **1.2** Reforest lands that have been deforested and afforest suitable lands **1.3** Increase the extent of forest cover within urban areas **1.4** Increase or implement agroforestry practices
Strategy 2: Sustain fundamental ecological functions
2.1 Reduce impacts on soils and nutrient cycling **2.2** Maintain or restore hydrology **2.3** Prevent the introduction/establishment of invasive plant species and remove existing invasives **2.4** Maintain or improve the ability of forests to resist pests and pathogens **2.5** Reduce competition for moisture, nutrients, and light
Strategy 3: Reduce carbon losses from natural disturbance, including wildfire
3.1 Restore or maintain fire in fire-adapted ecosystems **3.2** Establish natural or artificial fuel breaks to slow the spread of catastrophic fire **3.3** Alter forest structure or composition to reduce the risk, severity, or extent of wildfire **3.4** Reduce the risk of tree mortality from biological or climatic stressors in fire-prone systems **3.5** Alter forest structure to reduce the risk, severity, or extent of wind and ice damage
Strategy 4: Enhance forest recovery following disturbance
4.1 Promptly revegetate sites after disturbance **4.2** Restore disturbed sites with a diversity of species that are adapted to future conditions **4.3** Protect future-adapted seedlings and saplings **4.4** Guide species composition at early stages of development to meet expected future conditions
Strategy 5: Prioritize management of locations that provide high carbon value across the landscape
5.1 Prioritize low-vulnerability sites for maintaining or enhancing carbon stocks **5.2** Establish reserves on sites with high carbon density
Strategy 6: Maintain or enhance existing carbon stocks while retaining forest character
6.1 Increase structural complexity through retention of biological legacies in living and dead wood **6.2** Increase stocking on well-stocked or understocked forest lands **6.3** Increase harvest frequency or intensity because of greater risk of tree mortality **6.4** Disfavor species that are distinctly maladapted **6.5** Manage for existing species and genotypes with wide moisture and temperature tolerances **6.6** Promote species and structural diversity to enhance carbon capture and storage efficiency **6.7** Use seeds, germplasm, and other genetic material from across a greater geographic range
Strategy 7: Enhance or maintain sequestration capacity through significant forest alterations
7.1 Favor existing species or genotypes that are better adapted to future conditions **7.2** Alter forest composition or structure to maximize carbon stocks **7.3** Promote species with enhanced carbon density in woody biomass **7.4** Introduce species or genotypes that are expected to be adapted to future conditions

After Ontl et al. (2020).

TABLE 11.2 Impact of forest management practices on soil carbon stocks.

Positive	Neutral	Negative
Afforestation (particularly of former cropland)	Stand density management including thinning	Traditional removal of harvesting residues
Nitrogen addition and selection of nitrogen-fixing species		Soil disturbance during harvest or site preparation
Tree species diversity		Conversion of primary to secondary forest or plantation
Herbivore regulation		Litter and biomass removal for fodder, fuel, etc.

Based on Mayer et al. (2020).

opinions and reduce the risk of inappropriate decisions based on a simple majority view that may undervalue minority perspectives (Pommerening et al., 2020).

While forest management practices are largely aimed at increasing aboveground carbon, many will also have an impact on soil carbon, which may enhance or reduce net storage. Table 11.2 summarizes these effects for a number of management practices.

Active management of tree species diversity is one practice that can have a significant impact on forest resilience and carbon stocks. The relationship between biodiversity and ecosystem productivity—the biodiversity–productivity relationship or BPR—has been the subject of extensive ecological research over recent decades (see e.g., Naeem et al., 2012), and a positive relationship between forest productivity and tree species diversity has been identified across global forest ecosystems (Liang et al., 2016). Understanding the causal relationship between tree species diversity and forest productivity—whether diversity drives productivity or the reverse—is important to determine the circumstances in which species mixing can be an effective strategy to maximize forest carbon stocks. This question has been investigated by Madrigal-González and colleagues who found that the causal direction depends on the environmental conditions. In highly productive environments, stocks are more likely to be maximized by the establishment of forests with a variety of naturally cooccurring species as a result of ecological facilitation and/or niche differentiation (see Glossary), while in less productive situations stocks are more likely to be maximized by focusing on fewer species that are adapted to specific local stresses (Madrigal-González et al., 2020). In either case, an adaptive approach, taking account of the likely alteration in local environmental conditions resulting from climate change, will be beneficial.

Planting strategies that anticipate future climatic conditions—planting species that are currently less common or absent in a region but with niches that are predicted to become common under future climatic conditions—have also been studied, primarily for temperate or boreal forests, and can lead to improved resilience as well as higher net primary production (NPP) and carbon stocks (Hof et al., 2017). Although various studies indicate benefits from this approach, predicting the ecosystem-level impact of such interventions, including that on carbon stocks and fluxes, remains highly uncertain, and much further work is needed before firm conclusions can be drawn. Planting trials to provide model validation and assess factors that cannot yet be modeled, such as the impact of pests and pathogens, will be an essential part of this work.

Although less frequently included in the discussion of biomass production, bamboo forests can hold total carbon stocks of up to 390 tC/ha, considering above- and belowground biomass plus soil carbon. The high-standing biomass (up to 120 tC/ha) and rapid annual accumulation rates (8–15 tC/ha) show that bamboo forests can be efficient carbon sinks, as well as providing food, material for construction and other durable products, and additional ecosystem services (Yuen et al., 2017; Solomon et al., 2020).

11.2.4 Large-scale afforestation and reforestation projects

The recognition that forests have a major role in climate change mitigation has led to a number of ambitious political declarations, including the NYDF, although the increase in the rate of forest loss from 2014 to 2020 attests to the long and difficult road from political statements to effective action. Nevertheless, such declarations have encouraged the establishment of a number of international initiatives and projects targeting large-scale afforestation, reforestation, and forest protection (Table 11.3).

TABLE 11.3 Large scale afforestation, reforestation, and forest protection projects and initiatives.

AR initiative or project (Year established)	Objectives	Status
Bonn Challenge (2011)	To restore 150 Mha of deforested and degraded land by 2020, and a total of 350 Mha by 2030, potentially sequestering 15.7 $GtCO_2$.	43 countries, including Brazil, India, and China, have committed nearly 300 Mha of degraded land.
Three Norths Forest Shelterbelt (1978)	Afforestation of 410 Mha of arid and semiarid land in China by 2050, to counter expansion of the Gobi Desert and control dust storms.	35 Mha planted by 2013, but only 16 Mha reforested due to high mortality of planted trees, necessitating a reevaluation of planting strategies.
Pan-African Great Green Wall (Sahel Geen Barrier)	Addressing poverty and soil degradation by foresting a land area of 15 km × 7100 km (10.7 Mha) from Senegal to Djibouti.	Significant land restoration in Burkina Faso, Ethiopia, Niger, and Nigeria, and reforestation in Senegal.
Trillion Trees Campaign	Launched in 2018, building on the earlier Billion Tree Campaign, aiming to save from loss, better protect, and regrown a trillion trees by 2050.	Working on forest protection and restoration of over 400 Mha, impacting 20 billion trees.
WEF Trillion Trees Initiative	Launched in 2020 by the World Economic forum to support the global reforestation community during the UN Decade on Ecosystem Restoration (2021–30).	Pledges from 33 companies supporting the conservation and planting of 3.6 billion trees in 60 countries.

A recent global assessment suggests that existing forest canopy cover could be extended by 900 Mha globally (Bastin et al., 2019), and these initiatives and projects, if successful, would go a long way toward reaching that target. Several of the listed projects target arid and semiarid areas, with increased carbon stocks being a secondary objective alongside other social and environmental goals. Although these programs have resulted in significant areas being planted, the low survival rate of young trees in these challenging conditions indicates the need to adopt alternative strategies such as natural restoration and seminatural afforestation (Wang et al., 2020).

Large-scale reforestation and restoration projects have also faced opposition at local and international level, for example, due to the marginalization of local people in decision-making, and a focus in some cases on converting agricultural land into plantation monocultures, contributing to increased water stress and the loss of traditional livelihoods (O'Connor and Ford, 2014). Similar concerns are reflected in the 2021 update of the NYDF to strengthen recognition of the rights and contributions of indigenous peoples and local communities, and, based on early project experience, there is now widespread acceptance that the buy-in of communities and local decision-makers, and their inclusion in decision processes alongside governments, technical, and financial partners are essential for the long-term sustainability of these initiatives.

11.2.5 AR knowledge gaps and research priorities

Forestry has been practiced for sustainable timber production since at least the 13th century, and the scientific foundations of modern forestry were established in the 1750s, driven by the enlightenment rise of science in France and Germany. The emphasis on forests as carbon stores began in the 1980s and, although 4 decades of research have increased our understanding of these complex systems and the processes that affect their carbon stocks, the list of knowledge gaps (know-unknowns) has grown with that understanding.

Table 11.4 provides an overview of some of the main knowledge gaps, and an extensive recent review of the impact of forest management on carbon stocks, including knowledge gaps and research needs, can be found in Mayer et al. (2020).

Forest carbon monitoring, reporting, and verification

Accounted for carbon uptake resulting from land-use change activities such as AR, avoided deforestation, and other forest and land management practices require the measurement of deviations from a baseline—or "counterfactual" scenario—that describes how the land status and related uptake (or emissions) would have evolved in the absence of these activities. Without a credible counterfactual to support the additionality of the activity, measurement and monitoring using the approaches discussed below will be of little use in determining GHG benefits (Corbera and Friedli, 2012).

TABLE 11.4 AR knowledge gaps and research priorities.

Afforestation and reforestation
• Improved modeling of forest regrowth in global vegetation models • Establishment of internationally agreed performance criteria for forest carbon stock assessment, including remote sensing and local monitoring • Methods to determine site-specific optimal tree species and diversity, including (nitrogen fixing) understorey, to maximize soil carbon across the soil profile • Development of methods to evaluate the albedo effects of AR and other forest management practice in boreal regions and tools for albedo monitoring • Improved quantification of the impact of afforestation on carbon stocks in deep (>30 cm) and highly organic soils • Planting strategies to maximize productivity and carbon storage under future climatic conditions
Forest management practices
• Improved understanding of factors and processes (biotic and abiotic) controlling the impact of harvesting (and residue removal) and stand thinning on soil carbon at all depths (e.g., soil type, tree species, climatic conditions, harvesting/thinning practice, time since harvest, etc.) • Methods to determine optimal site-specific management strategy (preserve, (selective) harvest, etc.) • Establish the most carbon-efficient uses of wood, including site-dependent factors (fertility, regrowth rate, harvesting practices, local markets, etc.) • Improved understanding of forest dynamics following major disturbances (wildfire, drought, pests, site preparation) and influence on regeneration and carbon stocks • Improved modeling capability of postdisturbance recovery processes • Impact and potential for stump harvesting • Impact of herbivores and varying herbivore pressure on soil processes and soil carbon stocks
Wood products
• LCA approaches to establish substitution benefits, including counterfactual scenarios without increased production of wood products • Development of improved methods of timber treatment, new applications, and advanced fabrication techniques for timber use in construction • Development of technologies to increase reclamation and recycling of wood, including product design to enable deconstruction, separation, sorting, recovery, and reuse
Social and sustainability impacts
• Wider assessment (geographical, impact categories) of the direct and indirect socio-economic impacts of large-scale tree plantations on local communities • Assessment of the potential to improve the effectiveness of large-scale forest projects through greater emphasis on equitable approaches

Remote sensing techniques have been developed since the 1990s to quantify aboveground forest carbon stocks, using airborne and satellite imagery as well as light detection and ranging (LiDAR) and synthetic aperture radar (SAR, InSAR) surveys, calibrated using plot-level forest inventories and guided by empirical algorithms and, more recently, by machine learning methods (Powell et al., 2010; Gleason and Im, 2012; Li et al., 2021). These tools have been widely applied to national forest inventory programs and the assessment of aboveground biomass changes with time (Nguyen et al., 2020), and also enable accurate estimation of the potential to increase carbon storage through forest restoration (Ferraz et al., 2018).

Accounting methodologies are also well established in the IPCC's Guidelines for National Greenhouse Gas Inventories (IPCC, 2006, Chapter 4—Forest land), and forest monitoring using these techniques is available as a commercial service for project development and carbon verification, through platforms such as the Forestry Thematic Exploitation Platform (Forestry TEP).

For forest management activities seeking to build existing stocks or prevent carbon loss, accounting is a more significant challenge, since factors such as forest stand response to the legacy of historical management practices and natural disturbances will affect future net emissions, alongside newly introduced practices. The REDD+ accounting method for such activities under the Kyoto Protocol second commitment period (2013—20) was controversial since nonproject factors such as the assumed future implementation of national policies could be included in the forward-looking "forest reference level," potentially leading to an overstatement of project impacts (Grassi et al., 2018). For example, claimed emissions reductions for a number of such projects in the Brazilian Amazon have been shown to be overstated as a result of the

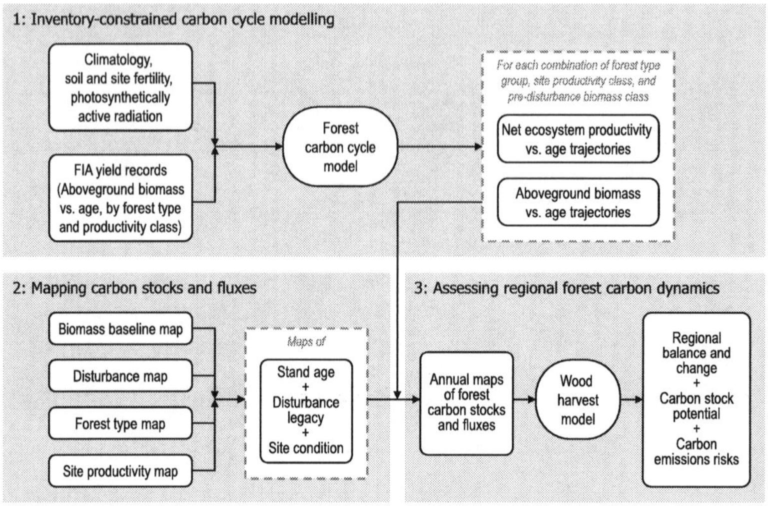

FIGURE 11.4 Forest carbon monitoring system. *After Zhou et al. (2021).*

adoption of a baseline that assumed higher deforestation than would have emerged absent the project activity (West et al., 2020), and various approaches have been proposed to developing baselines that more accurately reflect the true impact of such activities (Grassi et al., 2018; West et al., 2020).

Accounting systems for forest carbon have progressed significantly since the first Kyoto commitment period (2008–12), as methodological challenges have been addressed (see Krug, 2018 for an informative review of this "long road"), and further improvements are in progress, including the integration of inventory- and process-based models (Fig. 11.4) to assess the full forest carbon balance (Grassi et al., 2018; Zhou et al., 2021).

11.3 Biomass carbon removal and storage

The use of biomass for energy, with negative emissions achieved through the capture and storage of the CO_2 released during biomass conversion (BECCS) is discussed in Chapter 12. Biomass carbon removal and storage (BiCRS) was conceived as an alternative perspective on BECCS to recognize negative emissions methods that store carbon captured in biomass, with the added requirement that these approaches should not be detrimental to food security, rural livelihoods, or biodiversity, and ideally should promote these and other goals (Sandalow et al., 2021). The shift of focus from bioenergy production to carbon removal was motivated by Roger Aines' observation (labeled the "Aines Principle" by his coauthors) that, at a sufficient carbon price, the value of long-term storage of the carbon captured in biomass will exceed the value of its use for energy production. An emphasis on CDR and the avoidance of negative social and sustainability impacts results in a number of revisions and extensions to the standard BECCS concept.

Potential feedstocks for BiCRS largely overlap with those for BECCS, including wastes from agriculture, forestry, and industrial processes such as paper production, macro- and microalgae, and also including dedicated energy crops to the extent that these are not detrimental to other goals noted above. Suitable crops would include sugar cane, corn, rapeseed, palm oil, and soya that can contribute to food security, as well as woody biomass with a short restocking time, such as bamboo, eucalyptus, poplar, and willow.

The BiCRS perspective also recognizes a range of carbon storage alternatives, beyond the geological storage approach conventionally envisaged for BECCS, as summarized in Table 11.5.

These new carbon storage options, not linked to CO_2 capture from bioenergy production, are all at very early TRL stages but initial assessments of cost, scalability, and technical viability are encouraging for further development. Table 11.6 summarizes current knowledge gaps and research priorities for biomass carbon removal and storage.

11.4 Soil carbon sequestration

Global carbon stocks in soils total ~3600 GtC, with ~2050 GtC of soil organic carbon (SOC) and ~1550 GtC of soil inorganic carbon (SIC) in the upper 2 m (Lal, 2020). An estimated 130 GtC of SOC has been lost from the top 2 m of soil during 12,000 years of human land use, at a rate that reached 3 GtC/yr in the period from 2001 to 2015, and this "sink

TABLE 11.5 Alternative biomass related carbon storage options.

Carbon storage approach	Description
Wood harvesting and storage	Harvesting of biomass including wood (Scholz and Hasse, 2008; Zeng et al., 2013), leaves, or high yield crops such as switchgrass (Amelse and Behrens, 2021) and burial under anoxic conditions (such as dry landfill, peatlands, etc.) to prevent decomposition.
Biochar geological sequestration	Storage of biochar in the form of pellets or powder in former coal mines, salt mines, and other geological cavities, exploiting its long-term stability (Dufour, 2013).
Bioliquid injection	Geological storage of liquid products from biomass pyrolysis.
Deep ocean biomass sequestration	Deep ocean sequestration (abyssal dispatch) of crop residues (Metzger and Benford, 2001; Strand and Benford, 2009), macroalgae (Krause-Jensen and Duarte, 2016), or "biocoal" (Miller and Orton, 2021).

TABLE 11.6 BiCRS knowledge gaps and research priorities.

Topic	Research priorities
Biomass crops	• Location-specific selection of optimal plant species to grow on deforested land, to aid decisions on land reuse (biomass feedstocks vs. permanent reforestation) • Optimization of energy crops (including plant breeding) for life-cycle carbon removal, as well as energy production/conversion processability • Impact of biomass feedstock removal on carbon stocks in agricultural and forest soils
Sustainable potentials	• Regional assessment of sustainable potential from agricultural wastes (e.g., rice straw, oil palm residues, etc.) • Improved ecosystem modeling to improve estimates of sustainable potential • Methods to establish site-specific soil carbon capacity, to guide biomass removal practices consistent with long-term sustainability • Development of remote sensing and other assessment methodologies to enable compliance monitoring by carbon credit purchasers
Reducing land and water intensity	• Approaches to reducing the land and water intensity of biomass crop cultivation
Biomass conversion	• Development of feedstock handling and pretreatment systems, including systems that require no external fuel or power • Development of biomass supply chains that integrate hydrogen production and deliver all carbon to storage
Environmental impacts	• Improved understanding of the negative impacts of biomass cultivation, such as impact on water, and biodiversity, and ways to minimize these, providing input to optimal siting of future biomass cultivation and processing facilities • Development of systemic indicators to assess both expected and unintended impacts of large-scale biomass cultivation
Social and sustainability impacts	• Cross-disciplinary research to address the impacts of large-scale biomass production on food security, ecosystem services, and other sustainability goals • Development of sustainable biomass standards and practice guidance to ensure equitable biomass supply chains, minimizing risks, and maximizing benefits for local communities

potential" gives an indication of the scale of soil carbon sequestration (SCS) that might be achieved by restorative efforts. At the lower end of the scale, bottom-up estimates, accounting for social, economic, and technical limits on the uptake of best practices, suggest that between 10% and 30% of this sink (8–28 GtC) might be achievable in practice (Sanderman et al., 2017), while other estimates are as high as 178 GtC (650 $GtCO_2$; Lal, 2020).

Long-lived SOC is created in a process that starts with CO_2 removal from the atmosphere and photosynthetic conversion to biomass, followed by the transfer and conversion of plant biomass into soil organic matter (SOM), and its stabilization to achieve a long residence time (Lal et al., 2021). Fig. 11.5 shows a simplified model of these soil carbon

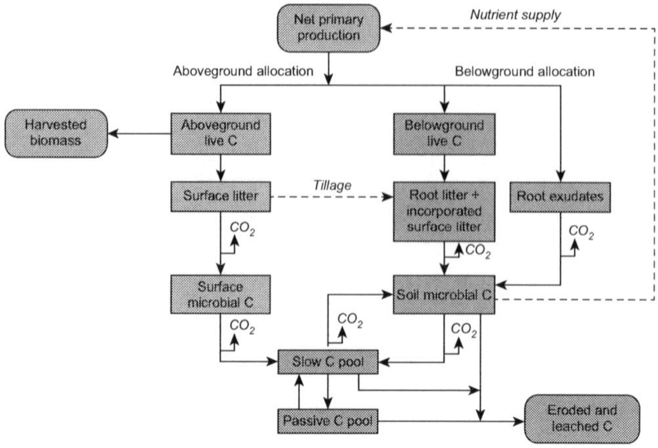

FIGURE 11.5 A simplified model of soil carbon pools, fluxes, and controls.

pools and processes. SCS, which aims to replenish depleted soil carbon stocks and to sustain them at higher levels, involves a range of approaches, summarized in Table 11.7, that influence these processes.

Soils that are managed to maximize SCS, for example through practices that maintain continuous vegetation cover and continuous root presence, are more fertile, less prone to erosion, and have a higher water holding capacity, reducing the need for artificial fertilization and irrigation, with the attendant energy costs and related GHG emissions (Lal et al., 2021).

Any activity that increases aboveground production will have an impact below ground, with complex implications for carbon stocks, and SCS can also be achieved through landscape-scale approaches such as afforestation and reforestation, and the reinstatement of habitats such as peatlands and coastal wetlands (Section 11.5).

TABLE 11.7 Soil carbon sequestration approaches.

SCS strategy	Specific approaches
Regenerative and conservation agriculture	Minimal soil disturbance
	Permanent soil cover (cover crops and crop residue mulches)
	Soil fertility management
	Crop rotations and associations (trees, livestock)
Increasing net primary production (NPP)	Use of high-yield crops
	Perennial planting, undercrops
Minimize and delay the return of carbon in plant products to the atmosphere	Carbonation of biomass residues as stable biochar
Increasing the fraction of NPP that finds its way into soil	Genetic manipulation of plants to increase belowground carbon allocation and recalcitrant carbon production
	Modifying soil chemistry to maximize belowground allocation
Manipulating soil processes to maximize carbon retention in slow and passive carbon pools	Use of soil amendments to maximize SOC physicochemical protection
	Manipulation of microbial genetics to maximize biological protection of SOC and production of biochemically recalcitrant microbial compounds

11.4.1 Agricultural practices

Agricultural carbon storage (ACS), or "carbon farming," refers to the suite of practices for managing vegetation and soils in agroecosystems with the aim, alongside agricultural production, of increasing carbon stocks and/or reducing GHG emissions (CO_2, N_2O, CH_4) from soil biochemical processes (Smith et al., 2008). Fig. 11.6 summarizes global GHG fluxes related to soils and vegetation.

Increasing the amount of crop residue available to be converted into humus can move long-term soil carbon stocks to a higher equilibrium level over a decadal time scale. Intensified cropping and conservation tillage (discussed below) are examples of agricultural practices that increase this availability. Biologically altered organic inputs, such as manure, can result in more sustainable increases in SOC due to the inclusion of material that decomposes less rapidly than unaltered crop residues. These practices—reduced tillage, crop residue retention, and increased carbon inputs—also result in increased levels of microbial biomass and retention in soils of microbial organic matter (MOM), further increasing soil carbon content.

Regenerative and conservation agriculture

Conservation agriculture (CA) aims to sustain crop yields by improving soils and crop environments, while regenerative agriculture (RA) extends these aims to include broader environmental concerns related to water quality, biodiversity, ecosystem health, and carbon sequestration (Elevitch et al., 2018). RA has been defined as "an approach to farming that uses soil conservation as the entry point to regenerate and contribute to multiple provisioning, regulating and supporting ecosystem services, with the objective that this will enhance not only the environmental, but also the social and economic dimensions of sustainable food production" (Schreefel et al., 2020). The improvement of soils is at the heart of both CA and RA, with SCS being the result of many of the underlying practices, although the sequestration benefits are not uncontested.

Conservation tillage

Tilling breaks down the aggregated structure of the soil, both at the macroaggregate level by physical action and at the microaggregate level by subsequent dehydration of binding clays (see Chapter 5). This allows previously protected labile SOM to be exposed to microbial decay and to loss by wind and water erosion. Soil disturbance through tillage also has a significant impact on the activity and make-up of the microbial community, and the retention of soil carbon, nitrogen, and undisturbed macroaggregated soil structure through a reduction in tillage intensity will increase fungal to bacterial biomass ratio through increased mycorrhizal colonization. Natural ecosystems have fungal-dominated microbial communities and are able to sustain the decomposition of organic material and recycling of nutrients without the supplementary inputs required in intensively managed agroecosystems. By avoiding the interruption of nutrient-recycling processes and maintaining the water-holding capacity of soils, practices that reduce soil disturbance therefore minimize the need for increased fertilization and irrigation to sustain soil productivity and soil carbon stocks (Lal et al., 2021).

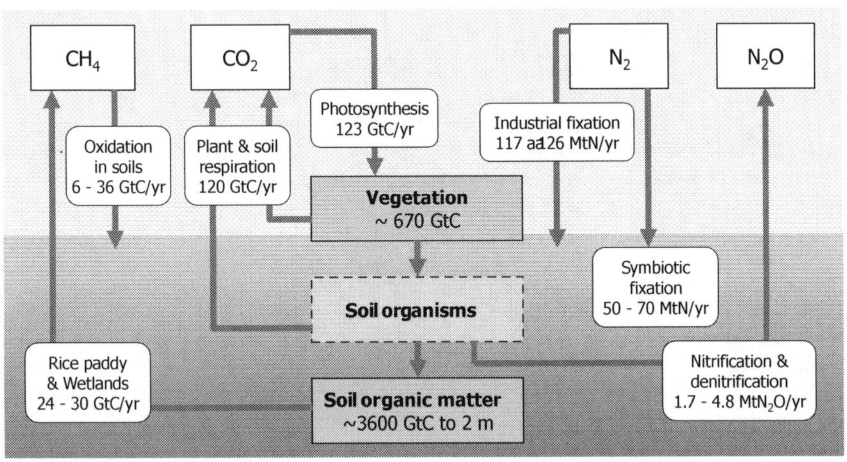

FIGURE 11.6 GHG fluxes and carbon stocks related to soils and vegetation.

The net effect on carbon stocks of this erosive loss of SOM as a result of tillage may be mitigated if the eroded material is deposited in downstream sediments, particularly if these are protected from decomposition. The interlinked processes of erosion and deposition and their impact on net carbon stocks are poorly understood and are the subject of ongoing research. Conservation tillage (CT) has the primary aim of reducing soil erosion and is defined by the IPCC as any tillage and planting method in which 30% or more of the crop residue remains on the soil after harvesting. These methods, which include no till, ridge till, minimum till, and mulch till as well as drill planting, minimize the disturbance of the soil. Crop residues either remain on the surface in the case of no-tillage or are partially incorporated into the soil surface layer for other CT practices.

Annual gains from reduced tillage of up to 0.5 tC/ha have been observed in experimental plots, depending on the soil type and climatic conditions, and a 10–50-year period to reach a new equilibrium would indicate significant cumulative potential. However, the overall impact of tillage practices on SOC stocks and fluxes, including the extent to which observed gains are attributable to improved SOM protection or to increased carbon input, for example from cover crops, remains an area of discussion (Chenu et al., 2019). At the other end of the spectrum from no-till, inverting the soil profile by deep plowing or deep tillage, potentially to a depth of up to 1.5 m, can bury high SOC surface soil and bring low SOC soils to the surface. Deep tillage has also been reported to enable additional accumulation in high SOC soils (Alcántara et al., 2016), although gains are highly dependent on other agricultural practices, and this is an area requiring further investigation across a range of soil and environmental conditions (Feng et al., 2020).

Crop selection, rotation, and intensified cropping

The amount of crop residue available to be converted into humus can be increased by a number of practices that intensify cropping, such as the elimination of fallow periods in seasonally cold or dry environments, the selection of crop varieties for high yield of aboveground and belowground biomass, and the selection, breeding, or genetic manipulation of crops for increased production of recalcitrant material, primarily lignin. Intensified cropping increases the amount of aboveground biomass available to contribute to SOC, and also increases belowground input to SOC through root and microbial biomass. When applied in conjunction with conservation tillage the resulting increased input availability will have maximum impact on SOC stocks.

Roots tend to decompose more slowly than leaf litter, possibly as a result of the higher proportion of complex organic compounds in roots that are more slowly transformed and therefore contribute to a longer SOC residence time. Selection, breeding, or genetic manipulation of crop varieties that result in higher belowground NPP allocation can therefore contribute to increasing soil carbon stocks (Jansson et al., 2010; Kell, 2012).

The establishment of perennial vegetation as a biomass crop, on previously tilled croplands, combines the advantages of intensified cropping and reduced tillage, while also potentially reducing soil erosion. Assessment of the impact of such crops on soil carbon storage is an ongoing research area.

Agroforestry systems, which integrate tree planting with agricultural practices (e.g., alley cropping, under-cropping, windbreaks, etc.) exploit the potential of forests to increase biomass production and soil carbon storage while maintaining an agricultural production system.

Managing soil biogeochemistry

The SOC pool is susceptible to the chemical conditions in the soil, and modification of soil chemistry can be used to influence the rate of accumulation or decomposition of humic material, as well as the redistribution of organic carbon into deeper subsurface layers. Sorption of organic molecules onto mineral surfaces protects these fragments from further mineralization and is aided by the presence in the soil of polyvalent cations. Addition of lime (CaO) or minerals containing iron and manganese oxides provides a source of these cations and can result in an increase in the sorption and protection of organic carbon. However, the shift in pH due to lime addition may also promote microbial decomposition of SOM. Alumino-silicates from basalts or clay wastes from mineral processing have also been proposed for stabilizing SOM (Gillman et al., 2002; Churchman et al., 2014).

Organic carbon remains susceptible to continued mineralization in the upper soil layers, particularly in the rhizosphere, where respiratory activity is high. Soil carbon stocks can therefore be increased by processes, including deep tillage (discussed above) that move organic carbon from the upper soil layers into deeper, less oxidative layers. Adsorbed organic carbon can be released into pore water by the addition of anions such as phosphate or sulfate that will compete for sorption sites on mineral surfaces. Under suitable hydrological conditions, vertical pore water flows can then carry dissolved organic carbon to deeper soil layers, where it can be adsorbed onto unsaturated mineral surfaces. As well as the primary aim of plant nutrition, fertilizer management may therefore achieve the additional objective of transporting carbon deeper into soils, increasing total carbon stocks.

Manipulation of microbial communities

Soil microbial communities facilitate some of the key biochemical processes that determine the fate of SOC as well as the emissions of GHGs (Fig. 11.6). These processes include the decomposition of animal and microbial biomass and the resulting supply of metabolites to soil organisms; mobilization of plant nutrients such as iron and phosphorus; nitrogen cycling, including fixation and denitrification; and removal of soil contaminants. While the biochemistry of these processes is increasingly known, their sensitivity to microbial community composition and dynamics is less well understood. Due to their different physiologies, the two main microbial groups—bacteria and fungi—interact differently with soil properties and processes, and community composition and activity therefore have an impact on soil carbon cycling and storage (Kallenbach et al., 2015).

Bacterial respiration mineralizes ~60% of the organic carbon metabolized from decaying plant material, while fungi respire only 30%—40%, which translates to a higher microbial growth efficiency (MGE), expressed as the weight fraction of metabolized carbon that is incorporated into microbial biomass and byproducts. A shift in the microbial community toward fungal dominance would therefore increase SOM by reducing the loss through respiration and increasing MOM. This shift toward a higher fungal-to-bacterial biomass ratio is one factor that can increase SOM under no-till cultivation, since tillage interrupts fungal growth by breaking up the network of mycelia. The incorporation of SOM into macroaggregates also provides physical protection, and the enhanced stability of aggregates in fungal-dominated soils reduces carbon mineralization by reducing bioavailability, leading to increased soil carbon accumulation.

A better understanding of these processes would open up the possibility of modification by direct or indirect means to maximize SCS and minimize GHG emissions. Direct methods could potentially include microbial inoculation of seeds (O'Callaghan, 2016), the use of aerobic composted biowaste as a soil amendment (Neklyudov et al., 2008), genetic engineering of microbes to control or modify specific functions (Doyle et al., 1995), while indirect means would include altering soil chemistry and pH to enhance existing functions.

The complexity of microbial communities involved in soil processes provides the opportunity to use the population dynamics of these diverse communities as indicators of the varying status of soil—for example as indicators of nutrient availability or SOC content (Sharma et al., 2010). Research studies using DNA sequencing techniques have confirmed correlations between soil status indicators, such as SOC, and the relative abundance of different microbial species, which respond differently to changing soil conditions.

11.4.2 Soil amendments: biochar and hydrochar

Biochar is the solid product of the dry pyrolysis of biomass when heated to 350—700°C in the absence of oxygen, typically accompanying the production of biooil and biogas (see Chapter 12 for a discussion of biomass conversion processes). Also referred to as pyrochar, biochar has been widely studied and applied as a soil amendment that can contribute to improved fertility and reduced soil GHG emissions through its influence on soil properties. Depending on the pyrolysis temperature and precursor material, a large proportion (>90%) of the carbon is resistant to microbial activity and its addition therefore augments the passive carbon pool in soils (Fig. 11.7).

The use of biochar for SCS was first proposed by Lehmann et al. (2006), building on the analysis of the Amazonian dark earths (*terra pretta*) by Sombroek et al. (2003). Recognizing the potential to create a significant, long-term terrestrial sink, with cobenefits for soil fertility and crop production, Lehmann and colleagues estimated that, if expected global bioenergy supply in 2100 was provided by biomass pyrolysis, returning the resulting biochar to the soil in croplands and temperate grasslands would sequester 5—10 GtC/yr, with a potential sink in the order of 400 GtC.

Multiple factors influence the response of soil carbon to biochar application, including biochar properties, application rate, soil type and properties, including initial carbon content, and climatic zone (Burrell et al., 2016; Chagas et al., 2022). Under suitable conditions, its application improves soil microaggregation and SOM protection (Weng et al., 2017) and reduces soil water loss (positively impacting water-use efficiency and microbial abundance), contributing to an increase in crop productivity by 10%—15%. In flooded soils such as paddy fields, CH_4 emissions and the loss of nitrogen through ammonia volatilization are reduced, and application can also mitigate increased N_2O and CH_4 emissions arising from the increased use of manures. Generally, biochar application improves most soil functions, particularly in degraded and low-fertility soils (El-Naggar et al., 2019; Blanco-Canqui, 2021), although the sorption of zinc and other micronutrients has been identified as a potentially significant in soils with low bioavailability (Wu et al., 2018). Impacts on soil organisms are similarly positive (Lehmann et al., 2011), and unraveling the many processes by which biochar application influences the abundance and composition of microbial communities, including the removal of pathogens and growth-inhibiting compounds, is an active research area.

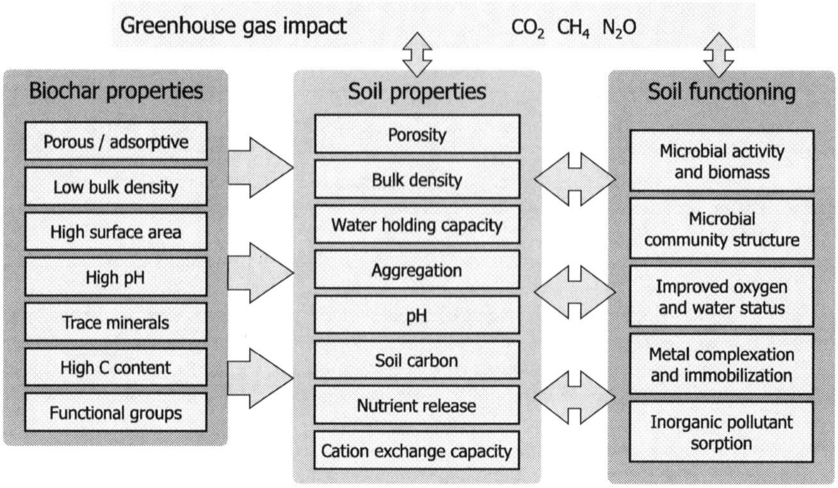

FIGURE 11.7 Biochar properties and impact on soil properties. *After Li et al. (2018).*

Biochar properties including bulk density, surface area, and adsorption behavior vary considerably depending on the type of biomass feedstock and pyrolysis conditions, and can be further modified by a variety of biological, chemical, and physical treatments. This flexibility leads to the possibility of engineered or "designer" biochars, which can be optimized for specific applications and conditions (Novak et al., 2009; Wang et al., 2017).

The residence time of increased soil carbon resulting from biochar application varies from decades to centuries, depending on biochar, soil, and climatic conditions, and the identification of an optimum biochar type and addition rate to maximize SCS will require soil-specific investigations.

Potential issues with large-scale application of biochars include a reduction in albedo for high application rates (Smith, 2016), leading to greater radiative heating of the soil surface during periods of high solar irradiance, the possibility of black carbon aerosol release during production, transportation, and application, causing reduced air quality and further exacerbating radiative heating of affected surfaces. These and other potential side effects point to the need for a comprehensive standardized assessment framework to guide the design of biochar projects (Azzi et al., 2021).

In contrast to biochar, hydrochar is the product of the hydrothermal conversion or carbonization (HTC) of wet or dry biomass (plant and other biowastes, sewage, etc.) into a carbon and nitrogen-rich slurry using hot, high-pressure water (175–295°C and 2–5 MPa pressure) as the reaction medium. Used as a soil amendment the effect of hydrochar differs from biochar, as the carbon is significantly more available for biological activity (i.e., less recalcitrant). Hydrochar application typically increases short-term CO_2 emission from soil respiration by stimulating microbial activity—an action known as priming. However, the priming effect can be avoided by feedstock conversion in oxygen-limited conditions, reducing the bio-availability of hydrochar carbon, and a number of investigations have demonstrated that hydrochar application can increase soil carbon storage on at least decadal timescales (Khosravi et al., 2022). With the global production of potential hydrochar feedstocks exceeding 10 Gt/yr, further assessment of the potential for sequestration over longer timescales is warranted.

The flexibility of biomass production and utilization to be implemented on small to medium scale to match locally available feedstocks means that this technology can also play a part in local development and energy independence. Distributed biomass pyrolysis plants, matched to locally available agricultural wastes, can provide heat and power and produce income from biochar SCS credits alongside cobenefits for soil fertility and food security (Pröll et al., 2017). When combined with local renewable energy production, biochar could be the basis for a negative emission hybrid renewable energy system generating income for communities and island nations (Li et al., 2019).

In pure bioenergy systems that do not produce biochar, 100% of the photosynthetically captured carbon is used to generate energy with, in the case of BECCS, recaptured carbon being geologically stored. In biomass conversion systems producing biochar, around half of the biomass carbon may be used for energy production while the remainder is returned to soils (Fig. 11.8). Bioenergy-biochar systems thus forego around half of the potential bioenergy production in favor of SCS and improved crop yields.

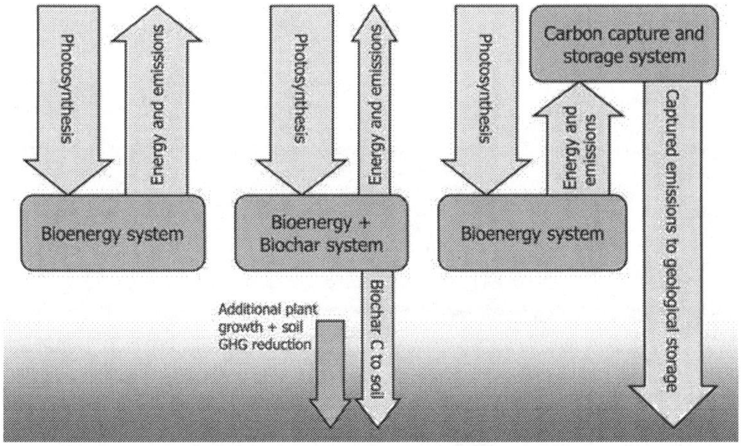

FIGURE 11.8 Carbon and energy flows in biochar and bioenergy systems.

From an overall life-cycle emissions perspective, the choice between these two options depends on the marginal carbon intensity of the energy source that would be replaced by additional bioenergy. If the local energy system still includes coal-fired generation with high carbon intensity then additional bioenergy production to reduce fossil fuel use would be preferred, while biochar production with SCS will result in lower overall emissions if the additional bioenergy would otherwise offset renewable sources (Roberts et al., 2010).

11.4.3 Major SCS projects and initiatives

The "4 per 1000" initiative was launched by the French Ministry of Agriculture in 2015, during COP21 in Paris, with the aim to increase the global soil carbon inventory by 0.4%/yr through agricultural, agroforestry, and other land management practices. Considering the ~700 GtC inventory in the top 30 cm of soils as the target for such practices, this would represent an additional sink of ~3 GtC/yr, reversing the recent rate of loss. Assessments suggest that this target is technically feasible (Lal et al., 2015; Minasny et al., 2017), although clearly ambitious given practical concerns such as scarcity of resources (e.g., manure availability) and the need for changes in regulations or farm subsidy frameworks to make SCS practices economic for farmers (Poulton et al., 2018).

In Europe, a Thematic Strategy for Soil Protection was first adopted in 2006, to promote the protection and sustainable use of soils, by preserving soil functions, preventing further degradation, and restoring already degraded soils. The strategy led to increased funding for research into soil issues as well as influencing EU policy on agriculture (e.g., the EU Carbon Farming Initiative) and rural development. In its current guise, the EU Soil Strategy for 2030 continues the theme of soil protection and sustainable use, aiming to significantly reduce nutrient losses and the use of chemical pesticides by 2030 and to achieve land-based climate neutrality (i.e., zero emissions from LULUCF) by 2035.

Globally, the FAO has launched a number of initiatives focusing on soils, including the Global Soils Partnership, the Intergovernmental Technical Panel on Soils, and the RECSOIL project, focusing on the recarbonization of global agricultural soils. The FAO's Soils Global Information System, which includes a global database and global information system (GIS) for soil organic carbon stocks and SCS potential, will be a valuable starting point for measurement and verification of the impact of SCS projects and initiatives (Amelung et al., 2020).

Through wide-ranging positive effects on soil functions, SCS can enable agricultural systems to adapt to climate change, while sustaining or improving productivity and other ecosystem services, and can also play a significant part in mitigating climate change. Given the necessary incentives, many of the approaches that enhance soil carbon can be quickly implemented, providing a critical bridge while other NETs are matured. Rapid scale-up of SCS-related initiatives on global, national, and local levels is critical to maximizing the impact of these approaches (Amelung et al., 2020).

11.4.4 SCS knowledge gaps and research priorities

Areas of ongoing research and field trials aiming to improve our knowledge of the many processes influencing SCS are summarized in Table 11.8. Useful recent reviews include those by Hartemink et al. (2014), Chenu et al. (2019), and Jansson and Hofmockel (2020); tracking citations of these reviews will be an effective guide to progress in these areas.

TABLE 11.8 Soil carbon sequestration research areas.

Soil carbon stocks
• Understanding and modeling the processes controlling accumulation, movement, distribution, and residence of carbon in the soil profile; impact of soil properties and carbon inputs • Improved understanding and modeling of the impact of soil erosion and deposition, and of hydrological conditions (e.g., peatlands) on net carbon stocks • Impact of tilling practices on microbial biomass, soil organic carbon, and nitrogen under various crops • Importance of soil disturbance, soil moisture content, and crop residue placement on fungal to bacterial biomass ratios • Synergistic effects of conservation tillage and cover crops on SOC pools and fluxes • Biotic regulation of SOM degradation, including the influence of plant secondary metabolites • Improved SOM stabilization using mineral amendments (e.g., basalt derived alumino-silicates, clay mine wastes)
Crop selection, rotation, and intensified cropping
• Role of nutrients (N, P) on SOC via primary production, above- and belowground allocation, litter decomposition, priming effects, and long-term SOC stabilization • Impact of cropping practices (cover crops, fertilization, etc.) and other CA/RA practices on microbial carbon use efficiency, and consequences for SOC stocks and fluxes; impact of enhanced CO_2 (eCO_2) and climate change on soil microbial communities; • Genetic controls of aboveground versus belowground biomass production • Plant–microbe interactions in the rhizosphere, influence on SOC inventory and longevity, and control through plant breeding or transgenic plant development • Influence of various crops, including perennials and other biomass crops, on SOC for varying climatic conditions and soil types
Biochar and hydrochar
• Optimal dosage rate and schedule for various soil types • Assessment of the varying properties of biochar and hydrochar derived from different biowaste feedstocks (designer biochars) • Identifying optimal biochar characteristics to maximum soil function and carbon sequestration benefits for specific soil and climatic conditions • Impact of pretreatment of feedstocks (e.g., torrefaction) and pyrolysis conditions on biochar properties • Basic and applied research on copyrolysis of biowastes with waste plastics, yielding pyrolysis oil, char, and gases • Product standards to support regulated large-scale deployment, e.g., ensuring product consistency and avoiding environmental issues such as toxic contaminants • Impact of biochar addition on wetland carbon stocks, microbial communities, hydrology, and vegetative recolonization • Influence of feedstock characteristics, processing steps, and environmental factors (soil conditions, hydrology, etc.) on biochar stability
Microbial processes in soils
• Understanding fungal and bacterial contributions to the decay pathways from above- and belowground plant residues to stable humified SOC • Understanding how microbial activity can be shifted toward increased SOC longevity for varying soil types and chemistry and for different plant residue types • Understanding the relative importance of soil disturbance, soil moisture, and residue placement on microbial community structure and microbial biomass • Understanding the influence of biochemical recalcitrance, physical protection in aggregates, and chemical protection by clay association in controlling the fate of microbial biomass and its contribution to soil carbon stocks • Understanding how eCO_2 and climate change affect microbial communities and soil processes
Monitoring and verification
• Further development of measurement standards for soil carbon • Full life-cycle GHG accounting in agricultural and other ecosystems, including off-site transfers • Further development of spectrometric and remote sensing techniques, methodologies and standards, as well as datasets for local calibration and validation • Development of indirect indicators for SOC stocks and potential (e.g., based on soil structural properties, fine mineral fraction, microbial biomass, etc.) • Establishment of new experimental sites, particularly in under-represented regions/soil types/climatic conditions, to provide long-term SCS model validation

Complementing these detailed investigations, the development of tools and techniques for the holistic management of soils (Savory and Butterfield, 2016), balancing the nutrient, water, and energy cycles, and the dynamics of soil communities, is also leading to an improved understanding of SCS (Hawkins et al., 2022).

Soil carbon monitoring, reporting, and verification

Credible and reliable MRV systems are a prerequisite for the funding of large-scale SCS projects, but compared to the relative simple task of "counting trees," monitoring of soil carbon stocks and accumulation through SCS activities poses a significant challenge. The high cost of direct measurement, particularly for deeper soil levels (>30−40 cm) and challenging ecosystems such as peatland swamps, and the complexity and knowledge gaps related to the diverse processes controlling SCS leads to high uncertainties, which may be compounded by a legacy of unrecorded historical land management practices. In addition, it is not currently feasible to verify sequestration rates that increase total SOC stocks by <1%/yr using direct soil measurements (Amelung et al., 2020), so that measurements in soils achieving the 4‰/yr target would still have an uncertainty in the order of ± 25% after a typical 10 years survey interval.

Direct soil sampling to quantify soil carbon fractions has more recently been complemented by CO_2 flux measurements using eddy covariance and surface chamber methods. Remote sensing techniques have been demonstrated using visible-near infrared and mid-infrared spectra (400−2500 nm), that reveal the absorption of light at specific wavelengths by organic molecular bonds and minerals in the soil (Chabrillat et al., 2019). These methods can also be used to measure soil water retention and water holding capacity (Baumann et al., 2022). Other measurement techniques under development for soil carbon assessment include laser-induced breakdown spectroscopy (LIBS), which can determine total, organic, and inorganic carbon, and bulk density of soil cores, and inelastic neutron scattering (INS), which can measure in situ soil carbon to 30 cm depth (Nayak et al., 2019). Indirect indicators for current and potential SOC storage from local to regional scale, also have potential for SOC monitoring, with fine mineral fraction, soil structural properties, and microbial biomass identified as promising indicators (Wiesmeier et al., 2019).

Further development of these new approaches will be needed to deliver a cost-effective SCS MRV system that achieves the accuracy required for verification and accounting. Such a system is likely to incorporate a range of different approaches, as illustrated schematically in Fig. 11.9.

Direct measurements, from field sampling and remote sensing, and other soil carbon indicators, most likely combined using machine learning algorithms, together with numerical models, validated by long-term data from field monitoring sites across the full range of soil types, climatic conditions, etc., will be the key components of such a system (see Campbell and Paustian (2015) for a recent review of the status of SOM modeling). Continued monitoring of field sites to extend the length of datasets is important to assess the fate of various SOC pools with residence times greater than the duration of currently available measurements.

Internationally agreed standards and standardized methods will also be required as a foundation of SCS MRV, and while standards are available covering basic soil parameter measurements, including C and N concentrations, no standards are yet in place for the assessment of SOC stocks, SCS potential, current GHG emissions and emission mitigation potential, and subsequently for verification of the effectiveness and permanence of the SCS actions (Bispo et al., 2017; Dynarski et al., 2020). One of the most comprehensive systems for soil carbon accounting is the Full Carbon Accounting Model (FullCAM), which includes estimates of carbon stocks and changes, including above and belowground biomass, standing and decomposing debris, and soils (Richards and Evans, 2004). FullCAM is used to establish Australia's national carbon inventory for the land sector, as well as estimates of carbon stock changes resulting from REDD+ type projects.

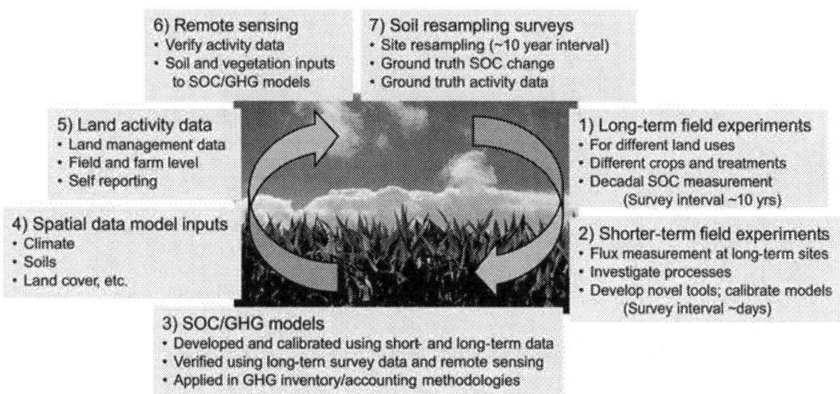

FIGURE 11.9 Components of a soil MRV system. *After Smith et al. (2020).*

11.5 Ecosystem restoration

In March 2019, the United Nations announced its Decade on Ecosystem Restoration (2021–30), a joint initiative by UNEP and FAO aiming to prevent, halt, and reverse the degradation of terrestrial and marine ecosystem worldwide. Alongside wide ranging social and sustainability benefits, ecosystem restoration can also help to preserve and enhance carbon stocks in soils and sediments, as well as increasing stocks in vegetation.

11.5.1 Coastal blue carbon

Vegetated coastal habitats (VCHs), notably mangrove and macroalgae forests, salt marshes, and seagrass meadows, are significant contributors to the ocean carbon sink as a result of their high primary productivity as well as their ability to trap "laterally imported" carbon originating in other ecosystems (Mcleod et al., 2011). By attenuating currents and waves, these habitats enable the settling of suspended particles and the build-up of shallow marine sediments, sequestering the incorporated carbon for millennia.

The relatively limited global extent of these habitats belies their significance as carbon sinks compared with both terrestrial ecosystems and the open ocean; covering just 0.2% of the ocean surface they deliver half of the global carbon flux into marine sediments (Duarte et al., 2005, 2013). Carbon burial rates in VCHs, typically 1.4–2.3 tC/ha/yr but ranging up to 17 tC/ha/yr, are some 40–50 times higher than in terrestrial forest ecosystems soils, and carbon stocks in coastal sediments do not reach saturation, since sediments will continue to accumulate in response to rising sea level (Mcleod et al., 2011). Global carbon sequestration in VCH sediments is estimated to approach 1 GtC/yr, with the potential to create significant negative emissions by increasing this flux through the range of methods discussed below.

The global extent of VCHs has been roughly halved due to human activities such as draining, dredging, and dyking, which alter hydrological conditions, and to excessive nutrient supply from agricultural run-off and sewage or water treatment effluent discharge. Excessive nitrogen and phosphorus supply can lead to community shifts in VCHs, increasing microbial activity and the production of more labile carbon, as well as hypoxia. Extreme events such as hurricanes and typhoons have also contributed to VCH loss as a result of the destruction of mangrove forests. There is also evidence that VCHs can in some cases be destabilized due to community shifts among organisms that rework sediments, such as crabs and marine worms, as a result of the loss of high-level predators (Atwood et al., 2015). Further research into the food webs in these habitats may open up the possibility of interventions, such as predator reintroduction or invasive species harvesting, with the aim of protecting carbon stocks (Macreadie et al., 2017).

VCHs also provide a range of other environmental services, including coastal protection and flood prevention, as well as supporting fisheries, and efforts to date to protect and restore these habitats have been largely driven by benefits in these areas, rather than by the potential for increased carbon storage, although mangrove conservation and afforestation projects are included within the scope of REDD+. Damaging human impacts can be controlled by improved fertilizer management and altering other land practices that reduce run-off and improve water quality, as well as by improving wastewater treatment (Macreadie et al., 2017). Practices that have been successfully implemented to restore VCHs include replanting in all habitats, and the construction and planting of new salt marshes and macroalgal habitats, while poor site selection (e.g., high wave or current energy, or poor water quality), or sites unsupportive of natural recruitment have been common causes of failed restoration projects (Bayraktrarov et al., 2016). Measures to restore hydrological conditions have also been successful, particularly for mangroves where the reinstatement of tidal flows can have the additional benefit of reducing methane emissions.

The effectiveness of VCH restoration for increased carbon storage may be offset by a number of secondary biophysical processes such as calcite precipitation and increases in other GHG emissions. The extent to which calcium carbonate accumulation in mangroves and seagrass meadows is the result of inorganic carbon import from other ecosystems versus local calcification is a research priority, since $CaCO_3$ precipitation in seawater results in the release of CO_2 which could offset up to 30% of the net carbon sequestration in these habitats (Howard et al., 2018; Saderne et al., 2019). Recent air–sea CO_2 exchange measurements over a seagrass meadow underlain by carbonate sediments demonstrated a consistent source of CO_2 to the atmosphere at an average rate far exceeding the organic carbon burial rate (Van Dam et al., 2021).

Increases in other GHG emissions due to increased availability of organic matter for metabolic conversion would also offset carbon gains, and for mangroves, an estimated 20% of carbon burial could be offset as a result of increased methane emissions (Rosentreter et al., 2018). An offset of ~10% was determined in a full experimental GHG accounting for a successful seagrass meadow restoration in Virginia, USA, where seagrass restoration was found to have removed 9600 tCO_2 over 15 years, but also released 950 tCO_2e as a result of enhanced CH_4 and N_2O production (Oreska et al., 2020).

11.5.2 Peatlands

Global peatlands, covering 4.2 million km^2 (420 MHa), just 2.8% of Earth's land surface, contain carbon stocks estimated at 390–450 GtC—roughly half the atmospheric inventory (Freeman et al., 2012), and losses or gains in these stocks can therefore have significant climate impact. The high soil carbon stocks of 260–300 tC/ha in peatlands occur as a result of the anoxic conditions, which limit the activity of phenol oxidase, the enzyme responsible for the breakdown of phenolic compounds, thereby inhibiting microbial decomposition of organic matter (Fenner and Freeman, 2011).

Drought conditions or the drainage of peatlands for forestry and agricultural uses, including peat mining, restores the conditions for microbial decomposition and can result in the loss of 50–60 tC/ha over a decadal timescale. Rewetting drained peatlands can "jump-start" the restoration of these ecosystems, with the potential to restore carbon stocks to former levels (Kareksela et al., 2015). Alongside rewetting and subsequent water management, which aims to restore hydrological function, other peatland restoration actions include revegetation and management to maximize vegetation cover, with the aim of reestablishing an active peat-forming ecosystem. The reestablishment of *Sphagnum spp*. is particularly important as these mosses have a symbiotic relationship with methanotrophic bacteria which oxidize much of the methane generated in peatlands, providing an additional carbon source for *Sphagnum* metabolism (Raghoebarsing et al., 2005).

Rewetting is however accompanied by an offsetting increase in methane emissions (Waddington and Day, 2007) and can also initiate further transient carbon losses as a result of increased nutrient and labile carbon availability, created during dry conditions, that can prime anaerobic decomposition on rewetting (Fenner and Freeman, 2011). Recent work has suggested that this positive feedback can be mitigated by shifts in vegetation type that inhibit decomposition (Wang et al., 2015). Further research is required, particularly long-term monitoring of GHG fluxes above restored peatlands and the impact of restoration practices on microbial communities, to develop a full understanding of the complex processes and timescales involved in peatland restoration, and the implications for carbon storage.

The incorporation of exogenous carbon-rich material, such as biochar and timber, in peatlands and other wetland soils has also been proposed as an approach to terrestrial carbon sequestration. As well as directly increasing long-lived carbon stocks, biochar addition influences carbon turnover in wetland soils in a number of ways, including increased microbial biomass and shifts in the microbial communities that reduce the mineralization of SOM and promote its incorporation and stabilization in soil aggregates (Zheng et al., 2018).

As noted above, the high level of carbon stocks in peatland is due to the inhibition of microbial decomposition and the possibility of exploiting this condition by either increasing the effectiveness of the inhibition or increasing the amount of carbon protected by it has been investigated by Freeman et al. (2012). Inhibition effectiveness could be increased by genetic engineering of *Sphagnum* mosses to increase the production of phenolic compounds and by localized water, pH, and nutrient management, while the storage of endogenous carbon could be increased by various practices to increase peatland NPP. Alternatively, exogenous carbon stocks such as harvested timber could be sequestered by "injection" into peatlands, with poplar, pine, and oaks being suitable tree species for temperate and boreal locations. Placement by vertical piling would be one injection method, the resulting pathway having the additional benefit of facilitating access for heavy vehicles during injection operations (Freeman et al., 2012).

11.5.3 Ecosystem restoration knowledge gaps and research priorities

As noted above, carbon sequestration has not generally been a key objective of ecosystem restoration projects progressed to date, and the complexity of processes influencing the rate, GHG impact, and longevity of carbon accumulation in wetland habitats suggests that significant further research will be required before "blue carbon" can be verified for carbon market inclusion. Table 11.9 provides an overview of the main knowledge gaps and research priorities in this area. The review by Mcleod et al. (2011) provides a comprehensive discussion of these topics.

Wetland ecosystem monitoring, reporting, and verification

In coastal and other wetlands, MRV of carbon stocks and the effectiveness of efforts to increase them share many challenges with SCS more generally, with the added complication of being often or continuously submerged, increasing the cost of direct measurements and precluding the use of some remote sensing techniques, such as those relying on infrared reflectance. Furthermore, carbon transport in terrestrial ecosystems, through soil erosion and runoff, plays a relatively small part in determining soil carbon stocks, but organic material in wetlands is far more mobile, with significant fluvial and tidal transport between ecosystems. The need in many cases to account for lateral flows of carbon, both particulate and dissolved is a further complication for wetland MRV (Maher et al., 2018; Akhand et al., 2021). High levels of carbon accumulation are also typically accompanied by high methane emissions, and these need to be measured and accounted for to assess net carbon equivalent accumulation.

TABLE 11.9 Ecosystem restoration knowledge gaps and research priorities.

Coastal blue carbon
• Relative importance of factors controlling carbon dynamics, including transport between ecosystems, to constrain VCH carbon budgets and guide restoration • Impact of land-use practices and other human activities on VCHs with high carbon sequestration potential • Food web influence on VCH carbon stocks and identification of the optimal target densities for key organisms • Development of tools and field evidence to determine the contribution of macroalgae to carbon burial rates and stocks in sites beyond source habitats • Impact of inorganic carbon cycling and non-CO_2 GHG emissions on VCH sequestration effectiveness • Impact of biochar on wetland native carbon stocks, microbial communities, hydrology, and vegetative recolonization
Social and sustainability impacts
• Quantification of the human and ecological benefits resulting from preservation and restoration of VCHs, in addition to carbon sequestration
Monitoring and verification
• Improved methods for quantification and verification of sediment carbon stocks, accumulation rates, and long-term storage fraction • Impact of ecosystem restoration practices on carbon stocks and other GHG balances • Methods to account for CH_4 and N_2O emissions that reflect the GWP of continuing and time varying emissions • Methods to assess the contribution of organic carbon inputs and exports, including carbonate precipitation • Baseline studies to determine the impact of climate change and human activity on carbon stocks • Methodologies for data integration (data fusion) including further development of machine learning techniques

Carbon stocks in wetlands are traditionally assessed using direct measurement of bulk density and soil carbon concentration on soil/sediment cores, together with assessment of soil/sediment accretion rate by measurement against a physical benchmark or using a variety of dating techniques, although there is at yet no standard methodology for this measurement (Villa and Bernal, 2018). Despite the challenges of occasional or continuous submergence, remote sensing methods have been successfully applied to wetlands presence/absence, cover, species, and biomass, including the use of machine learning algorithms to integrate optical, radar (SAR, InSAR) and LiDAR imagery, as well as multispectral and hyperspectral data (Pham et al., 2019).

Methodological guidance on the estimation of GHG emissions and carbon stocks in wetlands is included in the IPCC's guidelines for national GHG inventories (Hiraishi et al., 2014; see also Emmer et al. 2015), while methodologies for carbon offset estimation for coastal wetland conservation, restoration, and creation activities have recently been reviewed by Needelman et al. (2018) and by Sapkota and White (2020). However, the inclusion of wetlands in emissions reduction or removal programs remains problematic as there is as yet no standard methodology to include the global warming potential of continuing, time-dependent emissions of non-CO_2 GHGs (Villa and Bernal, 2018), and this requires further research.

11.6 Enhanced weathering

Rock weathering is a slow natural biogeochemical process that currently removes around 1 $GtCO_2$/yr from the atmosphere and transfers it, mostly as bicarbonate, into the ocean where its residence time is in excess of 100,000 years (Hartmann et al., 2009). Over a similar timescale, this process will also return atmospheric [CO_2] to its preindustrial equilibrium with the geosphere—unless large scale NET deployment does it first! The possibility of accelerating this process by applying ground minerals to the land surface was first proposed by Schuiling and Krijgsman (2006), who also recognized that this would counteract the effect of acid rain on forests, as well as improving soil quality. The enhanced weathering (EW) concept has since been extended to include the application of a range of mineral feedstocks, including ground mafic or ultramafic rocks, and basalts to beaches or to the open ocean.

Terrestrial EW (TEW) requires grinding these rocks to a micron-scale powder so that, when applied as a soil amendment, complete dissolution takes place within a few years. Depending on soil pore water chemistry, CO_2 would then be captured and stored either by the export of alkalinity in the form of dissolved bicarbonate via groundwater and rivers to the sea (reactions 11.1 and 11.3) or through the precipitation of secondary carbonate minerals in the soil (reactions 11.2 and 11.4). For the calcium silicate mineral wollastonite, the reactions are the following:

$$CaSiO_3 + 2CO_2 + 3H_2O \rightarrow Ca^{2+} + 2HCO_3^- + H_4SiO_4 \tag{11.1}$$

$$Ca^{2+} + 2HCO_3^- \rightarrow CaCO_3 + CO_2 + H_2O \tag{11.2}$$

and for olivine (forsterite);

$$Mg_2SiO_4 + 4CO_2 + 4H_2O \rightarrow 2Mg^{2+} + 4HCO_3^- + H_4SiO_4 \tag{11.3}$$

$$Mg^{2+} + 2HCO_3^- \rightarrow MgCO_3 + CO_2 + H_2O \tag{11.4}$$

In reactions 11.1 and 11.3, either 2 or 4 mol of CO_2 are converted to bicarbonate for each mole of silicate rock dissolved, equivalent to 0.76 tCO_2/t-rock (88 g/116 g) for wollastonite or 1.15 tCO_2/t-rock (176 g/153 g) for olivine, while in reactions 11.2 and 11.4, carbonate precipitation results in the release of 1 mole of CO_2, reducing this capture effectiveness by ½ and ¼, respectively.

If applied to beaches or to the open ocean, alkalinity released through dissolution would convert dissolved CO_2 to bicarbonate and carbonate ions, and the reduced CO_2 partial pressure in the surface mixed layer would drive an incremental air-to-sea CO_2 flux, effecting CDR. Silicic acid (H_4SiO_4) released through reactions 11.1 and 11.3 could have a further CDR benefit since this is the source of silica (Si) for diatom shell building, and is limiting in some parts of the ocean. Diatoms currently account for around half of the carbon export to deep ocean waters by the biological pump (Chapter 6), and this export could be enhanced by increased Si availability (Köhler et al., 2013).

When applied to beaches, a mineral sand (100–300 μm) would be preferred rather than a fine powder, although this involves a trade-off with dissolution rate due to the decline in a specific area with increasing particle diameter (Hangx and Spiers, 2009; Meysman and Montserrat, 2017). In the open ocean, a micron-scale powder would be necessary to maximize the dissolution of sinking particles within the mixed layer (Chapter 16), due to the long time delay before deeper waters are able to reequilibrate with the atmosphere (Köhler et al., 2013).

11.6.1 EW material and energy requirements

Feedstocks

Olivine-rich ultramafic rocks are ideal feedstocks and have been widely studied, with capture potentials typically in the range 0.8–0.9 tCO_2/t-rock, and up to 1.1 tCO_2/t-rock in the case of dunite, which contains >90% olivine. Basalt is also suitable, although with a lower capture potential of ~0.3 tCO_2/t-basalt, and many alkaline materials such as mine and other industrial wastes (e.g., slag) and recycled concrete (construction waste) can also be cost-effective feedstocks that are globally available in Gt quantities.

Heavy metals such as chromium and nickel are generally present in ultramafic rocks and to a lesser extent in basalts and may also be present in alkaline industrial wastes. Weathering will result in the release of these metals into the soil profile or ocean (Renforth et al., 2015; Beerling et al., 2018), and heavy metal content will therefore be an important criterion for feedstock selection, and may constrain maximum application rates to remain within environmental limits.

Energy requirements

EW requires the mining, grinding, and distribution of feedstock material, with grinding being the most energy intensive step. The required particle size it determined by the trade-off between energy input and dissolution time, so that faster dissolving minerals would incur less of an energy penalty. Measured mineral dissolution rates are generally lower in soils than under laboratory conditions, despite the fact that various biological processes, absent in laboratory studies, can increase in situ dissolution rates.

Based on measured dissolution rates for olivine, grinding and milling to ~1 μm or less would be required to ensure dissolution within 1–5 years in soils and to ensure dissolution of sinking particles within the mixed layer if applied in the open ocean. The energy required to achieve this fine particle size is in the range of 1.5–3.0 GJ/t-rock, which, in many cases, will exceed the solvent/sorbent regeneration energy requirement for CO_2 separation, for example in DAC (Kelemen et al., 2020; Strefler et al., 2018). For grid electricity with a carbon intensity of 0.25 tCO_2/MWhr, this would incur 0.1–0.2 tCO_2 emissions, reducing capture efficiency by around 10%–20%. The carbonation reactions of silicate rocks are spontaneous and exothermic, and finding ways to reduce external energy requirements by exploiting the chemical potential energy (~2 GJ/m^3 or 0.7 GJ/t in the case of olivine) is an important research priority (Kelemen et al., 2020).

11.6.2 Environmental and ecological impacts

Soil functioning

TEW is most suitably deployed in warm, humid regions where conditions hasten dissolution, and it is expected to have cobenefits for various soil functions, including restored or enhanced soil structure, and the protection of crops from pests and pathogens. Soil fertility will also increase through the release of nutrients such as silicon, phosphorous, and potassium, which would be beneficial in areas where plant growth is limited by their availability (Hartmann et al., 2013; Beerling et al., 2018).

Biological processes in soils can also accelerate mineral weathering—so called biologically accelerated weathering or bioweathering. Both fungi and soil bacteria direct impact mineral dissolution rates through a range of biochemical approaches to access minerals required for their metabolism (Bennett et al., 2001). As noted earlier in the case of SOC sequestration, manipulation of soil microbial communities may also be relevant for EW, with the aim of accelerating dissolution and minimizing biomineralization.

Biological processes in soils that contribute to SCS and to N_2O emissions will also be affected by EW, with complex consequences for overall GHG accounting. By increasing fertility, EW will result in increased NPP and consequent allocation to belowground carbon stocks, as well as improved stabilization of SOC through soil aggregation. Mineral addition will also affect N_2O emissions, particularly from agricultural soils with high N availability, due to pH increase favoring nitrification and other interactions with soil biota, notably mycorrhizal fungi and earthworms, that have a significant influence on N_2O emissions (Vicca et al., 2022).

Fluvial and estuarine biogeochemical processes will also be affected by pH increase downstream of the EW site, with as yet poorly understood consequences, although the eventual increase in alkalinity export to the ocean would help to mitigate the effects of ocean acidification, and could partially restore the ocean's buffering capacity—reducing the impact of further CO_2 uptake on ocean pH (Renforth and Campbell, 2021).

Heavy metal release

As noted above, many prospective EW feedstocks contain nickel and chromium, and although Ni can be beneficial for plant growth when present at low levels, in general, the build-up of these metals in the soil and entry into the food chain should be avoided. Various biological and physicochemical approaches could be used to manage heavy metals in soils, leading to potential synergies with other NETs. For example, some plants have adapted to high soil concentrations of heavy metals, either by accumulating them in biomass or by reducing their bioavailability through precipitation in the rhizosphere, and have been used for phytoremediation of contaminated soils (Suhrhoff, 2022). Crop rotation schemes that include such plants alongside other biomass crops could exploit the soil fertility benefits from EW to sustain biomass yields while exploiting the phytoremediation benefits to mitigate its heavy metals impact. Similarly, biochar can also immobilize heavy metals in soils (Amman and Hartmann, 2019), and coapplication with EW and biomass crops may also be beneficial.

The ecological impact of heavy metal release from coastal EW of olivine, including acute, chronic, and bioaccumulative toxicity, has been investigated for a set of model organisms as part of the foundational research being undertaken by Project Vesta (Moreau et al., 2021), and initial guidance for coastal olivine spreading, based on existing marine environmental quality standards, has been reported by Flipkens et al. (2021), while recognizing that further research is required to address knowledge gaps.

11.6.3 Enhanced weathering RD&D status

Field trial status

The earliest and most extensively studied field trial of TEW began in 1999 at the Hubbard Brook Experimental Forest (HBEF), New Hampshire, USA, with the one-off application of 3.4 t/ha of powdered and pelletized wollastonite ($CaSiO_3$) to an 11.8 ha forested watershed, including 631 kg deposited directly into the catchment's stream channel. Chemical weathering had been studied at HBEF since 1963, and the long-term loss of Ca (as well as Na, Mg, and K) from soils due to weathering by acid rain motivated the Ca addition trial (Peters et al., 2004). Initial analysis for the first 2 years focused on the fate of material added directly to the stream and demonstrated a 50% increase in Ca and Si dissolved in stream water, and an increase in soil pH from 3.9 to 4.4, mitigating ecosystem acidification. By the end of 2010, 11 years after application, 4% of the added Ca had been exported in stream water, with the remainder mostly transported to deeper soil layers as undissolved particles (Shao et al., 2015). The CDR implications of the trial were assessed after 15 years and demonstrated that the treatment had resulted in an increased GHG sink of between 8.5 and 12.2 tC/ha of which 80% was attributed to increased tree growth and 20% to a net reduction in soil respiration, most likely as a result of reduced labile carbon allocation to roots, although this was highly variable over the site, depending on both tree species and altitude (Taylor et al., 2021). Similar reductions in soil GHG emissions following silicate application have also been reported for a bamboo forest in China (Xu et al., 2020). Taylor and colleagues concluded from the HBEF trial that TEW could be an effective treatment for acid-impacted forests, with significant CDR benefits, but that site-specific studies would be needed to assess the suitability and effectiveness of the treatment and to optimize the rate and timing of mineral addition.

On agricultural soils, wollastonite has been used to remineralize nutrient-depleted soils, and a recent post hoc analysis of three field-scale applications in Ontario, Canada reported increased SIC storage, in the form of precipitated $CaCO_3$, equivalent to 6.0 tCO_2/ha in one field after addition of 6.2 t of wollastonite over three consecutive annual applications. Yields also increased for leafy vegetable and potato crops, but not for soy beans, and improved pest control was also reported, potentially reducing the need for pesticides (Haque et al., 2020). No assessment of soil respiration was undertaken in this instance.

A large number of trials on agricultural soils have recently been initiated, under crops including peas and sugar beet (UK), cotton (Project Carbdown—Greece), corn (Project Carbdown—Germany, and in the USA), miscanthus (USA) and oil palm (Malaysia). Results from these trials are just becoming available (Andrews et al., 2020, 2021). In the UK trials, with basalt application at 40 t/ha/yr, soil carbon increased significantly under peas but not under sugar beet, and a null result was also found in the early results from Project Carbdown for cotton fields in Greece (Smet et al., 2021).

Coastal EW is at a similar early stage in field trials and is being pioneered by Project Vesta, a research-driven nonprofit, established in 2019 to evaluate the technical feasibility and ecological impacts of coastal EW. Baseline surveying and monitoring of geochemistry, ecology, meteorology, and physical oceanography at the test and control sites in the Dominican Republic were completed in 2021, alongside social science studies and local community engagement. A parallel foundational research program included ecotoxicology studies on marine worms, modeling and batch shaker experiments on olivine dissolution, alkalinity, and trace metal release, as well as modeling the vertical sorting under wave action of olivine sand (3300 kg/m^3) within the lighter native quartz sand (2650 kg/m^3). A further 12 month monitoring program was initiated following the spreading of a 1 cm layer of olivine at the test site in early 2022 (Romaniello et al., 2021).

Knowledge gaps and research priorities

The initial TEW field trial results summarized above hint at the complexity of the biological and biogeochemical processes involved and changes induced by silicate addition to soils, with mineral application affecting crop yields (NPP), organic and inorganic soil carbon stocks, and soil emissions, and with these effects also being influenced by specific crop types and by crop related feedbacks. The frequency of unexpected results points to the many knowledge gaps, summarized in Table 11.10 and to significant challenges for monitoring and verification of carbon sequestration via TEW.

This extensive list of current knowledge gaps points to the need for a significant expansion and continuation of field trials, with a focus on understanding long-term impacts and developing measurement and monitoring capabilities to enable the verification of carbon removal. As observed in the 15 year HBEF trial, careful monitoring of the full ecosystem response to TEW treatment is essential to fully constrain the net carbon balance.

Of particular note is the observation in several field trials of increased SIC through carbonate precipitation, which will reduce the CDR efficiency by up to a factor of two compared to alkalinity export to the ocean. Because of its higher efficiency, ocean storage is the desired end result of TEW, but alkalinity export has yet to be demonstrated in field trials.

Development of a full system modeling capability will be essential (Taylor et al., 2017), both as a tool to minimize MRV costs and to enable incorporation into earth system and integrated assessment models to accurately reflect the impact of EW if deployed globally at multi-Gt scale.

11.7 Potential scale and costs

Estimates of the annual negative emissions rates, cumulative additions to carbon stocks, and unit costs of CO_2 removal potentially achievable through the global deployment of land-based NETs have been extensively published, with very broad ranges that reflect the various models and scenarios used, and the assumptions made to address current knowledge gaps. Table 11.11 summaries the results of recent assessments of scale and costs.

At the upper limit of the scale for AR plus forest restoration, Bastin et al. (2019) estimated a total storage potential of 205 GtC (750 $GtCO_2$), over 20 years worth of then current global emissions, and roughly the same as aboveground biomass in tropical forests (Fig. 11.1). This estimate has been contested—and defended—on a number of grounds relating

TABLE 11.10 Enhanced weathering knowledge gaps and research priorities.

Field trials
• Long-term effects on SOC, hydrology, soil function, soil organisms, GHG emissions, and crop yields • Impact of soil processes such as secondary mineral formation, and downward transport of mineral solids and solutes • Audited assessments of EW impact on biogeochemical cycles and of CDR effectiveness and permanence for different minerals across a range of crops, soil types, climatic conditions, and spatial scales • Ecosystem-scale feedbacks (e.g., increased terrestrial productivity leading to potentially impacts on the hydrologic cycle) • Investigation of phytoremedition to immobilize trace metals, including potential coapplication with urease • Developing dose–response relationships to quantify and optimize net carbon removal efficiency • Demonstration of downstream alkalinity response reaching the open ocean
Mineral dissolution
• Controls on mineral weathering kinetics in terrestrial and coastal systems (e.g., grain size, moisture variability (wetting/drying), saturation state, etc.) • Impact of soil biota and biological processes, including nutrient status, on weathering rate, SOC, and GHG emissions • Identification of high-weathering traits in plants • Opportunities to exploit the chemical potential energy of mineral dissolution
Ecological and environmental impacts
• Ecotoxicological effects of heavy metal release on terrestrial and marine biota • Ecosystem risks for large-scale EW in terrestrial, coastal, and open ocean environments • Impact on enhanced Si availability on crops (increased cell wall strength, resistance to pests and pathogens, etc.) • Impact of alkalinity and nutrients fluxes, and change in nutrient elemental ratios (N/Si, P/Si) on downstream ecosystems • Assessment of the environmental and social impact of extractive industry expansion and of mineral addition to terrestrial and coastal environments
MRV
• Processes controlling carbonate precipitation in soils, coastal sediments, and the open ocean • Developing full EW system modeling capability (including soil hydrological and biogeochemical processes, through to crop yields and crop-weathering feedbacks); validation using field trial results and incorporation into Earth system models • Developing techniques to accurately monitor mineral dissolution over large spatial scales • Developing effective methods for monitoring changes in biogeochemical fluxes on local to regional scales, including sampling and analysis protocols
Regional deployment potential
• Regional studies on EW potential, including distribution of mineral resources and additional benefits from increased biomass yields

to soil carbon impact, albedo effect, and the potential forestation of natural grasslands and savannah (Veldman et al., 2019, and others), but other assessments also suggest a potential of the same order for the sum of forest based methods.

When compared with the United Nations Strategic Plan for Forests—aiming to increase forest area by 3% (ca. 120 Mha) worldwide by 2030, roughly equivalent to a negative emissions rate of 0.5 GtC/yr (at a mid-range 4 tC/ha/yr)—these figures demonstrate how much more can be done in this area, and current estimates show that we are not even on course to deliver this plan (FAO and UNEP, 2020). Many countries will face significant challenges in meeting their voluntary forest-related commitments, particularly those where policies continue to encourage agricultural expansion and those that have a poor performance record on other sustainability goals (Fagan et al., 2020).

Soil carbon sequestration shows the second largest potential among land-based methods, and the estimated 130 GtC (480 $GtCO_2$) lost from soils (Sanderman et al., 2017) and 200 GtC (730 $GtCO_2$) total historical LULUCF emissions since the preindustrial era (Canadell and Raupach, 2008; Friedlingstein et al., 2022) provide a measure of scale, if not a rough upper limit for cumulative sequestration potential from such activities. While the estimate of 650 $GtCO_2$ (including wetlands and biochar application) by Lal et al. (2018) may seem unrealistic against these simple benchmarks, synergies from codeployment of NETs, for example with enhanced silicate weathering, are still to be taken into account and, with the *terra pretta* as an inspiration, the possibility of optimally engineered soil amendments could well take soil carbon stocks beyond preindustrial levels (Amman and Hartmann, 2019; Novak et al., 2009).

Global potential for TEW on agricultural soils has been estimated to be up to 95 $GtCO_2$/yr for dunite application at a cost of \$60/tCO_2 (Strefler et al., 2018), which would require the extraction of feedstocks on a similar scale to the global aggregate and raw material mining industries which together extract ~70 Gt-rock/yr (Beerling et al., 2018). Almost 75%

TABLE 11.11 Global potential and costs of land-based negative emissions.

Negative emissions method	Potential 2050 ($GtCO_2$/yr)	Potential 2100 ($GtCO_2$/yr)	Cumulative 2100 ($GtCO_2$)	Unit CO_2 removal costs ($/$tCO_2$)	References and notes
Afforestation and reforestation	0.5–3.6	1–12	80–260	5–50	Fuss et al. (2018)
	4.9		410	200	Doelmann et al. (2020)
			750		Bastin et al. (2019); 900 Mha, AR plus restoration
	2.7–17.9 (2030)				Griscom et al. (2017); 680 Mha
Avoided deforestation plus regrowth			400		Houghton and Nassikas (2018)
			55–108 (2050)	20–50	Houghton and Nassikas (2018)
	3.0–4.2 (2030)				Griscom et al. (2017)
Forest management	1.6–10 (2030)				Griscom et al. (2017), 1900 Mha
Soil carbon sequestration	2–5			0–100	Fuss et al. (2018)
	4.4–6.9 (2030)				Griscom et al. (2017); agricultural and grassland
		2.6–4.8 (max)		Negative	Smith (2016)
		5.5–12.5 (max)	195–420		Lal et al. (2018); excl. wetlands and biochar
		1.5–5.1 (max)	30–100		Sanderman et al. (2017)
Biochar	0.5–2			30–120	Fuss et al. (2018)
	0.6–1.5 (2030)				Griscom et al. (2017); 1.7 Gt/yr crop residues
	1–1.8		60–130		Woolf et al. (2010)
		1.6–3.4 (max)	95–205		Lal et al. (2018)
Enhanced weathering	2–4			20–200	Fuss et al. (2018)
		0.5–4			Beerling et al. (2018)
		0.2–1.0 (max)			Smith (2016)
		5–95 (max)		60–200	Strefler et al. (2018)
Wetlands	2.4–4.5 (2030)				Griscom et al. (2017); avoided and restoration
		1.28–3.9 (max)	130–260		Lal et al. (2018)
	0.8 (2060)	1.5			NASEM 2019; management/restoration of VCH's

Notes: Dates in brackets indicate alternative horizon for potentials and cumulatives.

of this potential could be realized in India, Brazil, South-East Asia, and China, where warm humid conditions favor rapid dissolution. However, the achievable potential for TEW still has very large uncertainties, particularly at a national/regional level, since many of the processes influencing weathering in soils are still the subject of ongoing research, and are yet to be incorporated into Earth system models.

These estimates summarized in Table 11.11 suggest that many land-based methods have the potential to deliver negative emissions on the multi-GtCO_2 scale by 2050, with 10 GtCO_2/yr being well within the range of estimates for the sum of forest-related approaches. Due to limited land availability, these figures are not simply additive, although codeployment opportunities such as TEW or biochar application together with biomass production would help to alleviate this constraint (Amann and Hartmann, 2019) as would improvements in biomass yields (food and biomass crops, and forests) through improved nutrient and water availability (de Oliveira Garcia et al., 2020).

References and resources

Akhand, A., Watanabe, K., Chanda, A., et al., 2021. Lateral carbon fluxes and CO_2 evasion from a subtropical mangrove-seagrass-coral continuum. Sci. Total Environ. 752, 142190.

Alcántara, V., Don, A., Well, R., Nieder, R., 2016. Deep ploughing increases agricultural soil organic matter stocks. Global Change Biol. 22 (8), 2939–2956.

Amann, T., Hartmann, J., 2019. Ideas and perspectives: synergies from co-deployment of negative emission technologies. Biogeosciences 16 (15), 2949–2960.

Amelse, J., Behrens, 2021. Sequestering biomass for natural, efficient, and low-cost direct air capture of carbon dioxide. Available at: https://ssrn.com/abstract=3861130.

Amelung, W., Bossio, D., de Vries, W., et al., 2020. Towards a global-scale soil climate mitigation strategy. Nat. Commun. 11 (1), 1–10.

Amiri, A., Ottelin, J., Sorvari, J., Junnila, S., 2020. Cities as carbon sinks—classification of wooden buildings. Environ. Res. Lett. 15 (9), 094076.

Andrews, G., Epihov, D., Pearce, C.R., James, R.H., Beerling, D.J., 2020. CO_2 sequestration by Enhanced Weathering of agricultural soils in Norfolk, UK. In: AGU Fall Meeting Abstracts, Vol. 2020, pp. B003–0003.

Andrews, M.G., Pearce, C.R., James, R.H., et al., 2021. Field trials of Enhanced Weathering in two contrasting climate zones. https://goldschmidtabstracts.info/2019/96.pdf.

Assmuth, A., Tahvonen, O., 2015. Continuous cover forestry vs. Clearcuts with optimal carbon storage. In: Paper Presented at BioEcon 2015, Cambridge, England, September 2015.

Atwood, T.B., Connolly, R.M., Ritchie, E.G., et al., 2015. Predators help protect carbon stocks in blue carbon ecosystems. Nat. Clim. Change 5 (12), 1038–1045.

Ayangbenro, A.S., Babalola, O.O., 2020. Reclamation of arid and semi-arid soils: the role of plant growth-promoting archaea and bacteria. Curr. Plant Biol. 25, 100173.

Azzi, E.S., Karltun, E., Sundberg, C., 2021. Assessing the diverse environmental effects of biochar systems: an evaluation framework. J. Environ. Manag. 286, 112154.

Bastin, J.F., Finegold, Y., Garcia, C., et al., 2019. The global tree restoration potential. Science 365 (6448), 76–79.

Baumann, P., Lee, J., Behrens, T., et al., 2022. Modelling soil water retention and water-holding capacity with visible–near infrared spectra and machine learning. Eur. J. Soil Sci. e13220.

Bayraktarov, E., Saunders, M.I., Abdullah, S., et al., 2016. The cost and feasibility of marine coastal restoration. Ecol. Appl. 26 (4), 1055–1074.

Becker, K., Wulfmeyer, V., Berger, T., Gebel, J., Münch, W., 2013. Carbon farming in hot, dry coastal areas: an option for climate change mitigation. Earth System Dyna. 4 (2), 237–251.

Beerling, D.J., Leake, J.R., Long, S.P., et al., 2018. Farming with crops and rocks to address global climate, food and soil security. Nature Plants 4 (3), 138–147.

Bellassen, V., Luyssaert, S., 2014. Carbon sequestration: managing forests in uncertain times. Nature News 506 (7487), 153.

Bennett, P.C., Rogers, J.R., Choi, W.J., Hiebert, F.K., 2001. Silicates, silicate weathering, and microbial ecology. Geomicrobiol. J. 18 (1), 3–19.

Betts, R.A., 2000. Offset of the potential carbon sink from boreal forestation by decreases in surface albedo. Nature 408 (6809), 187–190.

Bispo, A., Andersen, L., Angers, D.A., et al., 2017. Accounting for carbon stocks in soils and measuring GHGs emission fluxes from soils: do we have the necessary standards? Front. Environ. Sci. 5, 41.

Blanco-Canqui, H., 2021. Does biochar improve all soil ecosystem services? Bioener 13 (2), 291–304.

Bonan, G.B., 1997. Effects of land use on the climate of the United States. Climatic Change 37 (3), 449–486.

Branch, O., Wulfmeyer, V., 2019. Deliberate enhancement of rainfall using desert plantations. Proc. Natl. Acad. Sci. USA 116 (38), 18841–18847.

Brovkin, V., Ganapolski, A., Claussen, M., Kubatzki, C., Petoukhov, V., 1999. Modeling response to historical land cover change. Global Ecol. Biogeogr. 8, 509–517.

Burrell, L.D., Zehetner, F., Rampazzo, N., Wimmer, B., Soja, G., 2016. Long-term effects of biochar on soil physical properties. Geoderma 282, 96–102.

Campbell, E.E., Paustian, K., 2015. Current developments in soil organic matter modeling and the expansion of model applications: a review. Environ. Res. Lett. 10 (12), 123004.

Canadell, J.G., Raupach, M.R., 2008. Managing forests for climate change mitigation. Science 320 (5882), 1456–1457.

Chabrillat, S., Ben-Dor, E., Cierniewski, J., Gomez, C., Schmid, T., van Wesemael, B., 2019. Imaging spectroscopy for soil mapping and monitoring. Surv. Geophys. 40 (3), 361–399.

Chagas, J.K.M., de Figueiredo, C.C., Ramos, M.L.G., 2022. Biochar increases soil carbon pools: evidence from a global meta-analysis. J. Environ. Manag. 305, 114403.

Charney, J.G., 1975. Dynamics of deserts and drought in the Sahel. Q. J. R. Meteorol. Soc. 101 (428), 193–202.

Chenu, C., Angers, D.A., Barré, P., Derrien, D., Arrouays, D., Balesdent, J., 2019. Increasing organic stocks in agricultural soils: knowledge gaps and potential innovations. Soil Tillage Res. 188, 41–52.

Churchman, G.J., Noble, A., Bailey, G., Chittleborough, D., Harper, R., 2014. Clay addition and redistribution to enhance carbon sequestration in soils. In: Soil Carbon. Springer, Cham, Switzerland, pp. 327–335.

Claussen, M., Brovkin, V., Ganopolski, A., 2001. Biogeophysical versus biogeochemical feedbacks of large-scale land cover change. Geophys. Res. Lett. 28 (6), 1011–1014.

Corbera, E., Friedli, C., 2012. Planting trees through the clean development mechanism: a critical assessment. Ephemera 12 (1/2), 206–241.

de Oliveira Garcia, W., Amann, T., Hartmann, J., et al., 2020. Impacts of enhanced weathering on biomass production for negative emission technologies and soil hydrology. Biogeosciences 17 (7), 2107–2133.

Doelman, J.C., Stehfest, E., van Vuuren, D.P., et al., 2020. Afforestation for climate change mitigation: potentials, risks and trade-offs. Global Change Biol. 26 (3), 1576–1591.

Doyle, J.D., Stotzky, G., McClung, G., Hendricks, C.W., 1995. Effects of genetically engineered microorganisms on microbial populations and processes in natural habitats. Adv. Appl. Microbiol. 40, 237–287.

Duarte, C.M., Middelburg, J.J., Caraco, N., 2005. Major role of marine vegetation on the oceanic carbon cycle. Biogeosciences 2 (1), 1–8.

Duarte, C.M., Losada, I.J., Hendriks, I.E., Mazarrasa, I., Marbà, N., 2013. The role of coastal plant communities for climate change mitigation and adaptation. Nat. Clim. Change 3 (11), 961–968.

Duchelle, A.E., Simonet, G., Sunderlin, W.D., Wunder, S., 2018. What is REDD+ achieving on the ground? Curr. Opin. Environ. Sustain. 32, 134–140.

Dufour, A., 2013. Geological sequestration of biomass char to mitigate climate change. Environ. Sci. Technol. 47 (18), 10106–10107.

Dynarski, K.A., Bossio, D.A., Scow, K.M., 2020. Dynamic stability of soil carbon: reassessing the "permanence" of soil carbon sequestration. Front. Environ. Sci. 8, 218.

Dyson, F.J., 1977. Can we control the carbon dioxide in the atmosphere? Energy 2 (3), 287–291.

El-Naggar, A., Lee, S.S., Rinklebe, J., et al., 2019. Biochar application to low fertility soils: a review of current status, and future prospects. Geoderma 337, 536–554.

Elevitch, C.R., Mazaroli, D.N., Ragone, D., 2018. Agroforestry standards for regenerative agriculture. Sustainability 10 (9), 3337.

Fagan, M.E., Reid, J.L., Holland, M.B., Drew, J.G., Zahawi, R.A., 2020. How feasible are global forest restoration commitments? Conservation Letters e12700.

FAO, 2020. Global Forest Resources Assessment 2020—Key Findings. FAO, Rome.

FAO and UNEP, 2020. The State of the World's Forests 2020. Forests, Biodiversity and People. FAO and UNEP, Rome.

Feng, Q., An, C., Chen, Z., Wang, Z., 2020. Can deep tillage enhance carbon sequestration in soils? A meta-analysis towards GHG mitigation and sustainable agricultural management. Renew. Sustain. Energy Rev. 133, 110293.

Fenner, N., Freeman, C., 2011. Drought-induced carbon loss in peatlands. Nat. Geosci. 4 (12), 895–900.

Ferraz, A., Saatchi, S., Xu, L., et al., 2018. Carbon storage potential in degraded forests of Kalimantan, Indonesia. Environ. Res. Lett. 13 (9), 095001.

Flipkens, G., Blust, R., Town, R.M., 2021. Deriving nickel (Ni (II)) and chromium (Cr (III)) based environmentally safe olivine guidelines for coastal enhanced silicate weathering. Environ. Sci. Technol. 55 (18), 12362–12371.

Freeman, C., Fenner, N., Shirsat, A.H., 2012. Peatland geoengineering: an alternative approach to terrestrial carbon sequestration. Phil. Trans. Math. Phys. Eng. Sci. 370 (1974), 4404–4421.

Friedlingstein, P., O'sullivan, M., Jones, M., et al., 2022. Global carbon budget 2022. Earth Syst. Sci. Data 14 (11), 4811–4900.

Fuss, S., Lamb, W.F., Callaghan, M.W., et al., 2018. Negative emissions—part 2: costs, potentials and side effects. Environ. Res. Lett. 13 (6), 063002.

Gibbard, S., Caldeira, K., Bala, G., Phillips, T.J., Wickett, M., 2005. Climate effects of global land cover change. Geophys. Res. Lett. 32 (23).

Gillman, G.P., Burkett, D.C., Coventry, R.J., 2002. Amending highly weathered soils with finely ground basalt rock. Appl. Geochem. 17 (8), 987–1001.

Gleason, C.J., Im, J., 2012. Forest biomass estimation from airborne LiDAR data using machine learning approaches. Rem. Sens. Environ. 125, 80–91.

Grassi, G., Pilli, R., House, J., Federici, S., Kurz, W.A., 2018. Science-based approach for credible accounting of mitigation in managed forests. Carbon Bal. Manag. 13 (1), 8.

Griscom, B.W., Adams, J., Ellis, P.W., et al., 2017. Natural climate solutions. Proc. Natl. Acad. Sci. USA 114 (44), 11645–11650.

Hall, D.O., Mynick, H.E., Williams, R.H., 1991. Alternative roles for biomass in coping with greenhouse warming. Sci. Global Secur. 2 (2–3), 113–151.

Hangx, S.J.T., Spiers, C.J., 2009. Coastal spreading of olivine to control atmospheric CO_2 concentrations: a critical analysis of viability. Int. J. Greenh. Gas Control 3 (6), 757–767.

Haque, F., Santos, R.M., Chiang, Y.W., 2020. CO_2 sequestration by wollastonite-amended agricultural soils—An Ontario field study. Int. J. Greenh. Gas Control 97, 103017.

Hartemink, A.E., Gerzabek, M.H., Lal, R., McSweeney, K., 2014. Soil carbon research priorities. In: Soil Carbon. Springer, Cham, Switzerland, pp. 483–490.

Hartmann, J., Jansen, N., Dürr, H.H., Kempe, S., Köhler, P., 2009. Global CO_2-consumption by chemical weathering: what is the contribution of highly active weathering regions? Global Planet. Change 69 (4), 185–194.

Hawkins, H.J., Venter, Z.S., Cramer, M.D., 2022. A holistic view of Holistic Management: what do farm-scale, carbon, and social studies tell us? Agric. Ecosyst. Environ. 323, 107702.

Hiraishi, T., Krug, T., Tanabe, K., Srivastava, N., Baasansuren, J., Fukuda, M., Troxler, T.G., 2014. 2013 Supplement to the 2006 IPCC Guidelines for National Greenhouse Gas Inventories: Wetlands. IPCC, Switzerland.

Hof, A.R., Dymond, C.C., Mladenoff, D.J., 2017. Climate change mitigation through adaptation: the effectiveness of forest diversification by novel tree planting regimes. Ecosphere 8 (11), e01981.

Houghton, R.A., Nassikas, A.A., 2018. Negative emissions from stopping deforestation and forest degradation, globally. Global Change Biol. 24 (1), 350–359.

Houghton, R.A., Byers, B., Nassikas, A.A., 2015. A role for tropical forests in stabilizing atmospheric CO_2. Nat. Clim. Change 5 (12), 1022.

Howard, J.L., Creed, J.C., Aguiar, M.V., Fouqurean, J.W., 2018. CO_2 released by carbonate sediment production in some coastal areas may offset the benefits of seagrass "Blue Carbon" storage. Limnol. Oceanogr. 63 (1), 160–172.

IPCC, 2001. In: Metz, B., Davidson, O., Swart, R., Pan, J. (Eds.), Climate Change 2001: Mitigation. Contribution of Working Group III to the Third Assessment Report of the Intergovernmental Panel on Climate Change, Vol 3. Cambridge University Press, Cambridge, UK.

IPCC, 2006. In: Eggleston, H.S., Buendia, L., Miwa, K., Ngara, T., Tanabe, K. (Eds.), Agriculture, Forestry and Other Land Use. IPCC Guidelines for National Greenhouse Gas Inventories, vol. 4. IGES, Japan.

Jansson, J.K., Hofmockel, K.S., 2020. Soil microbiomes and climate change. Nat. Rev. Microbiol. 18 (1), 35–46.

Jansson, C., Wullschleger, S.D., Kalluri, U.C., Tuskan, G.A., 2010. Phytosequestration: carbon biosequestration by plants and the prospects of genetic engineering. Bioscience 60 (9), 685–696.

Kallenbach, C.M., Grandy, A.S., Frey, S.D., Diefendorf, A.F., 2015. Microbial physiology and necromass regulate agricultural soil carbon accumulation. Soil Biol. Biochem. 91, 279–290.

Kareksela, S., Haapalehto, T., Juutinen, R., et al., 2015. Fighting carbon loss of degraded peatlands by jump-starting ecosystem functioning with ecological restoration. Sci. Total Environ. 537, 268–276.

Kelemen, P.B., McQueen, N., Wilcox, J., Renforth, P., Dipple, G., Vankeuren, A.P., 2020. Engineered carbon mineralization in ultramafic rocks for CO_2 removal from air: Review and new insights. Chem. Geol. 550, 119628.

Kell, D.B., 2012. Large-scale sequestration of atmospheric carbon via plant roots in natural and agricultural ecosystems: why and how. Phil. Trans. Biol. Sci. 367 (1595), 1589–1597.

Kemena, T.P., Matthes, K., Martin, T., Wahl, S., Oschlies, A., 2017. Atmospheric feedbacks in North Africa from an irrigated, afforested Sahara. Clim. Dynam. 1–21.

Khosravi, A., Zheng, H., Liu, Q., Hashemi, M., Tang, Y., Xing, B., 2022. Production and characterization of hydrochars and their application in soil improvement and environmental remediation. Chem. Eng. J. 430, 133142.

Köhler, P., Abrams, J.F., Völker, C., Hauck, J., Wolf-Gladrow, D.A., 2013. Geoengineering impact of open ocean dissolution of olivine on atmospheric CO_2, surface ocean pH and marine biology. Environ. Res. Lett. 8 (1), 014009.

Krause-Jensen, D., Duarte, C.M., 2016. Substantial role of macroalgae in marine carbon sequestration. Nat. Geosci. 9 (10), 737–742.

Krug, J.H., 2018. Accounting of GHG emissions and removals from forest management: a long road from Kyoto to Paris. Carbon Bal. Manag. 13 (1), 1.

Lal, R., 2020. Managing soils for negative feedback to climate change and positive impact on food and nutritional security. Soil Sci. Plant Nutr. 66 (1), 1–9.

Lal, R., Negassa, W., Lorenz, K., 2015. Carbon sequestration in soil. Curr. Opin. Environ. Sustain. 15 (79–86).

Lal, R., 2018. Digging deeper: a holistic perspective of factors affecting soil organic carbon sequestration in agroecosystems. Global Change Biol. 24 (8), 3285–3301.

Lal, R., Smith, P., Jungkunst, H.F., Mitsch, W.J., et al., 2018. The carbon sequestration potential of terrestrial ecosystems. J. Soil Water Conserv. 73 (6), 145A–152A.

Lal, R., Monger, C., Nave, L., Smith, P., 2021. The role of soil in regulation of climate. Philos. Trans. Royal Soc. B 376 (1834), 20210084.

Lee, X., Goulden, M.L., Hollinger, D.Y., et al., 2011. Observed increase in local cooling effect of deforestation at higher latitudes. Nature 479 (7373), 384–387.

Lehmann, J., Gaunt, J., Rondon, M., 2006. Bio-char sequestration in terrestrial ecosystems—a review. Mitig. Adapt. Strategies Glob. Change 11 (2), 403–427.

Lehmann, J., Rillig, M.C., Thies, J., Masiello, C.A., Hockaday, W.C., Crowley, D., 2011. Biochar effects on soil biota—a review. Soil Biol. Biochem. 43 (9), 1812–1836.

Li, W., Ciais, P., Peng, S., et al., 2017. Land-use and land-cover change carbon emissions between 1901 and 2012 constrained by biomass observations. Biogeosciences 14 (22), 5053–5067.

Li, Y., Hu, S., Chen, J., et al., 2018. Effects of biochar application in forest ecosystems on soil properties and greenhouse gas emissions: a review. J. Soils Sediments 18 (2), 546–563.

Li, L., Yao, Z., You, S., Wang, C.H., Chong, C., Wang, X., 2019. Optimal design of negative emission hybrid renewable energy systems with biochar production. Appl. Energy 243, 233–249.

Li, X., Zhang, M., Long, J., Lin, H., 2021. A novel method for estimating spatial distribution of forest above-ground biomass based on multispectral fusion data and ensemble learning algorithm. Rem. Sens. 13 (19), 3910.

Liang, J., Crowther, T.W., Picard, N., et al., 2016. Positive biodiversity-productivity relationship predominant in global forests. Science 354 (6309) [aaf8957].

Macreadie, P.I., Nielsen, D.A., Kelleway, J.J., et al., 2017. Can we manage coastal ecosystems to sequester more blue carbon? Front. Ecol. Environ. 15 (4), 206–213.

Madrigal-González, J., Calatayud, J., Ballesteros-Cánovas, J.A., et al., 2020. Climate reverses directionality in the richness–abundance relationship across the World's main forest biomes. Nat. Commun. 11 (1), 1–7.

Maher, D.T., Call, M., Santos, I.R., Sanders, C.J., 2018. Beyond burial: lateral exchange is a significant atmospheric carbon sink in mangrove forests. Biol. Lett. 14 (7), 20180200.

Marland, G., Marland, S., 1992. Should we store carbon in trees?. In: Natural Sinks of CO_2. Springer, Dordrecht, pp. 181–195.

Mayer, M., Prescott, C.E., Abaker, W.E.A., Augusto, L., Cécillon, L., Ferreira, G.W.D., et al., 2020. Influence of forest management activities on soil organic carbon stocks: a knowledge synthesis. For. Ecol. Manag. 466, 118127.

McKinley, D.C., Ryan, M.G., Birdsey, R.A., Giardina, C.P., Harmon, M.E., Heath, L.S., et al., 2011. A synthesis of current knowledge on forests and carbon storage in the United States. Ecol. Appl. 21 (6), 1902–1924.

Mcleod, E., Chmura, G.L., Bouillon, S., et al., 2011. A blueprint for blue carbon: toward an improved understanding of the role of vegetated coastal habitats in sequestering CO_2. Front. Ecol. Environ. 9 (10), 552–560.

Metzger, R.A., Benford, G., 2001. Sequestering of atmospheric carbon through permanent disposal of crop residue. Clim. Change 49 (1), 11–19.

Meysman, F.J., Montserrat, F., 2017. Negative CO_2 emissions via enhanced silicate weathering in coastal environments. Biol. Lett. 13 (4), 20160905.

Miller, L.A., Orton, P.M., 2021. Achieving negative emissions through oceanic sequestration of vegetation carbon as Black Pellets. Clim. Change 167 (3), 1–23.

Minasny, B., Malone, B.P., McBratney, A.B., 2017. Soil carbon 4 per mille. Geoderma 292, 59–86.

Moreau, C., Montserrat, F., Green, T., et al., 2021. Progress toward small-scale field trials of coastal enhanced weathering of olivine. In: Goldschmidt2021 Virtual, 4–9 July.

Naeem, S., Duffy, J.E., Zavaleta, E., 2012. The functions of biological diversity in an age of extinction. Science 336 (6087), 1401–1406.

Nayak, A.K., Rahman, M.M., Naidu, R., et al., 2019. Current and emerging methodologies for estimating carbon sequestration in agricultural soils: a review. Sci. Total Environ. 665, 890–912.

Needelman, B.A., Emmer, I.M., Emmett-Mattox, S., et al., 2018. The science and policy of the verified carbon standard methodology for tidal wetland and seagrass restoration. Estuar. Coast 41 (8), 2159–2171.

Neklyudov, A.D., Fedotov, G.N., Ivankin, A.N., 2008. Intensification of composting processes by aerobic microorganisms: a review. Appl. Biochem. Microbiol. 44 (1), 6–18.

Nguyen, T.H., Jones, S., Soto-Berelov, M., Haywood, A., Hislop, S., 2020. Landsat time-series for estimating forest aboveground biomass and its dynamics across space and time: A review. Rem. Sens. 12 (1), 98.

Novak, J.M., Lima, I., Xing, B., et al., 2009. Characterization of designer biochar produced at different temperatures and their effects on a loamy sand. Ann. Environ. Sci 3 (2), 195–206.

O'Connor, D., Ford, J., 2014. Increasing the effectiveness of the "Great Green Wall" as an adaptation to the effects of climate change and desertification in the Sahel. Sustainability 6 (10), 7142–7154.

Ontl, T.A., et al., 2020. Forest management for carbon sequestration and climate adaptation. J. For. 118 (1), 86–101.

Oreska, M.P., McGlathery, K.J., Aoki, L.R., Berger, A.C., Berg, P., Mullins, L., 2020. The greenhouse gas offset potential from seagrass restoration. Sci. Rep. 10 (1), 1–15.

Ornstein, L., Aleinov, I., Rind, D., 2009. Irrigated afforestation of the Sahara and Australian Outback to end global warming. Clim. Change 97 (3), 409–437.

O'Callaghan, M., 2016. Microbial inoculation of seed for improved crop performance: issues and opportunities. Appl. Microbiol. Biotechnol. 100 (13), 5729–5746.

Peters, S.C., Blum, J.D., Driscoll, C.T., Likens, G.E., 2004. Dissolution of wollastonite during the experimental manipulation of hubbard Brook watershed 1. Biogeochemistry 67, 309–329.

Pham, T.D., Xia, J., Ha, N.T., Bui, D.T., Le, N.N., Tekeuchi, W., 2019. A review of remote sensing approaches for monitoring blue carbon ecosystems: Mangroves, seagrassesand salt marshes during 2010–2018. Sensors 19 (8), 1933.

Pommerening, A., Brill, M., Schmidt-Kraepelin, U., Haufe, J., 2020. Democratising forest management: applying multiwinner approval voting to tree selection. For. Ecol. Manag. 478, 118509.

Poulton, P., Johnston, J., Macdonald, A., White, R., Powlson, D., 2018. Major limitations to achieving "4 per 1000" increases in soil organic carbon stock in temperate regions: evidence from long-term experiments at Rothamsted Research, United Kingdom. Global Change Biol. 24 (6), 2563–2584.

Powell, S.L., Cohen, W.B., Healey, S.P., et al., 2010. Quantification of live aboveground forest biomass dynamics with Landsat time-series and field inventory data: A comparison of empirical modeling approaches. Rem. Sens. Environ. 114 (5), 1053–1068.

Pröll, T., Al Afif, R., Schaffer, S., Pfeifer, C., 2017. Reduced local emissions and long-term carbon storage through pyrolysis of agricultural waste and application of pyrolysis char for soil improvement. Energy Proc. 114, 6057–6066.

Raghoebarsing, A.A., Smolders, A.J., Schmid, M.C., et al., 2005. Methanotrophic symbionts provide carbon for photosynthesis in peat bogs. Nature 436 (7054), 1153–1156.

Renforth, P., Campbell, J.S., 2021. The role of soils in the regulation of ocean acidification. Philos. Trans. R. Soc. B 376 (1834), 20200174.

Renforth, P., von Strandmann, P.P., Henderson, G.M., 2015. The dissolution of olivine added to soil: Implications for enhanced weathering. Appl. Geochem. 61, 109–118.

Richards, G.P., Evans, D.M., 2004. Development of a carbon accounting model (FullCAM Vers. 1.0) for the Australian continent. Aust. For. 67 (4), 277–283.

Roberts, K.G., Gloy, B.A., Joseph, S., Scott, N.R., Lehmann, J., 2010. Life cycle assessment of biochar systems: estimating the energetic, economic, and climate change potential. Environ. Sci. Technol. 44 (2), 827–833.

Rockström, J., Williams, J., Daily, G., et al., 2017. Sustainable intensification of agriculture for human prosperity and global sustainability. Ambio 46 (1), 4–17.

Romaniello, S., Stopnitzky, S., Green, T., et al., 2021. Progress towards small-scale field trials of coastal enhanced weathering of olivine. In: EGU General Assembly 2021, Online, 19–30 Apr 2021, pp. EGU21–14112. https://doi.org/10.5194/egusphere-egu21-14112.

Rosentreter, J.A., Maher, D.T., Erler, D.V., Murray, R.H., Eyre, B.D., 2018. Methane emissions partially offset "blue carbon" burial in mangroves. Sci. Adv. 4 (6), eaao4985.

Saderne, V., Geraldi, N.R., Macreadie, P.I., et al., 2019. Role of carbonate burial in Blue Carbon budgets. Nat. Commun. 10 (1), 1–9.

Sandalow, D., Aines, R., Friedmann, J., McCormick, C., Sanchez, D.L., 2021. Biomass Carbon Removal and Storage (BiRCS) Roadmap (No. LLNL-TR-815200). Lawrence Livermore National Lab.(LLNL), Livermore, CA (United States).

Sanderman, J., Hengl, T., Fiske, G.J., 2017. Soil carbon debt of 12,000 years of human land use. Proc. Natl. Acad. Sci. USA 114 (36), 9575–9580.

Sapkota, Y., White, J.R., 2020. Carbon offset market methodologies applicable for coastal wetland restoration and conservation in the United States: a review. Sci. Total Environ. 701, 134497.

Savory, A., Butterfield, J., 2016. Holistic Management: A Commonsense Revolution to Restore Our Environment. Island Press, Washington, DC, USA.

Scholz, F., Hasse, U., 2008. Permanent wood sequestration: the solution to the global carbon dioxide problem. ChemSusChem: Chem. Sus. Energy Mater. 1 (5), 381–384.

Schreefel, L., Schulte, R.P.O., de Boer, I.J.M., Schrijver, A.P., van Zanten, H.H.E., 2020. Regenerative agriculture–the soil is the base. Global Food Secur. 26, 100404.

Schuiling, R.D., Krijgsman, P., 2006. Enhanced weathering: an effective and cheap tool to sequester CO2. Clim. Change 74 (1), 349–354.

Shao, S., Driscoll, C.T., Johnson, C.E., Fahey, T.J., Battles, J.J., Blum, J.D., 2015. Long-term responses in soil solution and stream-water chemistry at Hubbard Brook after experimental addition of wollastonite. Environ. Chem. 13 (3), 528–540.

Sharma, S.K., Ramesh, A., Sharma, M.P., et al., 2010. Microbial community structure and diversity as indicators for evaluating soil quality. In: Biodiversity, Biofuels, Agroforestry and Conservation Agriculture. Springer, Dordrecht, pp. 317–358.

Shuman, J.K., Shugart, H.H., O'Halloran, T.L., 2011. Sensitivity of Siberian larch forests to climate change. Global Change Biol. 17 (7), 2370–2384.

Smet, I., Evangelou, E., Tsadilas, C., et al., 2021. Field trials of enhanced weathering combined with cotton farming in Thessaly, Greece. In: Goldschmidt2021 Virtual 4–9 July.

Smith, P., 2016. Soil carbon sequestration and biochar as negative emission technologies. Global Change Biol. 22 (3), 1315–1324.

Smith, P., Martino, D., Cai, Z., et al., 2008. Greenhouse gas mitigation in agriculture. Phil. Trans. Biol. Sci. 363 (1492), 789–813.

Smith, P., Soussana, J.F., Angers, D., et al., 2020. How to measure, report and verify soil carbon change to realize the potential of soil carbon sequestration for atmospheric greenhouse gas removal. Global Change Biol. 26 (1), 219–241.

Solomon, T., Moon, H., Abebe, S., Minale, A.S., Teketay, D., 2020. Promoting bamboo-based agroforestry for enhancing ecosystem services from degraded lands. In: Agroforestry for Degraded Landscapes. Springer, Singapore, pp. 423–444.

Sombroek, W.I.M., Ruivo, M.D.L., Fearnside, P.M., Glaser, B., Lehmann, J., 2003. Amazonian dark earths as carbon stores and sinks. In: Amazonian Dark Earths. Springer, Dordrecht, pp. 125–139.

Strand, S.E., Benford, G., 2009. Ocean sequestration of crop residue carbon: recycling fossil fuel carbon back to deep sediments. Environ. Sci. Technol. 43 (4), 1000–1007.

Strefler, J., Amann, T., Bauer, N., Kriegler, E., Hartmann, J., 2018. Potential and costs of carbon dioxide removal by enhanced weathering of rocks. Environ. Res. Lett. 13 (3), 034010.

Suhrhoff, T.J., 2022. Phytoprevention of heavy metal contamination from terrestrial enhanced weathering: can plants save the day. Front. Climate 3, 820204.

Taylor, L.L., Beerling, D.J., Quegan, S., Banwart, S.A., 2017. Simulating carbon capture by enhanced weathering with croplands: an overview of key processes highlighting areas of future model development. Biol. Lett. 13 (4), 20160868.

Taylor, L.L., Driscoll, C.T., Groffman, P.M., Rau, G.H., Blum, J.D., Beerling, D.J., 2021. Increased carbon capture by a silicate-treated forested watershed affected by acid deposition. Biogeosciences 18 (1), 169–188.

Van Dam, B.R., Zeller, M.A., Lopes, C., et al., 2021. Calcification-driven CO_2 emissions exceed "Blue Carbon" sequestration in a carbonate seagrass meadow. Sci. Adv. 7 (51), eabj1372.

Veldman, J.W., Aleman, J.C., Alvarado, S.T., et al., 2019. Comment on "The global tree restoration potential". Science 366 (6463), eaay7976.

Vicca, S., Goll, D.S., Hagens, M., et al., 2022. Is the climate change mitigation effect of enhanced silicate weathering governed by biological processes? Global Change Biol. 28 (3), 711–726.

Villa, J.A., Bernal, B., 2018. Carbon sequestration in wetlands, from science to practice: an overview of the biogeochemical process, measurement methods, and policy framework. Ecol. Eng. 114, 115–128.

Waddington, J.M., Day, S.M., 2007. Methane emissions from a peatland following restoration. J. Geophys. Res.: Biogeosciences 112 (G3).

Wang, H., Richardson, C.J., Ho, M., 2015. Dual controls on carbon loss during drought in peatlands. Nat. Clim. Change 5 (6), 584–587.

Wang, B., Gao, B., Fang, J., 2017. Recent advances in engineered biochar productions and applications. Crit. Rev. Environ. Sci. Technol. 47 (22), 2158–2207.

Wang, F., Pan, X., Gerlein-Safdi, C., et al., 2020. Vegetation restoration in Northern China: a contrasted picture. Land Degrad. Dev. 31 (6), 669–676.
Weng, Z., Van Zwieten, L., Singh, B., et al., 2017. Biochar built soil carbon over a decade by stabilizing rhizodeposits. Nat. Clim. Change 7 (5), 371–376.
West, T.A., Börner, J., Sills, E.O., Kontoleon, A., 2020. Overstated carbon emission reductions from voluntary REDD+ projects in the Brazilian Amazon. Proc. Natl. Acad. Sci. USA 117 (39), p24188–p24194.
Wiesmeier, M., Urbanski, L., Hobley, E., et al., 2019. Soil organic carbon storage as a key function of soils—a review of drivers and indicators at various scales. Geoderma 333, 149–162.
Woolf, D., Amonette, J.E., Street-Perrott, F.A., Lehmann, J., Joseph, S., 2010. Sustainable biochar to mitigate global climate change. Nat. Commun. 1 (1), 1–9.
Wu, P., Cui, P.X., Fang, G.D., et al., 2018. Biochar decreased the bioavailability of Zn to rice and wheat grains: insights from microscopic to macroscopic scales. Sci. Total Environ. 621, 160–167.
Xu, L., Deng, X., Ying, J., Zhou, G., Shi, Y., 2020. Silicate fertilizer application reduces soil greenhouse gas emissions in a Moso bamboo forest. Sci. Total Environ. 747, 141380.
Yosef, G., Walko, R., Avisar, R., Tatarinov, F., Rotenberg, E., Yakir, D., 2018. Large-scale semi-arid afforestation can enhance precipitation and carbon sequestration potential. Sci. Rep. 8 (1), 996.
Yuen, J.Q., Fung, T., Ziegler, A.D., 2017. Carbon stocks in bamboo ecosystems worldwide: estimates and uncertainties. For. Ecol. Manag. 393, 113–138.
Zeng, N., King, A.W., Zaitchik, B., Wullschleger, S.D., Gregg, J., Wang, S., Kirk-Davidoff, D., 2013. Carbon sequestration via wood harvest and storage: an assessment of its harvest potential. Clim. Change 118 (2), 245–257.
Zheng, H., Wang, X., Luo, X., Wang, Z., Xing, B., 2018. Biochar-induced negative carbon mineralization priming effects in a coastal wetland soil: roles of soil aggregation and microbial modulation. Sci. Total Environ. 610, 951–960.
Zhou, Y., Williams, C.A., Hasler, N., Gu, H., Kennedy, R., 2021. Beyond biomass to carbon fluxes: application and evaluation of a comprehensive forest carbon monitoring system. Environ. Res. Lett. 16 (5), 055026.

Further reading

Busch, J., Engelmann, J., Cook-Patton, S.C., et al., 2019. Potential for low-cost carbon dioxide removal through tropical reforestation. Nat. Clim. Change 9, 463–466.
Emmer, I.M., Needelman, B.A., Emmett-Mattox, S., et al., 2015. Methodology for tidal wetland and seagrass restoration. In: VCS Methodology VM0033 (Verified Carbon Standard, Washington, D.C).
Lal, R., 2018. Digging deeper: a holistic perspective of factors affecting soil organic carbon sequestration in agroecosystems. Global Change Biol. 24 (8), 3285–3301.
Agnoletti, M., Dargavel, J., Johann, E., 2009. History of forestry. In: The Role of Food, Agriculture, Forestry and Fisheries in Human Nutrition, Vol. 2, pp. 1–28.
Batzer, D.P., Sharitz, R.R. (Eds.), 2014. Ecology of Freshwater and Estuarine Wetlands. University of California Press.
Biochar Journal; disseminating innovative research on economical and environmentally responsible uses for biochar: www.biochar-journal.org.
Brancalion, P.H., Holl, K.D., 2020. Guidance for successful tree planting initiatives. J. Appl. Ecol. 57 (12), 2349–2361.
Carbon 180. Forest Carbon Removal. Available at: https://carbon180.org/deep-dives.
Carbon Cycle Institute, 2020. Carbon farming. Available online at: https://www.carboncycle.org/carbon-farming/. (Accessed 26 June 2020). accessed.
Goffner, D., Sinare, H., Gordon, L.J., 2019. The great green wall for the Sahara and the Sahel initiative as an opportunity to enhance resilience in sahelian landscapes and livelihoods. Reg. Environ. Change 19 (5), 1417–1428.
Hartemink, A.E., McSweeney, K. (Eds.), 2014. Soil Carbon. Springer Science & Business Media.
Joseph, S., Cowie, A.L., Van Zwieten, L., et al., 2021. How biochar works, and when it doesn't: a review of mechanisms controlling soil and plant responses to biochar. GCB Bioener. 13 (11), 1731–1764.
Lorenz, K., Lal, R., 2009. Carbon Sequestration in Forest Ecosystems. Springer Science & Business Media, Cham, Switzerland.
Lorenz, K., Lal, R., 2018. Carbon Sequestration in Agricultural Ecosystems. Springer Science & Business Media, Cham, Switzerland.
McDonald, T., Gann, G.D., Jonson, J., Dixon, K.W., 2016. International Standards for the Practice of Ecological Restoration–Including Principles and Key Concepts. Society for Ecological Restoration, Washington, DC, USA.
McDonald, H., Frelih-Larsen, A., Lóránt, A., et al., 2021. Carbon farming—making agriculture fit for 2030. In: Study for the Committee on Environment, Public Health and Food Safety (ENVI). Policy Department for Economic, Scientific and Quality of Life Policies, European Parliament, Luxembourg.
Mbow, C., 2017. The great green wall in the Sahel. In: Oxford Research Encyclopedia of Climate Science.
Nellemann, C., Corcoran, E., Duarte, C.M., et al. (Eds.), 2010. Blue Carbon: The Role of Healthy Oceans in Binding Carbon. UNEP/GRID-Grid Arendal, Rapid Response Assessments Series.
Ok, Y.S., Tsang, D.C., Bolan, N., Novak, J.M. (Eds.), 2019. Biochar from Biomass and Waste: Fundamentals and Applications. Elsevier, Amsterdam, Netherlands.
Paustian, K., Lehmann, J., Ogle, S., Reay, D., Robertson, G.P., Smith, P., 2016. Climate-smart soils. Nature 532 (7597), 49–57.
Schmidt, H.P., Anca-Couce, A., Hagemann, N., et al., 2019. Pyrogenic carbon capture and storage. GCB Bioenergy 11 (4), 573–591.
Schwarzer, S., 2019. The Potential of Carbon Sequestration in the Soil. United Nations Environment Programme (Foresight Brief, 13).
Stafford, R., Chamberlain, B., Clavey, L., et al. (Eds.), 2021. Nature-based Solutions for Climate Change in the UK: A Report by the British Ecological Society. London, UK.

Projects and organizations

4-per-1000; International initiative aiming to increase carbon stocks in agricultural soils: https://4p1000.org.

AFR100; The African Forest Landscape Restoration Initiative: https://afr100.org/.

Blue Carbon Initiative; thebluecarboninitiative.org.

Bonn Challenge; Global effort to address land degradation and deforestation: www.bonnchallenge.org.

FAO; Global Forest Resources Assessments: www.fao.org/forest-resources-assessment/current-assessment/maps-and-figures GSOCseq programme:.

Forest Carbon Monitoring project; developing reliable and cost-efficient monitoring and accounting platform for forest carbon stocks: www.forestcarbonplatform.org.

Forest Stewardship Council; FSC Forest Carbon Monitoring Tool: https://fsc.org/en/for-forests/ecosystem-services/ecosystem-services-for-forest-managers.

Forest trends; Interactive database of forest projects: www.forest-trends.org.

Global Soil Sequestration Potential (GSOCseq) Map: www.fao.org/soils-portal/data-hub.

Global Soils Partnership: www.fao.org/global-soil-partnership.

Global Mangrove Alliance; Global community committed to protecting mangroves and securing their environmental role: www.mangrovealliance.org.

Great Green Wall; African-led movement to grow an 8,000 km "natural wonder of the world" across the Sahel: www.greatgreenwall.org/about-great-green-wall.

Holistic Management International (HMI); US not-for-profit organization promoting holistic management in agriculture: www.holisticmanagement.org.

RECSOIL: Recarbonization of global agricultural soils: www.fao.org/global-soil-partnership/areas-of-work/recarbonization-of-global-soils.

ICRLP; Institute for Carbon Removal Law and Policy, Biochar: An ICRLP explainer: https://www.youtube.com/watch?v=_DD6uZf4pyo, Enhanced weathering explainer: https://www.youtube.com/watch?v=p5RfvduMlzY.

Intergovernmental Technical Panel on Soils: www.fao.org/global-soil-partnership/itps.

International Biochar Institute; fostering good industry practices, and environmental and ethical standards to support safe and economic biochar systems; biochar-international.org.

IPCC, 2006. Guidelines for National Greenhouse Gas Inventories. www.ipcc-nggip.iges.or.jp/public/2006gl/.

IUCN; Focus area on building sustainable forest landscapes, protecting primary forests and advancing the rights of forest communities: www.iucn.org/theme/forests.

Mapping Ocean Wealth; data and decision support for mangrove restoration projects: oc https://oceanwealth.org/applications/mangrove-restoration/.

New York Declaration on Forests (NYDF); multi-stakeholder framework for forest action: forestdeclaration.org/.

Project Carbdown; progressing TEW field trials: www.carbon-drawdown.de/blog/2021-01-14-introducing-project-carbdown.

Project Vesta; Research driven start-up progressing coastal enhanced weathering: vesta.earth.

Trillion Tree Campaign; focussing on forest protection and restoration: www.trilliontrees.org.

UN Decade on Ecosystem Restoration; preventing, halting, and reverseing ecosystem degradation worldwide: www.decadeonrestoration.org.

US Biochar Initiative; a US not-for-profit organization promoting the sustainable production and use of biochar through research, policy, and technology: biochar-us.org .

WEF Trillion Tree Initiative; platform to support the global reforestation community: www.1t.org.

Chapter 12

Bioenergy with carbon capture and storage

Diarmaid Clery and Steve Rackley

12.1 Introduction

The photosynthetic production of biomass, discussed in Chapter 5, is the starting point for an increasingly wide range of approaches to achieving negative emissions. The incorporation of large-scale carbon dioxide removal (CDR) into integrated assessment models (IAMs), starting with IPCC's fifth assessment report (AR5; IPCC, 2014), encouraged an initial focus on bioenergy production with carbon capture and storage (CCS), but the field has since broadened to include such diverse options as the deep-sea burial of macroalgae or the injection of bio-oil into depleted hydrocarbon reservoirs. These methods can be collectively termed biomass carbon capture (or removal) and storage (BioCCS/BioCRS), or bioenergy with carbon capture and storage (BECCS) for the subset of methods that include bioenergy or biofuel production. All of these methods generate net negative emissions through the long-term storage of photosynthetically captured carbon that would otherwise largely return to the atmosphere as a result of biomass decomposition.

BECCS broadly refers to the conversion of biomass feedstocks through a range of processes including combustion, gasification, and fermentation, to produce thermal energy or biofuels, together with the capture and permanent storage of the CO_2 produced during the conversion process. This may involve the intermediate production of biofuels—energy-rich materials derived from plant, microalgal, or other microbial biomass, or produced from engineered or enhanced biological processes—that offer a potentially carbon-negative alternative to fossil fuels.

The combination of bioenergy with carbon capture and storage was first proposed in 1999 as a way of achieving negative carbon emissions (Rao, 1999), and is a key component of the NET portfolio (Keith, 2001; Obersteiner et al., 2001). Mitigation scenarios from IAMs give a central role to BECCS, particularly in scenarios that aim to meet the more stringent Paris goal of limiting global mean temperature rise to less than 1.5°C (Rogelj et al., 2015). While the current and projected costs of BECCS when compared to direct air capture (DAC; the other front-running NET discussed in Chapter 13) are clearly primary drivers, many other factors contribute to a narrower view of the portfolio than is likely to emerge, including land limitations, and other implementation challenges, and diverse energy system transition requirements, as well as social, environmental and political factors (Gambhir et al., 2019; Clery et al., 2021). These factors will define the real-world potential of BECCS and are not currently included in IAM scenarios (Fridahl, 2018), which could lead to an overreliance on negative emissions in the future and reduced demand for CO_2 emissions mitigation in the near term (Anderson and Peters, 2016; Stoddard et al., 2021).

The thermal energy produced in the combustion of biomass or biofuels may be used for power generation, including combined heat and power (CHP), but also has potential industrial applications in the production of pulp and paper, cement, and iron and steel (Möllersten et al., 2003; Mandova et al., 2019). The market niche that has enabled early commercial-scale deployment for BECCS has been one such application—in bioethanol production—and these early projects, discussed below, have also benefited from the sale of captured CO_2 for enhanced oil recovery (EOR); the full range of geological storage options is discussed in Chapter 15.

The use of sustainable biomass is critical to the successful implementation of BECCS as a negative emissions technology (Röder et al., 2015), and its effectiveness can be impacted by various project-specific conditions such as water usage, fertilizer type, and transport options. There are also trade-offs associated with BECCS facilities, such as land use, which will vary depending on the output prioritized by the facility, for example energy generation or CO_2 removal

Negative Emissions Technologies for Climate Change Mitigation. https://doi.org/10.1016/B978-0-12-819663-2.00017-4

(Fajardy et al., 2018; Smith et al., 2016). Dietary changes and other societal choices will also play an important part in determining the availability of land for biomass growth (Vaughan et al., 2018).

12.2 Biomass sources

The term "biomass" covers a variety of organic matter that can be used to generate useful energy, including trees, dedicated crops, trimmings, and organic wastes. Woody biomass is currently preferred for large-scale applications, although other types of biomass are an option for BECCS applications in the future. Herbaceous biomass (e.g., miscanthus and switchgrass) is faster growing and thus typically requires less land for energy production purposes, potentially making it the most sustainable biomass option (Fajardy and MacDowell, 2017).

"Biofuels" are energy-rich gases, liquids, or solids derived from biomass or generated through biological processes (Rodionova et al., 2017), and are classified according to the scheme shown in Table 12.1. The traditional use of wood, plant material, and animal waste are examples of primary biofuels.

12.2.1 Residual products from forestry and agriculture

Biomass residues are the waste products from biomass resources, including forestry off-cuts, sawmill residues, and pits or husks from fruit or cereal processing. Residues from the timber industry (forest bark, thinnings, branches, and sawdust) are currently used to supply the United Kingdom's Drax power station, which is planning to implement large-scale BECCS by 2030 (more detail in Section 12.6.2). The supply of biomass residues is linked to demand for primary products (e.g., timber, cereals, or fruit) but estimates suggest that residues could meet 7%–50% of global bioenergy demand by 2050, with residual waste products providing a cheaper alternative to bioenergy crops (Hanssen et al., 2020a).

The use of biomass residues provides the opportunity for bioenergy production with reduced land requirement and trade-offs compared to crops grown specifically for bioenergy, and CCS can be added to existing biomass residue applications, such as the pulp and paper industry, to achieve negative emissions (Santos et al., 2021). However, depending on the type of biomass residue being utilized, supply can be seasonal (Panoutsou et al., 2017), or impact on soil quality for future biomass growth (Thiffault et al., 2011).

12.2.2 Terrestrial bioenergy crops

Dedicated energy crops with a short rotation period (2/3 years) can be grown for BECCS, but will be in competition with agricultural land that could be used for food or farming unless grown on degraded or marginal land (The Royal Society, 2018; Mehmood et al., 2017). Crops such as miscanthus, wheat straw, and willow short rotation coppice (SRC) show promise for use in future BECCS applications due to their rapid growth, low water footprint (Fajardy and Mac Dowell, 2017; Freer et al., 2022), and potential benefits to biodiversity (Vanbeveren and Ceulemans, 2019).

Table 12.2 shows typical yields of bioenergy crops, where median values are used and the bracketed range shows the 10th and 90th percentiles. Various factors will influence the amount of biomass dry matter (DM) produced by a crop, such as species, location, soil, and environmental conditions, leading to the large ranges shown in the table, which are derived from field trials and measurements of globally grown lignocellulosic bioenergy crops (Li et al., 2018).

TABLE 12.1 Classification of biofuels.

Category	Generation	Examples
Primary		Combustion of woody or cellulosic plant material, or dry animal waste.
Secondary	First	Bioethanol produced from starchy food crops or crop residues, or biodiesel derived from waste animal fats.
	Second	Bioethanol produced from nonfood cellulosic biomass, or biodiesel derived from oil-rich plant seeds such as jatropha, palm, or soy.
	Third	Biofuels generated from micro- and macroalgae, cyanobacteria, and other microbial biomass.

TABLE 12.2 Typical of bioenergy crop yields.

Biomass type	Typical biomass yield (tonnes DM/ha/yr)
Miscanthus	11.5 (1–25)
Switchgrass	8.1 (3–16)
Willow	8.9 (3–14)
Poplar	7.2 (2–13)
Eucalypt	14.6 (5–27)

Source: Based on Li et al. (2018).

12.2.3 Aquatic and marine biomass sources

Aquatic biomass includes microalgae, macroalgae (seaweeds, kelp), and marine plants, and is responsible for approximately half of the oxygen released to the atmosphere through photosynthesis. Using aquatic biomass can dramatically reduce the water and land requirements associated with BECCS (Beal et al., 2018), although there are still many biological, economic, and technical constraints to producing large quantities of aquatic biomass.

Marine microalgae

Microalgae are unicellular, aquatic, photosynthetic plant-like organisms, part of the broader family of phytoplankton that play a significant part in marine organic carbon storage by using CO_2 dissolved in seawater as the carbon source for photosynthesis. Current research suggests that microalgae store 28 times more carbon than macroalgae (Raven, 2018). In the ocean, algae play an important role in transporting carbon from the euphotic zone to the ocean floor for biosequestration (Sengupta et al. 2017; see also Chapters 6 and 16). Microalgae have high photosynthetic efficiencies compared to terrestrial crops (Raheem et al., 2018) and can be used to convert CO_2 captured from point source emissions or directly from the atmosphere (Moreira and Pires, 2016).

To extract maximal value from microalgal biomass, cultivation is being optimized and processing technologies, such as hydrothermal liquefaction, are being used to produce fuels and high-value products (Pierobon et al., 2018). Third-generation biofuels (Table 12.1) can be derived from microalgae (as well as fungi and bacteria) in a number of ways, including;

- direct conversion of dried biomass by combustion or gasification,
- dark or photo fermentation of biomass carbohydrates by bacteria to produce biohydrogen, ethanol, and butanol, or
- lipid extraction and transesterification to produce biodiesel.

Ongoing research aims to improve biofuel yields through the genetic engineering of metabolic and other processes. Technological advances will be needed to achieve a positive energy balance in microalgae cultivation for biofuels (Slade and Bauen, 2013; Badar et al., 2018).

Microalgae can be grown in wastewaters or sea water, with growth requiring nutrients, light, and a carbon source—such as captured CO_2—with 1.83 $kgCO_2$ being required for the production of 1 kg of microalgae (Chisti, 2007). Nutrients, including phosphorous, nitrogen, iron, and sulfur, would typically be provided using fertilizers, although these contribute significantly to the cost of production.

Marine macroalgae

Macroalgae are large, aquatic, photosynthetic organisms, that also play a significant role in marine organic carbon storage. Most species of macroalgae are believed to have increased rates of photosynthesis and growth, in response to the increase in dissolved inorganic carbon, and increasing acidity, in the ocean during the industrial era (Koch et al., 2013), as shown from in situ experiments at natural carbon dioxide seeps (Celis-Pla et al., 2015).

Crude algae oil is an oil that can be extracted from algae, which is chemically similar to fossil crude oil. This is typically achieved using one of three methods: oil press, hexane extraction, or supercritical CO_2 extraction (Hannon et al., 2010), although these technologies are relatively expensive compared to fossil crude oil extraction. Alternatively, the cultivation of macroalgae for subsequent conversion, for example by anaerobic digestion to produce biogas, could be a more economically viable renewable energy source (Hughes et al., 2012).

Ocean macroalgal afforestation (OMA) can achieve negative emissions when combined with CCS (N'Yeurt et al., 2012), although many environmental, economic, and technical challenges remain as barriers to the large-scale implementation of macroalgae processing (Milledge and Harvey, 2016; Arenas and Vas-Pinto, 2014; Mac Monagail et al., 2017).

12.2.4 Municipal solid wastes

Municipal solid waste (MSW) is categorized as a form of biomass, although the negative emissions that can be delivered by burning such wastes depend on the ratio of biogenic to nonbiogenic material present. With significant biogenic content (or organic fraction), negative emissions can be achieved at an energy-from-waste (EfW) facility with CCS, where MSW is combusted to generate heat and electricity with CO_2 being captured and permanently stored.

12.2.5 Properties and composition of biomass

The most important properties of biomass from a bioenergy perspective are energy density and moisture content. Moisture is essential for biomass growth, and moisture content typically ranges from 6% to 12%. Higher moisture content increases the weight and volume of biomass, which increases transportation and storage costs so that, where possible, biomass is dried before being transported.

The composition varies considerably between different biomass types, with plant biomass comprising three major components: cellulose, hemicellulose, and lignin, and is therefore often referred to as "lignocellulosic biomass." The main components that make up the bulk of plant-derived biomass are shown schematically in Fig. 12.1. Cellulose and hemicellulose are also major components of marine macroalgal biomass, with lignin being absent as it is not required for structural support.

Cellulose, hemicelluloses (also known as polyoses), and lignin are the main constituents of the cell walls of plants, as shown in Fig. 12.2. Cellulose ($(C_6H_{10}O_5)_n$) and hemicelluloses are polysaccharides—polymers of sugars produced during photosynthesis (Chapter 5)—with cellulose consisting of hundreds to thousands of glucose ($C_6H_{12}O_6$) units. Hemicelluloses are a broad class of heteropolymers (copolymers) consisting of glucose and other sugars, of which xylose and mannose are the most common in shorter, heterogeneously structured molecules with a lower degree of polymerization compared to cellulose.

Lignins are a similarly broad class of heterogeneous copolymers, built up from the cross-linking of three alcohol components (lignols)—p-coumaryl alcohol, coniferyl alcohol, and sinapyl alcohol (Fig. 12.3), which provide structural strength in plants, particularly in wood, by bonding together the cellulose and hemicellulose fibers to produce lignocellulose.

Extractives are lower molecular weight, largely nonpolymeric substances including oil, protein, starch, and sugars that can be recovered by treatments such as solvent extraction.

12.3 Biomass conversion technologies

The high moisture content and low energy density of raw biomass are poor fuel characteristics, and pretreatments such as torrefaction and pyrolysis (see below) can be used to overcome this limitation and generate improved intermediate products for combustion, gasification, or biofuel production.

The main pathways for biomass conversion to energy are illustrated in Fig. 12.4, the three main conversion routes are thermochemical, biochemical, and physicochemical. The last two, biochemical conversion (microbial digestion, hydrolysis, and fermentation) and physicochemical conversion (lipid extraction and esterification), provide important routes to carbon-neutral biofuels, such as biodiesel, bioethanol, biomethanol, and other bio-oils (Stafford et al., 2017; Brown, 2019).

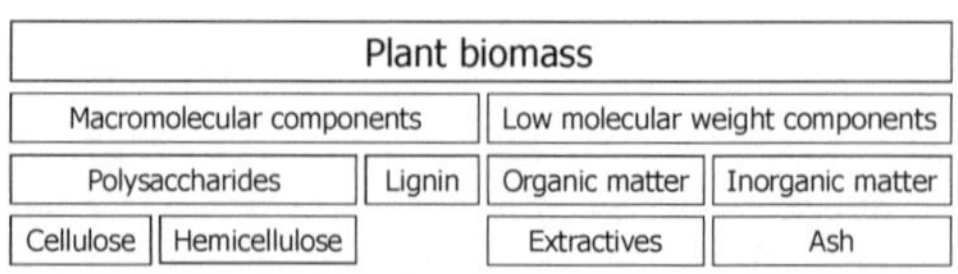

FIGURE 12.1 Main components of plant derived biomass. *After Mohan et al. (2006).*

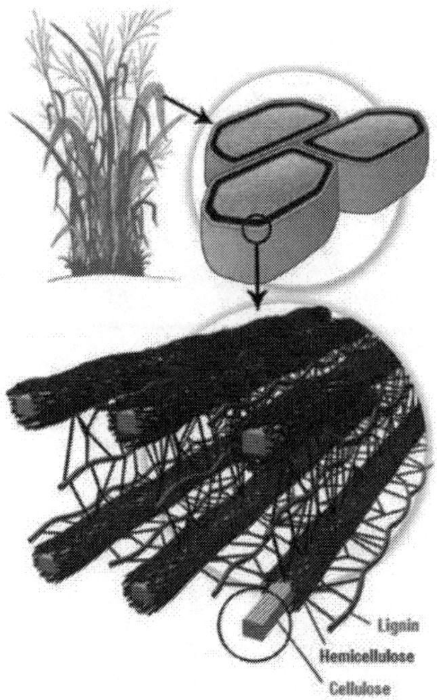

FIGURE 12.2 Spatial representation of cellulose, hemicellulose, and lignin in plant biomass. *Source: U.S. Department of Energy Genomic Science program, https://genomicscience.energy.gov.*

Biomass can also be used to produce biofuels through thermochemical processes (such as pyrolysis and gasification) or alternatively can be combusted to generate heat and electricity, with negative emissions achieved by applying CCS to the biomass conversion process.

12.3.1 Thermochemical decomposition of biomass

Biomass placed in a high-temperature environment will undergo different stages of thermal decomposition depending on the environmental conditions, leading to three main processes of thermochemical decomposition: pyrolysis, combustion, and gasification.

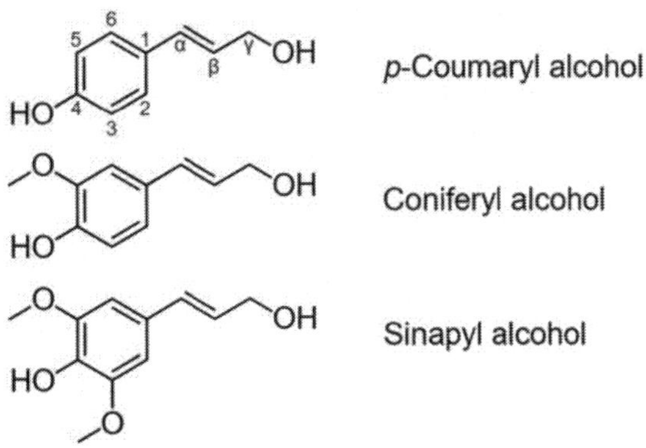

FIGURE 12.3 The three lignol components of lignin.

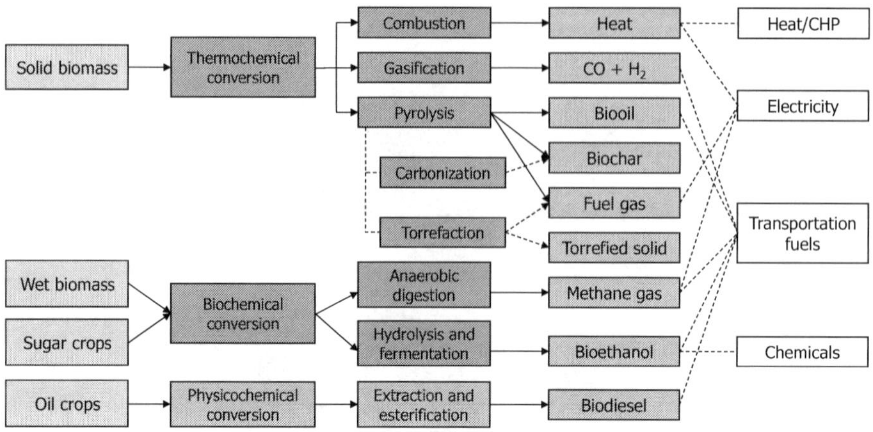

FIGURE 12.4 Biomass conversion pathways. *After Gent et al. (2017).*

Moisture is first released when a biomass particle is heated, with the other components being released as volatiles at various stages, or remaining in the solid fraction, depending on the conditions (Saddawi et al., 2010).

Biomass torrefaction and pyrolysis

As noted earlier, raw biomass typically has high volume and relatively low energy density, and its high moisture content and hygroscopic nature can lead to microbial degradation, as well as to difficulty grinding it into the small particles required in some applications. These drawbacks can be overcome through pretreatments such as torrefaction and pyrolysis—closely related thermal processes that decompose biomass into solid, liquid, and gaseous products in an inert or near-inert atmosphere to prevent or limit combustion. The processes are distinguished by the maximum temperature and the rate of heating, which leads to different degrees of decomposition and to different products.

Torrefaction takes place at a lower temperature, and typically involves slow heating over a period of several minutes to a maximum temperature in the range from 200 to 300°C, which is maintained for up to an hour. The aim of torrefaction—also known as mild pyrolysis—is to improve the properties of low-density, low-energy fibrous biomass feedstock such as straw and cotton waste, to produce a dry product with a higher energy content that is better suited to grinding and pelletization for subsequent combustion or gasification (Prins et al., 2006).

Initial heating to ~200°C removes all traces of moisture to produce a dry product, and further heating in the range from 200 to 270°C, corresponding to mild torrefaction, results in the onset of thermal decomposition, with the release of volatile products such as acetic acid, methanol, CO, and CO_2. From 270 to 300°C further decomposition releases additional gaseous products, and a temperature of around 300°C marks the onset of the release of tars from the thermal cracking of cellulose. This upper-temperature limit prevents energy loss through tar production and also minimizes the thermal decomposition of lignin, which is a natural binder that aids pelletization (Bergman, 2005).

Further slow heating results in carbonization in the temperature range from 400 to 500°C, with progressive and eventually complete release of tars and the production of chars—a range of solid products with high carbon and low to zero hydrocarbon content. Traditional charcoal production is one example of this carbonization process, with the released tars being combusted in an oxygen-limited atmosphere to heat the process. Chars have a wide variety of uses including as charcoal fuel, as a reducing agent to facilitate metal extraction from ores, as activated carbon with many environmental applications including as a sorbent for gas separation (Chapter 8), and as a soil amendment—commonly known as biochar—for carbon sequestration or soil remediation (Chapter 11).

In contrast to torrefaction, which is essentially a pretreatment before combustion or gasification, the aim of pyrolysis is to convert biomass to other products, principally biocrude that can be upgraded to produce liquid biofuels. The nature of the initial product depends principally on the rate of heating and the temperature reached, and the range of these operating conditions is summarized in Table 12.3. Residence time here refers to the time that released vapors are kept at the pyrolysis temperature; short residence times and rapid cooling (quenching) are necessary to maximize liquid yield, ensuring condensation occurs before breaking up the long polymeric chains and functional groups associated with cellulose and hemicellulose to produce volatile matter and char (Loo and Koppejan, 2008).

TABLE 12.3 **Characteristics and products of the range of pyrolysis processes.**

Final temp (°C)				
Heating rate	400	400–500	500–700	>700
Very high (>1000°C/s)			Fast pyrolysis 0.5–5 s residence Oil	Ultra pyrolysis <0.5 s residence Gas, chemicals
High		Hydropyrolysis <10 s residence Oil	Flash pyrolysis <1 s residence Oil, gas, chemicals	Methanopyrolysis <10 s residence Chemicals
Medium	Vacuum pyrolysis 2–30 s residence Oil			
Low (<100°C/min)			Conventional pyrolysis 5–30 min residence Oil, gas, char	

After Mohan et al. (2006).

Pyrolysis has received considerable research interest due to its ability to convert biomass directly into solid, liquid, and gaseous products, and pyrogenic carbon capture and storage (PyCCS) has wide potential application as a negative emissions technology (Schmidt et al., 2019), although it is currently represented largely by the biochar application. Negative emissions can also be achieved by the sequestration of liquid pyrolysis products, an approach being developed by Charm Industrial with the injection of pyrogenic bio-crude into depleted oil fields and salt caverns.

Biomass combustion

As a biomass particle combusts, it will undergo three main processes: moisture loss, devolatilization, and char combustion. Once enough moisture has been released the biomass particle can start to release volatile organic matter, which reacts with oxygen resulting in a visible flame (Mason et al., 2015). The temperature of the particle starts to increase during this stage, known as volatile combustion, resulting in the breakdown of chemical components and release of volatile compounds. Up to 80% of biomass composition is volatile organic matter, resulting in the majority of total heat output coming from this stage (Jenkins and Baxter, 1998). Volatile combustion begins when the surface of the particle reaches roughly 200°C. Hemicellulose decomposes first in the 200–260°C range, to form formaldehyde, acetic acid, CO, H_2, furfural (C_4H_3OCHO), and furan (C_4H_4O). Cellulose decomposes in the 240–350°C range, to produce hydroxylacetaldehyde, acetol, furfural, and CO as volatiles. Lignin is the last to decompose in the 280–500°C range. Lignin produces mostly char from devolatilization but also volatile aromatics are released (Desideri and Fantozzi, 2013).

Char combustion occurs after volatile combustion around the particle at temperatures in the 400–1000°C region. When the devolatilization is complete, oxygen can pass the surface of the particle and oxidize the char matter that remains, with CO and CO_2 diffusing out of the particle. Char combustion is typically the slowest stage of combustion and therefore the rate-determining step in the overall combustion process. After all carbon in the char has reacted with oxygen char combustion is complete, and only ash remains.

In large-scale combustion facilities, all combustion products with the exception of ash are vented to the atmosphere as flue gas, which passes through various forms of flue gas treatment processes to remove undesired components. Carbon capture from these flue gases is required to deliver negative emissions from BECCS, for which there are three main options: precombustion capture, postcombustion capture, and oxyfuel combustion.

Briefly, precombustion capture involves decarbonizing the biomass before combustion, typically through biomass gasification to produce a syngas mostly comprising hydrogen and carbon dioxide. CO_2 is then separated from this syngas leaving H_2 as a fuel for clean combustion. Postcombustion capture involves the separation of CO_2 from the flue gas products created during combustion, while oxyfuel combustion involves reengineering the combustion process to produce a sufficiently pure CO_2 product stream for subsequent utilization or storage, eliminating the need for a CO_2 separation stage.

These approaches are illustrated schematically in Fig. 12.5.

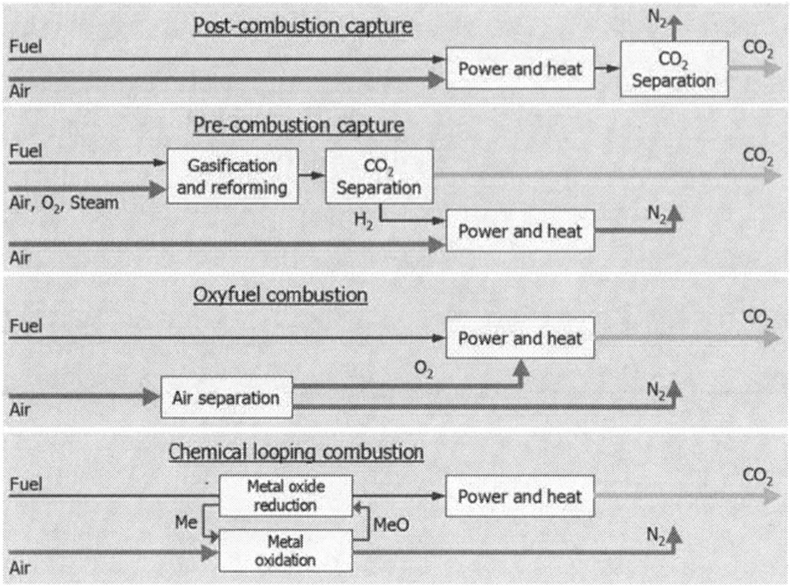

FIGURE 12.5 CO_2 capture options from biomass combustion for power generation.

As shown in the figure, the pre- and postcombustion approaches both require technologies to separate CO_2 from a gas mixture comprising $CO_2 + H_2$ or $CO_2 + N_2$, respectively. For oxyfuel combustion, the oxygen supply can be achieved either through a separation of O_2 from air ($O_2 + N_2 +$ trace gases) or by the delivery of oxygen to the combustion process in the form of a solid oxide (chemical looping). Some advantages and disadvantages of these options are summarized in Table 12.4.

The absorption, adsorption, chemical looping, and membrane technologies that can be applied in these approaches are discussed in Chapters 7–9, and the application of these technologies is discussed below. Biomass fuels are fundamentally different from coal and natural gas, and the differing composition of trace elements in the flue gas will lead to different challenges for these capture methods, such as different solvent degradation characteristics in postcombustion capture (Finney et al., 2019; Clery et al., 2021).

TABLE 12.4 Advantages and disadvantages of carbon capture options.

Capture option	Advantages	Disadvantages
Precombustion	Lower energy requirements for CO_2 capture and compression; fully developed technology, commercially deployed at the required scale in other industries	Temperature and efficiency issues associated with hydrogen-rich gas turbine fuel
Postcombustion	Fully developed technology, commercially deployed at the required scale in other industries	High parasitic power requirement for solvent regeneration; low CO_2 partial pressure impacts capture efficiency
	Opportunity for retrofit to existing power plants	High capital and operating costs for current absorption systems
Oxyfuel combustion	Mature air separation technologies available; very high CO_2 partial pressure simplifies capture process	Costly and energy-intensive air separation step; significant plant impact makes retrofit unattractive
Chemical looping combustion	Very high CO_2 partical pressure simplifies capture process	Immature technology, currently under development

Biomass cofiring

Cofiring of biomass with coal in the power generation sector was initially motivated by benefits from fuel diversity and to reduce emissions, particularly of oxides of nitrogen (NOx) and trace metals. It became more prominent in the 1990s, in the wake of the Kyoto Protocol, as a way to reduce the CO_2 intensity ($kgCO_2$/MWhr) of coal-fired power generation, often in anticipation of legislation introducing cap-and-trade carbon markets.

Biomass may be introduced at the start of fuel processing so that the mixed coal and biomass fuel is handled using the existing processing and fuel injection systems, or by the injection into the furnace of a separately processed biomass feed, either through the existing coal burners or through additional biomass burners. The premixing option has the advantage of rapid implementation and low capital cost, and is now well established at many large cofired plants, although it is generally limited to 5%−10% biomass on a heat input basis due primarily to limitation at the comilling stage.

This limitation can be overcome by indirect cofiring, where the biomass is processed separately, for example, by torrefaction (to improve grindability) or gasification (with the gas used to part-fuel the furnace), or alternatively by the use of a separate biomass furnace (parallel cofiring). Although the modification of existing coal burners for dedicated biomass injection has also been successfully applied, direct injection into existing coal firing systems has generally been preferred as a simple and lower-cost option.

The main technical issues arising from biomass cofiring are increased fire-side corrosion of boiler materials, particularly with biomass feeds containing potassium and chlorine, reduction of the ash fusion temperature with the addition of biomass ash, which can lead to increased maintenance costs, and impact on ash handling as well as NOx and particulate control systems (Tillman, 2000), although aluminosilicate additives, such as coal ash, are found to be effective at reducing these impacts (Clery et al., 2018).

While wood chips or pellets are most common, many different types of feedstock have been used in cofired plants, including agricultural residues (from soy, corn, rice, manioc, wheat, cotton, beans, sugarcane, and kernels from oil palm, olive, and cacao), energy crops (short rotation coppice and miscanthus), paper, and sewage sludge. Mechanical densification is a common pretreatment for lignocellulosic biomass to produce pellets, thereby increasing energy density and reducing volume, transportation, and handling costs.

Fluidized bed combustion (FBC) has been extensively investigated as an alternative to cofiring in conventional pulverized coal boilers. In FBC, a bed of coarse fuel particles, typically milled to ~5−10 mm, along with other inert bed particles, commonly crushed limestone or dolomite, is fluidized by an upward-flowing stream of air or oxygen. If the gas flow rate is sufficiently high the particles will be suspended in the circulating gas flow, resulting in a circulating fluidized bed (CFB). At lower gas rates the bed will resemble a bubbling fluid, called a bubbling fluidized bed (BFB). Fig. 12.6 illustrates the configuration of BFB and CFB combustors.

Fluidized beds have a number of advantages and disadvantages when compared to traditional pulverized coal combustion (PCC), the main ones being summarized in Table 12.5.

Fluidized bed combustors cover a wide range of capacities: bubbling beds typically range up to 50 MW, while circulating beds are common in the 100 to 500 MW range, with some designs up to 800 MW. CFB circulating velocities are typically 4−5.5 m/s, and 50−100 kg of solids are recycled through the bed for each kg of fuel burned.

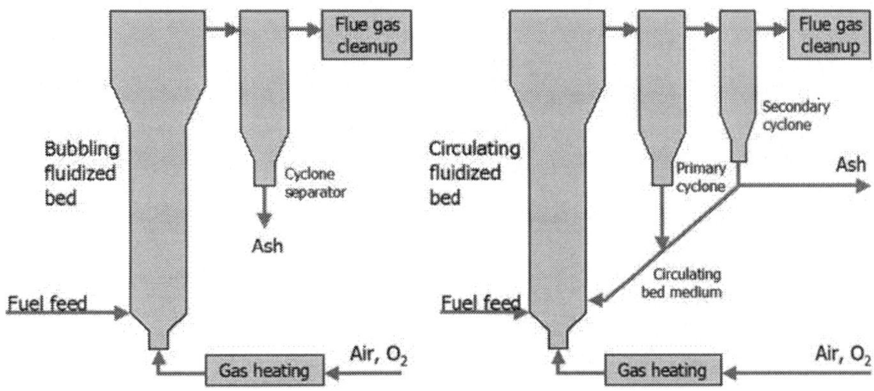

FIGURE 12.6 Fluidized bed combustion configurations.

TABLE 12.5 Main advantages and disadvantages of atmospheric fluidized bed combustion.

FBC advantages	FBC disadvantages
• Combustion at relatively low temperatures (800–900°C) with low NO_x formation	• Thermal efficiencies are ~2%–4% lower than conventional PCC due to higher fan power and heat losses from the cyclones, as well as heat loss from the removal of ash and sorbent (e.g., if used for SO_x control)
• Efficient heat transfer from the bed to immersed heat transfer surfaces	
• SO_x can be removed within the combustor at low cost by adding limestone particles to the bed	
• Fuel flexibility; FBC combustors can be designed to cater for a wide range of fuel types, including low-grade coals, biomass, and wastes	• In-bed heat transfer surfaces are subject to erosion
• High combustion efficiency due to turbulent mixing and longer fuel particle residence time	• N_2O formation is higher than in conventional PCC. Higher unburned fuel losses in CFBs, depending on fuel type

Pressurized bubbling and circulating fluidized beds (PBFB and PCFB), operating in the 1.0−1.5 MPa range, have also been developed and are similar to atmospheric systems except that the combustor and cyclones are housed in a pressure vessel. The advantage of pressurizing the bed is that the high-pressure combustion gas can be used directly to drive a gas turbine, significantly improving overall thermal efficiency. However, the system also becomes much more complicated due to the need to pressurize the fuel and air feeds, to depressurize the ash for removal, and possibly to filter the hot combustion gases to avoid erosion and deposition in the gas turbine system. Advanced PFBC combined cycle systems under development aim to achieve overall thermal efficiency well above 50%.

By 2017, there were around 230 cofired power and combined heat and power plants operating worldwide, with around two-thirds being located in Europe (Roni et al., 2017), largely as a result of the European Emissions Trading Scheme (ETS) and national policies supporting renewable energy and emissions reductions. However, as decarbonization of power generation continues, the progressive retirement of the coal-fired generation fleet means that cofiring will also decline and likely have ceased by mid-century, as BECCS deployment is peaking.

Postcombustion capture

In postcombustion capture, CO_2 is removed from the combustion reaction product steam—the flue gases—before emission to the atmosphere. Postcombustion capture is thus an extension of the common flue gas treatment for NO_x and SO_x removal, but with the need to remove higher quantities of CO_2 in the gas stream (typically 5%−15%, depending on the fuel being used).

Various technologies are available or under development to address the postcombustion separation of CO_2 (Table 12.6) the use of chemical solvents, such as MEA, being the most mature and widely deployed (Chapter 7).

TABLE 12.6 CO_2 separation technologies for postcombustion capture.

Technology area	Currently developed technologies	Example technologies under development
Absorption-based separation (Chapter 7)	Chemical solvents (e.g., MEA, chilled ammonia)	Novel solvents to improve performance (e.g., phase change solvents); improved design of processes and equipment
Adsorption-based separation (Chapter 8)	Zeolite and activated carbon molecular sieves	Carbonate sorbents; calcium/chemical looping
Membrane separation (Chapter 9)	Polymeric membranes	Immobilized liquid membranes; molten carbonate membranes
Cryogenic separation	CO_2 liquefaction	Hybrid cryogenic/membrane and hydrate-membrane processes

As well as technologies to address the gas separation challenge, several other aspects of the postcombustion process are also the focus of ongoing research and development, notably:

- Integration and optimization of the postcombustion capture process within the power plant or other industrial process
- Environmental impact, including life-cycle assessment (LCA), of the overall postcombustion capture process and of specific solvent use
- Procedures for optimal postcombustion process operation under varying plant conditions.

The world's first BECCS postcombustion capture pilot began capturing 1 tCO_2/d in February 2019 from a flue gas slipstream at the Drax Group's biomass-fired power plant in Selby, UK, as further discussed in Section 12.6.

Oxyfuel combustion

In oxyfuel combustion, the fuel is burned at ambient or elevated pressure using a combination of oxygen and recycled CO_2 rather than air to draft the combustion chamber, so that the flue gas is near-pure CO_2 rather than a mixture from which CO_2 would need to be separated. Oxygen may be delivered either as a gas stream, produced by the separation of O_2 from air or as a solid oxide in a chemical looping process (see below). The separation of oxygen from air for oxyfuel combustion can be achieved by a number of techniques, such as adsorption and membrane methods, although cryogenic systems are currently the most widely used for larger-scale applications (>200 tO_2/d).

Cryogenic air separation has been used for oxygen production since 1902, when the first air separation plant was constructed by Carl von Linde. This pioneering plant, with a capacity of 120 kgO_2/d, used Joule Thomson cooling together with countercurrent heat exchange to liquefy air and then separate oxygen and nitrogen in a distillation column. The process, as realized in a modern air separation unit (ASU), is illustrated in Fig. 12.7.

Filtered air is compressed, accompanied by heating, and cooled back to ambient temperature in a heat exchanger. Water vapor and CO_2 are removed after the first compression using molecular sieve units or absorption beds. A lower operating temperature for this air feed purification stage, achieved by additional cooling, improves CO_2 and water vapor removal and makes overall plant performance less sensitive to seasonal changes in ambient temperature. The cooled high-pressure air is fed through a countercurrent heat exchanger in which the returning end product from the cryogenic process is the heat sink. After this second cooling, the air is expanded through a valve and chilled to liquefaction temperature by Joule Thomson cooling.

The liquefied air is then fed into a tray or packed distillation column. Because nitrogen has a lower boiling point, the vapor above each tray will tend to be nitrogen-rich while the liquid will be oxygen-rich, the liquid cascading down the stack of trays while the vapor rises. A simple single-tower process yields a pure liquid oxygen bottom stream and a gaseous nitrogen top stream typically containing 5%–10% oxygen. An early enhancement of the technology was the addition of a second column (the Linde double column) to increase the purity of the nitrogen top stream.

As can be seen from Table 12.7, the operating temperature of the distillation tower will be in the range −196°C to −183°C, between the boiling points of nitrogen and oxygen. Helium and neon will not be liquefied in the process and will stay in the vapor phase with nitrogen, while krypton and xenon, with higher boiling points, will remain in the liquid phase

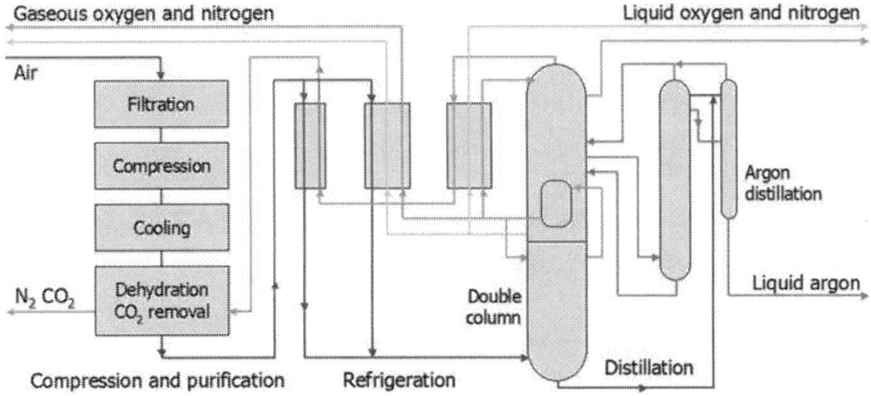

FIGURE 12.7 Air separation unit flow scheme.

TABLE 12.7 Composition of air and normal boiling points of its components.

Component	Boiling point (°C)	Volume %
Helium	−268.9	$5.0\ 10^{-4}$
Neon	−246.1	$1.8\ 10^{-3}$
Nitrogen	−195.8	78.1
Argon	−185.9	0.93
Oxygen	−183.0	20.9
Krypton	−153.2	$1.1\ 10^{-4}$
Xenon	−108.0	$9.0\ 10^{-6}$
Carbon dioxide	−78.5	$3.8\ 10^{-2}$

as impurities in the oxygen stream. Argon has a boiling point very similar to that of oxygen, and if a high-purity oxygen product is required (>95% pure) argon is removed at an intermediate point in the distillation column—the so-called argon belly—and can be separated and purified in additional distillation columns to yield a pure-argon product.

The principles of cryogenic air separation remain unchanged since Linde's time, but ASUs are now routinely constructed with single train capacities of ~7000 tO_2/d. The world's largest plant, at 30,000 tO_2/d, has been constructed by Linde AG for the Pearl Gas-to-Liquids project in Qatar using eight parallel 3750 tO_2/d trains.

Oxyfuel combustion has been developed as a clean coal technology since 2001, and a number of pilot plants at a 30 MW_{th} scale have been operated in Australia, China, and Europe, including the installation at the Schwarze Pumpe power plant in Germany (Liu and Zhang, 2018). Demonstration projects at a commercial scale (250−500 MW_e) were announced in China, Germany Korea, and the United Kingdom, although none of these has gone forward, in part due to national restrictions or resistance to further coal-fired power generation.

Oxyfuel combustion of biomass is less well advanced than postcombustion capture technologies, with bench-scale pilots up to 100 kW_{th} reported by a number of research groups (Kumar and Singh, 2016; Sher et al., 2018; Nguyen et al., 2020). Work to date has investigated cofiring biomass with coal as well as with sewage sludge (Kumar and Singh, 2017; Sung et al., 2018) in oxyfuel fluidized bed combustors.

Chemical looping combustion

Chemical looping combustion (CLC) is a form of oxyfuel combustion in which oxygen is introduced to the combustion reactor not as a gas but via a metal oxide carrier (Me_xO_y), which is either fully or partially reduced in the combustion reaction. The process is illustrated schematically in Fig. 12.8, applied to open cycle gas turbine power generation.

Considering carbon as the fuel, chemical looping combustion proceeds according to the reaction:

$$C_{(s)} + 2Me_xO_{y(s)} \rightarrow CO_{2(g)} + 2Me_xO_{y\text{-}1(s)} \tag{12.1}$$

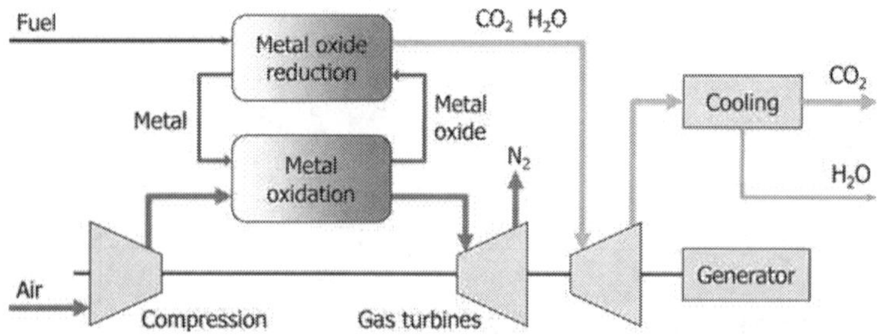

FIGURE 12.8 Chemical looping combustion applied in open cycle gas turbine generation.

This reaction may be applied either at high pressure in a gas turbine cycle or at atmospheric pressure in a steam cycle, while heat recovery from the CO_2 plus steam offgas from the fuel oxidation reactor can also be used to generate additional power. Depending on the fuel, steam can be condensed out of the offgas, leaving a high-purity CO_2 stream for further compression and storage.

Following this fuel oxidation/carrier reduction step, the spent carrier is circulated out of the combustion reactor into a carrier oxidation reactor and regenerated in the reaction:

$$2Me_xO_{y\text{-}1(s)} + O_{2(g)} \leftrightarrow 2Me_xO_{y(s)} \tag{12.2}$$

The metal oxidation reactor, which replaces the conventional gas turbine combustion chamber, captures oxygen directly from the air, eliminating the need for a separate air separation unit for oxygen supply. Oxides of the common transition metals (Fe, Cu, Ni, Mn, Ba) are possible carriers, with Fe_2O_3/Fe_2O_2, NiO/Ni, and BaO_2/BaO being extensively studied. The metal carrier oxidation step may be exothermic (for Cu, Mn, or Ba carriers) or endothermic (for Fe and Ni carriers), while the carrier reduction step is always exothermic.

In development work to date, the metal oxide carrier is typically in the form of particles with a diameter of 100–500 μm, and in some cases (e.g., Ni) it is carried on an alumina support. The reactions take place in two fluidized bed reactors, which achieve efficient heat and mass transfer. Recycling of solid carriers between the two reactors is an important control on the operating temperatures and overall heat balance between the two reactors as well as being a major technological challenge. For example, the BaO_2/BaO chemical loop requires 20 kgBaO_2/kgC, and applying this carrier to generate supercritical steam for a 500 MW_e power plant would require a carrier transport rate of 10^3 t/s (3.6 Mt/h).

In the case of biomass-derived fuels, solid-to-solid contact between oxygen carrier particles and fuel particles may not be sufficient to achieve the necessary reaction rates, so that a prior fuel gasification step or active bed particles to accelerate char gasification may be required. Alternatively, chemical looping with oxygen uncoupling (CLOU), a CLC variant in which gaseous oxygen is released from the metal oxide at high temperature, can be used for solid biomass-derived chars following torrefaction or pyrolysis pretreatments. A 1.5 kW_{th} CLOU proof-of-concept for solid fuel combustion was first reported in 2012 (Abad et al., 2012), and chemical looping combustion using biomass fuels has since been demonstrated in around 30 prototype facilities (recently reviewed by Lyngfelt et al., 2018 and Abuelgasim et al., 2020), the largest being a 4 MW_{th} demonstration reported by Vilches and colleagues, which used coarse wood pellets with illmentite ($FeTiO_3$) and manganese ore carriers (Vilches et al., 2017).

All the system components required for commercial-scale deployment of CLC/CLOU with biomass-derived fuels have been demonstrated; low-cost carrier materials have been tested over extended operations, with reasonable reactivity and lifetime performance; designs for commercial-scale systems have been completed; technoeconomic evaluations have confirmed technical feasibility with low CO_2 capture costs (Lyngfelt and Linderholm, 2017), and with negative emissions in the order of 0.5 tCO_2/MWh predicted for combined heat and power (CHP) applications (Peltola et al., 2020).

The inherent CO_2 capture during CLC of solid fuels significantly reduces capture costs compared to other combustion processes by avoiding the capital and energy costs of gas separation systems. SO_x and NO_x emissions can also be reduced to very low levels at a lower cost because they are concentrated in a smaller CO_2 flue gas stream (Lyngfelt et al., 2018).

The chemical looping concept is very flexible and, as well as fuel combustion, options have been demonstrated for fuel reforming to produce syngas (see next section), and for hydrogen production, either by reforming, WGS and CO_2 removal, or indirectly during carrier regeneration (Mendiara et al., 2018). Chemical looping can also be applied to capture CO_2 in a postcombustion application or in gasification (precombustion) systems, where it can be used in combination with a chemical loop to provide oxygen to the gasifier.

Precombustion capture—biomass gasification

An alternative to fully releasing the chemical energy of biomass or other fuel through complete combustion is to gasify the fuel to produce an intermediate synthesis gas or syngas product, which can then be oxidized as a power generation fuel or used as a feedstock for the production of liquid fuel, hydrogen, or methanol. Syngas is a mixture of CO and H_2 and results from the partial combustion of carbon and its high-temperature reaction in a gasifying stream of steam, oxygen, or air, with a wide range of reactions taking place, of which the most important are shown in Table 12.8.

Any CO_2 present under partial oxidation conditions will be converted to CO via the Boudouard reaction, particularly at high temperatures, which shift the equilibrium of this reaction to the right.

Precombustion CO_2 capture is then achieved by separating CO_2 and H_2, which can be done using a number of technologies, as shown in Table 12.9, of which the use of physical solvents (such as the Selexol process, described in Chapter 7) is currently the most commonly deployed at commercial scale. As well as the relatively higher CO_2 in the

TABLE 12.8 Carbon-based feedstock gasification processes and reactions.

Process	Reaction	Heat of reaction at 25°C
Partial oxidation	$C + 0.5O_2 \rightarrow CO$	−111 kJ/mol
	$CH_4 + 0.5O_2 \rightarrow CO + 2H_2$	−36 kJ/mol
Carbon–steam reaction	$C + H_2O \rightarrow CO + H_2$	+131 kJ/mol
Hydrogasification	$C + 2H_2 \leftrightarrow CH_4$	−75 kJ/mol
Water–gas shift (WGS) reaction	$CO + H_2O \leftrightarrow CO_2 + H_2$	−41 kJ/mol
Methanation	$CO + 3H_2 \leftrightarrow CH_4 + H_2O$	−206 kJ/mol
Steam methane reforming (SMR)	$CH_4 + H_2O \leftrightarrow CO + 3H_2$	+206 kJ/mol
Boudouard reaction	$C + CO_2 \leftrightarrow 2CO$	+172 kJ/mol

TABLE 12.9 CO_2:H_2 separation technologies for precombustion capture.

Technology area	Currently developed technologies	Example technologies under development
Absorption-based separation (Chapter 7)	Physical solvents (e.g., Selexol processes), chemical solvents	Novel solvents to improve performance; improved design of processes and equipment
Adsorption-based separation (Chapter 8)		Sorption enhanced water gas shift (SEWGS) process; elevated temperature pressure swing adsorption (ET-PSA)
Chemical looping systems (Chapter 8)		Chemical looping combustion, reforming and hydrogen production
Membrane separation (Chapter 9)		Metal and ceramic membrane WGS reactors; ion transport membranes
Cryogenic separation	CO_2 liquefaction	Hybrid cryogenic + membrane processes

syngas stream (>20% in the $H_2 + CO_2$ stream vs. 5%–15% in a postcombustion flue gas stream), CO_2:H_2 separation is easier than the postcombustion separation of CO_2 and N_2 due to the greater difference in molecular weights and molecular kinetic diameters.

Gasification reactors can take a variety of forms, the main ones being updraft and downdraft fixed bed gasifiers. Bubbling or circulating fluidized bed gasifiers are essentially updraft moving beds, and entrained flow gasifiers are downdraft moving beds. Table 12.10 summarizes their main characteristics and applications, and Fig. 12.9 illustrates the different conversion zones in up- and downdraft configurations.

In updraft gasifiers, the gasifying stream of steam, oxygen, or air entering the bottom of the bed first causes complete combustion of char, producing CO, CO_2, and H_2O, and raising the bed temperature to ~1000°C. Percolating upwards, the hot gases drive further reactions with char to form H_2 and CO, while cooling to ~750°C. Rising further, the gases pyrolyze the descending biomass that has been dried near the top of the reactor by the cooling gases. As released tars do not reach high enough temperatures for cracking in updraft gasifiers, the offgas typically has a high tar content of 10–20 wt%.

In downdraft gasifiers, the gasifying stream enters at a throated restriction part-way down the reactor where the dried and pyrolyzed biomass is combusted at temperatures of 1200–1400°C. Passing down over the hot char lower in the bed, the combustion gases are reduced, producing CO and H_2. Downdraft reactors produce a less tarry offgas as the tars released during pyrolysis are significantly cracked at high temperatures within the combustion and reduction zones.

Depending on the biomass feedstock and gasifier type, syngas will contain a variety of contaminants including particulate matter (ash and char), tars, nitrogen and sulfur compounds (NH_3, HCN, H_2S, COS), alkali metals (Na, K), and halides (HCl), with the potential for erosion, corrosion, catalyst poisoning, and NOx emissions in downstream processing. Some degree of gas cleaning is therefore required, depending on the syngas end-use. Synthesis pathways (methanol and Fischer–Tropsch synthesis) generally require more extensive cleaning than combustion applications, although tars are a significant problem for gas engines and gas turbines and require thermal or catalytic cracking to smaller, less problematic molecules such as CO, H_2, and CH_4.

TABLE 12.10 Characteristics and applications of gasification reactors.

Reactor type	Main characteristics	Applications
Fixed bed	Updraft and downdraft configurations; downdraft less suitable for high ash content biomass Gradual temperature increase creating biomass drying and pyrolysis zones Operating temperature up to 1400°C; high offgas temperature for downdraft reactors Lower syngas tar content from downdraft gasifiers	Small-scale (1–10 MW) biomass gasification (e.g., local-scale agricultural waste-to-power)
Fluidized bed	Lower operating temperature (<900°C) to prevent ash melting Excess air/gas flow reduces product calorific value Higher complexity, capital, and operating and maintenance costs Dual bed configuration separates gasification and char combustion	
Entrained flow	Top feed and side feed configurations Biomass pretreatment (drying/torrefaction/pyrolysis) outside the reactor Pretreated pulverized fuel (~75 μm), possibly introduced in a water slurry Typically operate at 1200–1500°C and 20–70 bar with oxygen as the gasifying medium High temperatures aid tar cracking	Most commonly used in IGCC power plants High-pressure syngas required for methanol and Fischer-Tropsch synthesis

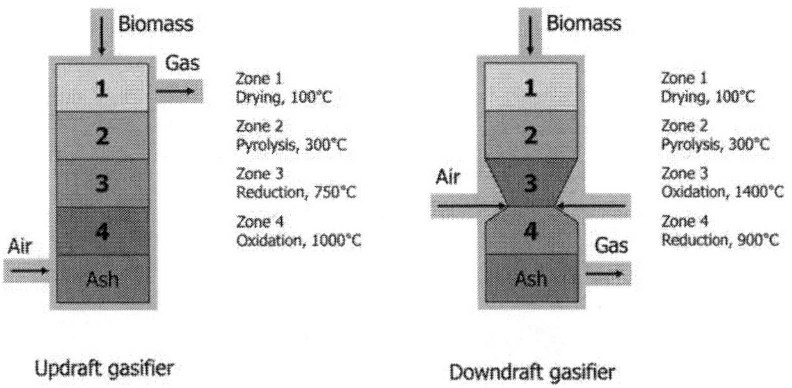

FIGURE 12.9 Updraft and downdraft biomass gasifier configurations.

Biomass integrated gasification combined cycle power generation

The application of biomass gasification in an integrated gasification combined-cycle power plant (BIGCC) is shown schematically in Fig. 12.10, where the gasification step typically includes pretreatment of the fuel by drying.

Combined cycle plants achieve high thermal efficiency by combining a hydrogen-fueled gas turbine (the top cycle) with a heat recovery steam generator (HRSG) using the gas turbine exhaust gases to generate steam (the bottom cycle), as shown in Fig. 12.11, which depicts the two cycles as ideal Carnot cycles.

In the top cycle (ABCD);

- A to B: the working fluid (air) is heated in the gas turbine combustion chamber
- B to C: the hot working fluid (air + combustion gas) is expanded through a turbine, doing work

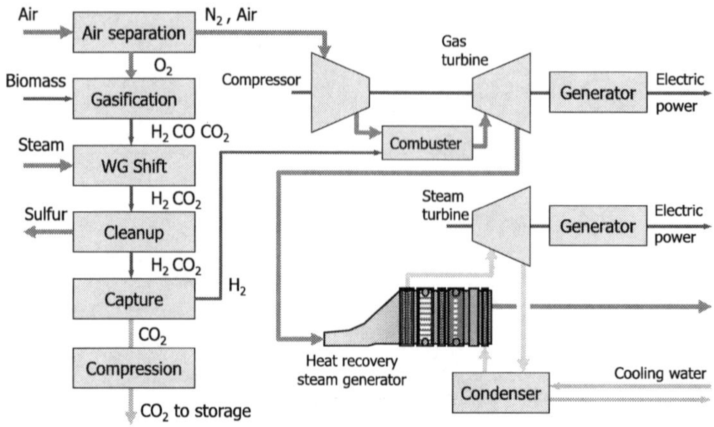

FIGURE 12.10 Process schematic of a biomass gasification combined cycle power plant.

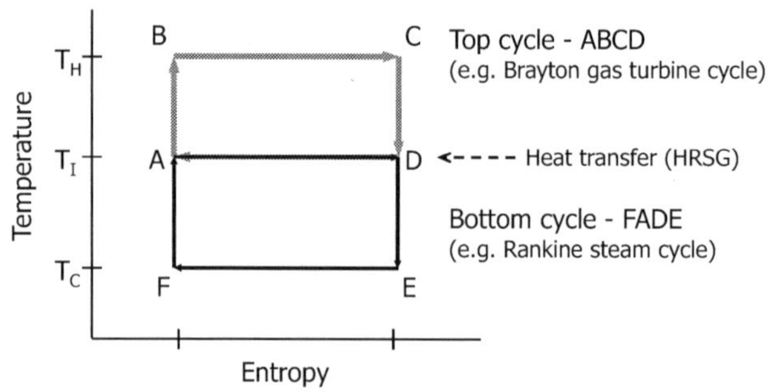

FIGURE 12.11 Combined cycle concept based on ideal Carnot cycles.

- C to D: the working fluid is cooled in the HRSG
- D to A: the working fluid is compressed to complete the cycle.

while in the bottom cycle (FADE);

- F to A: the working fluid (water) is heated in the HRSG producing steam
- A to D: the steam is expanded through a steam turbine doing work
- D to E: the working fluid is cooled and condensed
- E to F: the working fluid (water again) is pumped to the HRSG.

The ideal Carnot efficiency (η_{Carnot}) is the ratio of the work done to the heat input and is determined by the upper and lower temperatures of the cycle, with temperatures measured in Kelvin (K), so that for the top cycle;

$$\eta_{Top} = (T_H - T_I) / T_H = 1 - T_I / T_H \tag{12.3}$$

and is increased for the combined cycle to;

$$\eta_{Combined} = (T_H - T_L) / T_H = 1 - T_L / T_H \tag{12.4}$$

The Carnot efficiency represents the theoretical limit for perfectly reversible systems where all heat input and output are done at the hot, intermediate, and cold temperatures, that is, the combustion, HRSG, and cooling water temperatures. In practice, processes such as compression, expansion, and heat transfer are irreversible, and various other energy losses through friction, radiation, and leakage prevent this efficiency from being achieved in practice.

12.3.2 Biochemical conversion

Fermentation or anaerobic digestion are the biomass conversion processes used in the production of bioethanol and biomethane, with enzymatic hydrolysis being a common pretreatment to increase the biodegradability of lignocellulose or algal biomass. More detail on these processes is provided in this section.

Fermentation

Currently the only form of BECCS that is commercially viable is the production of bioethanol from the fermentation of sugar and starch crops, with carbon capture and storage by either saline aquifer injection or sale for enhanced oil recovery (EOR). This has provided an important early market opportunity for BECCS, with several commercial facilities starting up in the United States since 2009, the largest being located at the Archer Daniels Midland (ADM) plant near Decatur, IL, USA (see Section 12.6.1).

This process can be applied to different feedstocks such as sugar cane, beet, maize, and sorghum, following a pre-treatment step, usually enzymatic hydrolysis, in which enzymes such as cellulases convert cellulose to simple sugars that can be more easily fermented by bacteria or yeasts (Modenbach and Nokes, 2013). The fermentation process usually takes place in reactors held at 30°C for 48 h, in lightly acidic conditions (Khanal, 2011). Distillation is then typically required after fermentation to separate bioethanol from other liquid components, and a key focus of ongoing process improvement efforts is increasing overall energy efficiency through cogeneration systems and thermal integration.

CO_2 generated during fermentation can be captured using any of the separation methods discussed in Chapters 7–9, with amine-based solvent systems being most commonly applied. Brazil is a major producer of bioethanol from sugar cane and is a prime location for the implementation of this technology, having the potential to eliminate annual emissions of $\sim$30 $MtCO_2$ from bioethanol plants, around 5% of the country's emissions from energy production (Moreira et al., 2016).

Anaerobic digestion

Anaerobic digestion (AD) is the decomposition of organic material in the absence of oxygen and follows four main stages; hydrolysis, acidogenesis, acetogenesis, and methanogenesis, eventually resulting in the production of biogas (methane plus CO_2) and biofertilizer, the latter enabling the return of nutrients, minerals and fibrous material to the soil. The process is slow but can be optimized to improve the yield of methane and can also be used for hydrogen production, although this technology is less mature (Antoni et al., 2007).

Biogas is typically used to fuel CHP units without CO_2 separation, but CO_2 capture can be implemented by separating the biogas to produce biomethane and CO_2 streams, by postcombustion capture from CHP plant flue gases or by firing biogas combustion using oxygen rather than air (Li et al., 2017).

12.3.3 Hydrothermal liquefaction

Hydrothermal liquefaction converts biomass at low temperatures and under high pressure using a catalyst in the presence of hydrogen to obtain predominantly liquid products, often called bio-oil or bio-crude. Although this is an expensive process, producing a tarry substance which is not easily handled, it is nevertheless being researched further (Goyal et al., 2008; Toor et al., 2011).

As well as the conventional methods for capturing CO_2 from combustion of the liquid products, the subsurface sequestration of bio-oil is an alternative route under development to achieve negative emissions.

12.4 BECCS implementation

12.4.1 Technology readiness

Approaches for negative emissions through BECCS are advancing to the pilot and commercial scale around the world, and a number of projects are discussed in Section 12.6. Many of the individual components of a BECCS system are already at high technology readiness levels (TRL; see Chapter 2), for example, biomass combustion and chemical absorption of CO_2 for carbon capture (both are at TRL 9). However, with the exception of bioethanol production, the technologies are yet to be combined and tested at large-scale commercial facilities, so that most integrated BECCS options are at lower TRLs (Table 12.11).

TABLE 12.11 Technology readiness levels (TRLs) of different BECCS options.

BECCS option	TRL
BECCS Hydrogen	5—Lab scale system
BECCS Power	6—Pilot system
BECCS Industry	7—Demonstration system
BECCS Energy from waste	7—Demonstration system
BECCS Bioethanol	9—Commercial system

Niche markets are important enablers for the early development of new technologies, and markets for the sale of captured CO_2 for use in enhanced oil recovery (EOR) or to stimulate plant growth in greenhouses have been important stimuli for early demonstration plants. BECCS is most commonly discussed as an option for power plants, although opportunities have also been identified to apply BECCS in the cement, iron, and steel industries (Möllersten et al., 2003; Kuparinen et al., 2019; Mandova et al., 2019) and the decarbonizing of biofuel production and of Kraft pulp mills have been identified as market niches. There is also increased interest in combining BECCS with hydrogen production (Li et al., 2021), reflecting the growing role of hydrogen in the energy economy, and providing the opportunity to move toward biohydrogen as a carbon-negative industrial and transport fuel (Bello et al., 2020).

12.4.2 Implementation challenges

In competing for a major role in the NET portfolio, alongside other approaches that have been less commonly used in integrated assessment models (Gambhir et al., 2019), BECCS has the advantage of generating rather than consuming power alongside the removal of CO_2 from the atmosphere. Future energy systems may value this less, depending on the availability of renewable electricity, but biomass can provide cost-efficient baseload generation in net-zero power systems where most of the supply comes from wind and solar, supported by storage (Johansson et al., 2019).

The implementation of large-scale BECCS relies on the industrial-scale roll-out of CCS technology and the availability of land to produce the biomass feedstock, and achieving BECCS deployment at a climate impact scale will face a mix of technical, economic, and political challenges (Fridahl 2018; Fuss et al., 2018; Baik et al., 2018). Public acceptance will be important among these, for example, convincing landowners to convert their land to biomass is already facing challenges in some developed countries, as noted in Chapter 3, and levels of support for CO_2 storage can vary depending on the source of the CO_2 (Merk et al., 2022).

Negative environmental impacts can be associated with large-scale bioenergy crop cultivation, including emissions associated with feedstock transportation (Freer et al., 2022). Local-scale assessments of individual projects are critical to identify impacts on land use, water and air quality, biodiversity, and livelihoods (Creutzig et al., 2015; Fuss et al., 2014). Concerns have also been raised that large-scale BECCS implementation risks worsening other environmental impacts, including freshwater and marine toxicity and eutrophication, and ozone depletion and formation (Bello et al., 2020).

Appropriate policy instruments will be needed to financially support BECCS demonstration projects and the development of supply chains, which will also require global cooperation and careful carbon accounting. Emissions from biomass combustion are not reported in the energy sector to avoid double-counting, because carbon loss from harvesting is accounted in the land sector. This has led to some controversy over bioenergy being referred to as carbon neutral when net emissions from bioenergy can vary substantially depending on the feedstock, transport, and supply chain emissions (Röder et al., 2015).

Sustainability of bioenergy

The large-scale cultivation of bioenergy crops will face a number of sustainability challenges. Vast amounts of land will be required for biomass cultivation to enable BECCS deployment at gigatonne scale, with implications for water use, land rights, global shipping, and other ecological and societal risks; planning BECCS supply chains for sustainability is essential to avoid or minimize negative side effects for social equity and natural ecosystems (Pour et al., 2017). For example, the integration of biomass production into agricultural and forest landscapes can improve land and water use efficiency (Creutzig et al., 2015), as well as minimize emissions associated with land-use change, impacts on food security, and additional fertilizer use.

Sustainability is an overarching term and its definition is often contested (Reijnders, 2006). For biomass, it typically refers to environmental sustainability but can also consider economic sustainability, reduced inequalities, and other impacts in relation to the UN's Sustainable Development Goals (SDGs). BECCS sustainability impacts are case specific, depending, for example, on the type of biomass, location, supply chain, and other factors, and sustainability costs and cobenefits can vary considerably and require careful consideration in BECCS project planning (Donnison et al., 2020; Hunsberger et al., 2014).

Sustainability criteria for bioenergy systems have been published by the FAO under the Global Bioenergy Partnership initiative (GBEP, 2011) and by the International Organization for Standardization (ISO, 2015). A key metric for assessing biomass sustainability is the carbon payback period—the time taken for CO_2 removal through biomass regrowth to offset the emissions released during combustion. This time can vary from 1 or 2 years for dedicated energy crops, to hundreds of years for wood from natural forests with much slower growth rates. Life-cycle assessment (LCA), discussed in Chapter 2, is a common method of analysis for environmental sustainability and carbon payback. Biomass is typically considered to be carbon neutral, although this is dependent on the chosen LCA boundaries, and increasing wood harvests will inevitably impact on forest carbon stocks (Giuntoli et al., 2020). Other supply chain choices will also affect BECCS carbon removal effectiveness over the project lifetime (Fajardy and MacDowell, 2017).

Governance and policy issues

The rate of NETs implementation will need to be rapid in the coming decades, requiring significant policy support and regulation, and expert elicitation studies have identified societal support and governance structures as among the most significant implementation challenges (Vaughan and Gough, 2016). Key policy and governance issues include how best to fit BECCS into existing climate policies, for example, in relation to emissions reduction, and concerning the monitoring, reporting, and verification (MRV) of CO_2 removal from the atmosphere (Gough et al., 2018).

The introduction of policy instruments to incentivize deployment is expected to have a major influence on the pace of BECCS deployment, although commercial viability will also be aided by growth in the voluntary carbon market, as more and more businesses set targets for carbon neutrality and seek high-quality removal credits. Very few supportive policies currently exist, and policymakers will need to ensure that new instruments responsibly account for diverse societal values (Hunsberger et al., 2014; Bellamy et al., 2021). The type of policy instrument used to incentivize BECCS, and other NETs, can have a significant impact on public perceptions of the technology. Public support is higher for policies that reward companies for CO_2 removal than for those that pass on costs to consumers, for example, through guaranteed high prices. The use of standards and obligations are also well supported, as are persuasive approaches such as lobbying from governments, despite some concerns over the effectiveness of this approach (Bellamy et al., 2019; Babin et al., 2021).

It is likely that a portfolio of policies will be required to develop and evolve over time, using different instruments (e.g., price and quantity mechanisms) and different funding mechanisms (e.g., public and private), to aid in addressing the multiple challenges, including accounting and innovation barriers, facing BECCS implementation (Fridahl and Lehtveer, 2018). A whole systems approach will help to ensure that the wide range of factors are taken into account when policy choices are being made (Cross et al., 2021). These will include providing value to the economy, and fair risk and benefit sharing, as well as delivering confidence to investors.

12.4.3 Research and development agenda

As noted earlier, many of the system components required for BECCS implementation are already commercially deployed in other industrial contexts, so that the focus of further R&D is largely on issues of integration, efficiency, and effectiveness, as well as on sociopolitical aspects.

A recent review of the research priorities identified a range of technical issues requiring further study, including, life-cycle assessment (LCA) of BECCS value chains, improving the spatial and temporal resolution of land data, and the development of advanced biofuels (Stavrakas et al., 2018). Innovation and improvements in these and other technoeconomic areas will improve technical efficiency and economic viability, and pilot-scale demonstration projects will help to address some of the uncertainties that influence regulatory and policy design, including incentives (Lomax et al., 2015).

Further research is also needed into sociopolitical issues, which could prove to be a greater barrier to deployment. Public outreach is necessary to improve awareness of BECCS and address social concerns (Stavrakas et al. 2018; Babin et al., 2021), and further research into public engagement and the factors influencing acceptability is warranted (Bellamy et al., 2019), with the ultimate aim of improving public support and obtaining the essential social license to operate.

Research to improve the treatment of BECCS in IAMs is also needed, for example, to better reflect land-use competition and the impact of large-scale biomass cultivation on food security, water resources, biodiversity, and other ecosystem services (NASEM, 2019). Further development of national-level estimates of BECCS potential, accounting for local- to national-scale acceptability, country-specific business models, and techno-economic analyses, will help to provide a robust foundation for this work (Forster et al., 2020; Laude, 2020).

12.5 BECCS costs and global potential

12.5.1 Current and projected cost of BECCS

In one of the first techno-economic assessments of BECCS, Möllersten and colleagues estimated costs in the region \$30 to \$70/tCO_2 including CO_2 transportation and storage, with the potential to remove 165−220 MtCO_2/yr from global pulp production and 600 MtCO_2/yr from sugarcane-based ethanol production (Möllersten et al., 2003). In a more recent analysis, the global potential for pulp mills was estimated to be somewhat lower, at 137 MtCO_2/yr including emissions from lime kiln firing and other biomass combustion (Kuparinen et al., 2019). These and other published estimates of BECCS costs across a number of industrial sectors are summarized in Table 12.12.

The costs shown in Table 12.12 typically consider the respective capital and operating costs, net of any revenues generated from coproducts such as from the sale of electricity or biofuels, and therefore provide a comparison between the CDR cost-effectiveness of various BECCS applications (Richard et al., 2021). Nonfinancial costs and cobenefits, such as those arising from environmental or social impacts, are not included, but could shift these cost estimates up or down if they could be sensibly expressed in financial terms.

The lower CO_2 removal costs associated with certain applications are typically associated with processes, such as gasification, that can produce gas streams with high CO_2 concentration, leading to a lower capture cost when compared to postcombustion capture. Most techno-economic assessments of BECCS have focused on power applications and, apart from pulp and paper, there are relatively fewer assessments in other industrial settings. Applications such as in iron, steel, and cement production could also lead to low CO_2 removal costs if additional financial benefits accrue.

Many factors will affect the financial feasibility and speed of BECCS deployment, including economic and societal choices and industrial sector regulations. These will affect the appropriateness of different types of BECCS, for example, standards for fuel into an EfW plant may change—making it more or less financially attractive to investors. Competition between markets could also change the cost of assets, labor, and the development of international supply chains.

Carbon prices, providing a fixed revenue per tonne of CO_2 removed, will have a major impact on the financial viability of BECCS and other NETs, and will need to be in the \$50−−100/tCO_2 range by 2030 to make BECCS financially rewarding outside niche markets. Carbon prices currently remain below that range, so that incentives are needed to support short-term progress, although the social cost of carbon is likely already well above \$100/tCO_2 (Hope and Hope, 2013).

TABLE 12.12 Estimated BECCS costs in different sectors.

BECCS application	Cost (\$/tCO_2)	References
Pulp and paper mills	20−70	Möllersten et al. (2003) Fuss et al. (2018)
Bioethanol production	20−175	Fuss et al. (2018)
Biomass gasification	30−76	Sanchez and Callaway (2016), Fuss et al. (2018), Gough and Upham (2011)
Hydrogen production	70−160	Richard et al. (2021)
Energy from waste (EfW)	80−190	Richard et al. (2021)
Combustion	88−288	Fuss et al. (2018) Babin et al. (2021)

12.5.2 Global potential of BECCS

The global requirement for CO_2 removal through BECCS depends on a variety of factors such as the rate of innovation in other sectors and the emissions pathway assumed to net zero, which can vary dramatically depending on the approach taken, for example, to an overshoot in global mean surface temperatures. Pathways limiting warming to 1.5°C with limited overshoot suggest that BECCS will be used to remove up to 1, 8, and 16 $GtCO_2$/year by 2030, 2050, and 2100, respectively (de Coninck et al., 2018), although a systematic review concluded that the sustainable potential of BECCS is in the region of 0.5–5 $GtCO_2$/year by 2050, if severe adverse effects are to be avoided (Fuss et al., 2018). BECCS and afforestation are the two NETs that are currently relied upon in IAM scenario modeling (Köberle, 2019), but the gradual inclusion of other technologies such as DACCS, and of other options for short-term innovations in decarbonization, will likely result in a reduction in the quantities of BECCS required to meet climate targets.

Assessments of the physical global potential of BECCS indicate that this could be in the region of 300–700 $GtCO_2$ cumulative removal by 2100, requiring 100–500 Mha of land and producing roughly 10–100 EJ/year of power (The Royal Society, 2018). Global studies have also found that, although the required land area may be biophysically available, BECCS deployment should be limited to minimize land conversion and the resulting environmental impacts including biodiversity loss (Hanssen et al., 2020b). The availability of CO_2 storage capacity is not seen to be a barrier to achieving this quantity of removals (IEA, 2012), although there may be local or regional competition with other CCS applications (e.g., DACCS or fossil fuel CCS).

Studies of the United States have identified enough suitable land area to remove 110–120 $MtCO_2$ by 2020 and 360–630 $MtCO_2$ by 2040, in line with the energy system model and IAM predictions of US BECCS needs for a <2°C warming scenario (Baik et al., 2018). The United States is of particular interest for the BECCS deployment because it has a mature biomass production industry combined with substantial geological storage potential, located mostly in the south, west, and north-eastern parts of the country. Transportation, either of biomass to BECCS facilities located near storage sites or of CO_2 through long-distance pipelines will therefore be required, and the considerable level of opposition to pipeline construction in the United States highlights the importance and challenges likely to be faced in obtaining a social license for such projects (Lucky et al., 2015).

In terms of the scale of individual BECCS facilities, local assessments in the United Kingdom have found that 3 $MtCO_2$/yr of negative emissions could be delivered by a 500 MW power plant operating on local biomass resources, requiring approximately 2.33 Mt/yr of biomass for combustion (Donnison et al., 2020). Other studies of BECCS power stations at this scale in the United Kingdom have questioned whether BECCS plants will even be energy producers, highlighting the many supply chain decisions, with energy and emissions consequences, which could result in a net energy requirement if poorly managed (Fajardy and Mac Dowell, 2017).

12.6 BECCS projects

12.6.1 Bioethanol production

Four of the first five commercial-scale BECCS plants were bioethanol production facilities, of which the largest was the Illinois Industrial Carbon Capture and Storage (IL-ICCS) project, located at the Archer Daniels Midland (ADM) plant near Decatur, IL, USA, where CO_2 from the fermentation of corn is captured using an amine-based process (Chapter 7) and injected into the nearby Mt. Simon Sandstone saline aquifer formation. CO_2 injection at ~1.0 $MtCO_2$/yr commenced in 2017, although the ADM facility is not overall carbon-negative due to the emission of ca. 5 $MtCO_2$/yr from its power plant.

Some smaller-scale BECCS facilities include Arkalon Ethanol, USA (200,000 tCO_2/yr), Bonanza CCUS, USA (100,000 tCO_2/yr), and Husky Energy, Canada (90,000 tCO_2/yr). There are also a number of facilities that capture CO_2 from bioenergy but utilize the CO_2 for crop cultivation, rather than storing it underground. These include Organic Carbon Dioxide for Assimilation of Plants (OCAP), Netherlands (400,000 tCO_2/yr), Lantmännen Agroetanol, Sweden (200,000 tCO_2/yr), Calgren Renewable Fuels, USA (150,000 tCO_2/yr), Alco Bio Fuel (ABF), Belgium (100,000 tCO_2/yr) and Cargill Wheat Processing, UK (100,000 tCO_2/yr) (Babin et al., 2021).

12.6.2 Drax power plant (United Kingdom)

The world's first BECCS postcombustion capture pilot, using a proprietary solvent developed by Leeds University, UK spin-off C-Capture, began capturing 1 tCO_2/d in February 2019 from a flue gas slipstream at the Drax Group's biomass-fired power plant in Selby, UK. A second 300 $kgCO_2$/d pilot, using MHI's KM/CDR Process (Chapter 7), which uses a sterically hindered amine solvent (KS1) with a rate-promoting additive, was announced in June 2020. A front-end engineering design (FEED) study for full-scale BECCS implementation at the Drax plant is scheduled for 2022, and implementation, planned by 2030, is expected to result in negative emissions of 16 $MtCO_2$/yr.

Due to the economies of scale associated with CO_2 transportation and storage, it is likely that CCS at Drax will be delivered as part of a network of carbon capture projects feeding into a shared CO_2 transportation and storage system. This is currently being developed as part of *Zero Carbon Humber*, a partnership that has been successful in gaining UK Government funding to build the world's first net zero industrial region, and decarbonize the North of England.

12.6.3 Longship CCS project (Norway)

A similar regional industrial approach is being progressed in the Norwegian Longship project which, with government financial support, also aims to develop a full CCS supply chain incorporating a series of projects:

1. Capture of 400,000 tCO_2/yr at the Fortum Oslo Varme waste-to-energy plant.
2. Capture of up to 400,000 tCO_2/yr at the Norcem cement plant.
3. The Northern Lights project—a joint venture between Equinor, Shell, and Total to transport and permanently store CO_2 in the Norwegian North Sea.

The two carbon capture projects will use amine technology to separate CO_2, which will be transported in a liquefied state through a 110-km pipeline for permanent storage beneath the North Sea. The cement works currently cofires up to 30% biomass, and the waste-to-energy plant uses roughly 50% waste of biological origin, thus providing negative emissions once operational in 2024 (Fortum, 2022).

12.6.4 Mikawa power plant (Japan)

Mikawa power plant project in Fukuoka, Japan has retrofitted a 50 MW 100% biomass (primarily palm kernel shells) fired power plant with postcombustion capture. The site has been testing pilot-scale capture facilities since 2009, with the demonstration-scale facility becoming operational in 2020 and capturing 180,000 tCO_2/yr (Babin et al., 2021).

References and resources

Abad, A., Adánez-Rubio, I., Gayán, P., García-Labiano, F., Luis, F., Adánez, J., 2012. Demonstration of chemical-looping with oxygen uncoupling (CLOU) process in a 1.5 kW_{th} continuously operating unit using a Cu-based oxygen-carrier. Int. J. Greenh. Gas Control 6, 189–200.

Abuelgasim, S., Wang, W., Abdalazeez, A., 2020. A brief review for chemical looping combustion as a promising CO_2 capture technology: fundamentals and progress. Sci. Total Environ. 142892.

Anderson, K., Peters, G., 2016. The trouble with negative emissions. Science 354 (6309), 182–183.

Antoni, D., Zverlov, V.V., Schwarz, W.H., 2007. Biofuels from microbes. Appl. Microbiol. Biotechnol. 77 (1), 23–35.

Arenas, F., Vas-Pinto, F., 2014. Marine algae as carbon sinks and allies to combat global warming. In: Marine Algae: Biodiversity, Taxonomy, Environmental Assessment and Biotechnology. CRC Press, Boca Raton, USA.

Babin, A., Vaneeckhaute, C., Iliuta, M.C., 2021. Potential and challenges of bioenergy with carbon capture and storage as a carbon-negative energy source: a review. Biomass Bioenergy 146, 105968.

Badar, S.N., Mohammad, M., Emdadi, Z., Yaakob, Z., 2018. Algae and their growth requirements for bioenergy: a review. Biofuels 12 (3), 307–325.

Baik, E., Sanchez, D.L., Turner, P.A., Mach, K.J., Field, C.B., Benson, S.M., 2018. Geospatial analysis of near-term potential for carbon-negative bioenergy in the United States. Proc. Natl. Acad. Sci. USA 115 (13), 3290–3295.

Beal, C.M., Archibald, I., Huntley, M.E., Greene, C.H., Johnson, Z.I., 2018. Integrating algae with bioenergy carbon capture and storage (ABECCS) increases sustainability. Earth's Future 6 (3), 524–542.

Bellamy, R., Lezaun, J., Palmer, J., 2019. Perceptions of bioenergy with carbon capture and storage in different policy scenarios. Nat. Commun. 10 (1), 1–9.

Bellamy, R., Fridahl, M., Lezaun, J., et al., 2021. Incentivising bioenergy with carbon capture and storage (BECCS) responsibly: comparing stakeholder policy preferences in the United Kingdom and Sweden. Environ. Sci. Pol. 116, 47–55.

Bello, S., Galán-Martín, Á., Feijoo, G., Moreira, M.T., Guillén-Gosálbez, G., 2020. BECCS based on bioethanol from wood residues: potential towards a carbon-negative transport and side-effects. Appl. Energy 279, 115884.

Bergman, P.C.A., 2005. *Combined Torrefaction and Pellestisation.* ECN Report ECN-05-013. Energy Research Centre of the Netherlands.

Brown, R.C. (Ed.), 2019. Thermochemical Processing of Biomass: Conversion into Fuels, Chemicals and Power. John Wiley & Sons.

Celis-Pla, P.S., Hall-Spencer, J.M., Horta, P.A., et al., 2015. Macroalgal responses to ocean acidification depend on nutrient and light levels. Front. Mar. Sci. 2, 26.

Chisti, Y., 2007. Biodiesel from microalgae. Biotechnol. Adv. 25 (3), 294–306.

Clery, D.S., Mason, P.E., Rayner, C.M., Jones, J.M., 2018. The effects of an additive on the release of potassium in biomass combustion. Fuel 214, 647–655.

Clery, D.S., Mason, P.E., Barnes, D.C., et al., 2021a. The effect of biomass ashes and potassium salts on MEA degradation for BECCS. Int. J. Greenh. Gas Control 108, 103305.

Clery, D.S., Vaughan, N.E., Forster, J., Lorenzoni, I., Gough, C.A., Chilvers, J., 2021b. Bringing greenhouse gas removal down to earth: stakeholder supply chain appraisals reveal complex challenges. Global Environ. Change 71, 102369.

Creutzig, F., Ravindranath, N.H., Berndes, G., et al., 2015. Bioenergy and climate change mitigation: an assessment. GCB Bioenergy 7 (5), 916–944.

Cross, S., Welfle, A.J., Thornley, P., Syri, S., Mikaelsson, M., 2021. Bioenergy development in the UK & Nordic countries: a comparison of effectiveness of support policies for sustainable development of the bioenergy sector. Biomass Bioenergy 144, 105887.

de Coninck, H., Revi, A., Babiker, M., et al., 2018. Strengthening and implementing the global response. In: Global Warming of 1.5 °C: Summary for Policy Makers, pp. 313–443 (IPCC-The Intergovernmental Panel on Climate Change).

Desideri, U., Fantozzi, F., 2013. Biomass combustion and chemical looping for carbon capture and storage. In: Technologies for Converting Biomass to Useful Energy: Combustion, Gasification, Pyrolysis, Torrefaction and Fermentation. CRC Press, New York, pp. 129–167.

Donnison, C., Holland, R.A., Hastings, A., Armstrong, L.M., Eigenbrod, F., Taylor, G., 2020. Bioenergy with Carbon Capture and Storage (BECCS): finding the win–wins for energy, negative emissions and ecosystem services—size matters. GCB Bioenergy 12 (8), 586–604.

Fajardy, M., Mac Dowell, N., 2017. Can BECCS deliver sustainable and resource efficient negative emissions? Energy Environ. Sci. 10 (6), 1389–1426.

Fajardy, M., Chiquier, S., Mac Dowell, N., 2018. Investigating the BECCS resource nexus: delivering sustainable negative emissions. Energy Environ. Sci. 11 (12), 3408–3430.

Finney, K.N., Akram, M., Diego, M.E., Yang, X., Pourkashanian, M., 2019. Carbon capture technologies. In: Bioenergy with Carbon Capture and Storage. Elsevier, pp. 15–45.

Forster, J., Vaughan, N.E., Gough, C., Lorenzoni, I., Chilvers, J., 2020. Mapping feasibilities of greenhouse gas removal: key issues, gaps and opening up assessments. Global Environ. Change 63, 102073.

Fortum, 2022. A Full-Scale Carbon Capture and Storage (CCS) Project Initiated in Norway [online].

Freer, M., Gough, C., Welfle, A., Lea-Langton, A., 2022. Putting bioenergy with carbon capture and storage in a spatial context: what should go where? Front. Climate 26.

Fridahl, M., 2018. Bioenergy with Carbon Capture and Storage: From Global Potentials to Domestic Realities. European Liberal Forum.

Fridahl, M., Lehtveer, M., 2018. Bioenergy with carbon capture and storage (BECCS): global potential, investment preferences, and deployment barriers. Energy Res. Social Sci. 42, 155–165.

Fuss, S., Canadell, J.G., Peters, G.P., et al., 2014. Betting on negative emissions. Nat. Clim. Change 4 (10), 850–853.

Fuss, S., Lamb, W.F., Callaghan, M.W., et al., 2018. Negative emissions—Part 2: costs, potentials and side effects. Environ. Res. Lett. 13 (6), 063002.

Gambhir, A., Butnar, I., Li, P.H., Smith, P., Strachan, N., 2019. A review of criticisms of integrated assessment models and proposed approaches to address these, through the lens of BECCS. Energies 12 (9), 1747.

GBEP, 2011. The Global Bioenergy Partnership Sustainability Indicators for Bioenergy. FAO, Rome.

Gent, S., Twedt, M., Gerometta, C., Almberg, E., 2017. Introduction to thermochemical conversion processes. In: Theoretical and Applied Aspects of Biomass Torrefaction: For Biofuels and Value-Added Products. Butterworth-Heinemann, Elsevier, Oxford, UK.

Giuntoli, J., Searle, S., Jonsson, R., et al., 2020. Carbon accounting of bioenergy and forest management nexus. A reality-check of modeling assumptions and expectations. Renew. Sustain. Energy Rev. 134, 110368.

Gough, C., Upham, P., 2011. Biomass energy with carbon capture and storage (BECCS or Bio-CCS). Greenh. Gases: Sci. Technol. 1 (4), 324–334.

Gough, C., Garcia-Freites, S., Jones, C., et al., 2018. Challenges to the use of BECCS as a keystone technology in pursuit of 1.5 °C. Global Sustain. 1–9.

Goyal, H.B., Seal, D., Saxena, R.C., 2008. Bio-fuels from thermochemical conversion of renewable resources: a review. Renew. Sustain. Energy Rev. 12 (2), 504–517.

Hannon, M., Gimpel, J., Tran, M., Rasala, B., Mayfield, S., 2010. Biofuels from algae: challenges and potential. Biofuels 1 (5), 763–784.

Hanssen, S.V., Daioglou, V., Steinmann, Z.J.N., Doelman, J.C., Van Vuuren, D.P., Huijbregts, M.A.J., 2020a. The climate change mitigation potential of bioenergy with carbon capture and storage. Nat. Clim. Change 10 (11), 1023–1029.

Hanssen, S.V., Daioglou, V., Steinmann, Z.J., et al., 2020b. Biomass residues as twenty-first century bioenergy feedstock—a comparison of eight integrated assessment models. Clim. Change 163 (3), 1569–1586.

Hope, C., Hope, M., July 2, 2013. The social cost of CO_2 in a low-growth world. Nat. Clim. Change 3 (8), 722–724.

Hughes, A.D., Kelly, M.S., Black, K.D., Stanley, M.S., 2012. Biogas from macroalgae: is it time to revisit the idea? Biotechnol. Biofuels 5 (1), 86.

Hunsberger, C., Bolwig, S., Corbera, E., Creutzig, F., 2014. Livelihood impacts of biofuel crop production: implications for governance. Geoforum 54, 248–260.

IEA, 2012. Roadmap : Carbon Capture and Storage in Industrial Applications. IEA, Paris.

IPCC, 2014. Climate Change 2014. Working Group III: Mitigation of Climate Change. Cambridge University Press, Cambridge, UK.

ISO, 2015. ISO 13065: Sustainability Criteria for Bioenergy. International Organization for Standardization, Geneva.

Jenkins, B.M., Baxter, L.L., Miles Jr, T.R., Miles, T.R., 1998. "Combustion properties of biomass,". Fuel Process. Technol. 54, 17–46.

Johansson, V., Lehtveer, M., Göransson, L., 2019. Biomass in the electricity system: a complement to variable renewables or a source of negative emissions? Energy 168, 532–541.

Keith, D.W., 2001. Sinks, energy crops and land use: coherent climate policy demands an integrated analysis of biomass. Climatic Change 49 (1–2), 1.

Khanal, S.K., 2011. Anaerobic Biotechnology for Bioenergy Production: Principles and Applications. John Wiley & Sons.

Köberle, A.C., 2019. The value of BECCS in IAMs: a review. Curr. Sustain. Renew. Energy Reports 6 (4), 107–115.
Koch, M., Bowes, G., Ross, C., Zhang, X.H., 2013. Climate change and ocean acidification effects on seagrasses and marine macroalgae. Global Change Biol. 19 (1), 103–132.
Kumar, R., Singh, R.I., 2016. An investigation in 20 kW_{th} oxygen-enriched bubbling fluidized bed combustor using coal and biomass. Fuel Process. Technol. 148, 256–268.
Kumar, R., Singh, R.I., 2017. An investigation of co-combustion municipal sewage sludge with biomass in a 20 kW BFB combustor under air-fired and oxygen-enriched condition. Waste Manag. 70, 114–126.
Kuparinen, K., Vakkilainen, E., Tynjälä, T., 2019. Biomass-based carbon capture and utilization in kraft pulp mills. Mitig. Adapt. Strategies Glob. Change 24 (7), 1213–1230.
Laude, A., 2020. Bioenergy with carbon capture and storage: are short-term issues set aside? Mitig. Adapt. Strategies Glob. Change 25 (2), 185–203.
Li, H., Tan, Y., Ditaranto, M., Yan, J., Yu, Z., 2017. Capturing CO_2 from biogas plants. Energy Proc. 114, 6030–6035.
Li, W., Ciais, P., Makowski, D., Peng, S., 2018. A global yield dataset for major lignocellulosic bioenergy crops based on field measurements. Sci. Data 5 (1), 1–10.
Li, M., Lu, Y., Huang, M., 2021. Evolution patterns of bioenergy with carbon capture and storage (BECCS) from a science mapping perspective. Sci. Total Environ. 766, 144318.
Liu, Z., Zhang, T., 2018. Pilot and industrial demonstration of oxy-fuel combustion. In: Oxy-Fuel Combustion. Academic Press, pp. 209–222.
Lomax, G., Lenton, T.M., Adeosun, A., Workman, M., 2015. Investing in negative emissions. Nat. Clim. Change 5 (6), 498–500.
Loo, S. Van, Koppejan, J., 2008. The Handbook of Biomass Combustion and Cofiring. Routledge.
Lucky, J., Maciel, D., Harris, L., Satinsky, S., 2015. Lobos CO_2 Pipeline Health Impact Assessment. Report available from: https://humanimpact.org/wp-content/uploads/2015/06/CO2PipelineHIA_FinalReport.pdf.
Lyngfelt, A., Linderholm, C., 2017. Chemical-looping combustion of solid fuels–status and recent progress. Energy Proc. 114, 371–386.
Lyngfelt, A., Rydén, M., Linderholm, C., Mattisson, T., 2018. Chemical-looping combustion (CLC) of solid fuels (CLC-SF) - a discussion of operational experiences, costs, upscaling strategies and negative emissions (Bio-CLC). In: 14th Greenhouse Gas Control Technologies Conference. Melbourne 21-26 October 2018 (GHGT-14).
MacMonagail, M., Cornish, L., Morrison, L., Araújo, R., Critchley, A.T., 2017. Sustainable harvesting of wild seaweed resources. Eur. J. Phycol. 52 (4), 371–390.
Mason, P.E., Darvell, L.I., Jones, J.M., Pourkashanian, M., Williams, A., 2015. Single particle flame-combustion studies on solid biomass fuels. Fuel 151, 21–30.
Mandova, H., Patrizio, P., Leduc, S., et al., 2019. Achieving carbon-neutral iron and steelmaking in Europe through the deployment of bioenergy with carbon capture and storage. J. Clean. Prod. 218, 118–129.
Mehmood, M.A., Ibrahim, M., Rashid, U., et al., 2017. Biomass production for bioenergy using marginal lands. Sustain. Prod. Consum. 9, 3–21.
Mendiara, T., García-Labiano, F., Abad, A., et al., 2018. Negative CO_2 emissions through the use of biofuels in chemical looping technology: a review. Appl. Energy 232, 657–684.
Merk, C., Nordø, Å.D., Andersen, G., Lægreid, O.M., Tvinnereim, E., 2022. Don't send us your waste gases: public attitudes toward international carbon dioxide transportation and storage in Europe. Energy Res. Social Sci. 87, 102450.
Milledge, J.J., Harvey, P.J., 2016. Potential process 'hurdles' in the use of macroalgae as feedstock for biofuel production in the British Isles. J. Chem. Technol. Biotechnol. 91 (8), 2221–2234.
Modenbach, A.A., Nokes, S.E., 2013. Enzymatic hydrolysis of biomass at high-solids loadings–a review. Biomass Bioenergy 56, 526–544.
Mohan, D., Pittman Jr., C.U., Steele, P.H., 2006. Pyrolysis of wood/biomass for bio-oil: a critical review. Energy Fuels 20 (3), 848–889.
Möllersten, K., Yan, J., Moreira, J.R., 2003. Potential market niches for biomass energy with CO_2 capture and storage—opportunities for energy supply with negative CO_2 emissions. Biomass Bioenergy 25 (3), 273–285.
Moreira, D., Pires, J.C., 2016. Atmospheric CO_2 capture by algae: negative carbon dioxide emission path. Bioresour. Technol. 215, 371–379.
Moreira, J.R., Romeiro, V., Fuss, S., Kraxner, F., Pacca, S.A., 2016. BECCS potential in Brazil: achieving negative emissions in ethanol and electricity production based on sugar cane bagasse and other residues. Appl. Energy 179, 55–63.
NASEM, 2019. Negative Emissions Technologies and Reliable Sequestration: A Research Agenda. National Academies of Sciences, Engineering, and Medicine. The National Academies Press, Washington, DC, pp. 168–176.
Nguyen, H.K., Moon, J.H., Jo, S.H., et al., 2020. Oxy-combustion characteristics as a function of oxygen concentration and biomass co-firing ratio in a 0.1 MWth circulating fluidized bed combustion test-rig. Energy 196, 117020.
N'Yeurt, A.d.R., Chynoweth, D.P., Capron, M.E., Stewart, J.R., Hasan, M.A., 2012. Negative carbon via ocean afforestation. Process Saf. Environ. Protect. 90, 467–474.
Obersteiner, M., Mechler, R., Nilsson, S., et al., 2001. Managing Climate Risk. Interim Report. IR-01-051. International Institute for Applied Systems Analysis, Laxenburg, Austria.
Panoutsou, C., Perakis, C., Elbersen, B., Zheliezna, T., Staritsky, I., 2017. Assessing potentials for agricultural residues. In: Modeling and Optimization of Biomass Supply Chains. Academic Press, pp. 169–197.
Pierobon, S.C., Cheng, X., Graham, P.J., Nguyen, B., Karakolis, E.G., Sinton, D., 2018. Emerging microalgae technology: a review. Sustain. Energy Fuels 2 (1), 13–38.
Peltola, P., Saari, J., Tynjälä, T., Hyppänen, T., 2020. Process integration of chemical looping combustion with oxygen uncoupling in a biomass-fired combined heat and power plant. Energy 210, 118550.

Pour, N., Webley, P.A., Cook, P.J., 2017. A sustainability framework for bioenergy with carbon capture and storage (BECCS) technologies. Energy Procedia 114, 6044–6056.

Prins, M.J., Ptasinski, K.J., Janssen, F.J., 2006. More efficient biomass gasification via torrefaction. Energy 31 (15), 3458–3470.

Raheem, A., Prinsen, P., Vuppaladadiyam, A.K., Zhao, M., Luque, R., 2018. A review on sustainable microalgae based biofuel and bioenergy production: recent developments. J. Clean. Prod. 181, 42–59.

Rao, R.K., 1999. Eco-restructuring: Implications for Sustainable Development.

Raven, J., 2018. Blue carbon: past, present and future, with emphasis on macroalgae. Biol. Lett. 14 (10), 20180336.

Reijnders, L., 2006. Conditions for the sustainability of biomass based fuel use. Energy Pol. 34 (7), 863–876.

Richard, S., Mitchell, A., Evans, C., Whitaker, J., Thomson, A., Keith, A., 2021. Greenhouse Gas Removal Methods and Their Potential UK Deployment. Department for Business, Energy & Industrial Strategy, London.

Röder, M., Whittaker, C., Thornley, P., 2015. How certain are greenhouse gas reductions from bioenergy? Life cycle assessment and uncertainty analysis of wood pellet-to-electricity supply chains from forest residues. Biomass Bioenergy 79, 50–63.

Rodionova, M.V., Poudyal, R.S., Tiwari, I., et al., 2017. Biofuel production: challenges and opportunities. Int. J. Hydrogen Energy 42 (12), 8450–8461.

Rogelj, J., Luderer, G., Pietzcker, R.C., et al., 2015. Energy system transformations for limiting end-of-century warming to below 1.5 C. Nat. Clim. Change 5 (6), 519–527.

Roni, M.S., Chowdhury, S., Mamun, S., Marufuzzaman, M., Lein, W., Johnson, S., 2017. Biomass co-firing technology with policies, challenges, and opportunities: a global review. Renew. Sustain. Energy Rev. 78, 1089–1101.

Saddawi, A., Jones, J.M., Williams, A., Wojtowicz, M.A., 2010. Kinetics of the thermal decomposition of biomass. Energy Fuels 24 (2), 1274–1282.

Sanchez, D.L., Callaway, D.S., 2016. Optimal scale of carbon-negative energy facilities. Appl. Energy 170, 437–444.

Santos, M.P., Manovic, V., Hanak, D.P., 2021. Unlocking the potential of pulp and paper industry to achieve carbon-negative emissions via calcium looping retrofit. J. Clean. Prod. 280, 124431.

Schmidt, H.P., Anca-Couce, A., Hagemann, N., et al., 2019. Pyrogenic carbon capture and storage. GCB Bioenergy 11 (4), 573–591.

Sengupta, S., Gorain, P.C., Pal, R., 2017. Aspects and prospects of algal carbon capture and sequestration in ecosystems: a review. Chem. Ecol. 33 (8), 695–707.

Sher, F., Pans, M.A., Sun, C., Snape, C., Liu, H., 2018. Oxy-fuel combustion study of biomass fuels in a 20 kW_{th} fluidized bed combustor. Fuel 215, 778–786.

Slade, R., Bauen, A., 2013. Micro-algae cultivation for biofuels: cost, energy balance, environmental impacts and future prospects. Biomass Bioenergy 53, 29–38.

Stafford, W., Lotter, A., Brent, A., von Maltitz, G., 2017. Biofuels Technology: A Look Forward. WIDER Working Paper, No. 2017/87. The United Nations University World Institute for Development Economics Research (UNU-WIDER), Helsinki, ISBN 978-92-9256-311-0.

Stoddard, I., Anderson, K., Capstick, S., et al., 2021. Three decades of climate mitigation: why haven't we bent the global emissions curve? Annu. Rev. Environ. Resour. 46, 653–689.

Smith, P., Davis, S.J., Creutzig, F., et al., 2016. Biophysical and economic limits to negative CO_2 emissions. Nat. Clim. Change 6 (1), 42–50.

Stavrakas, V., Spyridaki, N.A., Flamos, A., 2018. Striving towards the deployment of bio-energy with carbon capture and storage (BECCS): a review of research priorities and assessment needs. Sustainability 10 (7), 2206.

Sung, J.H., Back, S.K., Jeong, B.M., et al., 2018. Oxy-fuel co-combustion of sewage sludge and wood pellets with flue gas recirculation in a circulating fluidized bed. Fuel Process. Technol. 172, 79–85.

The Royal Society, 2018. Greenhouse Gas Removal. Available at: https://royalsociety.org/-/media/policy/projects/greenhouse-gas-removal/royal-society-greenhouse-gas-removal-report-2018.pdf.

Thiffault, E., Hannam, K.D., Paré, D., et al., 2011. Effects of forest biomass harvesting on soil productivity in boreal and temperate forests—a review. Environ. Rev. 19 (NA), 278–309.

Tillman, D.A., 2000. Biomass cofiring: the technology, the experience, the combustion consequences. Biomass Bioenergy 19 (6), 365–384.

Toor, S.S., Rosendahl, L., Rudolf, A., 2011. Hydrothermal liquefaction of biomass: a review of subcritical water technologies. Energy 36 (5), 2328–2342.

Vaughan, N.E., Gough, C., 2016. Expert assessment concludes negative emissions scenarios may not deliver. Environ. Res. Lett. 11 (9), 095003.

Vaughan, N.E., Gough, C., Mander, S., et al., 2018. Evaluating the use of biomass energy with carbon capture and storage in low emission scenarios. Environ. Res. Lett. 13 (4).

Vanbeveren, S.P., Ceulemans, R., 2019. Biodiversity in short-rotation coppice. Renew. Sustain. Energy Rev. 111, 34–43.

Vilches, T.B., Lind, F., Rydén, M., Thunman, H., 2017. Experience of more than 1000 h of operation with oxygen carriers and solid biomass at large scale. Appl. Energy 190, 1174–1183.

Companies and projects

Charm industrial; underground storage of pyrogenic bio-crude: https://charmindustrial.com/.

Drax Group; UK BECCS project developer: https://www.drax.com/about-us/our-projects/bioenergy-carbon-capture-use-and-storage-beccs/.

Gassnova; Longship CCS Project: https://ccsnorway.com/.

Toshiba; Toshiba Starts Operation of Large-Scale Carbon Capture Facility: https://www.global.toshiba/ww/news/energy/2020/10/news-20201031-01.html.

Chapter 13

Direct air capture

Peter Styring, George Dowson and Steve Rackley

13.1 Introduction

The possibility of directly extracting CO_2 as a resource from ambient air was first proposed as a step in the production of carbon-neutral fuels from renewable energy (RE) (Bandi et al., 1995) and, with the recognition of the increasing need to reduce or mitigate emissions, it has been proposed as a way of capturing CO_2 from "a multitude of dispersed CO_2 emitters" (Lackner et al., 1999; Dubey and Ziock, 2001). Direct air capture (DAC) can therefore act as a valuable alternative, or supplement, to decarbonization of the dispersed emissions from road, sea, and air transport, given the difficulty of capturing carbon emissions directly from such dispersed sources. Additionally, DAC is the only engineering-based carbon capture method that can operate remotely from the emissions point source, and can therefore mitigate the CO_2 released by distant industries and regions that may be unable or unwilling to address their emissions at source. DAC can also be seen as the last resort to "clean up" legacy emissions and its deployment at scale, alongside other NETs, is increasingly seen as essential if climate goals are to be met (IEA, 2021).

The potential for operation of DAC remote from the emitters also confers several potential advantages over stationary point source emissions capture. Emissions from any and all sectors of the economy can be captured with equal ease, with the DAC process being located wherever is most convenient. This would include locations close to CO_2 storage sites (such as above saline aquifers or formations suitable for in-situ mineralization), which would reduce the requirement for CO_2 pipelines or other transportation infrastructure (Lackner et al., 1999), as well as areas with abundant low-carbon energy production, and areas with otherwise low economic value, such as deserts. When compared to point source carbon capture and storage (CCS) facilities, greater economies of scale may also be achieved, since capacity will not be limited by the requirement for the capture unit to be within reach of a single or small cluster of stationary emitters (Keith et al., 2006).

The concentration of carbon dioxide ($[CO_2]$) in the atmosphere now exceeds 420 ppm by volume, or 0.042% concentration (NOAA Global Monitoring, 2022), while point source emissions from power stations and certain industries mostly range from 5% to 15%, and up to 35% in carbon-intensive industries like iron and steel manufacture. This might suggest that the thermodynamic minimum energy costs for capturing CO_2 from such point sources would be many orders of magnitude lower than for capturing a similar quantity of CO_2 from ambient air. However, as discussed in the next section, this is not necessarily the case, and thermodynamic costs for air capture can be within a smaller factor of those for point source capture (Lackner, 2013), although the very low atmospheric $[CO_2]$ does unavoidably mean that capturing significant quantities of CO_2 requires an enormous volume of air to be processed. Each tCO_2 captured requires at least 1570 tonnes of air to be processed, so that any viable capture process needs to be extremely robust and efficient.

While a variety of technologies have been developed for performing gas separations in general, not all are well suited to the challenge of separating CO_2 from ambient air. The main candidate technologies (Fig. 13.1) involve some sorbent material that preferentially interacts with CO_2 at these concentrations until it is saturated, whereupon the CO_2 is released during sorbent regeneration. (The term "sorbent" is used here for brevity to include both solid and liquid, adsorbent and absorbent materials). Other capture methods like cryogenic condensation or membrane gas separation (see Chapter 9) have been shown in early assessments to require more energy to capture a unit mass of CO_2 than is released by the creation of that mass through fuel combustion (Lackner et al., 1999) and have therefore been typically ruled out as implausible for DAC (Keith et al., 2010), although investigations continue in these areas (see Section 13.8).

Negative Emissions Technologies for Climate Change Mitigation. https://doi.org/10.1016/B978-0-12-819663-2.00008-3

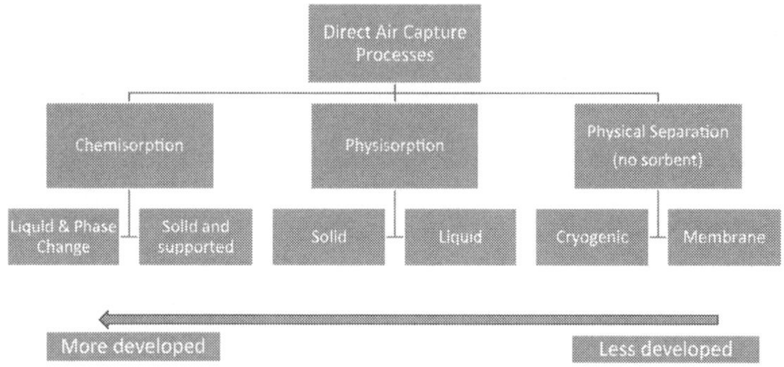

FIGURE 13.1 DAC approaches, indicating relative technology development levels.

Sorbents that are capable of capturing CO_2 at low concentration will also need to be cheap, highly selective toward CO_2, have a high loading capacity (expressed as molCO_2/kg-sorbent), fast recycle kinetics (reducing the sorbent inventory), and long operating lifetimes. All these factors will reduce process operating costs, especially by minimizing the amount of sorbent needed for a given capture rate. Solid sorbents must be physically robust, readily reusable, and highly tolerant of the other gas components in air. Liquid solvents must also have these features and additionally, to reduce losses, must not be readily evaporated.

Sorbent-CO_2 binding energy is a crucial factor in determining overall process energy costs. High binding energy will allow rapid CO_2 uptake, reducing sorbent inventory, but will necessarily increase the energy for sorbent regeneration, and hence process-energy costs. Conversely lower binding energy will typically result in slower CO_2 uptake, with less selectivity and lower capacity, increasing the sorbent inventory, although, with a lower regeneration energy requirement, it will more readily release the captured gas.

Fundamentally, sorbents can be classified by the type of interaction they have with CO_2; physical sorbents operate via the weak van der Waals interaction between the gas and the sorbent surface or involve gas dissolution (physisorption), whereas chemical sorbents will form much stronger chemical bonds and complexes between the sorbent surface or structure and one of the atoms in the CO_2 molecule (chemisorption). These two approaches occupy opposite ends of the balance between low binding energy (and hence slow uptake, low capacity, and regeneration energy) and high binding energy (and hence rapid uptake, high capacity, and regeneration energy).

Given the scale of the expected deployment of DAC, and therefore the importance of minimizing the sorbent inventory required for a given capture rate, the higher binding energy route, found in chemical sorbents, has typically been favored. These sorbents are considered to be most likely to lead to commercial air-capture systems, although development of alternative routes, which may have situational advantages, continues.

13.2 Thermodynamics of direct air capture

Individual DAC approaches may use thermal, electrical, or even mechanical forces to drive CO_2 separation, and when comparing them it is useful to keep in mind the fundamental thermodynamic minimum energy cost of gas separation, which informs the lower limit of energy use achievable in practice. Unlike the capture of CO_2 from postcombustion flue gas or syngas reprocessing, with CO_2 concentrations of ~15%, (equivalent to 15 kPa partial pressures), DAC requires a capture technology that can efficiently and effectively operate at CO_2 concentrations of ~420 ppm, corresponding at ambient pressure to a partial pressure of only ~42 Pa. This helps to set the theoretical minimum work required for the separation, assuming ideal gas behavior, by taking the negative value of the Gibbs free energy of mixing;

$$W_{\min} = -\Delta G_{\text{mix}} = -\text{RT}\ln\left(p/p_{CO_2}\right) \tag{13.1}$$

where R is the universal gas constant (8.314 J/mol K), T is the ambient absolute temperature and the logarithmic term is the ration of gas partial pressures. For ambient conditions (25°C = 298 K and $p_{CO_2} = 42$ Pa), $-\Delta G_{\text{mixo}} = 19.3$ kJ/mol. This is equivalent to 0.44 GJ/tCO_2 captured (1.61 GJ/tC) or 0.66 GJ/tCO_2 (2.42 GJ/tC) including compression to pipeline pressure. Note that this method also assumes no free energy change to the remaining air. From the logarithmic function in Eq. (13.1), it can be seen that the starting CO_2 concentration does not actually have a large effect on the minimum thermodynamic energy cost. Indeed, from a 15% CO_2 point source, the thermodynamic energy cost, calculated as earlier, reduces only to 4.7 kJ/mol, thus 0.11 GJ/tCO_2 (0.39 GJ/tC), despite the CO_2 stream being more than 350 times more concentrated.

However, approaching this thermodynamic minimum energy requirement in practice will be challenging. For comparison with existing CCS systems, benchmark amine-based processes designed for capture at concentrations of ~12% CO_2 from coal-fired electricity generation typically fall between 2.8 and 3.7 GJ/tCO_2 captured (Goto et al., 2013), where the thermodynamic minimum energy cost is 0.12 GJ/tCO_2. Additionally, if the thermodynamic limit for air capture at 420 ppm is compared to the energy released in the creation of the CO_2 (typically by combustion) then a limiting case can be calculated where the amount of energy used per mass of CO_2 being captured exceeds that which was released when the original fuel was burnt (note that this does not include thermal inefficiency of energy generation). For CO_2 from carbon, the standard free energy of formation is −394.4 kJ/mol or −8.96 GJ/tCO_2 created.

It can therefore be shown that the thermodynamic minimum energy required to remove the CO_2 equates to only ~4.9% of the energy released by the initial combustion of carbon. This observation motivated Lackner's early assessment of the potential feasibility of DAC (Lackner et al., 1999). This range, from 0.44 to 8.96 GJ/tCO_2, sets the idealized target range for DAC process energy consumption. Any energy used to power processes with energy consumptions in excess of this upper limit would, in principle, be better used in preventing further carbon emission by replacing any remaining carbon-based power generation, since prevention of emission has a lower energy cost than capture. However, there will be low- and zero-carbon energy sources including curtailed intermittent RE (that would otherwise go uncollected and unused), isolated power generation, and areas suitable for DAC where there may be no local carbon combustion emissions to avoid, and where DAC process above this threshold would still be sensible.

At the low end of the range, the thermodynamic limit implies a minimum binding energy of useable solvents or sorbents for capture from air, as this threshold of energy must be met and exceeded for the capture process to be even theoretically feasible. However, the energy penalty for release of the captured CO_2 and regeneration of the solvent/sorbent increases as the binding energy increases. If a comparison is made between the free energy of mixing (Eq. 13.1) and the overall system energy requirement (W_{system}), including solvent/sorbent regeneration energy, a figure for the overall efficiency (η) of the capture process can be reached;

$$\eta = W_{min}/W_{system} \tag{13.2}$$

Taking two examples of major types of DAC system that are under development, one using high-temperature liquid solvents with regeneration requiring temperatures of 800−900°C and one using solid sorbents with regeneration temperatures in the range 70−120°C, both types have regeneration energy requirements that, as might be expected, are well above the minimum given by Eq. (13.2). For example, the high-temperature sodium tri-titanate cycle (see Section 13.3.2) requires 90 kJ/mol for regeneration, equivalent to 2.04 GJ/tCO_2 (Sanz-Pérez et al., 2016), while the heat of calcination for $CaCO_3$ is 179 kJ/mol, equivalent to 4.07 GJ/tCO_2. Excluding other system energy requirements (discussed in more detail later), including for compression to storage pressure, these values exceed the minimum work requirement by factors of 4.6 and 9.2, respectively or, expressed in terms of the exergy or second law efficiency and Eq. (13.2), correspond to efficiencies of 22% and 11%, respectively.

The achievable efficiency is however not simply a function of the binding energy. As discussed earlier, a higher binding energy results in a higher and more rapid sorbate loading, so that the work applied for regeneration will be applied to a more concentrated product stream and less energy will be wasted thermally cycling unsaturated sorbent and inert sorbent mass. Therefore, the search for novel solvents, sorbents, and regeneration processes that approach the thermodynamic limit and balance the binding energy cost against sorption speed and capacity is a very active focus area for R&D work and is discussed in the following sections.

13.3 Liquid and phase-change chemisorption approaches

The fundamentals of absorption and the wide range of solvents and related capture and regeneration systems available, or under development, for gas separation applications are described in Chapter 7. Although these systems are mostly applicable to CO_2 capture at higher partial pressures, for example in relation to BECCS (see Chapter 12), liquid and solvent systems have also been developed for DAC. The key features of these systems are as follows: a CO_2-selective solvent, that readily reacts with CO_2 at air concentrations; a contacting system, typically either a packed absorber tower or cross-flow slab; and a solvent regeneration process that may involve solvent reboiling and CO_2 release in a stripping tower, flash separation of CO_2 (for high pressure input streams), CO_2 release through calcination of a precipitated inorganic carbonate, titanate, or other precipitated/phase change products, or through electrochemical methods.

13.3.1 Liquid amine sorbents

Of all of the carbon capture techniques that have been used at scale, chemical absorption using amine-based solvents in a packed absorber tower, also known as "wet scrubbing," is the most mature gas separation technology (Kearns et al., 2021). A number of different processes have been proposed and are being widely developed for CO_2 capture in power generation and other industrial applications, as discussed in more detail in Chapter 7. The general principle involves the reactivity of the amine functional group $-NH$ or $-NH_2$ reacting with the CO_2 to form carbamates salts (Fig. 13.2). These salts, upon moderate heating, return the purified CO_2 to the gas phase.

However, this technology is currently not well-suited to capture at very low CO_2 partial pressure, due both to the size of the absorber tower required to process large air volumes, and the high solvent circulation rate required for absorber operation versus the lower optimum rate for capture at low CO_2 (Kiani et al., 2020). Amine-based solvents also suffer from relatively slow reaction kinetics with CO_2 at air concentration, as well as losses due to oxidative and thermal degradation, which increase operating costs through solvent make-up requirements and the need to remove potentially harmful breakdown products (Vega et al., 2014). These factors lead to a high estimated cost for amine solvent-based DAC, despite the wide commercial availability of the underlying technology.

13.3.2 Alkaline solvents

Mineral hydroxides have been extensively investigated as solvents for CO_2 removal, both from flue gases and from air. These chemical sorbents have the advantage of reacting rapidly with CO_2 and, unlike amines, do not suffer from degradative or evaporative losses due to their inorganic structure, allowing far higher temperatures and more temperature cycles than can typically be tolerated by organic chemicals.

Calcium hydroxide $Ca(OH)_2$ was used as an exemplar in early proof-of-concept analyses of DAC (Lackner et al., 1999; Dubey and Ziock, 2001), but suffers the disadvantages of low solubility in water (1.73 g/L at 25°C, equivalent to 23.3 mmol/L), which limits the concentration of the solvent and hence the achievable maximum CO_2 loading to just 1.03 gCO_2/L. This in turn leads to a high energy requirement for solvent regeneration through calcination of the precipitated and dried $CaCO_3$ carbonation product at 850–950°C. Magnesium, another member of the alkaline earth metal series, can react similarly via $Mg(OH)_2$ to produce $MgCO_3$, which has a more moderate regeneration temperature of ~350°C (although higher temperatures are needed for complete release). This approach was also proposed by Dubey and colleagues for air capture (2001), but has the downside of an even lower hydroxide solubility, limiting the solvent to a maximum concentration in water of 6.4 mg/L at 25°C. At these concentrations, ~200 L of solution would be required to yield a single gram of separated CO_2.

A $Ca(OH)_2$ solution has nevertheless been investigated, using a spray–tower contactor and a calciner heated by concentrated solar power, with a total thermal energy input of 2.5 MJ/molCO_2 captured, which corresponds to 56.5 GJ/tCO_2, slightly more than 125 times the thermodynamic minimum energy (Nikulshina et al., 2006); this is explored in more detail in Section 13.4.3. In this case, and any method where the CO_2 capture energy exceeds 20 times the thermodynamic minimum energy cost (or 8.96 GJ/tCO_2), more energy would be spent in capturing CO_2 than was released when that CO_2 was generated.

Given these limitations of the rare earth elements, most subsequent work using caustic solutions has therefore focused on sodium hydroxide (NaOH), which in contrast can reach 25 M at 25°C, equivalent to 1 kg/L, allowing a maximum theoretical capture capacity of 550 gCO_2/L solvent (as two molecules of NaOH are required to form the carbonate Na_2CO_3). As such, this solvent has been used commercially for CO_2 capture from industrial flue gases as well as in early work on one of the most advanced DAC systems developed to date, by Keith and colleagues, firstly at the University of Calgary and later as Carbon Engineering (discussed in more detail below).

The absorption of CO_2 from air using sodium hydroxide was first investigated by Spector and Dodge (1946) and yields sodium carbonate (Na_2CO_3) as the reaction product, as shown in Fig. 13.3. Unlike the calcium and magnesium hydroxide routes, simple thermal cycling does not regenerate the NaOH and further reactions are required. To this end, two solvent regeneration routes have been investigated for DAC. The first, proposed by Zeman and Lackner (2004), operates via the

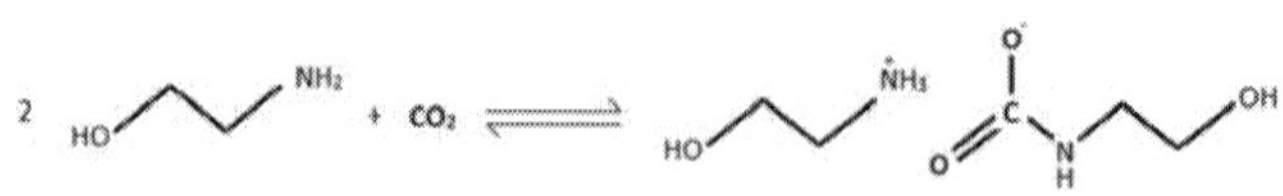

FIGURE 13.2 Reaction of monoethanolamine (MEA) with CO_2 to form carbamate salt.

CO_2 Capture

$$2\,NaOH + CO_2 \longrightarrow Na_2CO_3 + H_2O$$

Tri-titanate process, CO_2 release, 850 °C

$$7\,Na_2CO_3 + 5\,(Na_2O.3TiO_2) \longrightarrow 3(4Na_2O.5TiO_2) + 7\,CO_2$$

NaOH/Tri-titanate regeneration, 100 °C

$$3(4Na_2O.5TiO_2) + 7\,H_2O \longrightarrow 5(Na_2O.3TiO_2) + 14\,NaOH$$

FIGURE 13.3 Sodium hydroxide carbon capture followed by tri-titanate causticization regeneration process (Mahmoudkhani and Keith, 2009).

reaction of Na_2CO_3 with $Ca(OH)_2$ and is known as causticization, which is the general term for the regeneration of NaOH. Subsequent high-temperature calcination of the dried, precipitated $CaCO_3$ reaction product produces the CO_2 output and regenerates the $Ca(OH)_2$ reagent after the addition of water. This is an established process used for NaOH regeneration in the Kraft wood pulp process and the total energy cost is a minimum of 4.07 GJ/tCO_2 (Mahmoudkhani and Keith, 2009). This route forms the base case of the process being developed by Carbon Engineering (CE) and has an overall thermal energy costs of 8.1 GJ/tCO_2, plus an additional 1.78 GJ/tCO_2 of electricity, including compression to pipeline pressure (Socolow et al., 2011).

Although this route benefits from inexpensive chemicals, the high energy requirement for calcination is a significant disadvantage. The second approach, known as the titanate cycle, has been investigated by Keith and colleagues during the development of the CE DAC system (Mahmoudkhani and Keith, 2009). This is chemically more complex, involving poorly characterized reactions of sodium trititanate (initially formed by reaction of sodium carbonate with titanium dioxide, driven by decarbonization) with sodium carbonate, but is overall less energetically intensive than the causticization route. However, it requires that the sodium carbonate product of the carbon capture step is both dry and separated from unreacted sodium hydroxide, increasing process complexity. The total energy cost of this route is estimated to be approximately 2.04 GJ/tCO_2 (Sanz-Pérez et al., 2016).

The other common alkali metal, potassium, can also be used for carbon capture. Potassium hydroxide (KOH) has been used for CO_2 removal before cryogenic air separation since the 1950s and was investigated as an early DAC candidate by Bandi and colleagues, who investigated CO_2 capture with solvent regeneration by electrodialysis as the first step in a process to generate renewable fuels through electrochemical reduction of captured CO_2 (Bandi et al., 1995). The solvent was regenerated by reacting K_2CO_3 with H_2SO_4 releasing the captured CO_2, followed by regeneration of KOH and H_2SO_4 by separating the K_2SO_4 solution in a low-power two-compartment electrodialysis unit, using bipolar and cation exchange membranes (Fig. 13.4).

Due to the high energy requirement of this approach (9.7 GJ/tCO_2 captured) further work involving KOH has largely focused on chemical methods for solvent regeneration, although electrochemical regeneration methods have also attracted

CO_2 Capture

$$2\,KOH + CO_2 \longrightarrow K_2CO_3 + H_2O$$

CO_2 release using sulfuric acid

$$K_2CO_3 + H_2SO_4 \longrightarrow K_2SO_4 + CO_2 + H_2O$$

KOH regeneration by eletrodialysis

$$K_2SO_4 + H_2O \rightleftharpoons KOH + H_2SO_4$$

FIGURE 13.4 Capture using potassium hydroxide and sulfuric acid, followed regeneration by electrodialysis (Bandi et al., 1995).

attention with the aim of improving the energetics, as discussed later. Fig. 13.5 illustrates the more traditional approach using lime to regenerate KOH, which is the basis of the most recent proposal by Keith et al. (2018).

Similar to the NaOH cycle, discussed by Zeman and Lackner (2004) and others, the KOH caustic solvent is regenerated by reacting the carbonate-rich output of the air contactor with $Ca(OH)_2$, which is in turn regenerated by drying and calcination of $CaCO_3$ and hydration or "slaking" of CaO. A key component of the systems using the process shown in Fig. 13.5 is the fluidized bed pellet reactor, which represents a technical evolution from earlier approaches which involved precipitation of a lime mud of $CaCO_3$ and incurred significant handling costs for washing, filtering, and drying before calcination. In contrast, the pellet reactor, described by Burhenne and colleagues (2017), produces larger crystallized $CaCO_3$ pellets that are more easily handled for pretreatment ahead of calcination and is explored further in Section 13.3.5.

13.3.3 Phase change solvents

Phase change solvents are a special subclass of liquid CO_2 sorbents that incorporate the physical separation of the CO_2-rich state from the remaining unreacted or unsaturated sorbent. They exist both as biphasic liquid systems, where the CO_2-rich and CO_2-lean states are both liquid and immiscible, and precipitating systems where the CO_2-rich state exits the liquid or solution-based sorbent as a solid. They have been extensively investigated for postcombustion capture in view of their lower regeneration energy and their processing advantages compared to the benchmark amine solvents (see Chapter 7). Recently a number of research groups have investigated aqueous solvents based on iminoguanidine (CH_4N_4) that bind strongly to CO_2, forming carbonates and bicarbonates which crystallize due to their very low solubility. The product carbonated crystals can then be easily separated by filtration and the CO_2 released by mild heating to 80–120°C. Siepp and colleagues (2017) initially reported 2,6-Pyridine-bis(iminoguanidine), denoted as PyBIG, comprising two iminoguanidines bonded to a pyridine (C_5H_5N) ring (Fig. 13.6), however regeneration of PyBIG required a total energy input of

Carbon capture step:

$2\,KOH + CO_2 \longrightarrow K_2CO_3 + H_2O$ -95.8 kJ/mol

Causticization (KOH regenerated)

$K_2CO_3 + Ca(OH)_2 \longrightarrow CaCO_3 + 2\,KOH$ -5.8 kJ/mol

Carbon release (high temperature)

$CaCO_3 \longrightarrow CaO + CO_2$ +178.3 kJ/mol

Lime Slaking ($Ca(OH_2)$ regenerated

$CaO + H_2O \longrightarrow Ca(OH)_2$ -63.9 kJ/mol

FIGURE 13.5 Potassium hydroxide/lime direct air capture process.

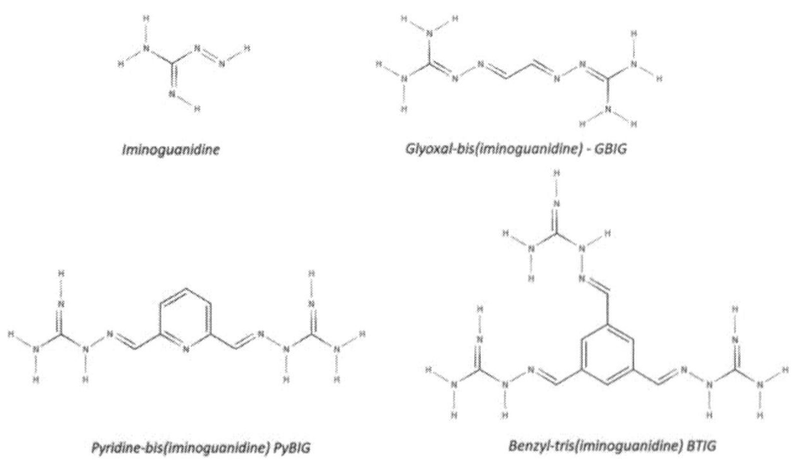

FIGURE 13.6 Structures of various iminoguanidine solvents for DAC.

223−288 kJ/molCO_2 (5.07−6.54 GJ/tCO_2) of which more than half (148 kJ/mol), was attributed to the sensible heat and dehydration of the crystals, and only 75 kJ/mol (equivalent to 1.70 GJ/tCO_2) for the subsequent CO_2 release (Custelcean, 2022). Comparing this latter figure with the thermodynamic minimum, calculated previously, illustrates an example of a suitably-low binding energy process for DAC that is let down by non-CO_2 related interactions, in this case with water, which raises the energy costs. However, this also suggests that if these additional sensible and dehydration energy costs can be avoided or mitigated by clever solvent design or process adjustments, these and similar sorbents can achieve the target energy requirements suitable for DAC described earlier in this chapter.

In later work using a simpler glyoxal ($C_2H_2O_2$) linked dimer (GBIG Fig. 13.6), a loading of 1.78 molCO_2/molGBIG, equivalent to 0.46 gCO_2/gGBIG, was demonstrated with a total regeneration energy of 152 kJ/molCO_2 or 3.45 GJ/tCO_2 (Williams et al., 2019). Other more complex iminoguanidines, such as the trimeric 1,3,5-benzene-tri(iminoguanidine) (BTIG; Fig. 13.6) have also been synthesized and investigated for DAC but have not yet matched the capacity or energy requirements of GBIG (Cai et al., 2020). In a related approach, these crystallizing solvents could also be used to regenerate other aqueous DAC solvents, such as K_2CO_3, Na_2CO_3, or simple amino acids such as glycine, by reacting the loaded solvent with the iminoguanidine to crystallize out the carbonate (Fig. 13.7), as an alternative to solvent regeneration in a reboiler/stripping tower (Seipp et al., 2017).

This would have the advantage that regeneration of the iminoguanidine occurs in the solid state, avoiding the need to heat the aqueous solution although, as noted earlier, the heating and dehydration of the crystals before CO_2 release also incurs an energy penalty. This regeneration approach has been demonstrated in a small-scale DAC unit based on a household humidifier and using an aqueous potassium glycine solvent (Brethomé et al., 2018).

13.3.4 Electrochemical solvent regeneration

As mentioned in Section 13.3.2, electrochemical regeneration of caustic solvents, after using sulfuric acid for the CO_2 release, is an alternative to the $Ca(OH)_2$ exchange−calcination−hydration approaches and, despite the high energy requirement of 9.7 GJ/tCO_2 found by Bandi and colleagues, further work has highlighted the possibility of significantly lower energy requirements and the easier integration of such methods with renewable energy (RE) sources, since no high-temperature thermal energy input is required.

One such electrochemical solvent regeneration method is bipolar membrane electrodialysis (BPMED) in which an alternating stack of ion-selective exchange membranes and water-dissociating bipolar membranes (see Chapter 9), is used to separate the loaded solvent into acidic and basic solutions. Captured CO_2 in the form of carbonate (CO_3^{2-}) and bicarbonate (HCO_3^-) ions in the loaded solvent is transported into the acidic solution, where the low pH drives the CO_2 hydration equilibrium toward $H_2CO_3^*$ (i.e., H_2CO_3 plus dissolved CO_2—see Section 6.2), and the low solubility of CO_2 in the acidic solution prompts the release of CO_2 into the gas phase (Fig. 13.8).

This method was first experimentally investigated by Nagasawa and colleagues who found a minimum solvent regeneration energy requirement of 92 kJ/molCO_2 (2.1 GJ/tCO_2), roughly half the energy required for $CaCO_3$ calcination (Nagasawa et al., 2009). A challenge with this method is the evolution of gas bubbles on the membrane surfaces, which reduces their effective area, increasing cell resistance and energy consumption and leading to localized hot-spots of high current density that can damage the membranes and reduce their operating life. A solution to this problem, proposed by Eisaman and colleagues, is to operate the electrolytic cell at high pressure (>5 bar), keeping the CO_2 dissolved until the pressure is released downstream. The energy penalty for pressurization amounted to an 8% reduction of energy requirements as a result of improved cell performance, under the experimental conditions investigated in this work (Eisaman et al., 2011). A techno-economic assessment by Sabatino and colleagues (2020), combining BPMED with the CE contactor described by Keith et al. (2018), indicated a total capture cost of \$773/tCO_2, three times higher than the CE high case of

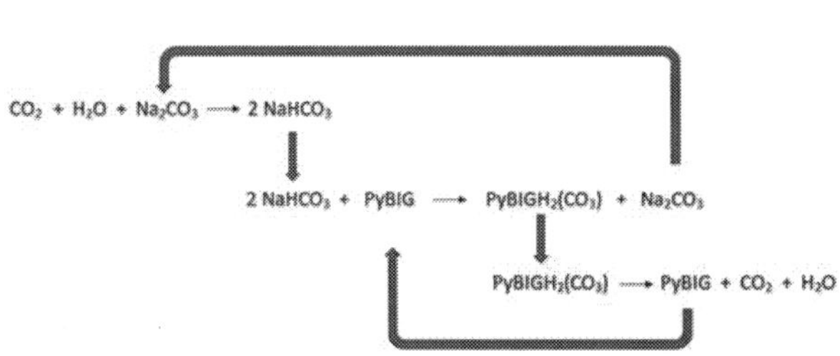

FIGURE 13.7 Aqueous carbonate solvent regeneration by reaction with iminoguanidine.

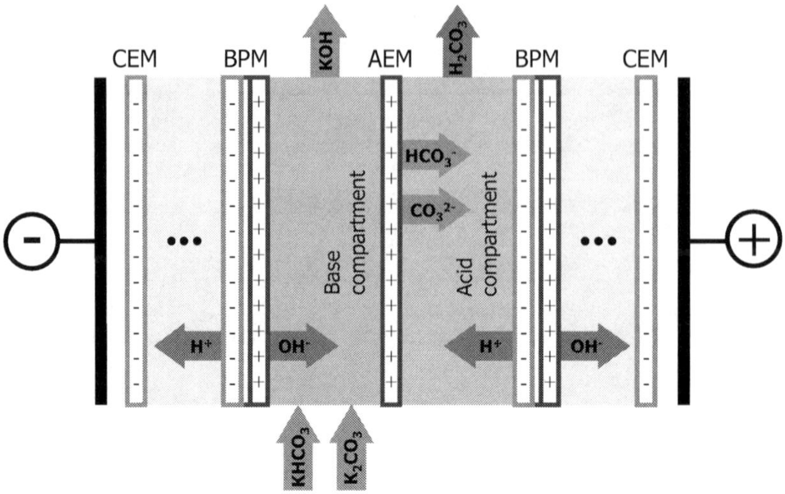

FIGURE 13.8 Electrochemical cell for caustic solvent regeneration. *After Sabatino et al. (2020).*

\$232/tCO_2, largely as a result of the expected membrane replacement cost (\$249/tCO_2 for a 3-year membrane life) and high electrical energy costs (\$364/tCO_2 at \$0.06/kWh), corresponding to an electrical energy demand of 21.8 GJ/tCO_2. Improved membranes, exhibiting greater thermal and mechanical stability and reduced costs, as well as reduced RE costs, would be needed to make such a system competitive with the baseline calcination route discussed in Section 13.3.2. Additionally, higher current densities are needed to shrink membrane surface requirements and there also needs to be a move away from platinum-group metal electrodes, especially for DAC scaling (Muroyama et al., 2020).

An alternative electrochemical solvent regeneration method was proposed by Shu and colleagues (2020) using an H_2-recycling electrochemical system (HRES), originally developed for ammonia recovery from municipal waste water, and shown schematically in Fig. 13.9. In this configuration, water splitting occurs at the cathode, evolving H_2 which is recycled to the anode compartment where it is oxidized, the protons being transported to the acidic compartment where (CO_3^{2-}) and (HCO_3^-) ions are converted to $H_2CO_3^*$. The solvent cations (Na^+ or K^+) are transported via a cation exchange membrane to the cathode compartment where they combine with OH^- to regenerate the caustic solvent. The energy requirement for NaOH regeneration was shown to be 374 kJ/mol (8.50 GJ/tCO_2) in lab-based experiments, with modeling results suggesting a minimum of 164 kJ/mol (3.73 GJ/tCO_2) after system optimization. These energy costs are similar to those reported for the BPMED, and regeneration costs are therefore likely to be comparable to those noted above.

13.4 Solid and supported chemisorption approaches

The fundamentals of adsorption and the range of sorbents and related capture and regeneration systems available or under development for gas separation applications are described in Chapter 8. The key features of these systems are a CO_2 selective chemical or physical adsorbent material, a contacting system that provides for periodic sealing or separation of the sorbent structure, and a sorption–desorption cycle with low energy requirements that may be driven by a temperature, pressure/vacuum or moisture swing. Rapid adsorption, driven by a sufficient binding energy, is also important for sorbent systems as it reduces the cycle time that in turn reduces the mass of sorbent required to achieve a target CO_2 output.

As discussed in Section 13.1, chemisorbents are generally thought to be better-suited to DAC conditions than physisorbents, considering their typically higher CO_2 capacity and selectivity. Additionally, in comparison with liquid-based systems, solid chemisorbents can have further improved capacity and selectivity through the absence of nonsorbent liquid, less likelihood of evaporative loss or degradation, and the ability to use surface structure to influence the CO_2-sorbent interaction. Particular examples include the rich structural and functional diversity of metal organic frameworks (MOFs), their tunability through pore geometry and pore chemistry modification by attachment of functional groups, and the potential for rapid, low-energy sorbent recycling, including charge-modulated processes, all of which motivates continued R&D work into their application as DAC sorbents.

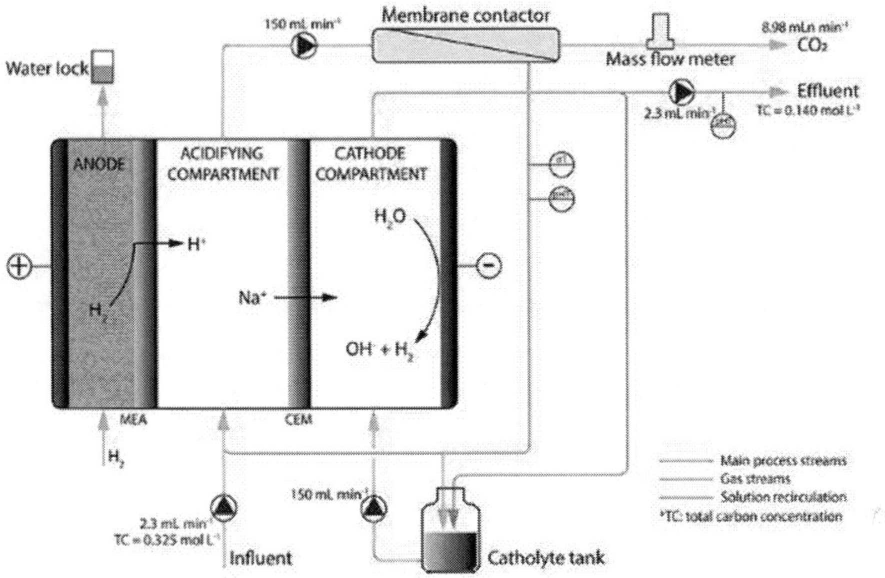

FIGURE 13.9 HRES electrochemical cell. *From Shu et al. (2020).*

One critical factor that solid sorbents encounter is the coadsorption of water on the solid surface. Removal of this water, especially in high-temperature regeneration processes can add significantly to the overall capture cost and process complexity, although Drechsler and Agar have demonstrated that good contactor design and energy recovery can mitigate this problem (Drechsler and Agar, 2019a). In other cases, explored in this section, water can act beneficially to increase uptake rate and capacity.

13.4.1 Supported amine sorbents

Given the general advantages of chemisorption and the potential additional advantages of solid sorbents mentioned above, early experimental work on sorbents for DAC has focused quite heavily on supported amine sorbents. These sorbents, also known as solid amines or amine solids, are prepared by combining amine or related organic functional groups and high specific area solid supports. As discussed in Chapter 7 and Section 13.3.1 in relation to absorption capture, amines capture CO_2 by forming carbamate salts under dry conditions:

$$2(RNH_2) + CO_2 \rightleftarrows RNHCO_2^- RNH_3^+ \tag{13.3}$$

or a carbonate or bicarbonate in the presence of water:

$$(RNH_2) + CO_2 + H_2O \rightleftarrows RNH_3^+ HCO_3^- \rightleftarrows RNH_3^+ CO_3^{2-} \tag{13.4}$$

Although many sorbents suffer reduced performance in humid conditions due to the competitive adsorption of H_2O alongside CO_2, this is not typically the case for supported amines. As the comparative stoichiometry of Reactions (13.3) and (13.4) shows, a mole of amine can adsorb a mole of CO_2 under humid conditions, due to the availability of OH^- ions, but only half a mole under dry conditions. Coadsorption of H_2O under humid conditions can also have a beneficial influence in the desorption step, since early desorption of H_2O provides an effective purge gas to drive air out from the sorbent before CO_2 desorption, helping to maximize the purity of the CO_2 product stream. The risk of oxidative amine degradation as the temperature rises is also reduced (Wurzbacher et al., 2016), and the preservation of sorbent outweighs the additional energy costs of drying the output gas stream.

A wide variety of porous materials have been investigated as amine supports, including alumina, carbon nanotubes, granular activated carbon, silica, resins, MOFs, and others. The support is not only a passive container for the sorbent molecule, but can also strongly influence its sorption capacity as well as the stability of the sorbent under regeneration. In experiments with polyethyleneimine (PEI) on mesoporous silica (MCM-41) Xu and colleagues showed an adsorption capacity of 0.22 gCO_2/g-PEI, double the capacity of PEI alone and more than 20 times the capacity of the bare silica support, attributing this increase to improved accessibility of the CO_2 affinity sites in surface bound molecules (Xu et al., 2002).

Amine solids are classified into three types, with different surface–amine bonding mechanisms resulting from different preparation methods, as illustrated in Fig. 13.10. Class 1 sorbents involve simple impregnation of support materials with free amines, Class 2 involve bonding siloamines or similar tethered amines chemically to the support surface, and Class 3 is an evolution of Class 2 where the chemically bound amines are extended and polymerized. In a scheme first proposed by Didas and colleagues (2015) hybrid Class 1–2 amine-solids have also been prepared, having both the covalent bonding of Class 2 materials as well as the high density of CO_2 affinity sites characteristic of Class 1.

Table 13.1 provides a comparison of the preparation and characteristics of these three classes.

Variations in temperature, humidity, and precise CO_2 concentration in the treated air can have dramatic effects on uptake capacity, depending on the class and specific nature of each supported sorbent. For those negatively affected by humidity, hydrophobic compounds can be included, as well as general stabilizers to prevent degradation. Additionally, some sorbents require a vacuum for CO_2 removal, rather than temperature swing alone, although this can also be carried out to reduce the required desorption temperatures and overall energy cost.

Branched amines such as PEI are commonly used in Class 1 sorbents and have the advantage of lower CO_2 binding energy compared to linear amines. PEI on silica-based support has the highest reported solid loading under DAC conditions (humid, ambient air at 400 ppm CO_2), reaching 7.31 mmolCO_2/g-PEI (0.32 gCO_2/g-PEI or 32 wt%) with 40 wt% PEI supported on pore expanding mesoporous silica MCM-41 (Sayari et al., 2016).

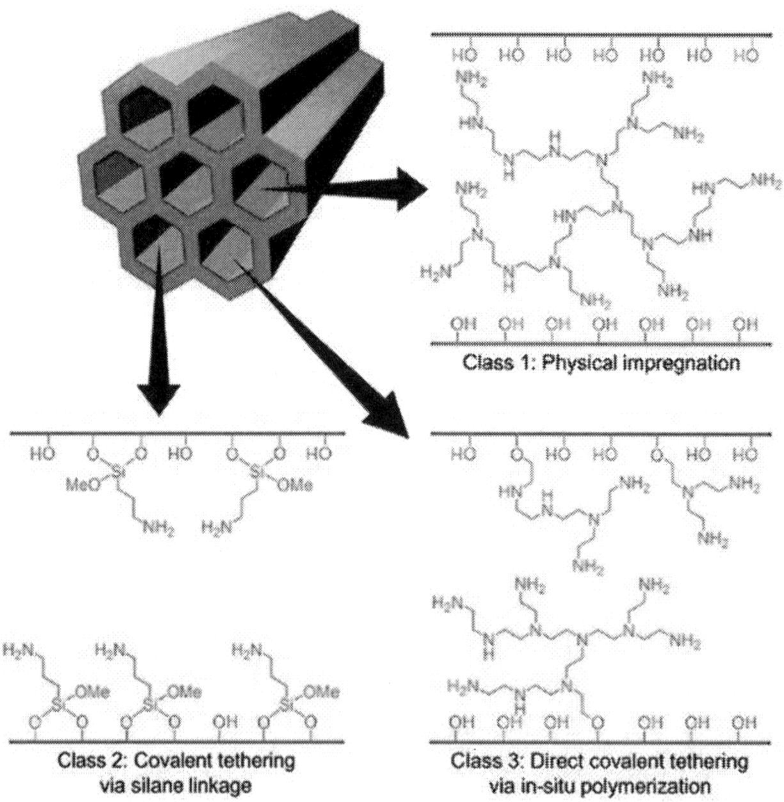

FIGURE 13.10 Illustration of the varying classes of solid-supported amines. *From Shi et al. (2020).*

TABLE 13.1 A comparison of supported amine classes.

Class	Preparation and bonding	Examples	Characteristics
1	Physical impregnation of presynthesized amine molecules, held in place by van der Waals forces	Simple amines such as MEA, DEA, and DIPA (ethanolamines) Low molecular weight Polyethyleneimine (PEI) M_w 400–25,000 Supports: micro- and mesoporous silica, fumed silica, porous alumina, and resins	Typically have the highest capacity and adsorption rate and are simplest to prepare but suffer from degradation and leaching, especially in the presence of water Polymeric and branched amines are used to reduce leaching of amines from the support surface, but this reduces sorption capacity and blocks sorbent pores Capacity: 0.8–2.3 mmolCO_2/g sorbent at 400 ppm
2	Amino-silane molecules grafted via a volatile solvent, leading to covalent bonding of silane to surface OH^- groups	Aminosilanes such as Aminopropyl silane (APS) on silica; Ethylenediamine on Mg_2(dobpdc) MOFs Also cellulose, alumina, and polymer supports	Typically lower uptake capacities than class 1 sorbents but tend to be more stable. High cost of sorbent materials can be an issue Capacity: 0.5–1.0 mmolCO_2/g sorbent at 400 ppm but higher capacities (3.9 mmolCO_2/g sorbent) have been reached under dry conditions (see later)
3	In situ polymerization of amine monomers, leading to direct covalent bonding	Covalently bound polymeric hyperbranched amines or amines polymerized onto support surface. Supports: Typically silica	Longer polymer chains and more amine sites increase sorbent capacity while the covalent bonding maintains the high stability of class 2. Capacity: Up to 1.5 mmolCO_2/g sorbent

An adsorption capacity of 2.83 mmolCO_2/g from ambient air has been demonstrated using a Class 2 MOF Mg_2(dobpdc) functionalized with ethylenediamine (Lee et al., 2014), while Liao and colleagues achieved a capacity of 3.89 mmolCO_2/g (0.17 gCO_2/g-sorbent) using the same MOF structure functionalized using the short diamine hydrazine (H_2N_4) (Liao et al., 2016), also at ambient conditions. The latter result is the highest CO_2 uptake capacity reported for MOF-based amine solids under dry DAC conditions and is attributed to the very high concentration of free amine groups (6.0 mmol/g) that can be incorporated into the MOF structure when using the smaller amine molecule.

Class 3 hyperbranched aminosilica materials, produced in situ polymerization of aziridines and azetidines on a support and is one of the approaches investigated by Global Thermostat for their air capture processes (Eisenberg and Chichilnisky, 2008). It is notable that there are few literature reports of regeneration energy costs for supported amine sorbents. Sorption isotherms are regularly reported between 50 and 100 kJ/molCO_2, which would imply a minimum energy cost for DAC using these sorbents of 1.1–2.3 GJ/tCO_2 (Varghese and Karanikolos, 2020), although actual desorption energies will be significantly higher.

13.4.2 Ion exchange resins and moisture swing capture

Closely-related to amine-based sorbents, ion exchange resins have an affinity for both CO_2 and H_2O. By changing the humidity of the environment, this can promote CO_2-sorption and desorption in a "moisture-swing" system. The sorbent in an anion exchange resin consists of a polystyrene or polypropylene resin sheet with amine ligands attached to the polymer backbone, exposing quaternary ammonium cations (NH_4^+), which are also found in many ionic liquids discussed in a later section. Amine supported on an amine-based anion exchange resin dispersed in a flat sheet of polypropylene was prepared in alkaline forms by Wang and colleagues (2011) to capture CO_2 from air. The resin, with ammonium salts attached to the polymer structure and hydroxide or carbonate groups as mobile counter ions, absorbs CO_2 when dry and releases it when wet (Fig. 13.11). In ambient air, the moist resin dries, spontaneously and subsequently absorbs a new portion of CO_2 for the next cycle which effectively raises the partial pressure from ~40 Pa to 5–10 kPa (5%–10% CO_2). Later studies have

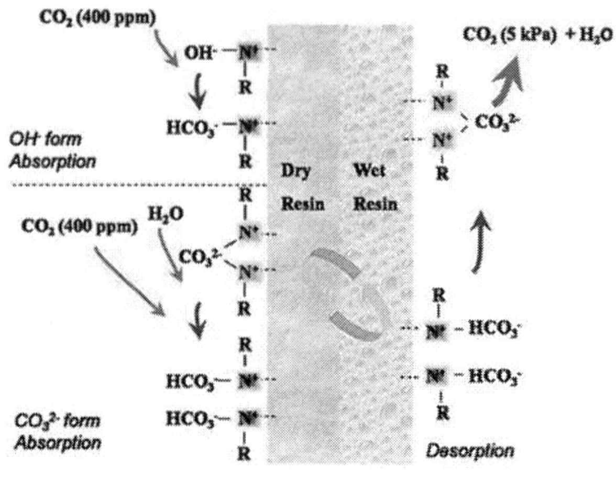

FIGURE 13.11 Moisture-swing sorption–desorption process on an ion exchange resin. *From Wang et al. (2011).*

shown that highly basic ion exchange resins, such as IRA900 demonstrate the highest CO_2 sorption capacities of up to 0.9 mmolCO_2/g sorbent with no short-term degradation over multiple cycles (Wang et al., 2020). Energy costs for this process, adjusted to produce liquid CO_2, have been estimated as approximately 50 kJ/mol CO_2 or 1.1 GJ/tCO_2 (Lackner, 2009).

A similar amine-functionalized polystyrene resin sorbent with an adsorption capacity from ambient air of 0.54 mmolCO_2/g sorbent (0.023 gCO_2/g sorbent) under dry conditions and 0.89 mmolCO_2/g (0.039 gCO_2/g sorbent) under humid conditions at 25°C has also been developed by Elfving and colleagues at the VTT Technical Research Centre in Finland (Elfving et al., 2017), but using a more traditional thermal vacuum swing (TVSA) cycle for sorbent regeneration.

13.4.3 Lime-based solid sorbents

The largest industrial application of lime-based materials is in the production of cement, and global cement production is responsible for ~7%–8% of global CO_2 emissions (Lehne and Preston, 2018). However, as described in Chapter 17, the slow carbonation of portlandite ($Ca(OH)_2$), produced during cement hardening by the hydration of allite (Ca_3OSiO_4) and bellite (Ca_2SiO_4), helps to reduce the net emissions from this industry on a decadal timescale as these cements "cure" after being poured and set. In one particular instance, involving a sealed environment designed to mimic long-term spaceflight, this curing process (which operated faster than expected) and the resulting drop in the internal atmospheric CO_2 threatened the experiment through a knock-on effect on the oxygen levels (Severinghaus et al., 1994). This carbonation reaction has been applied in a variety of configurations for CO_2 capture in the industrial sector, including chemical looping combustion as discussed in Chapters 8 and 12, but the disadvantage of portlandite as a potential DAC sorbent is the energy-intensive regeneration via the calcination of $CaCO_3$—as also discussed previously for NaOH solvent regeneration. Nevertheless, a number of routes have been investigated to address this challenge, mostly relying on integration with high-temperature heat sources to drive the sorbent regeneration.

The carbonation of lime (CaO), in the presence of water vapor to enhance reaction kinetics via calcium hydroxide, $Ca(OH)_2$, has been experimentally investigated for DAC by Nikulshina and colleagues (2009), using a fluidized bed reactor heated to 365–400°C by concentrated solar power. After carbonation the bed is switched to calcination mode by increasing the radiation flux to raise the bed temperature to 800–875°C (Fig. 13.12).

Assuming a 35% load factor and a CO_2 concentration of 400 ppm, a 500 tCO_2/year reactor, capturing approximately 1 molCO_2/s, would need to process 60 m^3/s of air using 286 kg of CaO sorbent and a solar thermal energy input of 2.5 MW (Nikulshina et al., 2009). Sorbent capacity and lifetime in fluidized bed applications can be improved by pelletizing lime with calcium aluminate cement binders, leading to the formation of mayenite ($Ca_{12}Al_{14}O_{33}$), which reduces sintering during repeated high temperature regeneration cycles and makes the sorbent pellets more resistant to attrition (Samari et al., 2019).

Heat integration with combined heat and power systems that use solid oxide fuel cells (SOFC) have also been investigated as a route to meet the energy requirements for calcination (Hanak and Manovic, 2018). This conceptual

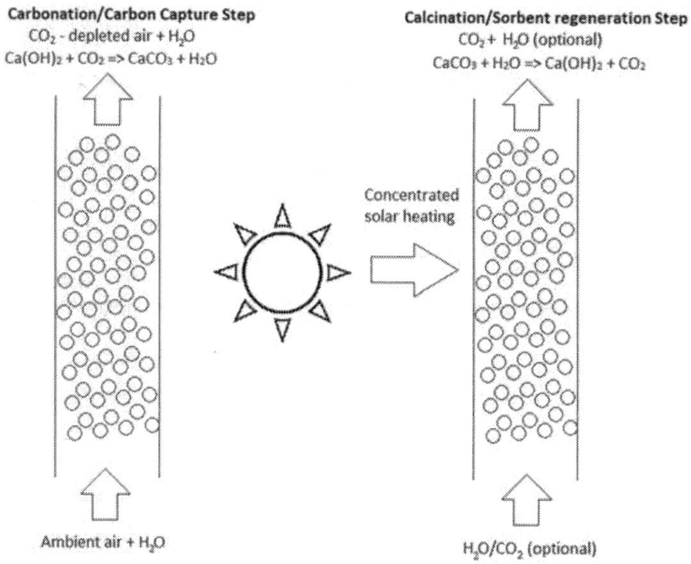

FIGURE 13.12 Schematic of a solar-heated fluidized-bed reactor for DAC using a CaO sorbent. *From Nikulshina et al. (2009).*

proposal combined a natural gas-fuelled SOFC delivering 23 MW_e with a calciner producing 300 tCaO/d, resulting in net negative emissions of 463.5 gCO_2/kWh_e assuming complete carbonation of the calcined material. In this system, carbonation was assumed to occur after distributing the calcined material (CaO) in an open environment, which would pose some problems for negative emissions verification and accounting. A 1 kW_e pilot of this system has been reported by Nabavi and colleagues (2019), and the concept is being further developed by the UK company Origen Power.

The open-air carbonation of porous portlandite structures, such as cladding tiles fabricated from pellets, has also been proposed as a passive approach to DAC (Abanades et al., 2020). Portlandite would be produced by calcination and hydration of limestone, with recovery of the heat of hydration significantly reducing the overall energy requirement (Criado et al., 2014). The estimated cost of this approach—\$140 to \$350/tCO_2, depending on the cost of fuel for limestone calcination, is within a factor of 2 of the latest estimates for large-scale DAC using liquid solvent technology. Abanades and colleagues consider the use of these tiles on residential buildings to be a viable option for individuals seeking to reduce their net emissions

13.4.4 Alkaline carbonate sorbents

As with the lime sorbents above, solid versions of the caustic CO_2 capture approaches explored in Section 13.3.2 have also been investigated, including the alkali metal carbonates Na_2CO_3 and K_2CO_3. As in the liquid phase, these are promising potential DAC sorbents that have the advantage of low cost, and low temperature regeneration under increased water vapor pressure via the formation of hydrated rather than anhydrous carbonates (the latter requiring a higher regeneration temperature). Once the bicarbonate ($NaHCO_3$) is formed with the assistance of water vapor, after drying it decomposes (Reactions 13.5 and 13.6) at a much lower temperature (approximately 100°C for $NaHCO_3$ and 100−120°C for $KHCO_3$) than when in aqueous solution, so that less high-grade heat is required for desorption (Toan et al., 2019).

$$2NaHCO_3 + 0.5H_2O_{(g)} \rightleftarrows Na_2CO_3 \cdot 1.5H_2O + CO_{2(g)} \quad (13.5)$$

$$2KHCO_3 + 0.5H_2O_{(g)} \rightleftarrows K_2CO_3 \cdot 1.5H_2O + CO_{2(g)} \quad (13.6)$$

Their disadvantage is slow uptake, limited by the rate of hydration of CO_2 to HCO_3^- (Reaction 6.2), and limited sorption capacity, being slightly higher for potassium-based than sodium-based sorbents. To improve this capacity, a wide variety of materials have been investigated to support potassium carbonate, with activated carbon, and alumina being the best performing to date (Okunev et al., 2000). Potassium carbonate supported on activated carbon has been investigated for CO_2 capture from flue gases (Shigemoto et al., 2006), and more recently, for DAC by Rodriguez-Mosqueda and colleagues, who achieved a peak capacity of 0.25 mmolCO_2/g-sorbent (0.011 g CO_2/g-sorbent), although salt loading was limited due to physical deterioration of the activated carbon support at higher salt loadings (Rodriguez-Mosqueda et al., 2018). Regeneration at temperatures as low as 50°C, using an air purge, gave a working capacity of just 0.1 mmolCO_2/g-sorbent (0.0044 gCO_2/g-sorbent), requiring therefore nearly 0.25 t-sorbent/kgCO_2 captured (Rodriguez-Mosqueda et al., 2019), and a product stream of 5000 ppm CO_2, making it potentially well-suited for CO_2 enhancement in greenhouses. Although not strictly a negative emissions application, this does contribute to emissions reduction by replacing enhancement systems that source CO_2 for greenhouse environments by natural gas or biomass combustion. The DAC potential for this approach led to a series of patents and the founding of the Dutch start-up company Antecy, which was acquired by Climeworks in 2019.

An alternative support, porous alumina (γ-Al_2O_3) has also been studied for DAC application. Chemisorption of CO_2 in the presence of water results in the production of crystalline potassium bicarbonate ($KHCO_3$) and potassium dawsonite ($KAlCO_3(OH)_2$) by reaction with the alumina support (Veselovskaya et al., 2013). A sorption capacity similar to supported amines has been demonstrated in lab experiments, 4.0–4.9 wt% corresponding to 0.91–1.1 mmolCO_2/g sorbent, but full regeneration requires heating to 250–300°C to decompose the dawsonite fraction and avoid significant capacity loss. Overall energy consumption is calculated at 333 kJ/molCO_2 or 7.57 GJ/tCO_2.

The enhanced weathering of alkaline minerals such as olivine distributed on the land was discussed in Chapter 11, and a similar approach that more directly delivers CO_2 as bicarbonate to its ultimate store in the oceans is discussed in Chapter 16. These methods require continuous mineral input, but alkaline mineral carbonation on the land surface, followed by regeneration, has also been considered as a DAC technology. Magnesia (MgO), produced by the "calcination" of magnesite ($MgCO_3$), has the advantage of a lower enthalpy of reaction for calcination of 118 kJ/mol (corresponding to 2.68 GJ/tCO_2 generated) compared to 179 kJ/mol for the calcium/lime route and at a lower calcination temperature range (typically 500°C or above), with lower temperatures leading to increased surface area and CO_2 uptake in the calcined product. McQueen and colleagues proposed the resulting magnesia as a distributed DAC sorbent, spreading the oxide on land plots for passive carbonation during a period of around 1 year, followed by collection and regeneration using oxy-fired natural gas fueled calcination with CCS (McQueen et al., 2020). The land requirement for this option, which captures 160 kgCO_2/m^2/y (1.6 ktCO_2/ha/year), is two orders of magnitude less than for biomass production, which typically removes at most 10 tCO_2/ha/year, and the techno-economic analysis gave an estimated cost of \$24–79/tCO_2 produced (including on-site capture from the calcination process) or \$46–159/tCO_2 avoided, accounting for grid supplied electricity costs and emissions. Application of this method at 1 GtCO_2 scale would require the extraction and processing of—about 25% of global magnesite resources (which total ca 8.5 Gt$MgCO_3$), but this could be supplemented by other mineral sources of Mg, such as anhydrous ultramafic rock (typically containing 40–45 wt% MgO), which are virtually unlimited (Kelemen et al., 2020).

13.5 Physisorption approaches

Solid physisorbents for CO_2 separation, such as activated carbon, zeolites, and metal-organic materials or frameworks (MOMs or MOFs), have been widely investigated for high-pressure applications such as postcombustion capture, but are less well-suited for DAC due to their lower selectivity to date, typical preferential adsorption of H_2O, and low CO_2 capacity. For example, Zeolite 13X under DAC conditions has a sorption capacity of just 3.5 μmolCO_2/g-sorbent, (0.15 mgCO_2/g-sorbent) while unadorned MOFs such as Mg-MOF-74 and SIFSIX-3-Zn have capacities below 0.2 mmolCO_2/g-sorbent. However, physisorption processes have the benefit of low to negligible desorption energies and require much lower temperatures than chemisorption processes. Liquid physisorption, otherwise known as simple dissolution, remains a niche area for CO_2 capture as a whole, and even more so for DAC. This route, using typically either ionic liquids or water itself, will usually only feature as part of a more complex capture process, often involving removal of the CO_2 after the capture step by a biological organism or via indirect chemisorption via carbonates.

13.5.1 MOFs as supports and sorbents

Metal-organic frameworks (MOFs), introduced in Chapter 8, have attracted considerable attention as potential DAC sorbents, in view of their great flexibility in structure and pore chemistry that allows them to be tuned for sorption capacity and selectivity. As described in the introduction, this balance of binding energy versus capacity is key to maximizing the performance of an air capture sorbent. The fine control that MOF structures can exert on this balance allows a degree of tuning of performance that is difficult to emulate with amines and other chemisorption processes. The selective adsorption process is essentially driven by the sieving nature of the MOF structures (Fig. 13.13). Pore sizes exclude other gases and

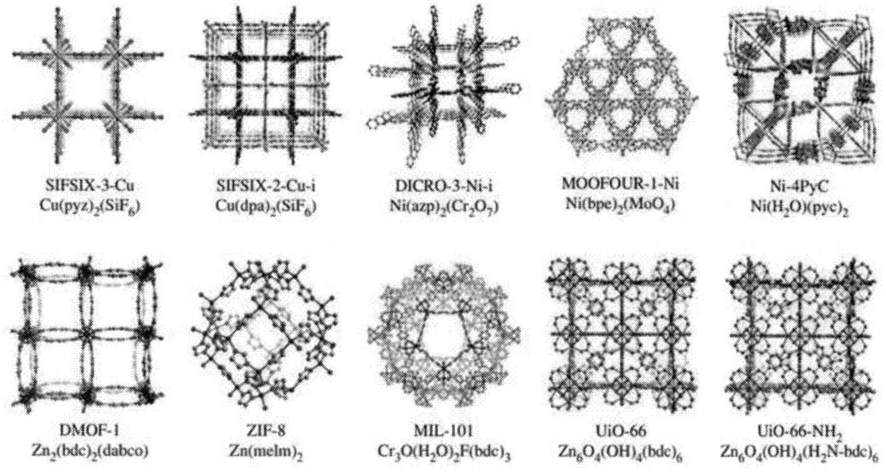

FIGURE 13.13 Structure of a range of MOM physisorbents. *From Madden et al. (2017).*

capture CO_2, despite it being larger than N_2, the major air component, as the CO_2 interacts with active sites, blocking the entry of undesirable gases (Li et al., 2020). At CO_2 concentrations of 15%, adsorption energy of these MOF structures can be as low as 16 kJ/molCO_2 (equivalent 0.36 GJ/tCO_2), approaching the thermodynamic limit at that concentration.

Desorption is usually performed using either temperature swing, vacuum swing, or both (in TVSA). Balancing these tunability advantages are the challenges of chemical, thermal, and mechanical stability that affect the performance under practical conditions where moisture, contaminants, and stress due to repeated thermal and pressure cycling can reduce capacity and compromise structural integrity (Ding et al., 2019a,b).

Among MOFs, Mg_2(dobpdc) (Mg_2-4,4′-dioxido-3,3′-biphenyldicarboxylate), based on the versatile MOF-74 structure, exhibits very high CO_2 capacity without amine functionalization (6.1 mmolCO_2/g-sorbent at 25°C and 15% CO_2 supply), but performs poorly under humid conditions, losing up to 80% of its adsorption capacity at 70% humidity. This is due to H_2O uptake on the exposed metal sites within the MOF that are responsible for the high CO_2 capacity.

Despite these issues, MOFs offer some of the lowest energy capture approaches. A fluoridated MOF, $NiNbOF_5$(-pyrazine)$_2$ also known as NbOFFIVE-1-Ni absorbed 1.3 mmolCO_2/g sorbent at 400 ppm and achieved complete desorption under vacuum and 55°C giving a corresponding energy consumption of just 54 kJ/mol or 1.23 GJ/tCO_2 (Guo et al., 2021). While this particular MOF is also resistant to water degradation, it is unfortunately highly crystalline, so difficult to use as a scale adsorbent, although with fluidized bed approaches it has demonstrated a capture rate equivalent to 2.1 kgCO_2/kg-sorbent per day of operation (see Section 13.7.2).

A compact DAC unit using a MOF polymer nanocomposite, termed Airthena, was shown to capture 140 gCO_2/d at 70%–80% output purity with a total system energy consumption of 2.3 kWh/kgCO_2 equivalent to 361 kJ/molCO_2 or 8.21 GJ/tCO_2 and a projected minimum energy cost of 1.6 kWh/kgCO_2 equivalent to 253 kJ/molCO_2 or 5.76 GJ/tCO_2. The system was run continuously for 2 months, producing a total of 8 kgCO_2, with plans to scale to a 6 kgCO_2/d system (2.2 tCO_2/year).

13.5.2 Liquid physisorption

Physisorption of CO_2 with liquids is a relatively niche area of carbon capture as a whole, and especially for DAC, with physisorption processes typically focusing on either low-cost mineral carbonates or tunable solid materials. However, in global terms, it is actually the second most effective form of atmospheric CO_2 removal after photosynthesis, as around 25% of the annual CO_2 drawdown in the global carbon budget is the result of physisorption into the ocean. While artificial processes cannot match the exposed surface area of the ocean, polar solvents may allow carbon capture at atmospheric partial pressures using temperature or pressure swing processes, with the degree of CO_2 solubility determined largely be solvent polarity. Ionic liquids (ILs), which are organic salts with low melting points (related to the anion exchange resins discussed in Section 13.4.2), are the most polar solvents known. Additionally, they have practically zero volatility at ambient temperatures, and their wide degree of synthetic flexibility—which enables variations in hydrophobicity, viscosity, and melting point—gives them a degree of tunability that cannot be found in other solvents.

While examples of moisture-swing polymeric ionic liquids (Wang et al., 2017) and MOF-encapsulated ILs (Lee et al., 2020) have been investigated, to date no liquid phase IL suitable for DAC has been identified, although the performance of supported ionic liquids, and the general variability of ionic liquid behavior, suggests that this is possible (Yang and Dai, 2021).

13.6 Systems configurations for liquid sorption DAC

Preconditioning of air for DAC, such as by heating, cooling, or significant pressurization, is not typically considered to be energetically feasible since, at 420 ppm, 2430 m^3 of air (3140 kg of air at STP) would need to be processed to expose a solvent or sorbent to an equivalent of 1 m^3 of CO_2 (1.95 kg at STP). Development work on air:solvent contactors has therefore typically focused on open systems requiring, at most, fanning of the input airstream. For reference, a 100 Pa (1 mbar) pressure drop would require 6.2 kJ/molCO_2 (0.14 GJ/tCO_2), roughly equivalent to the kinetic energy in air moving at 17 m/s (Lackner et al., 2012).

Early research into solvent-based DAC conducted by Keith and colleagues at the Energy and Environmental Systems Group at the University of Calgary, Canada, used an open contactor tower, with a large absorption area achieved using a fine spray rather than packing material which would reduce potential air flow (Stolaroff et al., 2006). While such a system reduces the air pressure drop and therefore any air pumping costs, solvent flow rate must also be constrained to avoid drop coalescence, reducing exposed surface area and limiting the CO_2 capture rate. The capital cost saving from eliminating packing material was found to be outweighed by the additional cost to increase the capture area.

A packed tower contactor configuration achieves an order of magnitude increase in the capture rate per unit area and was investigated in the experimental set-up shown in Fig. 13.14.

This 6 m high packed tower achieved a CO_2 reduction of 200 ppm (approximately 50% capture) with a solvent flow rate of 5 L/s and an air flow speed of 1.25 m/s and had a capture capacity of 15 tCO_2/year per m^2 of absorption tower cross-section. In these experiments, the capture performance of the tower was found to fall off only gradually with an intermittent solvent supply, reducing the energy required for solvent pumping. Capture performance was reduced by just 20% with a

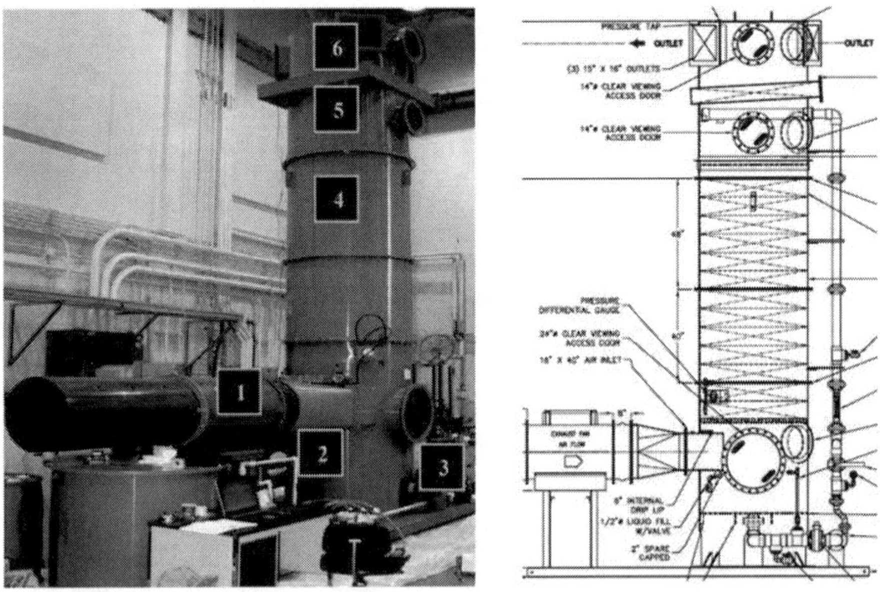

FIGURE 13.14 Experimental packed tower for DAC. Legend: 1: Fan/air drive, 2: Fluid basin, 3: Fluid pump, 4: Packed column, 5: Fluid distributor, 6: Air outlet and demister. *Photo courtesy University of Calgary, Canada; line drawing courtesy of Advanced Air Technologies Inc.*

5% solvent pump duty cycle (Mahmoudkhani et al., 2009), which reduced solvent pumping work by 90%. Including this optimization, the overall energy requirement for the capture system (excluding solvent regeneration, which is discussed in the previous sections) was 0.10 GJ/tCO_2 using NaOH and 0.09 GJ/tCO_2 using KOH.

Although packed towers are used commercially in many gas separation applications, the enormous air volumes required to be processed for DAC would lead to very large tower dimensions for a commercial-scale application. For example, a plant capturing 1 MtCO_2/year would need to process 5 km^3-air/d (assuming 420 ppm CO_2, 70% capture, and 95% system availability) requiring a tower cross-sectional area of 38,000 m^2 (equivalent to nearly 150 tennis courts) if drafted at an air speed of 1.5 m/s. Further development work has therefore focused on a cross-flow contactor configuration, as shown in Fig. 13.15, with horizontal airflow and vertical solvent flow, which also has the advantage of more straightforward wind-assisted airflow and reduces capital costs per unit capture capacity (\$/tCO_2.year) to roughly 20% when compared to a system based on vertical packed towers (Keith et al., 2018).

Latest contactor designs combine the elements shown in Fig. 13.16 into a slab some 20 m tall, 8 m deep, and 200 m long, and capable of capturing 100 ktCO_2/year. In a 1 MtCO_2/year capture facility, 10 such contacting units would be served by a central regeneration, compression, and processing plant (Holmes and Keith, 2012).

As with any open-air system, environmental conditions will be a crucial factor in determining the operational performance of a solvent:air contactor. Evaporative water loss in any aqueous-based system (particularly in dry environments), any freezing problems at low temperature, and reduced air density at altitude will all influence the selection of deployment locations. One alternative configuration that has been investigated experimentally for DAC is a hollow fiber membrane contactor, either with a solvent contained within the fibers absorbing CO_2 that diffuses through the membrane, or as a sorbent support, with the shell side holding a heat exchange medium (see Section 13.7.3). Stucki and colleagues

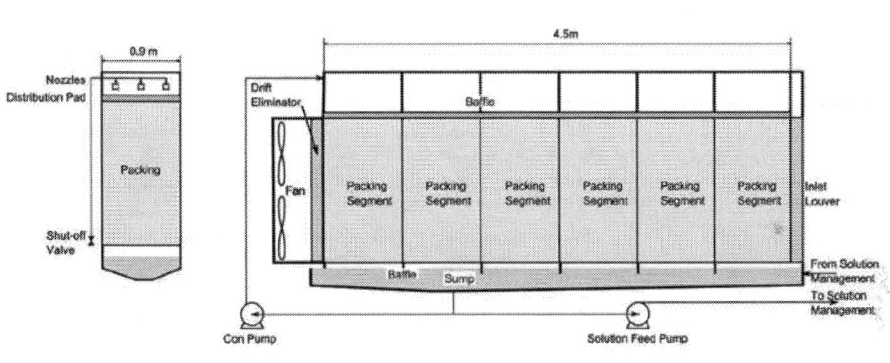

FIGURE 13.15 Schematic design of carbon engineering's cross-flow contactor. *From Holmes et al. (2013).*

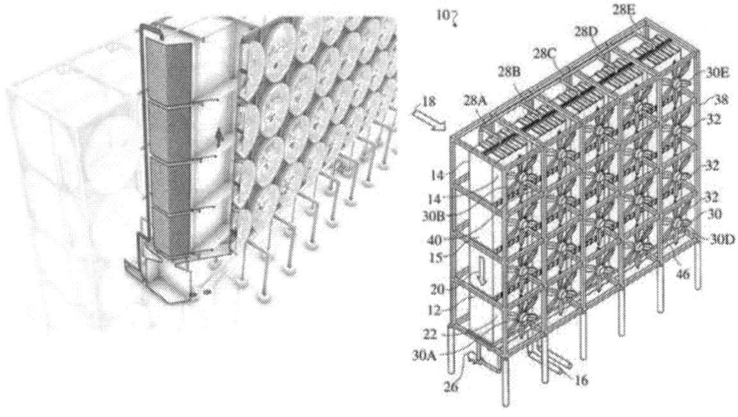

FIGURE 13.16 Schematic design of carbon engineering's slab contactor. *From Holmes and Kieth (2012).*

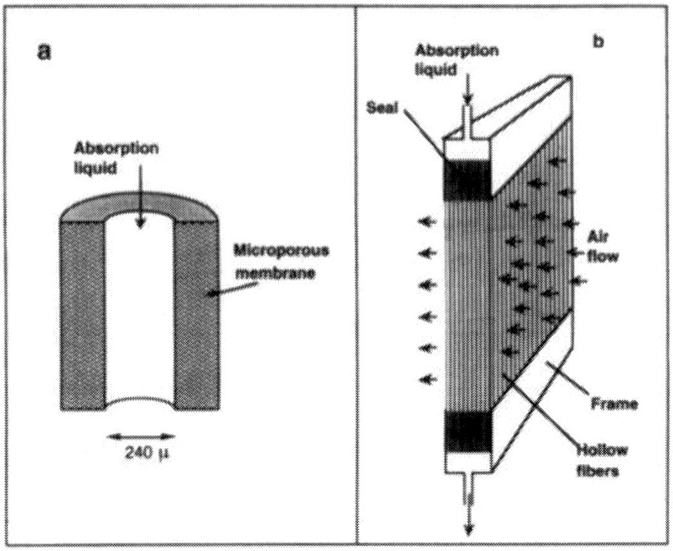

FIGURE 13.17 Hollow fiber air contactor for DAC. *From Stucki et al. (1995).*

investigated such an air contactor, shown schematically in Fig. 13.17, as a component of a plant to produce renewable methanol.

For an average wind velocity of 3 m/s, a hollow fiber air contactor reducing [CO_2] by 200 ppm would be able to capture 4.2 $kgCO_2/h$ (37 $tCO_2/year$) per m^2 of contactor cross-section, or 18 m^2 of membrane surface area (Stucki et al., 1995) —more than double the capture rate compared to an equivalently sized packed tower arrangement. Continuous electrolytic desorption of CO_2 and regeneration of the KOH solvent would require around 30 kW_e/m^2 of contactor cross-section, notionally provided by solar PV. Stucki and colleagues envisaged such a scheme operating in areas of steady wind conditions and high insolation, such as the west coasts of Africa, Australia, and the Americas. Although no economic analysis was given, a similar scheme has recently been considered by Bos and colleagues (2020) Using a supported amine sorbent in a packed bed absorber (discussed in Section 13.4.1) and powered by offshore wind. A 100 MW installation was estimated to produce renewable methanol from water and captured CO_2 at €800/t-methanol, with the full system (air capture, water electrolysis equipment, and methanol reactor) potentially installed in the base of the wind turbine tower, in line with its power generation capacity.

13.7 Sorption—desorption techniques for solid sorbents

In contrast to the liquid solvent systems discussed earlier, solid sorbents regenerated using temperature swing have the advantage that regeneration takes place at temperatures typically in the range from 70 to 120°C (compared to 800—900°C for liquid solvents) and can therefore take advantage of low-grade waste heat when integrated with other processes such as geothermal power generation (see Climeworks' Hellisheiði project, Section 13.9.2). Even lower heat demands can be realized when combining vacuum swing desorption processes with a small heat load (TVSA), provided the energy for evacuation is present (Guo et al., 2021). For processes that require some degree of vacuum swing for desorption, increased temperatures similarly reduce vacuum demands (Zhu et al., 2021), enabling the process to be customized to suit available energy sources. This flexibility of process energy requirements, in terms of both heat and vacuum, can give solid physisorbents a distinct advantage in some scenarios.

Both solvent-based and sorbent-based DAC systems face the challenge of constructing large air-contacting structures at low cost, while sorbent systems additionally require the sorbent material to be periodically sealed from the atmosphere during the desorption step. On a smaller scale, the effective design of a contactor system balances air-side limitations arising from the rate of CO_2 diffusion through a surface boundary layer, and sorbent-side limitations arising from the sorption reaction rate—for example, through the optimization of channel diameter in monolithic supports, or the surface area versus porosity in less structured systems (Lackner et al., 2012). On a larger scale, air flow considerations become more acute, with the use of any air blowers, or similar techniques to improve diffusion and reduce cycle times, adding to energy costs and this cost being proportional to the pressure drop experienced across the contacting structure (Wu et al., 2022). In the following section, some of the proposed architectures are explored and evaluated.

13.7.1 Packed beds

Laboratory investigations of sorbent properties have typically used packed beds, as they are the simplest to construct and test, but they present difficulties for large-scale applications and limit the dimensions of the air scrubbing units. A DAC air contactor using a temperature swing cycle needs to achieve efficient mass transfer during adsorption as well as fast and uniform heat transfer during desorption to minimize both cycle time (reducing sorbent demand) and energy consumption. The optimization of sorbent porosity, balancing the need to maximize sorbent volume and area against the need to minimize pressure drop, must also include heat transfer considerations. Achieving efficient heat transfer is challenging, particularly under a vacuum-assisted desorption, as the thermal conductivity of many sorbents is poor and convective heat transfer is negligible due to the reduced pressure. For example, Wurzbacher and colleagues (2016) reported a 2.5 h desorption time for a nanofibrillated cellulose sorbent bed with 72 × 72 mm cross-section, heated from its walls at 95°C. This, coupled with a long sorption period, significantly limits the number of cycles that can occur per day, increasing both sorbent and capital infrastructure costs (through enlarged sorption cross-section demands) per unit mass of CO_2 captured.

As with liquid-CO_2 contacting, presenting the sorbent surface as a thin film will accelerate sorption. While solid physisorbents are typically powders (at the macroscopic scale, at least), this thin-film-like approach can still be used by modifying the contact method. One example of modified packed beds includes using radial flow reactor design. These involve the gas flowing between an axial void down the length of a cylindrical column and a radial exhaust outlet (or inlet) path and have been demonstrated for air capture using supported amines (Yu and Brilman, 2020). This configuration would also be suitable for physisorption and, with some modifications, vacuum swing cycles, and can take advantage of reversed-flow procedures to enhance CO_2 output concentration (see Chapter 8).

Counter-current and cross-flow moving beds can perform similarly, again reducing required sorbent inventory by increasing uptake and desorption rate through reduced sorption distance (Yu and Brilman, 2017), although it faces hazards such as cavity formation and hydrodynamic "pinning," especially in cross-flow configurations where the gas travels perpendicularly to the solid (Schellevis et al., 2020). Moving belt systems may allow for high heat energy efficiency in thermal swing systems, where the cooling step of the cycle is carried out by heat exchange with a separate part of the belt that is undergoing heating, provided the process does not cause mechanical degradation of the solid sorbent (Drechsler and Agar, 2019b).

13.7.2 Fluidized beds

The use of moving solid sorbent surfaces to achieve the thin-film benefits of liquid solvents is also achieved in the case of a fluidized bed, which allows the shortest possible gas sorption distance by supporting the sorbent material on the gas stream itself. While this process will necessarily reduce the pressure drop compared to a similarly scaled packed bed and enable improved mass transfer of the CO_2, the fluidized sorbent will have lower effective density, thus requiring a larger vessel for the same CO_2 capture. Depending on the sorbent properties, it may also require a higher airflow, increasing presorption energy costs for fanning and pumping, especially when compared to monolithic approaches.

Zhang and colleagues proposed a bubbling fluidized bed (BFB) contactor using a silica-supported amine sorbent (reaction heat 90kJ/molCO_2 equivalent to 2.04 GJ/tCO_2). A conceptual 40 tCO_2/day BFB system, based on a bench-scale investigation of the sorbent performance, was anticipated to have an energy requirement of 6.6 GJ/tCO_2 and an operating cost of \$108/tCO_2 captured or \$152/tCO_2 avoided assuming 3.4 GJ/tCO_2 of electrical energy is provided by advanced NGCC generation with CCS and 3.2 GJ/tCO_2 of thermal energy from combustion of natural gas without CCS (Zhang et al., 2014). Similarly, Ghaemi and colleagues have demonstrated DAC using a fluidized bed of sodium hydroxide particles, illustrating the approach working across different chemisorption processes (Naeem et al., 2016).

Fluidized beds have also been shown to work with fast-cycling physisorbents. A bubbling fluidized bed of MOFs, supporting polyacrylates and using TSA, achieved only a modest total CO_2 loading of 0.063 gCO_2/g sorbent but this was countered by a dramatically shortened cycle time and increased working capacity, allowing 2.1 kCO_2/kg-sorbent to be captured per day at an energy cost of 54 kJ/molCO_2 (1.23 GJ/tCO_2), with only limited degradation of 1.3% after 2000 × 25 min cycles (Guo et al., 2021).

13.7.3 Monoliths and hollow fibers

In comparison with packed beds and even fluidized beds, monolithic and honeycomb-type sorbent structures have even lower pressure drop due to the many open channels and lower likelihood of blocking the gas flow. Many investigators envisage that practical implementation of adsorption-based DAC will employ a monolithic contactor of the type found in automotive catalytic converters, and the Global Thermostat LLC system (see Section 13.9.3) uses an amine sorbent supported on ceramic monoliths (Eisenberg and Chichilnisky, 2008). A wide variety of these monolithic sorbent architectures have been tested for DAC using TSA, TVSA, and VSA cycles. Examples include supported amine and alkaline chemisorbents, and physisorbents.

Kulikarni and Sholl modeled a monolithic contactor using an amino-modified silica to achieve unoptimized capture at 6.75 GJ/tCO_2 with a proposed capture unit capable of 1.1 tCO_2/d using a standard shipping container-sized unit and a proposed cost approaching \$100/tO_2 (Kulikarni and Sholl, 2012). Mukherjee and colleagues have demonstrated capture using physisorption monoliths (ultramicroporous nickel and silicon fluoride sorbent) at near atmospheric concentrations (1000 ppm) even in the presence of high relative humidity (Mukherjee et al., 2019).

Similar to monoliths in terms of free gas channels, hollow fiber sorbents (discussed in Section 13.6) prepared from silica/cellulose, or carbon nanotubes with dispersed silica or zeolite particles, among other materials, have also been proposed as an amine-solid support. When packed in a module, such a system would allow cooling water to flow on the shell side of the fibers to recover heat and reduce adverse temperature effects during sorption, and allow rapid thermal swing desorption by steam injection. Keller and colleagues have reported such a configuration, using carbon nanotubes with embedded silica particles formed into microtubes by filtration onto the bores side of polypropylene hollow fiber microfiltration membranes and then impregnated with PEI (Keller et al., 2019a). Sorption capacities of 1.1−1.9 mmolCO_2/g-sorbent were achieved using PEI loading from 5% to 20%, for a CO_2 partial pressure of 15 kPa, with further work under DAC conditions being planned. Temperature swing desorption with direct heat input to the sorbent from Joule heating has also been investigated for electrically conductive fibers (Keller et al., 2019b), although with a very high desorption energy requirement of 10−15 GJ/tCO_2.

13.8 Nonsorbent DAC methods

While the overwhelming majority of carbon capture and DAC approaches involve some kind of sorbent, there are alternative physical separation processes that allow CO_2 purification, even at ambient air concentration. While these approaches avoid binding energy, sorbent capacity, and stability issues, and as such are inherently scalable, the major drawback is that the separation process has to apply to all the gas components in the air, which amplifies energy costs dramatically.

13.8.1 Membrane-based DAC

Despite their potential for CO_2 capture from flue gases or other high CO_2 partial pressure applications (discussed in Chapters 9 and 12), membranes are not well suited to DAC due to the very low partial pressure of CO_2 at ambient conditions. This necessitates either a near vacuum on the product side and/or compression of immense air volumes on the feed side, at an energy cost of roughly 1.7 kJ/mol-air or 3.9 MJ/molCO_2 (88.6 GJ/tCO_2) for each doubling of CO_2 partial pressure.

However, this inherent penalty has not deterred research, and a number of membrane-based approaches have been investigated, encouraged by the process simplicity that such a system could offer. Amine facilitated transport membranes, discussed in Section 9.1.1, have been evaluated for DAC by Abd Rahaman and colleagues, using a PVA polymer layer functionalized by the addition of between 2% and 6% PEI and depositied by microfiltration onto a polymer hollow fiber support. At 1 bar feed pressure a permeance of 1.4 10^{-7} molCO_2/m^2/s/Pa or 415 GPU (gas permeance units, where 1 GPU = 7.5×10^{-12} m^3(STP)/m^2/s/Pa; CO_2 density at STP = 1.98 kg/m^3 or 44.9 mol/m^3). Permeance was found to be relatively insensitive to PEI content, although CO_2/N_2 selectivity increased from 40 to 300 when PEI content was increased from 2% to 6%. At 1 bar feed pressure, 55% CO_2 was recovered to a permeate stream containing 0.2% CO_2, and 40% recovery at 0.8% CO_2 with the feed pressure increased to 3 bar (Abd Rahaman et al., 2012). Feed pressurization to 3 bar incurs an energy penalty of 2.7 kJ/mol-air (6.2 MJ/molCO_2, 141 GJ/tCO_2), which, along with the modest enrichment illustrates the challenge faced by membranes for DAC—a challenge that led Keith and others to consider membrane-based DAC to be implausible (Keith et al., 2010).

After a long hiatus, the development of ultra-thin polymer membranes with CO_2 permeances of 30,000–40,000 GPU and CO_2/N_2 selectivity of 10–12 led Fujikawa and colleagues to reassess the feasibility of membrane-based DAC (Fujikawa et al., 2020). Process simulation for a four-stage membrane-based DAC system, shown schematically in Fig. 13.18, based on a membrane with 40,000 GPU CO_2 permeance and a CO_2/N_2 selectivity of 70, yielded a 42% CO_2 permeate stream requiring 14.2 kWh/kgCO_2 (51.1 GJ/tCO_2) to drive the vacuum pumps.

The energy requirement is many times higher than estimated for the latest generation of solvent-based systems, and for a wholesale electricity price of \$30–60/MWh would contribute \$480–960/tCO_2 to the overall capture costs of such a system. While this is still far from competitive, it represents a significant step forward and encourages further R&D on such membranes.

13.8.2 Direct CO_2 condensation/desublimation under antarctic conditions

The cryogenic separation of CO_2 from ambient air can in principle be achieved by cooling below −163°C (<135 K) resulting in condensation or "desublimation" to solid CO_2 (dry ice, referred to as CO_2 "snow"). The prodigious energy requirement of some 300 GJ/tCO_2 amply justifies the assessment by Keith et al. (2010) that such a scheme was implausible. However, widening the horizon to include potential locations where "ambient conditions" include a mean annual temperature below −50°C offers a different perspective. Agee and colleagues considered the possibility of locating a cryogenic DAC plant in Antarctica, using a liquid nitrogen cooling system as shown in Fig. 13.19.

In the capture unit, heat is captured from the ambient air (Q_{in}) by heat exchange with liquid nitrogen at 120 K (−153°C) resulting in the precipitation of CO_2 snow. The now "warmed" and evaporated nitrogen gas is compressed with adiabatic heating to 272 K (−1°C) at 600 bar. Heat is then extracted (Q_{out}) either for power or to heat buildings, liquefying the N_2 which is then depressurized across an expansion valve, with Joule–Thomson cooling back to 120 K. The notional system would be powered by coastal wind turbines, benefiting from strong katabatic winds around the pole, and would require a minimum of 617 GJ/tCO_2 (Agee and Orton, 2016). CO_2 snow would be excavated from beneath the desublimation heat exchanger and stored in insulated landfill sites, cooled to prevent sublimation at 190 K (−99°C), with the density of dry ice (1560 kg/m^3) resulting in a landfill volume of 6.4×10^8 m^3 per GtCO_2, or 40 m depth over a 4 × 4 km landfill area.

13.9 Technology readiness of DAC—example cases

Although many different sorbents have been investigated experimentally for CO_2 capture from dilute streams, few have been developed beyond the laboratory bench to the point of working capture systems. Most advanced are the systems developed by Carbon Engineering, Climeworks AG, and Global Thermostat, although several other companies are also advancing their processes to multi-tonne scale.

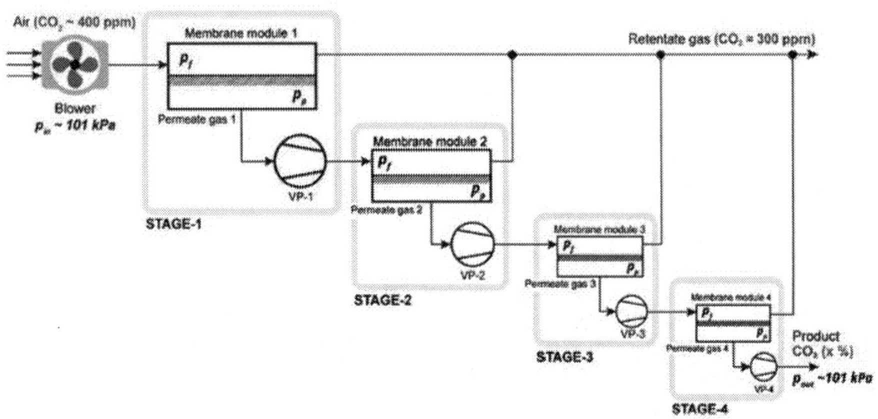

FIGURE 13.18 Schematic four-stage membrane-based DAC system. *From Fujikawa et al. (2020).*

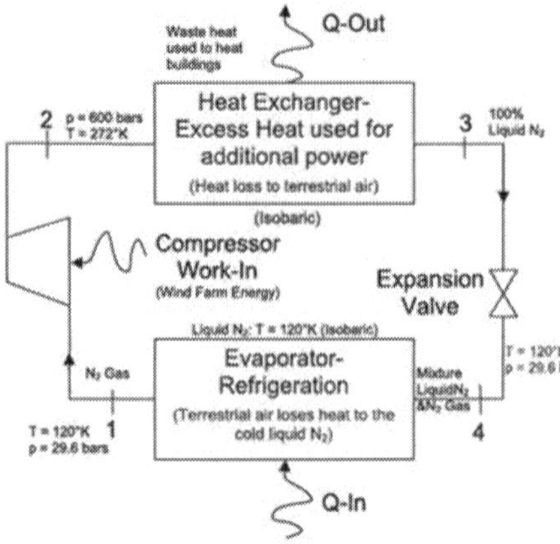

FIGURE 13.19 Closed-loop liquid−vapour cooling system for DAC by desublimation. *From Agee et al. (2013).*

13.9.1 Carbon Engineering

The NaOH/KOH system developed by Keith and colleagues at the University of Calgary and subsequently by Carbon Engineering is the only liquid solvent DAC system currently being actively developed toward commercial-scale deployment. The lab-scale investigations described in Section 13.3.2, which led to the founding of Carbon Engineering in 2009, were followed by a 1 tCO_2/d pilot plant, commissioned in 2015 (Fig. 13.20).

This pilot was designed to test the closed-loop operation of the process and assess operational risks such as contaminant build-up, as well as evaluating the performance under real operating conditions of the key system components, including the cross-flow air contactor (see Section 13.6), pellet reactor (described by Giacomin et al., 2020), oxy-fired calciner, and CaO-slaker, thereby demonstrating the technical feasibility and commercial viability of the technology. The pilot plant was upgraded in 2017 to produce liquid fuels from captured CO_2 (see Chapter 17), and the successful completion of the pilot brought this technology to TRL6, the threshold for commercialization.

Building on the results of this project, Carbon Engineering, Oxy Low Carbon Ventures, and other investors, formed the joint-venture company 1PointFive in 2019 to progress the engineering design and construction of a commercial scale 1$MtCO_2$/year DAC plant, essentially skipping demonstration and precommercial stages (TRL7/8). This plant, illustrated

FIGURE 13.20 Carbon Engineering 1 tCO_2/d pilot plant in British Columbia, showing air contactor (*blue, right*) and vertical calciner (*left*). *Courtesy: Carbon Engineering Ltd.*

FIGURE 13.21 Conceptual layout of the 1 $MtCO_2/d$ DAC plant under development by Carbon Engineering and partners. *Courtesy: Carbon Engineering.*

conceptually in Fig. 13.21, is expected to achieve capture costs in the range \$94−232/tCO_2 (Keith et al., 2018), depending on energy source (natural gas vs. natural gas + electricity), CO_2 delivery pressure, *n*th plant projected cost improvements, and cost of capital. Energy requirements are 8.81 GJ/tCO_2 for CO_2 delivery at 150 bar if fully gas fired, or 5.25 GJ/tCO_2 + 366 kWh/tCO_2 (total 6.57 GJ/tCO_2) for the same delivery pressure for a part-electric powered process, reducing to 5.52 GJ/tCO_2 for CO_2 delivery at 1 bar.

Pending a final investment decision, the plant is planned to be constructed in the Permian Basin oil region in Texas, with captured CO_2 being used for enhanced oil recovery; on planned completion around 2024, this will be the first DAC technology to reach TRL9. Further development of this process, with the aim of reducing capture costs and energy requirements below these estimates, includes a fully electric version of the calcium cycle, eliminating the need for natural gas input, and a liquid-phase regeneration system with no $CaCO_3$ crystallization, using an ion exchange membrane to remove CO_2 from the loaded solvent (Keith et al., 2018), with the aim of approaching the thermodynamic minimum regeneration energy for this route of 2.4 GJ/tCO_2.

13.9.2 Climeworks AG

Early research into DAC using amine-solid sorbents at the Swiss Federal Institute of Technology Zurich, included the investigation of nanofibrillated cellulose (NFC) as a novel support, typically produced by chemical or enzymatic treatment and mechanical delamination of wood pulp. The sorbent was synthesized by freeze-drying an aqueous suspension of NFC and an aminosilane such as 3-aminopropylmethyl-diethoxysilane or N-(2-aminoethyl)-3-aminopropylmethyldim ethoxysilane (Gebald et al., 2011). An air contactor comprising a packed bed of this filter material was developed, and the sorbent was regenerated by a TVSA swing at 70−95°C and 30−150 mbar, yielding an average sorption capacity of 0.9 mmolCO_2/g-sorbent (0.04 gCO_2/g-sorbent) (Gebald et al., 2013). A significant limitation of this prototype system was the long cycle time of ~5 h, roughly equally split between sorption and desorption, allowing just five cycles per day, although more recent results on NFC supporting PEI demonstrated shorter cycle times (Sehaqui et al., 2015).

Climeworks, founded in 2009 by Gebald and Wurzbacher, developed a prototype 50 tCO_2/year modular air contactor based on these results in 2014, and in 2017 became the first company to commission a commercial DAC plant, with 18 modular contactors capturing 900 tCO_2/year for supply to a nearby greenhouse (Fig. 13.22). The plant, located in Hinwil, Switzerland, requires 1.5−2.0 MWh_{th} and 200−300 kWh_e/tCO_2, equivalent to 6.12−8.28 GJ/tCO_2 overall.

Also in 2017, a Climeworks contactor module was installed to supply DAC CO_2 to the CarbFix2 project in Iceland (see Section 10.7.1), combining DAC with in situ mineralization of the captured CO_2 (Gutknecht et al., 2018). The Orca project has expanded this pilot installation, deploying second-generation contactors to achieve a 4 ktCO_2/year capacity. Construction was completed in September 2021 with the new plant adjacent to ON Power's Hellisheiði geothermal power plant, which will provide RE to power the project. Underground storage of captured CO_2 by in-situ mineralization will be performed by Carbfix.

FIGURE 13.22 Climeworks 900 tCO_2/year DAC plant in Hinwil, Switzerland, including the customer's greenhouse in the background. *Courtesy: Climeworks AG.*

Climeworks has also been a partner in DAC with carbon utilization power-to-liquids projects. In 2014, the CO_2 stream for an e-fuel pilot project completed by Audi AG and Sunfire GmbH was part-provided by Climeworks, and the company is also a partner in a project led by Norske e-Fuels to build a renewable aviation fuel plant in the Herøya Industrial Park, Porsgrunn, Norway. Plant construction started in 2022, with first production expected in 2024 at an initial capacity of 10,000 m^3/year of renewable fuel, equivalent to 25 $ktCO_2$/year avoided emissions through fossil fuel replacement. Expansion to 100,000 m^3/year is envisaged for 2026, although it is unclear whether the full 250 $ktCO_2$/year input stream would be provided through DAC.

While Climeworks DAC systems have been in commercial operation since 2017, installations to date have not exceeded 4000 tCO_2/d. Considering commercial DAC deployment at a climate-relevant scale ($\sim$1 $MtCO_2$/year) as TRL 9, commissioned projects reflect a prototype system (TRL 6) readiness level, including both in situ mineralization and e-fuels applications. The DAC component of the 25 $ktCO_2$/year Norske e-Fuels project may be considered a precommercial scale demonstration project, leading to TRL 7 in 2024. However, the long TSA cycle time is a fundamental challenge for the current configuration and, while later developments of amine-NFC sorbents have shown reduced cycle times, the 2019 acquisition of Antecy, which developed a low-temperature K_2CO_3 approach, may signal a change in the company's development direction (Fasihi et al., 2019).

13.9.3 Global Thermostat

Global Thermostat LLC, founded in 2010, is another leading developer of supported amine sorbents, using honeycomb ceramic monoliths supporting a proprietary amine, integrated with low-cost process heat or concentrated solar power delivered via an inert gas or steam purge at 85–100°C for sorbent regeneration. The prototype contactor system consisted of a cartridge containing 1280 alumina-coated cordierite ($(Mg,Fe)_2Al_4Si_5O_{18}$)-ceramic monoliths, with 30 m^2 total surface area, which is lowered into a "hot box" for regeneration, with a 16 min cycle time reported.

Among sorbent-based systems, Global Thermostat has reported the lowest energy demand at 150–260 kWh_{el}/tCO_2 (0.5–0.9 GJ/tCO_2) plus 1.2–1.4 MWh_{th}/tCO_2 (4.2–5.1 GJ/tCO_2) (Fasihi et al., 2019), with heat exchange between adjacent adsorbing and desorbing modules significantly reducing thermal energy input requirements.

In 2010, Global Thermostat with partners Corning (monoliths) and BASF (amines) commissioned a 500 tCO_2/year pilot plant at SRI International in Menlo Park, in California, which was upgraded to 1 $ktCO_2$/year in 2013. A number of plants with capacities up to 4 $ktCO_2$/year have since been promoted to provide CO_2 at industrial sites including for synthetic fuels, chemical and fertilizer production, and beverage carbonation. Modular systems designs, with capacity of $\sim$40 $ktCO_2$/year, have since been developed on the basis of the pilot performance. In 2019, Global Thermostat and ExxonMobil began working together to assess the scalability of Global Thermostat's DAC technology, and the joint venture was extended in 2020 and 2022. While Global Thermostat's DAC technology is clearly commercially viable at the scale appropriate for niche markets (TRL 8), systems below 10 $ktCO_2$/year can at best be considered prototypes (TRL 6) from the perspective of commercial deployment at climate impact scale.

13.9.4 Other DAC developers

Global Research Technologies/Silicon Kingdom Holdings

Research investigations by Lackner and colleagues at Los Alamos National Laboratory and Columbia University identified ion exchange resins as a promising DAC sorbent, since the desorption cycle is driven purely by the adsorption and evaporation of water, without any additional heat input (Lackner et al., 1999; Lackner, 2009). Further development of the concept was undertaken by Global Research Technologies and at Arizona State University (ASU), leading to a small-scale prototype system developed at ASU's Center for Negative Carbon Emissions (Arizona State University, 2019).

Capture columns (Mechanical Trees) contain 150 sorbent tiles that are saturated within 20 min of exposure to air before retracting into a sealed enclosure, which is partially evacuated before the humidity swing releases an initial 1% CO_2 product stream. The stream is concentrated by reacting with Na_2CO_3 to produce $NaHCO_3$, which is then heated to produce CO_2 plus steam; condensation of the steam yields a pure CO_2 stream that is compressed for delivery. Each column captures 85 $kgCO_2$/d (12 columns/tCO_2/d). Silicon Kingdom Holdings is planning a 12–24 column pilot project, capturing 1–2 tCO_2/d, with phased expansion to 30 tCO_2/d. A climate-impact scale DAC plant comprising 120,000 columns (4 $MtCO_2$/year) would occupy a land space of just 2 km^2.

Similar to the Climeworks projects in Iceland, an integrated project combining remote RE, humidity swing DAC, and in situ mineralization in basalt has also been considered by Goldberg and colleagues (Goldberg et al., 2013). Located in the French Kerguelen Islands in the Southern Ocean, the concept would exploit the region's exceptional wind resources but could equally well be geothermally powered, potentially at lower capital cost.

Hydrocell/Soletair

The amine-functionalized polystyrene resin sorbent developed by Elfving and colleagues at the VTT Technical Research Centre in Finland (see Section 13.4.2) has been applied in a small fixed bed DAC system by the Finnish company Hydrocell OY. The DAC unit, containing 30 kg-sorbent in each of eight beds, captures 3.8 $kgCO_2$/d, regenerated by a TVSA cycle with energy requirements of 7.6 $kWh_{th}/kgCO_2$ and 1.4–7.3 $kWh_e/kgCO_2$, depending primarily on sorption–desorption cycle times (Bajamundi et al., 2019). The Hydrocell DAC unit provides CO_2 to a power-to-liquid system developed in the SOLETAIR project, now being commercialized by the Finnish company also named Soletair. A pilot system, commissioned in 2018, combines the DAC unit with an electrolyzer for hydrogen production, and a synthesis reactor for hydrocarbon production, with an output of 6.2 kg of oil and wax per day.

Airthena

Airthena is a DAC unit using a MOF-polymer nanocomposite sorbent coated as a thin layer directly onto the vanes of a spiral wound resistive heater, developed by Sadiq and colleagues at Monash University and CSIRO in Australia (Sadiq et al., 2020). The pilot-scale unit has nine sorption chambers each containing 1 kg-sorbent with a 0.42 $molCO_2$/kg-sorbent capacity and can produce ~2 tCO_2/year with 60% purity. The unit has a power requirement of ~0.6 kW and requires 2.3 $kWh/kgCO_2$ (8.2 $MJ/kgCO_2$), although additional energy would be required to deliver a high-purity product stream.

13.10 Cost and potential of DAC

Given the low CO_2 concentration in air, it is not surprising that DAC is the most costly NET that has so far been demonstrated outside the laboratory, and techno-economic estimates of both currently achievable, theoretical, and eventual realistic target capture costs vary widely. Table 13.2 summarizes some recent cost estimates for various DAC systems configurations.

As can be seen in Tables 13.2A and 13.2B, wide and not always consistent ranges have been proposed for DAC costs, and these will continue to evolve and change over time. Although Lackner and colleagues (1999) suggested a cost as low as \$15/tCO_2 as a reasonable target price for DAC (assuming a $Ca(OH)_2$ sorbent regenerated through coal-fired calcination of $CaCO_3$, split roughly \$5 for CO_2 capture and \$10 for calcination), more recently, based on the cost of flue gas scrubbers, Lackner and (different) colleagues have argued that \$25/tCO_2 may be a plausible lower bound for sorbent regeneration costs in the long term (Lackner et al., 2012; \$32/tCO_2 if inflation adjusted to 2022), although no commercial process has reached or is likely to approach this figure in the near future. More recently, Azarabadi and Lackner (2019) demonstrated

TABLE 13.2A Estimated DAC costs across a variety of capture approaches for liquid solvents.

DAC configuration	Cost ($/t$CO_2$)	Comment	References
Liquid solvents			
Conceptual liquid solvent system	<136	Notional NaOH solvent, regenerated via addition of CaO and calcination of $CaCO_3$ to regenerate CaO.	Keith et al. (2006)
Generic; based on second law efficiency considerations	300 −2500	Cost per tCO_2 avoided; depending on second law efficiency and energy source (hydro at $\eta = 5\%$ to solar PV at $\eta = 2\%$)	House et al. (2011)
Conventional amine-based liquid solvent and absorber tower	676	Large absorber unit size, excessive solvent flow rate, and loss point to the need for optimized gas contactors	Kiani et al. (2020)
Liquid solvent with packed-slab contactor	336–389	Calculated contacting cost alone of \$60/tCO_2 in response to American Physical Society (APS) report suggesting \$240/tCO_2 (also contacting only)	Holmes and Keith (2012) Socolow et al. (2011)
Liquid solvent with packed-slab contactor	94–232	KOH solvent, regenerated via $Ca(OH)_2$ exchange, calcination, and hydration. Based on pilot plant performance	Keith et al. (2018)
Liquid solvent with packed-slab contactor	313–773	Carbon Engineering system (Keith et al., 2018) with electrochemical solvent regeneration	Sabatino et al. (2020)

TABLE 13.2B Estimated DAC costs across a variety of capture approaches for solid sorbents.

DAC configuration	Cost ($/t$CO_2$)	Comment	References
Solid sorbents			
Solid sorbent with LT regeneration	250–300	€222/268/tCO_2 with/without free waste heat declining to €54/71/tCO_2 by 2050 (2019 exchange rate 1.12 \$/€)	Fasihi et al. (2019)
Solid sorbent with LT regeneration	29–91	Calculations based on sorbent cost and lifespan	Azarabadi and Lackner (2019)
Solid sorbent with Joule heating regen.	45–350	MOF-polymer nanocomposite sorbent in small-scale capture unit	Sadiq et al. (2020)
Solid sorbent, no regeneration	140–340	Passive carbonation of large $Ca(OH)_2$ structures	Abanades et al. (2020)
MgO sorbent with calcination of magnesite	46–159	Passive carbonation of MgO spread on land, with regeneration by calcination of magnesite ($MgCO_3$)	McQueen et al. (2020)
LT solid sorbent	75–86	Calculated based on solar PV energy provision in North Africa, falling to \$32-\$58/tCO_2 under ideal conditions	Breyer et al. (2019)

that minimum costs for sorbent-based systems under various conditions range from \$29 and \$91/tCO_2. These figures are broadly supported by Fasihi and colleagues, who projected DAC costs to decline to between €38 and €54/tCO_2 (\$41−58/tCO_2) by 2050 for a range of implementation scale and learning curve assumptions and with access to free low-grade heat for regeneration rising to between €54 and €71/tCO_2 (\$58−76/tCO_2) without free heat provision (Fasihi et al., 2019). Along with several other estimates in the tables above, a consensus value for the lower bound of cost at approximately \$50−100/tCO_2 is indicated.

Looking at reported energy costs from the three largest operating DAC groups gives further information about current and target costs per tonne, although it should be noted that each company is using different processes and pursuing a different commercialization route. The Climeworks plants currently carry an estimated cost of \$600−1200/tCO_2 captured, with an ambition to reduce this figure to \$200−300 by the end of this decade and \$100−200 by the mid-2030s. Carbon Engineering reports current levelized costs of \$104−256/tCO_2, while Global Thermostat claims to have the lowest cost DAC system, delivering at below \$50/tCO_2. However, the latter figure seems at odds with the slow progress in scaling up the technology beyond 4 ktCO_2/year, since this price level would be competitive compared with other CO_2 sources for EOR.

Land and water requirements

Given the required future scale of DAC, cost calculations will eventually have to consider the land area needed for any given capture approach. When considering the land use requirements for DAC, Keith and colleagues draw an analogy with the requirements of wind power generation, because both depend on the rate of turbulent diffusion in the atmospheric boundary layer (Keith et al., 2006). For wind power, a spacing between turbines between 5 and 10 rotor diameters is needed to allow for the downward diffusion of momentum, while for DAC on a large scale, similar spacing between facilities will be needed to allow downward diffusion of CO_2 to ensure that each facility processes air with the highest ambient CO_2 concentration. It should be noted that 420 ppm is only the current global average CO_2 concentration in atmosphere and in some places it is consistently higher, especially downwind of heavy industrial processes. As it is desirable for the DAC units to be spread out to facilitate CO_2 capture, as in the case of large wind farms the actual footprint of the equipment is relatively small, and the land between remains available for other uses.

All carbon-neutral energy systems, from bioenergy to wind, occupy significant land areas, and a useful metric to compare them is the continuous power density in W/m^2 that can be achieved. For example, while energy from biomass represents less than 0.5 W/m^2, large wind turbines can deliver 1−4 W/m^2, while solar PV can achieve 50 W/m^2, taking account of typical load factors (30% and 25%, respectively). At scales in the order of 100s of kilometers, downward diffusion of CO_2 limits the air capture flux to around 1.5 ktCO_2/ha-year (400 tC/ha-year) (Elliott et al., 2001), equivalent to a power density of 41 W/m^2 if combined with natural gas fired power generation at the US average carbon intensity of 0.42 tCO_2/MWh (2019). This effective power density is comparable to current solar PV but, in the case of DAC, the equipment would occupy only a small fraction of the overall area (perhaps 4%, equivalent to a 30 ha site for a 1 MtCO_2/year DAC facility), while solar PV would have to cover 100% of this land area. Conversely, this indicates that to support a distributed DAC facility by low carbon energy alone will require significantly more land to supply the required power than for the DAC system footprint.

13.11 Future direction of DAC research

It is clear from the foregoing discussion that there is no shortage of technical options for DAC, and that this is an extremely active field of R&D work, in the search for improvements in solvent, sorbent, and overall system performance, as well as scaling approaches and ongoing commercialization.

For solvent-based systems taking the current Carbon Engineering concept (Keith et al., 2018) as state of the art, the key development challenges are process simplification, air contacting, and sorbent stability and performance, with the aim of reducing capital and operating costs. For Carbon Engineering, the current design features four major equipment items—the air contactor, pellet reactor, calciner-slaker, and air separation unit—to which must also be added compression, power plant, steam turbine, and fines filtration units. With capital recovery contributing over 50% of levelized capture costs, process simplification is a clear focus area.

The air contactor has the highest cost of the major equipment items, accounting for 30% of direct field costs, and is the least-tested part of DAC processes at scale for obvious reasons. Modeling, case studies, and field trials will be needed to optimize the layout and design of the air contactor, which is likely to vary in different locations due to local climatic conditions, average humidity, and availability of power.

Finally, the chemistry, stability, and performance of liquid and solid sorbents, of all types and using all approaches, over long timescales and realistic conditions, is going to be an ongoing process for the improvement of DAC performance as it advances toward full commercialization (TRL9) (Walton and Sholl, 2017). Given the urgent need for action to meet climate goals, a focus on practicality will be important to avoid time and effort being expended on capture processes that cannot reasonably be expected to lead to practical applications (Darunte et al., 2016).

However, advances in science and engineering are often the result of disruptive technologies that arise during R&D cycles and it is often the case that such disruptors appear roughly every 10 years, making it difficult to factor them into scenario planning (van der Veer, 2009). The DAC concept described in 2022 by the CO_2 Rail Company (Bachman et al., 2022) could be one such disruptor. In this proposal, the compact DAC unit, based on an electro-swing or pressure-swing cycle, is located in a moving railroad car and powered by the regenerative braking of the train. It may be that such a system, operating in a relatively higher-emissions environment such as on a railroad track, and considering the whole system, not just the single unit operation, could offer a novel solution to intensified DAC.

References

Abanades, J.C., Criado, Y.A., Fernández, J.R., 2020. An air CO_2 capture system based on the passive carbonation of large $Ca(OH)_2$ structures. Sustain. Energy Fuels 4 (7), 3409–3417.

Abd Rahaman, M.S., Zhang, L., Cheng, L.H., Xu, X.H., Chen, H.L., 2012. Capturing carbon dioxide from air using a fixed carrier facilitated transport membrane. RSC Adv. 2 (24), 9165–9172.

Agee, E.M., Orton, A., 2016. An initial laboratory prototype experiment for sequestration of atmospheric CO_2. J. Appl. Meteorol. Climatol. 55 (8), 1763–1770.

Agee, E., Orton, A., Rogers, J., 2013. CO_2 snow deposition in Antarctica to curtail anthropogenic global warming. J. Appl. Meteorol. Climatol. 52 (2), 281–288.

Arizona State University, 2019. Direct air capture machine. Video explainer of the prototype system at: https://www.youtube.com/watch?v=ikxj75mlHP8.

Azarabadi, H., Lackner, K.S., 2019. A sorbent–focused techno–economic analysis of direct air capture. Appl. Energy 250, 959–975.

Bachman, E., Tavasoli, A., Hatton, T.A., et al., 2022. Rail–Based Direct Air Carbon Capture: Laying the Tracks for Self–Powered 7+ Gigaton CO_2 Capture. Joule (in press).

Bajamundi, C.J.E., Koponen, J., Ruuskanen, V., et al., 2019. Capturing CO_2 from air: technical performance and process control improvement. J. CO_2 Util. 30, 232–239.

Bandi, A., Specht, M., Weimer, T., Schaber, K., 1995. CO_2 recycling for hydrogen storage and transportation—electrochemical CO_2 removal and fixation. Energy Convers. Manag. 36 (6–9), 899–902.

Bos, M.J., Kersten, S.R.A., Brilman, D.W.F., 2020. Wind power to methanol: renewable methanol production using electricity, electrolysis of water and CO_2 air capture. Appl. Energy 264, 114672.

Brethomé, F.M., Williams, N.J., Seipp, C.A., Kidder, M.K., Custelcean, R., 2018. Direct air capture of CO_2 via aqueous–phase absorption and crystalline–phase release using concentrated solar power. Nat. Energy 3, 553–559.

Breyer, C., Fasihi, M., Bajamundi, C., Creutzig, F., 2019. Direct air capture of CO_2: a key technology for ambitious climate change mitigation. Joule 3 (9), 2053–2057.

Burhenne, L., Giacomin, C., Follett, T., Ritchie, J., McCahill, J.S., Mérida, W., 2017. Characterization of reactive $CaCO_3$ crystallization in a fluidized bed reactor as a central process of direct air capture. J. Environ. Chem. Eng. 5 (6), 5968–5977.

Cai, H., Zhang, X., Lei, L., Xiao, C., 2020. Direct CO_2 capture from air via crystallization with a trichelating iminoguanidine ligand. ACS Omega 5 (32), 20428–20437.

Criado, Y.A., Alonso, M., Abanades, J.C., 2014. Kinetics of the $CaO/Ca(OH)_2$ hydration/dehydration reaction for thermochemical energy storage applications. Ind. Eng. Chem. Res. 53 (32), 12594–12601.

Custelcean, R., 2022. Direct air capture of CO_2 using solvents. Annual Rev. Chem. Biomol. Eng. 13 (10), 10.1–10.18.

Darunte, L.A., Walton, K.S., Sholl, D.S., Jones, C.W., 2016. CO_2 capture via adsorption in amine–functionalized sorbents. Curr. Opin. Chem. Eng. 12, 82–90.

Didas, S.A., Choi, S., Chaikittisilp, W., Jones, C.W., 2015. Amine–oxide hybrid materials for CO_2 capture from ambient air. Acc. Chem. Res. 48 (10), 2680–2687.

Ding, M., Cai, X., Jiang, H.L., 2019a. Improving MOF stability: approaches and applications. Chem. Sci. 10 (44), 10209–10230.

Ding, M., Flaig, R.W., Jiang, H.L., Yaghi, O.M., 2019b. Carbon capture and conversion using metal–organic frameworks and MOF–based materials. Chem. Soc. Rev. 48 (10), 2783–2828.

Drechsler, C., Agar, D.W., 2019a. Investigation of water co–adsorption on the energy balance of solid sorbent based direct air capture processes. Energy 192, 116587.

Drechsler, C., Agar, D.W., 2019b. Simulation and optimization of a novel moving belt adsorber concept for the direct air capture of carbon dioxide. Comput. Chem. Eng. 126, 520–534.

Dubey, M.K., Ziock, H., 2001. Extraction of Carbon Dioxide from the Atmosphere through Engineered Chemical Sinkage. (No. LA–UR–01–6473. Los Alamos National Laboratory.

Eisaman, M.D., Alvarado, L., Larner, D., Wang, P., Littau, K.A., 2011. CO_2 desorption using high–pressure bipolar membrane electrodialysis. Energy Environ. Sci. 4 (10), 4031–4037.

Eisenberg, P., Chichilnisky, G., 2008. System and Method for Removing Carbon Dioxide from Atmosphere and Global Thermostat using the Same. US Patent 12/124,864.

Elfving, J., Bajamundi, C., Kauppinen, J., 2017. Characterization and performance of direct air capture sorbent. Energy Proc. 114, 6087–6101.

Elliott, S., Lackner, K.S., Ziock, H.J., et al., 2001. Compensation of atmospheric CO_2 buildup through engineered chemical sinkage. Geophys. Res. Lett. 28 (7), 1235–1238.

Fasihi, M., Efimova, O., Breyer, C., 2019. Techno–economic assessment of CO_2 direct air capture plants. J. Clean. Prod. 224, 957–980.

Fujikawa, S., Selyanchyn, R., Kunitake, T., 2020. A new strategy for membrane–based direct air capture. Polym. J. 53 (1), 111–119.

Gebald, C., Wurzbacher, J.A., Tingaut, P., Zimmermann, T., Steinfeld, A., 2011. Amine–based nanofibrillated cellulose as adsorbent for CO_2 capture from air. Environ. Sci. Technol. 45 (20), 9101–9108.

Gebald, C., Wurzbacher, J.A., Tingaut, P., Steinfeld, A., 2013. Stability of amine–functionalized cellulose during temperature–vacuum–swing cycling for CO_2 capture from air. Environ. Sci. Technol. 47 (17), 10063–10070.

Giacomin, C.E., Holm, T., Mérida, W., 2020. $CaCO_3$ growth in conditions used for direct air capture. Powder Technol. 370, 39–47.

Goldberg, D.S., Lackner, K.S., Han, P., Slagle, A.L., Wang, T., 2013. Co–location of air capture, subseafloor CO_2 sequestration, and energy production on the Kerguelen plateau. Environ. Sci. Technol. 47 (13), 7521–7529.

Goto, K., Katsunori, Y., Higashii, T., 2013. A review of efficiency penalty in a coal–fired power plant with post–combustion CO_2 capture. Appl. Energy 111, 710–720.

Gutknecht, V., Snæbjörnsdóttir, S.Ó., Sigfússon, B., Aradóttir, E.S., Charles, L., 2018. Creating a carbon dioxide removal solution by combining rapid mineralization of CO_2 with direct air capture. Energy Proc. 146, 129–134.

Guo, M., Wu, H., Lv, L., Meng, H., Yun, J., Jin, J., Mi, J., 2021. A highly efficient and stable composite of polyacrylate and metal organic framework prepared by interface engineering for direct air capture. ACS Appl. Mater. Interfaces 13 (18), 21775–21785.

Hanak, D.P., Manovic, V., 2018. Combined heat and power generation with lime production for direct air capture. Energy Convers. Manag. 160, 455–466.

Holmes, G., Keith, D.W., 2012. An air–liquid contactor for large–scale capture of CO_2 from air. Phil. Trans. Math. Phys. Eng. Sci. 370 (1974), 4380–4403.

Holmes, G., Nold, K., Walsh, T., et al., 2013. Outdoor prototype results for direct atmospheric capture of carbon dioxide. Energy Proc. 37, 6079–6095.

House, K.Z., Baclig, A.C., Ranjan, M., et al., 2011. Economic and energetic analysis of capturing CO_2 from ambient air. Proc. Natl. Acad. Sci. USA 108 (51), 20428–20433.

IEA, 2021. Net zero by 2050: a roadmap for the global energy sector. In: Bouckaert, S., et al. (Eds.). IEA Publications, France.

Kearns, D., Liu, H., Consoli, C., 2021. Technology Readiness and Costs of CCS. Global CCS Institute Publications, Docklands, Australia.

Keith, D.W., Ha–Duong, M., Stolaroff, J.K., 2006. Climate strategy with CO_2 capture from the air. Clim. Change 74 (1–3), 17–45.

Keith, D.W., Heidel, K., Cherry, R., 2010. Capturing CO_2 from the atmosphere: rationale and process design considerations. In: Launder, B., Thompson, J.M.T. (Eds.), Geo–engineering Climate Change: Environmental Necessity or Pandora's Box. Cambridge University Press, Cambridge, UK.

Keith, D.W., Holmes, G., Angelo, D.S., Heidel, K., 2018. A process for capturing CO_2 from the atmosphere. Joule 2 (8), 1573–1594.

Kelemen, P.B., McQueen, N., Wilcox, J., Renforth, P., Dipple, G., Vankeuren, A.P., 2020. Engineered carbon mineralization in ultramafic rocks for CO_2 removal from air: review and new insights. Chem. Geol. 550, p119628.

Keller, L., Ohs, B., Abduly, L., Wessling, M., 2019a. Carbon nanotube silica composite hollow fibers impregnated with polyethylenimine for CO_2 capture. Chem. Eng. J. 359, 476–484.

Keller, L., Lohaus, T., Abduly, L., Hadler, G., Wessling, M., 2019b. Electrical swing adsorption on functionalized hollow fibers. Chem. Eng. J. 371, 107–117.

Kiani, A., Jiang, K., Feron, P., 2020. Techno–economic assessment for CO_2 capture from air using a conventional liquid–based absorption process. Front. Energy Res. 8, 92.

Kulikarni, A.R., Sholl, D.S., 2012. Analysis of equilibrium–based TSA processes for direct capture of CO_2 from air. Ind. Eng. Chem. Res. 51 (25), 8631–8645.

Lackner, K.S., 2009. Capture of carbon dioxide from ambient air. Eur. Phys. J. Spec. Top. 176 (1), 93–106.

Lackner, K.S., 2013. The thermodynamics of direct air capture of carbon dioxide. Energy 50, 38–46.

Lackner, K., Ziock, H.J., Grimes, P., 1999. Carbon Dioxide Extraction from Air: Is It an Option? (No. LA–UR–99–583. Los Alamos National Lab., NM (US).

Lackner, K.S., Brennan, S., Matter, J.M., Park, A.H.A., Wright, A., Van Der Zwaan, B., 2012. The urgency of the development of CO_2 capture from ambient air. Proc. Natl. Acad. Sci. USA 109 (33), 13156–13162.

Lee, W.R., Hwang, S.Y., Ryu, D.W., et al., 2014. Diamine–functionalized metal–organic framework: exceptionally high CO_2 capacities from ambient air and flue gas, ultrafast CO_2 uptake rate, and adsorption mechanism. Energy Environ. Sci. 7 (2), 744–751.

Lee, Y.–Y., Edgehouse, K., Klemm, A., Mao, H., Pentzer, E., Gurkan, B., 2020. Capsules of reactive ionic liquids for selective capture of carbon dioxide at low concentrations. ACS Appl. Mater. Interfaces 12 (16), 19184–19193.

Lehne, J., Preston, F., 2018. Making Concrete Change: Innovation in Low Carbon Cement and Concrete. Chatham House Policy Institute, London, UK.

Li, Z., Liu, P., Ou, C., Dong, X., 2020. Porous metal-organic frameworks for carbon dioxide adsorption and separation at low pressure. ACS Sustain. Chem. Eng. 8 (41), 15378–15404.

Liao, P.Q., Chen, X.W., Liu, S.Y., et al., 2016. Putting an ultrahigh concentration of amine groups into a metal–organic framework for CO_2 capture at low pressures. Chem. Sci. 7 (10), 6528–6533.

Madden, D.G., Scott, H.S., Kumar, A., et al., 2017. Flue–gas and direct–air capture of CO_2 by porous metal–organic materials. Phil. Trans. Math. Phys. Eng. Sci. 375 (2084), 20160025.

Mahmoudkhani, M., Heidel, K.R., Ferreira, J.C., Keith, D.W., Cherry, R.S., 2009. Low energy packed tower and caustic recovery for direct capture of CO_2 from air. Energy Proc. 1 (1), 1535–1542.

Mahmoudkhani, M., Keith, D.W., 2009. Low–energy sodium hydroxide recovery for CO_2 capture from atmospheric air—thermodynamic analysis. Int. J. Greenh. Gas Control 3 (4), 376–384.

McQueen, N., Kelemen, P., Dipple, G., Renforth, P., Wilcox, J., 2020. Ambient weathering of magnesium oxide for CO_2 removal from air. Nat. Commun. 11 (1), 1–10.

Muroyama, A., Patru, A., Gubler, L., 2020. CO_2 separation and transport via electrochemical methods. J. Electrochem. Soc. 167, 133504.

Mukherjee, S., Sikdar, N., O'Nolan, D., et al., 2019. Trace CO_2 capture by an ultramicroporous physisorbents with low water affinity. Sci. Adv. 5 (11).

Nabavi, S.A., Erans, M., Manović, V., 2019. Demonstration of a kW–scale solid oxide fuel cell–calciner for power generation and production of calcined materials. Appl. Energy 255, 113731.

Nagasawa, H., Yamasaki, A., Iizuka, A., Kumagai, K., Yanagisawa, Y., 2009. A new recovery process of carbon dioxide from alkaline carbonate solution via electrodialysis. AIChE J. 55 (12), 3286–3293.

Naeem, S., Ghaemi, A., Shahhosseini, S., 2016. Experimental investigation of CO_2 capture using sodium hydroxide particles in a fluidized bed. Kor. J. Chem. Eng. 33 (4), 1278–1285.

Nikulshina, V., Hirsch, D., Mazzotti, M., Steinfeld, A., 2006. CO_2 capture from air and co–production of H_2 via the $Ca(OH)_2$–$CaCO_3$ cycle using concentrated solar power–Thermodynamic analysis. Energy 31 (12), 1715–1725.

Nikulshina, V., Gebald, C., Steinfeld, A., 2009. CO_2 capture from atmospheric air via consecutive CaO–carbonation and $CaCO_3$–calcination cycles in a fluidized–bed solar reactor. Chem. Eng. J. 146 (2), 244–248.

NOAA Global Monitoring, 2022. Trends in Atmospheric Carbon Dioxide. Earth System Research Laboratories, Hawaii, USA.

Okunev, A.G., Sharonov, V.E., Aristov, Y.I., Parmon, V.N., 2000. Sorption of carbon dioxide from wet gases by K_2CO_3–in–porous matrix: influence of the matrix nature. React. Kinet. Catal. Lett. 71 (2), 355–362.

Rodríguez–Mosqueda, R., Bramer, E.A., Roestenberg, T., Brem, G., 2018. Parametrical study on CO_2 capture from ambient air using hydrated K_2CO_3 supported on an activated carbon honeycomb. Ind. Eng. Chem. Res. 57 (10), 3628–3638.

Rodríguez–Mosqueda, R., Rutgers, J., Bramer, E.A., Brem, G., 2019. Low temperature water vapor pressure swing for the regeneration of adsorbents for CO_2 enrichment in greenhouses via direct air capture. J. CO_2 Util. 29, 65–73.

Sabatino, F., Mehta, M., Grimm, A., et al., 2020. Evaluation of a direct air capture process combining wet scrubbing and bipolar membrane electrodialysis. Ind. Eng. Chem. Res. 59 (15), 7007–7020.

Sadiq, M.M., Batten, M.P., Mulet, X., et al., 2020. A pilot-scale demonstration of mobile direct air capture using metal-organic frameworks. Adv. Sustain. Syst. 4 (12), 2000101.

Samari, M., Ridha, F., Manovic, V., Macchi, A., Anthony, E.J., 2019. Direct capture of carbon dioxide from air via lime–based sorbents. Mitig. Adapt. Strat. Glob. Change 1–17.

Sanz–Perez, E.S., Murdock, C.R., Didas, S.A., Jones, C.W., 2016. Direct capture of CO_2 from ambient air. Chem. Rev. 116 (19), 11840–11876.

Sayari, A., Liu, Q., Mishra, P., 2016. Enhanced adsorption efficiency through materials design for direct air capture over supported polyethylenimine. ChemSusChem 9 (19), 2796–2803.

Schellevis, M., Jacobs, T., Brilman, W., 2020. CO_2 capture from air in a radial flow contactor: batch or continuous operation. Front. Chem. Eng. 2, 596555.

Sehaqui, H., Gálvez, M.E., Becatinni, V., et al., 2015. Fast and reversible direct CO_2 capture from air onto all–polymer nanofibrillated cellulose–polyethylenimine foams. Environ. Sci. Technol. 49 (5), 3167–3174.

Seipp, C.A., Williams, N.J., Kidder, M.K., Custelcean, R., 2017. CO_2 capture from ambient air by crystallization with a guanidine sorbent. Angew. Chem. 129 (4), 1062–1065.

Severinghaus, J.P., Broecker, W.S., Dempster, W.F., MacCallum, T., Wahlen, M., 1994. Oxygen loss in biosphere 2. Eos 75 (3), 33–37.

Shi, X., Xiao, H., Azarabadi, H., et al., 2020. Sorbents for the direct capture of CO_2 from ambient air. Angew. Chem. Int. Ed. 59 (18), 6984–7006.

Shigemoto, N., Yanagihara, T., Sugiyama, S., Hayashi, H., 2006. Material balance and energy consumption for CO_2 recovery from moist flue gas employing K_2CO_3–on–activated carbon and its evaluation for practical adaptation. Energy Fuels 20 (2), 721–726.

Shu, Q., Legrand, L., Kuntke, P., Tedesco, M., Hamelers, H.V., 2020. Electrochemical regeneration of spent alkaline absorbent from direct air capture. Environ. Sci. Technol. 54 (14), 8990–8998.

Socolow, R., Desmond, M., Aines, R., et al., 2011. Direct Air Capture of CO_2 with Chemicals. APS Technology Assessment Panel, Maryland, USA.

Spector, N.A., Dodge, B.F., 1946. Removal of carbon dioxide from atmospheric air. Trans. Am. Inst. Chem. Eng. 42 (5–6), 827–848.

Stolaroff, J., Keith, D., Lowry, G., June 2006. A pilot–scale prototype contactor for CO_2 capture from ambient air: cost and energy requirements. In: Proceeding of GHGT–8, the 8th International Conference on Greenhouse Gas Control Technologies, Trondheim, Norway.

Stucki, S., Schuler, A., Constantinescu, M., 1995. Coupled CO_2 recovery from the atmosphere and water electrolysis: feasibility of a new process for hydrogen storage. Int. J. Hydrogen Energy 20 (8), 653–663.

Toan, S., O'Dell, W., Russell, C.K., et al., 2019. Thermodynamics of $NaHCO_3$ decomposition during Na_2CO_3–based CO_2 capture. J. Environ. Sci. 78, 74–80.

van der Veer, J., 2009. My A to B: speeches, articles and letters. In: Shell International. ASIN: B00332OAPS.

Varghese, A.M., Karanikolos, G.N., 2020. CO_2 capture adsorbents functionalized by amine—bearing polymers: a review. Int. J. Greenh. Gas Control 96, 103005.

Vega, F., Sanna, A., Navarrete, B., Maroto–Valer, M.M., Cortes, V.J., 2014. Degradataion of ambine–based solvents in CO_2 capture process by chemical absorption. Greenhouse Gas Sci. Technol. 4 (6), 707–733.

Veselovskaya, J.V., Derevschikov, V.S., Kardash, T.Y., Stonkus, O.A., Trubitsina, T.A., Okunev, A.G., 2013. Direct CO_2 capture from ambient air using K_2CO_3/Al_2O_3 composite sorbent. Int. J. Greenh. Gas Control 17, 332–340.

Walton, K.S., Sholl, D.S., 2017. Research challenges in avoiding "showstoppers" in developing materials for large–scale energy applications. Joule 1 (2), 208–211.

Wang, T., Lackner, K.S., Wright, A., 2011. Moisture swing sorbent for carbon dioxide capture from ambient air. Environ. Sci. Technol. 45 (15), 6670–6675.

Wang, T., Hou, C., Ge, K., et al., 2017. Spontaneous cooling absorption of CO_2 by a polymeric ionic liquid for direct air capture. J. Phys. Chem. Lett. 8 (17), 3986–3990.

Wang, X., Song, J., Chen, Y., et al., 2020. CO_2 absorption over ion exchange resins: the effect of amine functional groups and microporous structures. Ind. Eng. Chem. Res. 59, 16507–16515.

Williams, N.J., Seipp, C.A., Brethomé, F.M., et al., 2019. CO_2 capture via crystalline hydrogen–bonded bicarbonate dimers. Chem 5 (3), 719–730.

Wu, X., Krishnamoorti, R., Bollini, P., 2022. Technological options for direct air capture: a comparative process engineering review. Ann. Rev. Chem. Biomol. Eng. 13, 13.1–13.22.

Wurzbacher, J.A., Gebald, C., Brunner, S., Steinfeld, A., 2016. Heat and mass transfer of temperature–vacuum swing desorption for CO_2 capture from air. Chem. Eng. J. 283, 1329–1338.

Xu, X., Song, C., Andresen, J.M., Miller, B.G., Scaroni, A.W., 2002. Novel polyethylenimine–modified mesoporous molecular sieve of MCM–41 type as high–capacity adsorbent for CO_2 capture. Energy Fuels 16 (6), 1463–1469.

Yang, Z., Dai, S., 2021. Challenges in engineering the structure of ionic liquids towards direct air capture of CO_2. Green Chem. Eng. 2 (4), 342–345.

Yu, Q., Brilman, D.W.F., 2017. Design strategy for CO_2 adsorption from ambient air using a supported amine based sorbent in a fixed bed reactor. Energy Proc. 114, 6102–6114.

Yu, Q., Brilman, W., 2020. A radial flow contactor for ambient air CO_2 capture. Appl. Sci. 10 (3), 1080.

Zeman, F.S., Lackner, K.S., 2004. Capturing carbon dioxide directly from the atmosphere. World Resour. Rev. 16 (2), 157–172.

Zhang, W., Liu, H., Sun, C., Drage, T.C., Snape, C.E., 2014. Capturing CO_2 from ambient air using a polyethyleneimine–silica adsorbent in fluidized beds. Chem. Eng. Sci. 116, 306–316.

Zhu, X., Ge, T., Yang, F., Wang, R., 2021. Design of steam–assisted temperature vacuum–swing adsorption processes for efficient CO_2 capture from ambient air. Renew. Sustain. Energy Rev. 137, 110651.

Chapter 14

Removal of methane and other non-CO_2 GHGs

Renaud de Richter, Wei Li, Steve Rackley and Tingzhen Ming

14.1 Introduction

Carbon dioxide removal (CDR) methods are negative emissions technologies (NETs), but not all NETs effect CDR. Greenhouse gas removal (GGR) is a synonym for NETs, which encompass the removal of all GHGs present in the atmosphere.

The natural elimination of non-CO_2 GHGs and organic compounds in the atmosphere, listed in Table 14.1, proceeds largely by reaction with sunlight-generated hydroxyl radicals (•OH) and chlorine atoms (Cl•), and through heterogeneous reactions occurring on mineral dust aerosols (Cwiertny et al., 2008; Chen et al., 2011; George et al., 2015). Several technologies for methane (CH_4) removal are described in this chapter, including enhancing the natural hydroxyl and chlorine sinks and through photocatalysis (de Richter and Caillol, 2011), and methods proposed to remove N_2O and halogenated gases are also discussed. A cost evaluation is provided to compare these NETs with those directed at CDR.

The degradation products of non-CO_2 GHGs will be at very low concentrations and can be left in the atmosphere without additional risks, even if the technologies discussed here were implemented at a climate-relevant scale. N_2O for instance, with a tropospheric concentration of ~335 ppb, is transformed into nitrogen and oxygen, the principal constituents of the air, while the oxidation of the whole atmospheric CH_4 inventory to CO_2 would increase [CO_2] by less than 0.5%, while substantially reducing radiative forcing. No issues of storage, sequestration, and monitoring for possible leakage would therefore arise with the large-scale implementation of these NETs.

The removal of CH_4 and other non-CO_2 GHGs is more challenging than for CO_2, and potentially more costly on a CO_2e basis, due to their lower concentrations (Lackner, 2020; Jackson et al., 2020). It is therefore even more important for these GHGs than for CO_2 that cost-effective mitigation through emissions reduction, such as addressing CH_4 emissions from landfill, oil and gas operations, and agriculture, should be implemented ahead of any attempts at large-scale removal (Lee et al., 2017; Nisbet-Jones et al., 2021).

Nevertheless, depending on the future prices of CO_2e removals, which should ideally be based on the 20 year global warming potential to reflect the 10 year lifetime of atmospheric CH_4, further technology development could lead to CO_2e removal options that are cost-competitive with CDR. It is with that possibility in mind that the various concepts that have been proposed are presented here.

14.2 Methane removal by enhancing atmospheric sinks

After CO_2, CH_4 is the GHG of second importance in terms of its radiative forcing, as shown in Fig. 14.1, which illustrates the contributions to present-day warming from all GHGs and other climate forcers.

The rationale for targeting CH_4 and transforming it into CO_2 is that the global warming potentials (GWPs) of these two GHGs are very different, both on 20 and 100 year time horizons (GWP_{20}, GWP_{100}). The GWP of CO_2 is always equal to 1 by definition, and CH_4 has a GWP of 86 over 20 years and 25–34 over 100 years, so that the complete oxidation of CH_4 would significantly reduce its climatic impact. Even if it is left as CO_2 in the atmosphere the climate benefit is high since their respective concentrations are very different; 1900 ppb for [CH_4] and ~420 ppm for [CO_2]. The current CH_4 stock in the atmosphere is around 5.3 Gt (460 $GtCO_2$e at GWP_{20}).

Negative Emissions Technologies for Climate Change Mitigation. https://doi.org/10.1016/B978-0-12-819663-2.00006-X

TABLE 14.1 Principal GHGs and their atmospheric concentrations in 1992 and 2021 (NOAA 2022) including reported ranges of GPW_{100} and tropospheric lifetime.

GHG	Year 1992	Year 2021	Unit	GWP_{100}	Lifetime (years)
CO_2	356.1	414.7	ppm	1	Variable
CH_4	1735.5	1895.6	ppb	25–34	12
N_2O	309.9	334.4	ppb	265–290	105–150
CFC-11	266.7	221.7	ppt	3500–4750	45–60
CFC-12	505.9	492.9	ppt	7300–10,900	84–130
CFC-113	80.2	68.4	ppt	4200	81–102
CCl_4	107.6	76.6	ppt	1300	41–54
CH_3CCl_3	135.2	1.2	ppt	100	6–37
HFC-23	8.6	35.1	ppt	12,400	222
HFC-125	0.14	35.42	ppt	2500–14,900	16–136
HFC-134a	0.27	118.9	ppt	1200–1300	13
HFC-143a	0.56	26.83	ppt	756–2900	4–41
HFC-152a	0.42	7.19	ppt	138–140	1.5–1.7
HCFC-22	101.9	248.9	ppt	1500–1810	12–15
HCFC-141b	0.35	24.63	ppt	440–580	8–11
HCFC-142b	3.2	21.5	ppt	1600	19–178
H-1211	2.89	3.03	ppt		10–14
H-1301	2.17	3.28	ppt		66–78
H-2402	0.36	0.4	ppt		14–16
SF_6	2.75	10.65	ppt	23,500	3200

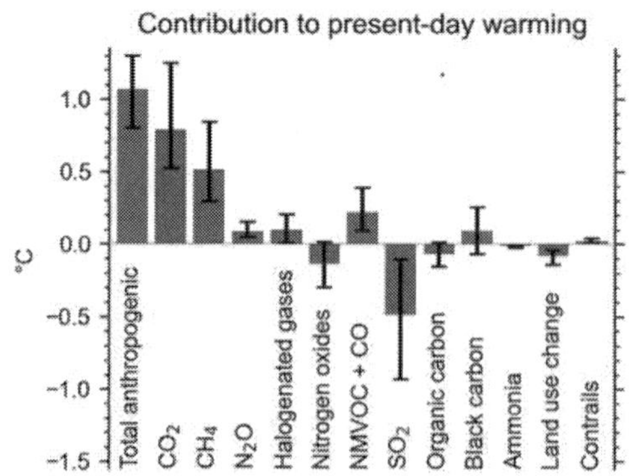

FIGURE 14.1 Contributions to current warming of different climate forcers. *Adapted from IPCC (2021).*

Targeting CH_4 can make an important contribution to keeping warming below 1.5°C, and the scenarios explored in IPCC AR6 require reductions of global CO_2 and CH_4 emissions of 48% and 34%, respectively, below 2019 levels by 2030, with net-zero emissions being achieved "in the early 2050s" (IPCC, 2022). Despite this need, both [CO_2] and [CH_4] continue to increase; the annual increases of tropospheric CH_4 in 2020 and 2021, respectively 15.3 and 17 ppb, broke all records since continuous monitoring started (NOAA, 2022). By way of comparison, the corresponding emissions of ~45 $MtCH_4$ are roughly equivalent to global LNG shipping capacity (Nisbet-Jones et al., 2021).

Developing NETs to remove gigatonnes of CO_2e from the atmosphere, therefore, appears to be essential to reach the Paris climate goals (Anderson, 2016). Currently, CH_4 is responsible for about 0.5°C of warming (UNEP, 2021), and climate models suggest that by the end of the century warming due to CH_4 emissions alone could range from 0.75 to 1.5°C with a median 0.9°C (Ocko et al., 2021).

"Natural" biogenic emissions of CH_4 are expected to rise as a result of global warming, in particular from wetlands and water bodies (lakes, hydro-electric reservoirs, etc.). There are also risks of increasing CH_4 emissions from thawing permafrost (O'Connor et al., 2010), and from the sudden destabilization of some submarine CH_4-hydrates deposits caused, for example, by earthquakes or submarine landsides in hydrate-rich zones. The risk may be small (Nisbet and Piper 1998), but in the early-Holocene the Storegga submarine landslide that took place on the continental slope west of Norway is thought to have released several gigatonnes of CH_4 (Paull et al., 2007).

Anthropogenic CH_4 emissions are also expected to rise, mainly from the energy sector and from agriculture (rice paddies, cattle breeding, and other livestock), driven by the needs of an increasing world population, expected to rise from 7.9 billion in 2021 to 9.8 billion by 2050, and to 11.2 billion in 2100 (UN, 2017).

At the 2021 Glasgow Conference of Parties (COP26) the global methane pledge was signed by more than 100 countries, committing to a reduction of anthropogenic CH_4 emissions by 30% compared to 2020 levels. Unfortunately, very large emitters such as India, Russia, Saudi Arabia, and Mexico did not sign the pledge. Also, NGOs (UNEP, 2021), as well as scientists (Ocko et al., 2021), think that a global reduction of 45% of anthropogenic CH_4 emissions is necessary by 2030 to have any chance of reaching the 1.5°C Paris goal. Deep reductions in future emissions from the oil and gas industry and from agriculture will be needed to prevent increasing warming due to CH_4.

As is the case for all NETs, removal is complementary to reduction, and CH_4 removal can help to prevent future warming increase, to deal with legacy emissions remaining in the atmosphere ($\sim$1900 ppb), and to reduce current warming if [CH_4] could be returned to its preindustrial level of $\sim$700 ppb (Abernethy et al., 2021). Two recent perspective papers (Jackson et al., 2021; Ming et al., 2022) review CH_4 removal methods, including those discussed below, and also discuss mitigation techniques targeting point sources, such as bio-covers on landfills (Lee et al., 2017).

The principal natural sinks of CH_4 are hydroxyl radicals (•OH) and chlorine atoms (Cl•) in the troposphere, as well as mineral soils (Lau et al., 2015), plants (Sundqvist et al., 2012), and trees (Price et al., 2004; Zhao et al., 2019), linked to the microorganisms they contain or are associated with (Dunfield, 2007). In the next subsections, methods aiming to enhance several of these sinks are described.

14.2.1 Enhancing the tropospheric chlorine atom methane sink

In the troposphere, about 2.5% of the total CH_4 is oxidized by Cl• (Hossaini et al., 2016), and in some polluted coastal areas, Cl• accounts for $\sim$10 to >20% of total CH_4 oxidation in the marine boundary layer (Thornton et al., 2010; Hossaini et al., 2016). Although •OH oxidizes about 90% of CH_4, it is interesting to target the Cl• sink because the speed of reaction with CH_4 is 16 times faster than is the case for •OH (Atkinson and Arey, 2003).

Chlorine atoms can be generated by UV photolysis of chlorine gas (Cl_2), which is produced at large scale by the chlor-alkali processes, involving the electrolysis of aqueous sodium chloride (NaCl) solution or of molten NaCl, with many applications in the chemical industry (Eurochlor, 2021). Over polluted regions where nitrogen oxides (NOx) are present, a heterogeneous autocatalytic Cl cycling involving marine aerosols is also possible (Lawler et al., 2009). In areas with less pollution, the reaction will not be catalytic but only stoichiometric, in which case atmospheric CH_4 removal at the scale required to reduce the tropospheric stock in about 15–20 years would require amounts of Cl_2 gas greater than the current global production. Increased production would be feasible, with the benefit of the coproduction of hydrogen and sodium hydroxide, the latter with potential DAC and marine CDR applications (Yuan et al., 2022; see also Chapters 13 and 16).

A natural catalytic process involving iron (Fe(III)/Fe(II)) and UV radiation at 365 nm also generates chlorine atoms (Lim et al., 2006). This catalytic photo-production involves the illumination of an acidified aqueous solution of Fe(III), which induces a photochemical reduction of Fe(III) to Fe(II) and the formation of Cl radicals. Within the marine boundary layer, the Fe(II) species formed are reoxidized and $FeCl_3$ is rapidly regenerated, making use of abundant seawater which contains NaCl and provides excess chloride ions (Wittmer et al., 2015a). By generating aerosols containing Fe_2O_3 + NaCl, using diluted HCl solutions at pH 2.6, the same authors were able to generate up to 78 chlorine atoms/hr per iron atom (Wittmer and Zetzsch 2017).

Oeste and colleagues have suggested that this natural mechanism might have occurred during the glacial ages, as Antarctic ice cores show a simultaneous decrease of CH_4 and CO_2 coincident with an increase in mineral dust (Oeste et al., 2017; Petit et al., 1999). This suggestion is based on the "Iron Dust hypothesis" for the glacial–interglacial changes in [CO_2] (Martin, 1990), with CO_2 drawn down by iron fertilization of the ocean leading to increased phytoplankton blooms, alongside an increase in the Cl• CH_4 sink, which can be assessed from $\delta^{13}CH_4$ ice core measurements (Levine et al., 2011). If this hypothesis is correct, increasing the amount of iron salt aerosols in the troposphere could fight global warming in several ways, the principal ones being by reducing the CH_4 concentration in the atmosphere and, after deposition in the ocean, by removing CO_2 from the atmosphere through a form of low dose ocean iron fertilization, that is, inducing phytoplankton blooms and enhancing oceanic productivity and the food web (Oeste et al., 2017).

It is worth noting that there are several anthropogenic sources of iron (Fig. 14.2), mainly from combustion processes in the iron and steel industry and from coal and other fossil fuels in power plants, as well as from bunker fuel used by marine vessels (Sedwick et al., 2007; Luo et al., 2008).

From the above, it is clear that anthropogenic emissions of iron salt aerosols (ISA) already cool the Earth by removing CH_4 and O_3 from the atmosphere (O'Grady, 2021), and also remove CO_2 by increasing oceanic and terrestrial primary productivity (Myriokefalitakis et al., 2015; Ito and Shi, 2015). Consequently, any increase in anthropogenic ISA emissions would enhance these natural GHGs removal mechanisms. Oeste and colleagues (2017) estimate that a doubling of current ISA emissions (an additional 300 kt/yr) would remove the equivalent of current annual anthropogenic CH_4 emissions, with costs for CH_4 removal through enhancement of the Cl• sink estimated to be in the order of or less than \$2/tCO_{2e} (Johnson, 2021). Two start-ups, Blue Dot Change and Atmospheric Methane Removal, are currently developing Cl• sink enhancement for the removal of atmospheric CH_4 (Petersen, 2022).

14.2.2 Enhancing the tropospheric hydroxyl radical methane sink

In the troposphere, the major oxidizing agent is the hydroxyl radical (•OH), which is generated naturally and is considered to be the detergent of the atmosphere, converting some 3.7 Gt of trace gases into CO_2 each year (Prinn et al., 2005). This includes several GHGs and many gases involved in stratospheric ozone depletion, such as man-made hydrofluorocarbons (HFCs), hydrochlorofluorocarbons (HCFCs), and volatile organic compounds (VOCs), as well as biogenic chloromethane and bromomethane, and urban air pollution. Accordingly, the •OH concentration [•OH] determines the atmospheric lifetimes of these gases.

The concentration of •OH depends on numerous factors including that of ozone ([O_3]), relative humidity, temperature, ultraviolet irradiance, emissions of VOCs, carbon monoxide (CO), nitrogen oxides (NOx), and some other factors (Nicely et al., 2020; He et al., 2021).

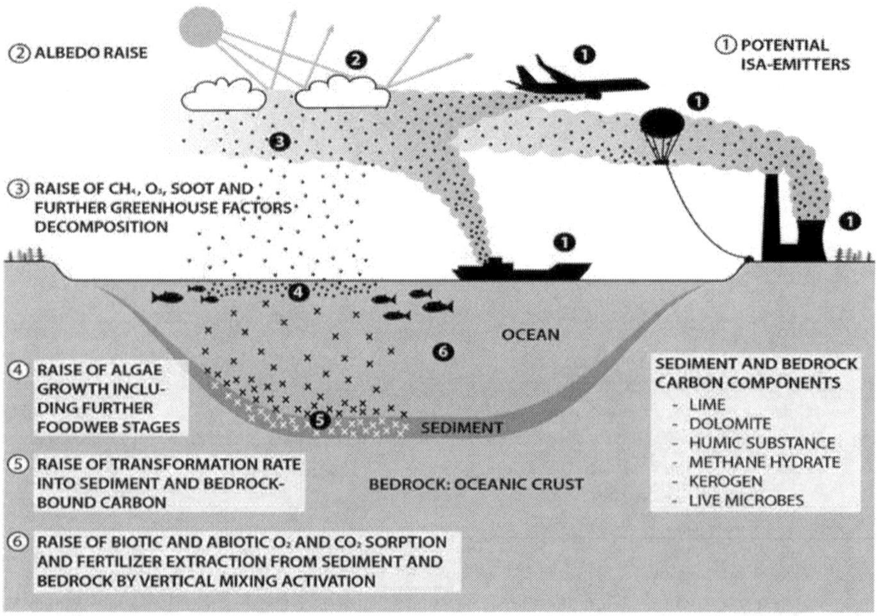

FIGURE 14.2 A schematic representation of current anthropogenic emissions of iron. *From Oeste et al. (2017).*

An important source for •OH during daytime is the UV photolysis (photodegradation) of ozone generating oxygen and exited atomic oxygen $O(^1D)$, which then reacts with water to produce two •OH radicals:

$$O_3 + h\nu\ (UV) \rightarrow O_2 + O(^1D) \tag{14.1}$$

$$O(^1D) + H_2O \rightarrow 2\ {\bullet}OH \tag{14.2}$$

In a variety of clean and moderately polluted environments, summertime field measurements demonstrated a high level of correlation between $O(^1D)$ and •OH (Heard and Pilling 2003, and references therein). These reactions show that ozone, water, and UV radiation in the wavelength range between 310 and 350 nm are essential in the troposphere to produce •OH (Matsumi et al., 2002; Hofzumahaus et al., 2004). As •OH generation is triggered by O_3, UV, and humidity, its concentration shows strong diurnal and seasonal cycles, as well as varying with height due to decreased water concentration with lower temperatures at altitude. The concentration of tropospheric •OH is highest in the tropics, where humidity is high and solar radiation intense.

UV levels vary mainly with the height of the sun in the sky, changing with the time of day, the season, and with latitude, as the distance traveled through the atmosphere determines the fraction of UV radiation absorbed. For the same reason, UV levels increase by approximately 10% with every 1 km increase in altitude. Coupled climate chemistry models indicate that about 85% of CH_4 oxidation by tropospheric •OH occurs between latitudes 40°S and 40°N (Holmes et al., 2012).

The IPCC AR4 WG1 (IPCC 2007, Chapter 7) underlines that Reactions 14.1 and 14.2 are the primary source of tropospheric •OH and that, under projected future global warming accompanied by increased amounts of water vapor, the tropospheric abundance of •OH, and of other oxidants such as HO_2•, is expected to increase. Lamarque and colleagues (2005) showed that under reduced aerosol emissions, a warmer and moister climate significantly increases [•OH], which illustrates the importance of water vapor concentration and its distribution. Some studies estimate that in a warmer climate with doubled tropospheric [CO_2], annual mean [•OH] concentration would increase by between 7% and 12.5% (Brasseur et al., 1998 and Johnson et al., 1999, respectively).

•OH generators are commercially available and have been proposed for the removal of some pollutants at point sources (Morneault, 2013; Engstrom, 2016). These generators are becoming cheaper, using new UV-LEDs with long lifetimes, and have seen rapid growth for UV-disinfection or sterilization, although such systems have not yet been suggested for large scale CH_4 removal.

There are, as yet, no publications proposing methods to enhance the tropospheric •OH sink of CH_4 or of other less concentrated GHGs but, from the above, it can be deduced that if humidity levels increase in hot and dry regions with high UV index, an increase in the amount of •OH generated can be expected. In some regions of Australia, Saudi Arabia, India, and the United States, anthropogenic humidity changes already occur, for instance where agriculture is irrigated using river flows, fossil groundwater, or desalinated seawater. Several studies have shown that irrigation can increase relative humidity by 9%–20%, and can have significant impacts on local meteorological conditions (Sorooshian et al., 2011). Irrigation in California increases humidity in the southwestern US states of Arizona, Colorado, New Mexico, Utah, and Wyoming (Lo and Famiglietti, 2013). Modeling studies in the early 2010s estimated that from 16.7 km^3 of irrigation water used in the summer months over the 52,000 km^2 of irrigated area in California's Central Valley, evotranspiration amounts to 14.7 km^3 which benefits the southwestern United States by water vapor export, strengthening the regional hydrological cycle. The moisture is blown over the Sierra Nevada, initiating an anthropogenic loop, with a 15% increase in summer precipitation in the other states, and much of the additional rain returning via a 28% increase in the flow of the Colorado River (Lo and Famiglietti, 2013). Continued expansion of irrigated land in developing countries will be needed to meet the needs of a growing global population, and if afforestation is employed at a large scale as a negative emissions method (Chapter 11), to avoid competition for fertile agricultural land some arid regions would need to be transformed, requiring irrigation and implying increasing evotranspiration and humidity.

A second example of man-made enhanced evaporation in dry regions arises from hydroelectric reservoirs. Several dams have been built in dry regions, one of the world's largest being Lake Nasser, almost 500 km long and covering about 6000 km^2, with a storage capacity of 162 km^3 of fresh water. The evaporated water loss is estimated to be 12–16 km^3 every year (Ebaid and Ismail, 2010; Abou El-Magd and Ali, 2012; Hassan et al., 2018). As temperatures and UV index are high, there can be little doubt that the increased humidity leads to an increase in •OH generation.

Various methods of increasing humidity in hot and dry regions to enhance •OH generation are possible (Wang et al., 2022a), one being the use of solar renewable energy to pump seawater to desert locations where shallow evaporation ponds with large surface area could be easily built.

14.3 Methane removal by catalysis and photocatalysis

A catalyst is a material or substance that accelerates chemical reactions while remaining unchanged by the chemical process, and usually also helps to reduce the reaction temperature and pressure. A photocatalyst, commonly a semiconductor metal oxide, is activated by light—such as sunlight—either UV or visible, and usually works at ambient temperature.

For more than a decade, it has been known that, at the laboratory scale, photocatalysis can transform almost all GHGs into harmless and climate-friendly products, or to CO_2 with much reduced warming potential (de Richter and Caillol, 2011).

Currently, many teams are researching the catalytic or photocatalytic partial oxidation of CH_4 (Sher Shah et al., 2020), with the aim of producing higher hydrocarbons, such as ethane and propane, or oxygenated compounds such as methanol and formaldehyde—value-added chemicals and fuels that are easier to handle and transport than CH_4. The selective and controlled oxidation of CH_4 is a difficult task and remains a great challenge at present, because the first hydrogen abstraction is difficult and slow, and overoxidation then frequently happens to produce CO_2, which is not the desired end product. When the aim is to transform CH_4 into useful chemicals, experiments that result in total oxidation are often considered failures and are not published, although they would be very valuable for those targeting atmospheric CH_4 removal. Fortunately, some results have been published and have confirmed that, at the laboratory scale, the total oxidation of CH_4 to CO_2 is possible and rapid, with good yields, both in batch photocatalytic reactors (PCRs) and in continuous-flow PCRs, using UV light and/or visible-light (Chen et al., 2016). A review of the different photocatalysts able to perform this oxidation has been published (de Richter et al., 2017) and proposes a method to perform large-scale atmospheric CH_4 oxidation using updraft solar chimneys (SCs) as air-moving devices and as PCRs. This is further described below and illustrated in Fig. 14.3.

A further method, using zeolites to capture atmospheric CH_4 followed by thermal catalytic oxidation has been proposed (Jackson et al., 2019), and is also described later. To remove CH_4 at atmospheric concentrations (~ 1.9 ppm), this method aims to profit from synergies with future direct air capture (DAC) infrastructure, including sharing the airflow.

14.3.1 Photocatalytic solar chimneys

Description of the solar chimney

A SC consists of a large greenhouse (GH) used as a solar collector where air is heated, and a tall tower or chimney to benefit from the buoyant rise of hot air (stack effect). An SC prototype has been built for air purification in the urban environment of Xian in China (Fig. 14.4).

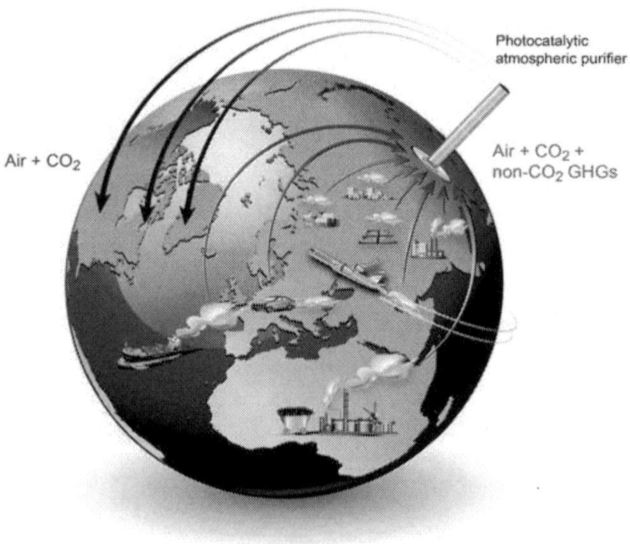

FIGURE 14.3 Artist's representation of GHG removal by photocatalytic SCs. *From de Richter et al. (2017).*

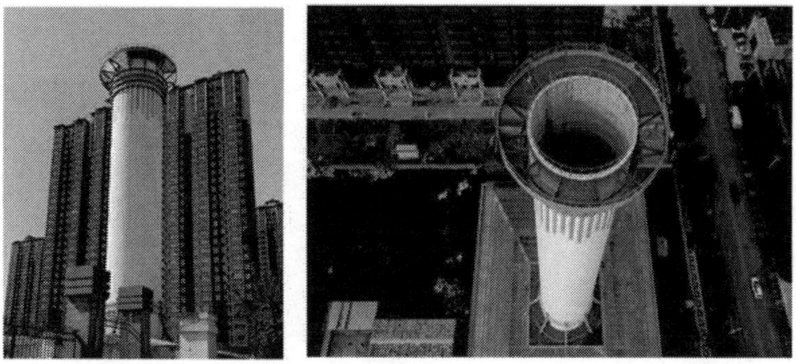

FIGURE 14.4 Solar chimney prototype in Xian, China. *Source: co-author pictures.*

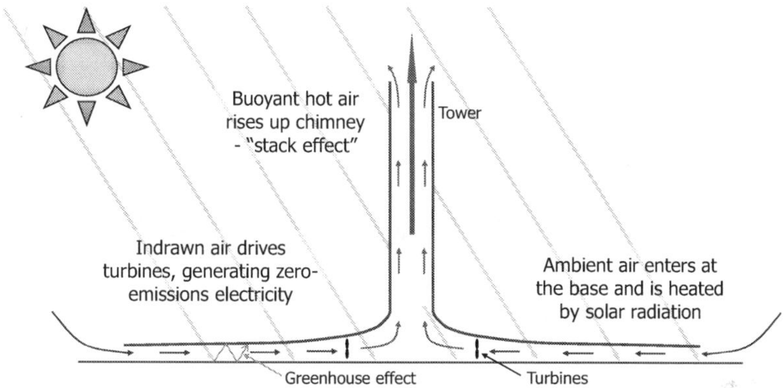

FIGURE 14.5 Representation of an SCPP and its principal components. *After Schlaich et al. (2005).*

If the solar collector is very large and the chimney is very tall, sufficient airflow can be created to allow the addition of turbines to generate electricity, resulting in a solar chimney power plant (SCPP; Fig. 14.5). An SCPP can operate continuously (24/7/365) if a thermal storage layer is added, for example comprised of 15 cm thick water bags (Ming et al., 2008). A prototype SCPP with a 195 m high chimney was built and tested in 1982–89 in Manzanares, 150 km south of Madrid, Spain, under the direction of German structural engineer Jörg Schlaich (Schlaich et al., 2005).

The typical dimensions of modeled SCPPs are 1 km chimney height and 38 km^2 GH area, and up to 1.5 km chimney heights have been studied (Pretorius, 2007; Schindelin, 2002). The energy return on energy invested (EROEI) of SCPPs has been compared to fossil fueled and renewable energy generation technologies in a full life cycle assessment (Zongker, 2013), and the CO_2 emissions due to construction and raw materials (cement, iron, glass) can be compensated in the first 16–20 months of operations. The lifetime of SCs and SCPPs is generally estimated to be 80–120 years, much longer than that of any other power plant. Comprehensive overviews of the SC technology have been published by Ming (2016) and by Kasaeian and colleagues (2017).

Rationale of using solar chimneys as air-moving devices

SCPPs would contact immense moving air volumes and receive similarly immense amounts of solar energy and, as a result, have been proposed to reduce the capital and operating costs of CO_2 DAC (de Richter et al., 2013).

Ming and colleagues (2017) gave a very complete overview of the possibilities of pollutant dispersion in the built environment, and SCs have been proposed to remove urban air pollution (Gong et al., 2017b). In December 2017, a 60 m high prototype SC was commissioned in Xian, China (Fig. 14.4), dedicated to the removal of particulate matter (PM2.5 and

PM10) using filters (Cao et al., 2018). As the generated pressure drop would be required for filtration, dedicated air-purification SCs would have no turbines to produce electricity.

Land footprint and costs of SCPPs

Removing Gt/yr of CO_2 or CO_2e using NETs will require the construction of huge infrastructures (Herzog, 2011), and SCs or SCPPs would not be exempt from this requirement. A 400 MW SC requires a chimney height of 1.5 km and diameter of 270 m and estimated costs are roughly $1 billion, excluding land acquisition (Hummel, 2010; Schindelin, 2002), although adding a 20 km^2 photocatalytic layer to the GH would increase the investment by only 1% (de Richter et al., 2017). The levelized cost of electricity produced by SCs, assuming capital repayment over 33 year, is typically below 0.10 €/kWh (Harte et al., 2013) and, for chimney heights >700 m and GH collector areas >20 km^2, the electricity generated can be cost-competitive with nuclear power, concentrated solar power, or offshore wind turbines.

The specific land area required for power generation by SCPPs at this scale (5 ha/MW = 2000 ha/400 MW) compares to ~400 ha/MW estimated for BECCS power plants in the United Kingdom (4.5 Mha for 22 × 500 MW plants; Donnison et al., 2020), and the SCPP requirement would be for arid or desert land rather than agricultural land suitable for food production.

Global electrical generation capacity (10 TW; IEA, 2018) could be provided by 25,000 SCPPs of 400 MW covering around 500,000 km^2 of arid or desert land (some 23% of Saudi Arabia, or 5% of the Sahara Desert).

As the daily air flow passing through a single 400 MW SCPP is 38 km^3 (48 Mt-air per day), 1 atmospheric volume (4.23×10^9 km^3, 5.2×10^{18} kg) could be processed in ~12 years by 25,000 SCPPs of 400 MW, costing $25 trillion to build. Other cost estimates have recently been presented by Li (2022). To put this construction cost in perspective, the International Renewable Energy Agency estimates that, by 2050, $131 trillion will be needed to transition to an energy system compatible with a 1.5°C Pathway (IRENA, 2021).

The capital cost for photocatalytic GGR could be reduced by building SCs and forgoing the energy production aspect of SCPPs. An air volume of 22–33 km^3/d (8000–12,000 km^3/yr) could be processed using a 500 m high SC (Cao et al., 2018), which, assuming an oxidation efficiency of 50%, could remove between 0.4 and 0.6 $MtCO_2e$/yr, based on a CH_4 GWP_{20} of 86. Combined with a 20 km^2 collector such a system could be constructed for $150–160 million, extrapolating the estimates of Harte and colleagues (Harte et al., 2013). The size and cost of the GH solar collector could also be reduced if low-grade waste heat was available to drive convection, for example from a thermal power plant or other industrial processes (Fathi et al., 2018), or from geothermal energy (Cao et al., 2014).

Deployment of photocatalytic SCPPs at a climate-relevant scale, within a diversified energy portfolio, could make a significant contribution to a carbon-negative economy, as SCPPs do not emit CO_2 during their operations, and remove CH_4, N_2O, and halocarbons from the atmosphere, generating CO_2e removal credits while producing rather than consuming electricity, in contrast to CO_2 DAC installations.

SC and SCPP synergy with CO_2 direct air capture

A 1 $MtCO_2$/yr DAC facility would require around 50,000 m^2 of frontal area, for example made up of five structures 1 km long × 10 m high, assuming 50% CO_2 capture and an airflow of 2 m/s (Socolow et al., 2011). The structures might be spaced 250 m apart, with the aim of preventing CO_2-depleted air from entering adjacent contactors, leading to a footprint of roughly 1.5 km^2 for the total system, including chemical regeneration and CO_2 export compression. A footprint of 1.5 km^2 for a 1 $MtCO_2$/yr removal capacity can be compared to 40 km^2/$MtCO_2$/yr for two 400 MW SCPPs described above.

Using waste or geothermal heat together with SCs would allow the area footprint of a DAC plant to be halved and the fans and associated energy consumption to be eliminated. As an example, assuming a pressure drop of 100 Pa across the DAC absorbers, using an SC with a circular GH collector of 0.8 km^2 area (~3.1 km circumference), an entrance height of 16 m results in a similar frontal area of 50,000 m^2. With a 550 m high chimney of 50 m radius, a difference of only 10°C between the ambient temperature outside the top of the chimney and the GH temperature at the chimney base would provide the required 2 m/s airflow to replace the fans and associated energy consumption. With a larger temperature difference, the height of the chimney could be reduced or the volume of air processed could be increased. If associated with a thermal power plant or another industrial process, the SC could also be used as a natural draft dry cooling tower for the plant (Bonnelle et al., 2010; Ghorbani et al., 2015), with improved heat capacity dissipation (Zou et al., 2013). Further synergies would arise from the release back inside the GH of residual waste heat following solvent or sorbent regeneration

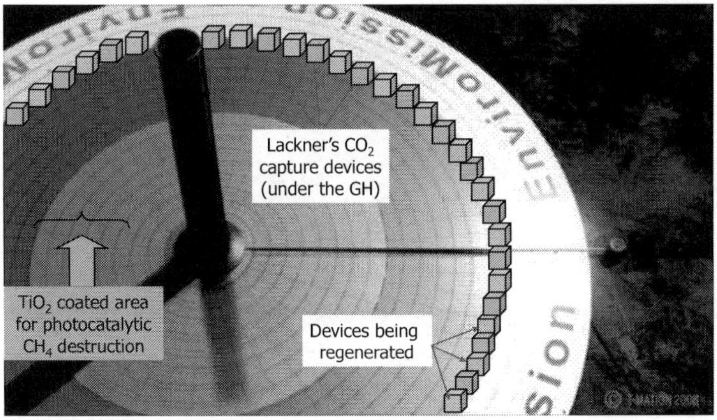

FIGURE 14.6 Possible configuration of an SCPP-DAC device to proceed to CDR and GGR. *From de Richter et al. (2013).*

in the DAC process. In this example, the SC would not be limited to tropical or equatorial locations and could potentially be operated in mid or even northern latitudes. Non-CO_2 GGR capability could be included in such an SC-DAC system by incorporating photocatalytic surfaces, as represented in Fig. 14.6.

In this example, the SC air flow is first intercepted by DAC devices that would cause some pressure drop. The PCs are located on the GH floor and on the lower surface of the GH glass, but could also be located inside the chimney using artificial UV light, although at the cost of additional electricity consumption.

14.3.2 Methane capture and removal by catalytic zeolites

The use of catalytic zeolites to transform CH_4 into CO_2, using air-moving devices similar to those used for CO_2 DAC has been investigated by a team from Stanford University (Jackson et al., 2019). Electric fans force air through catalytic zeolites that first trap CH_4 and then, when heated, oxidize CH_4 to CO_2 which is released back into the atmosphere (Fig. 14.7).

To be promising for CH_4 capture from dilute and medium-concentration sources, the zeolite materials need to have sufficient adsorption capacity as well as appropriate CH_4/N_2 and CH_4/CO_2 selectivity, properties found for a number of zeolite structures (Kim et al., 2013). This is illustrated in Fig. 14.8, where the adsorption sites are represented in blue, and for optimal interaction, each site is connected to three other sites (*yellow* arrows).

Recently, the capture of CH_4 at low concentrations was demonstrated using an abundant and inexpensive type of clay (mordenite), modified by ion exchange to incorporate Cu, although CH_4 oxidation required heating to 300−500°C

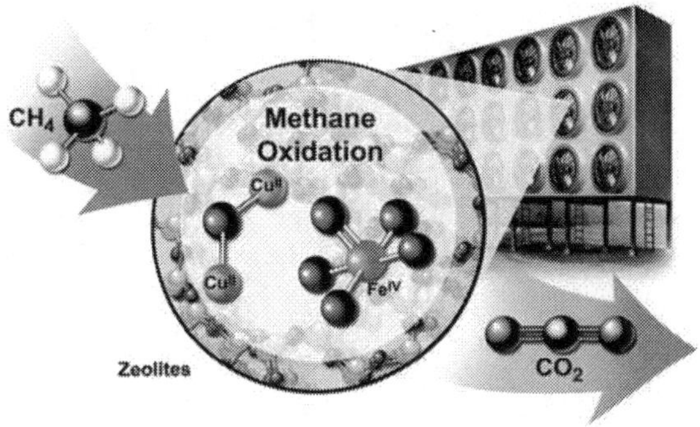

FIGURE 14.7 Conceptual device for atmospheric CH_4 removal. *From Jackson et al. (2019)*

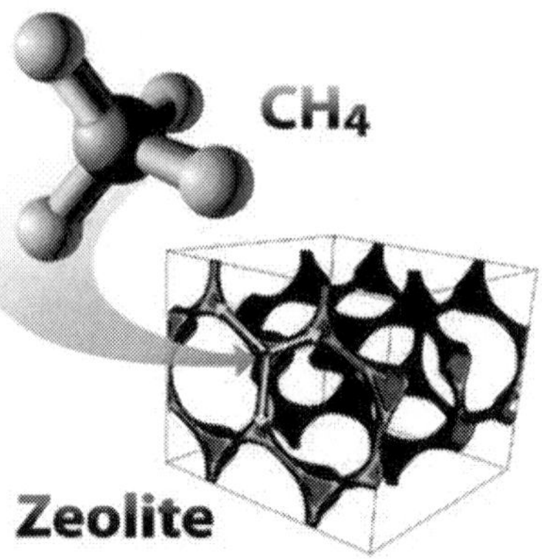

FIGURE 14.8 Capture of CH_4 in a Cu-zeolite. *From Kim et al. (2013).*

(Brenneis et al., 2021). A more direct route from CH_4 to methanol at nearly ambient temperature has been developed using bio-inspired zeolites containing iron and copper, mimicking enzymes used by organisms to oxidize hydrocarbons (Snyder et al., 2021).

Although the cost estimate for the process described by Jackson and colleagues (2019) has yet to be published, CH_4 removal in modified CO_2 DAC systems would share the costs of infrastructure and for moving the air with electrical fans. In view of the higher GWP of CH_4 compared to CO_2, the overall benefit of the combined process would be increased, despite the tropospheric [CH_4] being about 220 times lower than [CO_2].

14.4 Nitrous oxide and halogenated gases removal

As the transformation, destruction, or removal of all GHGs can be accelerated by several orders of magnitude using photocatalytic redox reactions (de Richter and Caillol, 2011), photocatalytic variants of SCPPs have been proposed to reduce atmospheric N_2O to N_2 and O_2 (Ming et al., 2016), and to destroy CFCs, HCFCs, and HFCs (de Richter et al., 2016). Table 14.2 lists some possible photocatalysts (PCs) for the removal of N_2O and some halogenated GHGs. The list is not exhaustive, and further details can be found in the review by de Richter and Caillol (2011). In addition to the removal of CH_4, SCPP-PCRs could be used to remove a range of GHGs, as illustrated in Fig. 14.9.

As an alternative to photocatalytic conversion, photolysis has also been proposed to address point sources of more concentrated N_2O, such as waste anesthetic gases, with estimated removal costs of \$49/tCO_2e, based on a 100 year GWP (Rauchenwald et al., 2020).

TABLE 14.2 Of the principal GHGs (Blasing, 2016) and examples of PCs for their degradation.

Target gas	GWP_{100} (AR5)	Concentration in 2021	Lifetime (years)	Photocatalysts for GHGs mineralization
Methane (CH_4)	28	1896 ppb	12	ZnO+0.1%Ag; Polytungstate dispersed on titania; TiO_2 supported on molybdena
Nitrous oxide (N_2O)	265	334 ppb	114	MgO, ZnO; TiO_2 + Ag or Cu; ZSM-5 zeolites + Cu, Ag, Ti, V, Mo, Cr
CFC-12 (CCl_2F_2)	10,200	493 ppt	100	TiO_2
CFC-11 (CCl_3F)	4660	222 ppt	45	TiO_2
HCFC-22 ($CHClF_2$)	1760	249 ppt	12	Zeolites
Tropospheric O_3	n.a.	n.a.	Days	TiO_2 + Cu, Mn, Fe; MnO_2

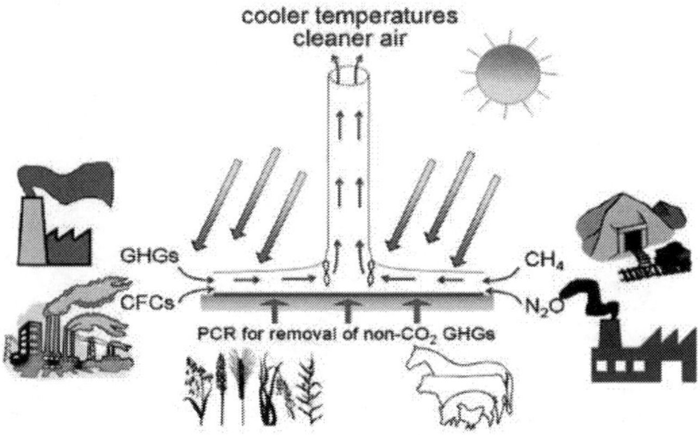

FIGURE 14.9 Schematic of an SCPP-PCR for GGR and renewable electricity generation. *From de Richter et al. (2017).*

14.5 Tropospheric ozone decomposition

Ozone (O_3) in the stratosphere protects all life on Earth from dangerous UV radiation, but tropospheric O_3, especially in the lower troposphere (surface boundary layer), is a significant GHG that is also dangerous for the health of all living organisms (Iriti and Faoro, 2008).

Tropospheric O_3 destruction by reactive halogen species, particularly chlorine, is well documented (Sherwen et al., 2016) and is responsible in the tropics for nearly 30% of tropospheric O_3 destruction (Read et al., 2008; Sommariva and von Glasow, 2012). During episodes called "tropospheric ozone depletion events" or "polar tropospheric ozone hole events," O_3 is completely destroyed in the lowest kilometer of the atmosphere on areas of several million km^2 and has a cooling effect (Roscoe et al., 2001). Enhancement of the Cl atom sink, discussed in Section 14.2.1 for CH_4 removal, has also been proposed as a possible method of large-scale surface O_3 removal (Ming et al., 2021). The use of iron salt aerosols could enhance this process of surface O_3 removal over the oceans (Wittmer, Bleicher and Zetzsch 2015b, 2016), but also globally, as methane removal reduces the ozone burden. Abernethy and colleagues (2021) have estimated that each $GtCH_4$ removed would result in a 1.0 ± 0.2 ppm surface [O_3] reduction, while also improving air quality (Fiore et al., 2008).

Many catalytic O_3 removal methods have been proposed (Namdari et al., 2021), and photocatalytic methods have been reviewed by de Richter and Caillol (2011). SCPPs or SCs could be used to decompose O_3 to O_2 at ambient temperature using TiO_2, MnO_2, or their derivatives as photocatalysts (Gong et al., 2017a; Jia and Zhang, 2018; Patzsch and Bloh, 2018; Namdari et al., 2019; Zhu et al., 2021; Wang et al. 2022b).

References and resources

References

Abernethy, S., O'Connor, F., Jones, C., Jackson, R., 2021. Methane removal and the proportional reductions in surface temperature and ozone. Philos. Trans. R. Soc. A 379, 20210104.

Abou El-Magd, I.H., Ali, E.M., 2012. Estimation of the evaporative losses from Lake Nasser, Egypt using optical satellite imagery. Int. J. Digit. Earth 5, 133–146.

Anderson, K., 2016. Duality in climate science. Nat. Geosci. 8, 898–900.

Atkinson, R., Arey, J., 2003. Atmospheric degradation of volatile organic compounds. Chem. Rev. 103, 4605–4638.

Blasing, T., 2016. Recent Greenhouse Gas Concentrations. Carbon Dioxide Information Analysis Center CDIAC. DOE—Oak Ridge. https://doi.org/10.3334/CDIAC/atg.032. http://cdiac.ornl.gov/pns/current_ghg.html. (Accessed 28 March 2018).

Bonnelle, D., Siros, F., Philibert, C., 2010. Concentrating solar parks with tall chimneys dry cooling. In: Solar Paces International Conference. Perpignan, France.

Brasseur, G.P., Kiehl, J.T., Müller, J.F., Schneider, T., Granier, C., Tie, X., Hauglustaine, D., 1998. Past and future changes in global tropospheric ozone: impact on radiative forcing. Geophys. Res. Lett. 25, 3807–3810.

Brenneis, R.J., Johnson, E.P., Shi, W., Plata, D.L., 2021. Atmospheric-and low-level methane abatement via an earth-abundant catalyst. ACS Environmental.

Cao, F., Li, H., Ma, Q., Zhao, L., 2014. Design and simulation of a geothermal—solar combined chimney power plant. Energy Convers. Manag. 84, 186—195.

Cao, Q., Shen, L., Chen, S.-C., Pui, D.Y., 2018. WRF modeling of PM 2.5 remediation by SALSCS and its clean air flow over Beijing terrain. Sci. Total Environ. 626, 134—146.

Chen, H., Navea, J.G., Young, M.A., Grassian, V.H., 2011. Heterogeneous photochemistry of trace atmospheric gases with components of mineral dust aerosol. J. Phys. Chem. 115, 490—499.

Chen, X., Li, Y., Pan, X., Cortie, D., Huang, X., Yi, Z., 2016. Photocatalytic oxidation of methane over silver decorated zinc oxide nanocatalysts. Nat. Commun. 7, 12273.

Cwiertny, D.M., Young, M.A., Grassian, V.H., 2008. Chemistry and photochemistry of mineral dust aerosol. Annu. Rev. Phys. Chem. 59, 27—51.

de Richter, R., Caillol, S., 2011. Fighting global warming: the potential of photocatalysis against CO_2, CH_4, N_2O, CFCs, tropospheric O_3, BC and other major contributors to climate change. J. Photochem. Photobiol. C Photochem. Rev. 12, 1—19.

de Richter, R.K., Ming, T., Caillol, S., 2013. Fighting global warming by photocatalytic reduction of CO_2 using giant photocatalytic reactors. Renew. Sustain. Energy Rev. 19, 82—106.

de Richter, R., Ming, T., Caillol, S., Liu, W., 2016. Fighting global warming by GHG removal: destroying CFCs and HCFCs in solar-wind power plant hybrids producing renewable energy with no-intermittency. Int. J. Greenh. Gas Control 49, 449—472.

de Richter, R., Ming, T., Davies, P., Liu, W., Caillol, S., 2017. Removal of non-CO_2 greenhouse gases by large-scale atmospheric solar photocatalysis. Prog. Energy Combust. Sci. 60, 68—96.

Donnison, C., Holland, R.A., Hastings, A., Armstrong, L.M., Eigenbrod, F., Taylor, G., 2020. Bioenergy with Carbon Capture and Storage (BECCS): finding the win—wins for energy, negative emissions and ecosystem services—size matters. GCB Bioenergy 12, 586—604.

Dunfield, P., 2007. The soil methane sink. In: Greenhouse Gas Sinks. CABI editions, 978 1 84593 189 6. 152.

Ebaid, H.M., Ismail, S.S., 2010. Lake Nasser evaporation reduction study. J. Adv. Res. 1, 315—322.

Engstrom, E.G., 2016. Hydroxyl Generation And/or Ozone Reduction System and Method. United States Patent Application 20160325004. US.

Eurochlor, 2021. Chlorine & Caustic Soda Production. https://www.eurochlor.org/about-chlor-alkali/how-are-chlorine-and-caustic-soda-made/. (Accessed 20 February 2022).

Fathi, N., McDaniel, P., Aleyasin, S.S., Robinson, M., Vorobieff, P., Rodriguez, S., de Oliveira, C., 2018. Efficiency enhancement of solar chimney power plant by use of waste heat from nuclear power plant. J. Clean. Prod. 180.

Fiore, A.M., West, J.J., Horowitz, L.W., Naik, V., Schwarzkopf, M.D., 2008. Characterizing the tropospheric ozone response to methane emission controls and the benefits to climate and air quality. J. Geophys. Res. Atmos. 113.

George, C., Ammann, M., D'Anna, B., Donaldson, D., Nizkorodov, S.A., 2015. Heterogeneous photochemistry in the atmosphere. Chem. Rev. 115, 4218—4258.

Ghorbani, B., Ghashami, M., Ashjaee, M., Hosseinzadegan, H., 2015. Electricity production with low grade heat in thermal power plants by design improvement of a hybrid dry cooling tower and a solar chimney concept. Energy Convers. Manag. 94, 1—11.

Gong, J.-Y., Chen, Y.-C., Yu, K.-P., 2017a. Photocatalytic decomposition of indoor ozone motivated by the white-light-emitting diode. Clean Technol. Environ. Policy 19, 2393—2404.

Gong, T., Ming, T., Huang, X., de Richter, R., Wu, Y., Liu, W., 2017b. Numerical analysis on a solar chimney with an inverted U-type cooling tower to mitigate urban air pollution. Sol. Energy 147, 68—82.

Harte, R., Höffer, R., Krätzig, W.B., Mark, P., Niemann, H.-J., 2013. Solar updraft power plants: engineering structures for sustainable energy generation. Eng. Struct. 56, 1698—1706.

Hassan, A., Ismail, S.S., Elmoustafa, A., Khalaf, S., 2018. Evaluating evaporation rate from high Aswan Dam Reservoir using RS and GIS techniques. Egypt. J. Remote Sens. Space Sci. 21, 285—293.

He, J., Naik, V., Horowitz, L.W., 2021. Hydroxyl radical (OH) response to meteorological forcing and implication for the methane budget. Geophys. Res. Lett. 48 e2021GL094140.

Heard, D.E., Pilling, M.J., 2003. Measurement of OH and HO_2 in the troposphere. Chem. Rev. 103, 5163—5198.

Herzog, H.J., 2011. Scaling up carbon dioxide capture and storage: from megatons to gigatons. Energy Econ. 33, 597—604.

Hofzumahaus, A., Lefer, B., Monks, P., Hall, S., Kylling, A., Mayer, B., Shetter, R., Junkermann, W., Bais, A., Calvert, J., 2004. Photolysis frequency of O_3 to O (1D): measurements and modeling during the international photolysis frequency measurement and modeling intercomparison (IPMMI). J. Geophys. Res. Atmos. 109.

Holmes, C., Prather, M., Søvde, A., Myhre, G., 2012. Future methane, hydroxyl, and their uncertainties: key climate and emission parameters for future predictions. Atmos. Chem. Phys. 12, 931.

Hossaini, R., Chipperfield, M.P., Saiz-Lopez, A., Fernandez, R., Monks, S., Feng, W., Brauer, P., Von Glasow, R., 2016. A global model of tropospheric chlorine chemistry: organic versus inorganic sources and impact on methane oxidation. J. Geophys. Res. Atmos. 121, 14271—14297.

Hummel, M., 2010. Can power from updraft and greentowers already compete against Conventional technologies? a profitability and risk analysis. In: SCPT 2010: Solar Chimney Power Technology International Conference. 28-30 September 2010, 355-360. Bochum, Germany.

IEA, 2018. Global Energy and CO_2 Status, 2017 Report. International Energy Agency, Paris.

IPCC, 2007. AR6 WG1 Climate Change 2021: The Physical Science Basis. https://www.ipcc.ch/report/ar4/wg1/. (Accessed 6 May 2022).

IPCC, 2021. AR6 WG1 Climate Change 2021: The Physical Science Basis. https://www.ipcc.ch/report/ar6/wg1/downloads/report/IPCC_AR6_WGI_Full_Report.pdf. (Accessed 4 May 2022).

IPCC, 2022. AR6 WG3 Climate Change 2022: Mitigation of Climate Change. https://report.ipcc.ch/ar6wg3/pdf/IPCC_AR6_WGIII_FinalDraft_FullReport.pdf. (Accessed 4 May 2022).

IRENA, 2021. World Energy Transitions Outlook: 1.5°C Pathway. International Renewable Energy Agency, Abu Dhabi. https://irena.org/publications/2021/Jun/World-Energy-Transitions-Outlook.

Iriti, M., Faoro, F., 2008. Oxidative stress, the paradigm of ozone toxicity in plants and animals. Water, Air, Soil Poll. 187, 285–301.

Ito, A., Shi, Z., 2015. Delivery of anthropogenic bioavailable iron from mineral dust and combustion aerosols to the ocean. Atmos. Chem. Phys. Discuss. 15, 23051–23088.

Jackson, R.B., Solomon, E., Canadell, J., Cargnello, M., Field, C., 2019. Methane removal and atmospheric restoration. Nat. Sustain. 2, 436–438.

Jackson, R.B., Solomon, E.I., Canadell, J.G., Cargnello, M., Field, C.B., Abernethy, S., 2020. Reply to: practical constraints on atmospheric methane removal. Nat. Sustain. 1–2.

Jackson, R., Abernethy, S., Canadell, J., et al., 2021. Atmospheric Methane Removal: A Research Agenda. Philosophical Transactions of the Royal Society A (in press).

Jia, J., Zhang, P., 2018. Catalytic decomposition of airborne ozone by $MnCO_3$ and its mechanism. Ozone Sci. Eng. 40, 21–28.

Johnson, C., Collins, W., Stevenson, D., Derwent, R., 1999. Relative roles of climate and emissions changes on future tropospheric oxidant concentrations. J. Geophys. Res. Atmos. 104, 18631–18645.

Johnson, M.S., 2021. Methane Removal and Emerging Technologies, Tuesday, 28 September, 2021 at Centre for Climate Repair. Cambridge, UK. At 2h02. https://www.youtube.com/watch?v=oFOnsY-_2zs. (Accessed 21 December 2021). Costs estimates at at 2h01'30".

Kasaeian, A., Molana, S., Rahmani, K., Wen, D., 2017. A review on solar chimney systems. Renew. Sustain. Energy Rev. 67, 954–987.

Kim, J., Maiti, A., Lin, L.-C., Stolaroff, J.K., Smit, B., Aines, R.D., 2013. New materials for methane capture from dilute and medium-concentration sources. Nat. Commun. 4, 1–7.

Lackner, K.S., 2020. Practical constraints on atmospheric methane removal. Nat. Sustain. 3, 357.

Lamarque, J.F., Kiehl, J., Hess, P., Collins, W., Emmons, L., Ginoux, P., Luo, C., Tie, X., 2005. Response of a coupled chemistry-climate model to changes in aerosol emissions: global impact on the hydrological cycle and the tropospheric burdens of OH, ozone, and NOx. Geophys. Res. Lett. 32.

Lau, M.C., Stackhouse, B., Layton, A.C., Chauhan, A., Vishnivetskaya, T., Chourey, K., Ronholm, J., Mykytczuk, N., Bennett, P., Lamarche-Gagnon, G., 2015. An active atmospheric methane sink in high Arctic mineral cryosols. ISME J. 9, 1880–1891.

Lawler, M., Finley, B., Keene, W., Pszenny, A., Read, K., Von Glasow, R., Saltzman, E., 2009. Pollution-enhanced reactive chlorine chemistry in the eastern tropical Atlantic boundary layer. Geophys. Res. Lett. 36.

Lee, E.-H., Moon, K.-E., Cho, K.-S., 2017. Long-term performance and bacterial community dynamics in biocovers for mitigating methane and malodorous gases. J. Biotechnol. 242, 1–10.

Levine, J.G., Wolff, E.W., Jones, A.E., Sime, L.C., 2011. The role of atomic chlorine in glacial-interglacial changes in the carbon-13 content of atmospheric methane. Geophys. Res. Lett. 38 (4).

Li, W., 2022. A Solar Energy Solution for Methane Removal. https://www.youtube.com/watch?v=kxegaWWvcL0. (Accessed 4 May 2021) (At Methane Action).

Lim, M., Chiang, K., Amal, R., 2006. Photochemical synthesis of chlorine gas from iron (III) and chloride solution. J. Photochem. Photobiol. Chem. 183, 126–132.

Lo, M.H., Famiglietti, J.S., 2013. Irrigation in California's Central Valley strengthens the southwestern US water cycle. Geophys. Res. Lett. 40, 301–306.

Luo, C., Mahowald, N., Bond, T., Chuang, P., Artaxo, P., Siefert, R., Chen, Y., Schauer, J., 2008. Combustion iron distribution and deposition. Global Biogeochem. Cycles 22.

Martin, J.H., 1990. Glacial-interglacial CO_2 change: the iron hypothesis. Paleoceanography 5, 1–13.

Matsumi, Y., Comes, F., Hancock, G., Hofzumahaus, A., Hynes, A.J., Kawasaki, M., Ravishankara, A., 2002. Quantum yields for production of O (1D) in the ultraviolet photolysis of ozone: recommendation based on evaluation of laboratory data. J. Geophys. Res. Atmos. 107. ACH 1-1-ACH 1-12.

Ming, T., Liu, W., Pan, Y., Xu, G., 2008. Numerical analysis of flow and heat transfer characteristics in solar chimney power plants with energy storage layer. Energy Convers. Manag. 49, 2872–2879.

Ming, T., 2016. Solar Chimney Power Plant Generating Technology. Academic Press, ISBN 978-0-12-805370-6.

Ming, T., de Richter, R., Shen, S., Caillol, S., 2016. Fighting global warming by greenhouse gas removal: destroying atmospheric nitrous oxide thanks to synergies between two breakthrough technologies. Environ. Sci. Pollut. Control Ser. 23, 6119–6138.

Ming, T., Peng, C., Gong, T., Li, Z., 2017. Pollutant Dispersion in Built Environment. Springer, Cham, Switzerland.

Ming, T., de Richter, R., Oeste, F.D., Tulip, R., Caillol, S., 2021. A nature-based negative emissions technology able to remove atmospheric methane and other greenhouse gases. Atmos. Pollut. Res.

Ming, T., Li, W., Yuan, Q., et al., 2022. Perspectives on removal of atmospheric methane. Adv. Appl. Energy 100085.

Wang, Y., Ming, T., Li, W., Yuan, Q., de Richter, R., Davies, P., Caillol, S., 2022a. Atmospheric removal of methane by enhancing the natural hydroxyl radical sink. Greenh. Gases: Sci. Technol. 12 (6), 784–795.

Morneault, G.J.E., 2013. Hydroxyl Generator, United States Patent Application 20130004381. US.

Myriokefalitakis, S., Daskalakis, N., Mihalopoulos, N., Baker, A., Nenes, A., Kanakidou, M., 2015. Changes in dissolved iron deposition to the oceans driven by human activity: a 3-D global modelling study. Biogeosciences 12, 3973–3992.

Namdari, M., Lee, C.-S., Haghighat, F., 2021. Active ozone removal technologies for a safe indoor environment: a comprehensive review. Build. Environ. int. 187, 107370.

Namdari, M., Lee, C., Haghighat, F., Bahloul, A., Huard, M., 2019. Reviewing MnOx-based catalysts for decomposition of indoor ozone. IOP Conf. Ser. Mater. Sci. Eng. 609, 042079.

Nicely, J.M., Duncan, B.N., Hanisco, T.F., Wolfe, G.M., Salawitch, R.J., Deushi, M., Haslerud, A.S., Jöckel, P., Josse, B., Kinnison, D.E., 2020. A machine learning examination of hydroxyl radical differences among model simulations for CCMI-1. Atmos. Chem. Phys. 20, 1341–1361.

Nisbet-Jones, P.B., Fernandez, J.M., Fisher, R.E., France, J.L., Lowry, D., Waltham, D.A., Woolley Maisch, C.A., Nisbet, E.G., 2021. Is the destruction or removal of atmospheric methane a worthwhile option? Philos. Trans. R. Soc. A 380, 20210108.

Nisbet, E.G., Piper, D.J., 1998. Giant submarine landslides. Nature 392, 329–330.

NOAA, 2022. Increase in Atmospheric Methane Set Another Record during 2021. https://www.noaa.gov/news-release/increase-in-atmospheric-methane-set-another-record-during-2021. (Accessed 4 May 2022).

O'Connor, F.M., Boucher, O., Gedney, N., Jones, C., Folberth, G., Coppell, R., Friedlingstein, P., Collins, W., Chappellaz, J., Ridley, J., 2010. Possible role of wetlands, permafrost, and methane hydrates in the methane cycle under future climate change: a review. Rev. Geophys. 48.

O'Grady, C., 2021. Methane removal seen as tool to slow warming. Science 374, 667–668.

Ocko, I.B., Sun, T., Shindell, D., Oppenheimer, M., Hristov, A.N., Pacala, S.W., Mauzerall, D.L., Xu, Y., Hamburg, S.P., 2021. Acting rapidly to deploy readily available methane mitigation measures by sector can immediately slow global warming. Environ. Res. Lett. 16, 054042.

Oeste, F.D., de Richter, R., Ming, T., Caillol, S., 2017. Climate engineering by mimicking natural dust climate control: the iron salt aerosol method. Earth System Dyn. 8.

Patzsch, J., Bloh, J.Z., 2018. Improved photocatalytic ozone abatement over transition metal-grafted titanium dioxide. J Catalysis Today 300, 2–11.

Paull, C., Ussler, W., Holbrook, W., 2007. Assessing methane release from the colossal Storegga submarine landslide. Geophys. Res. Lett. 34.

Petersen, O., 2022. Enhanced Atmospheric Methane Oxidation by Sending Iron-Salt Aerosols into the Atmosphere. At the Methane Emissions Technology Alliance. https://www.youtube.com/watch?v=z9gJd48ocAU. (Accessed 4 May 2022).

Petit, J.-R., Jouzel, J., Raynaud, D., Barkov, N.I., Barnola, J.-M., Basile, I., Bender, M., Chappellaz, J., Davis, M., Delaygue, G., 1999. Climate and atmospheric history of the past 420,000 years from the Vostok ice core, Antarctica. Nature 399, 429–436.

Pretorius, J.P., 2007. Optimization and Control of a Large-Scale Solar Chimney Power Plant. University of Stellenbosch, Stellenbosch.

Price, S.J., Sherlock, R.R., Kelliher, F.M., McSeveny, T.M., Tate, K.R., Condron, L.M., 2004. Pristine New Zealand forest soil is a strong methane sink. Global Change Biol. 10, 16–26.

Prinn, R., Huang, J., Weiss, R., Cunnold, D., Fraser, P., Simmonds, P., McCulloch, A., Harth, C., Reimann, S., Salameh, P., 2005. Evidence for variability of atmospheric hydroxyl radicals over the past quarter century. Geophys. Res. Lett. 32.

Rauchenwald, V., Rollins, M.D., Ryan, S.M., Voronov, A., Feiner, J.R., Šarka, K., Johnson, M.S., 2020. New method of destroying waste anesthetic gases using gas-phase photochemistry. Anesth. Analg. 131, 288–297.

Read, K.A., Mahajan, A.S., Carpenter, L.J., Evans, M.J., Faria, B.V., Heard, D.E., Hopkins, J.R., Lee, J.D., Moller, S.J., Lewis, A.C., 2008. Extensive halogen-mediated ozone destruction over the tropical Atlantic Ocean. Nature 453, 1232–1235.

Roscoe, H., Kreher, K., Friess, U., 2001. Ozone loss episodes in the free Antarctic troposphere, suggesting a possible climate feedback. Geophys. Res. Lett. 28, 2911–2914.

Schindelin, H., 2002. Entwurf eines 1500m hohen Turms eines Solar-Aufwindkraftwerks (Design of a 1500m high tower of a SUPP). PhD Thesis. University of Wuppertal, Germany.

Schlaich, J., Bergermann, R., Schiel, W., Weinrebe, G., 2005. Design of commercial solar updraft tower systems—utilization of solar induced convective flows for power generation. J. Sol. Energy Eng. 127, 117–124.

Sedwick, P.N., Sholkovitz, E.R., Church, T.M., 2007. Impact of anthropogenic combustion emissions on the fractional solubility of aerosol iron: evidence from the Sargasso Sea. G-Cubed 8.

Sher Shah, M.S.A., Oh, C., Park, H., Hwang, Y.J., Ma, M., Park, J.H., 2020. Catalytic oxidation of methane to oxygenated products: recent advancements and prospects for electrocatalytic and photocatalytic conversion at low temperatures. Adv. Sci. 7, 2001946.

Sherwen, T., Evans, M.J., Carpenter, L.J., Schmidt, J.A., Mickely, L., 2016. Halogen chemistry reduces tropospheric O_3 radiative forcing. Atmos. Chem. Phys. Discuss. 10.

Snyder, B.E., Bols, M.L., Rhoda, H.M., Plessers, D., Schoonheydt, R.A., Sels, B.F., Solomon, E.I., 2021. Cage effects control the mechanism of methane hydroxylation in zeolites. Science 373, 327–331.

Socolow, R., Desmond, M., Aines, R., Blackstock, J., Bolland, O., Kaarsberg, T., Lewis, N., Mazzotti, M., Pfeffer, A., Sawyer, K., 2011. Direct Air Capture of CO_2 with Chemicals: A Technology Assessment for the APS Panel on Public Affairs. American Physical Society.

Sommariva, R., von Glasow, R., 2012. Multiphase halogen chemistry in the tropical Atlantic Ocean. Environ. Sci. Technol. 46, 10429–10437.

Sorooshian, S., Li, J., Hsu, K.l., Gao, X., 2011. How significant is the impact of irrigation on the local hydroclimate in California's Central Valley? comparison of model results with ground and remote-sensing data. J. Geophys. Res. Atmos. 116.

Sundqvist, E., Crill, P., Mölder, M., Vestin, P., Lindroth, A., 2012. Atmospheric methane removal by boreal plants. Geophys. Res. Lett. 39.

Thornton, J.A., Kercher, J.P., Riedel, T.P., Wagner, N.L., Cozic, J., Holloway, J.S., Dube, W.P., Wolfe, G.M., Quinn, P.K., Middlebrook, A.M., 2010. A large atomic chlorine source inferred from mid-continental reactive nitrogen chemistry. Nature 464, 271–274.

UN, 2017. World Population Projected to Reach 9.8 Billion in 2050, and 11.2 Billion in 2100. https://www.un.org/development/desa/en/news/population/world-population-prospects-2017.html. (Accessed 4 May 2022).

UNEP, 2021. Global Methane Assessment: Benefits and Costs of Mitigating Methane Emissions, ISBN 978-92-807-3854-4. https://www.unep.org/resources/report/global-methane-assessment-benefits-and-costs-mitigating-methane-emissions. (Accessed 22 June 2021).

Wang, L., Guan, J., Han, H., Yao, M., Kang, J., Peng, M., Wang, D., Xu, J., Hao, J., 2022b. Enhanced photocatalytic removal of ozone by a new chlorine-radical-mediated strategy. Appl. Catal. B: Environ. 306, 121130.

Wittmer, J., Bleicher, S., Ofner, J., Zetzsch, C., 2015a. Iron (III)-induced activation of chloride from artificial sea-salt aerosol. Environ. Chem. 12, 461–475.

Wittmer, J., Bleicher, S., Zetzsch, C., 2015b. Iron (III)-induced activation of chloride and bromide from modeled salt pans. J. Phys. Chem. 119, 4373–4385.

Wittmer, J., Bleicher, S., Zetzsch, C., 2016. Report on the photochemical induced halogen activation of Fe-containing aerosols. J Climatol Weather Forecasting 4, 2.

Wittmer, J., Zetzsch, C., 2017. Photochemical activation of chlorine by iron-oxide aerosol. J. Atmos. Chem. 74, 187–204.

Yuan, Bo, Q.X., de Richter, R., Li, W., Quesada Cabrera, R., Ming, T., 2022. Atmospheric Methane Remediation by Possible Chlorine Atom Enhancement (To Be Submitted, in Progress).

Zhao, J.-F., Peng, S.-S., Chen, M.-P., Wang, G.-Z., Cui, Y.-B., Liao, L.-G., Feng, J.-G., Zhu, B., Liu, W.-J., Yang, L.-Y., 2019. Tropical forest soils serve as substantial and persistent methane sinks. Sci. Rep. 9, 1–9.

Zhu, G., Zhu, W., Lou, Y., Ma, J., Yao, W., Zong, R., Zhu, Y., 2021. Encapsulate α-MnO2 nanofiber within graphene layer to tune surface electronic structure for efficient ozone decomposition. Nat. Commun. 12, 1–10.

Zongker, J.D., 2013. Life Cycle Assessment of Solar Updraft Tower Power Plant: EROEI and GWP as a Design Tool. Wichita State University.

Zou, Z., Guan, Z., Gurgenci, H., 2013. Optimization design of solar enhanced natural draft dry cooling tower. Energy Convers. Manag. 76, 945–955.

Projects and organizations

Atmospheric Methane Removal (AMR); Start-up developing methane removal technology: www.amr.earth.

Blue Dot Change; Start-up developing methane removal technology: www.bluedotchange.com.

Global Methane Pledge: www.globalmethanepledge.org.

NOAA; global CH_4 data: www.esrl.noaa.gov/gmd/ccgg/trends_ch4.

Xian prototype solar chimney; New China TV video report: www.youtube.com/watch?v=WOIwHFszKYA.

Chapter 15

Geological carbon storage

Graham Andrews and Steve Rackley

15.1 Introduction

Among the engineered components of the NET portfolio, geological storage was the first technology to reach Technology Readiness Level 9 (TRL9; see Chapter 2), with individual projects being commissioned at the $MtCO_2$/yr scale starting in the 1990s (Bickle, 2009). Many technologies and methods that have been developed in the oil and gas industry are applicable with minor modifications to the challenges of geological storage, from the selection of suitable storage sites to the monitoring of CO_2 plume movement, and this contributed to the rapid implementation of early storage projects.

Geological storage was first proposed in the 1990s as the final step in the carbon capture and storage (CCS) process when it was recognized that "useless" saline aquifers were globally much more widely distributed than depleted oil or gas fields, where CO_2 injection for enhanced oil recovery (EOR) had been practiced since the 1970s (Koide et al., 1992). Although injection into saline aquifers and oil reservoirs for EOR are still considered the main contenders for large-scale storage, considerable project experience is also being gained on in situ mineralization—the injection of CO_2 into fractured basalt lavas and peridotite bodies, with storage via mineral carbonation (Kelemen and Matter, 2008; Oelkers et al., 2008; and Chapter 10). Although sharing many of the technical considerations with saline aquifer storage, storage related to EOR, enhanced recovery of gas (EGR), or coal bed methane (ECBM) is not discussed further in this chapter. A short introduction to these options can be found in Rackley (2017).

15.2 Storage in sedimentary basins

Hydrocarbon reservoirs and saline aquifers are characteristic of sedimentary basins where thick accumulations of sedimentary deposits host voluminous pore space. At their simplest, suitable reservoir rocks are highly porous and permeable (e.g., sandstone), and vertically confined by an overlying impermeable caprock layer (e.g., shale). Storage of CO_2 in a host formation is achieved through the pumping of CO_2 into injection wells that are drilled and completed in the selected formation. Both the capacity of a storage site and the security of containment are maximized if CO_2 is injected into the subsurface as a supercritical fluid ($scCO_2$), as in this state it has a liquid-like density and a significantly lower mobility through the pore space than it would have as a gas. This necessitates injection at a depth where the hydrostatic pressure exceeds the critical pressure ($P_c = 7.38$ MPa), which is typically reached at $\sim$800 m (the critical pressure depth). In the following discussion, all references to injected CO_2 refer to a supercritical fluid unless otherwise stated.

Once injected into the formation, the fluid will move under the influence of viscous, capillary, and gravity forces, its ultimate fate being determined by a number of trapping mechanisms upon which the storage capacity of the complex depends. For each trapping mechanism, a theoretical storage capacity can be determined that represents the physical limit of storage volume that could be achieved assuming that the entire pore space in the host formation is accessible and utilized. The classification of more practical capacities, derived by applying technical, regulatory, and commercial constraints to this theoretical maximum, is described in Section 15.4.

15.2.1 Structural and stratigraphic trapping

The buoyancy of injected CO_2 relative to the in situ pore fluid will result in its upward migration until an impermeable barrier or caprock is encountered (Fig. 15.1). The storage host formation, together with all formations that prevent upward or lateral migration, is collectively termed the storage complex or storage assessment unit (SAU). Subsequently, CO_2 may

Negative Emissions Technologies for Climate Change Mitigation. https://doi.org/10.1016/B978-0-12-819663-2.00014-9

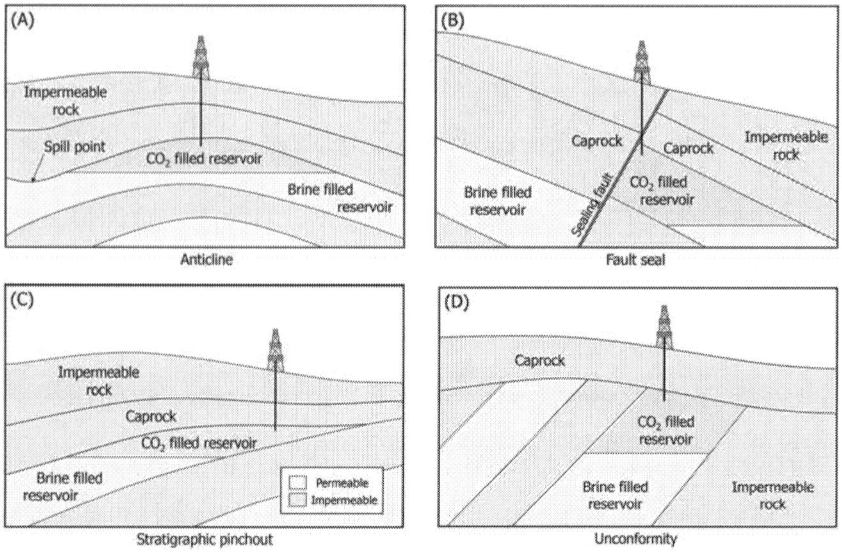

FIGURE 15.1 Stratigraphic (A, C, D) and structural (B) trap configurations.

migrate laterally under buoyant (i.e., gravity) or hydrodynamic forces, or it may "pond" beneath the caprock, if the geometry or other lateral changes in permeability prevent further migration. As this trapping mechanism is directly related to the buoyancy of CO_2 relative to formation brine, the term buoyant trapping is often used as an alternative to structural or stratigraphic trapping. Structural and stratigraphic traps are also known as static traps.

As migration of CO_2 continues, the trap will fill and the lower limit of the ponded CO_2 (the CO_2-water contact in an aquifer) will become deeper until one of the following occurs:

- The CO_2—water contact (CWC) drops to the spill point of the structure—the point where the structural or stratigraphic element causing the trap is on longer present, as indicated in Fig. 15.1A. Further migration into the trap will cause CO_2 to spill out at this point and migrate beyond the trap.
- The height of the trapped CO_2 column grows to the point where the CO_2 pressure at the crest of the structure exceeds the capillary entry pressure of the caprock (see next section), at which point CO_2 will start to leak into and possibly through the caprock.
- The pore pressure within the structure exceeds the fracture initiation pressure in the caprock or the reactivation pressure of any fault or fracture crossing the storage complex, at which point CO_2 may start to leak out of the storage complex.

Static trapping can also occur under smaller trapping elements, such as irregularities in a dipping planar caprock surface or small-scale impermeable layers within the storage formation (Fig. 15.10).

The timescale of structural trapping is determined by the time for CO_2 to migrate into the trap from the point of injection, and will typically occur in a few years to decades, after which the trapped CO_2 will more gradually disperse as a result of dissolution and gravity currents or geochemical reactions with the storage formation or caprock. Compared to the other trapping mechanisms described below, the risk of leakage is highest for CO_2 that is structurally trapped, as the fluid remains mobile and can migrate through the caprock if this is compromised by conductive faults, fractures, or permeable paths along wells, or fails at some point due to geochemical or geomechanical processes.

Theoretical storage capacity of structural and stratigraphic traps

The theoretical storage capacity is defined as the physical limit of storage volume or mass that can be accepted by a storage complex, assuming the entire pore space is filled and excluding any technical, engineering, or other constraints. When expressed as a mass of CO_2, the theoretical capacity of a structural or stratigraphic trap is given by;

$$m_{CO_2t} = A\,h\,\Phi\,(1 - S_{wirr})\rho_{CO_2} \tag{15.1}$$

where A, h, and Φ are the area, average thickness, and volume averaged effective porosity of the trap, S_{wirr} is the average irreducible water saturation within the trap pore volume, and ρ_{CO_2} is the average density of CO_2 in the trap at in situ conditions of temperature and pressure. The effective porosity excludes any disconnected porosity such as vugs in carbonate rocks and clay-bound water in shaly sandstones.

Eq. (15.1) assumes that the storage formation is open, for example, a deep saline aquifer at hydrostatic pressure that has a remote connection to the surface, and the increase in pore pressure of the system as a result of injection is not considered when estimating theoretical capacity. However, if injection is into a closed system then the increase in overall pore pressure must be considered in determining theoretical capacity. In this case, the theoretical storage capacity will be given by

$$m_{CO_2t} = V_{trap}\,\kappa_t\,\Delta P_{max}\rho_{CO_2} \qquad (15.2)$$

where V_{trap} is the total pore volume of the closed storage complex, κ_t is the total system compressibility—a combination of rock and fluid compressibility ($\kappa_t = \kappa_r + \kappa_f$), and ΔP_{max} is the maximum increase in pressure that can be sustained without leakage, for example as a result of mechanical failure of the caprock or the reactivation of existing faults or fractures. The pore pressure increase in a closed system can be relieved by pumping-off water, in which case the additional term $W_p\rho_{CO_2}$ would be added to the RHS of Eq. (15.2), where W_p is the volume at downhole conditions of the produced water. This and other reservoir management techniques for saline aquifer storage are discussed in Section 15.5.

15.2.2 Capillary and hydrodynamic trapping

Capillary trapping occurs when capillary forces immobilize lenses of CO_2 below small-scale, low permeability features, as illustrated in Fig. 15.2. Any estimate of the storage capacity of such small-scale elements will be extremely uncertain, even after detailed geological site assessment, as they are far below the resolution of the available surveying technologies. High-resolution 4D seismic monitoring (see Section 15.5.2) after a period of injection will be the most likely tool to provide insight into these mechanisms.

Storage can also take place in an open system where there is a continuous permeable flow path to the surface, where an active hydrodynamic flow opposes upward migration, for example, below a very gently dipping caprock (Fig. 15.3). Some mechanisms that result in hydrodynamic flows on a basin scale are illustrated in Fig. 15.4. If directed downdip, the hydrodynamic flow opposing updip buoyant movement of injected CO_2 would slow the migration, increasing the time available for dissolution or, in some cases, might halt or reverse the migration. Assuming a CO_2-brine density difference of 500 kg/m^3, a moderate hydraulic gradient of 5 m/km would balance the buoyant pressure gradient for a formation dipping at 0.6 degrees. Regional hydraulic gradients are therefore likely to have limited impact on plume dynamics except in near horizontal host formations and for hydraulic gradients at the top end of the typical range.

15.2.3 Residual trapping

As a plume of CO_2 rises through a water-saturated formation, the buoyant CO_2 will initially expel water from the pore space. This process is termed a drainage process as the saturation of the fluid that wets the grain surfaces (water in this case) is decreasing (Fig. 15.5).

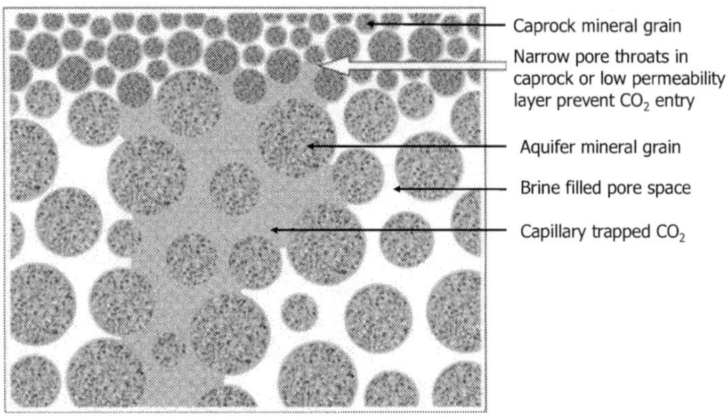

FIGURE 15.2 Capillary trapping mechanism.

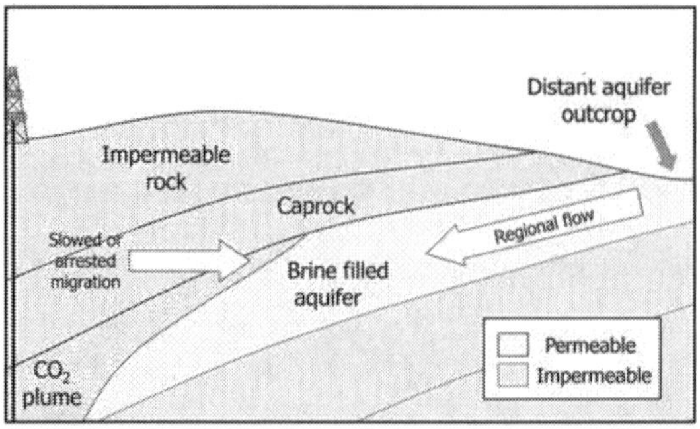

FIGURE 15.3 Hydrodynamic trapping—very slow movement in an open monocline.

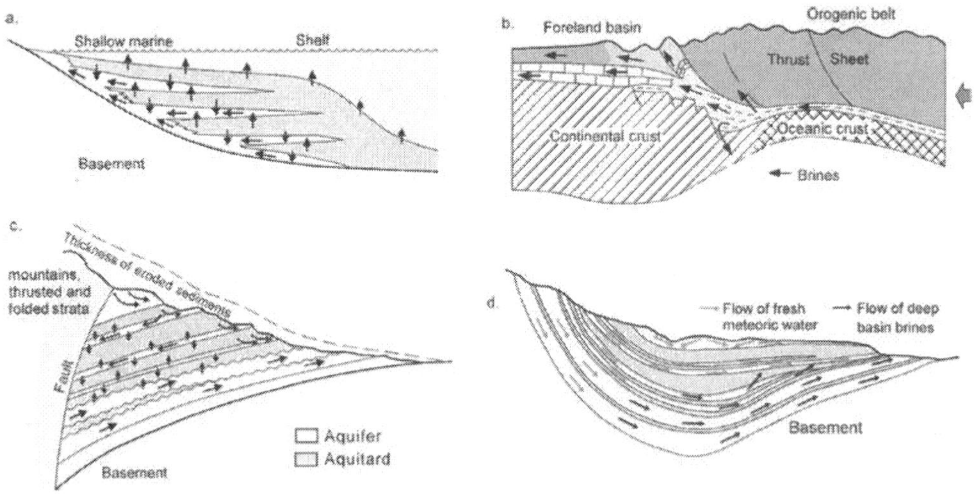

FIGURE 15.4 Basin scale aquifer flow-driving mechanisms: (A) compaction, (B) tectonic compression, (C) erosional rebound, and (D) gravity. *Source: From Bachu (2000) with permission.*

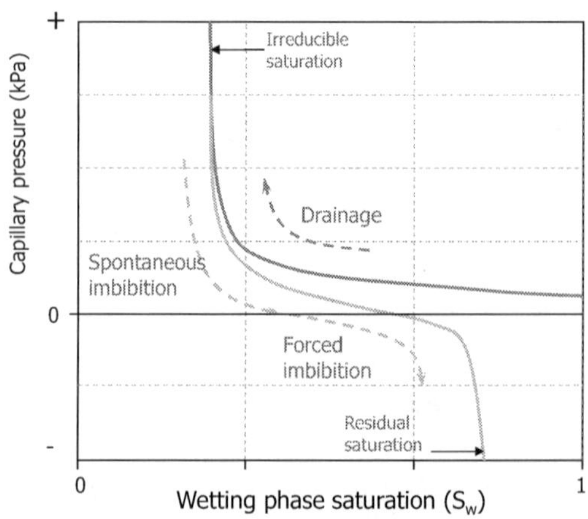

FIGURE 15.5 Schematic drainage-imbibition capillary pressure curves.

In the wake of the rising plume, water will again be drawn into the pores by spontaneous capillary imbibition. As the pore space saturation of CO_2 reduces, the permeability of the rock to the remaining CO_2 also declines, approaching zero at the residual saturation (Fig. 15.6).

At the plume front, this process reduces the water saturation to the residual water saturation (S_{wrc}), and later, in the wake of the plume, the CO_2 saturation will similarly be reduced to a residual value (S_{crw}). Here the suffix *irj* indicates that phase *i* is at residual saturation due to displacement by phase *j*. In this case, the whole rock volume swept out by the rising plume will retain a residual CO_2 saturation. Fig. 15.7 illustrates the residual CO_2 end point of this process on a pore scale.

Fig. 15.8 shows an example of this increasing and then decreasing CO_2 saturation behavior from the Frio brine pilot project. CO_2 saturation logging in the observation well (located 30 m away from the injection point) clearly shows the decline in CO_2 saturation toward a residual value as brine imbibition occurs behind the plume, following the 10-day injection period. The imbibition process is enhanced here by the relatively steep dip of the formation (11–16 degrees), which results in rapid updip migration of the plume. The dependence of CO_2 trapping on the maximum saturation reached during drainage is also demonstrated by these results; at ca. 5023 ft a peak CO_2 saturation in excess of 0.8 was reached leaving a residual (trapped) saturation of >0.3, while in the interval from 5030 to 5033 ft the peak CO_2 saturation was only 0.2 and trapped saturation was essentially zero.

Theoretical storage capacity for residual trapping

The theoretical storage capacity for residual trapping can be estimated from

$$m_{CO_2t} = A\,h\,\Phi\,S_{crw}\,\rho_{CO_2} \tag{15.3}$$

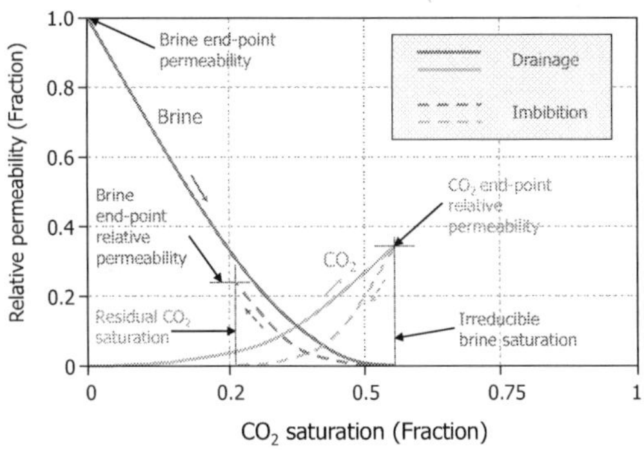

FIGURE 15.6 Relative permeability hysteresis for CO_2 injection.

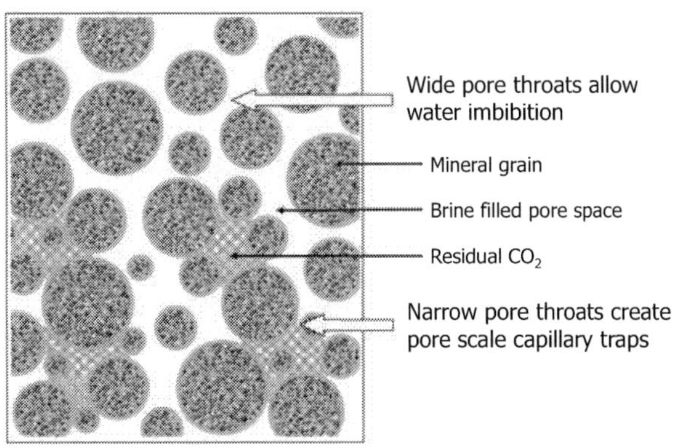

FIGURE 15.7 Residual trapping of CO_2 following plume migration through an aquifer.

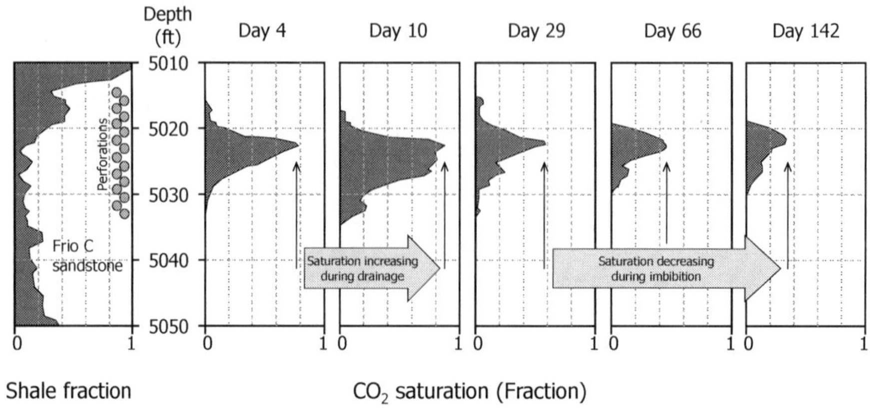

FIGURE 15.8 CO_2 saturation logging at the Frio Brine Pilot observation well. *Source: After Hovorka et al. (2006).*

with terms as defined earlier. This estimate of theoretical capacity assumes that injected CO_2 invades all of the effective pore space throughout the storage formation, a situation that could only be approached by applying various reservoir management techniques to maximize the gross rock volume swept by the CO_2 plume (see Section 15.5).

Residual-trapped CO_2 has a very low risk of leakage since the relative permeability of the residual phase is essentially zero. Over time, the residual trapped CO_2 will dissolve in the formation brine and mix with the surrounding unsaturated aquifer water.

15.2.4 Solubility trapping

Dissolution of CO_2 and convective mixing into the formation brine is the major long-term trapping mechanism in a saline aquifer. The solubility of CO_2 will determine the magnitude of solubility trapping and is dependent on the physical conditions of temperature and pressure as well as the ionic strength (salinity) and chemical composition of the formation brine. Fig. 15.9 illustrates the dependence of CO_2 solubility in pure water and in aqueous NaCl solutions on temperature, pressure, and solution ionic strength, as predicted by the model of Duan and Sun (2003).

CO_2 solubility for sites where storage in saline aquifers has been piloted (e.g., Frio, In Salah, Ketzin) or implemented at scale (e.g., Sleipner, Snøhvit), range from ~0.6 to ~1.0 mol/kg. The rapid decline in solubility with increasing salinity indicates a preference for host formations with lower salinity, if solubility trapping is to be maximized.

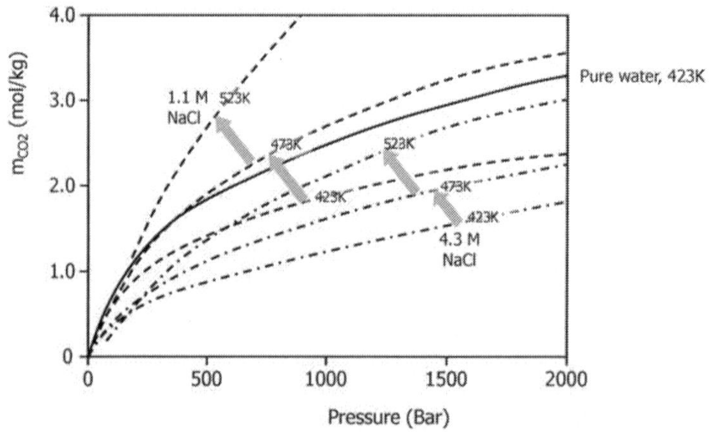

FIGURE 15.9 Solubility of CO_2 versus T and P in pure water and NaCl solutions (1.1 and 4.3 M).

Dissolution of residually trapped CO_2 will occur throughout the rock volume traced by the migrating plume, while dissolution of structurally trapped CO_2 will take place at the base of the trapped accumulation. Dissolved CO_2 will gradually disperse throughout the whole accessible volume by the two processes of diffusion and convection. Diffusion is a very slow process, driven by Brownian motion and the concentration difference between CO_2 saturated and unsaturated formation brine, while convection, driven by gravity and the brine density increase resulting from CO_2 dissolution (illustrated schematically in Fig. 15.10), will be several orders of magnitude faster, depending on the formation characteristics. In particular the vertical permeability of the formation will have a major influence on the speed of convective mixing as can be seen from the gravitational term in Darcy's equation;

$$q_v = -(k_v A/\mu)\,\Delta\rho\, g \tag{15.4}$$

where q_v is the vertical volume flow rate, k_v is the vertical permeability, and μ is the fluid viscosity. Simulation results, shown in Fig. 15.11, demonstrate that a 10-fold increase in k_v increases the rate of CO_2 dissolution by a similar factor. The onset of convective mixing arises from small-scale variability in brine density, and this will be affected by geochemical processes which consume dissolved CO_2, necessitating coupled fluid dynamic and geochemical modeling should these processes need to be investigated in detail.

Theoretical storage capacity for solubility trapping

The theoretical storage capacity for solubility trapping can be estimated from

$$m_{CO_2t} = A\,h\,\Phi\,(\rho_{bS}\,X_{SCO_2} - \rho_{b0}X_{0CO_2}) \tag{15.5}$$

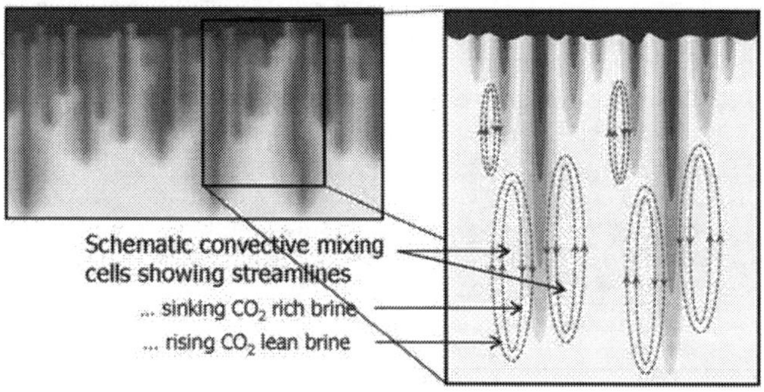

FIGURE 15.10 Convective mixing process leading to solubility trapping. *After Ennis-King and Paterson (2005).*

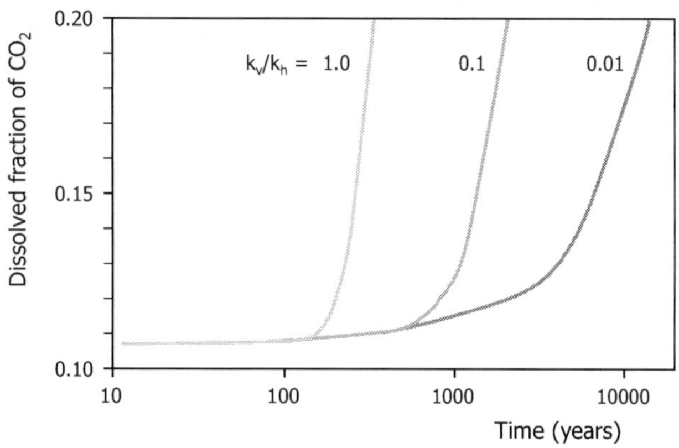

FIGURE 15.11 Impact of vertical permeability on rate of solubility trapping. *After Ennis-King and Paterson (2005).*

where ρ_b is the brine density, X^{CO_2} the mass fraction of CO_2 dissolved in the brine, subscripts 0 and S indicate initial and CO_2 saturated conditions, and other terms are as previously defined.

Solubility trapping presents a very low leakage risk, as the dissolved CO_2 will only come out of solution if there is a significant reduction in pressure, for example as a result of updip migration of saturated brine over long distances or leakage into a shallower formation.

Dissolution is effective on a centennial timescale, but can be considerably accelerated by applying specially designed injection strategies, as described in Section 15.5, or if hydrodynamic activity in the aquifer results in wide dispersion of the injected CO_2. Depending on the brine composition, pH, and the mineralogy of the storage complex, solubility trapping may be enhanced by the formation of bicarbonate or carbonate ions and other ionic complexes, resulting in additional ionic trapping.

15.2.5 Ionic trapping

Once CO_2 has dissolved in formation brine to form carbonic acid (H_2CO_3), successive deprotonations produce first the bicarbonate ion (HCO_3) and then the carbonate ion (CO_3^2), according to the sequence shown in Reaction 6.2. As shown in Fig. 6.3, above pH ~ 6, carbon will be predominantly in the form of bicarbonate or, above pH ~ 9, the carbonate ion. Ionic trapping, therefore, occurs for pH > 6, when the majority of the dissolved carbon is in an ionic state rather than as carbonic acid and aqueous CO_2.

Fig. 15.12 illustrates this pH dependence of solubility versus ionic trapping, and also shows that solubility and ionic trapping give way to mineral trapping—the most stable of the trapping mechanisms—with increased availability of Ca, Fe, or Mg cations.

15.2.6 Mineral trapping

By analogy with mineral carbonation, discussed in Chapter 10, the reaction of dissolved CO_2 with Ca^{2+}, Fe^{2+}, or Mg^{2+}-containing minerals in the rock matrix can result in the precipitation of carbonates in the pore space. This is the slowest potential trapping process, operating on a millennial timescale under subsurface conditions in sedimentary formations, but considerably faster in basalts (see Section 15.3).

Geochemical reactions need not be confined to the storage formation, and may also occur when CO_2 contacts the caprock above the injection formation. For example, geochemical analysis for the Sleipner project, discussed in Section 15.6, indicated that silicates present in the Utsira caprock would react with formation brine and CO_2 to form carbonates (Kaldi et al., 2011). In this case, the resulting increase in mineral bulk volume reduces porosity and permeability and leads to enhanced caprock integrity over time.

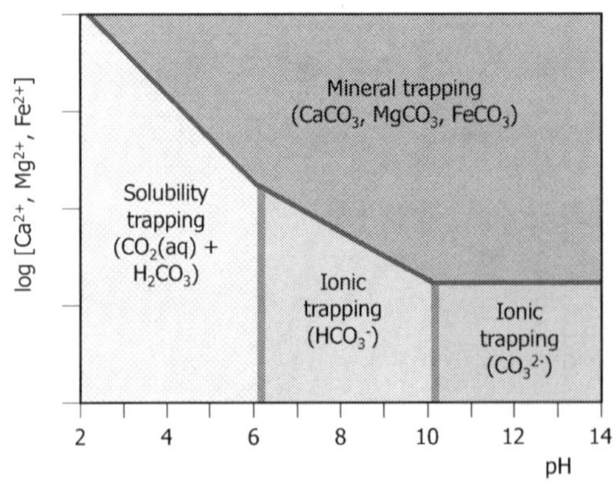

FIGURE 15.12 Trapping dependence on formation water pH and cation concentration. *Source: After Gunter et al. (2004).*

The theoretical storage capacity for mineral trapping is given by

$$m_{CO_2t} = Ah\Sigma_i c_i \chi_{iCO_2} M_{CO_2}/M_i \quad (15.6)$$

where c_i is the concentration of mineral i in the storage formation (kg m^3), χ_{iCO_2} is the stoichiometric fraction of CO_2 consumed in the carbonation reaction (molCO_2/mol-mineral i) and M_{CO_2}/M_i is the ratio of the molecular weights. This theoretical capacity assumes that dissolved CO_2 contacts all of the reactive minerals in the formation and remains in contact long enough for complete carbonation to occur.

15.2.7 Indicative specific capacities of saline aquifer trapping mechanisms

Table 15.1 gives an indication of the specific capacities of these trapping mechanisms, expressed in kgCO_2/m^3 of storage formation volume, as well as the timescale for trapping to occur.

The relevant formation volume to which these specific capacities would be applied, differs for each trapping mechanism, as shown in Fig. 15.13.

The relevant gross volume for structural trapping is the total volume beneath all structural trapping elements accessed by migrating CO_2, on all distance scales, limited either by the individual spill points or by the maximum buoyant column height that can be sustained given the geomechanical constraints. For residual trapping the relevant volume is the total gross rock volume swept out by migrating CO_2, while for solubility, ionic, and mineral trapping the latter volume plus the additional rock volume swept out by gravity currents or hydrodynamic flows of CO_2 saturated brine will apply. Consequently, with the exception of structural trapping, storage capacity will be increased by any feature or process that tends to increase the gross rock volume swept by the migrating plume. Examples of such features and processes include;

- Low vertical permeability
- Heterogeneous geological features that impede vertical flow
- Injection of water, concurrently or following CO_2 injection
- An active hydrodynamic regime
- Higher operating pressures in injection wells.

TABLE 15.1 Indicative specific capacities of saline aquifer trapping mechanisms.

Trapping mechanism	Indicative specific capacity (kgCO_2/m^3)	Effective timescale (years)
Structural trapping	50–200	10–1000
Residual trapping	10–50	10–1000
Solubility trapping	1–20	100–1000
Ionic trapping	1–20	100–1000
Mineral trapping	1–100	1000–10,000

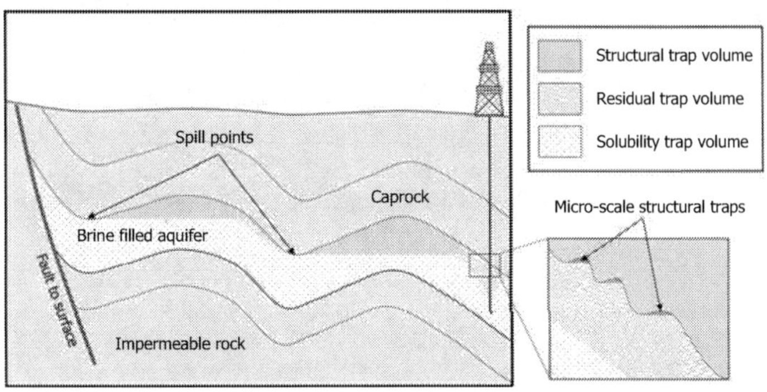

FIGURE 15.13 Storage volumes applicable to different trapping mechanisms.

15.2.8 Time dependence of saline aquifer trapping mechanisms

The trapping mechanisms described above operate on widely different timescales, as noted in Table 15.1 and illustrated in Fig. 15.14. Structural and residual trapping are the most important mechanisms on an operational timescale (years to decades), while solubility and mineral trapping largely determine the long-term fate of injected CO_2 (centuries to millennia). Because of the slower timescales, solubility and mineral trapping are often excluded in high-level (regional, national, etc.) estimates of storage capacity on the basis that they contribute at best a few extra percent over a realistic storage injection period.

15.3 In situ mineralization trapping

In contrast to saline aquifer storage, where trapping is a multi-stage process (e.g., primary structural followed by secondary solubility trapping), in basalt and ultramafic rocks trapping occurs in a single mineralization reaction, as described in Chapter 10. While the presence of a stratigraphic or structural trap provides additional security, it is not an essential part of the storage concept.

Adequate porosity and permeability are essential in both cases. In basalts, porosity and permeability are strongly affected by the degree of alteration that has occurred, primarily the deposition of secondary minerals, which increases with the depth (and therefore age) of the formation. The basalts in Iceland where the CarbFix project is located (Section 15.7.2) vary from 6% to almost 60% in porosity, with permeabilities from 0.1 to 6400 mD, the specific CarbFix target formation having lateral and vertical permeabilities of 300 and 1700 mD, respectively (Aradóttir et al., 2011). The deposition of secondary minerals is of course the aim of in situ mineralization, and this can cause a negative feedback through clogging of the pore space. Under certain conditions, this effect is countered by reaction-driven cracking that results from the pressure applied to the rock due to the increase in volume that accompanies the mineralization reaction (pressure of crystallization). In the case of mineralization in peridotite, such as the Samail Ophiolite Belt in Oman, sustaining the permeability of natural and induced fracture networks will be essential to maintain injectivity, and the conditions that control the reactive cracking process are an active area of research (Kelemen et al., 2018).

In the CarbFix pilot, initial monitoring results indicated that 95% of injected CO_2 was taken up along the flowpath between the injection and monitoring wells within 2 years (Matter et al., 2016). However, the subsequent microbiological assessment concluded that some fraction had been metabolized rather than mineralized, as a result of a bloom of iron-oxidizing bacteria (Trias et al., 2017). The CarbFix2 injection formation was at 2000 versus 800 m for CarbFix, and temperature of 200–260°C versus ~150°C, limiting the potential for subsurface biotic activity to affect injectivity and storage, and found similar rapid mineralization rates, with 50% of the injected CO_2 and 76% of the injected H_2S being mineralized in under 9 months (Gunnarsson et al., 2018).

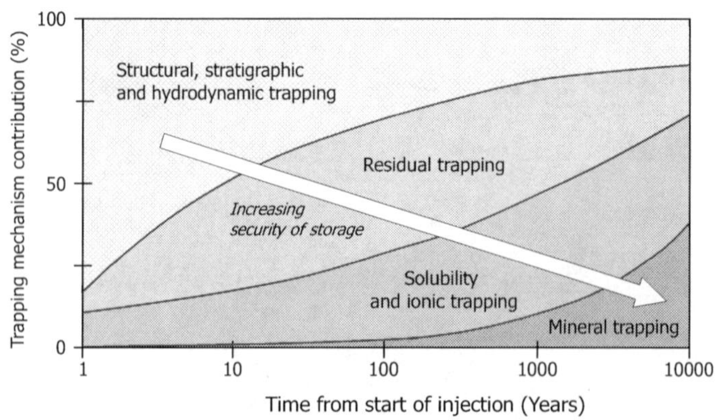

FIGURE 15.14 Relative importance of trapping mechanisms through time.

15.4 Storage capacity classification and estimation

Koide et al. (1992) estimated a global capacity for geological storage in saline aquifers of 320 $GtCO_2$ (87 GtC), on the assumption that 1% of the total area of all sedimentary basins would be utilized and that capacity would be determined by CO_2 solubility. More recent estimates suggest that this figure could be exceeded by a factor of 10 for the United States alone (Buursink et al., 2013), and that global storage capacity in saline aquifers could reach 25,000 $GtCO_2$ (Consoli and Wildgust, 2017) Storage capacity in flood basalts is estimated at 30,000 $GtCO_2$ for the Columbia River Basalt Group in the United States, alone (McGrail et al., 2006), and a global estimate including the ocean floors is probably orders of magnitude greater than the 1000s of $GtCO_2$ capacity that is likely to be required (see Chapter 1).

Geological storage capacity may be assessed on many scales, from country to basin or down to a specific site, and a wide range of assumptions and boundary conditions may be applied in individual estimates. Irrespective of the geographical scale of an estimate, consistency of definitions is essential to allow meaningful aggregation, and this begins with the classification of storage resource estimates.

15.4.1 Classification of storage capacity estimates

A number of systems have been developed to classify carbon storage capacity according to its technical maturity and economic viability, and the two main systems will be discussed here. The Carbon Sequestration Leadership Forum (CSLF) scheme, summarized in Table 15.2, focuses on the technical factors that determine how much of the theoretical capacity (discussed above for the various trapping mechanisms) is useable in practice.

This system is known as the Techno-Economic Resource-Reserve Pyramid for CO_2 Storage Capacity and is illustrated by the structure shown in Fig. 15.15, in which the theoretical capacity occupies the whole volume of the pyramid, while the progressively constrained volumes occupy reducing fractions of the total.

The committed storage capacity is an addition to the CSLF scheme by analogy with the concept of committing hydrocarbon gas reserves to long-term supply contracts, and will be relevant when geological storage is offered as a utility.

Economic factors are introduced in the step from effective to viable capacity, but two important aspects that are not explicit in the CSLF scheme are the potential for future economic conditions to increase the viable capacity and the capacity range (e.g., low, mid, and high values) as a result of remaining technical uncertainties. The scheme developed by the Society of Petroleum Engineers incorporates these two aspects in its Storage Resource Management System (SRMS) that is

TABLE 15.2 Classification scheme for GCS capacity.

Storage capacity category	Description
Theoretical storage capacity	The physical limit of storage volume in a geological system, assuming the entire pore space is accessible and utilized and that adsorption (where relevant) reaches 100% saturation within the entire volume.
Effective (or realistic) storage capacity	Derived from the theoretical storage capacity by applying appropriate geological and engineering cut-offs; dependent on site-specific or regional data, and time dependent as performance data and experience are acquired and new technologies become available.
Viable (or practical) storage capacity	Viable capacity is derived from the effective capacity by considering nontechnical factors, for example, economic or regulatory limits.
Matched storage capacity	Derived by matching the viable capacity with available sources that meet the technical requirements of a storage site, such as ability to deliver a specific supply rate. The difference between matched and viable capacity is termed stranded capacity, as it cannot be utilized due to a lack of sufficient economically viable or technically suitable CO_2 supplies.
Committed storage capacity	The committed capacity is that part of the matched capacity that has been contracted to specific CO_2 producers. The difference between matched and committed capacity is therefore available to be contracted to new producers or to the extension of existing storage contracts.

After CSLF (2007).

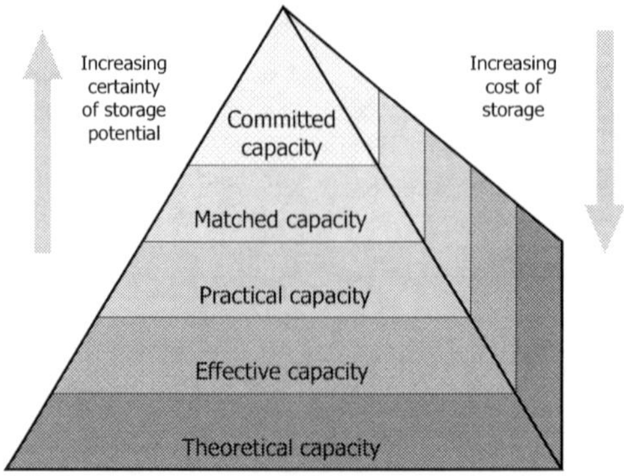

FIGURE 15.15 Techno-economic resource-reserve pyramid. *After CSLF (2007).*

modeled on the Petroleum Resource Management System (PRMS) widely used in the oil and gas industry. Fig. 15.16 illustrates the SRMS, along with the additional definitions in Table 15.3.

In the SRMS, that part of theoretical capacity that is not categorized as effective capacity (the bottom slice of the pyramid in Fig. 15.15) is further subdivided depending on its state of characterization, while the viable or practical storage resource is subdivided based on the likelihood of that capacity being achieved. The probabilistic assessment of capacity is discussed briefly below in Section 15.4.3.

15.4.2 Limitations on effective storage capacity

The ratio of effective to theoretical storage capacity is known as the storage efficiency, and results from the application of subsurface technical and engineering cut-offs, such as a geomechanical limit on well injection pressure or pore pressure increase. Some of the factors that determine effective storage capacity in relation to specific trapping mechanisms are summarized in Table 15.4 and the following sections. These factors also have a wide range of different distance and time scales, as illustrated in Figs. 15.17 and 15.18.

The cumulative volumetric effect of all factors is captured in the storage efficiency factor which represents the fraction of the total available pore space that will be occupied by CO_2. Storage efficiencies refer to specific trapping mechanisms and are designated B_{SE} for structural/buoyant, and R_{SE} for residual trapping. These factors address CO_2 present as a free phase within the pore space; as noted above solubility/ionic and mineral trapping are typically excluded from high-level storage capacity estimates. Various engineering approaches have however been proposed to accelerate CO_2 dissolution

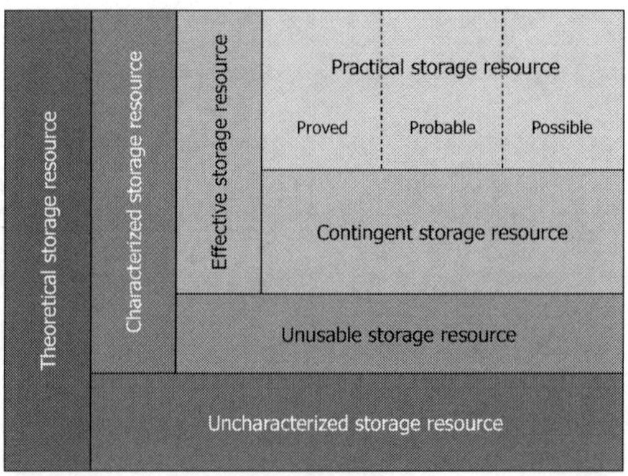

FIGURE 15.16 Storage resource management system. *After Gorecki et al. (2009).*

TABLE 15.3 Society of Petroleum Engineers (SPE) SRMS definitions.

Storage capacity category	Definition
Proved storage capacity	Proved storage capacity is the storage volume which, by analysis of geoscience and engineering data, can be estimated with reasonable certainty to be commercially achievable under current economic and commercial conditions.
Probable storage capacity	Probable storage capacity is the storage volume that analysis of geoscience and engineering data indicate is less likely to be achievable than proved capacity but more certain to be achieved than possible capacity. It is equally likely that the actual stored volume will be greater than or less than the sum of the estimated proved plus probable capacity.
Possible storage capacity	Possible storage capacity is the storage volume that analysis of geoscience and engineering data suggest is less likely to be achievable than probable capacity.
Contingent storage resource	The storage capacity in well-characterized sites and formations is technically feasible but requires improved future economic conditions to be commercially viable.

After Gorecki et al. (2009), and SPE (2017).

TABLE 15.4 Factors that determine effective geological storage capacity.

Limiting factor	Applicable trapping mechanisms	Description
Geomechanical	Structural and residual trapping	Maximum pressure increase that can be sustained without geomechanical failure of the storage formation or caprock.
Fluid flow	Residual trapping	Residual trapping can only occur within the gross volume swept by the migrating plume, which is controlled by factors such as vertical permeability and reservoir heterogeneity.
Rock–fluid interactions	Residual trapping	The residual saturation of CO_2 will be influenced by the maximum achievable injection rate.
	Structural trapping	The maximum column height in a structural trap may be limited by the capillary entry pressure of CO_2 into the caprock.
Fluid–fluid interactions	Structural trapping	The pore space occupied by irreducible formation brine becomes available for structural trapping on a time-scale reflecting its rate of dissolution into trapped CO_2.
Hydrological	Structural and residual trapping	Maximum pressure increase or plume extent that can be sustained without adverse impact on potable water resources.
Operational—pressure management	Structural and residual trapping	Potential for pressure management techniques to mitigate pore pressure increase and remove or relax the geomechanical limit.
Policy constraints	All	Regulatory requirements in a specific jurisdiction may further limit the extent to which the technical storage capacity can be effectively utilized.

which could substantially increase its contribution on an operational timescale (see Section 15.5). A full evaluation of these trapping processes is therefore essential for project-specific storage assessment and for evaluating the long-term fate of stored CO_2, and will also influence postclosure monitoring.

15.4.3 Probabilistic storage capacity estimation

It is clear from Table 15.4 that any storage capacity estimate is dependent on a wide range of factors, some of which may be determined with reasonable precision (e.g., ρ_{CO_2}), others with at least some degree of certainty (e.g., S_{wirr}, S_{crw}), while others, such as the spatial and size distribution of small scale trapping heterogeneities, will be very poorly determined.

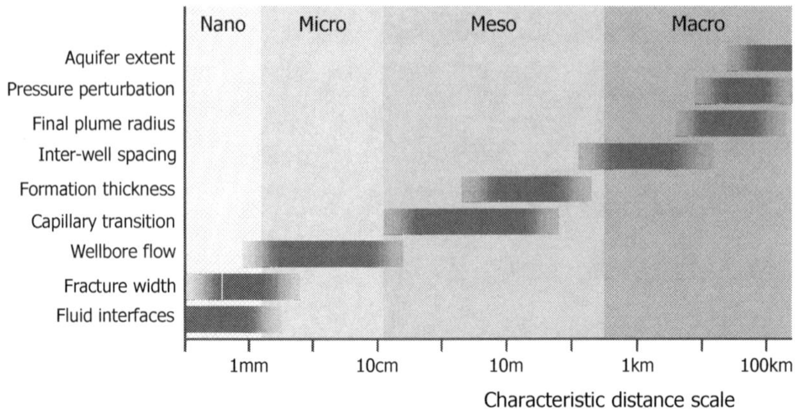

FIGURE 15.17 Characteristic distance scales of factors influencing storage capacity. *Source: After Celia and Nordbotten (2011).*

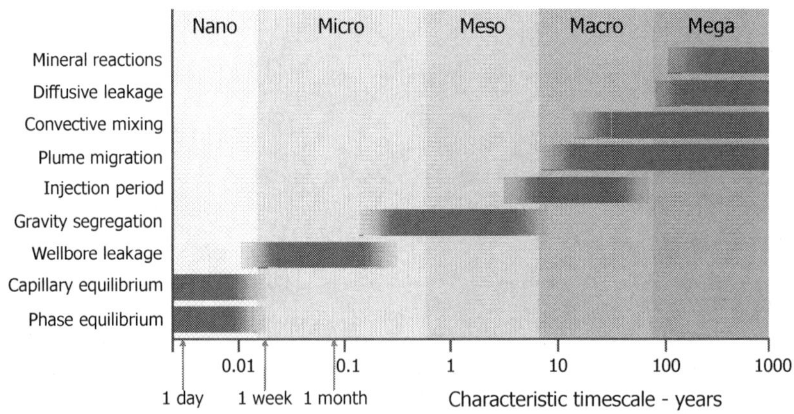

FIGURE 15.18 Characteristic time scales of factors influencing storage capacity. *Source: After Celia and Nordbotten (2011).*

Probabilistic estimation techniques can be used to ensure that these uncertainties are reflected in a storage capacity estimate. This practice is enshrined in the oil and gas industry's use of proved, probable and possible reserves, which reflect high, moderate, and lower levels of certainty that a resource volume will be economically producible. In similar vein, decision-makers—at all levels from individual project up to international policy—will need to understand the range of possible storage outcomes to make good decisions.

The probabilistic estimation method is illustrated in Fig. 15.19. Firstly, reference or "mid" case maps and estimates are made for each parameter in the capacity equation. Depending on data availability, mapping may be limited to a structural depth map to estimate applicable gross rock volume or, if there is sufficient well data, thickness, porosity, and permeability maps may also be generated. The range of uncertainty is then assessed for each input parameter, and represented as a probability density function (pdf). Common examples are a triangular distribution specified by a minimum possible, most likely, and maximum possible value, or a normal or log-normal distribution specified by a mean and standard deviation. For mapped inputs, low- and high-case maps would be generated to determine the pdf endpoints. Finally, these parameter uncertainty distributions are sampled using a Monte Carlo technique to generate many thousands of capacity estimates that are displayed as a density function or cumulative probability curve.

From these curves, the storage capacity at any confidence level can be determined. For example, a high confidence value such as a P(90) estimate—for which there is a 90% probability that actual capacity will exceed the estimate—may be used by a storage operator when committing volume to a CO_2 supplier. Sensitivity analysis, identifying which input parameters have the greatest impact on capacity uncertainty, may be used to define a data gathering or monitoring program that can have the greatest impact in narrowing the uncertainty range, potentially leading to larger committable volumes.

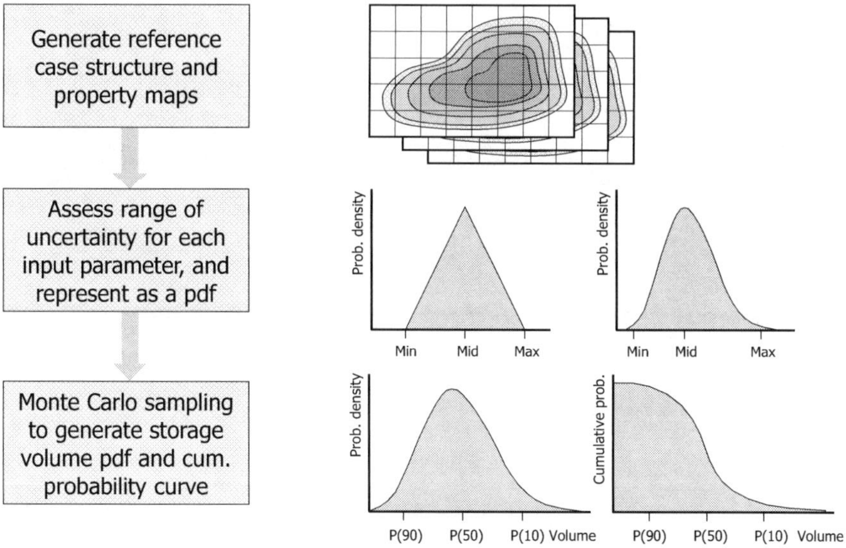

FIGURE 15.19 Probabilistic storage capacity estimation methodology.

In the Society of Petroleum Engineers (SPE) SRMS system, by analogy with oil and gas reserves reporting practice, for Proved (also termed "1P") storage capacity there should be at least a 90% probability that the quantities actually stored will equal or exceed the estimate (Peck et al., 2017), for Proved plus Probable (2P) there should be at least a 50% probability that the actual quantities stored will equal or exceed the estimate, while for Proved plus Probable plus Possible (3P) the required probability reduces to 10%.

15.5 Geological storage project planning

This section provides a brief overview of some of the activities involved in developing a geological storage project, with the aim of highlighting key processes such as risk management, monitoring and verification, and operational control. A fuller description of the project planning process and activities summarized here can be found in Senior et al. (2011), and in Rackley (2017).

15.5.1 Site identification and selection

At the start of the planning process, the regional location of the project will generally be a given, and existing basin and regional screening studies that have confirmed the presence of potentially suitable storage formations in the region will be the starting point for identifying possible storage sites for the project. The reviews of the United Kingdom and Norwegian North Sea storage potential and the US NETL Carbon Storage Program offshore CCS assessment are examples of such regional studies (see ETI, 2016; Allen et al., 2017; Rodosta et al., 2017).

Site screening and selection criteria

The initial screening and shortlisting of sites, following a process such as that shown in Fig. 15.20, will include the definition of a data gathering and assessment program for each potential site, and will typically require the issue of an exploration permit, depending on location-specific regulatory requirements.

The criteria for eventual site selection will be defined at this stage, based on the specific project requirements. Fig. 15.21 shows the site screening criteria used in the FutureGen project (FutureGen, 2007), while other examples of screening checklists and methodologies can be found in Oldenburg (2008) and NETL (2010).

Alongside the technical activities described below, public outreach activities will also start at this stage, particularly if the shortlist includes onshore sites. Experience from projects of all types shows that such engagements cannot start too early, and best practice guidance is available in NETL (2010), Brunsting et al. (2011), Lockwood (2017), and Aradottir and Hjalmarsson (2018).

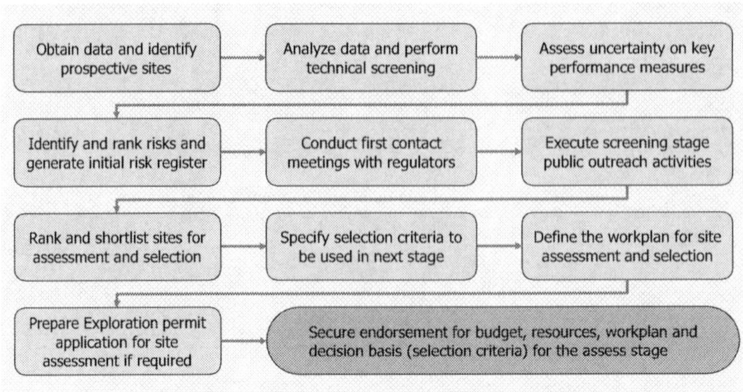

FIGURE 15.20 Site screening stage process steps.

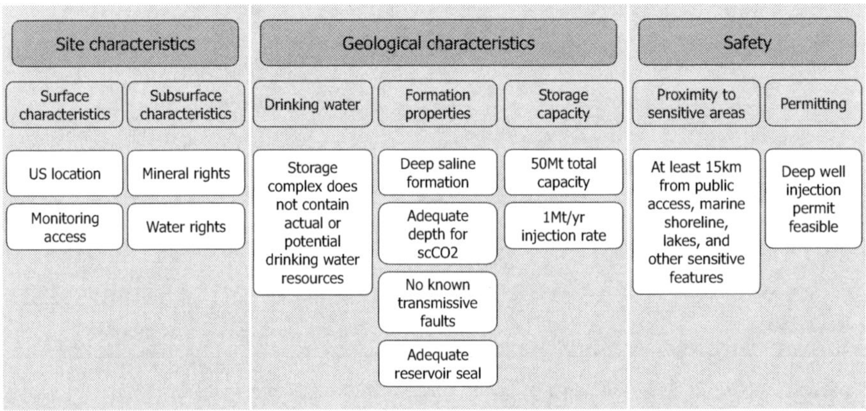

FIGURE 15.21 Site screening criteria used in the FutureGen project. *Source: After FutureGen (2007).*

Preliminary risk management activities

The identification and assessment of risks that could impact desired project performance, and the subsequent identification, assessment, and implementation of safeguards are activities that are revisited at each state of the project maturation process, within the overall risk management framework for the project (see e.g., Dean and Tucker, 2017). Fig. 15.22 shows schematically how such a framework underpins risk management activities through the project lifecycle.

At the site screening stage, the aim of risk analysis is to identify risks associated with each candidate site that could have a significant impact on achieving the project objectives. Risk assessment at this stage will be largely qualitative, except for sites that have comprehensive existing datasets such as depleted hydrocarbon fields, and a plan for additional data gathering and studies is an important output from this initial assessment. Informative reviews of risk management approaches for geological storage can be found in Condor et al. (2011), and Li and Liu (2016).

Subsurface modeling studies

Initial subsurface models will be constructed during the site assessment and selection stage, and these will be developed in subsequent project stages, as more data becomes available, to become the central tool for optimizing the project design and operating strategies. Table 15.5 summarizes typical elements of the storage system performance that will begin to be investigated at this stage.

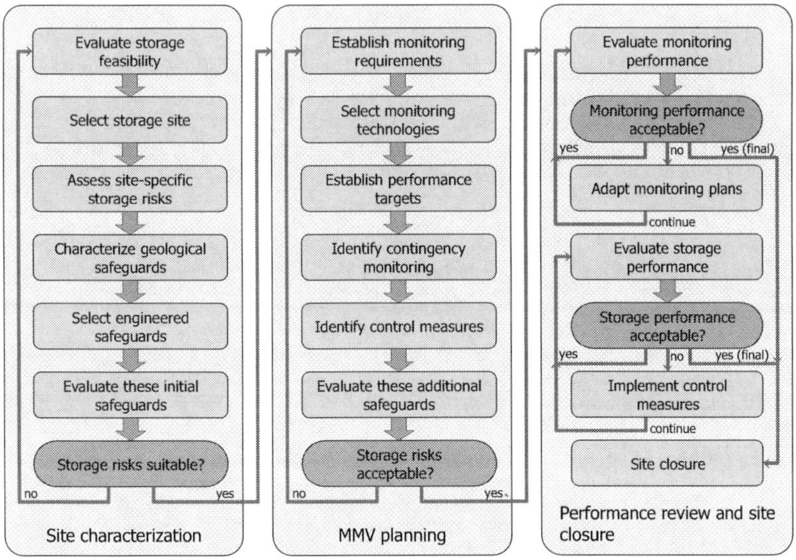

FIGURE 15.22 Schematic geological storage project risk management framework. *Source: After Shell Canada (2012).*

TABLE 15.5 Components of storage performance prediction during site assessment.

Prediction element	Description	Impacts
Areal and vertical extent of the CO_2 plume versus time	The CO_2 distribution (dissolved, trapped, mobile) in the storage formation, caprock, and any invaded overburden formations.	Storage capacity and evolving contribution of trapping mechanisms; areal extent of required containment assurance; necessity of remedial measures in the event of specific leakage risks such as faulted or fractured zones or regions with high legacy well risk.
Pressure behavior versus time	Prediction of both near well and far-field pressure behavior.	Potential pressure limitation on storage capacity; impact of pressure on well injectivity and well count; far-field hydrodynamic impact.
Injectivity performance versus time	Sustainable well injection rate.	Required well count to achieve target storage rate; possible need for additional wells in later storage life.
Leakage probability and rate versus time	Estimate of leak probability and rate for specific modeled leak paths (faults, existing wells, etc.)	Provides a basis for the risk-based monitoring plan and investigation of mitigation and remediation measures.
Effectiveness of potential remedial measures	Possibility to test remediation options such as injection or back-production.	Potential impact on well count, placement or timing; monitoring plan.

15.5.2 Measurement, monitoring, and verification planning

The measurement, monitoring, and verification (MMV) plan is perhaps the most important tool to manage uncertainty and risk during storage operations, to ensure safe and cost-effective storage, and, ultimately, allow straightforward site closure and postclosure liability transfer.

The monitoring plan identifies the type, location, and frequency of specified measurements as well as the measurement-specific performance requirements such as spatial resolution and accuracy that will be required to meet the monitoring objectives. The plan will be adaptive, incorporating a review and feedback mechanism, to enable updates throughout the project life in the light of field data, as well as the actual performance and emerging capabilities of the various monitoring technologies.

Fig. 15.23 illustrates the MMV design process adopted in the Shell Quest project (see Section 15.6.1) and shows the close links with the risk management framework of Fig. 15.22. Some further tools for monitoring and verification planning can be found in IEA GHG (2015), and Sijacic et al. (2014). Table 15.6 gives a brief overview of monitoring technologies for geological storage, and further details can be found in Jenkins et al. (2015) and Rackley (2017).

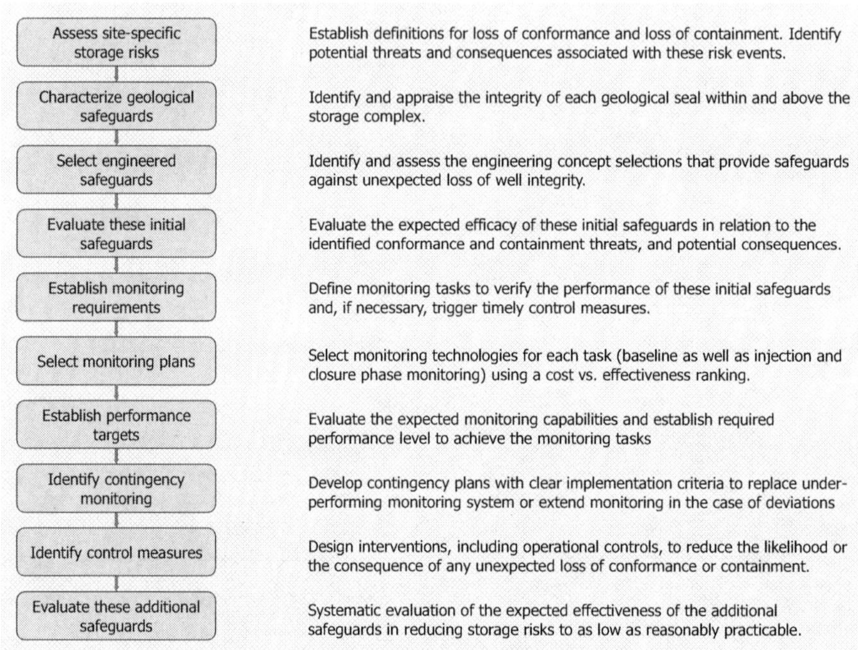

FIGURE 15.23 Shell Canada Quest MMV plan design process. *After Shell Canada (2012).*

TABLE 15.6 Monitoring techniques for geological storage.

Characterization or monitoring technique	Description
Plume location and movement	
Time lapse seismic	Typically acquired on a 3D grid, but also possible on 2D lines. May also be acquired using permanently installed geophones, particularly offshore.
Vertical seismic profiling	Seismic survey from a surface source to geophones located in a well, generating a one-way vertical profile that can be used to calibrate 3D data and provide near-well imaging of reservoir properties and CO_2 movement.
Crosswell seismic	Seismic survey with both source and geophones located in boreholes. Generates both reflection profiles and velocity tomography that can be used to calibrate 3D data and in time-lapse mode to image changes in reflectivity and velocity between wells.

TABLE 15.6 Monitoring techniques for geological storage.—cont'd

Characterization or monitoring technique	Description
Microseismic	Measurement of microseismic signal generated by fracturing events as a result of injection, using either surface or downhole-mounted geophones.
Electrical resistance tomography (ERT)	ERT images the subsurface resistivity with measurements between ground-based or borehole-mounted electrodes; long electrode ERT uses the casing of a well as an electrode to extend this technique to greater depth.
Crosswell resistivity	A variant of ERT using electrodes mounted or moving in two wells, and providing a 2D profile of the resistivity distribution between the wells.
Gravity surveys	Measurement of the local gravitational field using surface or airborne gravimetry, providing a low-resolution map of changes in pore fluid density due to plume movement.
Tracer injection	Marker chemicals co-injected with CO_2 and detectable at very low concentrations to enable early breakthrough detection.
Surface deformation monitoring (SDM)	
InSAR and DInSAR	Techniques for processing satellite-based radar images to detect surface deformation down to millimeter scale over timescales of months to years.
Tiltmeters	Shallow borehole mounted (<10 m) sensors that can detect microradian scale geomechanical deformation, equivalent to 1 mm deflection over a 1 km range.
CO_2 leak detection	
LIDAR	Laser monitoring techniques that can detect the location and movement of CO_2 in the atmosphere with ppb sensitivity.
Hyperspectral remote sensing	Imaging spectroscopy uses space-borne sensors to collect a spectrum at each image point, rather than an amplitude at a single wavelength. Can be applied to detect atmospheric gases and surface minerals, and to characterize ecological habitats.
Open-path laser gas detection	Detection of gases such as CO_2 and methane as a result of the attenuation of an infrared beam transmitted between a source and detector.
Eddy covariance measurement	A technique for determining the vertical flux of a gas species such as CO_2 by statistical analysis of the vertical velocity and fluctuating concentration of turbulent eddies.
Soil gas sampling	Collection and analysis of gas samples from the soil or shallow boreholes. Isotopic or tracer analysis can distinguish leakage of injected from biosphere-generated CO_2.
Fluid sampling	Collection and analysis of fluid samples from drinking water wells, and any monitoring wells located above the storage complex, to detect elevated CO_2 concentration.
Injection well monitoring	
Injection well pressure and temperature	Standard measurement of well pressures, volumes, and temperatures; used as input to dynamic reservoir modeling and to identify injection well problems.
Geochemical characterization	
Injection-production experiments	Injection of a limited volume of CO_2 into a host formation, followed by a soaking period (weeks to months) and a period of back production and sampling to assess the geochemical and other consequences of CO_2 injection.
Microbiological monitoring	
Fluid sampling and microbial analysis	Collection and analysis (geochemical and biological) of fluid samples from monitoring wells located in the host formation, to characterize microbial communities and activity following CO_2 injection.

15.5.3 Storage plan optimization

Following site selection, more detailed development planning work will commence and will typically have to address a number of optimization decisions, as shown in Fig. 15.24.

Primary injection optimization will aim to maximize the value of the storage project by minimizing the cost of injection wells, compression equipment, etc., subject to achieving the required storage volume and rate. The approach to this optimization will depend on whether the storage project is linked to a single CO_2 source with a well-defined delivery profile, or has more of the nature of a storage utility, in which case the project operator will want to keep open options to conclude additional storage agreements at future dates, as further insight is gained into the performance and capacity of the storage complex. In this situation, flexibility to accommodate these options will be an important consideration in project optimization.

Pressure control strategies

Preventative measures to control the pore pressure within the storage complex are likely to be a regulatory requirement, particularly in the case of a confined host formation, where storage capacity is determined by the maximum allowable pore pressure to prevent geomechanical failure or reactivation of faults or fractures (see e.g., European Commission Directive, 2009). This control can be achieved by the production of brine from the host formation via one or more wells drilled at suitable locations remote from the injection point, with the produced brine either being partially reinjected to enhance trapping (see below), disposed of to an unconstrained aquifer separate from the storage complex or, where environmentally acceptable, to the sea (McNemar et al., 2021). Chevron's Gorgon project is one example of a storage project employing pressure management wells, the eight injection wells being supplemented by four brine production wells ca. 5–6 km from the injection wells, and four reservoir monitoring wells (Chevron, 2005).

Disposal of produced brine into an aquifer overlying the storage complex could have a secondary benefit by creating a pressure gradient opposing potential leakage through faults, fractures, or a leaking wellbore. The use of such hydraulic barriers as a corrective pressure control measure is further discussed below.

Strategies to limit pressure build-up in the vicinity of injection wells generally rely on increasing the number and spacing of wells, and either operating all wells concurrently or rotating injection among wells (Tanaka et al., 2013). This has the disadvantage of increased total costs for well construction, but may be required in low-permeability host formations.

Trapping control strategies

As residual and solubility trapping are the two trapping mechanisms that can immobilize injected CO_2 on an operational timescale in sedimentary formations, measures that can enhance the effectiveness of these mechanisms will contribute to both storage optimization and leakage risk reduction.

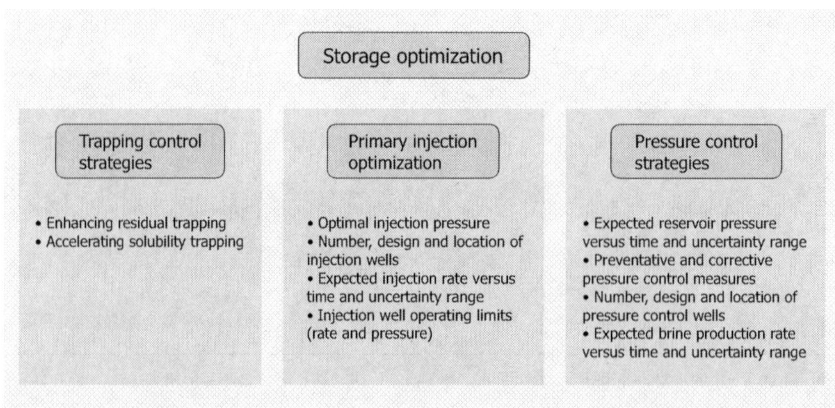

FIGURE 15.24 Elements of storage plan optimization.

Several approaches have been proposed to enhance residual trapping by brine injection, either concurrently with CO_2 injection or following the cessation of CO_2 injection. Concurrent injection of brine above the point of CO_2 injection has been shown in modeling studies to impede upward migration and increase residual trapping both as a result of an increase in the gross rock volume swept by the plume and by an increase and acceleration of imbibition following the cessation of injection (Nghiem et al., 2009). Water injection following the cessation of CO_2 injection has also been proposed to enhance residual trapping by breaking-up large connected and mobile CO_2 volumes (Spiteri et al., 2005). Combined brine injection both concurrently and following CO_2 injection has also been demonstrated in modeling studies to be very effective in rapidly immobilizing injected CO_2 as a residual phase (Qi et al., 2009; Huber et al., 2016).

Following the cessation of CO_2 injection, the production of brine from a region remote from the injection wells and its reinjection into the CO_2 plume has been proposed as a strategy to accelerate the dissolution of CO_2 (Leonenko and Keith, 2008; Hassanzadeh et al., 2009). The effectiveness of this approach requires good contact between the injected brine and the buoyant plume, and it is therefore most applicable in cases where the plume is structurally confined, for example beneath an anticlinal trap. Fig. 15.25 schematically illustrates the process in which CO_2 dissolves into CO_2-lean injected brine as it descends under gravity through the plume.

Modeling studies indicate that under ideal conditions this approach can raise the proportion of injected CO_2 trapped by dissolution in the 100 years after the start of injection from 5% or less to around 40%. Clearly, this is still a long-term process in relation to the operational life cycle of a storage project, so its application is likely to be limited to situations where it mitigates specific long-term risks, such as geochemical alteration of the caprock. Concurrent brine injection above the point of CO_2 injection, described above to enhance residual trapping (Nghiem et al., 2009), has also been shown to effectively enhance solubility trapping on an operational timescale for higher permeability formations ($k_h > 500$ mD).

15.5.4 Storage operations and monitoring

With the start of injection operations, performance monitoring becomes the main focus of project activity. Monitoring activities in the early injection period are important to establish conformance between predicted and observed movement of the injected CO_2 plume and indicate any early deviations, for example due to geological features that were not evident during initial site characterization (see Section 15.6.2).

Matching the observed and predicted plume behavior will enable updating of the subsurface simulation models and, if repeated time-lapse 3D seismic surveys are part of the monitoring program, seismic inversion studies will be important. In these studies, an initial static model is used to predict the expected seismic response, and a comparison between the predicted and actual seismic response aids in constraining subsurface uncertainties.

Well injectivity will also be closely monitored since any loss of injectivity, for example resulting from an injection of corrosion products from the CO_2 transportation system, could prevent injection at the required rate. The Hall plot (Fig. 15.26) is a frequently used tool for injectivity monitoring, where the product of injection wellhead pressure × injection duration is plotted against cumulative injection volume. A linear Hall plot indicates stable injectivity, while upward or downward deviations indicate a reduction or enhancement of injectivity, respectively, due for example to formation impairment or fracturing.

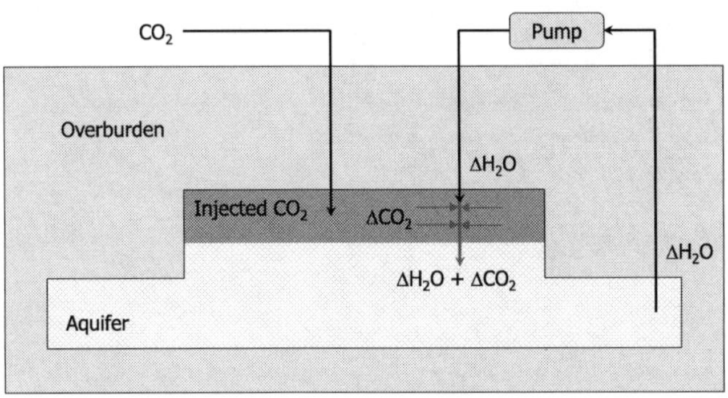

FIGURE 15.25 Schematic illustration of accelerated dissolution by brine injection. *Source: After Leonenko and Keith (2008).*

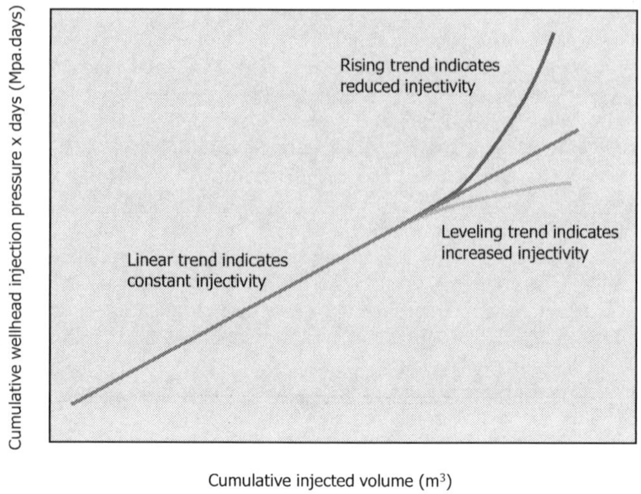

FIGURE 15.26 Schematic Hall plot showing diagnosis of injectivity impairment or enhancement.

Corrective measures and remediation

If monitoring results indicate a significant deviation from the expected subsurface performance, such as unexpected vertical or lateral plume movement, or elevated pressures in monitoring wells, corrective actions may be needed to bring performance back within permitted limits.

In Europe for example, the European Commission Directive (2009) on CO_2 storage requires project operators to submit a corrective measures plan, identifying measures to be taken in the event of significant irregularities and to have this plan approved by the relevant authority before the award of a storage permit. Operators are required to notify the authority of significant irregularities and to execute corrective measures as identified in the plan. Table 15.7 summarizes a range of such irregularities, together with possible response measures.

TABLE 15.7 Performance irregularities and possible corrective actions.

Performance irregularity	Description	Potential corrective measures
Decline in well injectivity	Loss of injectivity resulting in the need to increase injection pressure to sustain the required disposal rate	Well stimulation depending on diagnosed impairment mechanism; e.g., removal of corrosion products, dissolution of halite deposit, etc;
Results of monitoring surveys outside expected range	For example, unexpectedly high surface deformation or microseismic activity	Reduction of injection rate or cessation of injection;
Plume areal extent outside expected range	Plume migrates or is predicted to migrate outside the permitted area	Reduction of injection rate or cessation of injection; modification of the hydrodynamic regime to influence plume migration; modification of injection well pattern
Pore pressure rise outside expected range	Pore pressure exceeds or is predicted to exceed the maximum allowable pressure	Reduction of injection rate or cessation of injection; production of brine or back production of CO_2 from the host formation;
Detection of injected CO_2 at surface or in shallow aquifers	Leakage of CO_2 outside the storage complex, either through poorly sealing formations or along wells	Intervention into leaking boreholes to remediate leak paths; reduction of injection rate or cessation of injection; creation of a hydraulic barrier to arrest leakage

Remedial pressure control strategies

Pressure control strategies may be applied to address a number of irregularities, including unexpected pressure rise or unexpected plume movement, and are likely to be a major component of the corrective measures plan. A range of control strategies may be used to address pressure and flow anomalies on both local and storage complex scales.

a) ***Production of brine from the host formation***
The location of brine production will depend on whether pressures need to be reduced at a specific location, for example, to reduce local fault leak or reactivation risk, or more widely to reduce overall pressure build-up in the storage complex. Pressure relief could be achieved using predrilled production wells, the conversion of existing monitoring wells to producers, or the drilling of new dedicated pressure relief wells. Brine from the host formation could be produced to surface and disposed of, or possibly "dump flooded" into a permitted pressure relief formation.

b) ***Back-production of injected CO_2 from the host formation***
The volume of injected CO_2 that could be back-produced will depend strongly on its distribution within the host formation. For example, significant back-production may be possible from a structurally trapped CO_2 accumulation, but from a more dispersed CO_2 accumulation, for example as a result of internal geological heterogeneity (e.g., the Utsira aquifer at Sleipner—see Section 15.6.2), back production may require a significant number of additional wells.

c) ***Modification of the hydrodynamic regime***
The pressure field resulting from simultaneous production and injection in the region surrounding the plume can be used to modify the hydrodynamic regime, and this could be applied as a corrective measure to control the direction of plume movement, for example to prevent the plume approaching an area of increased leakage risk around a cluster of legacy wells. Brine injection within the well cluster would prevent plume migration into this area and could be a viable alternative to remedial work on a large number of wells.

d) ***Creation of a hydraulic barrier to remediate leakage***
Leakage along faults or wellbores can potentially be remediated by the injection of brine close to the point where the leak path emerges into the overlying permeable formation. The shallow injection establishes a pressure gradient opposed to the leakage direction, in effect creating a hydraulic barrier (Réveillère and Rohmer, 2011). If the injected brine is produced from the storage complex the effectiveness of this strategy will be increased as brine production from the complex will also reduce the driving pressure behind the leakage. Fig. 15.27 illustrates the concept and shows an example from modeling studies of the potential impact of a leaking injection wellbore. The subsurface configuration is shown on the left of the figure while, on the right, the solid curve shows the leak rate without corrective action and the dashed curves show the impact of a hydraulic barrier implemented at different times, which can effectively halt leakage after a few years.
Under the conditions described by Réveillère and Rohmer, the implementation of this corrective measure can reduce total leakage volume by >90%, depending primarily on the time delay between the cessation of primary injection and the initiation of remedial injection.
Brine injection at the upstream end of a leakage path (i.e., below the caprock leak path in the example of Fig. 15.27), has also been shown in modeling studies to be effective in leakage remediation by sweeping mobile CO_2 away from the leak path (Zahasky and Benson, 2016).

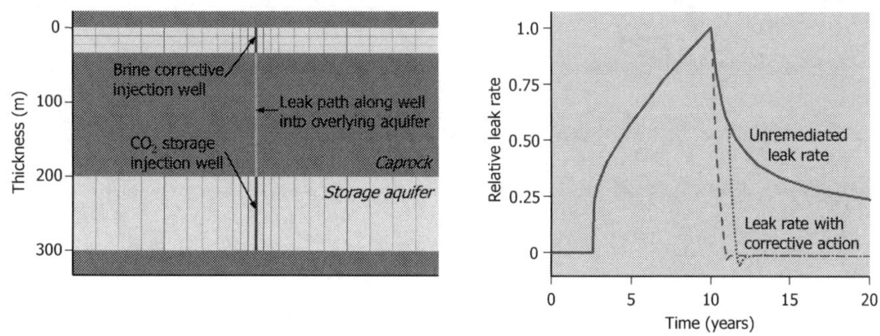

FIGURE 15.27 Impact of a hydraulic barrier on leakage rate. *Source: After Réveillère and Rohmer (2011).*

15.6 Saline aquifer storage case studies

To date, six large-scale saline aquifer storage projects have been implemented worldwide, just one being outside the oil and gas sector. A combined storage capacity of 7–8 $MtCO_2$/yr was reached following the start-up of injection at the Gorgon project in 2019, although actual injection averaged ~2.5 $MtCO_2$ from 2019 to 2021 (Table 15.8). With the commissioning of this project, saline aquifer storage has reached a facility scale that could accommodate the CO_2 produced by a single 500 MW_e thermal power station; 2000–3000 such facilities will be required to provide the 5–10 $GtCO_2$/yr of storage anticipated for BECCS plus DACCS.

The three case studies described below include two of these projects and have been selected because each offers unique insights into important aspects of saline aquifer storage project planning and operations:

- Shell Canada's Quest project is included for the wealth of information shared on the project's knowledge-sharing site, particularly on the initial stages of site selection, storage development planning, and MMV planning,
- Equinor's Sleipner project and the international multidisciplinary SACS research programs for the learning and best practices on storage performance monitoring and evaluation, and
- The Ketsin Storage Pilot as one of the few saline aquifer storage projects that have executed site closure and postclosure activities.

Together, they provide a useful learning set across the full life cycle of a saline aquifer storage project.

15.6.1 Saline aquifer storage case study 1: Quest

Shell's Quest project is the first commercial-scale CCS project operating in an industrial processing context—in this case, a bitumen upgrading plant at Scotford in Alberta, Canada. The project, which commenced injection in November 2015, captures up to 1.2 $MtCO_2$/y from a syngas stream using Shell's proprietary amine capture technology. The hydrogen product stream is used to upgrade bitumen mined from the Athabasca oil sands, while the captured CO_2 is compressed, dehydrated, and transported via an 84-km 12-in. dense-phase pipeline to be injected into the Basal Cambrian Sands at a depth of about 2.1 km. The injection facility includes three injection wells, providing redundancy to ensure high availability, and the pipeline has been sized for up to 3 $MtCO_2$/y to provide capacity for later expansion of the upgrading and capture project. The storage monitoring plan included the drilling of three deep monitoring wells and three shallow groundwater wells, at offset locations from each of the injection wells.

One unique aspect of this project is the open-access publication of major project documents. This approach was adopted to enable knowledge sharing and the documents provide a valuable knowledge base on which other storage project operators can build (see Shell Quest Knowledge Sharing, referenced below). Some key features of the Quest Storage Development Plan and MMV Plan are described later, while other useful project documents available online include the Environmental Assessment, independent project review of the storage component by Det Norske Veritas (DNV), and the Closure Plan.

TABLE 15.8 Large-scale saline aquifer storage projects.

Project operator	Location	Storage period	Storage rate $MtCO_2$/yr	CO_2 source
Sleipner equinor	Offshore Norway	1996–Present	1.0	Natural gas processing
In salah BP	Algeria	2004–2010	0.7	Natural gas processing
Snøhvit equinor	Offshore Norway	2008–Present	0.7	Natural gas processing
Decatur ICCS ADM[a]	Illinois, United States	2011–2014 2017–Present	0.3 1.0	Ethanol production
Quest Shell	Alberta, Canada	2015–Present	1.2	Hydrogen production
Gorgon Chevron	Western Australia	2019	2.0–4.0	Natural gas processing

[a]Archer Daniels Midland Company.

Quest storage development plan

The Quest storage development plan (SDP; Shell Canada, 2011) is a comprehensive example of such a document. The document structure and notable contents of each chapter are summarized in Table 15.9.

Quest measurement, monitoring, and verification plan

The Quest MMV plan (Shell Canada, 2012) provides an excellent example of the structure and content of such a plan. Starting from a statement of the purposes and design principles of MMV, the risks to containment and conformance before any MMV activities are then described, including an assessment of the natural safeguards in place that contribute to risk reduction (e.g., secondary seals, auxiliary storage formations, etc.). Monitoring technologies to address these risks are identified and a time schedule and coverage plan for data acquisition is constructed, together with performance targets for each selected monitoring technology.

TABLE 15.9 Summary of the structure and contents of the Shell Canada Quest project storage development plan.

No.	Chapter	Summary of contents
1	Executive summary	A high-level overview of the main project parameters, planning assumptions, and value (e.g., economics, if relevant).
2	The Quest opportunity	Describes the project at the next level of detail; what are the key objectives—what does success look like? What are the major components of the project? Overview of the plan to mature and execute the project. Summary of risk management and external stakeholder management plans—including regulatory and permitting requirements. Summary of project costs and economic value (where relevant).
3	HSSE and SD	Describes the main health, safety, security, and environmental (HSSE) aspects of the project, as well as its approach to sustainable development (SD). Describes the environmental and other regulatory requirements for permitting.
4	Capture site (facilities)	Describes the engineering concept to be applied at the CO_2 capture site, and the design basis of the major facility components, including delivery specification of product streams.
5	Storage site and well site selection	Describes the process of storage site selection including selection criteria and appraisal workplan execution and outcome. Similarly describes the process for selection of well locations for injection, deep monitoring, and shallow groundwater wells, both firm project wells and future contingency wells.
6	Pipeline	Describes the pipeline system, including pipeline design, metering and control systems, interfaces to capture system and wells, pipeline route selection, project construction schedule, and regulatory and land acquisition issues.
7	Integrated project design	Describes the overall operating limits of the system and how these integrate limits and requirements from subsurface (e.g., fracturing geomechanics), operational (e.g., flow assurance), and engineering design (e.g., compression and pipeline specification) perspectives.
8	Containment, storage capacity, injectivity, and conformance	Describes the storage complex and initial capacity and injectivity assessment. Describes in detail the main site-specific risks to each of these factors (containment, capacity, injectivity, and conformance) within the overall risk management framework for the project, as well as risk reduction and mitigation measures (safeguards).
9	Well engineering and production technology	Describes the design, construction (drilling and completion), and operation of each of the project well types. Considerations regarding well start-up, well interventions, and eventual abandonment are noted, although these activities will be subject to separate detailed planning.

Continued

TABLE 15.9 Summary of the structure and contents of the Shell Canada Quest project storage development plan.—cont'd

No.	Chapter	Summary of contents
10	Measurement, monitoring, and verification (MMV)	Provides a general description of the approach to MMV requirements and planning, including the link to the risk management framework, as well as a high-level summary of the MMV plan activities. Effectively an executive summary of the more detailed MMV plan.
11	New technology application	A summary of new technologies being applied across all aspects and phases of the project, from the capture facilities to the subsurface surveillance.
12	Asset management	Describes how the project facilities will be operated and maintained after start-up, including the organizational requirements
13	Project execution strategy	Describes how the project will be executed, including aspects such as project governance, budgeting and cost control, contracting and procurement, assurance, management of change, and risk management during the execution phase.
14	Start-up and commissioning	Summarizes specific considerations for the start-up of wells and other facilities, which will be covered by separate detailed plans,
15	Closure, postclosure, decommissioning, and abandonment	Summarizes considerations for site closure and later stages, including site closure performance targets. Detailed closure and postclosure plans will be prepared toward the end of the operating phase.

Additional contingency monitoring activities are identified, together with the triggers to initiate these activities. These activities would either be initiated in response to underperformance of the base plan monitoring technologies, in the case of safety critical containment monitoring, or by deviations from expected performance in the case of conformance monitoring. A time schedule and coverage plan for acquisition, as well as performance targets, are specified in the same way as for the primary MMV activities. The document then describes the residual risks after MMV activities, showing how these additional safeguards contribute to risk reduction (Shell Canada, 2012; Bourne et al., 2014).

15.6.2 Saline aquifer storage case study 2: Sleipner

In the Norwegian sector of the North Sea, Equinor (formerly Statoil) and partners have been producing natural gas from the Sleipner East and West fields since August 1996. The Sleipner gas contains 4.0%–9.5% CO_2, which had to be reduced to <2.5% to meet the EU pipeline gas delivery specification. In the absence of an economic EOR opportunity, the overlying Utsira aquifer was identified as a CO_2 disposal reservoir. Fig. 15.28 shows a schematic of the project.

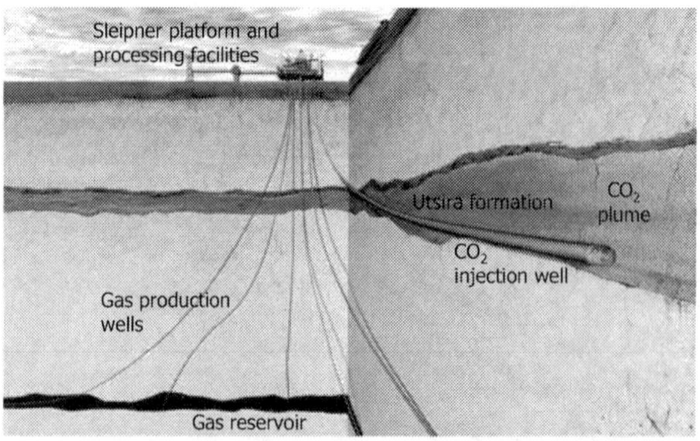

FIGURE 15.28 Schematic overview of the Sleipner CO_2 storage project. *Source: Courtesy, Equinor.*

An international multidisciplinary research program was established to collect and analyze data from the project and to leverage this to provide guidance for similar projects in the future. The Saline Aquifer CO_2 Storage (SACS) and subsequent SACS2 programs ran from 1998 to 2003 and culminated in the publication of a Best Practice Manual (Chadwick et al., 2008) based on the knowledge gained and lessons learned in this project.

Site characterization and selection

The Utsira formation is an unconsolidated water-bearing sandstone with high porosity (30%−42%) and permeability (1−3 Darcy). Initial characterization of the aquifer, which is > 400 km long and 50−100 km wide, was based on an extensive regional 2D seismic dataset, as well as data from some 300 well penetrations, of which 30 were within 20 km of the selected injection location.

At the Sleipner field, the Utsira formation lies at a depth of 800 m, deep enough for CO_2 to remain in a supercritical state, and forms an almost flat structure with a gentle dip to the south and southwest. This type of structure poses specific problems for predicting plume movement, as minor changes in the topography of the caprock can have a significant impact on the migration direction of injected CO_2. In this situation, accurate depth mapping of the top structure is essential, based on dense 2D or 3D surveys tied to adequate well data. This would be less of an issue in formations with higher-relief structural or stratigraphic traps, where plume movement will be more constrained by these geological features.

To minimize the risk from potential leak paths, the injection site was chosen to ensure that the predicted migration of the plume would take it away from the gas production wells, and a small structural trap in the aquifer was selected as the injection target. Preliminary reservoir modeling work indicated that a plume with a maximum lateral extent of 3 km would be formed after 20 years of injection.

Injection operation and monitoring

Capture of ~1 $MtCO_2$/yr using amine stripping and injection into the single disposal well commenced in September 1996. A 3D survey acquired in 1994 was used as the baseline for monitoring, and repeated time-lapse (4D) surveys acquired roughly every 2 years since 1999 have provided a detailed picture of the evolution of the plume.

These survey results (Fig. 15.29) show that the vertical movement of CO_2 was impeded by small-scale impermeable shale layers that were below the seismic resolution. These layers caused lateral spreading of the buoyant fluid, which then spilled at permeable boundaries and rose to the next thin layer. The 4D seismic results could be well-matched in the reservoir simulation once these fine-scale details were incorporated into the geological model.

In the longer term (>50 years), all mobile CO_2 is expected to migrate to the top of the formation, and the caprock topology will be the geological feature controlling further evolution of the plume (Ahmadinia and Shariatipour, 2020). These results demonstrated the central role of time-lapse seismic in monitoring geological storage operations, as well as the potential for plume dynamics and CO_2 trapping to be significantly affected by small-scale geological heterogeneities that may not be recognized or represented in the initial geological dataset.

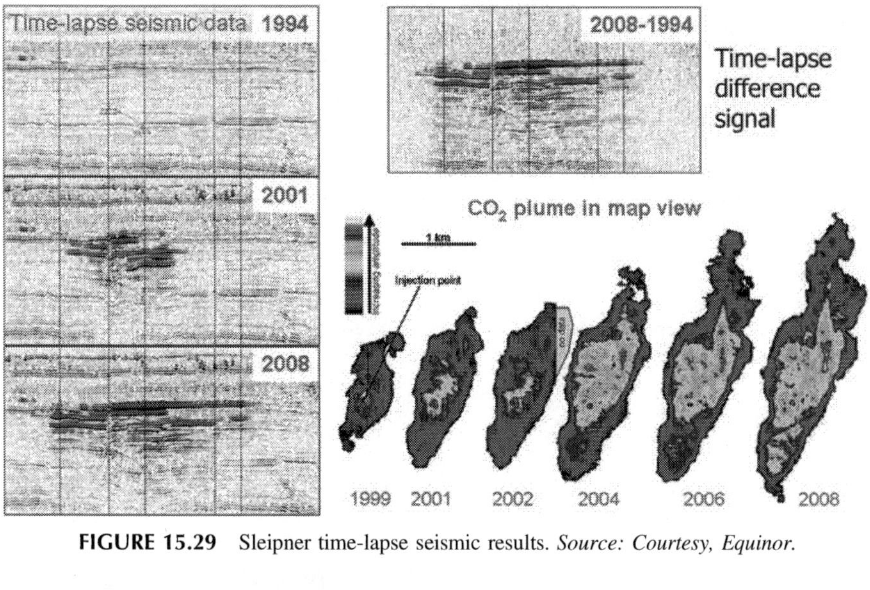

FIGURE 15.29 Sleipner time-lapse seismic results. *Source: Courtesy, Equinor.*

The Utsira aquifer, with an estimated storage capacity of 15 $GtCO_2$, has been identified as a potential host formation for a regional carbon storage facility that could accommodate injection of up to 300 $MtCO_2$/yr for 50 years. Effective subsurface modeling on this scale—incorporating sufficient resolution and process detail in a model extending over tens of thousands of square kilometers—is a significant R&D challenge (Gasda et al., 2017) that provides valuable guidance for similar basin-scale projects in other regions.

15.6.3 Saline aquifer storage case study 3: Ketzin pilot

The Ketzin pilot, located 25 km west of Berlin, Germany, was the first onshore saline aquifer storage pilot in Europe. Research activities, initially funded under the EU's CO_2SINK project, began at the site in 2004, with the drilling of one injection well and two observation wells. An extensive monitoring program, including the drilling of additional observation wells for above zone monitoring and to core the reservoir and caprock, was conducted during the injection period from June 2008 to August 2013, when some 67 $ktCO_2$ was injected into a saline aquifer at ~650 m depth.

In January 2014, the COMPLETE project was started at Ketzin, with the objective of capturing the knowledge and experience of the final phase in the life cycle of a saline aquifer storage site. The project scope covered three main themes:

- Operational activities involved in well abandonment and site decommissioning,
- Postclosure monitoring activities including a repeat 3D seismic survey, electrical, and surface CO_2 flux measurements, and
- Simulation model updates to support the postclosure monitoring campaigns.

The latter theme also included the integration of extensive project datasets into a single long-term archive. Some additional field experiments were also conducted, including an investigation of physicochemical properties of back-produced CO_2 and a brine injection test to assess residual gas saturation and to test brine injection as a possible remedial measure in the event of wellbore leakage (Wasch et al., 2013).

While it is less relevant for pilot-scale operations, regulatory requirements for site closure at a commercial scale site (>1 $MtCO_2$/yr) will include a demonstration that the agreed performance criteria for site closure have been met, as well as a formal assessment of postclosure risks and an update of the MMV plan to address those risks during the postclosure period, before the transfer of long-term responsibility to the relevant agency. An example of this closure plan documentation is available for the Shell Quest project (Shell Canada, 2014).

15.7 In situ mineralization case studies

15.7.1 Wallula injection pilot—in situ mineralization in flood basalt

The potential for carbon storage in continental flood basalts was pointed out by McGrail and colleagues (McGrail et al., 2006), and a pilot CCS project—the Wallula Basalt Sequestration Pilot Project—was planned by this group as part of the US Department of Energy's Big Sky Regional Carbon Sequestration Partnership. The initial pilot scope included the capture of up to 0.8 $MtCO_2$/yr from the power island of the Boise White Paper mill, located on the banks of the Columbia River in Washington state, and injection of $scCO_2$ into permeable flow units in the Columbia River Basalt Group (CRBG) which underlies over 200,000 km^2 of Washington, Oregon, and Idaho.

Following site assessment using 2D seismic, a pilot well was drilled in 2009 for site characterization and identified a suitable injection zone at ca. 850 m depth, with 28 m net thickness and an average permeability of 44–90 mD estimated from constant-rate pumping tests. Injection tests also confirmed the presence of three intervals that would act as flow barriers above the injection zone, with permeabilities of 0.01–0.1 μD. Unfortunately, the integrated CCS pilot was shelved in 2011 when DoE funding was not secured for the next phase. Funds were however available for an injection pilot and, in 2013, 977 tCO_2 was injected over 25 days in the world's first $scCO_2$ injection test into flood basalt.

Postinjection fluid sampling showed 100–1000-fold increase in Ca, Mg, Fe, and Mn cation concentrations compared to preinjection samples, as well as ^{13}C and ^{18}O isotopic shifts consistent with the onset of geochemical reactions. Geochemical analysis of sidewall core samples recovered before well abandonment in 2015 confirmed the rapid conversion of injected $scCO_2$ to carbonate minerals, which were visible as nodules of ankerite ($Ca(Fe,Mg,Mn)(CO_3)_2$) distributed within the basalt matrix (McGrail et al., 2017).

15.7.2 CarbFix—in situ mineralization in mid-ocean ridge basalt

Alongside continental flood basalts, mid-ocean ridges and submarine flood basalts also have the potential for climate impact scale carbon storage, with a potential global capacity estimated at 100,000–250,000 $GtCO_2$, and the CarbFix pilot project tested this concept in southwest Iceland. This project took a different approach from Wallula, where $scCO_2$ was injected. Instead, gaseous CO_2 and water were separately injected into the injection well and mixed at 350 m, at a hydrostatic pressure of 2.5 MPa. Small gas bubbles were thus entrained in the descending water stream and were fully dissolved at the injection depth of ca. 2000 m. Groundwater, available in plentiful supply, was used for the pilot with 22 kg-water required to fully dissolve each $kgCO_2$ at the operating pressure and temperature. Seawater could also be used but the required volume would increase to 31 $kg/kgCO_2$. Brine pH at the injection point was ~3.7 and, as can be seen from Fig. 6.3, the ~1 mol/kg total DIC would be almost entirely dissolved CO_2 plus H_2CO_3. Injection was into a porous basalt body of roughly 1 km^3 (2 km × 1 km x 500 m thick), which it is estimated can store 12 $MtCO_2$ on the basis of 10% of the pore space (~1% of the gross rock volume) being ultimately filled by carbonate precipitate.

In phase I of the pilot, conducted in early 2012, 175 tCO_2 were injected, while 73 t of a CO_2–H_2S gas mixture (55 tCO_2) was injected in phase II in mid-2012. Subsurface monitoring included conservative and reactive tracers, fluid sampling, and wireline coring in a number of monitoring wells, and showed rapid mineralization of the dissolved CO_2. Modeling of tracer breakthrough results indicated mineralization of >80% of injected CO_2 within 12 months and >95% within 24 months of injection (Matter et al., 2016).

Following the successful pilot project, the CarbFix injection system was integrated into the operations of the Hellisheidi geothermal power plant, the world's third largest geothermal plant, producing 303 MW_e plus 133 MW_{th} of hot water for district heating. Industrial scale injection began in 2014, and, after successive ramp-up stages, reached a level of 10 $ktCO_2$ and 6 ktH_2S in 2017, some 34% and 68%, respectively of the plant's previous annual emissions. Injection continues into the 80 km^2 Hellisheidi field, which has an estimated theoretical storage capacity of up 6 $GtCO_2$ (Sigfússon et al., 2018).

Building on the success of the CarbFix project (Clark et al., 2020), a follow-up project—CarbFix2—commenced in 2017 with the objective of integrating direct air capture (DAC) with in situ mineralization. A 50 tCO_2/yr DAC module developed by Climeworks was installed at the plant in 2017 (Gutknecht et al., 2018).

CarbFix is expanding at Hellisheidi under the name Project Silverstone, with a target capacity of 34 $ktCO_2$/yr and additional 4 $ktCO_2$/yr from the expansion of the DAC collaboration with Climeworks to 4 $ktCO_2$/yr under Project Orca. The same geological storage approach is also being developed at a new site on the Reykjanes Peninsula to store CO_2 shipped from continental Europe to the Coda Terminal near Reykjavik.

15.7.3 CarbonSAFE Cascadia project

The CarbonSAFE Cascadia project is a prefeasibility study, building on the Wallula pilot and CarbFix learnings, with the aim of developing an industrial-scale project capturing ~2.5 $MtCO_2$/yr from onshore sites, with transportation and injection into the Juan de Fuca basalts in the Cascadia basin, offshore Washington state, and British Columbia (Goldberg et al., 2018).

The scope of the study, completed in 2018, included the following:

- Evaluation of industrial CO_2 sources and potential modes of transportation
- Inventorization of existing geophysical and geological data in the area
- Laboratory analysis to measure mineralization rates in ocean basalt
- Initial reservoir modeling of the potential storage complex
- Preliminary review of offshore storage regulations in the United States and Canada
- Stakeholder analysis and evaluation of costs and potential economic incentives
- A preliminary project risk assessment.

The CarbonSAFE Cascadia project is part of the wider US DoE funded Carbon Storage Assurance Facility Enterprise (CarbonSAFE) initiative, which focuses on the development of geologic storage sites with capacity to store >50 $MtCO_2$ from industrial sources. The portfolio of CarbonSAFE projects aims to improve understanding of site screening and selection, characterization, and baseline MMV, as well as the design of injection and monitoring strategies for commercial-scale projects. The initiative is building toward the start-up of 50$MtCO_2$ scale storage sites by 2026.

15.8 Geological storage costs and global potential

The cumulative demand for geological storage capacity for CO_2 captured from BECCS and DACCS projects is expected to be 10 $GtCO_2$/yr to 2050 and 15–20 $GtCO_2$/yr through 2100 (Zahansky and Krevor, 2020). The estimated global potential for CO_2 storage in saline aquifers, ranging from 4000 to some 25,000 $GtCO_2$, carries a very wide range of uncertainty, largely because many of the sedimentary basins and formations where this capacity exists lack detailed geological characterization (Benson et al., 2012; Consoli and Wildgust, 2017). Global estimates essentially refer to theoretical capacity, although more detailed assessments of technically and economically viable capacity (e.g., reflecting pressure build-up limits and pressure management strategies) have been made in a few well-studied locations. Proximity to large CO_2 capture sources will also be an important factor in determining viable capacity; BECCS plant locations will be influenced by proximity to the biomass supply, to reduce carbon leakage in the supply chain, while DACCS plant siting will have greater flexibility to be close to storage sites.

The storage potential in basalts includes up to 7000 $GtCO_2$ in the most prospective active rift zone, covering 34,000 km^2 or roughly one-third of onshore Iceland plus 93,000 km^2 offshore within the Icelandic Exclusive Economic Zone (Snæbjörnsdóttir and Gislason, 2016). The global potential for storage in deep-sea basalt sites (below the buoyancy depth of liquid CO_2) has been estimated to be 30,000−150,000 $GtCO_2$ (Goldberg and Slagle, 2009), while ultramafic rocks like peridotite have about 10% of the potential storage volumes of onshore basalts (~10,000 $GtCO_2$). For comparison, ~4000 $GtCO_2$ would be emitted from the combustion of currently estimated global fossil fuel reserves, and ~2600 $GtCO_2$ (700 GtC, including from land-use change) has been emitted since the start of the Industrial Revolution (Frielingstein et al., 2022).

Estimated storage costs have a similarly wide range—from \$2 to \$28/tCO_2 (Rubin et al., 2015; Kali et al., 2022). Robust techno-economic analyses comparing storage in sedimentary basins and igneous rocks have yet to be conducted, but it seems likely that operational costs will be comparable. Volcanic stores will likely benefit from shallower drilling, and simpler and shorter monitoring than those in sedimentary basins, but will not benefit from the presence of legacy infrastructure.

15.9 Geological storage research and development agenda

Technologies developed in the oil and gas industry have enabled saline aquifer storage projects to proceed rapidly to the demonstration stage, but many of the geomechanical and geochemical processes, and monitoring requirements for these projects represent a step-out from oilfield practice and are the subject of ongoing R&D work. These and other R&D themes for saline aquifer storage are summarized in Table 15.10. Detailed discussion of these R&D themes can be found in Fuss et al. (2016), Bui et al. (2018), and NASEM (2019).

Many of the R&D topics listed in relation to saline aquifer storage are also relevant for in situ mineralization, while some additional mineralization-specific themes are listed in Table 15.11. Further discussion of these topics can be found in Trias et al. (2017), Daval (2018), Kelemen et al. (2018), and NASEM (2019).

TABLE 15.10 R&D themes for saline aquifer storage.

R&D theme	Description
Site characterization	Improved methods for site characterization for large-scale projects, including field testing techniques to quantify risks of pressure-induced slip along preexisting faults/fractures, and assess susceptibility to induced seismicity
Trapping mechanisms	Evolution of residual saturation after injection and linkage with solubility and mineral trapping; lab-scale experiments, field trials, and modeling studies to investigate trapping mechanisms and test trapping enhancement strategies, including factors controlling convective dissolution under in situ conditions; high-resolution scanning techniques (X-CT, PET, etc.) to investigate flow and trapping behavior in cores
Subsurface monitoring and imaging	Improvements in imaging technologies to quantify plume location, saturation, and volumetrics, monitor caprock integrity and enable leak detection; resolution improvements, e.g., imaging small-scale heterogeneities, spatial location of microseismic events; development of integrated monitoring systems to reduce cost and increase data quality for commercial scale projects
Geochemical processes	Reaction of CO_2 with reservoirs and seals - impact on caprock wettability, entry pressure, and self-sealing potential; processes in carbonate formations; mineral trapping, including in situ mineral carbonation; dissolution and mobilization of heavy metals or other potential contaminants; validation of reactive transport models

TABLE 15.10 R&D themes for saline aquifer storage.—cont'd

R&D theme	Description
Geomechanical processes	Mechanisms, risks, and consequences of pressure-induced slip along preexisting faults and associated seismicity; impact of injection pressures and temperatures on reservoir and caprock, faults, and fracture networks, including low-temperature injection from ship-borne CO_2; long-term impact of CO_2 on caprock mechanical properties; formation creep as a leak healing mechanism
Operational technologies	Development of efficient tools and automation technologies for real-time integration of subsurface and surface monitoring data, and comparison against baseline surveys; new techniques for leak detection; operational approaches to limit the potential for induced seismicity
Subsurface modeling	Incorporate improved process knowledge into mathematical simulation models from pore to basin scale; improved modeling of fault flow processes and in situ validation; simplified coupled hydrodynamic-geomechanical models to enable regional-scale studies; robust upscaling methods to parameterize fine-scale processes in coarse gridded models
Governance and outreach	Legal risks with respect to liability in the case of leakage from geological storage, including transnational aspects; further development of best practices for community engagement

TABLE 15.11 R&D themes for in situ mineralization.

R&D theme	Description
Mineralization reaction chemistry	Lab experiments and modeling studies on CO_2—basalt/peridotite reactions under in situ conditions; impact of injection water chemistry and other conditions on storage efficiency (promoting and controlling carbonate deposition, clogging of pore space, passivation of reactive surfaces, etc.)
Chemical and geomechanical processes	Lab investigation and modeling of chemical and rock mechanical processes and nano-scale rock characteristics (mineral—fluid surface energy, sorptivity, disjoining pressure) leading to reaction-driven cracking in peridotite and basalt formations
Biogeochemical processes	Impact of deep ecosystems and associated biogeochemical reactions on the fate of injected CO_2; relative contributions of microbial biomass and carbonates to storage in basalts; impact of biogeochemical reactions on passivation of mineral reactive surfaces
Peridotite field trials	Characterization of potential test sites leading to small- to medium-scale (~1 −100 $ktCO_2$/yr) field trials to investigate mineralization in peridotite; optimization (stimulation requirement, fluid type, pH, and other conditions) to achieve rapid mineralization while minimizing negative feedbacks
Governance	Legal risks with respect to liability in the case of leakage from geological storage, including transnational aspects

References and resources

Ahmadinia, M., Shariatipour, S.M., 2020. Analysing the role of caprock morphology on history matching of Sleipner CO_2 plume using an optimisation method. Greenhouse Gases: Sci. Technol. 10 (5), 1077–1097.

Allen, R., Nilsen, H.M., Andersen, O., Lie, K.A., 2017. Categorization of Norwegian continental shelf formations in terms of geological CO_2 storage potentials. Energy Proc. 114, 4583–4594.

Aradóttir, E.S.P., Hjalmarsson, E., 2018. CarbFix—public engagement and transparency. Energy Proc. 146, 115–120.

Aradóttir, E.S.P., Sigurdardóttir, H., Sigfússon, B., Gunnlaugsson, E., 2011. CarbFix: a CCS pilot project imitating and accelerating natural CO_2 sequestration. Greenhouse Gas Sci. Technol. 1 (2), 105–118.

Bachu, S., 2000. Sequestration of CO_2 in geological media: criteria and approach for site selection in response to climate change. Energy Convers. Manag. 41, 953–970.

Benson, S.M., Bennaceur, K., Cook, P., et al., 2012. Carbon dioxide capture and storage. In: Gomez-Echeverri, L., Johansson, T.B., Nakicenovic, N., Patwardhan, A. (Eds.), Global Energy Assessment: Toward a Sustainable Future. Cambridge University Press, Cambridge, UK.

Bickle, M.J., 2009. Geological carbon storage. Nat. Geosci. 2 (12), 815–818.

Bourne, S., Crouch, S., Smith, M., 2014. A risk-based framework for measurement, monitoring and verification of the Quest CCS Project, Alberta, Canada. Int. J. Greenh. Gas Control 26, 109–126.

Brunsting, S., de Best-Waldhober, M., Feenstra, C.Y., Mikunda, T., 2011. Stakeholder participation practices and onshore CCS: lessons from the Dutch CCS case Barendrecht. Energy Proc. 4, 6376–6383.

Bui, M., Adjiman, C.S., Bardow, A., et al., 2018. Carbon capture and storage (CCS): the way forward. Energy Environ. Sci. 11 (5), 1062–1176.

Buursink, M.L., Blondes, M.S., Brennan, S., et al., December 2013. National assessment of geologic carbon dioxide storage resources–trends and interpretations. In: AGU Fall Meeting Abstracts, Vol 2013. GC23A-0895.

Celia, M.A., Nordbotten, J.M., 2011. How simple can we make models for CO_2 injection, migration and leakage? Energy Proc. 4, 3857–3864.

Chadwick, A., Arts, R., Bernstone, C., May, F., Thibeau, S., Zweigel, P., 2008. Best Practice for the Storage of CO_2 in Saline Aquifers—Observations and Guidelines from the SACS and CO_2 STORE Projects. British Geological Survey Occasional Publication, Nottingham, UK, 14. British Geological Survey.

Chevron, 2005. Environmental Impact Statement/Environmental Review and Management Programme for the Gorgon Development (Chevron Australia).

Clark, D.E., Oelkers, E.H., Gunnarsson, I., et al., 2020. CarbFix2: CO_2 and H_2S mineralization during 3.5 years of continuous injection into basaltic rocks at more than 250°C. Geochem. Cosmochim. Acta 279, 45–66.

Condor, J., Unatrakarn, D., Wilson, M., Asghari, K., 2011. A comparative analysis of risk assessment methodologies for the geologic storage of carbon dioxide. Energy Proc. 4, 4036–4043.

Consoli, C.P., Wildgust, N., 2017. Current status of global storage resources. Energy Proc. 114, 4623–4628.

CSLF (Carbon Sequestration Leadership Forum), 2007. Estimation of CO_2 Storage Capacity in Geological Media. Phase 2 Report Prepared by the Task Force on CO_2 Storage Capacity Estimation for the Technical Group of the Carbon Sequestration Leadership Forum.

Daval, D., 2018. Carbon dioxide sequestration through silicate degradation and carbon mineralisation: promises and uncertainties. Nat. Partner J. Mat. Degrad. 2 (1), 11.

Dean, M., Tucker, O., 2017. A risk-based framework for Measurement, Monitoring and Verification (MMV) of the Goldeneye storage complex for the Peterhead CCS project, UK. Int. J. Greenh. Gas Control 61, 1–15.

Duan, Z., Sun, R., 2003. An improved model calculating CO_2 solubility in pure water and aqueous NaCl solutions from 273 to 533 K and from 0 to 2000 bar. Chem. Geol. 192, 257–271.

Ennis-King, J., Paterson, L., 2005. Role of convective mixing in the long-term storage of carbon dioxide in deep saline formations. Soc. Petrol. Eng. J. 10, 349–356 (SPE 84344).

ETI, 2016. A Summary of Results from the Strategic UK CO_2 Storage Appraisal Project. Energy Technology Institute, Pale Blue Dot Energy, Axis Well Technology, and Costain.

European Commission Directive, 2009. Directive 2009/31/EC of the European Parliament and of the Council of 23 April 2009 on the Geological Storage of Carbon Dioxide.

Friedlingstein, P., O'sullivan, M., Jones, M.W., Andrew, R.M., Gregor, L., Hauck, J., et al., 2022. Global carbon budget 2022. Earth Syst. Sci. Data 14 (11), 4811–4900.

Fuss, S., Jones, C.D., Kraxner, F., et al., 2016. Research priorities for negative emissions. Environ. Res. Lett. 11 (11), 115007.

FutureGen, 2007. FutureGen Initial Conceptual Design Report. Report PNWD-3760. FutureGen Alliance, Washington D.C., USA.

Gasda, S.E., Wangen, M., Bjørnarå, T.I., Elenius, M.T., 2017. Investigation of caprock integrity due to pressure build-up during high-volume injection into the Utsira formation. Energy Proc. 114, 3157–3166.

Goldberg, D., Slagle, A.L., 2009. A global assessment of deep-sea basalt sites for carbon sequestration. Energy Proc. 1 (1), 3675–3682.

Goldberg, D., Aston, L., Bonneville, A., et al., 2018. Geological storage of CO_2 in sub-seafloor basalt: the CarbonSAFE pre-feasibility study offshore Washington State and British Columbia. Energy Proc. 146, 158–165.

Gorecki, C.D., Holubnyak, Y., Ayash, S., et al., 2009. A new classification system for evaluating CO_2 storage resource/capacity estimates. In: Proceedings of the SPE International Conference on CO_2 Capture, Storage, and Utilization, November 2–4 2009, San Diego, CA, USA. SPE paper 126421.

Gunnarsson, I., Aradóttir, E.S., Oelkers, E.H., et al., 2018. The rapid and cost-effective capture and subsurface mineral storage of carbon and sulfur at the CarbFix2 site. Int. J. Greenh. Gas Control 79, 117–126.

Gunter, W.D., Bachu, S., Benson, S., 2004. The role of hydrogeological and geochemical trapping in sedimentary basins for secure geological storage of carbon dioxide. Geol. Soc., London, Special Publications 233 (1), 129–145.

Gutknecht, V., Snæbjörnsdóttir, S.Ó., Sigfússon, B., Aradóttir, E.S., Charles, L., 2018. Creating a carbon dioxide removal solution by combining rapid mineralization of CO_2 with direct air capture. Energy Proc. 146, 129–134.

Hassanzadeh, H., Pooladi-Darvish, M., Keith, D.W., 2009. Accelerating CO_2 dissolution in saline aquifers for geological storage—mechanistic and sensitivity studies. Energy Fuel. 23 (6), 3328–3336.

Hovorka, S.D., Benson, S.M., Doughty, C., et al., 2006. Measuring permanence of CO_2 storage in saline formations: the Frio experiment. Environ. Geosci. 13 (2), 105–121.

Huber, E.J., Stroock, A.D., Koch, D.L., 2016. Analysis of a time dependent injection strategy to accelerate the residual trapping of sequestered CO_2 in the geologic subsurface. Int. J. Greenh. Gas Control 44, 185–198.

IEA GHG, 2015. Review of Offshore Monitoring for CCS Projects. IEAGHG Report 2015/02. IEAGHG, Cheltenham, UK.

Jenkins, C., Chadwick, A., Hovorka, S.D., 2015. The state of the art in monitoring and verification—ten years on. Int. J. Greenh. Gas Control 40, 312–349.

Kaldi, J., Daniel, R., Tenthorey, E., et al., 2011. Caprock Systems for CO_2 Geological Systems. International Energy Agency. Report 2011/01 IEA/CON/10/179.

Kali, R.F.S., Fan, D., Hazel, N., Striolo, A., 2022. Techno-socio-economic analysis of geological carbon sequestration opportunities. Environ. Sci. J. Integr. Environ. Res.: Advances 2022 (1), 138–155.

Kelemen, P.B., Matter, J., 2008. In situ carbonation of peridotite for CO_2 storage. Proc. Natl. Acad. Sci. USA 105, 17295–17300.

Kelemen, P.B., Aines, R., Bennett, E., et al., 2018. In situ carbon mineralization in ultramafic rocks: natural processes and possible engineered methods. Energy Proc. 146, 92–102.

Koide, H., Tazaki, Y., Noguchi, Y., Nakayama, S., Iijima, M., Ito, K., Shindo, Y., 1992. Subterranean containment and long-term storage of carbon dioxide in unused aquifers and in depleted natural gas reservoirs. Energy Convers. Manag. 33 (5–8), 619–626.

Leonenko, Y., Keith, D.W., 2008. Reservoir engineering to accelerate the dissolution of CO_2 stored in aquifers. Environ. Sci. Technol. 42 (8), 2742–2747.

Li, Q., Liu, G., 2016. Risk assessment of the geological storage of CO_2: a review. In: Geologic Carbon Sequestration. Springer, Cham, Switzerland, pp. 249–284.

Lockwood, T., 2017. Public Outreach Approaches for Carbon Capture and Storage Projects. IEA Clean Coal Centre. Report CCC/276.

Matter, J.M., Stute, M., Snæbjörnsdottir, S.Ó., et al., 2016. Rapid carbon mineralization for permanent disposal of anthropogenic carbon dioxide emissions. Science 352 (6291), 1312–1314.

McGrail, B.P., Schaef, H.T., Ho, A.M., Chien, Y.J., Dooley, J.J., Davidson, C.L., 2006. Potential for carbon dioxide sequestration in flood basalts. J. Geophys. Res. Solid Earth 111 (B12).

McGrail, B.P., Schaef, H.T., Spane, F.A., et al., 2017. Wallula basalt pilot demonstration project: post-injection results and conclusions. Energy Proc. 114, 5783–5790.

McNemar, A., Myer, L., Damiani, D., McKoy, M., Bromhal, G., 2021. Developing and validating pressure management and plume control strategies. In: Proceedings of the 15th Greenhouse Gas Control Technologies Conference, pp. 15–18. March 2021.

NASEM (National Academies of Sciences, Engineering, and Medicine), 2019. Negative Emissions Technologies and Reliable Sequestration: A Research Agenda. National Academies Press, Washington, DC, USA.

NETL, 2010. Best Practices for Site Screening, Selection, and Initial Characterization for Storage of CO_2 in Deep Geologic Formations. National Energy Technology Laboratory report DOE/NETL-401/090808.

Nghiem, L., Yang, C., Shrivastava, V., Kohse, B., Hassam, M., Card, C., 2009. Risk mitigation through the optimization of residual gas and solubility trapping for CO_2 storage in saline aquifers. Energy Proc. 1 (1), 3015–3022.

Oelkers, E., Gíslason, S.R., Matter, J., 2008. Mineral carbonation of CO_2. Elements 4 (5), 333–337.

Oldenburg, C.M., 2008. Screening and ranking framework for geologic CO_2 storage site selection on the basis of health, safety, and environmental risk. Environ. Geol. 54, 1687–1694.

Peck, W.D., Azzolina, N.A., Ge, J., Gorecki, C.D., Gorz, A.J., Melzer, L.S., 2017. Best practices for quantifying the CO_2 storage resource estimates in CO_2 enhanced oil recovery. Energy Proc. 114, 4741–4749.

Qi, R., LaForce, T.C., Blunt, M.J., 2009. Design of carbon dioxide storage in aquifers. Int. J. Greenh. Gas Control 3 (2), 195–205.

Rackley, S.A., 2017. Carbon Capture and Storage. Elsevier, Oxford, UK.

Réveillère, A., Rohmer, J., 2011. Managing the risk of CO2 leakage from deep saline aquifer reservoirs through the creation of a hydraulic barrier. Energy Proc. 4, 3187–3194.

Rodosta, T., Bromhal, G., Damiani, D., 2017. US DOE/NETL carbon storage program: advancing science and technology to support commercial deployment. Energy Proc. 114, 5933–5947.

Rubin, E.S., Davison, J.E., Herzog, H.J., 2015. The cost of CO_2 capture and storage. Int. J. Greenh. Gas Control 40, 378–400.

Senior, B., Bradshaw, J., Chikkatur, A., Wright, M., 2011. Planning saline reservoir storage developments—the importance of getting started early. Energy Proc. 4, 4575–4582.

Shell Canada, 2011. Quest Storage Development Plan. Available at: https://open.alberta.ca/dataset/quest-carbon-capture-and-storage-project-report-2011.

Shell Canada, 2012. Quest Carbon Capture and Storage Project, Measurement, Monitoring and Verification Plan. Available at: https://open.alberta.ca/publications/quest-carbon-capture-and-storage-project-report-2012.

Shell Canada, 2014. Quest CCS Project Closure Plan- 3rd Year Update. Available at: https://open.alberta.ca/publications/quest-carbon-capture-and-storage-project-report-2013.

Sigfússon, B., Arnarson, M.þ., Snæbjörnsdóttir, S.Ó., Karlsdóttir, M.R., Aradóttir, E.S., Gunnarsson, I., 2018. Reducing emissions of carbon dioxide and hydrogen sulphide at Hellisheidi power plant in 2014–2017 and the role of CarbFix in achieving the 2040 Iceland climate goals. Energy Proc. 146, 135–145.

Sijacic, D., Wildenborg, T., Steeghs, P., 2014. TNO Monitoring plan development tool. Energy Proc. 63, 4834–4840.

Snæbjörnsdóttir, S.Ó., Gislason, S.R., 2016. CO_2 storage potential of basaltic rocks offshore Iceland. Energy Proc. 86, 371–380.

SPE (Society of Petroleum Engineers), 2017. Petroleum Resources Management System. Available at: www.spe.org/industry/docs/Petroleum-Resources-Management-System.pdf.

Spiteri, E., Juanes, R., Blunt, M.J., Orr, F.M., 2005. Relative-permeability hysteresis: trapping models and application to geological CO2 sequestration. In: Society of Petroleum Engineers 2005 Annual Technical Conference and Exhibition, Dallas, Texas, USA (SPE 96448).

Tanaka, K., Vilcáez, J., Sato, K., 2013. Improvement of CO_2 geological storage efficiency by injection and production well design. Energy Proc. 37, 4591–4597.

Trias, R., Ménez, B., le Campion, P., et al., 2017. High reactivity of deep biota under anthropogenic CO_2 injection into basalt. Nat. Commun. 8 (1), 1–14.

Wasch, L.J., Wollenweber, J., Tambach, T.J., 2013. Intentional salt clogging: a novel concept for long-term CO_2 sealing. Greenhouse Gases: Sci. Technol. 3 (6), 491–502.

Zahasky, C., Benson, S.M., 2016. Evaluation of hydraulic controls for leakage intervention in carbon storage reservoirs. Int. J. Greenh. Gas Control 47, 86–100.

Zahasky, C., Krevor, S., 2020. Global geologic carbon storage requirements of climate change mitigation scenarios. Energy Environ. Sci. 13 (6), 1561–1567.

Further reading

Akhurst, M., Kirk, K., Neele, F., Grimstad, A.A., Bentham, M., Bergmo, P., 2021. Storage Readiness Levels: communicating the maturity of site technical understanding, permitting and planning needed for storage operations using CO_2. Int. J. Greenh. Gas Control 110, 103402.

Benson, S.M., Hepple, R., Apps, J., Tsang, C.F., Lippmann, M., 2002. Lessons Learned from Natural and Industrial Analogues for Storage of Carbon Dioxide in Deep Geological Formations. Lawrence Berkeley National Laboratory report LBNL #51170.

Blondes, M.S. (Ed.), 2013. National Assessment of Geologic Carbon Dioxide Storage Resources: Methodology Implementation (No. 2013-1055). US Geological Survey.

Castaneda-Herrera, C.A., Stevens, G.W., Haese, R.R., 2018. Review of CO_2 leakage mitigation and remediation technologies. In: Vialle, S., Ajo-Franklin, J., Carey, J.W. (Eds.), Geological Carbon Storage: Subsurface Seals and Caprock Integrity. John Wiley & Sons, Hoboken, NJ, USA. Geophysical Monograph 238.

Dean, M., Blackford, J., Connelly, D., Hines, R., 2020. Insights and guidance for offshore CO_2 storage monitoring based on the QICS, ETI MMV, and STEMM-CCS projects. Int. J. Greenh. Gas Control 100, 103120.

Duong, C., Bower, C., Hume, K., Rock, L., Tessarolo, S., 2019. Quest carbon capture and storage offset project: findings and learnings from 1st reporting period. Int. J. Greenh. Gas Control 89, 65–75.

Frailey, S.M., Tucker, O., Koperna, G.J., 2017. The genesis of the CO_2 storage resources management system (SRMS). Energy Proc. 114, 4262–4269.

Gerdes, K.F., 2003. Carbon Dioxide Capture for Storage in Deep Geologic Formations – Results from the CO_2 Capture Project. In: Gerdes, K.F. (Ed.), CCS Technology Development and Demonstration Results (2015-2022), 5. CO_2 Capture Project, UK.

IEA, G.H.G., 2009. A Review of the International State of the Art in Risk Assessment Guidelines and Proposed Terminology for Use in CO_2 Geological Storage. IEA Greenhouse Gas R&D Programme, Cheltenham, UK.

Heidug, W., 2013. Methods to Assess Geologic CO2 Storage Capacity, IEA Workshop Report. Available at: www.iea.org/publications/freepublications.

Mortezaei, K., Amirlatifi, A., Ghazanfari, E., Vahedifard, F., 2018. Potential CO_2 Leakage from Geological Storage Sites: Advances and Challenges. Environmental Geotechnics, pp. 1–25.

Preston, C., Monea, M., Jazrawi, W., et al., 2005. IEA GHG Weyburn CO_2 monitoring and storage project. Fuel Process. Technol. 86, 1547–1568.

Shell UK, 2015. Peterhead CCS Project, Storage Development Plan. Available at: assets.publishing.service.gov.uk/government/uploads/system/uploads/attachment_data/file/531016/DECC_Ready_-_KKD_11.128_Storage_Development_Plan.pdf.

UK Carbon Capture and Storage knowledge sharing site: Knowledge collected from UK CCS projects. www.gov.uk/government/collections/carbon-capture-and-storage-knowledge-sharing.

US DOE National Energy Technology Laboratory: reportBest Practices for Public Outreach and Education for Carbon Storage Projects, Report DOE/NETL-2017/1845. www.netl.doe.gov/node/5828.

Projects and organizations

CarbonSAFE; US DoE funded project aiming to mature industrial scale geological storage sites: www.netl.doe.gov/coal/carbon-storage/storage-infrastructure/carbonsafe (see also Solid Carbon.).

GECO project; European project, building on the CarbFix project, to advance zero-emissions geothermal energy: www.carbfix.com/geco and geco-h2020.eu.

Global CCS Institute; CCS project database: co2re.co/FacilityData.

Porthos project; CCS transport and storage infrastructure project led by the Port Authority of Rotterdam: www.porthosco2.nl.

Shell Quest Knowledge sharing; https://open.alberta.ca/dataset?tags=Quest+CCS+project.

SINTEF CO_2 storage data consortium; sharing data from CO_2 storage projects: www.sintef.no/en/projects/co2-storage-data-consortium-sharing-data-from-co2-storage-projects/.

Solid Carbon; Follow-up of the CarbonSAFE project, working to develop and demonstrate in-situ mineral carbonation: www.solidcarbon.ca.

STEM-CCS; Strategies for Environmental Monitoring of Marine CCS, Including Controlled Seabed Release Experiments: stemm-ccs.Eu.

Zero Carbon Humber; Project led by DRAX, Equinor, and others to develop a zero-carbon industrial hub in NE England: www.zerocarbonhumber.co.uk/.

Chapter 16

Ocean storage and ocean CDR methods

Steve Rackley and Michael Tyka

16.1 Introduction

The ocean contains an estimated 39,000 GtC (143,000 $GtCO_2$) of dissolved inorganic carbon (DIC) and is playing a critical role in mitigating climate change, having taken up around one-quarter (~680 from ~2570 $GtCO_2$) of anthropogenic CO_2 emissions (including E_{LUC}) during the industrial era (Khatiwala et al., 2013; Friedlingstein et al., 2022). The scale of anthropogenic CO_2 uptake into the oceans has been increasing with the rise in CO_2 partial pressure over the past 200 years and averaged ~2.9 GtC/yr (10.6 $GtCO_2$/yr) in the decade to 2021. (Friedlingstein et al., 2022).

On a 100 kyr timescale, limited by the exposed surface area and slow kinetics of silicate rock dissolution (Archer, 2005), the ocean will be the long-term sink of all anthropogenic CO_2 emissions, as a result of processes such as the biological and solubility pumps (see Chapter 6), reaction with carbonate sediments, and the surface run-off of alkalinity from rock weathering (Lord et al., 2016; Penman et al., 2020).

Given its immense inventory and role as a natural sink, the ocean is a prime candidate for the storage of captured CO_2 and for the enhancement of natural processes to accelerate drawdown from the atmosphere (Caldeira et al., 2005; Rau, 2014). This chapter describes the many options that have been proposed, including direct storage of captured CO_2 as a liquid or hydrate, enhancement of the biological pump, and storage in the form of enhanced ocean alkalinity.

16.2 Ocean storage of captured CO_2

Approaches to increase the ocean's carbon inventory by storing CO_2 captured from industrial sources, or from BECCS or DAC projects, are summarized in Table 16.1.

TABLE 16.1 Options for ocean storage of captured CO_2.

Physical concept	Technological scheme
Accelerated weathering of limestone (AWL)	Flue-gas captured by accelerated weathering of limestone with ocean discharge of alkaline product stream
Direct CO_2 injection	Rising plume
	Neutral buoyancy (isopycnal) spreading plume
	Sinking plume
Liquid or hydrate	Piped feed to seabed lake
CO_2 isolation	Sinking liquid plus CO_2 hydrate slurry
	Constrained or contained liquid or hydrate accumulation
	Injection into ocean-floor sediments

Negative Emissions Technologies for Climate Change Mitigation. https://doi.org/10.1016/B978-0-12-819663-2.00003-4

16.2.1 Accelerated weathering of limestone

Carbonate rock weathering is the result of the reaction with carbonic acid, produced by CO_2 dissolution in rainwater, according to the reaction:

$$CO_2 + CaCO_3 + H_2O \rightarrow Ca^{2+} + 2HCO_3^- \quad (16.1)$$

and results in the addition of ~2 Gt/yr of dissolved calcium bicarbonate to the ocean, equivalent to ~0.3 GtC/yr or ~1 $GtCO_2$/yr. The kinetics of this process can be enhanced by increasing the partial pressure of CO_2 in the solution, thereby reducing the pH. Dissolving carbonate minerals in seawater equilibrated with CO_2-rich flue gases or direct air-captured CO_2 was first proposed by Rau and Caldera (1999) (see also Caldera and Rau (2000)). The process is shown in Fig. 16.1, including the two-step reactor design of Chou et al. (2015).

According to Reaction 16.1, sequestering 1 tCO_2 would require 2.3 $tCaCO_3$ plus 0.3 tH_2O, and would produce 2.8 $tHCO_3^-$. However, since the reaction takes place only on the wetted reactive mineral surface, a significantly greater amount of mineral would need to be present in the reactor, depending on the size of the crushed or ground particles. Rau and Caldera estimated that a 45 m^3 reactor containing 100 t of mineral crushed to a 1 cm radius would be required to capture 1 tCO_2/d. A significant challenge for this process is the water requirement, since CO_2 capture at 6 mmol/kg would require 3.8 kt-seawater to be pumped per tCO_2 sequestered (Rau, 2011; Renforth and Henderson, 2017). Coastal power plants typically use ~2 kt-seawater/tCO_2 emitted for cooling, so that around 57% of emissions could be captured using this water alone (Rau, 2011, quoting California power plant performance figures). The initial product stream also contains a high concentration of dissolved CO_2 (in equilibrium with the input pCO_2) and would have to be degassed and diluted with 10–100 parts of seawater before release, to prevent outgassing and calcite reprecipitation (Rau et al., 2007; Chou et al., 2015). Other potential issues include reduced oxygen content in the discharge stream and the potential for discharge of impurities from the flue gas or from the mineral input.

Pulverized limestone wastes produced by the cement and stone industries could provide a zero or negative-cost mineral supply if suitably located, and Rau et al. (2007) estimated that, if limestone transportation costs to coastal power plants can be minimized, AWL could achieve sequestration costs below \$30/tCO_2.

At TRL4—bench scale system validation—AWL is the most advanced approach to ocean CO_2 storage. Laboratory validation (Rau et al., 2007; Rau, 2011) has been followed by a feasibility case study for the Hoping coal-fired power station in northeastern Taiwan, including reactor design optimization and assessment of the impact on seawater chemistry in the discharge area (Chou et al., 2015). More recently, a pilot-scale AWL reactor was reported, processing a 200 m^3/h flue gas slipstream from a coal-fired power plant at Wilhelmshaven, Germany (Kirchner et al., 2020a). AWL can be a low-cost carbon capture option for coastal power plants and other emitters, although keeping alkalinity and mineral impurity perturbations within regulatory limits may pose a challenge, depending on the hydrography of the discharge location (Kirchner et al., 2020b).

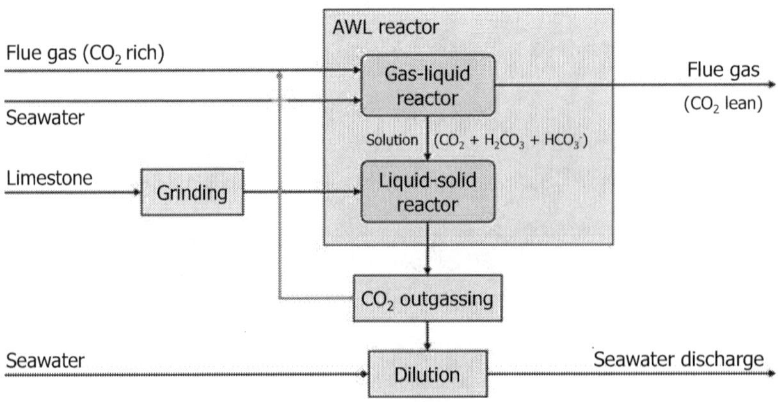

FIGURE 16.1 Schematic AWL process.

An alternative AWL approach, proposed by Righi (2020) and denoted enhanced pressurized weathering of limestone (EPWL), uses a pipeline to a deep sea discharge point as the reaction vessel, thereby benefiting from the increased $CaCO_3$ solubility at high pressure and reducing the water requirement by a factor of ~15.

16.2.2 Direct CO_2 injection

The behavior of CO_2 released directly into seawater depends primarily on the pressure (*P*) and temperature (*T*) of the water into which it is released. The important properties are the following (Chow, 2014):

- The solubility of CO_2 and its dependence on (*P*, *T*) (Fig. 15.9)
- The liquefaction pressure
- The variation of CO_2 liquid density with *P*, which determines its buoyancy in seawater
- The (*P*, *T*) of CO_2 hydrate formation.

At temperatures of ~10°C, CO_2 liquefies at 4–5 MPa, corresponding to water depths of 400–500 m, with a liquid density of 860 kg/m^3 at 10°C (Fig. 16.2). At these depths, liquid CO_2 will therefore be positively buoyant, and free liquid droplets will rise and evaporate into gas bubbles as pressure drops below the saturation pressure.

Fig. 16.3 shows the densities of CO_2 and seawater versus depth for a range of conditions. Liquid CO_2 is more compressible than water and becomes neutrally buoyant at depths of 2500–3000 m. Below this depth range, released liquid droplets will sink to the ocean floor.

The potential for ocean waters to carry a higher DIC load is illustrated in Fig. 16.4, which shows the approximate DIC profile for mid-Pacific and mid-Atlantic Ocean locations. An increase in DIC of 50 μmolC/kg, ~2% of current levels, for the body of water from 1000 to 3000 m depth, represents an increase in the carbon inventory of >80 GtC, equivalent to ~7 years of global anthropogenic CO_2 emissions at recent levels.

Direct dissolution below the mixed layer (ML) is one option to make use of this storage capacity. Liquid CO_2 release at depths >500 m would be accompanied by the immediate formation of a hydrate shell around individual liquid droplets (Linga et al., 2007).

Hydrate formation and decomposition

Gas hydrates, also known as clathrate hydrates or clathrates, were discovered by Humphry Davy in 1811 and are a crystalline phase of water in which a guest molecule is trapped within a cage of hydrogen-bonded water molecules. In the case of CO_2 hydrates, the cage of water molecules takes the form of a 12- or 14-faced polyhedron as illustrated in Fig. 16.5.

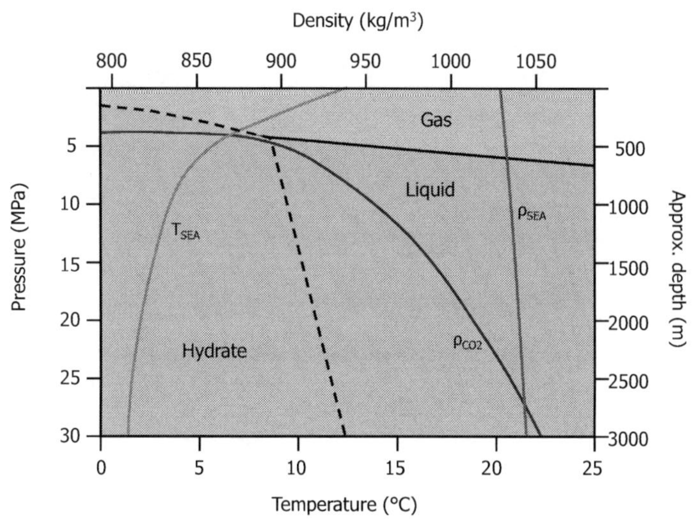

FIGURE 16.2 CO_2 phase behavior in seawater.

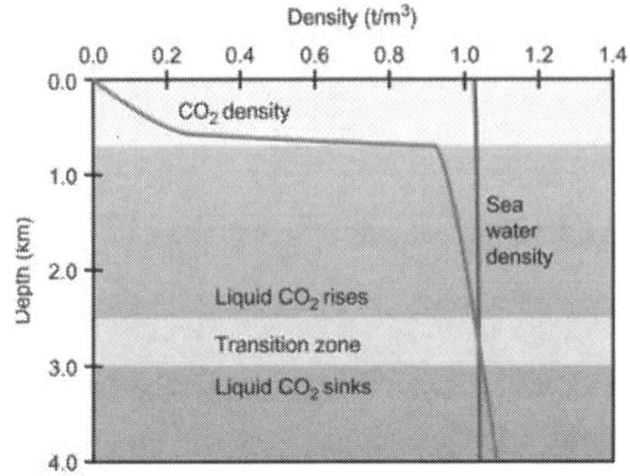

FIGURE 16.3 CO_2 and seawater density versus depth.

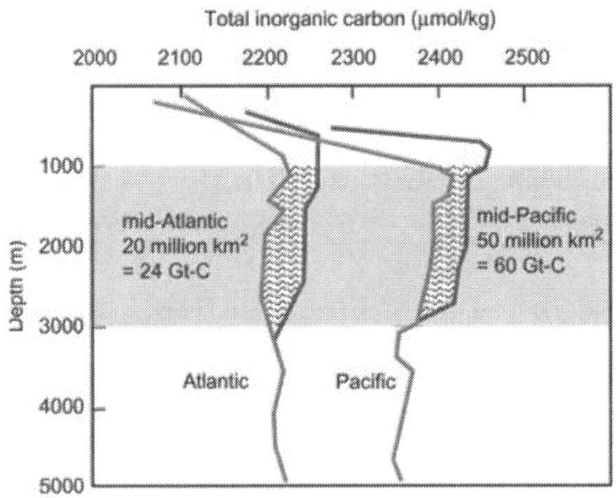

FIGURE 16.4 Ocean storage capacity for dissolved CO_2. *After Ohsumi (2003).*

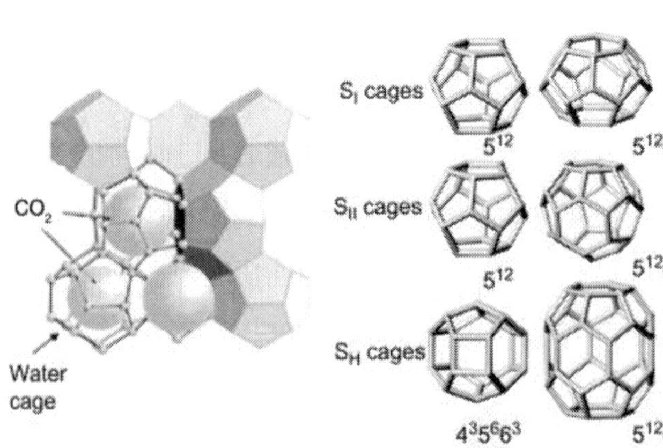

FIGURE 16.5 CO_2 hydrate cage structures. *Graphic elements from Yang et al. (2011) and Ma et al. (2016).*

Hydrates are formed under low-temperature, high-pressure conditions, in an exothermic "enclathration" process, that has an enthalpy of formation of ~65 kJ/molCO_2 for CO_2 hydrate. CO_2 hydrates can form in seawater deeper than ~400 m, as long as there is sufficient CO_2 present, and as shallow as 150 m in polar regions where the water temperature is close to 0°C (Fig. 16.2). A CO_2 gas bubble or liquid drop at hydrate-stable P and T will rapidly form a crystalline hydrate skin, which will slow but not halt CO_2 dissolution. Naturally occurring hydrates have been observed forming around CO_2-rich gas bubbles venting from hydrothermal vents at ~1500 m depth in the Okinawa Trough.

The density of solid CO_2 hydrate is 1110 kg/m^3, and it will therefore sink in seawater, which has a density of ~1030 kg/m^3. However, the density of a hydrate mass will depend on the gas or liquid density, as well as the skin thickness and droplet size, which determine the particle's gas:liquid:hydrate ratio. Depending on this "average" particle density, the neutral buoyancy depth in seawater may be between 1000 and 3000 m.

Hydrate particles can dissolve at hydrate-stable conditions if the surrounding water is not CO_2 saturated and can also decompose by dissociation if the particle moves outside the hydrate-stable region.

Field trials of CO_2 dissolution and hydrate formation

Laboratory and in situ experiments, mathematical modeling studies, and small-scale field trials have been conducted to investigate the properties, behavior, and ecological impact of CO_2 released into seawater, generally as liquid droplets, larger liquid masses, or liquid–hydrate–water composites.

Experiments conducted at the Monterey Bay Aquarium Research Institute (MBARI), typically involving the release and observation of liquid droplets by remotely operated vehicles (ROVs), have yielded insights into the dissolution rate of individual hydrate-enclosed droplets and their rising or sinking rate, depending on release depth (Brewer et al., 2002; Aya et al., 2004). Experimental releases of larger liquid CO_2 volumes have also been performed to investigate the use of sonar as a tracking method (Brewer et al., 2006). Hydrate-enclosed CO_2 droplets were found to be easily detectable using high-frequency sonar (38 and 675 kHz) as a result of the high acoustic velocity difference between liquid CO_2 and seawater.

Other small-scale release experiments have investigated the behavior of CO_2 liquid–hydrate–seawater composite material formed by the injection and vigorous mixing of seawater into a liquid CO_2 stream (Tsouris et al., 2007). Unlike hydrate-enclosed liquid CO_2 droplets, which become negatively buoyant at ~2500 m, this composite material contains a higher proportion of hydrate and becomes negatively buoyant at ~1000 m.

These small-scale experiments have given insight into the fate of individual droplets and small hydrate masses but are unable to address larger-scale effects that would determine the dynamic behavior of a large plume. Mathematical simulations indicate that if a large mass of hydrate pellets was released, the resulting negatively buoyant stream would sink significantly faster and further than individual pellets, as a result of the entrainment of water into a plume and the increasing density of entrained water as CO_2 dissolves into the plume. A plume containing a mass flux of 100 kgCO_2/s was predicted to descend ~500 m before the pellets were fully dissolved, compared to 10–70 m for individual pellets. Production of hydrate particles with different sizes and densities would therefore assist in vertical spreading of an injected plume, as a result of differing settling velocities (Chow et al., 2009).

Ecological impact of ocean storage of captured CO_2

The drawdown of anthropogenic CO_2 from the atmosphere has resulted in a surface ocean pH reduction of ~0.1, and this acidification will continue while atmospheric [CO_2] continues to rise, with increasingly severe consequences for calcifying organisms (Raven et al., 2005). The capture and direct injection of CO_2 into the deep ocean could avoid the adverse impact of further acidification on shallow ecosystems, by transferring the impact to the deep ocean. The local impact of direct CO_2 injection on marine ecosystems has been studied experimentally in CO_2 enrichment experiments—discussed in the next section—while the global impact has been the subject of simulation studies.

Reith and colleagues considered a base case in which unmitigated emissions follow the RCP4.5 pathway and a scenario in which CO_2 capture and direct injection into the deep ocean reduced emissions to a level consistent with maximum warming of 1.5°C (Reith et al., 2019). This required the injection of 1430 GtCO_2 by the year 2100 and ~5700 GtCO_2 by the year 3020, at the end of the simulations, the 3020 figure including ~2200 GtCO_2 of leakage that had to be recaptured and injected to achieve the warming target in the long-term. Compared to the unmitigated base case, the pH of the shallow oceans (to 130 m depth) was increased by 0.13–0.18, and the areal extent of potential coral reef habitats was significantly increased. This upper ocean improvement came at the cost of significant acidification in the intermediate and deep ocean, with pH dropping by as much as 2.37 around injection locations, with severe implications for ocean floor ecosystems. Other simulation studies have assessed the local ecological impact of direct injection from a moving ship, either as sinking hydrate particles or as buoyant liquid droplets (Israelsson et al., 2010). Results indicate that sufficient dilution can be achieved to avoid acute, local impacts, and the authors concluded that direct injection cannot be immediately rejected on these grounds.

Free ocean CO_2 enrichment experiments

By analogy with the free air CO_2 enrichment (FACE) experiments performed to test the effects of elevated [CO_2] in terrestrial ecosystems (Chapter 11), free ocean CO_2 enrichment (FOCE) systems have been constructed, most notably at MBARI, to study the effects on different marine ecosystems of increased [CO_2] in seawater—from the deep-sea floor to near-shore sand flats and seagrass meadows (Kline et al., 2012; Kirkwood et al., 2014; Srednick et al., 2020).

In these systems, seafloor currents carry a parcel of CO_2-enriched seawater through a series of baffles within a rectangular flume. The CO_2-enriched parcel mixes and reacts with the surrounding seawater and the system is actively controlled to achieve a consistent elevated [CO_2] at seafloor level in a test area of $\sim 1\ m^2$ in the middle of the flume (Fig. 16.6).

The FOCE system is typically operated to achieve a pH offset of −0.1 to −0.5 from ambient conditions and is designed to be left in place for weeks to months to investigate the impact of reduced pH on such factors as foraging success and feeding efficiency in biological communities including crustaceans, fishes, and mollusks (Barry et al., 2004, 2014), and the survivability of microorganisms such as foraminifera (Bernhard et al., 2009). Further developments of the system included oxygen and temperature controls within the test area, to enable a wider variety of in situ experiments into the impact of climate change on these ecosystems (Gattuso et al., 2014).

Sequestration effectiveness of direct CO_2 injection

A number of studies have looked at the effectiveness of direct CO_2 injection as a climate change mitigation measure, using Earth system models to assess the impact of feedbacks from changing climate and the response of terrestrial carbon stocks and sinks. Reith et al. (2016) observed that feedbacks such as rapid outgassing of injected CO_2, and reduced ocean-surface and varying terrestrial CO_2 uptake reduced the sequestration effectiveness so that for every 1 tC injected, only around 0.84 tC was retained at the end of a 100-year injection period, for injection at depths of 1500−3000 m.

Long-term sequestration effectiveness was also found to be dependent on injection site location, as a result of interaction with global ocean circulation. Multicentury timescales are possible in many locations, and up to 1000 years in the deep North Pacific (Ridgwell et al., 2011; Siegel et al., 2021), although weakening of the Atlantic meridional overturning circulation (AMOC) could result in North Atlantic sites becoming more effective (Jain and Cao, 2005).

Ridgwell and colleagues also observed that ocean sequestration became more effective for high future emissions scenarios, or in the case of a high equilibrium climate sensitivity (ECS; see Glossary and Chapter 1), as a result of a greater

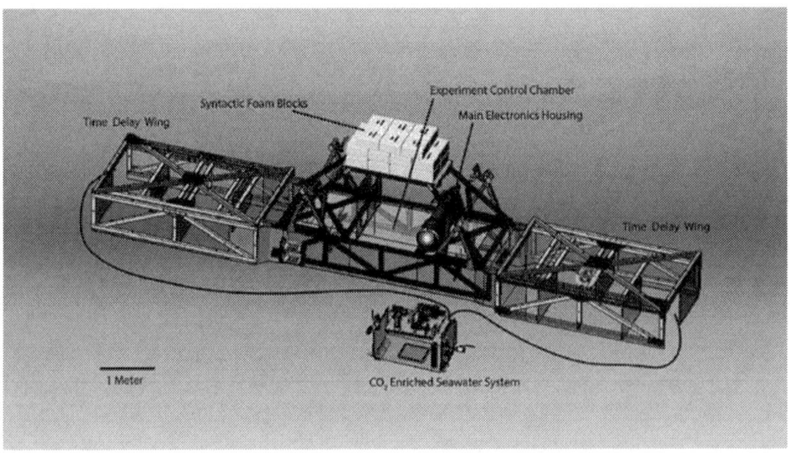

FIGURE 16.6 MBARI FOCE flume configuration. *From Kirkwood et al. (2014), with permission.*

fraction of emitted CO_2 being retained in the atmosphere. This occurs under higher emissions scenarios due to the more rapid depletion of the ocean's carbonate buffer, and under high climate sensitivity due to a higher mean surface temperature leading to reduced CO_2 solubility and increased soil respiration.

16.2.3 Ocean floor liquid CO_2 injection

The negative buoyancy of liquid CO_2 below ~2500 m, and of CO_2 hydrates below ~1000 m, opens up the possibility of sequestering CO_2 as a gravity-stable deposit on the ocean floor—either as a liquid lake in an ocean-floor depression or deep trough, as a contained liquid or hydrate accumulation, or below ocean-floor sediments.

Ocean-floor CO_2 lakes

An ocean-floor lake could be formed either by releasing liquid CO_2 directly into an ocean-floor depression or by releasing negative-buoyancy droplets or hydrated particles close to the bottom. Deep ocean trenches, such as the Indonesian Sunda trench and the Puerto Rico trench, have been proposed as suitable locations (Goldthorpe, 2017), and these two trenches alone would have a capacity to store in excess of 40,000 $GtCO_2$ at depths greater than 6–7 km. Contained deep sea basins, such as occur in the Mediterranean and Arabian Seas, could also be suitable, with potential capacities in the 100 $GtCO_2$ range.

Naturally occurring ocean-floor CO_2 lakes have been observed in the vicinity of hydrothermal vents, as a result of the liquefaction and pooling of CO_2-rich gases released from these vents. These naturally occurring CO_2 lakes are typically held in place by overlying sediments and by hydrate crusts or cement in these sediments.

Small-scale ocean-floor experiments have been performed to assess the impact of liquid CO_2 pools on deep-sea ecosystems. Small pools were created by releasing ~20 L of liquid into open-topped cylindrical "corrals" on the sea bed at 3600 m depth. At this depth, the excess liquid density keeps the mass constrained within the corral, but slow dissolution results in a dense CO_2-rich, low-pH plume being swept downcurrent across the ocean floor. The results of these experiments show that significant reductions in pH will occur in the close vicinity of an ocean-floor CO_2 lake, resulting in high mortality rates in the affected marine ecosystem (Barry et al., 2004), and attempts to perform mesoscale trials have consequently been met with considerable environmental opposition.

Numerical studies have also been performed to assess the dissolution rate from a CO_2 lake, both under static conditions and under the influence of ocean-floor currents. These studies show that vertical mixing above such a lake is reduced as a result of the formation of a high-density boundary layer. In an environment that is not disturbed by ocean-bottom currents, the dissolution rate from the lake would be reduced as a result of the surface hydrate layer, as well as by the stratification resulting from denser, CO_2-rich water filling the depression above the lake surface. A lifetime in excess of 10,000 years would be possible for a 50 m deep lake with a dissolution rate of <5 mm/year (Ohsumi, 1995), although these residence times would be reduced 25–500 fold in the presence of moderate or stronger ocean-floor currents (Fer and Haugen, 2003).

Investigations around hydrothermal vents at ~1400 m depth in the Okinawa Trough, located off Taiwan in the East China Sea, have identified flourishing benthic microbial communities in the 20–40 cm sediment layer overlying lakes of CO_2. Genetic and biochemical analyses show that the dominant organisms in these communities are well adapted to the conditions and are assimilating CO_2 from the pools (Inagaki et al., 2006). Injection of CO_2 into seabed sediments (discussed below) may therefore have less ecological impact around hydrothermal vents, in view of this preexisting adaptation.

Contained ocean-floor storage

Adverse ecological impacts could also be avoided by isolating the liquid or hydrate, for example within a synthetic geomembrane (Palmer et al., 2007). One envisaged containment system is a sausage-like envelope, 500 m in circumference and perhaps 10 km in length, each of which could contain 16 $MtCO_2$. Membrane material would need to be resistant to the chemical effects of CO_2 and have low permeance to prevent osmosis, but the requirement for mechanical strength would not be great if deployed close to the neutral buoyancy point, in view of the small density difference. Polymer geomembranes have been used in dam, reservoir, and land-fill construction since the 1960s, with 1000-year lifetimes projected. A similar containment system has also been proposed for storing CO_2 in the form of hydrates (Capron et al., 2013). Although this would require additional volume to be stored, it would have the advantage that leakage would likely be much slower in the event of a membrane rupture, since both heat and unsaturated water are required for hydrate dissolution.

Casserini et al. (2017) have proposed containment within glass-ceramic capsules, nominally 0.3 m diameter × 0.6 m in length, each containing 30 L of liquid CO_2 and suitably weighted with sand to control the rate of descent to the sea floor. End-to-end LCA, including raw material extraction, capsule manufacture, filling, and transportation indicated an emissions impact of 0.1 tCO_2e/tCO_2 stored, with an average levelized cost of $\$17/tCO_2$ stored. Potential storage areas in the Mediterranean (Tyrrhenian) and Black Seas, at depths from 1500 to 3000 m and <200 km from shore, were assessed as having storage potentials of 14 and 36 $GtCO_2$, respectively.

16.2.4 Storage in ocean-floor sediments

If CO_2 was injected below the ocean floor, the temperature rise, in line with the local geothermal gradient, would eventually result in hydrate dissociation and a reduction in density to the point where, even if below the open water neutral buoyancy depth, liquid CO_2 would become buoyant in the pore fluid. Depending on the water depth and temperature gradient, the hydrate formation zone (HFZ), and also the negative buoyancy zone (NBZ), can extend several hundred meters into ocean-floor sediments, as shown in Fig. 16.7. The NBZ, from ~2500 m down to its base in ocean-floor sediments, provides a physical (gravitational) barrier that would prevent upward movement of injected CO_2, opening up the possibility of storing CO_2 in a gravitationally and thermodynamically stable state below unconsolidated ocean-floor sediments.

Storage under seabed sediments was first proposed by Koide et al. (1997), and the concept was later modified (House et al., 2006) with a proposal to inject liquid CO_2 below the HFZ at a location where this was deeper than the NBZ (see Fig. 16.8). Injected liquid CO_2 would rise slowly due to its low buoyancy and high viscosity (Teng and Zhang, 2018), and start to form hydrates at the bottom of the HFZ, which would inhibit further upward movement and cause lateral spreading (Fig. 16.8A). Any CO_2 that did penetrate this hydrate cap would eventually reach the base of the NBZ at which point it would cease to be buoyant, while dissolution and advection would carry CO_2 in the aqueous phase away from the HFZ (Fig. 16.8B). Ultimately, on a timescale of up to 10^6 years, all of the CO_2 would dissolve, the aqueous phase mixing with the ambient fluid until it became neutrally buoyant (Fig. 16.8C).

House and colleagues estimated storage capacity within the 200 mile US EEZ to exceed 10,000 $GtCO_2$ while subsequent analysis suggests a global capacity of between 1000 and 28,000 $GtCO_2$. The main uncertainty in this approach is whether these ocean-floor sediments, which are typically carbonate or clay rich, have sufficient intrinsic permeability to permit injection at the required rates, and the extent to which hydraulic fracturing could be applied without risking the release of (albeit negatively buoyant) liquid CO_2 to the ocean floor. These issues have been investigated by Dai et al. (2017) in modeling studies that establish thresholds for sediment thickness, porosity, and permeability to achieve necessary injection rates for sediments in the Gulf of Mexico setting. The modeling work also showed that highly heterogeneous sediments could effectively trap injected CO_2 at around 1200 m, less than half the neutral buoyancy depth, significantly expanding the storage capacity of seabed sediments (Dai et al., 2018). This conclusion is borne out by the earlier observation of CO_2 lakes adjacent to hydrothermal vents at ~1400 m depth in the Okinawa Trough, which were contained by as little as 20–40 cm of sediments (Inagaki et al., 2006).

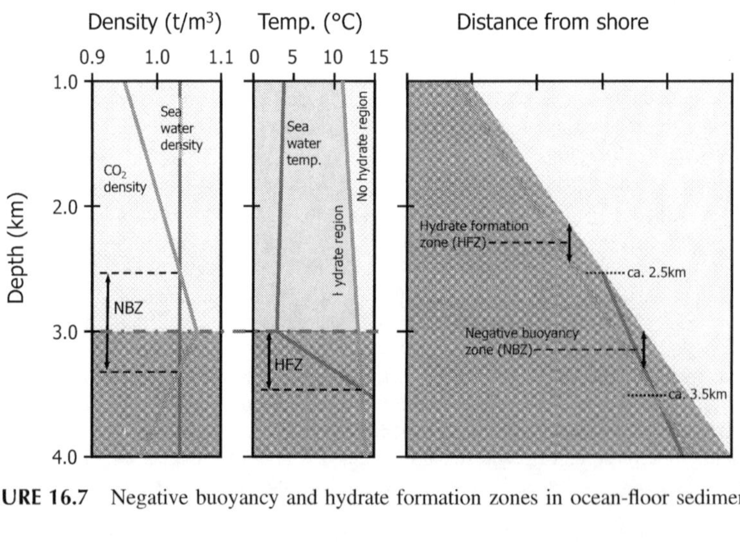

FIGURE 16.7 Negative buoyancy and hydrate formation zones in ocean-floor sediments.

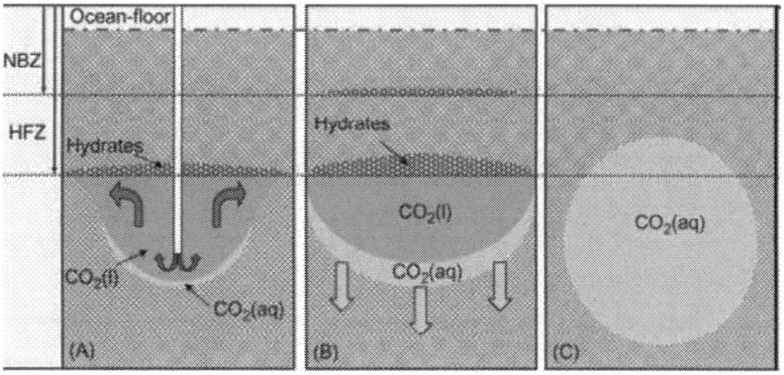

FIGURE 16.8 Evolution of CO_2 injected into deep ocean-floor sediments. *After House et al. (2006).*

16.2.5 Mesoscale direct ocean storage field trials

The only mesoscale ocean storage field trial to reach an advanced planning stage was the CO_2 Ocean Sequestration Field Experiment, which was proposed under the International Collaboration on CO_2 Ocean Sequestration signed by Japan, Norway, and the United States during the 1997 UNFCCC COP3 in Kyoto, Japan.

The first series of field experiments designed under this agreement aimed to:

- Investigate CO_2 droplet plume dynamics
- Clarify the effects of hydrates on the dissolution of CO_2 droplets
- Trace the evolution of the CO_2-enriched seawater by performing 3D mapping of velocity, pH, and DIC
- Assess the potential biological impact of changes in seawater pH by quantifying changes in bacterial biomass, production, and growth efficiency both in the water column and on the ocean floor.

Acute and chronic environmental impacts and subsequent recovery were the focus of further experiments planned under the agreement, which were to be conducted over a time scale of at least a year, comparable to the lifetimes of affected organisms.

The final project design envisaged the release of 20–40 tCO_2 over a period of 1–2 weeks using a 4 cm diameter coiled steel tubing deployed from a ship down to the injection depth at 800 m. The buoyant plume of CO_2 droplets, injected from a diffuser assembly on the ocean floor at a rate of ~1 kg/s, was expected to rise ~100 m before dissolving.

The experiment was initially planned to be hosted by the Natural Energy Laboratory of Hawaii Authority (NELHA) and to take place in the summer of 2001 in the ocean research corridor operated by NELHA at Keahole Point, Kona, on the west coast of Hawaii (Herzog et al., 2000). However, as a result of strong public opposition the project was unable to secure the required permits and had to abandon the Hawaii experiment. Subsequently, a scaled-down version of the experiment was planned in 2002, to release and monitor 5.4 tCO_2 off the coast of Norway (Drange et al., 2001), but failed to secure approval from the Norwegian environment ministry. As Keeling has noted (2009), it is ironic that these experiments were stopped on environmental grounds, given that current fossil fuel CO_2 emissions result in the disposal of ~2.8 $GtCO_2$/yr to the ocean.

16.3 Marine CDR—enhancing ocean alkalinity

In contrast to the various ocean storage approaches described above, marine CDR (mCDR) methods seek to enhance the ocean's ability to drawdown CO_2 from the atmosphere, by exploiting the air:solvent contactor that covers over 70% of Earth's surface. The chemical, electrochemical, and biological approaches to mCDR are described in the following sections.

The natural weathering of silicate and carbonate minerals currently accounts for the removal of ~0.5 $GtCO_2$/yr from the atmosphere, resulting in an increase in the alkalinity of the ocean, while countering the acidification that results from the formation of carbonic acid, and chemical mCDR methods aim to enhance this process. Storage of all anthropogenic carbon emissions in this form would increase the ocean's DIC inventory by ~0.025% per annum, at recent emissions levels (~10 GtC/yr).

The possibility of increasing CO_2 storage through increased ocean alkalinity was first proposed by Kheshgi (1995) and the various techniques to achieve this, collectively referred to as ocean alkalinity enhancement (OAE), are summarized in Table 16.2.

16.3.1 Physical and chemical fundamentals of OAE

Seawater alkalinity

Alkalinity is a measure of the excess of proton acceptors—bases and weak acids that can react with and neutralize stronger acids by accepting H^+ ions—versus H^+ ions plus strong acids that can donate protons in neutralizing reactions. The

TABLE 16.2 A taxonomy of OAE methods.

Category	Method	Mineral input	Mineral treatment	References
(1) Addition of reactive bases	(1.1) Limestone dissolution in upwelling regions (LDURs)	Limestone $CaCO_3$	Grinding to rock powder	Harvey (2008)
	(1.2) Ocean liming (OL)	Soda ash (Na_2CO_3)	Grinding	Kheshgi (1995)
(2) Addition of reactive bases (OH^-)	(2.1) Ocean liming	Limestone $CaCO_3$/Lime CaO	Grinding, calcination with CCS	Kheshgi (1995)
		Brucite ($Mg(OH)_2$) or brine	Electrolysis of desalination plant reject brine; grinding of brucite precipitate before dispersal	Davies et al. (2018), Tan et al. (2019)
	(2.2) Mineral carbonation—ocean liming (MC-OL)	Silicates, commonly Olivine $(Mg,Fe)_2SiO_4$	Grinding, carbonation, calcination with CCS, regrinding of MgO calcination product to ~10 μm	Renforth et al. (2013), Renforth and Kruger (2013)
	(2.3) Open ocean enhanced weathering (OOEW)	Silicates, commonly Olivine $(Mg,Fe)_2SiO_4$	Grinding to rock powder	Schuiling and De Boer (2011), Hartmann et al. (2013)
	(2.4) Electrochemical enhanced weathering (EEW)	Enstatite $MgSiO_3$; other silicates, commonly Olivine $(Mg,Fe)_2SiO_4$	Grinding, reaction with HCl produced, along with NaOH, by brine or seawater electrolysis	House et al. (2007)
	(2.5) Electrochemical splitting (ES)	Limestone $CaCO_3$	Grinding, splitting using anolyte acidity in electrochemical cell and reaction to produce $Ca(OH)_2$	Rau et al. (2018)
		Silicates, commonly Enstatite $MgSiO_3$ or Wollastonite $CaSiO_3$	Grinding, electrochemical cell anolyte acidity splitting, reaction to produce $Ca(OH)_2$ or $Mg(OH)_2$	Rau et al. (2013)
(3) Seawater treatment to increase alkalinity	(3.1) Indirect ocean capture (IOC)	None	Seawater acidification by electrodialysis, CO_2 stripping with CCS	de Lannoy et al. (2018), Eisaman et al. (2012), Willauer et al. (2017)
	(3.2) Hybrid seawater desalination-carbon capture (SD-CC)	None	NaOH production during seawater battery discharge, carbonation of resulting hydroxide	Bae et al. (2019)

principle proton acceptors in seawater are bicarbonate and carbonate ions, and borate ($B(OH)_4$). A simple expression for total alkalinity, excluding less abundant species, is given by Zeebe and Wolf-Gladrow (2001):

$$A_T = [HCO_3^-] + 2[CO_3^{2-}] + [B(OH)_4^-] + [OH^-] - [H^+] \quad (16.2)$$

For seawater at pH ≈ 8.2, HCO_3^- contributes ~71%, CO_3^{2-} ~28%, and other species less than 1% to A_T. When seawater alkalinity is increased, the equilibrium distribution of DIC between dissolved CO_2, bicarbonate, and carbonate ions (Fig. 6.3) shifts toward carbonate as a result of the reactions:

$$CO_2 + OH^- \rightarrow HCO_3^- \quad (16.3)$$

$$HCO_3^- + OH^- \rightarrow CO_3^{2-} + H_2O \quad (16.4)$$

By converting dissolved CO_2 to bicarbonate (Reaction 16.3), alkalinity addition enables further CO_2 uptake (to restore partial pressure equilibrium), although some of the added alkalinity will be consumed in converting bicarbonate to carbonate ions (Reaction 16.4). The actual uptake efficiency (η) will therefore be less than 1 $molCO_2/molA_T$ and is given by the following:

$$\eta_{CO_2} = \delta DIC/\delta A_T \quad (16.5)$$

The value of η_{CO_2}, which varies in the range of 0.6–0.85, depends on the location-specific DIC, A_T, atmospheric pCO_2, salinity, and other seawater parameters, and can be determined by solving the full carbonate system (Burt et al., 2021; Tyka et al., 2022).

Storing carbon as ocean bicarbonate is ~60% more efficient than storage as solid carbonate minerals. For example, the mineralization reaction of $Ca(OH)_2$ with CO_2 stores 1 molC/molCa:

$$Ca(OH)_2 + CO_2 \rightarrow CaCO_3 + H_2O \quad (16.6)$$

whereas the dissolution of 1 $molCa(OH)_2$ raises ocean alkalinity by two moles, with a subsequent uptake of $2\eta_{CO_2} \approx 1.6$ $molCO_2$.

In-depth treatments of the seawater carbonate system can be found in Zeebe and Wolf-Gladrow (2001) and Dickson (2011).

Mineral saturation state

The saturation state of a mineral such as $CaCO_3$ (denoted Ω_{CaCO_3}) reflects the potential for the mineral to either dissolve or precipitate and is given by the following:

$$\Omega_{CaCO_3} = [Ca^{2+}][CO_3^{2-}]/K_{sp} \quad (16.7)$$

where the solubility product K_{sp} is the product of the concentrations of the two ions at equilibrium, when the mineral is neither precipitating nor dissolving.

Under normal conditions, the shallow ocean is supersaturated with respect to calcium carbonate ($\Omega_{calcite}$ ~3–4, $\Omega_{aragonite}$ ~1–2), but $CaCO_3$ does not spontaneously precipitate due to the inhibitory action of magnesium, which is five times more abundant than calcium in seawater (Sun et al., 2015). This inhibition eventually breaks down as rising pH increases $[CO_3^{2-}]$ and precipitation occurs, removing one unit of DIC as well as two units of alkalinity:

$$Ca^{2+} + 2HCO_3^- \rightarrow CaCO_3 + CO_2 + H_2O \quad (16.8)$$

In the deeper ocean, mineral dissolution will occur below the depth where $\Omega = 1$ (the saturation horizon), which generally varies from 0.5–1.0 km for the more soluble aragonite and 0.7–3.0 km for the less soluble calcite.

Ocean liming with $Ca(OH)_2$ has been shown in the lab to induce significant seawater $CaCO_3$ precipitation (Moras et al., 2021), potentially precipitating more alkalinity than was originally added. This occurs even after dilution to moderate excess alkalinity because $CaCO_3$ seeds introduced in the added material can nucleate further precipitation even as low as $\Omega \cong 5$. Even slowly dissolving alkalinity may be problematic due to the presence of nucleation sites, and the CO_2 drawdown potential of olivine dissolution may be reduced by a factor of 5 due to secondary precipitation (Fuhr et al., 2022).

To avoid a pH spike at the addition site, it will be critical for alkalinized seawater to disperse and dilute more quickly than the spontaneous nucleation time of calcium carbonate, and these kinetic considerations will therefore influence alkalinity addition strategies (Renforth and Henderson, 2017, He and Tyka, 2023).

The carbonate buffer

As discussed in Chapter 6, the dissolution of CO_2 in seawater results in the production of carbonic acid, but the acidifying effect is moderated or buffered by the availability of carbonate ions, which convert the excess H^+ to bicarbonate:

$$H_2CO_3 \rightarrow H^+ + HCO_3^- \tag{16.9}$$

$$H^+ + CO_3^{2-} \rightarrow HCO_3^- \tag{16.10}$$

the equilibrium carbonate ion concentration being ultimately restored through the dissolution of carbonate minerals on the ocean floor. The overall result is that the availability of carbonate ions and of carbonate minerals in deep ocean sediments mitigates the acidification that would otherwise result from increasing atmospheric $[CO_2]$. OAE methods enhance this process by providing additional base ions to neutralize carbonic acid (Lackner, 2002).

Air:sea gas exchange

Gas exchange across the air:sea interface is driven by a difference in partial pressures between the surface few millimeters of seawater and the atmosphere, and by a transfer velocity that is strongly dependent on the turbulent state of the sea surface, and therefore on wind speed (Liss and Merlivat, 1985). For CO_2, the resulting gas flux can be expressed as follows (Jacobs et al., 1999):

$$F = K_0\,(pCO_{2AIR} - pCO_{2SEA})\,k_v \tag{16.11}$$

where K_0 is the solubility coefficient (Henry's constant) for CO_2 in seawater (mol/m^3/μatm), F is the flux (mol/m^2/s), k_v (m/s) is the wind-speed-dependent gas transfer velocity, pCO_{2AIR} and pCO_{2SEA} (μatm) are the respective CO_2 partial pressures. Various correlations, based on tracer and eddy covariance measurements, have been published for k_v as a function of wind speed, and for CO_2 a simplified expression due to Wanninkhof (2014) is:

$$F = 7.7 \times 10^{-4} <U_{10}^2> \Delta pCO_2 \tag{16.12}$$

where F has the units of mol/m^2/yr, $<U_{10}^2>$ is the mean value of the square of the wind speed at 10 m height in (m/s)^2, and the CO_2 partial pressure difference (ΔpCO_2) is in μatm. Table 16.3 shows typical values for NW Europe coastal waters, assuming $pCO_{2AIR} = 420$ μatm.

The CO_2 partial pressure equilibration time (τ_{CO_2}) is in the order of weeks to months for most of the ocean surface, but up to a year in some regions, dictated primarily by the ML depth as well as wind speeds (Jones et al., 2014). Addition of quickly dissolving alkalinity would therefore need to be performed in a careful manner so as to avoid excessive ocean-surface pH rise, until CO_2 uptake can neutralize the excess alkalinity (He and Tyka, 2023).

The long equilibration time also poses significant challenges for mCDR MRV, since CO_2-depleted seawater needs to stay in contact with the atmosphere for an extended period, and any direct quantification of CO_2 influx will require measurement over large spatial and temporal scales (Bach et al., 2021).

Alkaline minerals

Many alkaline minerals have been proposed for use in OAE, as summarized in Table 16.4. The CO_2 removal effectiveness of each mineral is determined from the moles of A_T added, less the moles of DIC added, with, in the final column, a factor

TABLE 16.3 Air:sea CO_2 flux for NW Europe coastal conditions.

	Units	Winter	Summer
$<U_{10}^2>$	(m/s)^2	95	40
CO_{2SEA}	μatm	400	345
ΔpCO_2	μatm	20	75
F	molCO_2/m^2/yr	1.46	2.31
	tCO_2/km^2/yr	64	102

TABLE 16.4 Reactions and efficiencies of alkaline minerals for OAE.

Mineral/Compound	Reaction	Mol weight	DIC/mol	Alk/mol	$gCO_2//gMineral$	Eff. $gCO_2/gMineral$[2]
Oxides and hydroxides						
Periclase	$MgO + H_2O \rightarrow Mg^{2+} + 2OH^-$	40.3	0	2	2.18	1.86
QuickLime	$CaO + H_2O \rightarrow Ca^{2+} + 2OH^-$	56.1	0	2	1.57	1.33
Brucite	$Mg(OH)_2 \rightarrow Mg^{2+} + 2OH^-$	58.3	0	2	1.51	1.28
Portlandite	$Ca(OH)_2 \rightarrow Ca^{2+} + 2OH^-$	74.1	0	2	1.19	1.01
Lye	$NaOH \rightarrow Na^+ + OH^-$	40.0	0	1	1.10	0.91
Carbonates						
Magnesite[1]	$MgCO_3 + H_2O \rightarrow Mg^{2+} + CO_{2(g)} + 2OH^-$	84.3	1	2	0.52	0.37
Dolomite[1]	$CaMg(CO_3)_2 + 2H_2O \rightarrow Ca^{2+} + Mg^{2+} + 2CO_{2(g)} + 4OH^-$	184.4	2	4	0.48	0.33
Calcite[1]	$CaCO_3 + H_2O \rightarrow Ca^{2+} + CO_{2(g)} + 2OH^-$	100.1	1	2	0.44	0.31
Soda ash[1]	$Na_2CO_3 \rightarrow 2Na^+ + CO_2 + 2OH^-$	106.0	1	2	0.42	0.29
Nahcolite[1]	$NaHCO_3 \rightarrow Na^+ + CO_2 + OH^-$	84.0	1	1	0	−0.08
Silicates						
Forsterite[3]	$Mg_2SiO_4 + 4H_2O \rightarrow 2Mg^{2+} + SiO_2 + 4OH^-$	140.7	0	4	1.25	1.06
Olivine, Fo92[4]	92% forsterite, 8% fayalite	145.7	0	3.68	1.11	0.94[5]
Olivine, Fo88[4]	88% forsterite, 12% fayalite	148.3	0	3.52	1.04	0.88[5]
Olivine, Fo80[4]	80% forsterite, 20% fayalite	153.3	0	3.20	0.92	0.78[5]
Fayalite[4]	$Fe_2SiO_4 + \frac{1}{2}O_2 + 3H_2O \rightarrow 2Fe(OH)_3 + SiO_2$	203.8	0	0	0	0

Notes: (1) Reactions are written to emphasize the net alkalinity as pure OH^-. (2) Ratio includes an assumed ocean uptake efficiency $\eta_{CO_2} = 0.85$. (3) The silicon oxide products of weathering are represented as anhydrate (SiO_2) rather than silicic acid (H_4SiO_4). (4) The hydroxide-consuming oxidation of Fe^{2+} to Fe^{3+} (Griffioen, 2017) is included in the calculation. Thus fayalite does not provide net alkalinity when added to the ocean. (5) Uptake efficiency calculated based on different molar fractions of forsterite and fayalite.

$\eta_{CO_2} = 0.85$ also applied to the A_T increment. Effectiveness is reduced for carbonate and bicarbonate minerals due to the incremental DIC added and for iron(II) containing silicates due to the alkalinity consuming oxidation and precipitation of iron(III) hydroxide.

16.3.2 Marine CDR by carbonate base addition

Kheshgi's original OAE proposal discussed the dissolution of naturally occurring soda ash (Na_2CO_3), but ruled this out for large-scale application due to the limited global reserves, and also ruled out limestone due to its low solubility (Kheshgi, 1995). Harvey (2008) proposed to overcome this limitation by applying powdered limestone to upwelling regions where the saturation depth is relatively shallow (250–500 m), and where the upwelling velocity is relatively high (30–300 m/yr). Such an approach could therefore start to enhance surface alkalinity within a few years to decades and would also be applicable to areas where surface waters are under-saturated due to freshwater runoff (Chierici and Fransson, 2009).

Operational optimization of limestone dissolution in upwelling regions (LDUR), for example, optimal particle size and dose rate for location- and time-specific upwelling conditions, will require the development of systems for routine monitoring of upwelling rates (Kadko and Johns, 2011; Haskell et al., 2015), including seasonal variability (Black et al., 2011), and of climate change and wind-driven upwelling intensification (Garcia-Reyes et al., 2015).

16.3.3 Marine CDR by oxide or hydroxide base addition

The dispersal of lime (CaO), produced by limestone calcination, was the main focus of Kheshgi's (1995) proposal. Each mole of CaO dissolved in the ocean would react according to:

$$CaO + 1.79\ CO_2 + 0.79\ H_2O \rightarrow Ca^{2+} + 1.62\ HCO_3^- + 0.17\ CO_3^{2-} \qquad (16.13)$$

the fractionation between carbonate and bicarbonate reaction products depending on the local ML pH. A schematic of this ocean liming (OL) process is shown in Fig. 16.9, and similar schematics for other OAE methods discussed later can be found in Renforth and Henderson (2017).

The energy requirement for calcination (~3 GJ/tCaO) is a significant challenge for OL, although improvements over the current state-of-the-art (reviewed by Renforth et al., 2013) are being progressed through integration with low carbon energy sources such as solar thermal energy (Gil et al., 2010), fuel cell heat (Hanak et al., 2017), or hot syngas from biomass gasification (Caserini et al., 2019).

Techno-economic assessment of OL for a range of feedstock and calcination options, based on the process schematic shown in Fig. 16.9, indicates costs in the range of \$20–100/tCO_2 (Renforth and Henderson, 2017), with the OL + BECCS concept falling mid-way in this range when H_2 revenues are included (Caserini et al., 2019).

Mineral carbonation with ocean liming

The carbonation of silicate minerals and calcination of the carbonation product has been proposed as an alternative form of OL. In this variant (Fig. 16.10), magnesium carbonate, produced by the carbonation and slaking of forsterite

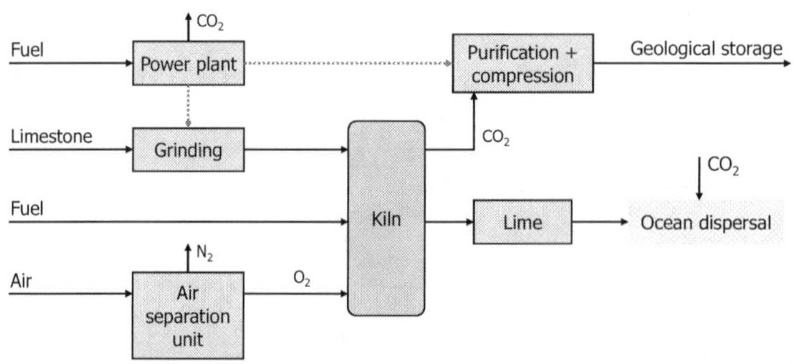

FIGURE 16.9 Schematic process for ocean liming. *After Renforth and Henderson (2017).*

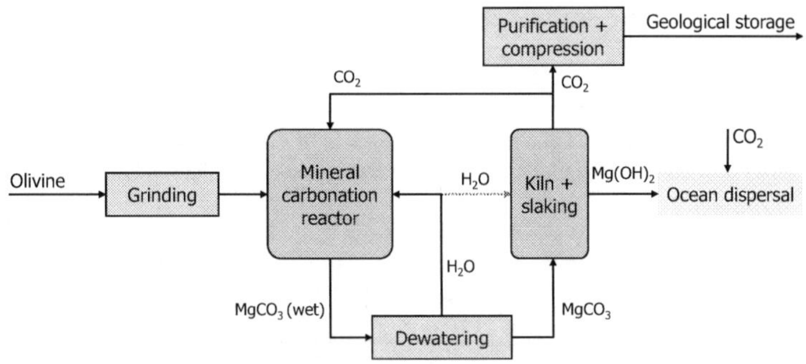

FIGURE 16.10 Schematic process for coupled mineral carbonation and ocean liming. *After Renforth and Henderson (2017).*

(Mg_2SiO_4), is calcined to produce brucite ($Mg(OH)_2$). This approach benefits from the lower calcination temperature of $MgCO_3$ versus $CaCO_3$, and the recycling of CO_2 from calcination output to carbonation input, although this gain is offset by the need to dewater the carbonation product slurry (Renforth and Kruger, 2013).

Alternative approaches to producing $Mg(OH)_2$ include acid extraction (leaching) from olivine or other ultramafic rocks using HCl or H_2SO_4, followed by electrolysis of the $Mg(Cl)_2$ or $MgSO_4$ solution to precipitate $Mg(OH)_2$ and regenerate the acid (Scott et al., 2021), or solar thermal decomposition of $Mg(Cl)_2$ from desalination brine, with the coproduced HCl sequestered in geological formations or by contact with other alkaline materials (Davies, 2015).

Enhanced silicate mineral weathering

Enhanced weathering (EW), discussed in Chapter 11, can be extended to coastal waters or to the open ocean (OOEW), where mineral dissolution releases reactive bases to neutralize carbonic acid and increase CO_2 uptake (Renforth, 2012; Hartmann et al., 2013). Olivine ($(Mg_xFe_{1-x})_2SiO_4$)—as the primary constituent of the earth's upper mantle—is the most common silicate mineral, and OOEW is usually exemplified by the reaction of forsterite—the Mg end-member ($x = 1$) of the olivine solid solution:

$$Mg_2SiO_4 + 4CO_2 + 4H_2O \rightarrow 2Mg^{2+} + 4HCO_3^- + H_4SiO_4 \quad (16.14)$$

In this reaction, the orthosilicate ion (SiO_4^{4-}), released on dissolution, is protonated to form orthosilicic acid, yielding four hydroxyl ions. Although the stoichiometry of Reaction 16.14 implies the consumption of 4 mol of dissolved CO_2 per mole of mineral, thermodynamic studies (Griffioen, 2017) and bench experiments (Montserrat et al., 2017) have shown that the effectiveness is significantly reduced if secondary minerals are precipitated, for example;

$$Mg_{1.6}Fe_{0.4}SiO_4 + 1.6CO_2 + 0.1O_2 + 2.6H_2O \rightarrow 1.6MgCO_3 + 0.4Fe(OH)_3 + H_4SiO_4 \quad (16.15)$$

where iron(III) oxide-hydroxide production also consumes dissolved oxygen. This reduction in CDR effectiveness may be countered by an increase in the dissolution rate of iron-containing olivine in natural environments due to the effect of sideophores secreted by microorganisms to increase the bioavailability of nutrient iron (Torres et al., 2019).

Mineral grinding requirements for OOEW are much more stringent than for coastal EW or LDUR, since only grains dissolving in the ML will contribute to CDR on a relevant time scale. Although a 10 μm grain size would be optimal for coastal EW (Feng et al., 2017), Köhler et al. (2013) have shown that 20% of applied olivine would be lost below the ML for 1 μm grains (55 m ML depth, 25°C SST), increasing to 40% for 30 m ML depth, to 60% for SST reduced to 10°C, and to over 95% for grain size increased to 10 μm (single factor sensitivities). Fine grinding is energetically inefficient and introduces significant additional costs (Li and Hitch, 2015), although there would be an opportunity to optimize grinding costs for location- and season-specific ML depth and temperature. Interaction with particulate matter would also result in accelerated sinking and limited dissolution within the ML, even for very fine particle sizes (Fakhraee et al., 2023), which would pose a significant challenge for CDR accounting.

16.4 Marine CDR—electrochemical methods

16.4.1 Electrochemical accelerated weathering

The central idea of electrochemical approaches to mCDR, first proposed by House et al. (2007), is to split a solution, such as desalination reject brine, into H^+ and OH^- ions using electrolysis. The basified stream is released into the ocean while the acidified stream, typically hydrochloric acid, is neutralized by contact with alkaline materials—effectively accelerating the natural weathering process.

The separation of these two streams has the advantage that the rock-dissolution products are not discharged into the ocean, enabling the use of rocks with higher heavy metal content, or rocks with a lower proportion of alkaline minerals, especially if the neutralizing reaction occurs underground by means of acid injection.

House's original proposal used Chlor-alkali electrolysis of NaCl brine to generate NaOH (the base), H_2 (a valuable by-product), and Cl_2. Chlorine has no market at the scale required for significant negative emissions, so in this proposal the two gases were recombined in a fuel cell to recover energy and generate hydrochloric acid.

Brine electrolysis generates Cl_2 preferentially over O_2, although this can be overcome using special oxygen-selective anodes (Bennett, 1980), or by preventing chloride ions from contacting the anode. In that case, O_2 is evolved and protons are generated directly at the anode, acidifying the solution. This acidity cannot be allowed to diffuse to the cathode where it would neutralize the desired base product, and a number of approaches to this have been proposed and tested.

The acidity will be neutralized as soon as it is created if the anode is directly packed and surrounded by powdered alkaline material such as calcium carbonate (Rau, 2009) or silicate rock (Rau et al., 2013). The coproduced H_2 can either be sold as a green fuel or recombined with the evolved oxygen in a fuel cell to recover energy (Rau, 2009).

A more energy-efficient alternative to electrolysis is bipolar membrane electrodialysis (Tanaka, 2007; Oener et al., 2020), where the production of gases is avoided altogether by using a special polymer membrane capable of directly splitting H_2O into H^+ and OH^- under the influence of an electric field (Eisaman et al., 2012; Tyka et al., 2022). The evolution of gases can also be avoided by reusing the produced H_2 directly on the anode side using a gas-diffusion electrode—in a sense combining the electrolysis and fuel cell energy recovery in a single process (Xie et al., 2014; Davies et al., 2018).

The acidity generated by these methods can be neutralized either in situ by alkaline rocks following underground injection or using an alternative source of alkalinity such as red mud from aluminum smelting (Qi, 2021) or alkaline mine tailings (Hamilton et al., 2020), industrial wastes available at the 10–100 Mt scale.

16.4.2 Electrochemical stripping of CO_2 from seawater

Electrochemistry can also be used to extract CO_2 directly from seawater, and several approaches have been proposed. Bipolar membrane electrodialysis can be used to acidify seawater (Eisaman et al., 2012; de Lannoy et al., 2018; Digdaya et al., 2020), which shifts the carbonate equilibrium toward CO_2, increasing the partial pressure and allowing CO_2 to be extracted under a weak vacuum. After CO_2 removal the acidified and basified streams are recombined and discharged to the sea, where the increased alkalinity results in CO_2 drawdown from the atmosphere.

Another option is to use electrolysis, together with a pair of cation exchange membranes, to strip CO_2 under acidic conditions and also produce H_2. This has been investigated as a way of generating hydrocarbon fuels onboard nuclear-powered ships (Willauer et al., 2017).

Capture costs for these electrochemical methods have been estimated in the range of \$373–604/tCO_2 (Eisaman et al., 2012; Digdaya et al., 2020), largely due to the high electrical energy requirements both for electrolysis and for pumping large quantities of seawater.

16.5 Marine CDR—biological methods

Biological mCDR methods aim to enhance the biological pump, discussed in Chapter 6, increasing photosynthetic production in the shallow ocean and the export of organic carbon to the ocean floor.

16.5.1 Ocean afforestation

Ocean macroalgal afforestation (OMA) involves the cultivation of kelp or other seaweeds over large areas of shallow, euphotic waters, with the advantage when compared to terrestrial biomass that it would not compete with food production for land and water use. China and Indonesia are the dominant producers of seaweed, which is used mainly for food and fertilizer production. Global production grew from 10 Mt/yr in 1998 to ~36 Mt/yr in 2019 (Cai, 2021), and this is likely to accelerate as its potential as a third/fourth-generation biofuel feedstock is realized (Buschmann et al., 2017).

Macroalgal productivity is exceptionally high, with CO_2 uptake reaching ~15 tCO_2/ha/yr (~1.5 $ktCO_2/km^2$/yr), comparable to the most productive terrestrial biomass crops, and can bring environmental and ecological co-benefits including local increases in fish stocks, mitigation of ocean acidification (OA) and deoxygenation, and shoreline protection by damping wave energy (Duarte et al., 2017). Harvested crops can be used as bioenergy feedstocks (see Chapter 12), either with or without CCS, or could be sunk to the deep ocean floor where the embodied carbon would be sequestered in line with the ocean ventilation time (N'Yeurt et al., 2012; Raven, 2017).

The use of macroalgae as a biofuel feedstock was first proposed in the 1970s, both as an additional energy source and as a climate change response (Wilcox, 1977). Research into industrial-scale macroalgal cultivation for biofuels was conducted in the US DOE-funded Marine Biomass Program from 1979 to 1985, although this effort was subsequently abandoned when natural gas price deregulation in the United States led to the discovery and development of vast new gas resources.

Biomass conversion options were discussed in Chapter 12, and methods such as anaerobic digestion or fermentation are particularly suitable for macroalgae due to the high water and carbohydrate content (Milledge et al., 2014; Rajak et al., 2020). High conversion efficiency has been achieved by genetic engineering of microorganisms to metabolize macroalgal carbohydrates (alginate polysaccharides) and directly synthesize bioethanol. In a pilot-scale demonstration, commenced in 2002 by the Tokyo Gas Company, 0.6 t/d of seaweed was anaerobically digested and the offgas (~20 m^3/t, 60% CH_4, 40% CO_2) was combined with natural gas to power a 10 kW gas engine (Aizawa et al., 2007). A review of more recent developments can be found in Kostas et al. (2021).

The deep ocean disposal of macroalgal biomass is also being investigated (Williamson et al., 2021; Wu et al., 2022), following an earlier proposal for the deep ocean disposal of crop residues (Strand and Benford, 2009). The ecological impact of this approach is largely unknown and likely complex, and it would also face significant carbon accounting challenges—both topics discussed further below. Despite these challenges, a number of companies are exploring this proposal, including the US start-up Running Tide, operating in the Gulf of Maine, and Dutch start-up KelpBlue, operating in Namibia.

16.5.2 Ocean iron fertilization

The concept of ocean iron fertilization (OIF) as a carbon sequestration strategy arises from the observation that primary production is limited in some so-called high-nutrient, low-chlorophyll (HNLC) regions, particularly in the Southern Ocean and in the North Pacific (Boyd, 2013). English biologist Joseph Hart first suggested in the 1930s that this might be due to a local deficiency of iron—an essential micronutrient—and this was experimentally confirmed in the 1980s by oceanographer John Martin (Hart, 1942; Martin et al., 1990).

Unlike macronutrients (nitrate, phosphate, and silicic acid) that are mostly delivered into the euphotic zone by the upwelling of nutrient-rich deep ocean waters, iron is delivered as a component of wind-borne (aeolian) dust, carried offshore from arid land masses. The low level of primary production in the Southern Ocean and North Pacific results from the lack of upwind dust sources. Analysis of sea-bed cores from the Southern Ocean has confirmed that increased dust-borne iron from Patagonia to the Subantarctic Atlantic resulted in a drop in atmospheric [CO_2] of some 40 ppm during the most recent glacial maximum (Kumar et al., 1995).

Iron fertilization of these HNLC ocean areas would increase primary production and potentially increase the flux of particulate organic carbon (POC) into the deep ocean. Since the mid-1990s, 13 OIF trials have been conducted, 11 in HNLC areas (7 in the Southern Ocean, 3 in the subarctic North Pacific, and 2 in the equatorial East Pacific), and one combined Fe + P study in the LNLC subtropical North Atlantic (Aumont and Bopp, 2006; Yoon et al., 2018). The ranges of parameters in these mesoscale experiments are shown in Table 16.5.

Tracers such as SF_6 and thorium-234 (^{234}Th) are used for monitoring. SF_6 can be detected in minute quantities and is used to map the location of the iron-enriched patch as it is advected by ocean currents. The naturally occurring radioactive isotope thorium-234 has a high affinity for particles, and a declining concentration indicates the rate at which particles sink from the enriched patch.

These experiments have confirmed that OIF does affect the rate of primary production, causing a bloom of phytoplankton. However, they have also demonstrated other changes in the ecology and biogeochemistry of the enriched patch and revealed that many factors affect the fate of the bloom and the extent to which POC is exported to the deep ocean (Table 16.6). One example of an adverse impact is increased generation of nitrous oxide (N_2O; atmospheric lifetime ~130 years, GWP_{100} ~280) as a result of microbial nitrification of ammonia from the decaying bloom.

TABLE 16.5 Parameters of mesoscale OIF experiments.

Parameter	Range	Description
Iron quantity	350–2800 kg	Commonly as $FeSO_4$ solution
Patch area	50–100 km^2	
ML thickness	10–100 m	
Tracers	SF_6, ^{234}Th	Tracing patch movement and particle export

TABLE 16.6 Iron fertilization: impacts and controlling factors.

Impacts	Controlling factors
Characteristics of the phytoplankton bloom	Initial chemical and optical conditions
	Preexisting phytoplankton communities
	Dilution rate of enriching solution
	Iron supply rate and duration
	Macronutrient availability
Local ecological impact of the bloom	Response of phytoplankton grazers to increased food availability (increased growth and reproduction)
	Changes in community compositions and food web structure
	Depletion of macronutrient supply
	Increased bacterial activity and biogenic gas production

The effectiveness of induced blooms in increasing POC export remains highly uncertain and has been demonstrated in only two of the mesoscale experiments. Estimates of the required rate of iron addition to the rate of carbon export range from 2 to 2000 μmolFe/molC, the lower value assumes 100% biomass export while the upper value reflects a more conservative estimate of net export reaching the ocean floor.

Skepticism on the effectiveness of OIF has been expressed since Martin and colleagues' (1990) confirmation of the iron deficiency hypothesis (Broecker, 1990; Joos et al., 1991). Zeebe and Archer (2005) estimated that, based on the results of the Southern Ocean iron enhancement experiment (SoFeX), fertilization of 20% of the ocean surface, 15 times per year until 2100 would reduce [CO_2] by just 15 ppm below a 700 ppm business-as-usual scenario. Their conclusion, that large—scale OIF appeared infeasible as a mitigation approach, was echoed in the IPCC's Sixth Assessment report which expressed "low confidence" in this method.

Lauderdale et al. (2020) have suggested that, on a global level the ocean is not iron deficient, but contains "just enough" iron, regulated by a microbial feedback that influences iron bioavailability. Under this hypothesis, OIF in HNLC regions like the Southern Ocean would locally consume nutrients that would otherwise be exported and sustain productivity in ecosystems such as the North Atlantic ("nutrient stealing"), so that OIF would potentially have a negative overall impact when looked at globally. This demonstrates the importance of taking a global, holistic perspective when assessing the impact of interventions in complex systems—a characteristic of all mCDR methods.

16.5.3 Ocean macronutrient fertilization

Ocean macronutrient fertilization (OMF) has also been investigated both as an mCDR approach and as a means of increasing global food production. Early modeling studies (Jones and Caldeira, 2003; Matear and Elliott, 2004) indicated significant potential for OMF via phosphate and/or nitrate addition, although with a number of compensating GHG-related side effects including increased calcification, N_2O outgassing, and surface acidification, as well as increased areas of deep water anoxia.

Mesoscale fertilization experiments have subsequently been conducted in macronutrient-poor waters, such as the phosphate-deficient eastern Mediterranean Sea (Krom et al., 2005) and eastern subtropical North Atlantic (Dixon, 2008), with both yielding unexpected results. The EU-funded CYCLOPS project was the largest of these trials, and involved the addition of phosphoric acid (H_3PO_4), buffered with $NaHCO_3$ and including SF_6 as the inert tracer, to a 4 × 4 km area south of Cyprus, with the aim of investigating phosphorus cycling in the Eastern Mediterranean. The expected increase in primary production and chlorophyll *a* was, however, not observed, despite the added phosphate being rapidly taken up by the microbial communities. Based on a microcosm experiment it was concluded that the phytoplankton community had been both nitrogen and phosphate limited while the bacterial and micrograzer communities had only been phosphate limited. Fertilization had resulted in an increase in predation and a decrease in photosynthetic activity (Krom et al., 2005). A similar result was found in the subtropical Atlantic trial.

OMF studies have generally considered phosphorous addition, but nitrogen fertilization, making use of excess phosphate present outside iron-limited areas, has also been assessed to have significant mCDR potential (Lawrence, 2014), although no nitrogen addition field trials have yet been reported (Williamson et al., 2021).

16.5.4 Artificial upwelling

The upwelling of deep ocean waters is the main source of nutrients to the surface ocean, supporting Earth's most productive marine ecosystems (Anderson and Lucas, 2009), and artificial upwelling (AU) seeks to enhance this process by lifting cool, nutrient rich water from beneath the ML to increase primary productivity and CO_2 drawdown into the ocean. Although the primary aim of AU is generally the stimulation of phytoplankton, at the base of the food web, it has also been investigated as a means to provide additional nutrients for seaweed cultivation (Fan et al., 2020) and other types of mariculture (Heilweck and Moore, 2021). The cooling effect of large-scale AU can also be significant, and model results predict that reduced terrestrial respiration due to lower air temperatures can account for ~80% of the carbon sequestered (Oschlies et al., 2010), posing a significant challenge for monitoring and verification.

A range of engineering approaches to AU have been investigated, including RE-powered air-lift (Yang et al., 2017), current induced (Xu et al., 2019), temperature and salinity-driven, and wave-driven systems.

The earliest self-powered approach was proposed by Stommel et al. (1956) and relies on the lower salinity, and therefore density, of deeper water to drive a "perpetual salt fountain". The concept, illustrated in Fig. 16.11, uses a vertical pipe filled with cool, low salinity deep water. As this contained water equilibrates in temperature with the surrounding water it becomes buoyant, and rises, resulting in a sustained inflow of deep water at the base of the pipe.

Maruyama and colleagues reported a 3-day test of this system in the Mariana Trench area of the Pacific Ocean, using a 280 m long, 0.3 m diameter, pipe made from reinforced PVC sheet, achieving an upwelling velocity of 212 m/d - (Maruyama et al., 2004). This was followed by a 34-day free-floating open-ocean experiment in the Philippines Sea, in

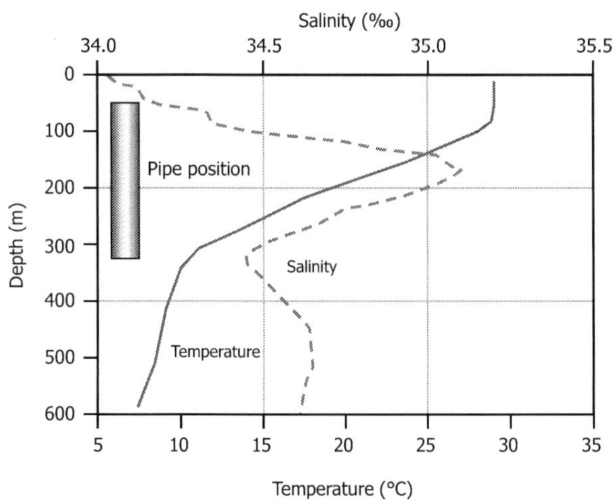

FIGURE 16.11 AU based on Stommel's perpetual salt fountain. *After Maruyama et al. (2011).*

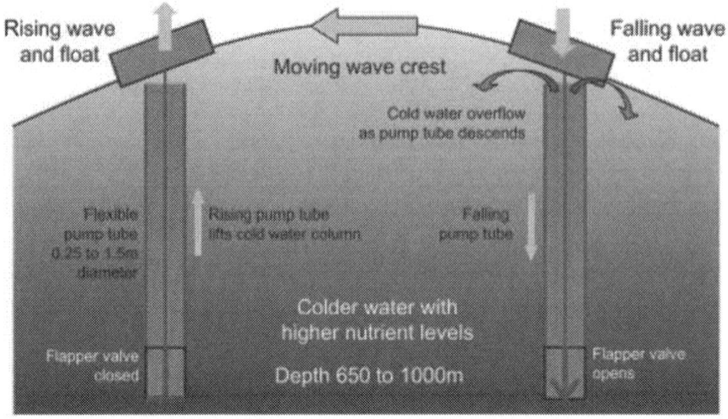

FIGURE 16.12 Wave-driven ocean upwelling system.

which an increase in chlorophyll *a* was observed via satellite remote sensing, demonstrating an increase in primary production (Maruyama et al., 2011).

Wave-driven AU, originally proposed by Isaacs and colleagues (Isaacs et al., 1976), has attracted considerable attention (Fig. 16.12). Each pumping unit comprises a vertical tube, sealed at its base by a flapper valve that only allows water to pass into the tube, and attached to a surface float that lifts the water-filled tube on each wave crest. As the float and tube start to descend into the next trough, the flapper valve opens, the tube drops, and the water in the upper section of the tube is released into the surface ocean (Kirke, 2003).

An initial field trial of wave-driven AU—the Ocean Productivity Perturbation Experiment (OPPEX)—took place in 2008 and was designed to evaluate the feasibility of inducing a phytoplankton bloom in the HNLC area of the North Pacific. The trial used a 300 m polyethylene tube (0.75 m in diameter) and although the system demonstrated significantly faster uplift velocity than the salt fountain (>150 m/h vs. 212 m/d), catastrophic failure of the pump after 2 h precluded any evaluation of the bloom hypothesis (White et al., 2010). Recent research on mechanical systems has focused on a self-powered air-lifted approach, as described in (Pan et al., 2016), which also reviews other AU trails. Ongoing research on the ecological impacts is discussed below.

One practical problem with powered AU is that cold deepwater will tend to sink due to its higher density before temperature equilibration (Kirke, 2003), potentially carrying nutrient-rich water back below the ML. However, simulation studies indicate that nutrient enhancement still occurs in the ML, and can be improved by designing the pump outflow to maximize surface spreading (Yao et al., 2020; Kemper et al., 2022). More importantly, the outgassing of CO_2 from cool, DIC-rich deep waters that are warmed and brought to the ML would reduce CDR effectiveness (Bauman et al., 2014), and the lower pH of these waters would also exacerbate surface acidification (Keller et al., 2014).

16.6 Environmental and ecological impacts of mCDR

Large-scale deployment of any mCDR method will have wide-ranging environmental and ecological implications, many of which remain largely unknown and are topics of ongoing research (Section 16.8). The main impacts are briefly described in this section, and comprehensive reviews can be found in Bach et al. (2019), Boyd et al. (2022), and Gore et al. (2019).

16.6.1 Environmental impacts

Impact on surface water pH and the carbonate system

The impact on pH and carbonate chemistry of mCDR methods that increase alkalinity will depend on how much alkalinity is added, how quickly treated water mixes with the surrounding waters, and how quickly the treated seawater equilibrates with the atmosphere (Ilyina et al., 2013; Bach et al., 2019, He and Tyka, 2023). Keller et al. (2014) investigated OAE in a high emissions scenario (RCP 8.5) at an addition rate of 0.25 Pmol/yr (equivalent to 7 GtCaO/yr) from 2020 and observed an increase in ocean carbon uptake of 166 GtC by 2100, and average increases in pH of 0.06 and in $\Omega_{calcite}$ of 0.48 by 2100, offsetting 20% and 30%, respectively of the expected declines under this scenario. OAE impact in a low emissions case has been modeled by Lenton et al. (2018) using similar alkalinity addition rate, resulting in a similar increases in global ocean carbon uptake, average pH (0.07), and $\Omega_{aragonite}$ (0.5), while global mean A_T increased by 105 μmol/kg (~5%).

These and other studies suggest that, with a modest increase in surface seawater alkalinity, large-scale deployment could significantly reduce the expected future decline in pH and Ω (Bach et al., 2019), although these impacts are not uniform, with the Arctic and tropical oceans displaying greater pH and Ω sensitivity (González and Ilyina, 2016). OMA can also mitigate the expected decline in pH and Ω, with increases in pH and $\Omega_{aragonite}$ of up to 0.1 being observed in Chinese seaweed farms (Xiao et al., 2021).

Terrestrial and other environmental impact

All mCDR methods that rely on mineral input face the challenge to minimize energy requirements and land-based environmental impacts of mining, grinding, land transport, and in some cases calcination. Mineral extraction and comminution poses a number of environmental hazards, including the generation and spreading of rock dust, and the disturbance of ecosystems, including soil removal (Renforth et al., 2013). OAE at the $GtCO_2$ scale would require mineral extraction on the same scale as the current global cement industry or roughly 10% of global rock extraction for aggregate (Gagern et al., 2019), with corresponding globally distributed environmental impacts. Methods for assessing, mitigating, and remediating the environmental impact of such operations are, however, well established (Berry and Pistocchi, 2003).

Increased production of dimethyl sulfide (DMS) was observed in some OIF field trials in the Southern Ocean (Boyd, 2013), and other mCDR methods that promote phytoplankton blooms, such as AU and OMF, could be expected to lead to a similar increase. DMS is a precursor of sulfate aerosols that contribute to the formation of sunlight-reflecting clouds (Lovelock and Rapley, 2007), so that fertilization methods could lead to increased albedo and reduced radiative forcing in areas such as the Southern Ocean, where cloud formation is limited by a lack of condensation nuclei (Boyd, 2013).

16.6.2 Ecological impacts

Developing an adequate understanding of ecological impacts to inform decision-making is one of the major challenges facing mCDR (Gattuso et al., 2018). In contrast to the extensively studied ecological impact of OA, the impact on marine biota and ecosystems of increasing alkalinity and other methods has been investigated in only a limited, albeit rapidly growing, number of studies.

An increase in alkaline metal (Ca and Mg) concentration resulting from large-scale OAE is unlikely to have a significant ecological impact on a global level, as mineral application at the 10 Gt/yr scale would only increase the natural concentrations by 1%–2% over a century (Bach et al., 2019). Perturbations would be significantly higher at dispersal points, before wider mixing, and local ecological impacts would be expected. Alkalinity manipulation is routinely used to counter acidification in aquaria and increasingly to protect shellfish in open marine settings, and OAE has been proposed as an intervention to protect coral reefs from the impact of OA. Increased biocalcification in response to increased alkalinity and aragonite saturation state has been demonstrated in a limited number of studies (Albright et al., 2016; Feng et al., 2016; Gore et al., 2019). For example, Feng and colleagues showed that enhanced calcification following regional ocean liming in the Caribbean, South China Sea, and over the Great Barrier Reef could protect these reefs from the effects of ocean acidification, and would require the application of around 4 GtCaO/yr spread across these three regions.

Dissolved silicate (DSi) is an essential nutrient for silicifying organisms such as diatoms and is the nutrient that generally limits the extent and duration of diatom blooms. The natural input of DSi to the oceans would be doubled by the application of between 0.7 and 2.0 Gt/yr of olivine (Bach et al., 2019), potentially shifting the balance of competition in favor of silicifiers over calcifiers (Köhler et al., 2013). In ocean carbon cycle model studies, Hauch and colleagues (2016) found that the CDR effectiveness of OAE by dunite dissolution was increased by ~6% as a result of silicic acid fertilization. Silicate-based OAE could therefore limit the advantage that calcifiers would gain from increased alkalinity, limiting any increase in biocalcification (and subsequent loss of alkalinity from surface waters) and potentially increasing total ballast—silicate plus carbonate—available to drive the biological pump, both of which constitute positive feedbacks for CDR effectiveness (Bach et al., 2019). An intriguing possibility arising from these competing effects is the ecosystem-dependent optimization of CDR effectiveness by combining $Ca(OH)_2$ and $MgSiO_4$.

In the case of AU, the mode of operation can also influence the relative impact on different communities (autotrophs vs. heterotrophs). Ortiz and colleagues found in a 37-day mesoscale (44 m^3) experiment in the oligotrophic waters of the Canary Islands, Spain, that a recurring supply of nutrients favored an autotrophic system and was therefore more efficient at increasing primary production, while a single upwelling pulse of equal volume and with the same nutrient load favored a more heterotrophic system (Ortiz et al., 2022).

Some marine ecosystems experience significant variability in pH and other carbonate chemistry parameters on diurnal and longer timescales (Hoffmann et al., 2011; Hernández et al., 2015), and organisms have developed a range of response mechanisms (Kroeker et al., 2020). Ferderer and colleagues investigated the impact of OAE on a coastal phytoplankton community and found noticeable but moderate impacts on the various groups in the community for an alkalinity enhancement of ~500 μmol/kg alkalinity, equivalent to a 21% increase in DIC. They concluded that the impact on phytoplankton communities of OAE at this level was justified, given the climate benefit that would result from an increase in the ocean carbon inventory in the order of 1000s of GtC (Ferderer et al., 2022).

However, studies have also demonstrated a range of negative effects from enhanced alkalinity. Recent single-species impacts have been reported by Gore et al. (2019), Comeau et al. (2017), and earlier work has been reviewed by Renforth and Henderson (2017), Bach et al. (2019), and Gagern et al. (2019). OAE via open ocean dispersal of slow-dissolving alkaline minerals, such as brucite, olivine, or basalt, may have a significant impact on zooplankton food quality and grazing efficiency, with potentially complex implications for the marine food web and biological pump (Fakhraee et al., 2023). Gim et al. (2018) found species-specific negative impacts from increased [HCO_3^-], and proposed an ecotoxicological limit (EC50-DIC) of 11 mM. This is more than four times the current open ocean DIC and, although it may occur locally in the case of direct CO_2 injection or seabed storage, is unlikely to arise in the case of mCDR.

OMA can affect offshore ecosystems in a number of ways, for example, through nutrient competition, canopy shading, the absorption of carbon, and the creation of new habitats for native and invasive species, as well as for diseases and pathogens (Campbell et al., 2019). Small-scale investigations (from patch scale up to ~2 ha) have typically found positive impacts, including increased biodiversity, limited or no impact on nutrient availability, increasing mean pH-aiding biocalcification, and the prevention of harmful algal blooms (Wahl et al., 2018; Visch et al., 2020; Xiao et al., 2021). Large-scale OMA would reduce mid-layer oxygen depletion, but create new ocean floor oxygen minimum zones (Wu et al., 2022).

Trace metal impact

Trace metals present in mineral sources for OAE may impact marine biota as nutrients or toxins and may accumulate across foodwebs. In the case of the application of olivine, for example, Fe and Ni have been highlighted as significant (Bach et al., 2019), primarily through fertilization effects if applied in Fe- or Ni-limited areas.

As discussed above, the impact of such fertilization on primary production, carbon export, and OAE CDR effectiveness remains uncertain. Hauck and coworkers simulated OAE using an ocean carbon cycle model and found that iron fertilization through the dissolution of dunite (olivine with Mg/Fe > 9) could increase the CDR effectiveness by more than a third in a high emissions scenario (RCP8.5). The increase was largely due to increased productivity in the Southern Ocean, and the subduction of nutrient-depleted water would subsequently have a negative impact on productivity in upwelling regions on a decade-to-century timescale (Hauck et al., 2016). Where primary production is colimited by Fe and N, iron fertilization would favor N_2-fixing cyanobacteria, potentially leading to a further increase in primary production and CDR effectiveness, as well as radiating impacts on the foodweb (Bach et al., 2019).

A similar fertilization effect would be expected as a result of an increase in dissolved nickel (DNi) from olivine dissolution, Ni being an essential micronutrient for the predominant N_2-fixing cyanobacteria *Trichodesmium* (Bach et al., 2019). However, DNi near dispersal locations could reach concentrations several orders of magnitude higher than natural levels and would be toxic to many organisms (Montserrat et al., 2017; Bach et al., 2019).

Some trace metals potentially present in alkaline minerals, such as chromium (Cr) and cadmium (Cd), are subject to bio-accumulation and large-scale application could pose a health risk. As pointed out by Bach et al. (2019) trace metal toxicity could be the decisive factor in determining the acceptability of specific mineral source rocks or alkaline industrial by-products for use in OAE.

16.7 Marine CDR effectiveness, permanence, potential, and cost

The ocean is a physically, chemically, and biologically complex environment, and all mCDR methods face multiple, often poorly understood, challenges regarding their effectiveness. The resulting uncertainties lead to wide ranges in estimated CDR potential and removal costs, as discussed in the following sections.

16.7.1 mCDR effectiveness and permanence

At the global level, the effectiveness of all mCDR methods would be reduced by various carbon cycle linkages such as a back-flux from untreated ocean areas (Oschlies et al., 2010), and from terrestrial carbon pools (Keller et al., 2018a), and consideration of these effects will be crucial in assessing the effectiveness of large-scale deployment. It is less clear whether back-fluxes should also be included in mCDR accounting, for example in the sale of carbon removals, and ignoring back-fluxes for this purpose would be consistent with the way in which emissions are currently accounted (Keller et al., 2018a).

Side effects remote from the area of application are likely to be a feature of mCDR methods, and will be a major challenge in determining effectiveness and in carbon accounting. In addition to the possible "nutrient stealing" effect of OIF in the Southern Ocean discussed earlier, Krumhardt et al. (2020) have shown that increased calcification in the Southern Ocean—a potential side effect of OAE or OF—would reduce alkalinity leakage from the Antarctic to other ocean regions, thereby reducing the ocean's ability to draw down CO_2.

Nutrient reallocation on a more local scale could also have a significant impact on the additionality of OMA (Bach et al., 2021), unless this was supplemented, for example by artificial upwelling. Other factors potentially limiting OMA effectiveness are the impact on ocean albedo and radiative forcing (Raven, 2017; Bach et al., 2021), and the release of ozone-destroying halocarbons ($CHBr_3$, CH_2Br_2, and others), particularly for tropical locations (Leedham et al., 2013).

Of the methods discussed above, OAE is considered the most effective to increase ocean CO_2 uptake (Gattuso et al., 2021), largely because the chemical and physical processes are governed by the well-understood chemistry of seawater, and the physical meteorology of air:sea gas exchange. Regional variations in the marine carbonate system and meteorological and physical oceanographic conditions will therefore influence OAE effectiveness (Burt et al., 2021).

Impact of calcification and abiotic carbonate precipitation

mCDR effectiveness can be affected by any change in calcification, and both positive and negative potential feedback mechanisms have been identified (Riebesell et al., 2009; Bach et al., 2019). An increase in calcification would reduce the effectiveness of OAE, since the formation of 1 mole of $CaCO_3$ reduces alkalinity by 2 mol and releases 1 mole of CO_2 (House et al., 2007; Montserrat et al., 2017), and this has been observed to occur when reactive alkaline solids (e.g., $Mg(OH)_2$) lead to a sufficiently high A_T increase (>600 μmol/kg) (Hartmann et al., 2022). On the other hand, precipitated $CaCO_3$ could act as ballast to accelerate the sinking of particulate organic matter (POM), strengthening the biological pump and the effectiveness of fertilization methods. Increased calcification may also limit the effectiveness of OMA, which provides a habitat for calcifying epibionts such as barnacles (Bach et al., 2021). A broad assessment of the impact on calcifying organisms and related biogeochemical processes will therefore be important to fully understand the effectiveness of mCDR.

Abiotic carbonate precipitation is often considered to require $\Omega_{CaCO_3} > 18$ for unseeded nucleation in seawater at 25°C, and up to $\Omega = 25$ when nucleation is hindered by the presence of other ions (Rau, 2011; Renforth et al., 2013). However, House et al. (2007) consider "flash precipitation" at the point of discharge to be inevitable, unless steps are taken to reduce the Ω perturbation by additional mixing. For example, dilution in a ship's wake may proceed faster than the homogenous nucleation time of $CaCO_3$ (He and Tyka, 2023). The potential for OAE to initiate precipitation that removes more alkalinity than was originally added, demonstrated in the lab by Moras et al. (2021), also requires further investigation under ocean conditions.

Mineral dispersal and dissolution

For methods that rely on milled mineral dispersal (other than LDUR), dissolution must take place within the ML, and effectiveness is, therefore, dependent on a range of operational parameters and environmental conditions (particle size, application rate, temperature, and ML depth). Maximum effectiveness will require operational controls optimized for varying dispersal conditions (Harvey, 2008).

Permanence and termination effects

Given the diversity of mCDR methods, a wide range of storage lifetimes is to be expected. Modeling studies suggest that AU, OIF, and OMF will only have a short-term influence on atmospheric $[CO_2]$, with ~70% of the exported carbon being transported back into the ML within 50 years, while deep ocean storage of biomass, similar to deep ocean direct CO_2 injection, would be expected to have a storage lifetime of centuries to a millennium, depending on storage location (Siegel et al., 2021). OAE has highly predictable and assured permanence, so long as no other process subsequently reduces ocean alkalinity (Ilyina et al., 2013). In any event, it is likely that an OAE effectiveness close to 0.8 $molCO_2/molA_T$ addition would persist for the next century when the climate impact will be most important (House et al., 2007). Recent simulations have shown that this effectiveness is dependent on the location of OAE deployment, with some areas losing a significant fraction of the induced CO_2 deficit to downwelling before equilibration can occur (He and Tyka, 2023).

A number of studies have investigated the effect of terminating mCDR deployment, particularly for OAE. If applied on a regional level, for example as a means of protecting vulnerable ecosystems from the effects of OA, termination results in a return to surrounding conditions due to rapid dilution and dispersion (Feng et al., 2016). This rapid return also occurs for global deployment of AU and other mCDR methods with short storage lifetimes (Ochlies et al., 2010) although not in the case of global OAE deployment (Keller et al., 2014; Ilyina et al., 2013). González et al. (2018) found that, under a high emissions scenario (RCP8.5), termination of global OAE would be followed by a short period (~ 10 years) in which ocean pH and $\Omega_{calcite}$ decline more rapidly than in the base case, although still remaining above base case values by ~0.1 and ~1.0, respectively. Such a rapid change in seawater chemistry could be a significant stressor for ocean ecosystems.

mCDR potential and costs

The capacity of the oceans to absorb additional anthropogenic CO_2 is effectively unlimited. As Lackner (2002) has observed, the oceans could take up 10,000 GtC as increased alkalinity with changes in pCO_2, $[CO_3^{2-}]$, and $[H^+]$ that are of the same order as have resulted from the historical drawdown of emissions. Seawater chemistry, therefore, does not impose a meaningful constraint on mCDR potential.

Table 16.7 summarizes published estimates for mCDR potentials and costs. In contrast to the relatively low estimates of the potential for deepwater macroalgae disposal, N'Yeart et al. (2012) also considered the option of macroalgae cultivation as a biomethane feedstock, and envisage a potential approaching 6 $GtCO_2$/yr for each 1% of the ocean surface devoted to OMA, while also replacing ~10% of global fossil energy. Several mCDR methods have the potential to deliver at a 1 $GtCO_2$/yr scale or greater and it seems very feasible that a portfolio of such methods could collectively deliver a 1 GtC/yr (3.7 $GtCO_2$/yr) climate stabilization wedge (Pacala and Socolow, 2004).

OAE costs are largely determined by the cost of energy—either for calcination or electrolysis—as can be seen from the spread of estimates given for electrochemical enhanced weathering by Rau et al. (2018), ranging from \$3/tCO_2 removed using low-cost hydro at \$0.01/kWh to \$161/tCO_2 using high cost solar at \$0.29/kWh. Given this strong dependency, cost reduction efforts will be largely aimed at reducing the energy cost, for example by integration with energy production or other processes (Caserini et al., 2019), or the use of stranded energy resources (Renforth and Kruger, 2013).

TABLE 16.7 mCDR estimated potentials and costs.

mCDR method	Potential ($GtCO_2$/yr)	Cost (\$/tCO_2)	Comment; reference
OL (CaO)	3.7 feasible	\$72–159	Oxy-fuel flash calciner; Renforth et al. (2013)
OL (CaO or MgO)	6–23 in 2100		Moderate-emissions scenario; Fakhraee et al. (2023)
OL ($Mg(OH)_2$)	1.0 by 2050		Desalination reject brine electrolysis; Davies et al. (2018)
LDUR	0.6 after 50 years, 1.05 after 200 years	\$40–44	4 $GtCaCO_3$/yr over 8% of the ocean surface. Costs assume 4 t-mineral/tCO_2 and are undiscounted. Harvey (2008)
EW	0.8		3 Gt-olivine/yr; Köhler et al. (2013)
	10		17–56 Gt-olivine/yr; Feng et al. (2017)
	0.5	\$3–161	Electrochemical EW with H_2 production; Rau et al. (2018)
OIF		\$18–58	Optimistic case for favorable locations; Williamson et al. (2021)
OMF	>3.7	~\$20	Jones (2014)
OMA	<1.0	<\$100	Deep ocean disposal; N'Yeurt et al. (2012)
	12–20*		Deep ocean disposal, * coupled with AU; Wu et al. (2022)
	19–53		Biofuel production; N'Yeurt et al. (2012)
AU	0.9	\$100–150	Ochlies et al. (2010), N'Yeurt et al. (2012)
	~0.1		Koweek (2022).

16.8 Marine CDR research and development themes

The main barriers to mCDR deployment are the need for improved understanding of ecological impacts and CDR effectiveness, the development of MRV approaches and deployment systems, and the establishment of an enabling ocean governance framework. This section highlights the most important R&D themes in each of these areas, while further discussion can be found in Gaggern et al. (2019), and N'Yeurt et al. (2012).

16.8.1 Ecological impact assessment

Although the ecological assessment of mCDR is in its infancy, the methodologies and expertise developed through OA research can enable rapid progress. In particular, the assessment of seawater chemistry, trace metal cycling and toxicology are well-established research areas, while expertise and established protocols from OA research can facilitate the design and execution of experiments to investigate ecological impacts (Riebesell et al., 2011; Andersson et al., 2015; Cornwall and Hurd, 2016; Boyd et al., 2018; Bach et al., 2019).

The observations of natural systems which display conditions relevant to mCDR such as occur in the Black Sea (an analog for OAE), and on volcanic coasts (sites of intense basalt weathering) can improve our understanding of ecological impacts (Andersson et al., 2015), while paleo-oceanographic studies of periods when ocean conditions were similar to those expected under large scale mCDR may also provide insights (Bach et al., 2019).

Key areas for further research on ecological impacts include the following:

- Field trials to investigate biogeochemical and ecological impacts, from single-species, to multispecies interactions and ecosystem levels, in coastal and open ocean locations (Renforth et al., 2013; EFI, 2020).
- Assessing the synergistic or antagonistic effects of mCDR alongside other stressors such as warming, including potential cessation effects (Gaggern et al., 2019).
- Further investigation of biocalcification response to increased alkalinity and saturation state, for more species and at an ecosystem level (Gore et al., 2019).
- Ecological impact (toxicity and fertilization) of trace metals including iron, nickel, and chromium (Hartmann et al., 2013; Bach et al., 2019).
- Extension of FOCE experiments to test the impact on benthic communities and ecosystems over longer durations (Stark et al., 2019)
- OMA impacts, including genetic depression of natural algal populations, facilitation of algal diseases, alteration of hydrodynamic regimes, artificial habitat creation, and canopy shading (Buschmann et al., 2017; Campbell et al., 2019)

16.8.2 CDR effectiveness and process efficiency

Mineral dissolution and secondary precipitation

Current understanding of the processes and controls on biotic and abiotic (re)precipitation of $CaCO_3$ and other secondary minerals under conditions of increasing A_T is insufficient to quantify their impact on mCDR effectiveness (Henderson et al., 2008; Bach et al., 2019). Further work is required in the following areas:

- Investigation of the rate of dispersion/dilution and fate of added mineral particles under in situ conditions (Henderson et al., 2008; Meysman and Montserrat, 2017), including methods to monitor dissolution in and transport through the ML.
- Investigation of the potential for onset of calcite dissolution at depths shallower than the saturation horizon (Chung et al., 2003; Feely et al., 2004).
- Investigation of controls on secondary mineral precipitation and flash precipitation under field conditions, and impact on OAE effectiveness.
- Development of standardized engineered materials, mineral specifications, and operational guidelines to avoid or minimize any loss in effectiveness (Hartmann et al., 2022).
- Optimization of mineral dispersal (e.g., particle size and dosage for location and time/season), to minimize ecological impact and maximize CDR effectiveness.

Process efficiency

- Novel methods to improve energy efficiency for mineral comminution and calcination, including cross-industry synergies.
- Optimization of seawater electrolysis to address fouling, ohmic losses, impact of impurities, suppression of ozone-depleting compounds (CH_3BR, CH_3Cl), etc.
- AWL reactor design optimization and operating procedures (Chou et al., 2015); potential for catalytic reaction enhancement (e.g., immobilized CA) to enable process equipment size and cost reduction.

16.8.3 Measurement, reporting, and verification

Marine CDR faces some unique measurement and monitoring challenges, including the large temporal and spatial scales separating the operating location and the point of eventual CO_2 drawdown, terrestrial side effects (e.g., for AU; Ochlies et al., 2010), the strong background variability of ocean chemical and biological conditions (Hofmann et al., 2011), and the difficulty of measuring changes in air-sea CO_2 flux (Bach et al., 2021). The rapid turnover of biomass and the potential for particulate and dissolved organic carbon export to other systems are further challenges for OMA, OIF, OMF, and AU, which aim to increase primary production (Hurd et al., 2022).

The development of internationally recognized MRV and accounting systems is a prerequisite for mCDR deployment and for mCDR removals to be eligible for participation in carbon markets and other incentive schemes (Peters and Geden, 2017).

Measurement and monitoring

The extensive and ongoing R&D effort on global climate observation and particularly OA monitoring, including sensor development and calibration protocols, satellite remote-sensing, data QC and processing algorithms, and best practice guidance (Martz et al., 2015), provides a valuable foundation for the rapid development of mCDR MRV systems, standards, and protocols. Fig. 16.13 outlines such a framework for mCDR methods involving mineral dispersal.

For monitoring the near field impact on seawater chemistry, sensors capable of continuous autonomous measurement of A_T have recently been demonstrated, with a precision of 2–10 μmol/kg (ca. 0.1%–0.5% of seawater A_T) depending on signal averaging (Briggs et al., 2020). Moored, free drifting, and autonomous systems have been developed that will be applicable for mCDR monitoring.

The effectiveness of mCDR methods is likely to be observed as a change in ocean chemistry well before any climate impact is detectable (Fröb et al., 2020), and a range of different systems have been developed that can support regional monitoring of ocean chemistry and air–sea gas exchange (Bushinsky et al., 2019). These could eventually be linked to regional and global assessments of air–sea CO_2 exchange (Laruelle et al., 2014) and to existing observation networks such as Argos SOCONET and ICOS-Oceans (Wanninkhof et al., 2019).

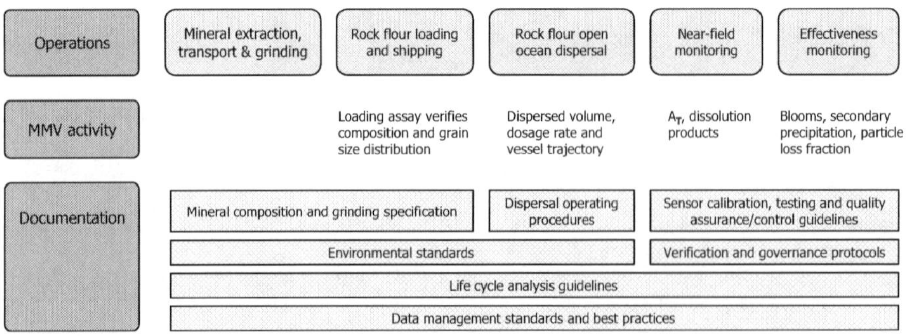

FIGURE 16.13 Prototype MMV framework for mCDR methods using open ocean mineral dispersal.

Satellite remote sensing products such as sea surface temperature and color, related to chlorophyll content, are also increasingly used in combination with locally derived empirical relationships to assess surface ocean chemistry (Land et al., 2019) and will be applicable to regional mCDR monitoring.

Other measurement and monitoring systems development needs include the following:

- Further development of monitoring systems to enable cost-effective MRV over large spatial and temporal scales (Fröb et al., 2020), including remote sensing techniques, advanced sensors, and autonomous ocean vehicles (EFI, 2020).
- Development of techniques to monitor the fate (ML lateral transport or subduction) of CO_2-depleted seawater and to directly measure seawater CO_2 uptake over large spatial and temporal scales (Bach et al., 2021; Hurd et al., 2022; Freeman, 2022).
- Development of techniques to measure organic and inorganic carbon losses (e.g., respiration, calcification, grazing) and exports in nearshore and offshore ecosystems (Hurd et al., 2022).
- Basic research to establish an improved baseline for natural carbon cycling against which ocean carbon removal can be compared (EFI, 2020).

Modeling

Pending further progress in the previously listed areas, the verification of CO_2 uptake in early mCDR projects will inevitably rely on modeling, as direct measurement of CO_2 flux additionality is currently impossible. Measurement and monitoring of the relevant flow field and other ocean parameters to calibrate models and thereby indirectly verify mCDR is a likely path forward.

The regional and global impacts of mCDR have been assessed in a number of modeling studies (Section 16.5.1), and specific cases have been proposed for inclusion in the CDR model intercomparison project (CDRMIP; Keller et al., 2018b). While the impacts on seawater chemistry can be modeled with some confidence, the wide-ranging ecological impacts and the consequences for effectiveness require further investigation and model development work.

Further development of modeling and planning tools that accurately predict mCDR outcomes will aid the identification of optimal locations for testing and deployment, and inform decision- and policymaking (Gaggern et al., 2019; EFI, 2020).

16.8.4 Governance, policy, and outreach

A governance framework for ocean-based NETs is currently lacking, but is an essential enabler for field trials that could have substantial ecological, transboundary, or other impacts (Russell et al., 2012). Ideally, this should provide for stage-gated decision-making involving all relevant stakeholders, should take account of all potential direct and indirect impacts, and should ensure careful control and monitoring, with open publication of results (EFI, 2020).

Given the uncertainties inherent in the portfolio of ocean-based methods, governance and policy frameworks should allow for iterative learning and regular review, as the impacts, benefits, and costs of each method become more fully understood. Alternative approaches that break from the current fragmented international ocean governance regime, such as demand-based governance (Jinnah, 2018) and objective-based frameworks to align multilevel governance efforts (Harrould-Kolieb and Hoegh-Guldberg, 2019), warrant further investigation.

Further social science research is also needed to stimulate governance and policy development and to ensure effective public engagement in decision-making as ocean-based methods progress toward deployment.

Governance, policy, and outreach topics needing further research include:

- Development of an effective governance framework (e.g., extending the LC/LP Assessment Framework; Dixon et al., 2014; Boettcher et al., 2019; EFI, 2020).
- Development of risk trade-off approaches so that the risks of deployment can be quantified and weighed against the risks of taking no action (EFI, 2020).
- Research and consultation on policy options to enable demonstration and early deployment projects, including mechanisms for mCDR participation in carbon markets (EFI, 2020).
- Research and consultation to identify the questions that policymakers need to answer, for example, regarding quantification, permanence, and accounting issues, and to provide the basis for those answers (Lomax et al., 2015).
- Social science research into public acceptability (e.g., under what conditions will demonstration and deployment be acceptable?)
- Development of a knowledge base and expertise on wider societal issues including potential socioeconomic and geopolitical barriers (Boettcher et al., 2019; EFI, 2020).

References and resources

References

Aizawa, M., Asaoka, K., Atsumi, M., Sakou, T., 2007. Seaweed bioethanol production in Japan—the ocean sunrise project. In: OCEANS 2007. IEEE, pp. 1–5.

Albright, R., Caldeira, L., Hosfelt, J., Kwiatkowski, L., Maclaren, J.K., Mason, B.M., et al., 2016. Reversal of ocean acidification enhances net coral reef calcification. Nature 531 (7594), 362–365.

Anderson, T.R., Lucas, M.I., 2009. Upwelling ecosystems. In: Jørgensen, S.E. (Ed.), Ecosystem Ecology. Elsevier, Amsterdam.

Andersson, A.J., Kline, D.I., Edmunds, P.J., et al., 2015. Understanding ocean acidification impacts on organismal to ecological scales. Oceanography 28 (2), 16–27.

Archer, D., 2005. Fate of fossil fuel CO_2 in geologic time. J. Geophys. Res.: Oceans 110, C09S05.

Aumont, O., Bopp, L., 2006. Globalizing results from ocean in situ iron fertilization studies. Global Biogeochem. Cycles 20 (2).

Aya, I., Kojima, R., Yamane, K., Shiozaki, K., Brewer, P.G., Peltzer III, E.T., 2004. In situ experiments of cold CO_2 release in mid-depth. Energy 29 (9–10), 1499–1509.

Bach, L.T., Gill, S., Rickaby, R., Gore, S., Renforth, P., 2019. CO_2 removal with enhanced weathering and ocean alkalinity enhancement: potential risks and co-benefits for marine pelagic ecosystems. Front. Climate 1, 7.

Bach, L.T., Tamsitt, V., Gower, J., Hurd, C.L., Raven, J.A., Boyd, P.W., 2021. Testing the climate intervention potential of ocean afforestation using the Great Atlantic Sargassum Belt. Nat. Commun. 12 (1), 1–10.

Bae, H., Park, J.S., Senthilkumar, S.T., Hwang, S.M., Kim, Y., 2019. Hybrid seawater desalination-carbon capture using modified seawater battery system. J. Power Sources 410, 99–105.

Barry, J.P., Buck, K.R., Lovera, C.F., et al., 2004. Effects of direct ocean CO_2 injection on deep-sea meiofauna. J. Oceanogr. 60, 759–766.

Barry, J.P., Lovera, C., Buck, K.R., et al., 2014. Use of a free Ocean CO_2 enrichment (FOCE) system to evaluate the effects of ocean acidification on the foraging behavior of a deep-sea urchin. Environ. Sci. Technol. 48, 9890–9897.

Bauman, S.J., Costa, M.T., Fong, M.B., et al., 2014. Augmenting the biological pump: the shortcomings of geoengineered upwelling. Oceanography 27 (3), 17–23.

Bennett, J.E., 1980. Electrodes for generation of hydrogen and oxygen from seawater. Int. J. Hydrogen Energy 5 (4), 401–408.

Bernhard, J.M., Barry, J.P., Buck, K.R., Starczak, V.R., 2009. Impact of intentionally injected carbon dioxide hydrate on deep-sea benthic foraminiferal survival. Global Change Biol. 15 (8), 2078–2088.

Berry, P., Pistocchi, A., 2003. A multicriterial geographical approach for the environmental impact assessment of open-pit quarries. Int. J. Surf. Min. Reclamat. Environ. 17 (4), 213–226.

Black, B.A., Schroeder, I.D., Sydeman, W.J., Bograd, S.J., Wells, B.K., Schwing, F.B., 2011. Winter and summer upwelling modes and their biological importance in the California Current Ecosystem. Global Change Biol. 17 (8), 2536–2545.

Boettcher, M., Chai, F., Cullen, J., et al., 2019. High level review of a wide range of proposed marine geoengineering techniques. In: Boyd, P., Vivian, C. (Eds.), GESAMP Reports & Studies Series. IMO, London, p. 98.

Boyd, P.W., 2013. Ocean fertilization for sequestration of carbon dioxide from the atmosphere. In: Geoengineering Responses to Climate Change. Springer, New York, NY, pp. 53–72.

Boyd, P.W., Collins, S., Dupont, S., et al., 2018. Experimental strategies to assess the biological ramifications of multiple drivers of global ocean change—a review. Global Change Biol. 24 (6), 2239–2261.

Boyd, P.W., Bach, L.T., Hurd, C.L., Paine, E., Raven, J.A., Tamsitt, V., 2022. Potential negative effects of ocean afforestation on offshore ecosystems. Nat. Ecol. Evol. 6 (6), 675–683.

Brewer, P.G., Peltzer, E.T., Friederich, G., Rehder, G., 2002. Experimental determination of the fate of rising CO_2 droplets in seawater. Environ. Sci. Technol. 36 (24), 5441–5446.

Brewer, P.G., Chen, B., Warzinki, R., et al., 2006. Three-dimensional acoustic monitoring and modeling of a deep-sea CO_2 droplet cloud. Geophys. Res. Lett. 33 (23), L23607.

Briggs, E.M., De Carlo, E.H., Sabine, C., Howins, N., Martz, T.R., 2020. Autonomous, ISFET-based Total Alkalinity and pH measurements on a barrier reef of Kāne'ohe Bay. ACS Earth and Space Chemistry 4 (3), 355–362.

Broecker, W.S., 1990. Comment on "Iron deficiency limits phytoplankton growth in Antarctic waters" by John H. Martin et al. Global Biogeochem. Cycles 4 (1), 3–4.

Burt, D.J., Fröb, F., Ilyina, T., 2021. The sensitivity of the marine carbonate system to regional ocean alkalinity enhancement. Frontiers in Climate 2021, 68.

Buschmann, A.H., Camus, C., Infante, J., et al., 2017. Seaweed production: overview of the global state of exploitation, farming and emerging research activity. Eur. J. Phycol. 52 (4), 391–406.

Bushinsky, S.M., Takeshita, Y., Williams, N.L., 2019. Observing changes in ocean carbonate chemistry: our autonomous future. Curr. Clim. Change Rep. 5 (3), 207–220.

Cai, J., 2021. Global Status of Seaweed Production, Trade and Utilization. FAO Seaweed Innovation Forum, Belize.

Caldeira, K., Rau, G.H., 2000. Accelerating carbonate dissolution to sequester carbon dioxide in the ocean: geochemical implications. Geophys. Res. Lett. 27 (2), 225–228.

Caldeira, K., Akai, M., Brewer, P., et al., 2005. Ocean storage. In: Metz, B., Davidson, O., de Coninck, H.C., Loos, M., Meyer, L.A. (Eds.), IPCC Special Report on Carbon Dioxide Capture and Storage. Cambridge University Press, Cambridge, UK.

Campbell, I., Macleod, A., Sahlmann, C., et al., 2019. The environmental risks associated with the development of seaweed farming in Europe-prioritizing key knowledge gaps. Front. Mar. Sci. 6, 107.

Capron, M.E., Stewart, J.R., Rowe, R.K., 2013. Secure seafloor container CO_2 storage. In: 2013 OCEANS-San Diego. IEEE, pp. 1–8.

Caserini, S., Barreto, B., Lanfredi, C., Cappello, G., Ross Morrey, D., Grosso, M., 2019. Affordable CO_2 negative emission through hydrogen from biomass, ocean liming, and CO_2 storage. Mitig. Adapt. Strategies Glob. Change 24 (7), 1231–1248.

Chierici, M., Fransson, A., 2009. Calcium carbonate saturation in the surface water of the Arctic Ocean: undersaturation in freshwater influenced shelves. Biogeosciences 6 (11), 2421–2431.

Chou, W.C., Gong, G.C., Hsieh, P.S., et al., 2015. Potential impacts of effluent from accelerated weathering of limestone on seawater carbon chemistry: a case study for the Hoping power plant in northeastern Taiwan. Mar. Chem. 168, 27–36.

Chow, A.C., Adams, E.E., Israelsson, P.H., Tsouris, C., 2009. Carbon dioxide hydrate particles for ocean carbon sequestration. Energy Proc. 1 (1), 4937–4944.

Chow, A., 2014. Ocean carbon sequestration by direct injection. In: Morgado, C.R., Esteves, V. (Eds.), CO_2 Sequestration and Valorization. IntechOpen, London, UK.

Chung, S.N., Lee, K., Feely, R.A., et al., 2003. Calcium carbonate budget in the Atlantic Ocean based on water column inorganic carbon chemistry. Global Biogeochem. Cycles 17 (4).

Comeau, S., Tambutté, E., Carpenter, R.C., Edmunds, P.J., Evensen, N.R., Allemand, D., et al., 2017. Coral calcifying fluid pH is modulated by seawater carbonate chemistry not solely seawater pH. Proc. Biol. Sci. 284 (1847), 20161669.

Cornwall, C.E., Hurd, C.L., 2016. Experimental design in ocean acidification research: problems and solutions. ICES (Int. Counc. Explor. Sea) J. Mar. Sci. 73, 572–581.

Dai, Z., Zhang, Y., Stauffer, P., et al., 2017. Injectivity evaluation for offshore CO_2 sequestration in marine sediments. Energy Proc. 114, 2921–2932.

Dai, Z., Zhang, Y., Bielicki, J., et al., 2018. Heterogeneity-assisted carbon dioxide storage in marine sediments. Appl. Energy 225, 876–883.

Davies, P.A., 2015. Solar thermal decomposition of desalination reject brine for carbon dioxide removal and neutralisation of ocean acidity. Environ. Sci. J. Integr. Environ. Res.: Water Res. Technol. 1 (2), 131–137.

Davies, P.A., Yuan, Q., de Richter, R., 2018. Desalination as a negative emissions technology. Environ. Sci. J. Integr. Environ. Res.: Water Res. Technol. 4 (6), 839–850.

de Lannoy, C.-F., Eisaman, M.D., Jose, A., et al., 2018. Indirect ocean capture of atmospheric CO_2: Part I. Prototype of a negative emissions technology. Int. J. Greenh. Gas Control 70, 243–253.

Dickson, A.G., 2011. The carbon dioxide system in seawater: equilibrium chemistry and measurements. In: Riebesell, U., Fabry, V.J., Hansson, L., Gattuso, J.P. (Eds.), Guide to Best Practices for Ocean Acidification Research and Data Reporting. Office for Official Publications of the European Communities.

Digdaya, I.A., Sullivan, I., Lin, M., et al., 2020. A direct coupled electrochemical system for capture and conversion of CO_2 from oceanwater. Nat. Commun. 11 (1), 1–10.

Dixon, J.L., 2008. Macro and micro nutrient limitation of microbial productivity in oligotrophic subtropical Atlantic waters. Environ. Chem. 5 (2), 135–142.

Dixon, T., Garrett, J., Kleverlaan, E., 2014. Update on the London Protocol—developments on transboundary CCS and on geoengineering. Energy Proc. 63, 6623–6628.

Drange, H., Alendal, G., Johannessen, O.M., 2001. Ocean release of fossil fuel CO_2: a case study. Geophys. Res. Lett. 28 (13), 2637–2640.

Duarte, C.M., Wu, J., Xiao, X., Bruhn, A., Krause-Jensen, D., 2017. Can seaweed farming play a role in climate change mitigation and adaptation? Front. Mar. Sci. 4, 100.

EFI (Energy Futures Initiative), 2020. Uncharted Waters: Expanding the Options for Carbon dioxide Removal in Coastal and Ocean Environments.

Eisaman, M.D., Parajuly, K., Tuganov, A., Eldershaw, C., Chang, N., Littau, K.A., 2012. CO_2 extraction from seawater using bipolar membrane electrodialysis. Energy Environ. Sci. 5 (6), 7346–7352.

Fakhraee, M., Li, Z., Planavsky, N., Reinhard, C., 2023. Environmental impacts and carbon capture potential of ocean alkalinity enhancement. Res Square. Available from:. https://doi.org/10.21203/rs.3.rs-1475007/v1. Submitted for publication.

Fan, W., Zhang, Z., Yao, Z., et al., 2020. A sea trial of enhancing carbon removal from Chinese coastal waters by stimulating seaweed cultivation through artificial upwelling. Appl. Ocean Res. 101, 102260.

Feely, R.A., Sabine, C.L., Lee, K., et al., 2004. Impact of anthropogenic CO_2 on the $CaCO_3$ system in the oceans. Science 305 (5682), 362–366.

Feng, E.Y., Keller, D.P., Koeve, W., Oschlies, A., 2016. Could artificial ocean alkalinization protect tropical coral ecosystems from ocean acidification? Environ. Res. Lett. 11 (7), 074008.

Feng, E.Y., Koeve, W., Keller, D.P., Oschlies, A., 2017. Model-based assessment of the CO_2 sequestration potential of Coastal Ocean Alkalinization. Earth's Future 5 (12), 1252–1266.

Fer, I., Haugan, P.M., 2003. Dissolution from a liquid CO_2 lake disposed in the deep ocean. Limnol. Oceanogr. 48 (2), 872–883.

Ferderer, A., Chase, Z., Kennedy, F., Schulz, K.G., Bach, L.T., 2022. Assessing the influence of ocean alkalinity enhancement on a coastal phytoplankton community. Biogeosci. Discuss. 1–36.

Freeman, S., 2022. Tools for climate solutions: developing techniques for marine carbon dioxide removal measurement, reporting and validation. J. Acoust. Soc. Am. 151 (4), A242.

Friedlingstein, P., O'sullivan, M., Jones, M.W., Andrew, R.M., Gregor, L., Hauck, J., et al., 2022. Global carbon budget 2022. Earth System Science Data 14 (11), 4811–4900.

Fröb, F., Sonntag, S., Pongratz, J., Schmidt, H., Ilyina, T., 2020. Detectability of artificial Ocean Alkalinization and stratospheric aerosol injection in MPI-ESM. Earth's Future 8 (10) e2020EF001634.

Fuhr, M., Geilert, S., Schmidt, M., et al., 2022. Kinetics of olivine weathering in seawater: an experimental study. Frontiers in Climate 4, 39.

Gagern, A., Rau, G., Iglesias-Rodriguez, M.D., et al., 2019. Ocean Alkalinity Enhancement. Current state of knowledge and potential role of philanthropy. In: Meeting Proceedings, September 9th, 2019, Half Moon Bay, California.

García-Reyes, M., Sydeman, W.J., Schoeman, D.S., et al., 2015. Under pressure: climate change, upwelling, and eastern boundary upwelling ecosystems. Front. Mar. Sci. 2, 109.

Gattuso, J.P., Kirkwood, W., Barry, J.P., et al., 2014. Free-ocean CO_2 enrichment (FOCE) systems: present status and future developments. Biogeosciences 11 (15), 4057–4075.

Gattuso, J.P., Magnan, A.K., Bopp, L., et al., 2018. Ocean solutions to address climate change and its effects on marine ecosystems. Front. Mar. Sci. 337.

Gattuso, J.P., Williamson, P., Duarte, C.M., Magnan, A.K., 2021. The potential for ocean-based climate action: negative emissions technologies and beyond. Front. Climate 2, 575716.

Gil, A., Medrano, M., Martorell, I., Lázaro, A., Dolado, P., Zalba, B., Cabeza, L.F., 2010. State of the art on high temperature thermal energy storage for power generation. Part 1—Concepts, materials and modellization. Renew. Sustain. Energy Rev 14 (1), 31–55.

Gim, B.M., Hong, S., Lee, J.S., et al., 2018. Potential ecotoxicological effects of elevated bicarbonate ion concentrations on marine organisms. Environ. Pollut. 241, 194–199.

Goldthorpe, S., 2017. Potential for very deep Ocean Storage of CO_2 without ocean acidification: a discussion paper. Energy Proc. 114, 5417–5429.

González, M.F., Ilyina, T., 2016. Impacts of artificial ocean alkalinization on the carbon cycle and climate in Earth system simulations. Geophys. Res. Lett. 43 (12), 6493–6502.

González, M.F., Ilyina, T., Sonntag, S., Schmidt, H., 2018. Enhanced rates of regional warming and ocean acidification after termination of large-scale ocean Alkalinization. Geophys. Res. Lett. 45 (14), 7120–7129.

Gore, S., Renforth, P., Perkins, R., 2019. The potential environmental response to increasing ocean alkalinity for negative emissions. Mitig. Adapt. Strategies Glob. Change 24 (7), 1191–1211.

Griffioen, J., 2017. Enhanced weathering of olivine in seawater: the efficiency as revealed by thermodynamic scenario analysis. Sci. Total Environ. 575, 536–544.

Hamilton, J.L., Wilson, S.A., Morgan, B., et al., 2020. Accelerating mineral carbonation in ultramafic mine tailings via direct CO_2 reaction and heap leaching with potential for base metal enrichment and recovery. Econ. Geol. 115 (2), 303–323.

Hanak, D.P., Jenkins, B.G., Kruger, T., Manovic, V., 2017. High-efficiency negative-carbon emission power generation from integrated solid-oxide fuel cell and calciner. Appl. Energy 205, 1189–1201.

Harrould-Kolieb, E.R., Hoegh-Guldberg, O., 2019. A governing framework for international ocean acidification policy. Mar. Pol. 102, 10–20.

Hart, T.J., 1942. Phytoplankton periodicity in the Antarctic surface waters. Discov. Rep. 21, 261–356.

Hartmann, J., West, A.J., Renforth, P., et al., 2013. Enhanced chemical weathering as a geoengineering strategy to reduce atmospheric carbon dioxide, supply nutrients, and mitigate ocean acidification. Rev. Geophys. 51 (2), 113–149.

Hartmann, J., Suitner, N., Lim, C., et al., 2022. Stability of alkalinity in Ocean Alkalinity Enhancement (OAE) approaches – consequences for durability of CO_2 storage. In: Biogeosciences Discussion (in press).

Harvey, L.D.D., 2008. Mitigating the atmospheric CO_2 increase and ocean acidification by adding limestone powder to upwelling regions. J. Geophys. Res. 113, 1669.

Haskell II, W.Z., Hammond, D.E., Prokopenko, M.G., 2015. A dual-tracer approach to estimate upwelling velocity in coastal Southern California. Earth Planet Sci. Lett. 422, 138–149.

Hauck, J., Köhler, P., Wolf-Gladrow, D., Völker, C., 2016. Iron fertilisation and century-scale effects of open ocean dissolution of olivine in a simulated CO_2 removal experiment. Environ. Res. Lett. 11 (2), 024007.

He, J., Tyka, M.D., 2023. Limits and CO_2 equilibration of near-coast alkalinity enhancement. Biogeosciences 20 (1), 27–43.

Heilweck, M., Moore, D., 2021. Saving the planet with appropriate biotechnology: 3. The high seas solution. Mexican J. Biotechnol. (1).

Henderson, G.M., Rickaby, R.E.M., Bouman, H., 2008. Decreasing Atmosphere CO_2 by Increasing Ocean Alkalinity. Available at: https://www.earth.ox.ac.uk/~gideonh/reports/Cquestrate_report.pdf.

Hernández, C.A., Clemente, S., Sangil, C., Hernández, J.C., 2015. High-resolution ocean pH dynamics in four subtropical Atlantic benthic habitats. Biogeosci. Discuss. 12 (23), 19481–19498.

Herzog, H., Adams, E., Akai, M., et al., 2000–2001. Update on the international experiment on CO_2 ocean sequestration. In: Williams, D., Durie, B., McMullan, P., Paulson, C., Smith, A. (Eds.), Proceedings of the Fifth International Conference on Greenhouse Gas Control Technologies. CSIRO Publishing, pp. 399–404. Greenhouse Gas Control Technologies.

Hofmann, G.E., Smith, J.E., Johnson, K.S., et al., 2011. High-frequency dynamics of ocean pH: a multi-ecosystem comparison. PLoS One 6 (12), e28983.

House, K.Z., Schrag, D.P., Harvey, C.F., Lackner, K.S., 2006. Permanent carbon dioxide storage in deep-sea sediments. Proc. Natl. Acad. Sci. USA 103, 12291–12295.

House, K.Z., House, C.H., Schrag, D.P., Aziz, M.J., 2007. Electrochemical acceleration of chemical weathering as an energetically feasible approach to mitigating anthropogenic climate change. Environ. Sci. Technol. 41 (24), 8464–8470.

Hurd, C.L., Law, C.S., Bach, L.T., et al., 2022. Forensic carbon accounting: assessing the role of seaweeds for carbon sequestration. J. Phycol. 58 (3), 347–363.

Ilyina, T., Wolf-Gladrow, D., Munhoven, G., Heinze, C., 2013. Assessing the potential of calcium-based artificial ocean alkalinization to mitigate rising atmospheric CO_2 and ocean acidification. Geophys. Res. Lett. 40 (22), 5909–5914.

Inagaki, F., Kuypers, M.M., Tsunogai, U., et al., 2006. Microbial community in a sediment-hosted CO_2 lake of the southern Okinawa Trough hydrothermal system. Proc. Natl. Acad. Sci. USA 103 (38), 14164–14169.

Isaacs, J.D., Castel, D., Wick, G.L., 1976. Utilization of the energy in ocean waves. Ocean Eng. 3 (4), 175–187.

Israelsson, P.H., Chow, A.C., Adams, E.E., 2010. An updated assessment of the acute impacts of ocean carbon sequestration by direct injection. Int. J. Greenh. Gas Control 4 (2), 262–271.

Jacobs, C.M., Kohsiek, W.I.M., Oost, W.A., 1999. Air–sea fluxes and transfer velocity of CO_2 over the North Sea: results from ASGAMAGE. Tellus B 51 (3), 629–641.

Jain, A.K., Cao, L., 2005. Assessing the effectiveness of direct injection for ocean carbon sequestration under the influence of climate change. Geophys. Res. Lett. 32 (9), L09609. https://doi.org/10.1029/2005GL022818.

Jinnah, S., 2018. Why govern climate engineering? A preliminary framework for demand-based governance. Int. Stud. Rev. 20 (2), 272–282.

Jones, I.S., 2014. The cost of carbon management using ocean nourishment. Int. J. Climate Change Strategies Manag. 6 (4), 391–400.

Jones, I.S., Caldeira, K., 2003. Long-term ocean carbon sequestration with macronutrient addition. In: Proceedings of the 2nd Annual Conference on Carbon Sequestration, Washington, USA.

Jones, D.C., Ito, T., Takano, Y., Hsu, W.C., 2014. Spatial and seasonal variability of the air-sea equilibration timescale of carbon dioxide. Global Biogeochem. Cycles 28 (11), 1163–1178.

Joos, F., Sarmiento, J.L., Siegenthaler, U., 1991. Estimates of the effect of Southern Ocean iron fertilization on atmospheric CO_2 concentrations. Nature 349 (6312), 772.

Kadko, D., Johns, W., 2011. Inferring upwelling rates in the equatorial Atlantic using ^{7}Be measurements in the upper ocean. Deep Sea Res. Oceanogr. Res. Pap. 58 (6), 647–657.

Keller, D.P., Feng, E.Y., Oschlies, A., 2014. Potential climate engineering effectiveness and side effects during a high carbon dioxide-emission scenario. Nat. Commun. 5, 3304.

Keller, D.P., Lenton, A., Littleton, E.W., Oschlies, A., Scott, V., Vaughan, N.E., 2018a. The effects of carbon dioxide removal on the carbon cycle. Curr. Clim. Change Rep. 4 (3), 250–265.

Keller, D.P., Lenton, A., Scott, V., et al., 2018b. The carbon dioxide removal model intercomparison project (CDRMIP): rationale and experimental protocol for CMIP6. Geosci. Model Dev. (GMD) 11 (3), 1133–1160.

Kemper, J., Riebesell, U., Graf, K., 2022. Numerical flow modeling of artificial ocean upwelling. Front. Mar. Sci. 1932.

Khatiwala, S., Tanhua, T., Mikaloff Fletcher, S., et al., 2013. Global ocean storage of anthropogenic carbon. Biogeosciences 10 (4), 2169–2191.

Kheshgi, H.S., 1995. Sequestering atmospheric carbon dioxide by increasing ocean alkalinity. Energy 20, 915–922.

Kirke, B., 2003. Enhancing fish stocks with wave-powered artificial upwelling. Ocean Coast Manag. 46 (9–10), 901–915.

Köhler, P., Abrams, J.F., Völker, C., Hauck, J., Wolf-Gladrow, D.A., 2013. Geoengineering impact of open ocean dissolution of olivine on atmospheric CO_2, surface ocean pH and marine biology. Environ. Res. Lett. 8 (1), 014009.

Kostas, E.T., Adams, J.M., Ruiz, H.A., Durán-Jiménez, G., Lye, G.J., 2021. Macroalgal biorefinery concepts for the circular bioeconomy: a review on biotechnological developments and future perspectives. Renew. Sustain. Energy Rev. 151, 111553.

Koweek, D.A., 2022. Expected limits on the potential for carbon dioxide removal from artificial upwelling. Front. Mar. Sci. 9, 841894.

Kirchner, J.S., Berry, A., Ohnemüller, F., Schnetger, B., Erich, E., Brumsack, H.J., Lettmann, K.A., 2020a. Reducing CO_2 emissions of a coal-fired power plant via accelerated weathering of limestone: carbon capture efficiency and environmental safety. Environ. Sci. Technol. 54 (7), 4528–4535.

Kirchner, J.S., Lettmann, K.A., Schnetger, B., Wolff, J.O., Brumsack, H.J., 2020b. Carbon capture via accelerated weathering of limestone: modeling local impacts on the carbonate chemistry of the southern North Sea. Int. J. Greenh. Gas Control 92, 102855.

Kirkwood, W.J., Walz, P.M., Peltzer, E.T., et al., 2014. Design, construction, and operation of an actively controlled deep-sea CO_2 enrichment experiment using a cabled observatory system. Deep Sea Res. Oceanogr. Res. Pap. 97, 1–9.

Kline, D.I., Teneva, L., Schneider, K., et al., 2012. A short-term in situ CO_2 enrichment experiment on Heron Island (GBR). Nature Scientific Reports 2, 413.

Koide, H., Shindo, Y., Tazaki, Y., et al., 1997. Deep sub-seabed disposal of CO_2—the most protective storage. Energy Convers. Manag. 38, S253–S258.

Kroeker, K.J., Bell, L.E., Donham, E.M., et al., 2020. Ecological change in dynamic environments: accounting for temporal environmental variability in studies of ocean change biology. Global Change Biol. 26 (1), 54–67.

Krom, M.D., Thingstad, T.F., Brenner, S., et al., 2005. Summary and overview of the CYCLOPS P addition Lagrangian experiment in the Eastern Mediterranean. Deep Sea Res. Part II Top. Stud. Oceanogr. 52 (22–23), 3090–3108.

Krumhardt, K.M., Long, M.C., Lindsay, K., Levy, M.N., 2020. Southern Ocean calcification controls the global distribution of alkalinity. Global Biogeochem. Cycles 34 (12) e2020GB006727.

Kumar, N., Anderson, R.F., Mortlock, R.A., et al., 1995. Increased biological productivity and export production in the glacial Southern Ocean. Nature 378 (6558), 675–680.

Lackner, K.S., 2002. Carbonate chemistry for sequestering fossil carbon. Annu. Rev. Energy Environ. 27 (1), 193–232.

Land, P.E., Findlay, H.S., Shutler, J.D., et al., 2019. Optimum satellite remote sensing of the marine carbonate system using empirical algorithms in the global ocean, the Greater Caribbean, the Amazon Plume and the Bay of Bengal. Remote Sens. Environ. 235, 111469.

Laruelle, G.G., Lauerwald, R., Pfeil, B., Regnier, P., 2014. Regionalized global budget of the CO_2 exchange at the air-water interface in continental shelf seas. Global Biogeochem. Cycles 28 (11), 1199–1214.

Lauderdale, J.M., Braakman, R., Forget, G., Dutkiewicz, S., Follows, M.J., 2020. Microbial feedbacks optimize ocean iron availability. In: Proceedings of the National Academy of Sciences. www.pnas.org/cgi/doi/10.1073/pnas.1917277117.

Lawrence, M.W., 2014. Efficiency of carbon sequestration by added reactive nitrogen in ocean fertilisation. Int. J. Glob. Warming 6 (1), 15–33.

Lawrence, M.G., Schäfer, S., Muri, H., Scott, V., Oschlies, A., Vaughan, N.E., et al., 2018. Evaluating climate geoengineering proposals in the context of the Paris Agreement temperature goals. Nat. Commun. 9 (1), 3734.

Leedham, E.C., Hughes, C., Keng, F.S.L., Phang, S.M., Malin, G., Sturges, W.T., 2013. Emission of atmospherically significant halocarbons by naturally occurring and farmed tropical macroalgae. Biogeosciences 10 (6), 3615–3633.

Lenton, A., Matear, R.J., Keller, D.P., Scott, V., Vaughan, N., 2018. Assessing carbon dioxide removal through global and regional ocean alkalization under high and low emission pathways. Earth System Dynamics 9, 339–357.

Li, J.J., Hitch, M., 2015. Ultra-fine grinding and mechanical activation of mine waste rock using a high-speed stirred mill for mineral carbonation. Int. J. Minerals, Metallurgy, Materials 22 (10), 1005–1016.

Linga, P., Kumar, R., Englezos, P., 2007. The clathrate hydrate process for post and pre-combustion capture of carbon dioxide. J. Hazard Mater. 149 (3), 625–629.

Liss, P.S., Merlivat, L., 1985. Air-sea gas exchange rates: introduction and synthesis. In: Buat-Ménard, P. (Ed.), The Role of Air-Sea Exchange in Geochemical Cycling. NATO ASI Series C: Mathematical and Physical Sciences, Vol 185. D. Reidel, Dordrecht.

Lomax, G., Workman, M., Lenton, T., Shah, N., 2015. Reframing the policy approach to greenhouse gas removal technologies. Energy Pol. 78, 125–136.

Lord, N.S., Ridgwell, A., Thorne, M.C., Lunt, D.J., 2016. An impulse response function for the "long tail" of excess atmospheric CO_2 in an Earth system model. Global Biogeochem. Cycles 30 (1), 2–17.

Lovelock, J.E., Rapley, C.G., 2007. Ocean pipes could help the Earth to cure itself. Nature 449 (7161), 403-403.

Ma, Z.W., Zhang, P., Bao, H.S., Deng, S., 2016. Review of fundamental properties of CO_2 hydrates and CO_2 capture and separation using hydration method. Renew. Sustain. Energy Rev. 53, 1273–1302.

Martin, J.H., Fitzwater, S.E., Gordon, R.M., 1990. Iron deficiency limits phytoplankton growth in Antarctic waters. Global Biogeochem. Cycles 4 (1), 5–12.

Martz, T.R., Daly, K.L., Byrne, R.H., Stillman, J.H., Turk, D., 2015. Technology for ocean acidification research: needs and availability. Oceanography 28 (2), 40–47.

Maruyama, S., Tsubaki, K., Taira, K., Sakai, S., 2004. Artificial upwelling of deep seawater using the perpetual salt fountain for cultivation of ocean desert. J. Oceanogr. 60 (3), 563–568.

Maruyama, S., Yabuki, T., Sato, T., et al., 2011. Evidences of increasing primary production in the ocean by Stommel's perpetual salt fountain. Deep Sea Res. Oceanogr. Res. Pap. 58 (5), 567–574.

Matear, R.J., Elliott, B., 2004. Enhancement of oceanic uptake of anthropogenic CO_2 by macronutrient fertilization. J. Geophys. Res.: Oceans 109 (C4).

Meysman, F.J., Montserrat, F., 2017. Negative CO_2 emissions via enhanced silicate weathering in coastal environments. Biol. Lett. 13 (4), 20160905.

Milledge, J.J., Smith, B., Dyer, P.W., Harvey, P., 2014. Macroalgae-derived biofuel: a review of methods of energy extraction from seaweed biomass. Energies 7 (11), 7194–7222.

Montserrat, F., Renforth, P., Hartmann, J., Leermakers, M., Knops, P., Meysman, F.J., 2017. Olivine dissolution in seawater: implications for CO_2 sequestration through enhanced weathering in coastal environments. Environ. Sci. Technol. 51 (7), 3960–3972.

Moras, C.A., Bach, L.T., Cyronak, T., Joannes-Boyau, R., Schulz, K.G., 2021. Ocean Alkalinity Enhancement—avoiding runaway $CaCO_3$ precipitation during quick and hydrated lime dissolution. Biogeosci. Discuss. 1–31.

N'Yeurt, A.R., Chynoweth, D.P., Capron, M.E., Stewart, J.R., Hasan, M.A., 2012. Negative carbon via ocean afforestation. Process Saf. Environ. Protect. 90 (6), 467–474.

Oener, S.Z., Foster, M.J., Boettcher, S.W., 2020. Accelerating water dissociation in bipolar membranes and for electrocatalysis. Science 369 (6507), 1099–1103.

Ohsumi, T., 1995. CO_2 storage options in the deep sea. Mar. Technol. Soc. J. 29 (3), 58–66.

Ohsumi, T., 2003. Ocean storage, including costs and risks. In: Paper Prepared for IPCC Workshop on CO_2 Capture and Storage, 2003.

Ortiz, J., Arístegui Ruiz, J., Taucher, J., Riebesell, U., 2022. Artificial upwelling in singular and recurring mode: consequences for net community production and metabolic balance. Front. Mar. Sci. 8, 73105.

Oschlies, A., Pahlow, M., Yool, A., Matear, R.J., 2010. Climate engineering by artificial ocean upwelling: channelling the sorcerer's apprentice. Geophys. Res. Lett. 37 (4).

Pacala, S., Socolow, R., 2004. Stabilization wedges: solving the climate problem for the next 50 years with current technologies. Science 305 (5686), 968–972.

Palmer, A., Keith, D., Doctor, R., January 2007. Ocean Storage of carbon dioxide: pipelines, risers and seabed containment. In: International Conference on Offshore Mechanics and Arctic Engineering, vol 4269, pp. 495–505.

Pan, Y., Fan, W., Zhang, D., et al., 2016. Research progress in artificial upwelling and its potential environmental effects. Sci. China Earth Sci. 59 (2), 236–248.

Peters, G.P., Geden, O., 2017. Catalysing a political shift from low to negative carbon. Nat. Clim. Change 7 (9), 619.

Penman, D.E., Rugenstein, J.K.C., Ibarra, D.E., Winnick, M.J., 2020. Silicate weathering as a feedback and forcing in Earth's climate and carbon cycle. Earth Sci. Rev. 209, 103298.

Qi, Y., 2021. The neutralization and recycling of red mud–a review. J. Phys.: Confer. Series 1759 (1), 012004. IOP Publishing.

Rajak, R.C., Jacob, S., Kim, B.S., 2020. A holistic zero waste biorefinery approach for macroalgal biomass utilization: a review. Sci. Total Environ. 716, 137067.

Rau, G.H., 2009. Electrochemical CO_2 capture and storage with hydrogen generation. Energy Proc. 1 (1), 823–828.

Rau, G.H., 2011. CO_2 mitigation via capture and chemical conversion in seawater. Environ. Sci. Technol. 45, 1088–1092.

Rau, G.H., 2014. Enhancing the ocean's role in CO_2 mitigation. Atmosphere 100, 760.

Rau, G.H., Caldeira, K., 1999. Enhanced carbonate dissolution: a means of sequestering waste CO_2 as ocean bicarbonate. Energy Convers. Manag. 40, 1803–1813.

Rau, G.H., Knauss, K.G., Langer, W.H., Caldeira, K., 2007. Reducing energy-related CO_2 emissions using accelerated weathering of limestone. Energy 32, 1471–1477.

Rau, G.H., Carroll, S.A., Bourcier, W.L., Singleton, M.J., Smith, M.M., Aines, R.D., 2013. Direct electrolytic dissolution of silicate minerals for air CO_2 mitigation and carbon-negative H_2 production. Proc. Natl. Acad. Sci. USA 110 (25), 10095–10100.

Rau, G.H., Willauer, H.D., Ren, Z.J., 2018. The global potential for converting renewable electricity to negative-CO_2-emissions hydrogen. Nat. Clim. Change 8 (7), 621–625.

Raven, J., Caldeira, K., Elderfield, H., et al., 2005. Ocean Acidification Due to Increasing Atmospheric Carbon Dioxide. Royal Society Policy document 12/05. The Royal Society, London, UK.

Raven, J.A., 2017. The possible roles of algae in restricting the increase in atmospheric CO_2 and global temperature. Eur. J. Phycol. 52 (4), 506–522.

Reith, F., Keller, D.P., Oschlies, A., 2016. Revisiting ocean carbon sequestration by direct injection: a global carbon budget perspective. Earth System Dyna. 7, 797–812.

Reith, F., Koeve, W., Keller, D.P., Getzlaff, J., Oschlies, A., 2019. Meeting climate targets by direct CO_2 injections: what price would the ocean have to pay? Earth System Dyna. 10, 711–727.

Renforth, P., 2012. The potential of enhanced weathering in the UK. Int. J. Greenh. Gas Control 10, 229–243.

Renforth, P., Henderson, G., 2017. Assessing ocean alkalinity for carbon sequestration. Rev. Geophys. 55 (3), 636–674.

Renforth, P., Jenkins, B.G., Kruger, T., 2013. Engineering challenges of ocean liming. Energy 60, 442–452.

Renforth, P., Kruger, T., 2013. Coupling mineral carbonation and ocean liming. Energy Fuels 27 (8), 4199–4207.

Ridgwell, A., Rodengen, T.J., Kohfeld, K.E., 2011. Geographical variations in the effectiveness and side effects of deep ocean carbon sequestration. Geophys. Res. Lett. 38 (17).

Riebesell, U., Fabry, V.J., Hansson, L., Gattuso, J.P., 2011. Guide to Best Practices for Ocean Acidification Research and Data Reporting. Office for Official Publications of the European Communities, European Commission, Brussels, Belgium.

Riebesell, U., Körtzinger, A., Oschlies, A., 2009. Sensitivities of marine carbon fluxes to ocean change. Proc. Natl. Acad. Sci. USA 106 (49), 20602–20609.

Righi, D., 2020. Analysis of enhanced pressurized weathering of limestone as a technology for CO_2 storage. Politechnico Di Milano. www.politesi.polimi.it/bitstream/10589/166552/3/Tesi_Davide_Righi.pdf.

Russell, L.M., Rasch, P.J., Mace, G.M., et al., 2012. Ecosystem impacts of geoengineering: a review for developing a science plan. Ambio 41 (4), 350–369.

Schuiling, R.D., De Boer, P.L., 2011. Rolling stones; fast weathering of olivine in shallow seas for cost-effective CO_2 capture and mitigation of global warming and ocean acidification. Earth Syst. Dyn. Discuss. 2, 551–568.

Scott, A., Oze, C., Shah, V., et al., 2021. Transformation of abundant magnesium silicate minerals for enhanced CO_2 sequestration. Commun. Earth Environ. 2 (1), 1–6.

Siegel, D.A., DeVries, T., Doney, S.C., Bell, T., 2021. Assessing the sequestration time scales of some ocean-based carbon dioxide reduction strategies. Environ. Res. Lett. 16 (10), 104003.

Srednick, G., Bergman, J.L., Doo, S.S., et al., 2020. Shallow coral reef free ocean carbon enrichment: Novel in situ flumes to manipulate pCO_2 on shallow tropical coral reef communities. Limnol Oceanogr. Methods 18 (3), 116–128.

Stark, J.S., Peltzer, E.T., Kline, D.I., et al., 2019. Free Ocean CO_2 Enrichment (FOCE) experiments: scientific and technical recommendations for future in situ ocean acidification projects. Prog. Oceanogr. 172, 89–107.

Stommel, H., Arons, A.B., Blanchard, D., 1956. An oceanographical curiosity: the perpetual salt fountain. Deep Sea Res. 3 (2), 152–153.

Strand, S.E., Benford, G., 2009. Ocean sequestration of crop residue carbon: recycling fossil fuel carbon back to deep sediments. Environ. Sci. Technol. 43 (4), 1000–1007.

Sun, W., Jayaraman, S., Chen, W., Persson, K.A., Ceder, G., 2015. Nucleation of metastable aragonite $CaCO_3$ in seawater. Proc. Natl. Acad. Sci. USA 112 (11), 3199–3204.

Tan, R.R., Aviso, K.B., Foo, D.C., Lee, J.Y., Ubando, A.T., 2019. Optimal synthesis of negative emissions polygeneration systems with desalination. Energy 187, 115953.

Tanaka, Y., 2007. Bipolar membrane electrodialysis. Membr. Sci. Technol. 12, 405–436.

Teng, Y., Zhang, D., 2018. Long-term viability of carbon sequestration in deep-sea sediments. Sci. Adv. 4 (7), eaao6588.

Torres, M.A., Dong, S., Nealson, K.H., West, A.J., 2019. The kinetics of siderophore-mediated olivine dissolution. Geobiology 17 (4), 401–416.

Tsouris, C., Szymcek, P., Taboada-Serrano, P., et al., 2007. Scaled-up ocean injection of CO_2–hydrate composite particles. Energy Fuels 21 (6), 3300–3309.

Tyka, M.D., Van Arsdale, C., Platt, J.C., 2022. CO_2 capture by pumping surface acidity to the deep ocean. Energy Environ. Sci. 15 (2), 786–798.

Visch, W., Kononets, M., Hall, P.O., Nylund, G.M., Pavia, H., 2020. Environmental impact of kelp (Saccharina latissima) aquaculture. Mar. Pollut. Bull. 155, 110962.

Wahl, M., Schneider Covachã, S., Saderne, V., et al., 2018. Macroalgae may mitigate ocean acidification effects on mussel calcification by increasing pH and its fluctuations. Limnol. Oceanogr. 63 (1), 3–21.

Wanninkhof, R., 2014. Relationship between wind speed and gas exchange over the ocean revisited. Limnol Oceanogr. Methods 12 (6), 351–362.

Wanninkhof, R., Pickers, P.A., Omar, A.M., et al., 2019. A surface ocean CO_2 reference network, SOCONET and associated marine boundary layer CO_2 measurements. Front. Mar. Sci. 6, 400 (and see citations).

White, A., Björkman, K., Grabowski, E., et al., 2010. An open ocean trial of controlled upwelling using wave pump technology. J. Atmos. Ocean. Technol. 27 (2), 385–396.

Wilcox, H.A., 1977. The U.S. Navy's Ocean food and energy farm project. In: Ocean Energy Resources (Ed.), Proceedings of the Energy Technology Conference, Houston, Texas, 18–23 September 1977. American Society of Mechanical Engineers, New York, USA.

Willauer, H.D., DiMascio, F., Hardy, D.R., Williams, F.W., 2017. Development of an electrolytic cation exchange module for the simultaneous extraction of carbon dioxide and hydrogen gas from natural seawater. Energy Fuels 31 (2), 1723–1730.

Williamson, P., Boyd, P., Harrison, D., Reynard, N., Mashayekhi, A., 2021. Biologically-based negative emissions in the open ocean and coastal seas. In: Bui, M., Mac Dowell, N. (Eds.), Negative Emission Technologies. Royal Society of Chemistry, London.

Wolf-Gladrow, D.A., Zeebe, R.E., Klaas, C., Körtzinger, A., Dickson, A.G., 2007. Total alkalinity: the explicit conservative expression and its application to biogeochemical processes. Mar. Chem. 106 (1–2), 287–300.

Wu, J., Keller, D.P., Oschlies, A., 2022. Carbon dioxide removal via macroalgae open-ocean mariculture and sinking: an Earth system modeling study. Earth System Dynamics Discussions 1–52.

Xiao, X., Agustí, S., Yu, Y., et al., 2021. Seaweed farms provide refugia from ocean acidification. Sci. Total Environ. 776, 145192.

Xie, H., Wang, Y., Chu, W., Ju, Y., 2014. Mineralization of flue gas CO_2 with coproduction of valuable magnesium carbonate by means of magnesium chloride. Chin. Sci. Bull 59, 2882–2889.

Xu, Z., Fan, W., Xiao, C., Yao, Z., Qiang, Y., Chen, Y., 2019. Experimental and numerical study of current-induced artificial upwelling. Appl. Ocean Res. 87, 26–37.

Yang, D., Le, L.A., Martinez, R.J., Currier, R.P., Spencer, D.F., 2011. Kinetics of CO_2 hydrate formation in a continuous flow reactor. Chem. Eng. J. 172 (1), 144–157.

Yang, J., Zhang, D., Chen, Y., Fan, W., Liang, H., Tan, M., 2017. Feasibility analysis and trial of air-lift artificial upwelling powered by hybrid energy system. Ocean Eng. 129, 520–528.

Yao, Z., Fan, W., Zhang, Z., Pan, Y., Di, Y., Chen, Y., 2020. The hydrodynamic study of artificial upwelling plume in density-stratified ocean. Appl. Ocean Res. 103, 102341.

Yoon, J.E., Yoo, K.C., Macdonald, A.M., et al., 2018. Reviews and syntheses: ocean iron fertilization experiments—past, present, and future looking to a future Korean Iron Fertilization Experiment in the Southern Ocean (KIFES) project. Biogeosciences 15 (19), 5847–5889.

Zeebe, R.E., Wolf-Gladrow, D., 2001. CO_2 in Seawater: Equilibrium, Kinetics, Isotopes. Elsevier Oceanography Series, Vol 65. Elsevier, Amsterdam.

Zeebe, R.E., Archer, D., 2005. Feasibility of ocean fertilization and its impact on future atmospheric CO_2 levels. Geophys. Res. Lett. 32 (9).

Further reading

Albright, R., Cooley, S., 2019. A review of interventions proposed to abate impacts of ocean acidification on coral reefs. Reg. Stud. Mar. Sci. 29, 100612.

Caserini, S., Dolci, G., Azzellino, A., et al., 2017. Evaluation of a new technology for carbon dioxide submarine storage in glass capsules. Int. J. Greenh. Gas Control 60, 140–155.

Harrison, D.P., 2017. Global negative emissions capacity of ocean macronutrient fertilization. Environ. Res. Lett. 12 (3), 035001.

Keeling, R.F., 2009. Triage in the greenhouse. Nat. Geosci. 2 (12), 820.

Krause-Jensen, D., Duarte, C.M., 2016. Substantial role of macroalgae in marine carbon sequestration. Nat. Geosci. 9 (10), 737–742.

Projects and Organizations

Atmocean; Wave energy and artificial up/downwelling system developer: https://atmocean.com.

Blue Carbon Initiative; thebluecarboninitiative.org.

Carboocean IP; EU-funded assessment of marine carbon sources and sinks, focusing on the Atlantic and Southern Oceans over the past 200 and next 200 years: www.carboocean.org.

Coastal ESW (Enhanced Silicate Weathering) Project; Antwerp University project to advance enhanced silicate weathering: coastalesw.com.

Cquestrate; An "open source" project to develop ocean liming: www.cquestrate.com.

Desarc-Maresanus project; Politecnico Milano project researching OAE by slaked lime dispersal in the open ocean: www.desarc-maresanus.net.

Ebb Carbon; Electrochemical indirect capture: www.ebbcarbon.com.

Global Mangrove Alliance; Global community committed to protecting mangroves and securing their environmental role: www.mangrovealliance.org.

GLODAP (Global Ocean Data Analysis Project): cdiac.esd.ornl.gov/oceans/glodap.

MBARI Free Ocean Carbon Dioxide Enhancement (FOCE): www.mbari.org/at-sea/cabled-observatory/mars-science-experiments/foce.

Ocean artUP; Research project investigating artificial upwelling primarily for food production; https://ocean-artup.eu.

Origen Carbon Solutions; Decarbonizing lime production: https://origencarbonsolutions.com.

Project Vesta; https://vesta.earth.

SmartStones; The Olivine Foundation: http://smartstones.nl.

Woods Hole Oceanographic Institution: www.whoi.edu.

Chapter 17

Carbon dioxide utilization

Peter Styring, Stephen McCord and Steve Rackley

17.1 Introduction

Carbon dioxide utilization (CDU) is the name given to a broad, somewhat ambiguously defined group of processes that are underpinned by a single commonality—namely the "use" of CO_2 in the production of a product or the delivery of a service. Utilization processes are transformative; CO_2 is transformed into a new chemical species.

CDU is often conflated with CCS, with the term "carbon capture, utilization, and storage" (CCUS) used to refer to this amalgamation of CO_2-driven processes. While this is logical to a degree, common definitions of CCUS and discussions around the topic fail to capture the nature of CDU, and where it differs from CCS. Recent developments to redefine the technologies have taken place and the term "conversion" has been adopted as a natural transition toward a universal description.

There are many complete textbooks, some even stretching to two volumes, that have been published in recent years to highlight CDU technologies (Aresta, 2019; Aresta and Dibenadetto, 2021; Styring et al. 2011, 2014; Centi and Peranther, 2014; North and Styring, 2019a, 2019b), and it is clearly beyond the scope of a single chapter to give an in-depth analysis of the field. With that in mind, this chapter aims to highlight the general concepts, the current state of the art, recent advances, and some of the socio-political thinking behind the use of CDU as a climate change mitigation technology that can also provide an additional and significant income stream. One area that is often neglected in CDU textbooks is biological conversion processes. This is partly due to a lack of maturity in such systems and partly to land use issues, particularly relating to the complex land-food-energy nexus, where competing priorities exist in different societies and locations. This chapter includes a deeper analysis of biological CDU processes.

The best form of mitigation is to avoid the production of waste in the first place (Lansink, 2018), which means, in this case, avoiding new fossil carbon entering the supply chain. In essence, this equates to a move away from a linear economic approach of continuous extraction, use, and disposal, to a circular economy where carbon is used as a regenerable feedstock. Much emerging climate legislation focuses on permanent CO_2 storage as the end goal of carbon dioxide removal (CDR), and a 100-year storage lifetime has been adopted as a somewhat arbitrary benchmark of permanence, looking only at the current linear economy and ignoring the possible circularity of a new carbon economy.

Meeting this permanence requirement may seem problematic for CDU technologies such as the production of synthetic fuels, discussed in Section 17.3. However, as these fuels will continue to be produced using captured CO_2, the net result compared to the fossil-carbon-based "extract-use-emit" alternative is accumulating negative emissions. Looked at from a mass balance perspective, the circular carbon economy effectively recycles the carbon emitted from the use of these fuels, and if the cycle is sustained for 100 years the permanence requirement will have been met. During the transition, when fossil-carbon is being progressively replaced by recycled carbon, CDU is functionally equivalent to direct air capture (DAC) using RE or the production of biomass, and, while the cycle persists, carbon is effectively sequestered within a new net-zero economic carbon cycle. Unlike DAC however, CDU has an additional benefit in that its products have economic value rather than being a waste that incurs additional cost for disposal.

To appreciate the mitigation impact of CDU, as with biomass production, it is therefore necessary to consider the whole (repeated) lifecycle, as a linear mindset will not give the complete picture. Moving away from a linear mindset to one of carbon accounting, using recycled carbon with known provenance, reveals the full impact of CDU achieved through the avoidance of further virgin fossil carbon use.

Negative Emissions Technologies for Climate Change Mitigation. https://doi.org/10.1016/B978-0-12-819663-2.00005-8

17.1.1 Carbon sequestration versus utilization

CO_2 sequestration and utilization are different functions with a common aim of reducing or avoiding greenhouse gas (GHG) emissions:

- *Sequestration* as a function seeks to reduce net GHG emissions through the long-term storage of CO_2, achieving carbon neutrality in the case of fossil carbon storage (i.e., CCS) or negative emissions in the case of biogenic carbon (as in BECCS) or DAC
- *Utilization* as a function seeks to (re)use CO_2 as a carbon feedstock, achieving negative emissions relative to a counterfactual case in which the product or service is delivered using virgin fossil carbon.

In some cases, depending on the lifetime of the product, both of these functions may be fulfilled by a single product, process, or service, an example being CO_2 mineralization for building materials.

The IPCC's Sixth Assessment Report (IPCC, 2022) now firmly includes CDU as a key mitigation technology, and waste avoidance is an important tool in the mitigation arsenal that can reduce and avoid fossil carbon emissions while sustaining a vital supply chain. Carbon-containing materials play a key role in the world today, and moving away from this reliance on carbon is difficult in some sectors (e.g., long-distance freight or air transport), and impossible in others (e.g., parts of the chemical industry, especially organic chemicals). Replacing this function, previously provided by fossil carbon, is a key driver for CDU.

Renewable energy sources (RE) and hydrogen (H_2) are prerequisites for CDU in most cases, mineralization processes being an exception. CDU itself is not primarily about producing energy, although producing fuels is a major goal, but about transforming existing energy supplies and carbon into something with more value at a time and location required by the end user. The term "consumer" is avoided deliberately here, recognizing the need for society to move away from a mindset of carbon consumption to one of carbon custodianship, in which the "end user" has a duty of care to ensure its sustainable (re)use. Alongside this, in the move to a more circular economy, consideration needs to be given to the concept of "carbon provenance," in the same way that the Soil Association guarantees the provenance of organic products within the food supply chain (Links to projects and organizations referred to in the text are provided in Section 17.10.2).

Neither sequestration nor conversion are stand-alone solution, and both will be necessary, together with other approaches such as reduced or avoided consumption, to meet the goals of the Paris Agreement.

17.1.2 Defining utilization within the confines of CDU

Processes in which CO_2 is used as an input, can be broadly split into two categories—transformative and nontransformative.

Nontransformative processes cover many diverse applications, ranging from carbonating drinks, which have been produced since the mid-18th century (Priestley, 1774), decaffeination of coffee and tea, use as a dry-cleaning fluid, and enhanced oil and gas recovery (EOR, EGR). The key aspect of all of these technologies is that the chemical nature of the CO_2 molecule does not change during its use; CO_2 is effectively a working fluid and provides a service. These processes cannot be considered CDU as they do not meet the functional requirement of reducing emissions.

The inclusion within utilization approaches of CO_2 as a working fluid has been contentious, and the use of different terminologies (CCU, CCUS, CCS, CDU) has also caused confusion, highlighting the need for a unified terminology (Olfe-Kraeutlein et al., 2022). Progress in resolving these issues has recently been made with the adoption within the US DoE's Carbon Utilization Program (NETL, 2022) of the term "Carbon Conversion," differentiating between utilization in its widely used sense, and the use of CO_2 as a working fluid, which is now included within NETL's carbon storage pathway (Fig. 17.1).

Carbon conversion encompasses the transformative uses of CO_2—the breaking of the O=C=O structure—and the formation of new bonds to create a range of new products. Transformative CDU, which positions CO_2 at the center of a new circular carbon economy, can be split into three broad categories:

- *Chemical and catalytic conversion*: where CO_2 is used as a carbon source for the direct production of a range of molecules including bulk and fine chemicals, synthetic fuels, and polymers. Chemical conversion covers production pathways such as Fischer–Tropsch, direct hydrogenation, plasmolysis, electrochemical conversion, and photochemical conversion. Also in this category are atom-efficient CO_2 insertion reactions to make carbonates, carbamates, polycarbonates/carbamates, urea, and urea derivatives.

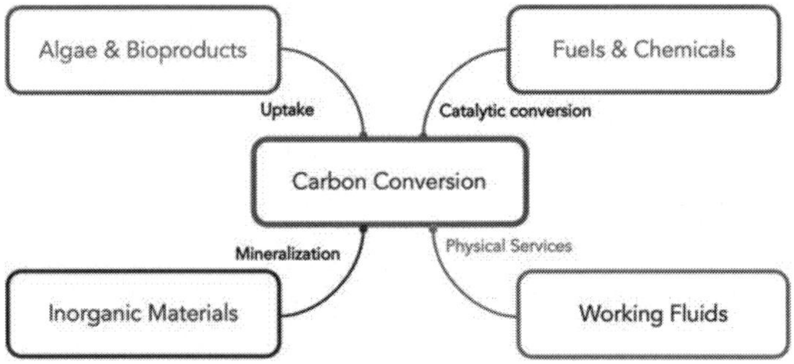

FIGURE 17.1 US DoE definition of carbon conversion in the utilization landscape. *From NETL (2022).*

- *Accelerated mineralization and inorganic materials*: where CO_2 is used to produce inorganic products, including building materials such as aggregates and cementitious materials, typically by reaction with alkaline salts and silicates. These processes are highly exothermic and occur spontaneously and stoichiometrically on mixing.
- *Biotechnological conversion*: where CO_2 is used as a carbon feedstock for natural photosynthesis and bacterial or other microbial systems. Algae production in particular is a research area of much interest within the scope of CDU. It should be noted however that biological reactions often occur at much lower rates than chemical reactions.

CO_2 is the product of the combustion of carbonaceous materials such as hydrocarbons, wood (lignocellulose), or other plant-based materials. As energy is released in combustion, the CO_2 molecule has a low Gibbs energy (−394 kJ/mol; see Fig. 10.1). Inorganic carbonates have a still lower Gibbs energy than CO_2, so that carbonation reactions are exothermic. For example, the carbonation reaction of calcium oxide (CaO) with CO_2 to produce calcium carbonate ($CaCO_3$) has an enthalpy of reaction of −179 kJ/mol. Therefore, mineralization is an energy-producing process that not only makes carbon capture relatively simple, but also produces value-added products with considerable economic potential. All other products in a CO_2 economy are higher energy species, so that energy needs to be added to the production process. In the case of fuel production, the high energy comes typically from hydrogen which, with the help of catalysts, reacts with CO_2 to form energy-dense C−H bonds. The importance of mineralization and fuel production within the CDU portfolio is reflected in a number of recent assessments (GCI, 2016; Sick et al., 2022), which highlight accelerated mineralization (aggregates and cement curing) and fuels production as having a combined GHG mitigation potential of ~7 $GtCO_2$/y, and a potential market size of ~\$0.8−1.0 trillion/y by 2050.

Fig. 17.2 shows schematically the supply chains that can be addressed through switching from fossil-based feedstocks to CDU processes, starting with the input of low carbon energies such as solar, wind, hydro, and nuclear. To ensure low emissions from CDU processes, hydrogen as an input should be produced by water electrolysis (green hydrogen) or by reforming biomethane, with CO_2 captured and recycled back into the biosphere or otherwise sequestered. Processes involving the electro- or photocatalytic reduction of CO_2 can also lead to the production of low-carbon hydrogen as an intermediate product. Emissions can also be minimized by recovering the reaction heat from accelerated mineralization and using this elsewhere in the CDU ecosystem.

The other product lines will require added energy. These include polymers and advanced materials, such as carbon fibers and graphene, synthetic fuels (synfuels or e-fuels), fertilizers such as urea and other agritech products, and a wide range of personal care, home care, and health care products, including surfactants and pharmaceuticals. Treating CO_2 as a valuable feedstock, rather than a waste attracting disposal costs, and developing an income generation approach to CO_2 mitigation from both high- and low-volume products, will help to accelerate the commercial uptake of CDU.

The concept of fossil carbon avoidance is key to developing a circular economy in the chemicals sector. The UK National Interdisciplinary Centre for Chemical Circular Economy (NIC3E) is spearheading the replacement of fossil carbon in a drive toward "nonpetro" chemical manufacture. Fig. 17.3 shows how, by removing fossil carbon from the front end of the supply chain, CO_2 can become the primary feedstock, together with water and renewable or low-carbon energy, in a new e-chemical supply chain.

The second level e-chemicals would be hydrogen and carbon monoxide (CO) with CO_2, CO, and H_2 enabling the production of further products as the supply chain expands. Mineral carbonates are also part of the second product level. The higher levels include hydrocarbons (alkanes and alkenes), heteroatom products (alcohols, ethers, esters, carbamates), polymers, and advanced carbon materials (fibers, graphene). Aromatic compounds have been a challenge, although a

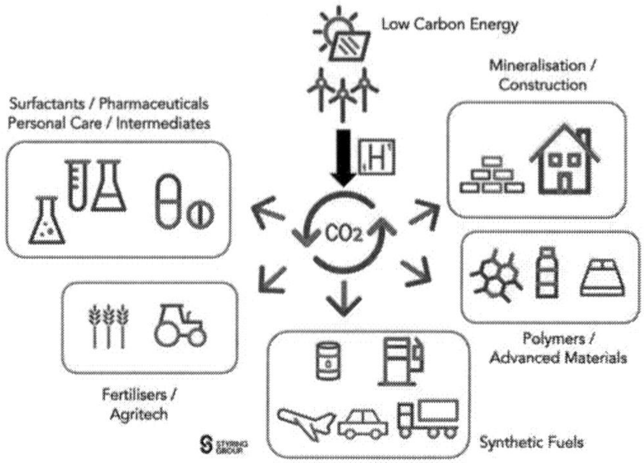

FIGURE 17.2 Potential for CDU in a wide range of manufacturing sectors.

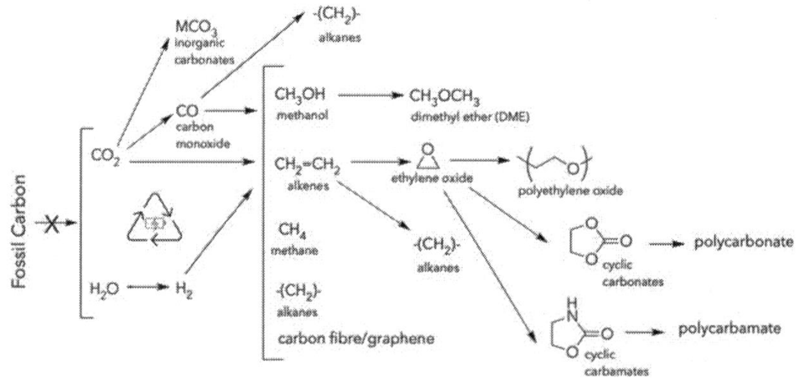

FIGURE 17.3 Transitioning from a petrochemical to a CO_2-based e-chemical supply chain.

recent publication by Miller et al. (2022) shows the potential for the production of benzene derivatives from CO_2. However, in the short term, it is likely that aromatics will still need to come from the petrochemicals industry, or feasibly from lignin-rich feedstocks (Faruk and Sain, 2015).

In other sectors, urea production accounts for the largest use of CO_2 by mass with the global market accounting for around 230 $MtCO_2$/yr IEA, (2019). While this is small in comparison with global emissions of around 40 $GtCO_2$/yr, it is still an order of magnitude higher than the CO_2 currently injected into the subsurface (Global CCS Institute, 2021), predominantly for EOR and with geological storage other than EOR accounting for an average of just 2–5 $MtCO_2$/yr from 2019 to 2021 (Chapter 15).

17.1.3 Determining GHG reduction potential

Another area in which utilization differs from sequestration is in how the reduction in GHG is calculated. As stated above, sequestration as a function intends to store CO_2 long term, and the net emissions reduction is determined as the amount of CO_2 stored less the amount of CO_2 produced in the life cycle of the sequestration process.

Thus, the net negative emissions achieved in a BECCS or DACCS project are determined using a life cycle approach that inventorizes and quantifies all emissions arising upstream and downstream of the capture and storage process (see Section 2.4).

In seeking primarily to avoid fossil GHG emissions, CDU requires a different approach, using a baseline or a benchmark to compare the life cycle emissions of the CDU process with those of the alternative process based on fossil carbon usage, the so-called counterfactual, a lower value for the CDU process or product indicating that emissions have been avoided. Selecting a benchmark that ensures a fair comparison between product systems can be difficult, particularly since whole life cycle emissions need to be calculated for both cases.

The focus on the whole product life cycle is one attribute of an approach called "life cycle thinking" (LCT), which goes beyond the traditional focus on the production phase, to consider upstream (e.g., resource extraction, transformation, and transportation) and downstream (e.g., use and disposal) phases in greater detail. A focus on the whole life cycle allows for a clearer assessment and analysis of the impact a product has on the economy, society, and the environment.

Within the LCT toolkit, a range of techniques allow the assessment of different aspects of sustainability, such as economic and social impact (including life cycle costing (LCC), and social life cycle assessment (S-LCA)), in addition to the more common focus on assessing environmental impact. The effectiveness of CDU in reducing net fossil GHG emissions requires such an assessment of environmental impact, using standard life cycle assessment (LCA) approaches, as discussed in Chapter 2. Further resources on LCT, LCA, and associated tools can be found in general overviews (e.g., Toniolo et al., 2020), ISO standards (ISO, 2006), detailed guidance (ILCD, 2010), academic papers (Zimmermann et al., 2020; Müller et al., 2020), and beyond. Guidance on conducting LCA and other impact assessments can also be found (Mutchek et al., 2022) to help deal with specific methodological pitfalls that can arise when assessing CDU technologies and products.

CDU is a key mitigation technology in the fight against climate change; avoiding the use of virgin fossil carbon mitigates emissions, and the reuse of captured carbon to produce materials and fuels creates a circular economy that sustains this benefit.

17.2 Introduction to CDU technologies

The widespread deployment of modern CDU technologies remains a challenge, not least because of the low availability of captured and purified CO_2. Despite this, there are a growing number of examples of developed CDU processes and products becoming available, and a number of plants are currently under construction. These are summarized in the Zero Emissions Platform Database (Section 17.11.2) and will be briefly mentioned in the relevant sections below. CDU is not new, and in some industries, the utilization of CO_2 has a long history at a commercial scale. Salicylic acid, an intermediate in the production of aspirin, has been synthesized from CO_2 since the 19th century, and, in the 1920s, the Meisner–Bosch process used CO_2 in the synthesis of urea fertilizer—one of the most important global chemicals that enhanced global food production and hence economic and population growth, especially in developing nations.

Early modern CDU adoptions have typically occurred in situations with specific economic incentives (e.g., an availability of low-cost, low-carbon energy, or the existence of waste processing fees that can be avoided by using the waste material as a feedstock) or where advantageous use can be made of thermodynamic conditions (e.g., mildly exothermic carbonization or carbamation reactions). In some cases, both of these conditions may be present, for example, in the development of CDU-based building aggregates by companies such as Carbon 8 and others.

As noted earlier, CDU processes for the production of aggregates, cement, and synthetic or e-fuels have the greatest mitigation and income generation potential (GCI, 2016), with polymers (low volume-high value) and methanol (low volume-medium value) at significantly lower levels. Methanol, however, can also be included in the fuels category as it is an intermediate in many processes, including methanol to olefin or gasoline (Maitlis and de Klerk, 2013), and dimethyl ether (DME) synthesis (Styring and Dowson, 2021; Styring et al., 2021a). As these areas show the greatest promise, these will be the focus of the following discussion. Biological processes such as algae cultivation have had limited success for a number of reasons, including reactor design to maximize light penetration into the bulk, high water usage, algae stability, and the vast land area needed for large-scale production. While algal synthesis is fast compared to most biological processes, it is nevertheless far slower than catalyzed chemical reactions (Supasitmongkol and Styring, 2014).

A barrier to commercialization in many instances is the need for low fossil hydrogen for the chemical reduction of CO_2. Hydrogen is typically produced by steam methane reforming (SMR) of natural gas—a fossil-based technology with associated GHG emissions (Section 8.2.2). Recent roadmaps and policies for the production of clean hydrogen have included so-called "blue hydrogen" from SMR of natural gas combined with CCS to sequester the CO_2, or "green hydrogen" from water electrolysis, although a recent assessment of the life-cycle emissions of blue hydrogen suggests it can make a little contribution toward net-zero goals, primarily as a result of fugitive methane emissions from natural gas production (Howarth and Jacobson, 2021). Green hydrogen is also challenged, in this case by high costs, which would be reduced with the continuing decline in the costs of PV cells (Verlinden, 2020). Alongside RE, green hydrogen also requires clean water, which will put further stress on global water resources, although efforts are continuing to create commercially viable electrolysis processes that can tolerate brackish or saline water (Tong et al., 2020).

17.3 Fuels

Of the high-energy processes, the conversion of CO_2 to fuels, either using excess RE or by artificial photosynthesis (Chartier et al., 2016), holds out the possibility of true impact scale fossil fuel substitution. Biofuel production through CO_2 utilization in algal biomass production or by bioenergy crops, remains more of a challenge, as biomass takes considerably longer to produce than synthetic fuels through thermo-catalytic methods.

The major chemical and fuel production options are discussed in the following sections, while new routes continue to be developed. However, while there is a rich chemistry that can be applied to fuel production, care must be taken when assessing the potential of these reactions for commercial e-fuel production. In particular, catalysts need to be based on Earth-abundant metals and other elements and must take into account social impacts, linked to geo-political availability and ethical supply routes. In the EU, for example, restrictions preclude the use of iridium and cobalt (European Commission, 2020), while high costs limit the potential use of exotic ligand architectures, especially for soluble (homogeneous) catalysts.

17.3.1 Methane

The Sabatier reaction (Müller et al., 2013) is the catalytic reduction of CO_2 to produce methane, as shown in Reaction 17.1.

$$CO_{2(g)} + 4H_{2(g)} \rightarrow CH_{4(g)} + 2H_2O_{(l)} \qquad 17.1$$

The reaction has both positive and negative aspects, resulting in its limited commercial use. The positive aspect is that the e-methane produced is chemically the same as natural gas (NG) and can therefore be used directly in the existing NG energy infrastructure without modification of the grid or of end-user appliances. However, the methane produced has a higher global warming potential than the CO_2 feed, so it is not necessarily a mitigation option, especially as fugitive methane emissions from the process are difficult to avoid. Secondly, the reaction consumes valuable hydrogen gas which, if combusted directly, does not emit greenhouse gases. Thirdly, the process is wasteful, both economically and environmentally, since 50% of the hydrogen feedstock reacts to produce water.

Catalysts have been developed by a number of companies for high- and low-temperature methanation of CO_2 to produce synthetic natural gas (SNG), using a mixed gas stream of carbon monoxide, carbon dioxide, and hydrogen resulting, for example, from biomass gasification.

17.3.2 Methanol

Methanol production is a major focus of deployed and developing CDU technologies, in part as an intermediate in the production of many common chemicals. Carbon Recycling International (CRI), based in Iceland, opened the world's first industrial "e-methanol" plant in 2012, converting CO_2 to methanol using electrolysis-derived hydrogen. The plant had a production capacity of 1.3 kt-methanol/yr, and a second plant opened in 2015 with a capacity of 4 kt-methanol/yr. A third plant at a much larger scale of 110 kt-methanol/yr was opened in China in 2021 to utilize a hydrogen by-product stream. A fourth 116 kt-methanol/yr plant, recycling 160 ktCO_2/yr, is planned to be opened in Norway in 2024 and will be the largest of its kind in operation. While these plants remain relatively small in comparison to their fossil counterparts, the decade following the start-up of the first plant has seen the deployed scale increase by a factor of nearly 100. The creation of small-scale local plants also highlights the benefit of distributed synthesis, as production of methanol at the point of use can further reduce end-to-end emissions.

Production of e-methanol is also planned in a number of net zero hubs under development in the EU. In the "Power to Methanol" project in Antwerp, Belgium, a consortium of seven partners will produce methanol from CO_2 and H_2 derived from RE-powered water electrolysis. Construction of an 8 kt/yr demonstration plant is expected to commence in 2022. A similar project is also under development in Ghent, Belgium, where the North CCU Hub (consisting of 16 partners and a further 15 supporters) plans to start up the "North-C-Methanol" demonstration plant by 2024. This plant will produce 43 kt-methanol/yr making use of captured CO_2 and H_2 from a 63 MW electrolyzer. The plant will be upgraded to full commercial scale, with a 300 MW electrolyzer, producing both methanol and ammonia by 2028, with a further doubling of the electrolyzer size and production of formic acid planned by 2030.

Another example of a multiproduct CDU project is the Westküste 100 project in Schleswig–Holstein, Germany, where a cross-industry partnership aims to produce a range of CDU products including cement, methanol, and various e-fuels for the aviation and motor industries.

A novel compact electrified steam reforming (eSMR) process to produce syngas has recently been developed by Haldor Topsøe and partners (Wismann et al., 2019). The process uses methane from biogas and RE, and 40% of the CO_2 generated in the eSMR reaction is used together with the syngas to produce methanol. A demonstration plant with an annual capacity of 10 m^3-methanol, opened in 2022 at Aarhus University, Foulum, Denmark, and, as well as being CO_2-neutral, the process is expected to be cost-competitive with fossil-derived methanol.

Methanol is a major component of the petrochemicals supply chain and is an important intermediate in the production of synthetic fuels such as DME and hydrocarbons. The importance of synthetic fuel production was highlighted by the European Union in the Clean Planet for All publication (European Commission, 2018), which considered several scenarios for both CCS and CDU. A concerning aspect of the analysis was the prediction of little growth until around 2035, well short of what would be required to achieve net zero by mid-century; real impact was seen post-2050 with continued growth until 2070. It was highlighted in the report that CCS and CCU were unlikely to be effective in a fossil carbon energy scenario due to the difficulty, complexity, and economic cost of producing CO_2 above 90% purity (IEA GHG, 2012). The analysis predicted that the only viable negative emissions scenarios were those where BECCS/U or DAC was deployed and where synthetic or eFuels are produced. However, this only resulted in the capture of 400–600 $MtCO_2$/yr by 2070 for different scenarios. Approximately 46% of the CO_2 captured would most likely be geologically sequestered with the remainder converted into materials, especially synthetic fuels.

17.3.3 Dimethyl ether

The use of DME (CH_3OCH_3) as a fuel is seen as one of the major tools in driving the defossilization of transport, particularly long-haul commercial road traffic. DME is a drop-in diesel replacement that has lower tailpipe emissions despite having only half the energy density of the fossil alternative. DME has also been shown to have a higher combustion efficiency than diesel—more of the chemical energy being delivered as useful heat. DME can be produced thermo-catalytically from methanol (which may itself be produced either directly from CO_2 or via the partially reduced carbon monoxide (CO) intermediate, in a simple condensation reaction) or directly from CO_2 in processes that are typically electrochemically driven (Fig. 17.4).

One of the first such processes to be commercialized was developed by Oberon Fuels, a Californian-based company that distributes a DME-propane blend to off-grid users in the United States, in partnership with Suburban Propane.

Research by Ford Motors in Aachen, Germany, has shown that DME derived from RE and captured CO_2 is a much cleaner fuel in terms of tailpipe emissions (Willems, 2018), and other studies have shown it to have lower lifecycle emissions than equivalent battery electric vehicles (BEVs) (Ellingsen, 2016; Styring et al., 2021b). The latter is due to the higher emissions during production and end-of-life disposal of BEVs compared to internal combustion engine (ICE) vehicles, and because of variable renewable grid mixes in different countries. Further emissions reductions can be achieved by colocating RE generation, carbon capture, and DME production, and this approach will be encouraged by regulatory issues regarding the allocation of emission, particularly in the EU.

When considering emissions reduction, a further benefit of DME production is the selectivity of the synthesis process, unlike the production process for fossil- or synthetic diesel that generates intermediate products with distributions of carbon chain lengths that incur additional processing costs through the need for cracking and fractional distillation (Styring and Dowson, 2021).

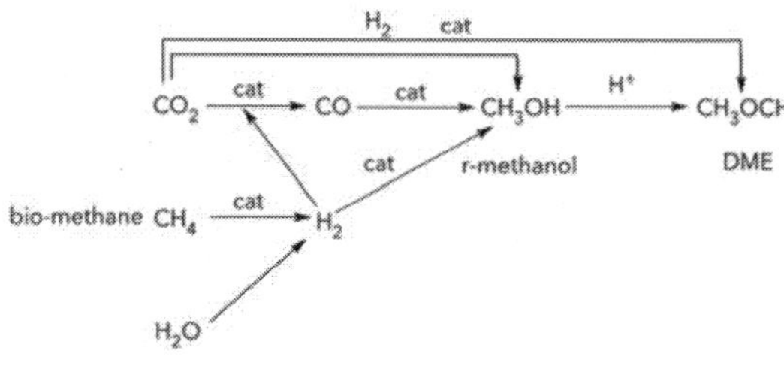

FIGURE 17.4 Pathways to DME synthesis form CO_2. *From Styring and Dowson (2021), with permission.*

17.3.4 Hydrocarbons

The production of hydrocarbons from CO_2 and CO is not new, the Fischer–Tropsch (FT) process being developed by SASOL in the wake of the Second World War to produce transport fuels that were otherwise embargoed (Maitlis and de Klerk, 2013). The drive to produce synthetic hydrocarbon fuels has previously been accelerated at times of crisis, and the climate emergency has focused attention on the need for fossil feedstocks to be reduced in a managed transition, eventually eliminating virgin fossil-carbon use. The reason behind a focus on hydrocarbons is that they can directly replace fossil derivatives in legacy equipment, with little or no modification. Synthetic fuels, can also enhance engine performance, evident, for example, from flame tests that show that Neste's MY Renewable Diesel^{TM} burns with a much cleaner flame, producing substantially lower particulate emissions.

Extensive research has been undertaken into producing hydrocarbons from CO_2 and reviews are plentiful (Jiang et al., 2010; North and Styring, 2019b; Styring and Dowson, 2021), while new methods are continuously evolving. Other notable recent publications include an extensive review of solar fuels from CO_2 by Gong et al. (2022), with an emphasis on synthetic hydrocarbons, and seminal work on solar fuels by Loh et al. (2021). The ultimate aim of research into hydrocarbon synthesis is selectivity toward a specific molecular weight or carbon chain distribution. In broad terms, C-8 hydrocarbons are sought after for gasoline engines, C-12 for kerosene/jet engines, and C-18 for diesel engines. The problem is that while there has been considerable progress in the development of catalysts and processes, there tends to be limited opportunity for selectivity. There is a balance between reaction temperature, pressure, and catalyst choice that can direct the selection of chain length distribution, but this selectivity is often binary. Low-temperature Fischer–Tropsch (LTFT) cobalt-based catalysts tend to show greater selectivity toward methane, one of the less desired hydrocarbons, and short *n*-alkenes, whereas high-temperature (HTFT) iron-based catalysts favor high molecular weight waxes, typically paraffins, that need to be cracked to produce a range of more desirable shorter chain products (Maitlis and de Klerk, 2013). A major advance in selectivity has, however, been recently reported by Yao et al. (2020) using an iron-based catalyst system that produces a narrow distribution of alkenes centered on octene ($CH_2CHC_6H_{13}$, or C_8H_{16}). Reactor design also plays a significant role in the products formed, with HTFT reactions typically being two-phase iron-catalyzed reactions over fluidized beds, and LTFT reactions over cobalt being performed in three-phase slurry or fixed-bed tubular reactors (Steynberg and Dry, 2004).

The German company Sunfire has led the way in synthetic hydrocarbon production, with CO_2 being obtained from point source capture as well as DAC in collaboration with Swiss company Climeworks (see Chapter 13). The Sunfire process involves coelectrolysis using a solid oxide electrolysis cell (SOEC) system to yield syngas ($CO + H_2$).

Direct solar conversion of CO_2 into fuels is also gaining significant attention, with "drop-in" aviation fuels produced from CO_2 and water, both extracted from air, converted using solar energy as the reaction driving force (Schäppi et al., 2022). A 5 kW_{th} demonstration plant has been built to produce transport fuels such as methanol and kerosene operating under field conditions. The solar redox reaction is carried out using a ceria-based (CeO_2) catalyst to produce syngas, which is then converted to methanol using a gas-to-liquid (GTL) reactor modified to produce Fischer-Tropsch liquids.

Shell, in collaboration with KLM, used a Hydrogenics PEM electrolyzer to produce hydrogen at the Shell Pernis refinery, which reacts with CO_2 captured from the refinery and from agricultural waste fermentation. Reduction in a reverse water–gas shift (RWGS) reaction produces CO, which is then converted in an FT process to a wax, which, in turn, is hydrocracked to produce kerosene. In an initial demonstration, 0.5 m^3 of syn-kerosene was produced, blended with 50% Jet A (the current maximum permissible synthetic content) and used to fuel a passenger aircraft. Blending is seen as part of the transition from fully fossil-carbon fuels to a fully synthetic kerosene, with the interim use of blue hydrogen (i.e., from fossil methane) until enough RE is available to produce green hydrogen at the required scale. As part of the Clean Skies for Tomorrow Coalition, KLM aims for 10% of its fuel use to be sustainable aviation fuel (SAF) by 2030. The Jet Zero Consultation (UK Gov, 2021) is a UK-based approach to SAF production, using a variety of sustainable feedstocks, building on the Royal Society (2019) review of synthetic transport fuels.

17.3.5 Electrochemical CO_2 conversion

Lackner and colleagues (Lackner et al., 1999) noted that, once RE had become cost competitive with fossil fuels, it could be used to convert CO_2 captured from the atmosphere into hydrocarbons, essentially "refilling the carbon with energy."

In electrochemical processes, an input of electrical energy drives conversion reactions in an electrolytic cell in which CO_2 is reduced on the cathode while oxygen evolves at the anode. CO_2 can be converted to a range of fuels and fuel-precursors by this means including CO, formic acid (HCOOH), methane (CH_4), and ethylene (C_2H_4). The cathode typically includes a catalytic layer designed to improve the kinetics of the reduction reaction and also to be selective for a specific reaction product, for example, a copper (Cu) electrode for methane production or silver (Ag) for CO production (Fig. 17.5).

A major drawback of electrochemical conversion is the cost of electricity, and this process will only be economically viable if the energy input comes at near-zero cost, for example, by consuming RE during periods when generation exceeds demand. Electrochemical CO_2 reduction can then be seen as a method of storing RE in a form that can be used in sectors not normally amenable to RE, such as aviation.

17.4 Biofuels

Although biofuels can be produced from terrestrial biomass crops, the aqueous cultivation of microalgal biomass as a feedstock, using captured CO_2 as a carbon source, has a number of advantages that can lead to higher productivity per unit land usage, as summarized in Table 17.1.

These advantages are balanced by the cost of systems to supply CO_2, and for continuous cultivation and harvesting, which are significant due to low biomass intensity, resulting from the low concentration and microscopic nature of algal cells (Benemann, 1997).

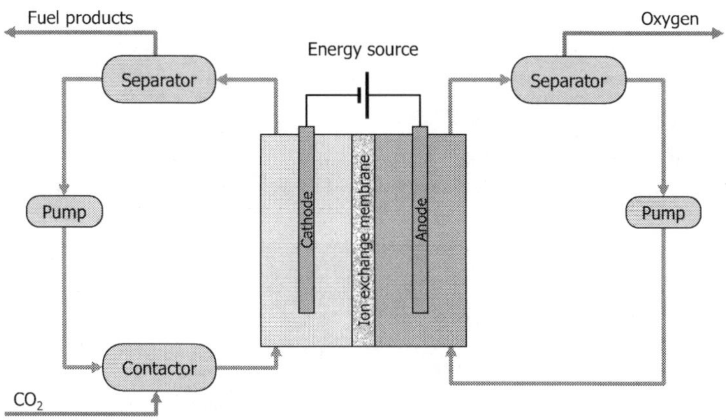

FIGURE 17.5 Schematic CO_2 electrolyzer system.

TABLE 17.1 Advantages of aqueous microalgae cultivation for biomass production.

System feature	Aqueous cultivation advantage
Photosynthetic efficiency	Higher photosynthetic efficiency leads to carbon fixation rates 10–50 times higher in microalgae compared to terrestrial plants
Land usage	With efficient access to water, CO_2, and nutrients, microalgae can lead to higher productivity per unit of land usage compared to terrestrial crops
Process simplicity	Simpler system with fewer process variables than terrestrial agroecosystem crops, making process engineering for higher yield simpler and more generic
Continuous production and harvesting	Avoids the unproductive period during which new terrestrial crops are becoming established, maximizing photosynthetic production period
Nutrient control	Enables continuous optimization of levels of CO_2 and nutrients (nitrogen and phosphorus) to maximize lipid production
Specific productivity	Photosynthetic productivity is not "wasted" in constructing roots, stems, etc., which have low or zero yield of end product (e.g., compared to oilseed crops)
Land and water quality	Aqueous cultivation of algae can use poor-quality land and water that is unsuitable for conventional crop production

17.4.1 Algal biomass production systems

Algal cultivation in shallow open constructed ponds has been considered as a CO_2 capture method for power plant flue gases, and open systems such as this are typically able to convert 1%–2% of total incident solar energy into stored chemical energy under the high-intensity levels typical of full sunlight. This compares to a typical 0.1%–0.2% of sunlight converted into biomass for terrestrial crops grown under similar outdoor conditions, although some crops, such as irrigated sugar cane, can achieve productivities similar to algae cultures. To capture the CO_2 from a power plant, the algae cultivation ponds would therefore have to be large enough to capture solar energy equal to at least 20 and possibly up to 100 times the output of the power plant.

Algae farms based on open ponds, as illustrated schematically in Fig. 17.6, are used commercially to produce algal strains such as spirulina for use as "nutraceuticals." Systems of this type are operated by algae producers in the United States, including Earthrise Farms, CA, and Cyanotech Corp, HI, and in several other countries, principally China and India.

For use in a large-scale capture project, individual ponds would range up to 10 hectare (ha) in area, with a total area of >10,000 ha being required to capture a maximum of 30% of the CO_2 released from a 500 MW coal-fired power plant. This type of open pond system was demonstrated in the late 1980s on a 0.1 ha scale at the US Department of Energy site in Roswell, NM, achieving a 3.5% visible light conversion efficiency (roughly half of this level based on total solar input) and producing an extrapolated 70 t-biomass/ha/yr during the summer months.

A microalgal capture pilot project developed by GreenFuel Technologies and Debacsa, a subsidiary of the Spanish group Aurantia, commenced treating flue gas from a cement plant in Jerez, Spain in December 2007 using an open pond system. A second development phase utilizing a 100 m^2 prototype vertical thin-film closed photobioreactor was initiated in late 2008, but plans for a demonstration-scale follow-up project using microalgal-derived biofuel to power a 30 MW cogeneration plant were terminated in 2009.

Sapphire Energy currently operates a 120-ha open pond cultivation system at Columbus, New Mexico, that produces 100 barrels/day of green crude for refining into transport fuels (White and Ryan, 2015).

The challenges of open pond systems and closed photobioreactors, described in the following section, include (Kumar et al., 2015):

- Poor light conversion efficiency
- Incomplete capture due to the inability to sustain a high CO_2 concentration
- Contamination of the culture by unwanted algal species and algal predators
- Heat losses or overheating, requiring heating or cooling and/or limiting site suitability
- High cost of harvesting and processing.

Contamination may be particularly problematic in open systems if an unwanted local species outcompetes a species that has been specifically selected for some characteristic such as high lipid production. However, contamination is also an issue in closed systems, which may be more difficult to clean once contaminated.

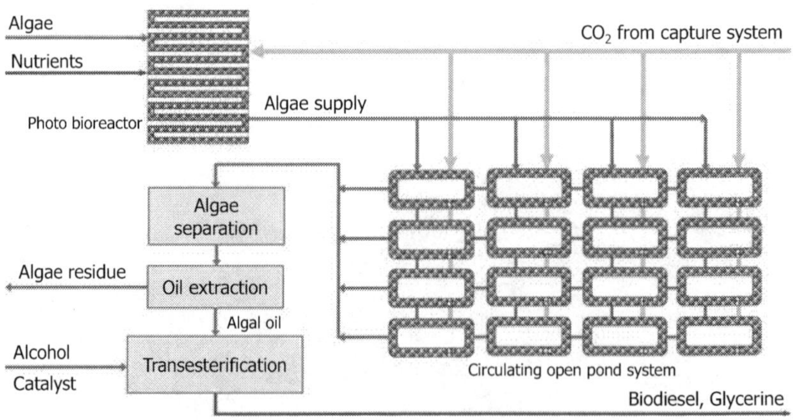

FIGURE 17.6 Algae farm using bioreactor algae supply and open circulating ponds.

17.4.2 Light conversion efficiency and saturation

Although algae are able to photosynthesize under very low light levels, their ability to convert incident light saturates at a light intensity roughly equivalent to a bright overcast day (about one-tenth of full direct sunlight). At higher intensities, overall light-conversion efficiency is reduced and incident light above this saturation intensity (IS) is wasted as heat. For the roughly 10-fold increase in light intensity from IS up to a full sunlit day, only a 2–3-fold increase in photosynthetic conversion will occur in a static aqueous culture, so that 70%–80% of solar energy is wasted as a result of saturation.

Incident light above the saturation intensity can still be efficiently used if the duration of illumination is very short (<1 ms) and the dark period is about five times longer. Such "flashing light" effects have been shown to increase photosynthetic rates in algal mass cultures (Abu-Ghosh et al., 2016). Several solutions that have been proposed and tested to overcome this saturation limitation and increase light conversion efficiency are summarized in Table 17.2.

One aspect of the genetic engineering of algal strains is the possibility of reducing the concentration in microalgae of so-called antenna pigments, often primarily chlorophyll, to improve the utilization of photons captured for photosynthesis (Stephenson et al., 2011). This would reduce the amount of nonproductive photon capture that occurs at the surface while increasing irradiance deeper in the culture, thus increasing overall productivity.

17.4.3 Development and demonstration of closed algae production systems

A number of companies are developing closed photobioreactors, typically using one or more of these approaches to increase light conversion efficiency beyond that achievable in open ponds. Temperature and nutrient levels can also be more easily and continuously controlled in closed reactor systems, further improving productivity, although O_2 saturation also needs to be managed to avoid growth inhibition due to oxygen toxicity.

One example is Green Fuel's 3D Matrix System (3DMS), which increases the area-to-volume ratio more than 500-fold compared to open ponds, and achieved productivity of up to 1.7 t-biomass/ha/d (ash-free, dry biomass weight) in field trials capturing flue gas CO_2 from power plants, including a trial at the APS Redhawk 1.06 GW gas-fired combined-cycle plant near Phoenix, AZ (Pulz, 2007; Iglina et al., 2022).

In 2014, Pond Biofuels (now Pond Technologies Inc.) began operating an algal production demonstration facility, using raw flue gas from Votorantim's St. Marys Cement plant in Ontario, Canada, as the carbon source. Supplementing the moderate sunlight levels and shorter daylight hours at this latitude, the 20 m^3 bioreactor was illuminated by custom-designed flashing red LED arrays. The long-term goal of the project was to establish a commercial-scale facility producing 225 kt/yr of algal biomass and 29,000 m^3/yr of algal-derived oil.

A two-stage algae production system that combines closed photobioreactors with open ponds has been developed by Cellana (previously HR BioPetroleum) in Hawaii. The closed reactor was used to produce a pure feed culture, which was then transferred into open ponds for batch cultivation (Huntley and Redalje, 2007). This approach achieves the main advantage of the closed photobioreactor in eliminating contamination, as residence time in the open ponds is reduced, while also minimizing cost through the use of the open pond system. A small pilot system, consisting of several closed reactors and open ponds with a capacity of 600 m^3, started operation at Kona in Hawaii in 2004 and was expanded to 2.5 ha in a joint venture with (then Royal Dutch) Shell, which ran from 2007 to 2011. Although Shell pulled out of the

TABLE 17.2 Methods to increase light conversion efficiency in microalgae cultivation.

Approach	Description
Turbulent mixing	Increasing the turbulence of the culture under illumination results in increased light-conversion efficiency since under illuminated culture is frequently brought up to the saturation intensity level
Light redistribution	Use of optical systems, such as waveguides, to conduct high-intensity incident sunlight and distribute more uniformly throughout the bioreactor
Reactor geometry	Vertical reactor geometry to increase the surface-to-volume ratio and achieve more uniform illumination close to the saturation intensity
Selection and genetic engineering of algal species	Variations in light sensitivity of different species of algae provide the opportunity to optimize light conversion efficiency through species selection and engineering (e.g., reducing the "antenna" pigment)

venture in 2011, Cellana continues to develop the technology and is evaluating potential sites for a 1000 ha demonstration-scale facility to produce 100 kt/yr of algal biomass.

17.4.4 Photobioreactor carbon delivery options

The photobioreactor approach typically involves bubbling CO_2 through the reactor, with carryover gas containing any excess CO_2 that is not absorbed by the culture being emitted to the atmosphere. CO_2-rich flue gas can also be bubbled through the culture liquid external to the reactor, in a separate "bioscrubber." CO_2 loading in the reactor can be increased by using microbubbles (diameter 10–50 μm, which dissolve faster than larger bubbles due to their high surface-to-volume ratio, and can also fully dissolve while rising in shallow ponds due to their lower buoyant velocity (Huang et al., 2017). Hollow-fiber membranes have also been investigated for gaseous CO_2 delivery in photobioreactors (Pires et al., 2012).

As an alternative to gaseous CO_2, sodium or ammonium bicarbonate, byproducts of NaOH and aqueous ammonia-based capture systems (see Chapter 7) have also been used as carbon sources for microalgae cultivation. A further novel approach, proposed by the Netherlands Organisation for Applied Scientific Research (TNO) (Könst et al., 2017), is to integrate algae production in a more traditional absorber-regenerator configuration, as illustrated in Fig. 17.7. In this scheme, which has been demonstrated at a laboratory bench scale, carbon is delivered to the photobioreactor dissolved in the rich solvent, with lean, regenerated solvent recirculated to the absorber. Absorption is preceded by flue-gas scrubbing to remove SOx and NOx, which can inhibit algal growth.

Although details of the solvent used in the TNO demonstration have not been published, similar patented systems propose carbonate solutions (e.g., Na_2CO_3), with the bicarbonate-rich solvent transported as the carbon source to the bioreactor, where the carbonate is regenerated (Schipper et al., 2013).

17.4.5 Fuel production from algal biomass

Although algal biomass can be used without further processing as a feedstock for fish farming (aquaculture), or as a solid fuel for power generation, the main organic components can also be processed into high-value products: proteins can be used as animal feed, carbohydrates can be fermented to produce bioethanol, and lipids (oils and fats produced for energy storage) can be used for biodiesel production.

Algae are potentially well suited as a feedstock for biodiesel production, as some strains can produce lipid levels up to 60% of the dry biomass weight when certain nutrients are limited (Sharma et al., 2012); active nutrient control can therefore be used to maximize lipid production, although it has yet to be demonstrated that algae with a high lipid content (>30% dry weight) can also be productive in mass cultures.

Lipids can be extracted from algal biomass by several processes, which may include drying or the use of solvents, depending on the algal species, strain, and growing conditions. Two potential processes are; cell disruption followed by emulsification with recycled oil and centrifuging, or cold press extraction of 75%–80% of the oil, followed by solvent extraction using hexane (C_6H_{14}) or diethyl ether ($C_2H_5OC_2H_5$), the solvents then being separated and recovered by distillation (Enamala et al., 2018). Extraction can also be achieved using a supercritical fluid such as CO_2 or methanol (de Jesus and Maciel Filho, 2020).

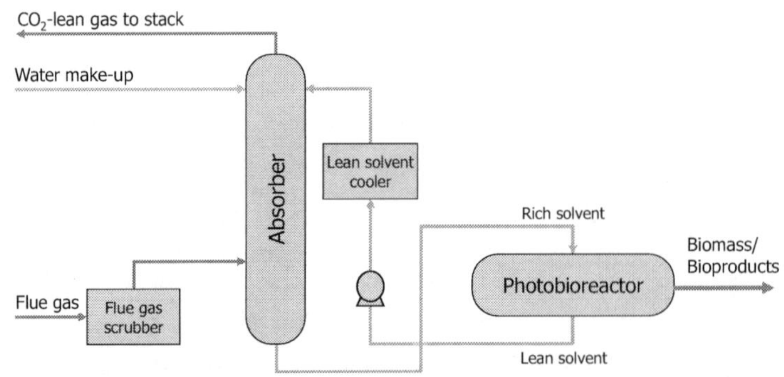

FIGURE 17.7 Integration of absorption capture with algal solvent regeneration.

Extracted lipids are mostly in the form of triglycerides, in which a long-chain fatty acid is attached to each of the three hydroxyl (OH) groups of a glycerol molecule ($C_3H_5(OH)_3$). Conversion to biofuels (diesel or jet fuel) involves the process of transesterification, shown in Fig. 17.8, in which the triglyceride is catalytically reacted with a simple alcohol, such as methanol or ethanol, transforming the triglyceride into three methyl or ethyl esters of the fatty acids, and releasing the glycerol molecule (Nagle and Lemke, 1990).

The catalyst for the reaction is a strong base such as sodium or potassium hydroxide (NaOH, KOH). Separation of the reaction products is by settling due to density differences, and the light biofuel fraction is then finished by washing and drying.

17.4.6 Research focus in algae and biodiesel production

Current R&D work aims to increase algal productivity toward a target level of 100 t-biomass/ha/yr with an oil content target of 40% triglycerides, to minimize land usage and reduce the cost of production, harvesting, and processing to the point where algal biofuels become cost competitive with traditional fuels. The following areas are being addressed by ongoing R&D and pilot-scale demonstration programs (Pierobon et al., 2018; Assunção and Malcata, 2020; Park et al., 2022):

- Optimization of photobioreactor configuration to overcome light saturation effects and achieve other productivity benefits
- Alternative bioreactor designs for microalgae cultivation (e.g., growth of biofilms in porous support bioreactors (Podola et al., 2017)
- Protection of the open pond cultures against invasion by other algal strains
- Genetic engineering to reduce antenna pigment concentration and increase oil yield
- Optimization and control systems for photobioreactors (irradiance, nutrients, acidity, etc.) to maximize product yield
- Development of low-cost harvesting systems for specific algal strains (e.g., centrifugation, filtration, floatation, and flocculation).
- Processing of harvested algal biomass to extract oil and yield high-value residues
- In-situ transesterification of lipids without prior extraction from algal biomass.

17.5 Accelerated mineralization to inorganic carbonates

This CDU subclass includes the production of metal carbonates, typically from alkaline minerals sources, and enhanced cementation processes. While the production of fuels requires energy input, mineralization processes lead to products with Gibbs-free energies that are lower than CO_2, so that the reactions are exothermic and heat generated can potentially be used in other processes. While many carbonation reactions occur without the addition of a catalyst, processes can be enhanced if a catalyst is added.

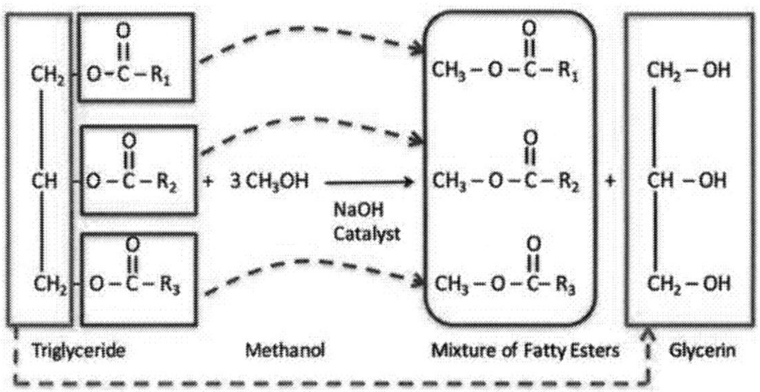

FIGURE 17.8 Transesterification of triglyceride to produce biodiesel.

17.5.1 Precipitated calcium carbonate production

Alkaline wastes from a number of industrial processes, such as ash from coal combustion or municipal waste incinerator (MWI), or slag from steelmaking are potential feedstocks for mineral carbonation. The potential for generating high-value products such as precipitated calcium carbonate (PCC) from these wastes, while also capturing CO_2, is an additional economic incentive. A number of options for CCS projects that produce PCC have been proposed or are under active commercial development, including mineral carbonation of alkaline wastes, and calcium or magnesium carbonate precipitation from seawater, either of which may be combined with DAC or point source capture from flue gases.

Whether these uses can be considered to constitute storage of CO_2 will depend on the longevity of the end-product and on whether CO_2 is also captured from the eventual recycling of the product, although, if the alternative of calcium carbonate extraction is avoided by reuse then there will be net storage compared to this basecase.

The use of steel slag or mining waste (or other alkaline industrial wastes) in a DAC carbonation scheme is a possible early-stage demonstrator project that is being pursued by several companies. This relatively simple, option, illustrated in Fig. 17.9, consists of a number of large basins (each $\sim 10{,}000\ m^2$) into which the prepared waste material is loaded. Water is sprayed onto the basin as a fine mist, absorbing CO_2 as the drops fall through the air. The solution dissolves calcium hydroxide ($Ca(OH)_2$) as it trickles through the bed and is recycled, dissolving more CO_2 and $Ca(OH)_2$, and precipitating $CaCO_3$ on each pass through the system.

Water consumption, as a result of evaporation and droplet loss through entrainment in the wind, would be a significant proportion of the overall operating cost of such a system. Stolaroff and colleagues have estimated a sequestration cost of \$8/tCO_2 sequestered for a scheme that would sequester 32 ktCO_2/yr in 140 kt of steel slag or 680 kt of concrete waste (Stolaroff et al., 2005).

Similar integration opportunities exist in other industries, such as the pulp and paper industry, where the capture of CO_2 from a paper plant could be achieved by producing an otherwise energy-intensive feedstock (PCC), while also consuming industrial wastes. Global consumption of PCC as a feedstock for the paper, pharmaceutical, and plastics industries is a relatively modest 10 Mt/yr, but this could be expanded by one or two orders of magnitude by increasing the use of PCC in the ~4 Gt/yr global cement industry.

Capture of CO_2 from power plant flue gas by absorption into seawater and precipitation of calcium and magnesium carbonates for use as cement additives has also been proposed. Seawater contains 0.01 mol-Ca/kg and 0.05 mol-Mg/kg, and precipitation of the carbonates can therefore capture 1 tCO_2 in 370 t-seawater, producing 2.1 t-carbonates, comprising 81% $MgCO_3$ and 19% $CaCO_3$. Precipitation requires a pH of ~10, so the addition of an alkali such as NaOH is necessary for a seawater feed with a pH of ~8. Of course, a full LCA of the process must also include the economic burden and environmental impact of producing the NaOH, although this is often neglected in comparative studies.

Carbon8 Systems have successfully produced construction materials from air pollution control residues (APCR) through an accelerated mineralization process, commissioning the first commercial (30 kt-APCR/yr) plant in 2012. In addition to three commercial plants operating in the UK, demonstrations have been carried out in Ontario, Canada, and a further commercial plant has opened recently in France. Other companies producing PCC and precipitated magnesium carbonate (PMC) by CO_2 mineralization as feedstock for construction materials include Carbon Capture Machine, a recent Carbon XPRIZE finalist, and Fortera (previously Calera Corp), which started up a demonstration plant producing up to 10 t/d of carbonates in Monterey County, CA, in 2008, and commissioned its first commercial-scale plant at Redding, CA, in 2022.

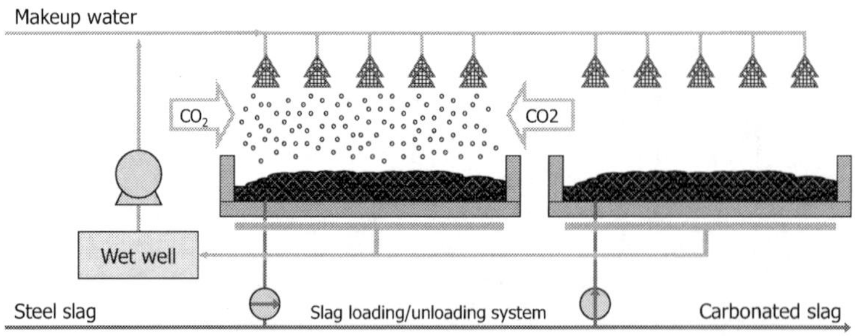

FIGURE 17.9 Direct air capture of CO_2 by steel slag carbonation.

17.6 CO_2 curing of concrete products

In addition to the opportunities for capture at cement plants that have been discussed in previous chapters, a significant storage opportunity exists where cement finds its end use, in the curing of concrete products. The use of MgO-based rather than CaO-based cements also offers the opportunity for reduced emissions on production, and additional capture during product curing.

17.6.1 Accelerated CO_2 curing of Portland concrete products

When concrete is produced by mixing cement, water, and sand or other aggregate materials, the hardening process occurs as a result of the hydration of the silicate and aluminate compounds in the cement, such as allite (Ca_3OSiO_4) and bellite (Ca_2SiO_4), according to reactions:

$$2Ca_3OSiO_4 + 6H_2O \rightarrow 3CaO.2SiO_2.3H_2O + 3Ca(OH)_2 \quad 17.1$$

$$2Ca_2SiO_4 + 4H_2O \rightarrow 3CaO.2SiO_2.3H_2O + Ca(OH)_2 \quad 17.2$$

These hydration processes initially form an open microcrystalline structure, which is progressively filled in and strengthened by further hydration products as the process continues over a time scale of hours to weeks. Hydration will continue as long as water is present, and poured concrete is typically cured under a layer of water or covered by an impermeable membrane for 1–2 weeks to ensure maximum strength is achieved (Stark, 2011).

Precast concrete products can also be cured for 12–24 h in a steam kiln, at carefully controlled temperatures in the range of 55–75°C, to accelerate early strength gain, although steam production adds an additional energy and emissions cost to the process. Curing in a CO_2-rich atmosphere for around 1 h at 20°C has been demonstrated to be a viable and faster alternative to steam curing, which eliminates this energy cost and offers a route for the cement industry to significantly reduce its net emissions (Liu et al., 2021). Aramco has developed such a curing process for precast concrete building elements that utilize CO_2 captured from cement plants or other industrial processes.

In the curing process, CO_2 reacts with the cement hydration products, principally calcium hydroxide ($Ca(OH)_2$), to produce calcium carbonate. This carbonation reaction, which results in CO_2 absorption from the atmosphere, has traditionally been considered to be undesirable. Although limited carbonation increases strength and makes concrete less permeable and more resistant to shrinkage and cracking, excessive carbonation reduces the pH of water in the cement pores, potentially leading to corrosion of steel reinforcing bars. This is not a concern for many precast concrete products, such as concrete masonry blocks. In principle, 1 t-cement is capable of absorbing 1 tCO_2 during the hydration process if all hydration products are carbonated, while 0.5 tCO_2/t-cement is considered a more reasonable uptake limit. In practice, uptake is limited by the formation of a carbonated crust in the first 2–10 mm of the product, in which rapid carbonate infilling of pores reduces permeability by three to five orders of magnitude, preventing deeper CO_2 ingress. RD&D work in the concrete industry is ongoing, focusing on increasing CO_2 uptake and penetration into the product (Ashraf, 2016). Mirroring this process, the carbonation of fabricated porous portlandite ($Ca(OH)_2$) structures, such as tiles for cladding buildings, has also been proposed by Abanades et al. as a passive DAC approach (Abanades et al., 2020).

17.6.2 Magnesium silicate cement

Compared to conventional Portland cement (PC), produced using a limestone feedstock, the use of magnesium oxide (MgO) as the intermediate cement product, based on a magnesium silicate (talc) feedstock, results in a significant reduction in the CO_2 footprint. Emissions due to production are reduced from $\sim$1.0 tCO_2 to 0.2–0.4 tCO_2/t-cement produced, while absorption during hydration and curing is increased from $\sim$0.3 tCO_2 to 1.0 tCO_2/t-cement, giving an overall negative emission of $\sim$0.7 tCO_2/t-cement produced and used, or an improvement of $\sim$1.4 tCO_2/t-cement produced and used compared to a PC alternative.

When cured in a high-CO_2 atmosphere, the strength of pure MgO-based concrete blocks can exceed that of equivalent PC blocks. Alternatively, the use of MgO as an additive in PC products increases the permeability of the concrete product, leading to increased CO_2 uptake during curing (Naqi and Jang, 2019). Commercial trials of MgO cements are being progressed by several companies spun off from recent R&D projects, including Novacem in the UK (subsequently acquired by Calix), and TecEco in Australia.

17.7 Low volume products

While fuels and mineral products represent the largest CDU opportunity, other high-value chemicals, while not produced at the volume of the previous examples, provide an economic niche that can help to finance a transition away from fossil carbon. Many companies are following the decarbonization path that replaces fossil carbon with captured CO_2, and a few examples from a range of sectors are briefly described here.

In 2019, the Air Company in New York City were awarded the Nova Institute prize for their innovative production of ethanol from CO_2 which was captured from the urban air in NYC. Hydrogen from water hydrolysis was then reacted with the CO_2 to produce pure ethanol. The original product, "Air Vodka," made significant impact due to its novelty, although it was the move by the company at the start of the Covid-19 pandemic to produce hand sanitizer which was distributed free to NYC hospitals that really cemented their position in the sustainability market. Since then, the company has moved on to produce the high-value "Air Eau de Parfum," which generates income to support their CDU efforts.

In a commitment to remove fossil carbon from their products by 2039, Unilever has defined a "Carbon Rainbow" approach in their Clean Future program for homecare products, in which they will increasingly use green carbon (from biological sources), blue carbon (from marine sources), gray carbon (from plastic wastes), and purple carbon (from CDU). In 2021, the company reported the production of nonionic surfactants from steelmaking off-gas in a collaboration with LanzaTech and India Glycol, the resulting cleaning products being released initially in the Chinese, German, and South African markets.

Polymers are another example of low volume-high value products, and the German company Covestro (formerly Bayer MaterialScience) has led the way in the production of polyurethane polyols that have been used in the production of sustainable foams. Life cycle assessment of these materials shows that emissions reductions of close to 20% can be achieved by incorporating CO_2-derived intermediates, eliminating less sustainable intermediates from the supply chain (von der Assen et al., 2013, 2014).

17.8 CDU development and deployment: opportunities and challenges

While the foregoing sections have shown the broad potential for CDU arising from a range of chemistries, commercial deployment requires processes to be simple and cheap and to not create further environmental burden—requirements that many of the technologies discussed here and reported in the literature have yet to meet.

17.8.1 Sustainable and ethical raw materials

Catalysts and coreactants in particular need to be derived from sustainable and socially responsible sources (Douglas et al., 2009), with emphasis placed on the use of Earth-abundant elements, avoiding those that are on scarcity lists, including rhodium, ruthenium, and iridium, or are obtained from ethically compromised sources, such as cobalt (Sovacool, 2021). The EU maintains a list of critical raw materials for which there is a concern regarding their continued use (European Commission, 2020).

The chemical reduction of CO_2 using hydrogen also raises sustainability concerns, as water produced in the removal of two oxygen atoms reduces the economic and environmental benefits of employing this reaction in a circular economy. One way to circumvent this is through novel technologies where molecular hydrogen is not used and water production is avoided. Plasma chemistry is one such technology (Bogaerts and Centi, 2020) that may present an advantage, especially coplasmolysis to produce syngas. Plasma chemistry is driven by radical reactions and, in the case of CO_2 plasmolysis, one oxygen is removed to produce CO with the oxygen radical either combining with another to give molecular O_2 or reacting with nitrogen to give nitrogen oxides which can be separated. While this is a good way of producing CO, much work is needed to increase efficiency as a function of yield, although this reaction does not produce hydrogen (Moss et al., 2017). Coplasmolysis reactions such as the dry reforming of methane from biogas mixtures ($CH_4 + CO_2$) have the benefit of producing syngas according to Reaction 17.3 (North and Styring, 2019a).

$$CH_4 + CO_2 \rightarrow 2CO + 2H_2 \quad 17.3$$

17.8.2 Cost competitiveness

The cost of CDU products compared to the fossil carbon-derived alternative is often considered to be a barrier to commercial viability. Recent IEA GHG reports predict that, in the long-term, production costs for middle distillates such as gasoline and kerosene will be approximately 30% more than fossil alternatives if the CO_2 comes from industrial emissions capture, but only 12% more if it comes from biomass combustion (IEA GHG, 2021a). The cost of hydrogen is a major factor in these production costs and is currently a barrier to widespread e-fuels production (IEA GHG, 2021b). These cost differentials assume the setting of a sensible carbon price, for example in emissions trading schemes, but do not include the removal of current fossil fuel subsidies. However, the US DoE's Energy Earthshot—Hydrogen, launched in 2021, aims to reduce the cost of clean hydrogen by 80% to \$1 for 1 kg within 1 decade ("111") and, if this can be achieved, it would have a significant impact on e-fuel competitiveness.

17.8.3 Product compatibility

The two IEA GHG studies referenced earlier looked at the viability of CDU pathways and the role of hydrogen generation within those pathways and included an assessment of the mitigation potential, possible market penetration and profitability, and the potential to scale for a range of synthetic fuels (IEA GHG, 2021a, 2021b). All the fuels considered—hydrocarbons (middle distillates), methane, methanol, and DME—gave good mitigation potential but also had some less positive aspects (Fig. 17.10). The "other" category in the figure encompasses a number of variables including current engine compatibility, chemistry limitations, and social impact. Engine compatibility is an important criterion that needs to be considered in fuel selection, as legacy vehicles will persist in society as the transition to an electric vehicle environment progresses (Fig. 17.11). Although this is just a snapshot in time and will likely change with further R&D progress, it suggests that, while each fuel has a positive mitigation potential, methanol and methane suffer compatibility issues, and hydrocarbons and DME show the greatest overall potential, with almost identical profiles.

A similar analysis was made for the accelerated mineralization options for cured concrete and aggregates (Fig. 17.12). While both are shown to offer benefits over all aspects of the analysis, CO_2-cured concrete shows outstanding potential. If this can be carried out in controlled environments, then the mitigation and economic potential are considerable and will be enabled by the growing use of preformed concrete structures in modular construction.

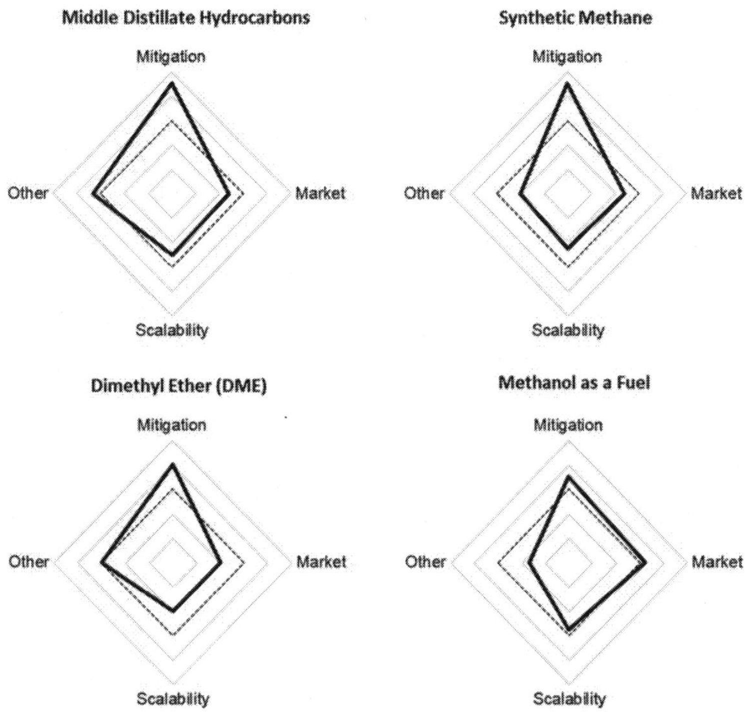

FIGURE 17.10 Analysis of the e-fuels landscape for power-to-x products from CO_2. *From IEA GHG (2021b), with permission.*

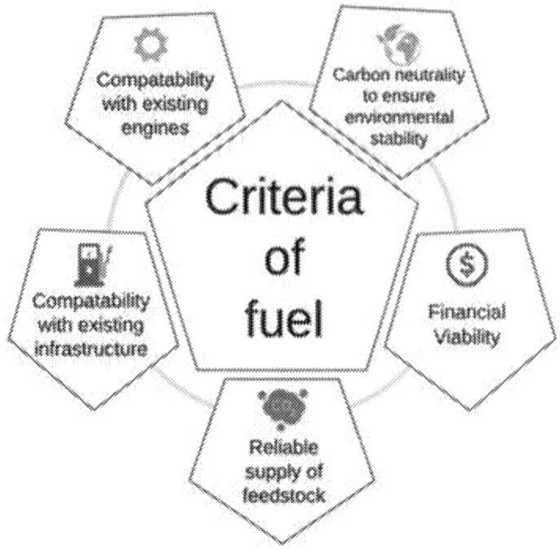

FIGURE 17.11 Defining criteria for synthetic and low-carbon fuels for an energy transition in transport vehicles. *From Styring et al. (2021b), with permission.*

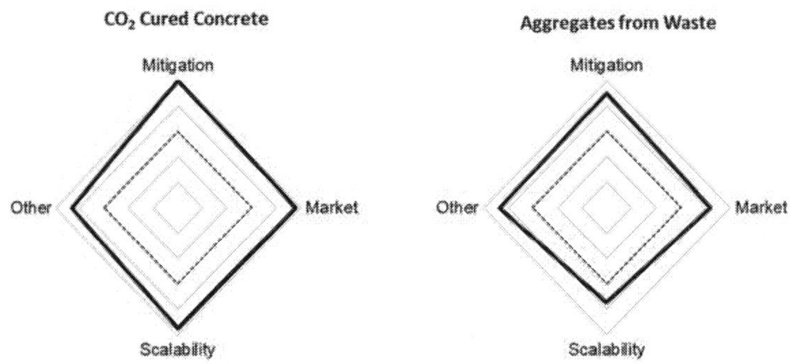

FIGURE 17.12 Analysis of the accelerated mineralization landscape for power to X products from CO_2. *From IEA GHG (2021b), with permission.*

17.8.4 The personal transportation transition

The path to net zero by 2050 will be strongly influenced by policy decisions. In many jurisdictions, fossil fuels are still subsidized to ensure jobs and security of energy supply, although an important transport policy enabling the transition away from fossil carbon is the ban on the sale of new ICE vehicles, from 2030 or 2035 in some European countries, and from 2040 in 30 others—the United States, Germany, and China being notable exceptions. However, although the widely accepted alternative—battery electric vehicles (BEVs)—have zero tailpipe emissions, this is not the case over the full lifecycle, as it ignores emissions related to the source of the electric power as well as those from construction and decommissioning, which are considerably higher for BEVs than for ICE vehicles. Recent research has shown that, when vehicle production and end-of-life disposal are taken into account, the perceived benefits of electrification are negated in all but the case of small EVs (Ellingson et al., 2016; Styring et al., 2021a).

A sustainable synthetic carbon fuel will still be required during that transition in view of the need to fuel legacy ICE vehicles, possibly for up to 20 years after the ban. A Theory of Change approach has been used by Styring et al. (2021a) to show how synfuels can be used to service such a transition, and if DME is used as a low emissions fuel, then BEVs may turn out not to be the preferred solution for personal cars and other light-duty vehicles (Fig. 17.13). BEVs require new infrastructure to deliver power to the point of need, while synfuels can "drop-in" to the existing fuel infrastructure and supply chain. Synfuels are also an effective store for renewable energy generated at times when it is surplus to demand and would otherwise be curtailed.

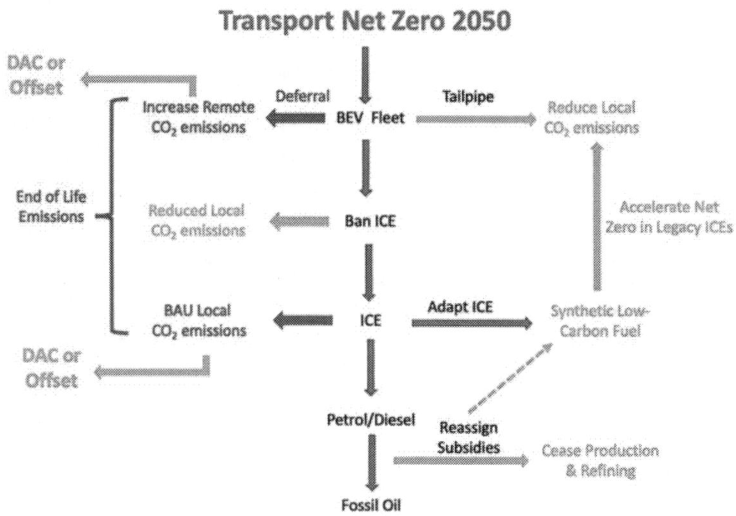

FIGURE 17.13 Theory of change approach to accelerated transport defossilization. *From Styring et al. (2021a), with permission.*

In both BEVs and synfuel ICE scenarios, energy availability will be key in the personal transport transition, and this will be time and weather dependent as well as depending on the geography of the region, with some areas being RE poor while others have a surplus. CO_2 availability will also be an issue, with the highest capture potential being in industrial and urban areas whereas rural areas and developing nations may have limited access to CO_2 point sources. DAC will be important in this respect, despite the challenges associated with it (see Chapter 13), and the \$3.5 billion program to encourage DAC development recently announced by the US DoE will provide much-needed impetus for its further development (NETL, 2022). The wide availability of low-cost CO_2, either from point source capture or from DAC, will therefore be essential for its use at scale as an economical industrial feedstock.

Care must also be taken to ensure that any legislatively imposed transition does not create societal transport poverty and a transport underclass where lower-income groups cannot afford to buy and use private transport. While this is a problem in urban areas, where the cost of a 10-year-old BEV can be 10 times that of an equivalent diesel vehicle, the impact is amplified considerably in rural areas where there is poor public transport.

17.9 Concluding remarks and future perspectives

Carbon dioxide utilization, or perhaps more accurately carbon conversion, is an important area of chemistry and engineering, and part of the suite of technologies that can contribute to achieving net zero emissions. Most net-zero strategies tend to concentrate on the energy sector and how this can be decarbonized or, more accurately, defossilized, although almost 10% of carbon from oil and gas is used in the petrochemicals sector to create the raw materials used by various industries to meet society's needs.

As a society we need "things"; things are energy, they store energy in the chemical bonds that hold them together. Most things incorporate carbon and, in current supply chains, this comes from fossil sources. CDU allows us to reuse that carbon to make new chemicals and so new things; to move from a linear economy to a circular economy of things.

CDU offers a new perspective on mitigating CO_2 emissions, which is not limited to burying used CO_2. Mitigation encompasses improving, lessening, and relieving, according to various dictionary definitions. Ad Lansink's waste hierarchy makes this very clear; the best way to mitigate emissions is not to create them in the first place. Therefore, if recycled carbon is used instead of virgin fossil carbon, future emissions will be mitigated, and the more carbon is recycled, the greater the mitigation. Capturing CO_2 and reusing it in a circular economy is predicted to have a mitigation potential of $\sim$7 $GtCO_2$/y and a global economic potential approaching \$1 trillion/yr (GCI, 2016), generating income that will boost economies locally and globally while maintaining a secure supply of sustainable carbon.

The greatest opportunities for mitigation lie in accelerated mineralization to give sustainable aggregates and enhanced concrete curing ($\sim$5 $GtCO_2$/y) and in synthetic fuels ($\sim$2 $GtCO_2$/y). Mineralization gives obvious long-term storage potential while synthetic fuels offer substantial avoidance potential while maintaining a sustainable supply of high energy density fuels, enabling a transition to a sustainable transport system while keeping available important legacy transport options while that transition progresses (Royal Society, 2019).

Mineralization reactions are exothermic and so offer an opportunity to produce sustainable construction materials while harnessing an additional energy stream through heat recovery. With fuels and organic chemicals, the cost and availability of sustainable hydrogen is a challenge, and this area of CDU offers the greatest potential for a paradigm technology shift, with coelectrolysis and coplasmolysis offering the possibility for reduction chemistry to be carried out without the need to produce and isolate molecular hydrogen. Catalysts play an essential role here and need to be highly active and selective, as well as being derived from Earth-abundant elements to ensure they are both environmentally sustainable and ethically sourced.

The current state of Earth's climate and environment demands a move to a more sustainable and circular approach to carbon chemistry, and CDU can be an important step in achieving a new equilibrium. CDU is happening now and in many cases is ready to be deployed at even faster rates, growing rapidly to make a significant contribution to emissions mitigation. Barriers to deployment need to be quickly lifted so that ambition can be replaced by action.

A narrow approach to systems analysis, considering only a small part of the supply chain, has led some to claim that CDU cannot contribute to climate change mitigation. This misconception is avoided when CDU is correctly seen as a key component of a much broader and complex ecosystem, incorporating the principles of a waste hierarchy. In fact, the definition of a waste also needs scrutiny, as the "waste" from one process will increasingly be used as the feedstock for another. Most of society's material needs include carbon in some form—organic or inorganic—and meeting those needs by reusing carbon, as in CDU, creates a circular economy and avoids perpetuating an unsustainable linear one.

References and resources

References

Abanades, J.C., Criado, Y.A., Fernández, J.R., 2020. An air CO_2 capture system based on the passive carbonation of large $Ca(OH)_2$ structures. Sustain. Energy Fuels 4 (7), 3409–3417.

Abu-Ghosh, S., Fixler, D., Dubinsky, Z., Iluz, D., 2016. Flashing light in microalgae biotechnology. Bioresour. Technol. 203, 357–363.

Aresta, M., 2019. Carbon dioxide utilization: the way to the circular economy. Greenhouse Gases: Sci. Technol. 9 (4), 610–612.

Aresta, M., Dibenedetto, A., 2021. The Carbon Dioxide Revolution: Challenges and Perspectives for a Global Society. Springer, Switzerland.

Ashraf, W., 2016. Carbonation of cement-based materials: challenges and opportunities. Construct. Build. Mater. 120, 558–570.

Assunção, J., Malcata, F.X., 2020. Enclosed "non-conventional" photobioreactors for microalga production: a review. Algal Res. 52, 102107.

Benemann, J.R., 1997. CO_2 mitigation with microalgae systems. Energy Convers. Manag. 38, S475–S479.

Bogaerts, A., Centi, G., 2020. Plasma technology for CO_2 conversion: a personal perspective on prospects and gaps. Front. Energy Res. 8, 111.

Centi, G., Perathoner, S. (Eds.), 2014. Green Carbon Dioxide: Advances in CO_2 Utilization. John Wiley & Sons, New York, NY, USA.

Chartier, O., Baker, P., Oberč, B., 2016. Artificial Photosynthesis: Potential and Reality. European Commission, Directorate-General for Research & Innovation, Brussels. Available at: https://policycommons.net/artifacts/227746/artificial-photosynthesis/889034.

de Jesus, S.S., Maciel Filho, R., 2020. Recent advances in lipid extraction using green solvents. Renew. Sustain. Energy Rev. 133, 110289.

Douglas, D., Papadopoulos, G., Boutelle, J., 2009. Citizen Engineer: A Handbook for Socially Responsible Engineering. Prentice Hall, Upper Saddle River, NJ, USA.

Ellingsen, L.A.W., Singh, B., Strømman, A.H., 2016. The size and range effect: lifecycle greenhouse gas emissions of electric vehicles. Environ. Res. Lett. 11 (5), 054010.

Enamala, M.K., Enamala, S., Chavali, M., et al., 2018. Production of biofuels from microalgae-a review on cultivation, harvesting, lipid extraction, and numerous applications of microalgae. Renew. Sustain. Energy Rev. 94, 49–68.

European Commission, 2018. A Clean Planet for All—a European Strategic Long-Term Vision for a Prosperous, Modern, Competitive and Climate Neutral Economy, COM. European Commission, p. 773 (2018). Available at: https://eur-lex.europa.eu/legal-content/EN/TXT/?uri=CELEX%3A52018DC0773.

European Commission, 2020. Study on the EU's List of Critical Raw Materials Final Report 2020. European Commission, Brussels, Belgium.

Faruk, O., Sain, M., 2015. Lignin in Polymer Composites. William Andrew/Elsevier, Oxford.

GCI, 2016. Global Roadmap for Implementing CO_2 Utilization. Global CO_2 Initiative. https://deepblue.lib.umich.edu/bitstream/handle/2027.42/150624/CO2U_Roadmap_FINAL_2016_12_07%28GCI%29.pdf.

Global CCS Institute, 2021. Global Status of CCS in 2021. Available at: www.globalccsinstitute.com/resources/global-status-report/.

Gong, E., Ali, S., Hiragond, C.B., et al., 2022. Solar fuels: research and development strategies to accelerate photocatalytic CO_2 conversion into hydrocarbon fuels. Energy Environ. Sci. 15, 880–937.

Howarth, R.W., Jacobson, M.Z., 2021. How green is blue hydrogen? Energy Sci. Eng. 9, 1676–1687.

Huang, Y., Zhao, S., Ding, Y.D., Liao, Q., Huang, Y., Zhu, X., 2017. Optimizing the gas distributor based on CO_2 bubble dynamic behaviors to improve microalgal biomass production in an air-lift photo-bioreactor. Bioresour. Technol. 233, 84–91.

Huntley, M.E., Redalje, D.G., 2007. CO_2 mitigation and renewable oil from photosynthetic microbes: a new appraisal. Mitig. Adapt. Strategies Glob. Change 12, 573–608.

IEA, 2019. Putting CO_2 to Use: Creating Value from Emissions. Technology Report—September 2019. Available at: www.iea.org/reports/putting-co2-to-use.

IEA GHG, 2012. CO_2 Capture at Gas Fired Power Plants. IEA GHG Report 2012(8). Available at: https://ieaghg.org/docs/General_Docs/Reports/2012-08.pdf.

IEA GHG, 2021a. CO_2 Utilization: Hydrogenation Pathways. IEA GHG Report 2021-03, November 2021. Available at: https://ieaghg.org/ccs-resources/blog/new-ieaghg-report-co2-utilisation-hydrogenation-pathways.

IEA GHG, 2021b. CO_2 as a Feedstock: Comparison of CCU Pathways. IEA GHG Report 2021-02, November 2021. Available at: https://ieaghg.org/ccs-resources/blog/new-ieaghg-report-2021-02-co2-as-a-feedstock-comparison-of-ccu-pathways.

Iglina, T., Iglin, P., Pashchenko, D., 2022. Industrial CO_2 capture by algae: a review and recent advances. Sustainability 14 (7), 3801.

ILCD, 2010. International Reference Life Cycle Data System (ILCD) Handbook - General Guide for Life Cycle Assessment - Provisions and Action Steps. EUR 24378 EN. Publications Office of the European Union, Luxembourg.

IPCC, 2022. In: Shukla, P.R., Skea, J., Slade, R., A., et al. (Eds.), Climate Change 2022: Mitigation of Climate Change. Contribution of Working Group III to the Sixth Assessment Report of the Intergovernmental Panel on Climate Change. Cambridge University Press, Cambridge, UK.

ISO, 2006. ISO 14044:2006 - Environmental Management - Life Cycle Assessment - Requirements and Guidelines. Available at: www.iso.org/iso/catalogue_detail?csnumber=38498.

Jiang, Z., Xiao, T., Kuznetsov, V.Á., Edwards, P.Á., 2010. Turning carbon dioxide into fuel. Phil. Trans. Math. Phys. Eng. Sci. 368 (1923), 3343–3364.

Könst, P., Mireles, I.H., van der Stel, R., van Os, P., Goetheer, E., 2017. Integrated system for capturing CO_2 as feedstock for algae production. Energy Proc. 114, 7126–7132.

Kumar, K., Mishra, S.K., Shrivastav, A., Park, M.S., Yang, J.W., 2015. Recent trends in the mass cultivation of algae in raceway ponds. Renew. Sustain. Energy Rev. 51, 875–885.

Lackner, K., Ziock, H.J., Grimes, P., 1999. Carbon Dioxide Extraction from Air: Is it an Option? (No. LA-UR-99-583). Los Alamos National Lab. (LANL), Los Alamos, NM, USA.

Liu, B., Qin, J., Shi, J., Jiang, J., Wu, X., He, Z., 2021. New perspectives on utilization of CO_2 sequestration technologies in cement-based materials. Construct. Build. Mater. 272, 121660.

Loh, J.Y., Kherani, N.P., Ozin, G.A., 2021. Persistent CO_2 photocatalysis for solar fuels in the dark. Nat. Sustain. 4 (6), 466–473.

Lansink, A., 2018. Challenging changes—connecting waste hierarchy and circular economy. Waste Manag. Res. 36 (10), 872–872.

Maitlis, P., de Klerk, A., 2013. Greener Fischer-Tropsch Processes for Fuels and Feedstocks. Wiley-VCH, Weinheim, Germany.

Miller, D., Armstrong, K., Styring, P., 2022. Assessing methods for the production of renewable benzene. Sustain. Prod. Consum. 32, 184–197.

Moss, M., Reed, D.G., Allen, R.W., Styring, P., 2017. Integrated CO_2 capture and utilization using non-thermal plasmolysis. Front. Energy Res. 5, 20.

Müller, K., Fleige, M., Rachow, F., Schmeißer, D., 2013. Sabatier based CO_2-methanation of flue gas emitted by conventional power plants. Energy Proc. 40 (2013), 240–248.

Müller, L.J., Kätelhön, A., Bachmann, M., Zimmermann, A., Sternberg, A., Bardow, A., 2020. A guideline for life cycle assessment of carbon capture and utilization. Front. Energy Res. 15.

Mutchek, M., Krynock, M., Kumar, A., et al., 2022. Carbon Dioxide Utilization Life Cycle Analysis Guidance for the US DOE Office of Fossil Energy and Carbon Management, Version 2.0 (No. DOE/NETL-2022/3774). National Energy Technology Laboratory, USA.

Nagle, N., Lemke, P., 1990. Production of methyl ester fuel from microalgae. Appl. Biochem. Biotechnol. 24 (1), 355–361.

Naqi, A., Jang, J.G., 2019. Recent progress in green cement technology utilizing low-carbon emission fuels and raw materials: a review. Sustainability 11 (2), 537.

NETL, 2022. National Energy Technology Laboratory (NETL) Carbon Utilization Program. https://netl.doe.gov/carbon-management/carbon-utilization.

North, M., Styring, P. (Eds.), 2019a. Carbon Dioxide Utilization: Volume 1 Fundamentals. de Gruyter, Berlin.

North, M., Styring, P. (Eds.), 2019b. Carbon Dioxide Utilization: Volume 2 Transformations. de Gruyter, Berlin.

Olfe-Kraeutlein, B., Armstrong, K., Mutchek, M., Cremonese, L., Sick, V., 2022. Why Terminology Matters for Successful Rollout of Carbon Dioxide Utilization Technologies. Frontiers in Climate, p. 89.

Park, W.K., Min, K., Yun, J.H., et al., 2022. Paradigm shift in algal biomass refinery and its challenges. Bioresour. Technol. 346, 126358.

Pierobon, S.C., Cheng, X., Graham, P.J., Nguyen, B., Karakolis, E.G., Sinton, D., 2018. Emerging microalgae technology: a review. Sustain. Energy Fuels 2 (1), 13–38.

Pires, J.C.M., Alvim-Ferraz, M.C.M., Martins, F.G., Simões, M., 2012. Carbon dioxide capture from flue gases using microalgae: engineering aspects and biorefinery concept. Renew. Sustain. Energy Rev. 16 (5), 3043–3053.

Podola, B., Li, T., Melkonian, M., 2017. Porous substrate bioreactors: a paradigm shift in microalgal biotechnology? Trends Biotechnol. 35 (2), 121–132.

Priestley, J., 1774. Directions for impregnating water with fixed air: in order to communicate to it the peculiar spirit and virtues of Pyrmont water, and other mineral waters of a similar nature. In: Experiments and Observations on Different Kinds of Air. Published for J. Johnson, St. Paul's Churchyard, London, UK.

Pulz, O., 2007. Performance Summary Report: Evaluation of GreenFuel's 3D Matrix Algae Growth Engineering Scale. Unit IGV Institut für Getreideverarbeitung GmbH.

Royal Society, 2019. Sustainable Synthetic Transport Fuels. Royal Society, London. Available at: https://royalsociety.org/-/media/policy/projects/synthetic-fuels/synthetic-fuels-briefing.pdf.

Schipper, K., van der Gijp, S., van der Stel, R., Goetheer, E., 2013. New methodologies for the integration of power plants with algae ponds. Energy Proc. 37, 6687–6695.

Schäppi, R., Rutz, D., Dähler, F., et al., 2022. Drop-in fuels from sunlight and air. Nature 601, 63–68.
Sharma, K.K., Schuhmann, H., Schenk, P.M., 2012. High lipid induction in microalgae for biodiesel production. Energies 5 (5), 1532–1553.
Sick, V., Stokes, G., Mason, F.C., 2022. CO_2 utilization and market size projection for CO_2-treated construction materials. Front. Climate 96.
Sovacool, B.K., 2021. When subterranean slavery supports sustainability transitions? power, patriarchy, and child labor in artisanal Congolese cobalt mining. Extr. Ind. Soc. 8 (1), 271–293.
Stark, J., 2011. Recent advances in the field of cement hydration and microstructure analysis. Cement Concr. Res. 41 (7), 666–678.
Stephenson, P.G., Moore, C.M., Terry, M.J., Zubkov, M.V., Bibby, T.S., 2011. Improving photosynthesis for algal biofuels: toward a green revolution. Trends Biotechnol. 29 (12), 615–623.
Steynberg, A.P., Dry, M., 2004. Fischer-Tropsch Technology. Vol. 152 of Studies in Surface Science and Catalysis. Elsevier, Amsterdam.
Stolaroff, J.K., Lowry, G.V., Keith, D.W., 2005. Using CaO-and MgO-rich industrial waste streams for carbon sequestration. Energy Convers. Manag. 46 (5), 687–699.
Styring, P., Jansen, D., De Coninick, H., Reith, H., Armstrong, K., 2011. Carbon Capture and Utilization in the Green Economy: Using CO_2 to Manufacture Fuel, Chemicals and Materials. CO_2Chem Media & Publishing, Sheffield, UK.
Styring, P., Quadrelli, E.A., Armstrong, K. (Eds.), 2014. Carbon Dioxide Utilisation: Closing the Carbon Cycle. Elsevier, Amsterdam.
Styring, P., Dowson, G.R., 2021. Oxygenated transport fuels from carbon dioxide: driving towards net zero. Johnson Matthey Technol. Rev. 65 (2), 170–179.
Styring, P., Duckworth, E.L., Platt, E.G., 2021a. Synthetic fuels in a transport transition: fuels to prevent a transport underclass. Front. Energy Res. 9, 707867.
Styring, P., Dowson, G.R., Tozer, I.O., 2021b. Synthetic fuels based on dimethyl ether as a future non-fossil fuel for road transport from sustainable feedstocks. Front. Energy Res. 9, 240.
Supasitmongkol, S., Styring, P., 2014. A single centre aluminium (III) catalyst and TBAB as an ionic organo-catalyst for the homogeneous catalytic synthesis of styrene carbonate. Catal. Sci. Technol. 4 (6), 1622–1630.
Tong, W., Forster, M., Dionigi, F., et al., 2020. Electrolysis of low-grade and saline surface water. Nat. Energy 5 (5), 367–377.
Toniolo, S., Tosato, R.C., Gambaro, F., Ren, J., 2020. Life cycle thinking tools: life cycle assessment, life cycle costing and social life cycle assessment. In: Life Cycle Sustainability Assessment for Decision-Making. Elsevier, Amsterdam, pp. 39–56.
UK Gov, 2021. Jet Zero Consultation A Consultation on Our Strategy for Net Zero Aviation. Department for Transport, UK Government, Crown Copyright, 2021.
Verlinden, P.J., 2020. Future challenges for photovoltaic manufacturing at the terawatt level. J. Renew. Sustain. Energy 12 (5), 053505.
von der Assen, N., Jung, J., Bardow, A., 2013. Life-cycle assessment of carbon dioxide capture and utilization: avoiding the pitfalls. Energy Environ. Sci. 6 (9), 2721–2734.
von der Assen, N., Bardow, A., 2014. Life cycle assessment of polyols for polyurethane production using CO_2 as feedstock: insights from an industrial case study. Green Chem. 16 (6), 3272–3280.
White, R.L., Ryan, R.A., 2015. Long-term cultivation of algae in open-raceway ponds: lessons from the field. Ind. Biotechnol. 11 (4), 213–220.
Willems, W., 2018. Sustainable Fuels for Future Mobility: The DME Opportunity for CI-Engines. In *8th International DME Conference, Sacramento, CA, USA, 10th–12th September.*
Wismann, S.T., Engbæk, J.S., Vendelbo, S.B., et al., 2019. Electrified methane reforming: a compact approach to greener industrial hydrogen production. Science 364 (6442), 756–759.
Yao, B., Xiao, T., Makgae, O.A., et al., 2020. Transforming carbon dioxide into jet fuel using an organic combustion-synthesized Fe-Mn-K catalyst. Nat. Commun. 11 (1), 1–12.
Zimmermann, A.W., Wunderlich, J., Müller, L., et al., 2020. Techno-economic assessment guidelines for CO_2 utilization. Front. Energy Res. 5.

Projects and organizations

Air Company; producer of high value products from captured CO_2: aircompany.com.
Aramco; CO_2-curing process for precast concrete elements: www.aramco.com/en/sustainability/climate-change/managing-our-footprint/carbon-curing.
Calix; sustainable cement manufacture; https://calix.global/industries/cement.
Carbon8; mineralization of industrial wastes: www.c8s.co.uk.
Carbon Capture Machine; building materials from CO_2 mineralization: https://ccmuk.com/technology.
Cellana; biofuel production from algae: www.cellana.com.
CircularChem; See NIC3E.
Clean Skies for Tomorrow Coalition; enabling a transition to sustainable aviation fuels and carbon-neutral flying. www.weforum.org/projects/clean-skies-for-tomorrow-coalition.
Climeworks; DAC technology developer. www.climeworks.com.
Covestro; Production of sustainable polymers from plant waste; www.solutions.covestro.com/.
CRI; Methanol production from CO_2; www.carbonrecycling.is.
Cyanotech; Microalgae production: www.cyanotech.com.
Earthrise Farms; Microalgae production: www.earthrise.com.
European Union; Critical raw materials initiative; https://ec.europa.eu/growth/sectors/raw-materials/areas-specific-interest/critical-raw-materials_en.
Greenfuel Technologies; biofuel production from microalgae and captured fossil-carbon emissions: https://greenfueltechnologies.com/.

Green NCAP; independent initiative promoting the development of clean, energy efficient, and environmentally unharmful cars: www.greenncap.com.
H. Topsøe; Syngas methanation: www.topsoe.com/processes/sng/methanation.
Neste; MY Renewable Diesel™ developer; www.neste.com/products/all-products/renewable-road-transport/neste-my-renewable-diesel.
NIC3E; transforming the UK chemical industry to fossil-independence: www.circular-chemical.org.
North CCU Hub; Regional CCU hub centered on the North Sea Port area in Belgium: https://northccuhub.eu/.
Oberon; ultra low-carbon or carbon-negative fuel producer: https://oberonfuels.comv.
Pond Technologies Inc.; closed-loop algal biomass production systems: www.pondtech.com/blog/algae-biofuels/.
Power to Methanol Antwerp B.V.; CDU for e-methanol production https://powertomethanolantwerp.com.
Sapphire Energy; biofuels from microalgae: www.harrisgroup.com/projects/sapphire-energy.
Soil Association; UK organic and sustainability certifier across food, farming, forestry, and other sectors: www.soilassociation.org.
Suburban Propane; DME-propane distributor: www.suburbanpropane.com/suburban-propane-experience/suburban-renewables/oberon/.
Sunfire; Renewable hydrogen, syngas, and e-fuel producer: www.sunfire.de.
TecEco; Sustainable cement manufacture: www.tececo.com.au.
Unilever; Cleaning and laundry products manufacturer aiming for 100% renewable or recycled carbon: www.unilever.com/news.
US DoE; Earthshot—Hydrogen "111": www.energy.gov/eere/fuelcells/hydrogen-shot.
Westküste 100; Cross-industry green hydrogen and CDU hub: www.westkueste100.de.
Zero Emissions platform; Database of CDU projects: Zeroemissionsplatform eu/about-ccs-ccu/css-ccu-projects/.

Units and NET-related acronyms

A compendium of units, conversion factors, acronyms, and a glossary of common terms related to CCS technologies.

Units and conversion factors

Prefixes

G	=	giga	=	10^9
M	=	mega	=	10^6
k	=	kilo	=	10^3
m	=	milli	=	10^{-3}
μ	=	micro	=	10^{-6}
n	=	nano	=	10^{-9}

Quantity	SI unit
Permeability	mol/m/s/Pa
Permeance	mol/m^2/s/Pa

Conversion factors

1 Å	=	10^{-10} m	=	0.1 nm
1 bar	=	100 kPa		
1 barrer	=	3.4×10^{-16} mol/m/s/Pa		
	=	10^{-10} cm^3(STP)/cm/s/cmHg		
1 GJ	=	277.8 kW h	=	0.278 MW h
1 tonne (t)	=	1000 kg	=	1 Mg
1 tC	=	3.66 tCO_2		
1 tCO_2	=	0.273 tC		

NET-related acronyms

AABW antarctic bottom water
ACS agricultural carbon sequestration
AEM anion exchange membrane
ALARP as low as reasonably practicable
ALW accelerated limestone weathering
AMOC atlantic meridional overturning circulation
AOA artificial ocean alkalinization
APCR air pollution control residues
AR afforestation/reforestation
AR4/5 ../n IPCC Fourth/Fifth ../nth Assessment Report
ARD afforestation, reforestation, and deforestation
ARW amine reclaimer waste
A_T total alkalinity

BAT best available technology
BAU business as usual (the "Do nothing" case)
BAW biologically accelerated weathering
BCA benefit–cost analysis; below-ground carbon allocation
BCM biologically controlled mineralization
BECCS biomass energy with carbon capture and storage (or BECS)
BiCRS biomass carbon removal and storage
BIGCC biomass integrated gasification combined cycle
BIM biologically induced (or influenced) mineralization
BPM biologically passive mineralization
[CO_2] CO_2 concentration
CA conservation agriculture
CaL calcium looping
CCM CO_2 concentrating mechanism
CCS carbon capture and storage (or sequestration)
CCU carbon capture and utilization
CDI capacitive deionization
CDM Clean Development Mechanism
CDR carbon dioxide removal
CEM cation exchange membrane
CERs certified emission reductions
CFBC circulating fluidized bed combustion
CFME cell-free metabolic engineering
CGS carbon geological storage
CLC chemical looping combustion
CLG chemical looping gasification
CLOU chemical looping with oxygen uncoupling
CMI carbon mitigation initiative
CS-SSGS carbon storage in sub-seabed geological structures
CSLF Carbon Sequestration Leadership Forum
CSM carbon storage by mineralization
CT conservation tillage
DACCS (U) direct air capture with carbon storage (or utilization)
DGVM dynamic global vegetation model
DIC dissolved inorganic carbon
DMEPG dimethyl ethers of polyethylene glycol
DoE US Department of Energy
DSF deep saline formation
DSGE dynamic stochastic general equilibrium (models)
eCO_2 enhanced CO_2 (concentration or partial pressure)
ECS equilibrium climate sensitivity
EfW energy from waste
EMAR electrochemically mediated amine regeneration
EPS extracellular (or exo) polymeric substances (biofilm)
ESA electrical swing adsorption
ESM Earth system model
eSMR electrified steam methane reforming
ESW enhanced silicate weathering
ET-PSA elevated temperature pressure swing adsorption
ETS emissions trading scheme
EW enhanced weathering
FACE free air carbon dioxide enrichment
FAR First Assessment Report (IPCC, 1990)
FBC fluidized bed combustion
FCCC Framework Convention on Climate Change
FOAK first of a kind
FOCE free ocean carbon dioxide enrichment
FT Fischer–Tropsch process
GCS geological carbon storage

GFBCC gasification fluidized-bed combined cycle
GHG greenhouse gas
GMST global mean surface temperature
GS geological storage
$GtCO_2$ Gigaton CO_2 (10^9 metric tonnes = 10^{12} kg)
GTL gas-to-liquids
GTP global temperature potential
GW gigawatt
GWP global warming potential
H enthalpy
HAS hyperbranched aminosilica
HBGS hydrate-based gas separation
HDAC humidity-swing direct air capture
HNLC high nutrient, low chlorophyll
HRSG heat recovery steam generator
HSE health, safety, and environment (also SHE or HE)
HTC hydrothermal conversion or carbonization
IAM integrated assessment model
IEA International Energy Agency
IGCC integrated gasification combined cycle
IGFC integrated gasification fuel cells
IL ionic liquid
ILM ionic liquid membrane, immobilized liquid membrane
InSAR interferometric synthetic aperture radar
IPCC Intergovernmental Panel on Climate Change
ITM ion transport membrane
IUPAC International Union of Pure and Applied Chemistry
kPa kilopascal
kW kilowatt
LCA life cycle analysis
LiDAR light detection and ranging; laser imaging, detection, and ranging
LM liquid membrane
LULUCF land use, land use change, and forestry
MAOM mineral-associated organic matter
MCDI membrane capacitive deionization
mCDR marine carbon dioxide removal
MEA monoethanolamine
MECC mixed electron and carbonate-ion conductor
MECS microencapsulated carbon sorbents
MFC microbial fuel cell, magnetic framework composite
MGA membrane gas absorption
MGE microbial growth efficiency
MIC(C)P microbially induced calcite (or calcium carbonate) precipitation
MISA magnetic induction swing adsorption
MOCC mixed oxide-ion and carbonate-ion conductor
MOF metal-organic framework
MOM metal-organic material, microbial organic matter
MPa megapascal
MSW municipal solid waste
$MtCO_2$ megaton CO_2 (10^6 metric tonnes = 10^9 kg)
MW molecular weight
MW_e megawatts electric power
MW_{th} megawatts thermal power
NADW North Atlantic deep water
NEOTEC negative emissions ocean thermal energy conversion
NEP net ecosystem production
NET negative emissions technology
NOAA National Oceanic and Atmospheric Administration (US Department of Commerce)
NOAK *N*th-of-a-kind

OA ocean acidification
OAE ocean alkalinity enhancement
OF ocean fertilization
OIF ocean iron fertilization
OMA ocean macroalgal afforestation
OMF ocean macronutrient fertilization
OTEC ocean thermal energy conversion
OTM oxygen transport membrane
P_c critical pressure
PCFBC pressurized circulating fluidized bed combustion
pCO_2 partial pressure of CO_2
PEM proton exchange membrane
PFBC pressurized fluidized bed combustion
PIC particulate inorganic carbon
POC particulate organic carbon
POM particulate organic matter
POX partial oxidation
ppb parts per billion (10^{-9})
ppm parts per million (10^{-6})
PSA pressure swing adsorption
PU porosity unit (1 PU = 1% porosity)
QRA quantitative risk assessment
RA regenerative agriculture
RAM risk assessment matrix
RCP representative concentration pathway
RD&D (RD3) research, development, demonstration (and deployment)
REDD reducing emissions from deforestation in developing countries
REDD+ reducing emissions from deforestation and forest degradation in developing countries, and the role of conservation, sustainable management of forests, and enhancement of forest carbon stocks in developing countries
RTIL room temperature ionic liquid
RWGS reverse water–gas shift
S entropy
S-TSA steam-assisted temperature swing adsorption
SAF sustainable aviation fuel
SAR synthetic aperture radar
SC supercritical
scCO_2 supercritical CO_2
SCM soil carbon management
SCS soil carbon storage
SIC soil inorganic carbon
SILM supported ionic liquid membrane
SLM supported liquid membrane
SMB simulated moving bed
SMBC soil microbial biomass carbon
SOC soil organic carbon
SOFC solid oxide fuel cell
SOM soil organic matter
SPS switchable polarity solvents
SRB sulfate-reducing bacteria
SRES Special Report on Emissions Scenarios (IPCC)
SSA supercapacitive swing adsorption
SSGS sub-seabed geological storage (or structures)
STP standard temperature and pressure (IUPAC; 0°C, 100kPa)
T temperature
TAB threshold avoidance budget
T_{ALK} **(or** T_A**)** total alkalinity
TAR Third Assessment Report (IPCC, 2001)
TBCA total belowground carbon allocation

TBS thermomorphic bi-phasic solvents
T_c critical temperature
T_C total carbon content
tCDR terrestrial carbon dioxide removal
T_{CO_2} total CO_2 content
TCR transient climate response (to cumulative carbon emissions)
TDS total dissolved solids
TEB threshold exceedance budget
TEW terrestrial-enhanced weathering
THC thermohaline circulation
TIC total inorganic carbon
TOA trade-off analysis
TOC total organic carbon
TOR transfer of responsibility; terms of reference
TRL technology readiness level
TSA temperature swing adsorption
TSIL task-specific ionic liquid
TVSA temperature-vacuum swing adsorption
UIC underground injection control
UNFCCC United Nations Framework Convention on Climate Change
USDW underground sources of drinking water
VCH vegetated coastal habitat
VOI value of information
VPSA vacuum pressure swing adsorption
WACC weighted average cost of capital
WCRP World Climate Research Programme
WHS wood harvest and storage
ZET zero emissions technologies
ZIF zeolitic imidazolate framework

NET-related glossary

Absorption Physical or chemical absorption is a process in which atoms or molecules of a gas or liquid enter the bulk phase of another material (liquid or solid) and are taken up within the volume, for example, gas adsorption into a liquid, or liquid into a solid. See *Adsorption*.

Adsorption A physical or chemical process in which molecules of a gas or liquid adhere to the surface of a solid adsorbent material. See also *Pressure-swing adsorption, Temperature-swing adsorption, Electrical-swing adsorption*.

Aquifer A porous and permeable geological formation containing water. Saline aquifers contain water, that is, nonpotable without desalination.

Autotrophs Autotrophs are plants and some bacteria that use photosynthesis to produce organic compounds or substrates directly from inorganic carbon. See also *Heterotrophs*.

Azeotrope An azeotrope is a mixture of two or more chemicals (e.g., CO_2 and methane) such that the ratio of constituents in the vapor phase is the same as in the liquid phase. As a result of this constant ratio, the components in the azeotrope cannot be separated by distillation unless an additive is introduced to break the azeotrope.

Barrer The Barrer, named after Richard Barrer, a New Zealand chemist and pioneer in the measurement of gas permeation, is a CGS unit of gas permeability commonly used to measure the permeance of membrane materials. The unit is used to express membrane permeability in gas separation applications.

Biochar Biochar is a charcoal resulting from the dry pyrolysis of biomass—either naturally occurring as a result of wildfires or as a result of biomass pyrolysis for bioenergy production. Applied as a soil amendment, it is expected to sustainably sequester carbon while also improving soil functioning. See also *Hydrochar*.

Biofuels Biofuels are any fuels (solid, liquid, or gas) derived from biomass or other biological material and are characterized by "generation" according to the type of feedstock; first-generation—carbohydrate feedstocks such as sugars and starch that have a direct alternative use as foods; second-generation—nonfood carbohydrate feedstocks such as agricultural wastes and short rotation crops; third-generation—algae; fourth-generation—direct secretion of biofuel by metabolic engineering of photosynthetic microorganisms.

Brayton cycle A thermodynamic cycle that describes the operation of internal combustion engines, including gas turbines. The cycle comprises compression of the working fluid, heating by combustion with an injected fuel, expansion against a piston or turbine, and cooling to return the working fluid to the initial state.

Carbonate buffering A geochemical effect that stabilizes the pH of ocean water. An increase in ocean pH (basification) favors the deprotonation of H_2CO_3 and subsequently of the bicarbonate ion (HCO_3^- to CO_3^{2-}) resulting in the release of H^+ and a stabilizing drop in pH. The resulting carbonate ion is then available for precipitation as calcium carbonate ($CaCO_3$). If ocean pH drops (acidification), carbonate ions recombine with H^+, reducing H^+ availability and resulting in a compensatory increase in pH. While carbonate ions are available, the impact of CO_2 dissolution on ocean acidity is thus reduced as a result of this buffering. Once the carbonate ion concentration becomes depleted, further CO_2 addition results in unbuffered acidification.

Carbonic anhydrase Carbonic anhydrases are a family of enzymes that catalyze the reversible conversion of CO_2 and water into bicarbonate and protons. In animals, the enzyme helps to maintain the acid–base balance in the blood and other tissues and to remove CO_2 from tissues. This hydration reaction is also central to amine and carbonate-based absorption systems, which can be enhanced using these enzymes as catalysts.

Carnot cycle The Carnot cycle is an idealized reversible thermodynamic cycle that consists of a succession of four processes operating on the working fluid: an isothermal expansion at a temperature T_1, a reversible adiabatic expansion to temperature T_2, an isothermal compression at temperature T_2, and finally a reversible adiabatic compression to the original state of the working fluid to complete the cycle. The French physicist Sadi Carnot first proposed and studied this idealized cycle in the 1820s, in establishing the first theoretical framework to describe the steam engine.

Chemical absorption An absorption process in which the sorbate chemically combines with the sorbent.

Chemical looping combustion Chemical looping combustion (CLC) is a circulating combustion process in which the oxygen required for combustion is delivered to the combustion reactor as a metal oxide rather than as free oxygen. Reduction of the oxide releases oxygen for combustion of the fuel, and the reduced metal oxide is conveyed to a second reactor where it is reoxidized, completing the loop.

Clathrate A clathrate is a molecular complex in which a guest molecule is trapped within the crystal or lattice framework of host molecules. Clathrates in which the host framework is composed of water molecules, linked by their hydrogen bonds, are also known as hydrates or clathrate hydrates and are the basis of hydrate-based CO_2 capture (HBC).

Clean Development Mechanism The Clean Development Mechanism (CDM) is a project-based mechanism under the Kyoto Protocol to the UN Framework Convention on Climate Change (UNFCCC) that enables the generation and issue of Certified Emission Reductions (CERs) from eligible projects.

Combined cycle When heat recovered from the working fluid of one thermodynamic cycle is used to heat the working fluid of another cycle, the cycles are described as combined. Most commonly applied to the heat recovery from Brayton cycle (gas turbine) exhaust gas to drive Rankine (steam) cycle power generation. Either cycle may be a general industrial process, and the add-on cycle is generically known as a topping cycle if at higher temperature or a bottoming cycle if at lower temperature.

Dendrimer A molecule composed of multiple identical branches (dendrons) around a central core, forming a tree-like structure and a typically spherical symmetric geometry.

Directed evolution Directed evolution is a genetic engineering technique that mimics the process of evolution by natural selection and is used to evolve organic molecules expressed by genes (such as proteins or nucleic acids) toward a desired property. This is achieved by subjecting a gene to successive rounds of mutation, thereby creating a so-called "library" of gene variants, expressing these gene variants and isolating those that show an improvement toward the desired function, and then using these "selected" variants as the starting point for the next cycle. The technique has been used to evolve the absorption catalyst carbonic anhydrase, in sulfate-reducing bacteria, toward greater temperature and alkalinity tolerance.

Dissolved inorganic carbon Dissolved inorganic carbon (DIC) is the sum of all inorganic carbon species in a solution, including CO_2, carbonic acid, and bicarbonate and carbonate anions. The average DIC in the world's oceans is $\sim 2 \times 10^{-3}$ mol-C/kg.

Earth system model An Earth system model (ESM) is a numerical simulation used to assess the global and regional response of the Earth system to future climate change, particularly to different GHG emissions scenarios. The biological, chemical, and physical processes occurring in the atmosphere, ocean, land, and biosphere are represented by algorithmic components of the ESM. An important aim of ESMs is to capture the feedback from these processes that can either enhance or mitigate climate change.

Ecological facilitation Interactions between species that result in a mutual net benefit without harming either are termed facilitative. Examples in a forest ecosystem include interactions that counteract pathogens, lead to nutrient enrichment, or influence microclimatic conditions.

Eddy covariance measurement A technique for determining the vertical flux of a gas species such as CO_2 by statistical analysis of the vertical velocity and fluctuating concentration of turbulent eddies.

Electric-swing adsorption Electric-swing adsorption (ESA) refers to an adsorption–desorption cycle in which sorbent regeneration is achieved using an applied electric field to heat the sorbent and release the sorbate.

Endothermic An endothermic reaction requires an input of energy in the form of heat to enable the reaction to proceed. The enthalpy change of reaction ΔH is positive in an endothermic reaction ($\Delta H > 0$). See also *Exothermic*.

Enthalpy Enthalpy is a thermodynamic property of a system, measured in J/mol, and is equal to the internal energy of the system (U) plus the product of pressure (p) and volume (V) of system ($H = U + pV$). Enthalpy change (ΔH) is related to the change in entropy (ΔS) by $\Delta H = T\Delta S + V\Delta p$.

Entropy Entropy is a thermodynamic quantity, measured in kJ/K-mole. It is a measure of the degree of disorder within a system, with higher entropy corresponding to a greater degree of disorder.

Equilibrium climate sensitivity The equilibrium climate sensitivity (ECS) is the change in global mean surface temperature that would result once climate equilibrium is reached following a doubling of atmospheric [CO_2], for example, from a preindustrial 290 to 560 ppm.

Euphotic The euphotic (or photic) zone is the well-lit upper layer of a body of water, reaching down to the euphotic depth where light intensity has reduced to 1% of the intensity at the water surface. Depending on water turbidity, the euphotic depth can be as much as 200 m in the open ocean.

Exergy Exergy is a thermodynamic property of a system, for given conditions of its external environment, and is equal to the energy (in Joules) available to perform useful work in a process that brings the system into equilibrium with that environment. Unlike energy, exergy can be destroyed as a result of any increase in total entropy due to any irreversibility in the work-generating process.

Exothermic An exothermic reaction releases energy, typically in the form of heat as the reaction proceeds. The enthalpy change of reaction ΔH is negative in an exothermic reaction ($\Delta H < 0$). See also *Endothermic*.

Facilitated transport membrane In a facilitated transport membrane, the permeate is transported across the membrane attached to a carrier.

Fischer–Tropsch process The Fischer–Tropsch process is a chemical reaction, most commonly catalyzed by iron or cobalt, in which a mixture of carbon monoxide and hydrogen is converted into a synthetic liquid hydrocarbon. Feedstock for the process is commonly natural gas (gas to liquids (GTL)) or syngas from coal or biomass gasification.

Fluidized bed reactor Fluidization of a bed of solid particles is achieved by the controlled injection of gas into the bed, causing the particles to be suspended in the flowing gas by viscous forces. The efficient heat and mass transfer that occurs in such a bed make this an effective medium for various reactions such as solid fuel combustion and gasification.

Functional membrane Functionalization of a membrane is the process of incorporating an active chemical group (the functional group) into the matrix or onto the pore surface of a membrane to enhance the separation of a specific species through surface interactions with the functional group. An example of a functional membrane is the amine-modified silica membrane for CO_2 separation. Membrane separation properties are strongly influenced by the density and the surface chemistry of the functional group.

Gas turbine cycle See *Brayton cycle*

Gasification Gasification is the high-temperature conversion of carbonaceous fuels, such as municipal wastes, biomass, petroleum residues, or coal, into a mixture of carbon monoxide and hydrogen by partial oxidation ($C \rightarrow CO$) or steam reforming ($C \rightarrow H_2 + CO$). The resulting gas (syngas) may be used directly as a fuel or as an intermediate product for the production of liquid fuel or hydrogen.

Glass transition temperature The glass transition temperature is the temperature (T_g) below which an amorphous material exhibits a brittle, glassy structure. Above T_g, bonds within the material weaken and it becomes rubbery (soft and deformable without fracturing). The glassy versus rubbery distinction is important for polymeric membranes.

Global temperature change potential The global temperature change potential (GTP) of a greenhouse gas (GHG) is an estimate of the rise in global mean surface temperature, over a given time period, resulting from heat absorption by the GHG, relative to the temperature rise that would result from heat absorption by the same mass of CO_2. The time period, in years, is indicated by a suffix, for example, GTP_{20} or GTP_{100}.

Global warming potential The global warming potential (GWP) of a GHG is the amount of heat (i.e., infrared radiation) absorbed by a gas in the atmosphere, over a certain period of time, relative to that absorbed by the same mass of carbon dioxide (CO_2). The time period, in years, is indicated by a suffix, for example, GWP_{20} or GWP_{100}, so that GHGs with a short lifetime compared to CO_2 have $GWP_{20} > GWP_{100}$.

Gross primary production Primary production refers to the production by living organisms of chemical energy in the form of organic compounds, almost exclusively as a result of photosynthesis. Part of this energy is used by the producing organism for its own growth and maintenance; Net Primary Production (NPP) excludes this "own use," while gross primary production (GPP) includes it. GPP/NPP is typically measured in tC/ha-yr. See also *Net ecosystem production*

Heterotroph In contrast to autotrophs, which produce organic compounds or substrates directly from inorganic carbon, heterotrophs obtain organic carbon by feeding on autotrophs or other heterotrophs. Plants and some bacteria use photosynthesis to produce organic substrates and are thus autotrophs, while bacteria, fungi, and animals that feed on plant matter or plant products such as root exudates are heterotrophs.

Hydrochar Hydrochar is the product of the low-temperature, high-pressure thermochemical conversion of wet biomass (plant and other bio-wastes, sewage, etc.) into a carbon and nitrogen-rich slurry, in a process known as hydrothermal carbonization (HTC). Similar to *Biochar* (q.v), this can be used as a soil amendment to enhance biomass growth and increase soil carbon content.

Hydrogen embrittlement When in contact with a metal surface, hydrogen atoms diffuse into the metal and recombine to form hydrogen molecules or combine with carbon in alloy steels to produce methane, creating stress within the metal's atomic structure. This stress can increase to levels where the metal fails due to reduced ductility, toughness, or tensile strength. Factors such as high temperatures, which increase the solubility of hydrogen in metals, will increase the risk of hydrogen embrittlement.

Hydrothermal conversion Hydrothermal conversion uses hot, high-pressure (but subcritical) water as a reaction medium to convert wet or dry biomass to carbonaceous products. High-pressure operation reduces water and energy consumption in the conversion process.

Hypersorption Hypersorption refers to an adsorption process using a moving or simulated moving bed of sorbent. The moving bed concept was initially patented in 1922, as a method to separate syngas components from coal gasification, although the hypersorption name was only applied in the 1940s.

Intergovernmental Panel on Climate Change Established by the World Meteorological Organization (WMO) and the United Nations Environment Programme (UNEP) in 1988, the aim of the Intergovernmental Panel on Climate Change (IPCC) is to assess the state of knowledge of all aspects of the climate system and of climate change, whether due to natural variability or resulting from human activities.

Ion exchange Ion exchange is a process in which ions are exchanged between two electrolytes or between an electrolyte and an insoluble solid. The replacement of calcium and magnesium ions by sodium ions is an example of ion exchange that takes place during water softening.

Ionic strength Ionic strength measures the total concentration of ions in solution, weighted by the square of the charge number (valency) of each species; $I = \Sigma_i c_i z_i^2$ where c_i and z_i are the molar concentration (mol/L) and charge number of ionic species i. For ionic strength close to zero, interactions between ions can be ignored and the solution is called an ideal solution, by analogy with an ideal gas. As ionic strength increase, concentration c_i is replaced by activity $a_i = f_i c_i$, where f_i is the I-dependent activity coefficient, to properly reflect the impact of interactions between ions.

Isochoric An isochoric or constant-volume process is a thermodynamic process in which the volume of the system under consideration remains constant.

Isosteric heat (enthalpy) of adsorption The quantity of heat (enthalpy) generated when a differential quantity of a sorbate is adsorbed onto a solid surface at constant pressure.

Joule Thomson cooling When a gas expands, work must be done to separate molecules against the weak attractive van der Waal's forces between molecules. If the expansion occurs without heat being exchanged with the surrounding environment, the thermal energy of the gas is expended to do this work, and the gas is cooled as a result. The effect was discovered in 1852 by physicists J.P. Joule and W. Thomson (Lord Kelvin) and is used in many industrial processes requiring cooling, including the liquefaction of gases.

Labile carbon See *Recalcitrant carbon*

Macroporous A porous medium having pores with a diameter >50 nm

Mafic See *Ultramafic*

Mechanical vapor recompression Mechanical vapor recompression (MVR) is a technique to concentrate a solution by evaporation in which the heat generated by compressing vapor from the solution is used to heat the solution. One interesting application under development is the use of MVR for small-scale water desalination, as an energy-efficient alternative to methods such as reverse osmosis.

Mesoporous A porous medium having pores with a diameter in the range of 2−50 nm.

Metal-organic framework Metal-organic frameworks (MOFs) are typically three-dimensional molecular structures in which metal ions, either singly or in clusters, are interlinked by organic ions or molecules known as ligands.

Microporous A porous medium having pores with a diameter <2 nm.

Miscibility Two substances are miscible if they can be mixed in all proportions to form a homogeneous solution—that is, there is no limit to the solubility of either component in the other.

Moiety A specific fragment of a molecule (e.g., an amine moiety), typically a charge-carrying species.

Net ecosystem production Net ecosystem production (NEP) is the difference between gross primary production and total ecosystem respiration (i.e., from all organisms). NEP is thus the total amount of organic carbon produced in an ecosystem, that is, available for subsequent storage, export, or nonbiological oxidation to CO_2 from fires or UV radiation.

Net primary production See *Gross primary production*

Niche differentiation When competing species use the resources in their shared environment in different ways, creating their own ecological niche, they relieve the competitive pressure that would otherwise drive one to dominance and the other to extinction. Shade tolerance (i.e., differentiated use of sunlight) is an example of an evolved trait that enables niche differentiation in a forest environment.

Nutrient mineralization Nutrient mineralization is the conversion of a nutrient from an organic to an inorganic form and occurs when organic material (such as soil organic matter, plant residues, manure, or other biologically derived solid material) is decomposed by soil micro-organisms, releasing the nutrient for uptake by plants.

Peridotite Peridotite, an igneous rock that makes up most of the earth's upper mantle, is a mixture of olivine ($(Mg, Fe)_2SiO_4$) and pyroxenes, which comprise a wide range of chain silicate minerals with the general formula $((Ca, Na, Fe^{+2}, Mg), (Cr, Al, Fe^{+3}, Mg, Mn, Ti, V))Si_2O_6$.

Perovskite A class of crystalline minerals with the general chemical formula ABO_3, in which A is a lanthanoid element and B is a transition metal. Metal cations A and B are typically arranged in two offset cubic structures linked together by oxide anions. The first perovskite discovered was $CaTiO_3$. Mixed ionic electronic conductors (MIECs) are perovskite-type materials in which one or both of the A and B lattices are doped with other cations to improve stability and performance; examples are so-called LSFO ($La_{0.5}Sr_{0.5}FeO_{3-\delta}$), LSCFO ($La_{0.6}Sr_{0.4}Co_{0.8}Fe_{0.2}O_{3-\delta}$), and LSFCO ($La_{0.2}Sr_{0.8}Fe_{0.8}Cr_{0.2}O_{3-\delta}$), where δ represents the proportion of oxide vacancies in the structure. These materials have important applications in oxygen transport membranes and solid oxide fuel cells.

Pervaporation Pervaporation is a liquid separation process in which one component of a liquid mixture permeates through a membrane and evaporates into a vapor phase on the permeate side of the membrane. The word, like the process, is a combination of permeation and evaporation.

Physical absorption An absorption process in which the sorbate does not chemically combine with the sorbent but is held within the sorbate by physical forces such as Van der Waals force.

Potential temperature In oceanography, the potential temperature of a water parcel is the temperature that the parcel would be at if brought to the surface adiabatically. The correction from in situ temperature, therefore, deducts the small temperature increase resulting from the work done in compressing a sinking parcel.

Pressure-swing adsorption Pressure-swing adsorption (PSA) refers to an adsorption–desorption cycle, that is, driven by the drop in sorbent carrying capacity with decreasing pressure. Sorbent regeneration is achieved using a low-pressure or vacuum to release the sorbate.

Pyrolysis Pyrolysis is the thermal conversion of organic matter in the absence of oxygen at temperatures between 500 and 800°C.

Recalcitrant carbon Recalcitrance in the context of soil carbon refers to a resistance to microbial decomposition. In the spectrum of susceptibility to decomposition, recalcitrant carbon occupies the middle ground between labile (easily changed) and inert material. Resistance to decomposition is here due to physical or chemical protection, for example, as a result of adsorption onto mineral surfaces (physical recalcitrance) or the blocking of catabolic pathways due to the absence of essential enzymes or oxidants (chemical recalcitrance), or other environmental factors.

Soil taxonomy (USDA) A hierarchical classification of soil types according to their properties and the conditions under which they were formed. Taxonomic differentiation is based on a wide range of characteristics including physical structure (particle size, aggregation, layering), organic versus mineral content, and chemical and electrical properties.

Solid oxide fuel cell A solid oxide fuel cell (SOFC) is a device that generates electricity directly from the electrochemical oxidation of the fuel by oxygen ions (O^{2-}) that diffuse through a solid oxide electrolyte. In the case of hydrogen as fuel, electrochemical oxidation at the anode of the cell produces water and releases two electrons from the oxygen ion, which are conducted back to the cathode through the external circuit.

Specific area The total surface area of a material per unit mass (m^2/kg). The rate of chemical reactions that take place on the surface of solid particles can be increased by increasing the specific area of a material, for example, by grinding it into smaller particles.

Steric hindrance Steric hindrance occurs when the presence and position of chemical groups within a large molecule prevent or hinder chemical reactions that proceed more easily in smaller molecules. Steric hindrance is exploited in the amine absorption reaction to increase the absorption capacity of the solvent, but has the associated disadvantage that the sorption rate is very low.

Stratification In ocean waters, stratification occurs when layers of differing salinity and temperature form a stable layering, with less dense layers overlaying denser layers. Stratification impedes vertical mixing and overturning of surface waters that are critical *inter alia* for the oxygenation of deep ocean waters and the recycling of nutrients to the euphotic zone.

Substrate In the context of biochemical processes in soils, a substrate is an organic compound such as a root exudate or biomass decomposition product that is available for metabolism by a heterotroph, typically through the action of an extracellular enzyme (exoenzyme).

Syngas Syngas is the product of a gasification or steam reforming process and consists of a mixture of hydrogen and carbon monoxide.

Temperature-swing adsorption Temperature-swing adsorption (TSA) refers to an adsorption–desorption cycle that is driven by the drop in sorbent carrying capacity with increasing temperature. Sorbent regeneration is achieved by heating to drive off the sorbate.

Thermocline The thermocline is the boundary layer that separates the warmer surface layers of the ocean (the epilimnion) from colder deeper water (the hypolimnion). In tropical oceans, the water temperature can drop from 25°C at the surface to ca. 10°C below the thermocline, at 100–200 m below sea level.

Thermodynamic cycle A thermodynamic cycle is a sequence of changes in the state of a system such that at the end of the sequence all properties of the system (e.g., pressure, temperature, entropy) are returned to the values they had at the initial state of the cycle.

Thermosiphon A thermosiphon is a mechanism of fluid circulation in which the driving force is either fully or partially provided by the relative buoyancy of the circulating fluid when warm relative to the same fluid when cold.

Third-generation biofuels See *Biofuels*

Threshold avoidance budget Cumulative CO_2 emissions over a given time period that ensures warming remain below a given threshold, with a given probability, takings into account the impact of non-CO_2 driven warming at peak warming.

Threshold exceedance budget Cumulative CO_2 emissions at the time a given warming threshold is exceeded, with a given probability, taking into account the impact of non-CO_2 driven warming at the time the threshold is exceeded.

Transesterification The process used in biofuel production where triglycerols present in the lipids extracted from biomass feedstock, such as oilseeds crops or algae, are converted to methyl or ethyl esters by reaction with methanol or ethanol, typically in the presence of a strong basic catalyst, such as NaOH.

Transient climate response Transient climate response (TCR) is the global mean surface temperature (GMST) rise at the time of atmospheric carbon dioxide doubling, both relative to preindustrial levels, in ESM simulations for the specific case where [CO_2] increases at 1% per year.

Ultramafic rocks Mafic or ultramafic rocks are igneous rocks that are relatively poor in silica and are composed predominantly or, in the case of ultramafic rocks, almost entirely of iron and magnesium minerals. Typically, although not exclusively, ultramafic rocks contain <45% silica and are a prospective source material for CO_2 sequestration by mineral carbonation.

Water–gas shift In the water–gas shift (WGS) reaction, steam reacts with the carbon monoxide to produce CO_2 and hydrogen. Separation of CO_2 from the WGS off-gas then yields a pure hydrogen stream.

Zeolites Zeolites from the Greek for "boiling stone" are a group of naturally occurring and synthetic microporous minerals composed of hydrated sodium, potassium, calcium, or magnesium aluminosilicate. Zeolites are able to trap other molecules, such as water or gases, within their open crystal structure, and as a result, they have a wide range of applications as sorbents and catalysts, for example, for amine solvent regeneration and as molecular sieves for gas separation.

Index

Note: 'Page numbers followed by '*f*' indicate figures and '*t*' indicate tables.'

D

E

T

U

V

W

Z

Made in the USA
Monee, IL
13 September 2026

63138382R00245